Integrating Architecture, Processes, and AI on SAP® BTP

Kumar Piraviperumal

Integrating Architecture, Processes, and AI on SAP® BTP

Designing an AI-Ready Execution Layer Using SAP Signavio, SAP LeanIX, and WalkMe

Rheinwerk Publishing

Editor Meagan White
Acquisitions Editor Hareem Shafi
Copyeditor Rachel Gibson
Cover Design Abigail Damke
Photo Credit iStock: 183241249/ © Bluberries
Layout Design Vera Brauner
Production Kelly O'Callaghan, Hannah Lane
Typesetting SatzPro, Germany
Printed and bound in the United States of America, on paper from sustainable sources

ISBN 978-1-4932-2880-5
1st edition 2026

Rheinwerk Publishing, Inc.
2 Heritage Drive, Suite 305
Quincy, MA 02171
USA
info@rheinwerk-publishing.com
+1.781.228.5070

Represented in the E. U. by:
Rheinwerk Verlag GmbH
Rheinwerkallee 4
53227 Bonn
Germany
service@rheinwerk-verlag.de
+49 (0) 228 42150-0

Library of Congress Cataloging-in-Publication Control Number: 2026027637

Contents at a Glance

Contents

PART V The Data Spine on SAP BTP

Preface

A transformation program does not fail due to a lack of vision from its architects or a lack of discipline from its process teams. It fails because the artifacts produced by those teams never come together in a place trusted by machines, auditors, and AI systems. This book is about building that place—and governing it well enough for an enterprise to finally let its AI reason about how it actually operates.

Most enterprises that run on SAP already own all the tools they need to describe how they operate. They have a process repository that documents how work is supposed to be performed. They have an architecture inventory that records which applications exist and how they depend on one another. They have an adoption platform that watches what users actually do. They have an integration and extension platform on which delivery happens, and they are now being asked to expose all of it to generative and agentic AI. The tools are not the problem. The problem is that these descriptions live in separate systems and were never anchored to one another—so the process model says one thing, the configured system does another, the architecture inventory records a third, and no single artifact can be trusted as the authoritative account of how the enterprise really runs.

That gap was tolerable for a long time. A misaligned process model produced inefficient handoffs and avoidable manual work, but the system still ran, the close still happened, and the audit still passed with enough effort. AI changes the economics of the gap completely. An AI agent that consumes a fragmented landscape does not merely produce inefficient suggestions; it produces confident, fluent, and structurally incorrect recommendations at the speed of an API call, referencing deprecated applications and drawing from process versions that were never approved. AI is not a neutral consumer of enterprise data. It is an amplifier. When the landscape is coherent, AI compounds that coherence into faster reasoning and clearer explanations. When the landscape is fragmented, AI compounds the fragmentation just as efficiently. The tolerance for semantic drift collapses at precisely the moment an enterprise decides to put AI to work.

This book is a practitioner's answer to that problem. It describes, end to end, how to integrate SAP Signavio, SAP LeanIX, WalkMe, SAP Build, and SAP Business Technology Platform (SAP BTP) into a single, governed execution layer that can support SAP AI safely and at scale. The integrating idea is a canonical data spine on SAP BTP: A governed semantic layer that does not replace the upstream systems of record but harmonizes their identity, versioning, ownership, and cross-system relationships so that execution services and AI all read from one stable, auditable account of the enterprise. The argument can be stated in a single sentence, and the rest of the book supports it: process and architecture data, when stored and governed in isolation, cannot support trustworthy automation or execution-grade AI—but governed together through a canonical layer, they can.

Who This Book Is For

This book is written for the people who have to make the integration real rather than merely diagram it: The enterprise architect who owns the application landscape, the process architect who governs the model; the SAP BTP engineer who builds the spine; the integration specialist who creates the flows; and the governance, controls, and audit professionals who must defend the result. It is equally written for the business-transformation lead and the finance or controlling professional who is accountable for the program's outcomes and must explain to an auditor why an AI-influenced change can be trusted. Throughout, the material is grounded in a controller's instinct for evidence: Every design choice is examined not only for whether it works, but for whether the resulting change can be reconstructed, attributed, and defended after the fact. We assume that you are familiar with the SAP landscape in broad terms; the book supplies the integration discipline that connects the parts.

What You Will Learn

The book moves deliberately from problem to foundation, to consumption, and then to operation. It opens by establishing the structural problem and the reference architecture, then specifies each source system precisely enough to integrate it; next, it builds and implements the canonical spine on SAP BTP; then layers AI and adoption on top of that spine; and finally, turns to running the platform as a governed, audited operation. Rather than treating the five technologies as separate products, it treats them as one execution loop—design, delivery, telemetry, improvement—and shows exactly where each tool fits in that loop and how the loop is closed.

Chapter 1 and Chapter 2 establish the thesis and the reference integration architecture: Why isolated process and architecture repositories cannot ground AI, and what a coherent target landscape looks like, including its security and authorization model. Chapter 3 through Chapter 8 specify the source systems. They define the SAP Signavio process data model at L4 and L5 depth and how to extend it for downstream consumption; the SAP LeanIX data model and how to map architecture to process semantics so that applications and interfaces bind to the activities that execute them; and how WalkMe is designed for process-aware adoption and how its telemetry becomes a governed signal rather than a private analytics silo.

Chapter 9 through Chapter 12 are the structural heart of the book. They design the canonical data spine—its entities, keys, versioning, lineage, and ownership rules—then implement it on SAP BTP with concrete persistence, schema, API, and governance patterns. These chapters then build the ingestion and reconciliation flows from SAP Signavio and SAP

LeanIX, and expose the process-aware execution services that let any application resolve process context, enrich it, and analyze change impact against one consistent source of truth.

Chapter 13 through Chapter 17 bring AI and the user experience onto the governed spine. They introduce the SAP AI architecture—SAP AI Core, SAP AI Launchpad, lifecycle governance, and Joule as the interaction layer—then specify how data is prepared for AI so that grounding is stable and hallucination and drift are prevented by platform properties rather than hoped-for model behavior. They show how AI generates process suggestions that are scored, explained, traceable, and always subject to mandatory human validation; how SAP Build Process Automation turns an accepted suggestion into a governed workflow on top of the spine; and how SAP Build Work Zone surfaces process-driven navigation, role-based task views, and AI insights without letting the user interface become an ungoverned back door into canonical data.

Chapter 18 and Chapter 19 close the book by discussing where a transformation program actually lives: in operations. They define the operating model of the integrated platform—monitoring, observability, data-quality management, version-drift resynchronization, and scaling across programs—and then turn to security, controls, and auditing: data access and authorization, control effectiveness, audit trails and evidence, and the compliance considerations that make the whole layer defensible. The final chapter draws the threads together into a single picture of the integrated, AI-ready enterprise.

A Single Worked Example, Followed Throughout

The book does not develop its ideas in the abstract. A single, concrete scenario runs through the chapters: a cost center governed in SAP Master Data Governance and activated in controlling on SAP S/4HANA, referenced by a position created in SAP SuccessFactors during headcount planning. That one cross-system relationship is small enough to hold in the mind and rich enough to exercise every hard problem in the book—stable identity across systems, approved versions, execution ownership, change impact resolution, and the audit trail that joins a decision to its evidence. By the final chapters, the same example is what lets AI propose a defensible change and lets a controller point to the exact row that explains why the change was made.

How to Use This Book

Each chapter is built to be worked through, not merely read. Architectural arguments are carried by figures drawn in a consistent visual language, so the same shapes recur and accumulate meaning as the book progresses. Implementation chapters include real schema, configuration, and code listings—persistence definitions, API contracts, security scopes, and integration logic—that can be adapted rather than admired. Every chapter closes with a set of practitioner features: a common design pitfalls section that names the antipatterns that recur in the field and the remedy for each, a set of review questions that test whether the structural argument landed, and student companion notes with hands-on exercises. The chapters can be read in sequence for a complete methodology, or consulted individually once the canonical spine is in place; the cross-references are explicit so that you can start with the AI chapters if you wish and follow the thread back to the foundation that makes those chapters work.

Above all, this book takes a position. It argues that AI proposes, governance approves, and the system of record records—and that the value of an integrated SAP landscape is not the cleverness of any single tool but the trust that survives across all of them. By the time you finish the book, you should be able to build that trust deliberately and defend it under audit. The first chapter begins where every honest transformation story begins: with the structural reasons most programs drift, and what it takes to keep this one from drifting too.

The chapters that follow assume nothing has been integrated yet. By the end, everything is.

PART I

Technical Context and Integration Principles

Chapter 1
Why Enterprise Architecture and Business Process Architecture Data Must Be Integrated for AI Execution

Transformation programs do not fail because architects lack vision or because process teams lack discipline. They fail because the artifacts those teams produce never meet in a place that machines, auditors, and AI systems can trust. Integration is not a connector problem. It is a problem of preserving meaning across the design, delivery, telemetry and improvement of an enterprise.

This book describes how to integrate SAP Signavio, SAP LeanIX, WalkMe, and SAP Business Technology Platform (SAP BTP) into a single, governed execution layer that is capable of supporting SAP AI safely and at scale. We move from technical context through tool-specific configuration into the canonical data spine on SAP BTP, and then into AI consumption, SAP Build adoption, and operational governance. This chapter establishes the technical problem the rest of the book addresses.

The argument of this chapter can be stated in a single sentence: Process and architecture data, when stored and governed in isolation, cannot support trustworthy automation or execution-grade AI. This chapter develops that claim from four directions: the limitations of isolated process repositories, the limitations of isolated enterprise architecture repositories, the requirement for unified semantic context that SAP AI imposes on any landscape it consumes, and the architectural role that SAP BTP plays as the data spine, integration layer, and extension layer that holds these systems together. We close by explaining what AI must not be permitted to automate, even when the underlying technology becomes capable of doing so.

You should leave this chapter with three things:

- A working mental model of the enterprise execution pyramid, in which process is the executable language of enterprise alignment
- A clear understanding of the design-to-delivery feedback loop that SAP Signavio, SAP LeanIX, WalkMe, and SAP BTP collectively close
- Recognition that AI amplifies existing coherence or fragmentation rather than creating coherence on its own

1.1 The Structural Failure of Most Transformation Programs

After two decades of SAP transformation work across industries, a recurring pattern emerges. Programs do not fail at go-live. They fail two or three years after go-live, when the documented process diverges from the configured system, the configured system diverges from the actual user behavior, and no single artifact can be trusted as the authoritative description of how the enterprise actually operates. By that point, the process repository says one thing, the enterprise architecture inventory says another, the integration platform has accumulated point-to-point logic that nobody owns, and the user community has built local workarounds that nobody has documented. This is not a tooling problem. It is a structural problem. The artifacts of design, delivery, telemetry, and improvement live in separate systems and were never anchored to one another.

The cost of this structural failure used to be tolerable. A misaligned process model meant inefficient handoffs and manual workarounds, but the underlying SAP configuration still ran. With enough effort, the close still happened, and the audit was passed. Generative and agentic AI changes that. An AI agent that consumes a fragmented landscape does not merely produce inefficient suggestions; it produces confident, fluent, and structurally wrong recommendations at the speed of an application programming interface (API) call. Tolerance for semantic drift collapses at exactly the moment enterprises are being asked to expose more of their operating model to AI.

In the following sections, we'll introduce you to the design-to-delivery feedback loop and the enterprise execution pyramid, and we'll discuss how adding AI to the loop can both amplify coherence and fragmentation.

1.1.1 Design, Delivery, Telemetry, and Improvement

Every enterprise that runs on SAP closes the same loop, whether deliberately or by accident. Process and architecture are designed. Design becomes delivery through configuration, integration, and extension. Delivery generates telemetry through user interaction, transactional data, and adoption signals. Telemetry feeds improvement, which reenters design. The four stages are inseparable. When design is not connected to delivery, configuration drifts from intent. When delivery is not connected to telemetry, adoption gaps remain invisible. When telemetry is not connected to improvement, the loop never closes, and transformation becomes a one-time event rather than a continuous capability.

SAP's portfolio is now organized around closing this loop. SAP Signavio governs design. SAP LeanIX governs the architecture context within which delivery occurs. SAP BTP and SAP S/4HANA host delivery itself. WalkMe captures the telemetry of how users interact with the product being delivered. SAP AI Core and Joule reason across the loop and propose improvements. None of these tools is sufficient on its own. The integration discipline that connects them is what determines whether the loop closes or merely appears to close. Figure 1.1 shows the design-to-delivery feedback loop. The loop closes only when stable identifiers, approved versions, and explicit ownership are preserved across every stage.

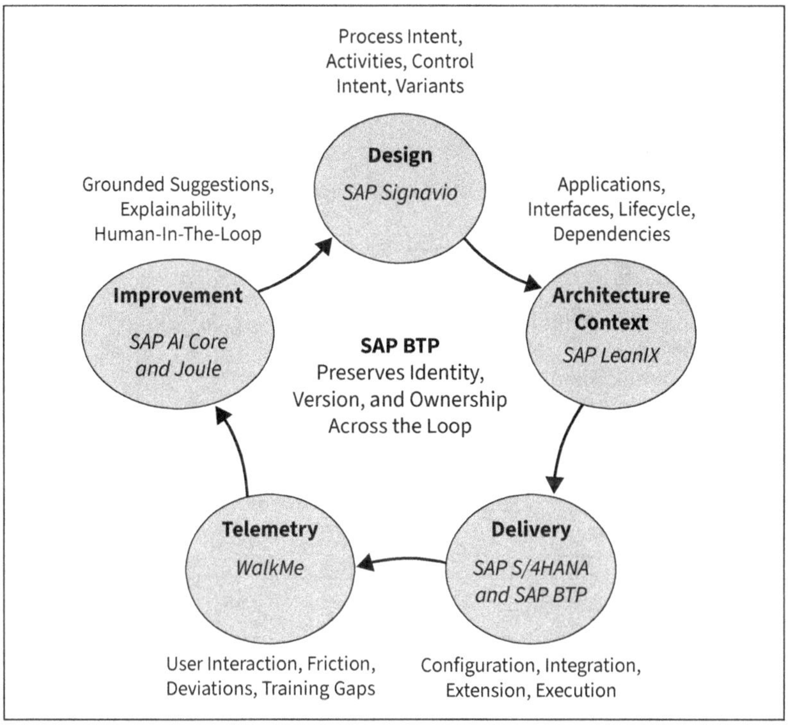

Figure 1.1: Design-to-Delivery Feedback Loop

1.1.2 The Enterprise Execution Pyramid

Behind the design-to-delivery loop sits a deeper structural model. The enterprise execution pyramid expresses the dependency chain that every operating model relies on, whether it is documented or not. Figure 1.2 shows the enterprise structural execution model. *Vision* describes the long-term direction of the enterprise. *Mission* translates that direction into purpose. *Value streams* are the outcomes the enterprise must produce to fulfill that purpose. *Capabilities* are what the enterprise must be able to do, repeatably, to deliver those outcomes. *Strategies* are the choices about how to develop and deploy those capabilities. *Resources* are the people, capital, and partnerships allocated to those strategies. *Data and systems* are the technical substrate on which everything above ultimately runs.

It is tempting to read the pyramid as a static hierarchy, but it is actually a dependency chain. A change at any layer propagates through every layer below and constrains every layer above. Process is the only artifact in the enterprise that explicitly traverses all layers. A purchase-to-pay process simultaneously expresses value stream intent, capability execution, strategic choice (centralized versus local procurement), resource allocation (which approvers, which buyers), and data and system behavior (which application performs the three-way match, which interface posts the invoice). For this reason, process becomes the executable language of enterprise alignment. Without it, the pyramid is descriptive. Process makes the pyramid executable.

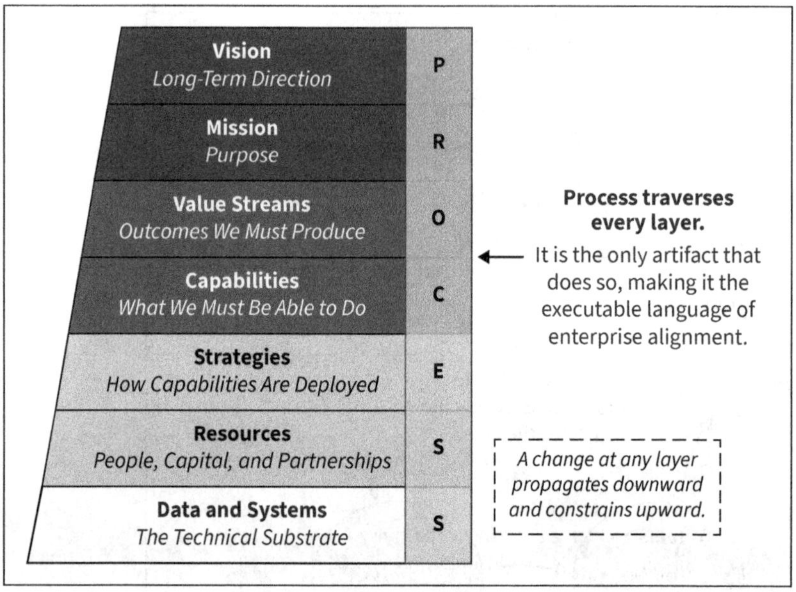

Figure 1.2: Enterprise Structural Execution Model

The enterprise structural execution model is the foundation this book builds on. SAP Signavio corresponds to the process layer, while SAP LeanIX corresponds to the data and systems layer. WalkMe describes the user interaction within the data and systems layer. SAP BTP is the layer at which all three are reconciled. SAP's AI solutions comprise the layer that reasons across that reconciled view. None of this works if the layers are not connected.

1.1.3 Why AI Amplifies Both Coherence and Fragmentation

AI is not a neutral consumer of enterprise data. It is an amplifier. When the underlying landscape is coherent—when process semantics, architecture context, configuration data, and adoption telemetry share stable identifiers and approved versions—AI compounds that coherence into faster reasoning, cleaner explanations, and more reliable recommendations. When the underlying landscape is fragmented, AI compounds that fragmentation just as efficiently. It produces fluent recommendations based on inconsistent semantics, plausible workflows that reference deprecated applications, and confident answers that no human owner can defend.

This is why we are starting this book with a chapter about integration discipline rather than a chapter about AI models. The model is the easy part. The semantic foundation that the model consumes is the hard part. The remainder of this chapter develops that semantic foundation in four movements, beginning with the limitations of isolated process repositories.

1.2 Limitations of Isolated Process Repositories

SAP Signavio is the system of record for process semantics. It is the tool in which activity-level intent, roles, control intent, application references, and process variants are modeled and governed. When used well, SAP Signavio gives an enterprise something it usually does not have: a single, versioned, auditable description of how work should be performed. The discipline of modeling at L4 and L5 depth, with stable identifiers and explicit variants, is what transforms SAP Signavio from a documentation tool into a system of record.

Business Process Mapping Levels

Before we continue, here is a brief primer on business process mapping levels for anyone who has not encountered them before. Business process mapping levels describe how detailed a process map is and where it sits in the overall process hierarchy. SAP Signavio can map up to seven levels, but in this book we go only to L5, as that depth reflects the needs of most companies. L1 and L2 sit above our scope, so we will focus on L3–L5:

- L3 covers subprocesses, major workflows, and larger stages. An example would be creating a sales order.
- L4 covers the workflow of a specific task. An example would be routing and approving a sales order in SAP S/4HANA.
- L5 provides specific instructions for how the end user accomplishes the task. An example would be documentation that details which fields to fill out and which buttons to press to create a sales order.

SAP Signavio is far less effective when used in isolation. A process repository that is not anchored to architecture, configuration, telemetry, and governance produces artifacts that are internally consistent but operationally untrustworthy. In practice, the following limitations recur most often:

- **Process models without execution ownership**
  A process diagram in SAP Signavio describes what should happen. It does not describe which application initiates the activity, which application executes it, and which downstream applications consume its output. In a single-system landscape, this ambiguity may be tolerable. However, in most SAP landscapes—where a single hire-to-retire process spans SAP SuccessFactors, SAP Master Data Governance, SAP S/4HANA, SAP Concur, and frequently a non-SAP identity provider—ambiguity is fatal. Two modelers in two business units will produce two diagrams that look semantically identical but execute against entirely different system boundaries.

  A modeler cannot solve this from inside SAP Signavio alone. SAP Signavio does not control the application catalog; it can reference application names only as text. Those text references drift the moment SAP LeanIX retires an application, renames it, or rehosts it from on-premise to cloud. Execution ownership must be modeled outside SAP Signavio in a layer that owns architecture identity. Anything else is convention rather than control.

- **Process models without configuration traceability**
  Imagine that a process step says "Post goods receipt." An SAP S/4HANA configuration specialist knows that this single sentence corresponds to dozens of configuration decisions: movement type determination, account assignment, tolerance limits, output determination, partner role determination, and the corresponding batch and serial number behavior. None of those configuration decisions is visible inside SAP Signavio, and none of them is traceable from the process model to the configured system unless an external structure connects them.

  When configuration is not traceable to process, two things happen. First, configuration review becomes archaeology rather than engineering; teams reverse engineer business intent from technical settings. Second, change impact analysis becomes guesswork. When a process changes, no automated mechanism identifies which configuration objects must be revisited. AI cannot rescue an enterprise from this condition because AI itself depends on the same traceability that the enterprise lacks.

- **Process models without AI grounding stability**
  AI grounding is the practice of supplying an AI service with the structured context it needs to produce defensible answers. For an SAP-centric AI service, that context includes approved versions of the relevant process definition, the relevant activities, and the relevant applications. An AI service grounded on a process definition that exists in three concurrent draft revisions inside SAP Signavio will produce three concurrent recommendations, all of them defensible in isolation and none of them defensible together. Without external version discipline—the separation of identity from version state—AI grounding becomes unstable.

  It is worth being precise about what version discipline requires. It requires that every process and every activity carries a stable identifier that does not change across revisions, and a separate version state that captures whether the revision is a draft, validated, approved, published, deprecated, or archived. SAP Signavio supports versioning natively, but the discipline that determines which version is exposed to AI must be enforced outside SAP Signavio in a publication boundary that the AI service consults rather than assumes.

- **Process models without lifecycle discipline**
  A mature SAP Signavio repository contains thousands of process and activity objects. Without lifecycle discipline, that volume becomes its own liability. Deprecated processes

remain searchable. Draft variants are exported alongside published variants. Localizations diverge from their global parents and eventually become independent models with no lineage back to the original. The repository slowly transforms from a system of record into a system of artifacts.

Lifecycle discipline is not an SAP Signavio configuration setting. It is a governance decision about which version states are eligible for which downstream consumption. A draft variant may be visible to a process modeler, and a validated variant may be visible to a peer reviewer, but only a published variant should be visible to integration services, execution services, and AI grounding. SAP BTP helps enforce those boundaries technically rather than procedurally. The lifecycle layer of the canonical data spine, which we introduce in Chapter 9, is what makes that enforcement possible.

The cumulative effect of these four limitations is that an isolated process repository, however well modeled, cannot be the foundation of an AI-ready execution layer. It is a necessary input but not a sufficient one. The next section examines this same problem on the architecture side.

1.3 Limitations of Isolated Enterprise Architecture Repositories

SAP LeanIX is the system of record for enterprise architecture. It describes applications, interfaces, dependency chains, and lifecycle states. When used well, SAP LeanIX gives an enterprise a stable, governed inventory of what exists in the landscape and how the components depend on one another. It is the architectural memory of the enterprise, and that memory is essential for any change-impact, modernization, or sovereignty analysis.

The limitation is that SAP LeanIX describes systems rather than execution. An application object in SAP LeanIX records that the SAP Ariba system exists, that it is in production, that it interfaces with SAP S/4HANA, and that it is owned by the procurement function. It does not record which activities of which processes execute in SAP Ariba versus in SAP S/4HANA, nor does it record under which process variants that distribution holds. Architecture inventories without process binding become structurally similar to organizational charts: accurate and necessary but insufficient.

1.3.1 Architecture Without Execution Semantics

An architecture team that operates without execution semantics tends to optimize for the wrong variables. Without process binding, the temptation is to rationalize the application portfolio based on cost, vendor consolidation, or architectural elegance. Each of these optimizations is locally defensible, but none of them is globally safe. An application that appears redundant on the architecture diagram may, at process level, host a critical control activity for a regulated variant in a single country, executed by a small number of users for whom no alternative path has been designed. Retiring such an application is a process decision disguised as an architecture decision.

The reverse problem is equally common. Architecture teams accept the proliferation of applications that, in process terms, do nothing meaningfully different from those already in the landscape, because the absence of process binding makes the duplication invisible. In both directions, the missing connection is the same. Without an explicit linkage from architecture artifacts to L4 and L5 process activities, architectural reasoning becomes disconnected from operational reality.

1.3.2 Middleware Drift and Disconnected Integration Logic

A second pattern of failure occurs in the integration layer itself. SAP Integration Suite, custom middleware, and event-driven integration platforms accumulate transformation logic over years of project work. That logic encodes implicit semantic assumptions about which applications produce which data, which fields are mandatory, and which downstream consumers tolerate which variations. SAP LeanIX cannot typically model these assumptions because it captures only the existence of the interface, not the semantics it carries.

The result is a phenomenon that integration architects know well: The landscape operates correctly only because a small number of senior engineers retain undocumented knowledge of how the integration layer interprets edge cases. When those engineers leave the company, the landscape becomes fragile in ways that only surface during incidents. AI-assisted integration cannot remediate this condition because AI needs the very semantics that the middleware never recorded. That information existed, but it was lost the moment it was implemented as code instead of being captured as a governed artifact.

1.3.3 Architecture Describes Systems; Process Describes Execution

The most economical way to express the limitation of isolated architecture repositories is by contrast. SAP LeanIX answers the question “What exists?” SAP Signavio answers the question “What should happen?” Neither answers the question “Where does what should happen execute against what exists?” This is what SAP AI must be able to answer to ground its recommendations. This requires an explicit, governed mapping between L4 and L5 process steps and the SAP LeanIX applications and interfaces those steps execute against. Chapter 6 is devoted to that mapping, which is the natural outcome of the structural argument made here.

Until that mapping is in place, an enterprise has two systems of record describing the same operational reality from different angles, with no enforced connection between them.

Process knows the intent and architecture knows the substrate, but neither knows what the other knows. AI inherits that gap and amplifies it.

1.4 Why SAP AI Requires Unified Semantic Context

Section 1.2 and Section 1.3 established that process and architecture in isolation are insufficient foundations for execution. This section argues something stronger: SAP AI, as it is being delivered in 2026 and beyond, makes the integration of those foundations a hard requirement rather than a desirable improvement. To make that argument precisely, you need a working model of what SAP AI actually refers to. The phrase has become so overloaded in vendor communications that we must first clarify its structure before making the integration argument.

1.4.1 SAP AI Is Layered, Not Monolithic

There is no single tool called SAP AI. There is an AI operating stack, in which distinct technology layers run simultaneously, each with different governance needs and different appetites for semantic context. The stack, ordered from foundation to frontier, runs through rule-based and expert systems, industrial control systems, neural networks, classical machine learning, predictive analytics, the Internet of Things (IoT), deep learning, transformers, edge AI, generative AI, and agentic AI. The frontier above that—artificial general intelligence (AGI) and beyond—remains speculative.

What matters for an SAP transformation architect is not that this list is comprehensive, but that none of the lower layers has been replaced by the higher ones. Rule-based systems still execute the SAP tax determination logic and the materials requirement planning (MRP) run. Industrial control systems still operate plants in real time, irrespective of whether a Joule agent is reasoning about scheduling. Classical machine learning still powers most demand forecasting and predictive maintenance models in production today. Generative AI did not displace these layers; it stacked on top of them and depends on them. Figure 1.3 shows all thirteen layers of the AI operating stack in dependency order.

SAP's AI investments make the layered nature of the stack explicit. SAP AI Core executes models. SAP AI Launchpad catalogs them. The generative AI hub provides curated access to a portfolio of foundation models, including SAP's own models. SAP-RPT-1 is a tabular foundation model designed for the rows-and-columns data that lives natively in enterprise resource planning (ERP) systems. SAP-ABAP-1 is a foundation model fine-tuned on ABAP, the language in which most SAP custom logic is written. Joule provides a unified way for end users, developers, and consultants to use SAP's AI capabilities. Joule agents build on this by taking actions on their own, using tools and executing tasks autonomously. Each of these components has a different role, and each of them imposes different requirements on the underlying semantic context.

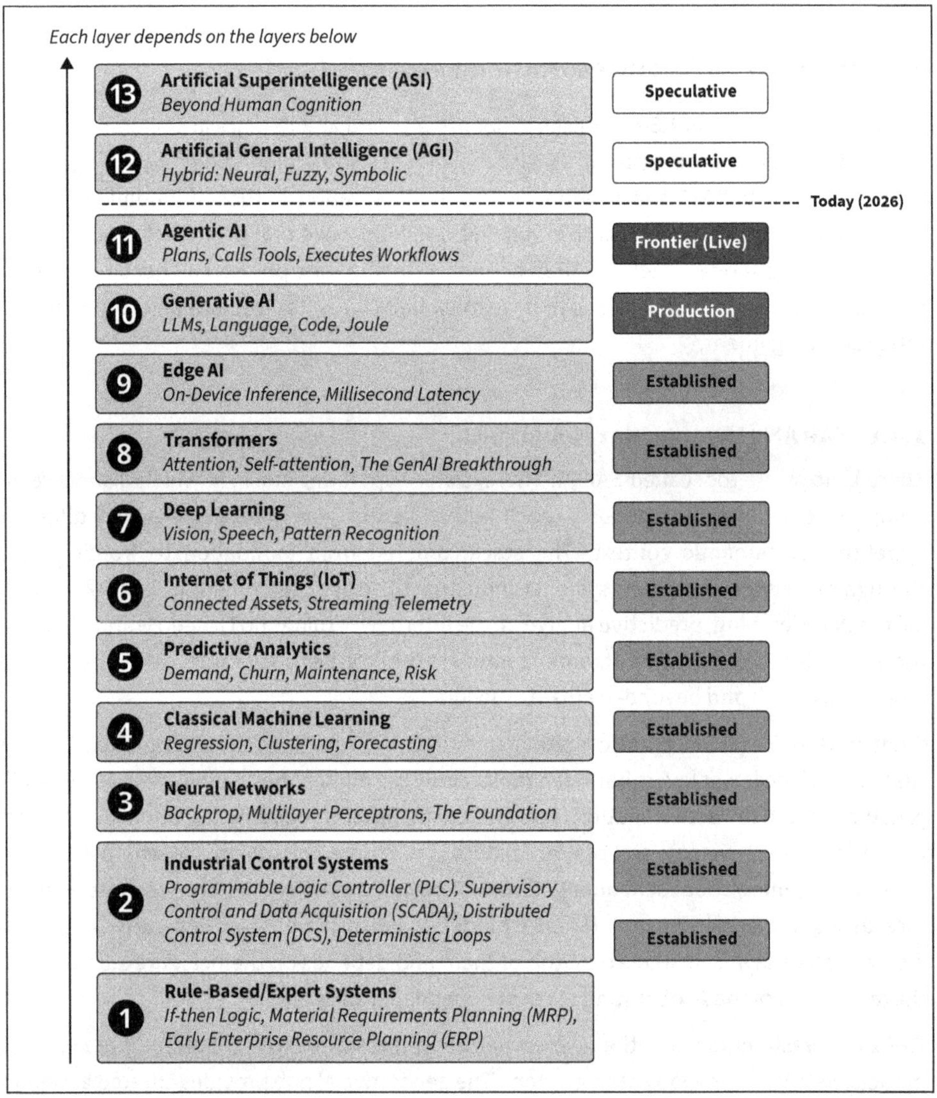

Figure 1.3: The AI Operating Stack

1.4.2 What Each Layer Requires from the Semantic Foundation

Different layers of the AI operating stack require different things from the underlying semantic foundation. Rule-based systems require complete and consistent configuration data. Classical machine learning models require clean, well-governed training data with stable feature definitions. Predictive analytics requires reliable operational signals correlated with stable identifiers. Generative AI requires grounded context: structured data, scoped documents, and stable references that allow the model to retrieve rather than invent. Agentic AI requires all of the above as well as an explicit, permissioned action surface, audit trails, and human-in-the-loop checkpoints for actions that affect systems of record.

These requirements are all about identity, version, and ownership. Every layer of the stack assumes that the artifacts it consumes can be referred to by stable identifiers, that the version of those artifacts is unambiguous, and that the ownership of those artifacts is technically enforced rather than conventionally agreed. When an enterprise cannot supply those three guarantees, every layer of the stack degrades, but generative and agentic AI do so most visibly and most expensively.

1.4.3 The Specific Requirements of Generative and Agentic AI

Generative AI introduces a particular sensitivity to ambiguous semantics. A foundation model is, in effect, a confident pattern matcher. Given two plausible references to the same business object—one current, one deprecated—the model has no intrinsic mechanism to choose the current one. Given three concurrent draft versions of the same process, it has no intrinsic mechanism to identify the authoritative one. *Hallucination* is what happens when a model is forced to produce an output without sufficient grounding. The remedy for hallucination is better grounding, not a better model. Better grounding means stable identifiers, approved versions, and explicit ownership flowing from the integration layer into the AI runtime.

Agentic AI raises the stakes by an order of magnitude. An agent is a generative model that has been given tools, permissions, and the autonomy to plan a multistep task. Joule agents and the agents built in the Joule Studio agent builder operate this way. Each agent is a named actor in the SAP landscape, and each action an agent takes must be auditable to the same standard as a human user. This is not an aspirational standard; it is what auditors, regulators, and internal control frameworks already require. The agent that drafts a goods receipt or proposes a configuration change must be able to explain, in production, which process version it consulted, which application it acted against, which approved data it used, and which human authority approved its proposal.

The agent can't answer any of those questions by itself. They can only be answered by the integration layer that supplied the agent with its context in the first place. If the integration layer cannot answer them, the agent's output is unauditable. In any regulated environment, unauditable output is unusable, no matter how fluent it sounds.

1.4.4 AI Grounding Bundles and the Role of the Canonical Layer

The practical mechanism by which structured context flows into SAP AI is the AI grounding bundle. A grounding bundle is a curated, versioned, role-scoped package of data and metadata that the AI service is permitted to reason over for a specific purpose. It typically contains the relevant published process definition, the associated activities and their execution context, the SAP LeanIX application and interface references, and any in-scope master data references. A grounding bundle should not be assembled by the AI service but by the canonical integration layer that owns the underlying semantics and that enforces the publication boundary.

The reason this matters is that the canonical integration layer is the only place in the architecture where the boundaries between systems of record can be enforced consistently. SAP Signavio cannot enforce SAP LeanIX boundaries. SAP LeanIX cannot enforce SAP Signavio boundaries. Neither can enforce master data boundaries that belong to SAP Master Data Governance, SAP S/4HANA, or SAP SuccessFactors. The canonical layer—implemented on SAP BTP and described in detail in Chapter 9—is where these boundaries are reconciled and where the rules that govern AI consumption are enforced.

The argument of this section can therefore be summarized as follows: SAP AI does not require a unified semantic context for marketing purposes. It requires one because the operating model of generative and agentic AI, applied to systems of record, is structurally incompatible with fragmented identity, ambiguous versions, and implicit ownership. The next section examines how SAP BTP discharges this requirement.

1.5 The Role of SAP BTP as the Data Spine, Integration Layer, and Extension Layer

SAP BTP is frequently introduced as middleware. That description is technically not wrong, but it is incomplete. Middleware, in the traditional sense, moves data between applications, while a data spine preserves meaning across them. This distinction is important because moving data is a transport concern and preserving meaning is a semantic concern. SAP BTP capable of doing both, but only if it is designed and governed as a semantic layer rather than as a set of integration flows. This section walks you through how SAP BTP fulfills these roles.

1.5.1 Three Architectural Roles of SAP BTP

SAP BTP plays three distinct architectural roles, and they should be kept conceptually separate even when they are physically co-located. As the data spine, SAP BTP hosts the canonical entities and relationships that normalize process semantics from SAP Signavio, architecture semantics from SAP LeanIX, master data references from SAP S/4HANA and SAP SuccessFactors, and adoption telemetry from WalkMe. As the integration layer, it hosts the ingestion flows, validation logic, and publication services that move governed semantics in and out of the spine. As the extension layer, it hosts the execution services, AI grounding services, and human-in-the-loop workflows that consume the spine on behalf of business users and AI agents.

These three roles must be kept conceptually separate because they have different governance regimes. The data spine is governed by lifecycle rules: which version of which entity is published, who approved it, and which downstream consumers are entitled to see it. The integration layer is governed by data-quality rules: which ingestion is reconciled, which validation has passed, and which errors require human intervention. The extension layer is governed by application-level rules: which execution service performs which action, with which authority, and with which audit trail. Treating all three roles of SAP BTP as if they were a single kind of middleware is the most common reason architectures fail to scale.

Figure 1.4 shows SAP BTP as the semantic integration layer. This is the only layer where process, architecture, master data, and telemetry are reconciled into a single governed view. AI consumes only published, validated, and role-scoped grounding bundles assembled from the data spine.

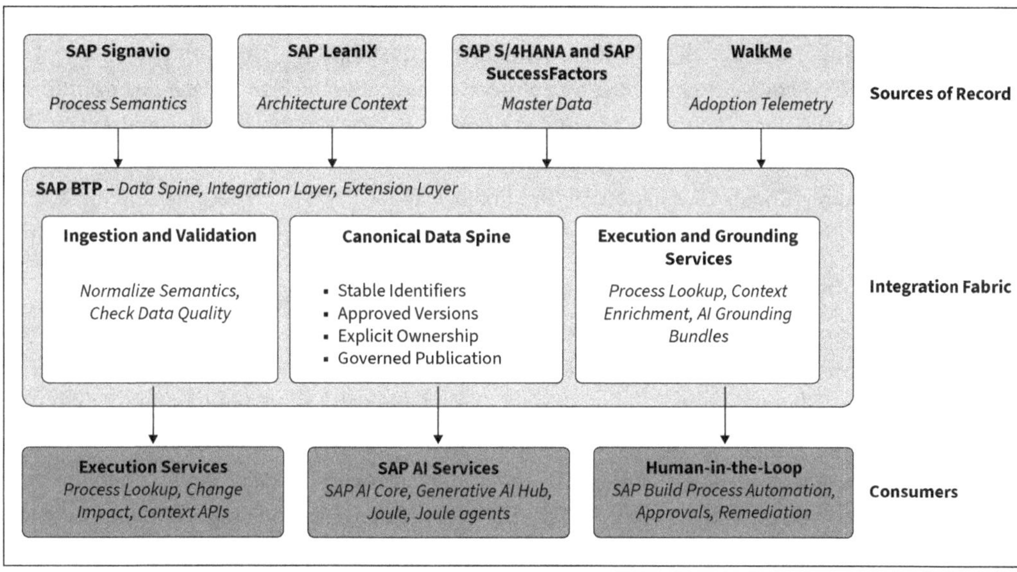

Figure 1.4: SAP BTP as the Semantic Integration Layer

1.5.2 Why Canonical Semantics Live on SAP BTP

The canonical entities on SAP BTP—processes, process versions, activities, applications, interfaces, execution contexts, master-data references, adoption artifacts, and governance overlays—are deliberately smaller than any of the upstream systems they reconcile. The data spine does not duplicate SAP Signavio, SAP LeanIX, or SAP S/4HANA's master data tables. It captures only the minimum semantics required to make execution and AI consumption safe. This minimalism is not a cost-saving measure; it is a governance decision. Every additional attribute the data spine duplicates is an additional attribute that can drift from its system of record.

Chapter 9 develops the canonical entity model in full detail, including table designs, key relationships, validation logic, and lifecycle rules. Our discussion in the current chapter is deliberately kept at the architectural level. You should leave this chapter with a clear understanding that the canonical layer exists, where it sits, and why it is structurally necessary. The implementation discipline will come later.

1.5.3 Validation, Publication, and the Role of Boundary Enforcement

Two specific responsibilities of the SAP BTP integration layer deserve emphasis because they reappear throughout the book. The first is *validation*. Before any canonical entity is exposed to execution services or AI grounding, it must be validated for structural integrity: stable identifiers must resolve, version state must be consistent, execution context must be unambiguous, and master-data references must be valid against their systems of record. Validation is a deterministic function. It either passes or fails, and the failure modes must be both observable and actionable.

The second responsibility is *publication*. Validation determines correctness, but publication determines authority. Only entities that have been explicitly published by an authorized governance role are eligible for downstream consumption. This boundary cannot be enforced by convention; it must be enforced technically, through service-layer authorization that prevents any path of consumption from reaching an unpublished entity. The architectural pattern that enforces this boundary is introduced here and developed in detail across Chapter 9, Chapter 10, and Chapter 14.

1.5.4 SAP BTP as the Coordinator of the Design-to-Delivery Loop

Our discussion of the design-to-delivery loop in Section 1.1 can now be made precise. SAP BTP coordinates that loop because it is the only layer that brings all three inputs together: It takes design from SAP Signavio, architecture from SAP LeanIX, and adoption telemetry from WalkMe, and then provides a governed, published view of all three to execution and AI services. Without SAP BTP in this role, the design layer and the delivery layer remain in different conversations. With SAP BTP in this role, the conversation closes and improvement becomes data-driven rather than experience-driven.

Keep in mind that SAP AI Core and the generative AI hub are the AI platforms, not SAP BTP. The role of SAP BTP is to make those AI platforms safe to use. It is the layer at which semantic discipline is enforced, version boundaries are protected, and the AI consumption audit trail is anchored. Programs that try to deploy SAP AI without first establishing a governed SAP BTP foundation will discover why this distinction matters, usually after their first audit cycle.

1.6 Non-Goals: What AI Must Not Automate

Every architectural argument about AI eventually reaches the question of where AI should not be permitted to act. This is sometimes treated as a matter of ethics or policy. In an SAP execution context, it is more useful as a matter of architecture. The boundaries that AI must respect are the same boundaries that protect the integrity of the canonical layer, the authority of the systems of record, and the auditability of the operating model. They are not optional.

Five non-goals set the direction for the rest of this book. We introduce them here and reinforce them in later chapters:

- **AI must not mutate canonical truth**
  The canonical entities on SAP BTP—processes, versions, activities, applications, execution contexts, and master data references—are the structural memory of the enterprise. AI services may read from them, propose changes against them, and surface anomalies in them. AI services must not write to them. Modification of canonical truth remains the responsibility of the relevant system of record and the human or service account authorized by that system. This boundary is enforced by service-layer authorization, not by convention.
- **AI must not bypass governance**
  The validation and publication discipline described in Section 1.4 applies to AI consumers exactly as it applies to human consumers. An AI service that consumes a draft process version, an unvalidated execution context, or a deprecated application reference is consuming ungoverned data, and its output inherits the risk. AI services must consume only published, validated, and entitled artifacts. The temptation to expose a draft to AI in the name of agility must be resisted; it almost always produces output that cannot be defended in audit.
- **AI must not autopublish versions**
  An AI service may detect that a process is out of date, that a control activity is missing, or that an architectural relationship has drifted. It may propose the remediation, but it must not unilaterally publish the new version. Publication is a governance act and must remain bound to a human or service account explicitly authorized for that act. The same pattern appears throughout this book: AI proposes, governance approves, and the system of record records.
- **AI must not directly execute configuration**
  Generating a configuration specification is not the same as executing a configuration change. The boundary between the two is exactly where SAP Build Process Automation, SAP Solution Manager's Change and Request Management (ChaRM), and the relevant transport mechanisms operate. AI may produce a candidate configuration artifact. The artifact must enter the same review, transport, and validation path as any human-authored configuration. Direct execution of AI-generated configuration in production environments is not an automation gain; it is an audit failure waiting to occur.
- **AI must not bypass separation of duties (SoD) controls**
  SoD is the structural protection that prevents a single actor from executing every step of a sensitive process. Whether the actor is human or AI-driven, the protection must hold. An agent that creates a vendor master record must not also approve the first invoice paid to that vendor. An agent that drafts a configuration change must not also transport it to production. Modeling AI agents as named actors in the SAP authorization model, and applying the same SoD rules that apply to human users, is the architectural answer to this requirement. This control is part of the integration layer, not the AI service itself.

The unifying principle behind these five non-goals is that AI is a powerful augmenter of governance and a poor substitute for authority. Authority, in an SAP execution context, is the legally and operationally meaningful capacity to publish a process version, release a configuration, approve a payment, or retire an application. That authority is held by human roles and service accounts explicitly delegated to act on behalf of those roles. AI may inform, propose, accelerate, and explain, but it does not, in this architecture, hold authority. This book is consistent about this point because the audit, regulatory, and operational consequences of any other design are unacceptable.

1.7 Summary

This chapter established the technical problem this book addresses. When governed in isolation, process and architecture data cannot support the trustworthy execution or AI grounding that SAP-anchored enterprises now require. We introduced the design-to-delivery loop that SAP Signavio, SAP LeanIX, WalkMe, and SAP BTP collectively close, and the enterprise execution pyramid that situates process as the executable language of enterprise alignment. We explained why isolated process repositories lack execution ownership, configuration traceability, AI grounding stability, and lifecycle discipline; why isolated architecture repositories lack execution semantics and accumulate middleware drift; why SAP AI imposes hard requirements on identity, version, and ownership across the layered AI operating stack; why SAP BTP is the natural and necessary layer at which canonical semantics, validation, publication, and execution governance are enforced; and finally, what AI must not be permitted to do, regardless of its capability.

The chapters that follow develop these arguments into implementation discipline. Chapter 2 defines the reference integration architecture, including security, identity, and auditability principles. Chapter 3 through Chapter 8 cover SAP Signavio, SAP LeanIX, and WalkMe as configured systems of record. Chapter 9 and Chapter 10 develop the canonical data spine on SAP BTP. Chapter 11 and Chapter 12 cover ingestion flows and execution services. Chapter 13 through Chapter15 cover SAP AI architecture, data preparation, and process suggestions. Chapter 16 and Chapter 17 cover SAP Build adoption. Chapter 18 and Chapter 19 cover operations, security, and audit. The online appendices, available at *https://www.sap-press.com/6352*, document the attribute models, table designs, and APIs that practitioners will need to implement the patterns described in the body of the book.

1.7.1 Key Learning Points

- Process is the executable language of enterprise alignment. The enterprise execution pyramid expresses a dependency chain from vision through data and systems, and process is the only artifact that traverses every layer.

- The design-to-delivery feedback loop closes only when stable identifiers, approved versions, and explicit ownership are preserved across design, delivery, telemetry, and improvement.
- Isolated process repositories produce internally consistent artifacts that are operationally untrustworthy because they lack execution ownership, configuration traceability, AI grounding stability, and lifecycle discipline.
- Isolated enterprise architecture repositories describe what exists but not what executes. The missing connection is the explicit mapping from L4 and L5 process steps to the SAP LeanIX applications and interfaces those steps execute against.
- SAP AI is layered. Rule-based systems, classical machine learning, deep learning, transformers, generative AI, and agentic AI run simultaneously, and none of them replace the others. Each layer requires stable identifiers, approved versions, and explicit ownership from the integration fabric.
- Generative and agentic AI are particularly sensitive to fragmented semantics. Hallucination is a grounding problem, not a model problem. The remedy is a governed canonical layer, not a better foundation model.
- SAP BTP plays three distinct architectural roles—data spine, integration layer, and extension layer—and these roles have different governance regimes. Treating them as a single type of middleware is the most common reason architectures fail to scale.
- AI augments governance but does not hold authority. Modification of canonical truth, publication of versions, execution of configuration, and circumvention of SoD controls should remain outside the scope of AI services, regardless of their capability.

1.7.2 Common Design Pitfalls

- A frequent pitfall is treating SAP BTP as connectivity middleware rather than as a semantic layer. Programs that take this view eventually accumulate point-to-point logic in the integration platform itself, and the resulting drift is both hard to discover and expensive to remediate.
- Another pitfall is exposing draft process versions to AI services in the name of agility. The cost of this shortcut is rarely seen during development; it only becomes visible during the first audit cycle, and by then the AI-generated artifacts have propagated into operational data.
- Programs frequently underestimate the modeling discipline required for execution context. Assuming that the system referenced in an SAP Signavio diagram is the same system that executes the step in production is the most common source of configuration misalignment in cross-application landscapes.
- Architecture teams that rationalize the application portfolio without process binding routinely retire applications that host critical activities for regulated variants. The resulting remediation cost typically exceeds the savings claimed at the time of retirement.

- Finally, granting AI services direct modification authority over canonical entities, even with the intent of accelerating routine maintenance, undermines the structural integrity of the entire integration layer. AI proposes; governance approves; the system of record records. Any deviation from this pattern eventually produces outcomes that cannot be defended in audit.

1.7.3 Review Questions

1. Why is process described as the executable language of enterprise alignment, and what does that imply for the role of SAP Signavio relative to SAP LeanIX in an AI-ready landscape?
2. What are the four limitations of isolated process repositories, and which of them is most consequential for AI grounding?
3. Why does an isolated enterprise architecture repository, however well governed, remain insufficient as a foundation for execution?
4. What does a *layered AI operating stack* mean in concrete SAP terms, and how does each layer's requirement for semantic context differ?
5. How does the AI grounding bundle relate to the canonical data spine, and why is the bundle assembled by the integration layer rather than by the AI service itself?
6. What are the three architectural roles that SAP BTP plays in this book, and why must they be governed separately?
7. What are the five non-goals of AI in an SAP execution context, and what is the underlying principle that unifies them?

1.7.4 Student Companion Notes

Students approaching this chapter through an SAP University Alliance program or an enterprise architecture curriculum should focus on the structural relationships rather than the tools themselves. The tools will continue to evolve, but the structural argument—that process and architecture must be integrated through a governed canonical layer in order to support AI execution—will not. We recommend the following three exercises to help you see how the structural argument plays out in real systems.

First, take a familiar end-to-end process, such as hire-to-retire or order-to-cash, and explicitly enumerate the systems that participate in it. For each activity, identify the trigger system, the executor system, and the downstream consumer. Notice how often these are different systems and how often the difference is implicit in current documentation.

Second, draft a one-page argument for why a particular AI use case in your own organization, or in a published SAP customer story, would or would not be safe to deploy without a canonical integration layer. Identify the specific failure modes that the absence of stable identity, approved versions, or explicit ownership would introduce.

Third, sketch the design-to-delivery feedback loop for one process, naming the specific SAP Signavio artifacts, SAP LeanIX artifacts, WalkMe interactions, and SAP BTP services that would be required to close the loop. This exercise is most valuable if you can identify exactly where the loop currently breaks in your own environment.

The remaining chapters of this book build on the conceptual foundation established here. Chapter 2 follows by showing what the reference integration architecture looks like in practice.

Chapter 2
Reference Integration Architecture

Architecture is what survives the program. It is the structure that an enterprise will carry into its next reorganization, its next compliance regime, and its next platform decision. A reference architecture is not just a slide in a presentation deck; it is a contract about where authority lives, where data is allowed to move, and how the consequences of every action are recorded.

Chapter 1established the technical problem this book addresses: When governed in isolation, process and architecture data cannot support trustworthy execution or AI grounding. This chapter turns that problem into a reference integration architecture. It defines the systems of record, the layers of SAP Business Technology Platform (SAP BTP), the consumption surfaces, and the security disciplines that make the architecture safe to operate at scale. It also defines the boundaries—data ownership, write authority, identity propagation, and audit obligation—that the rest of the book will configure, integrate, and govern.

The chapter is organized in two halves. The first half, Section 2.1 through Section 2.5, defines the logical architecture: the systems of record, the data ownership and write boundaries between them, the integration and extension responsibilities of SAP BTP, and the execution and consumption layers that sit on top of the canonical data spine. The second half, Section 2.6 through Section 2.11, treats security as a first-class architectural concern: trust zones, identity and authorization, role-based exposure of the data spine, the security boundaries between execution services and AI services, audit and non-repudiation, and the explicit allocation of security responsibilities across SAP Signavio, SAP LeanIX, WalkMe, SAP BTP, and the consuming applications.

Throughout the chapter, four figures anchor the discussion. Figure 2.1 presents the logical view of the reference integration architecture and is referenced from every subsequent section. Figure 2.2 makes the data ownership and write boundaries explicit as a matrix between systems and entities. Figure 2.3 introduces the trust zone model that organizes security obligations across the integrated landscape. Figure 2.4 shows the identity and authorization topology that carries authenticated principals from the corporate identity provider to the resource servers on SAP BTP, including the AI services. These figures are intended to be referenced repeatedly when you implement the patterns described in later chapters.

You should leave this chapter with three things:

- An internalized model of the reference architecture as a set of named layers with explicit responsibilities

- A clear picture of where authoritative writes happen, where derived views are computed, and where consumption is permitted to read
- A concrete understanding of the security architecture that protects this structure: the trust zones, the identity flow, the authorization model, and the audit trail

Subsequent chapters develop the configuration, integration, and operation of each layer in detail. This chapter sets the contract that those chapters implement.

2.1 Logical Architecture Overview

The reference architecture is best understood as four bands stacked vertically with a security and audit rail running across all of them. Figure 2.1 shows this organization. The top band contains the systems of record. The middle band, which is by far the largest, contains SAP BTP, organized internally into four layers: ingestion and validation, the canonical data spine, execution services, and AI services. The bottom band contains the consumption surfaces: SAP Build Work Zone, Joule, SAP Build Process Automation, and AI agents. The right-hand rail represents identity and audit, which apply uniformly to every layer rather than being concentrated in any one of them.

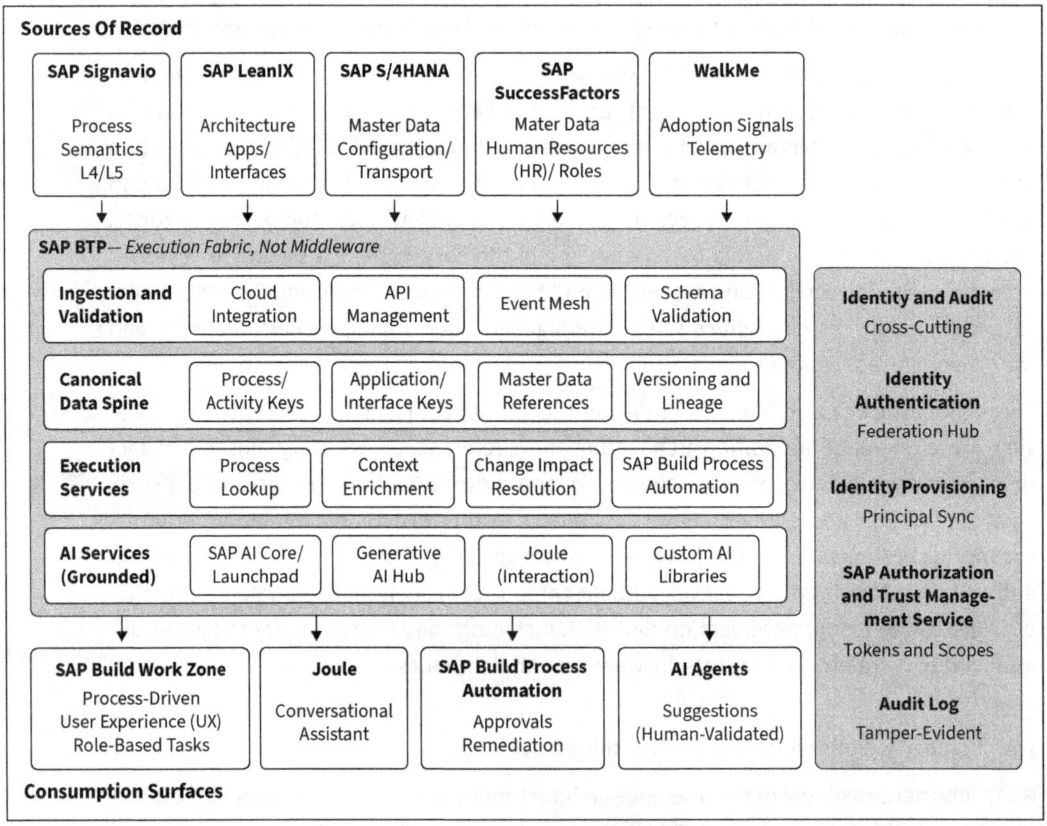

Figure 2.1: Reference Integration Architecture (Logical View)

Business Process Fabric

The full plate suite for this architecture—the consolidated view together with the execution, adoption, and AI-consumption path diagrams—is provided as Appendix H in the online supplements at *https://www.sap-press.com/6352*. Taken together, SAP Signavio, SAP LeanIX, WalkMe, Joule, SAP Build, and our custom AI libraries—woven on SAP BTP, developed in Joule Studio 2.0, and bound by stable process identity—form what we call the business process fabric: the design-layer counterpart of the canonical data spine.

Each band has a different governance regime. The systems of record are governed by their own application-level controls: SAP Signavio owns process publication; SAP LeanIX owns architecture lifecycle; SAP S/4HANA and SAP SuccessFactors own master data; WalkMe owns adoption telemetry. SAP BTP is governed by canonical lifecycle rules, validation discipline, and service-layer authorization. The consumption surfaces are governed by application-level rules and the role-based scope of the tokens they carry. Treating these as a single integration platform ignores governance regimes that must remain distinct in order to scale and audit.

Figure 2.1 is the master reference diagram for the chapter. The downward arrows from the systems of record into SAP BTP represent ingestion flows rather than direct data exposure. Each arrow is a controlled, audited path subject to schema validation and version checks. Inside SAP BTP, the four horizontal layers are arranged in the order in which data passes through them: ingestion and validation receive raw data from the sources, the canonical data spine normalizes and persists it, execution services use it to answer business questions, and AI services consume the data spine according to grounding rules described in Chapter 14. The arrows from SAP BTP to the consumption surfaces represent governed application programming interfaces (APIs) that provide published, validated artifacts to users and AI agents.

Two structural points in Figure 2.1 deserve emphasis because they recur throughout the book. First, no system of record writes directly to a consumption surface. Every write transits SAP BTP, where canonical semantics are preserved and authorization is enforced. Second, AI services do not have privileged access to the canonical data spine. They consume the same APIs as any other service, with role-scoped tokens that constrain exactly which entities they can read. These two structural choices, made consistently across the architecture, are what make audit and compliance workable in the chapters that follow.

Architecture diagrams are easy to draw and hard to defend. The reference architecture in Figure 2.1 makes a few key structural commitments, and each one rules out an alternative that organizations often fall back on under pressure. A reference architecture earns the label by being prescriptive about these choices. Naming the rejected alternatives is how the prescriptiveness becomes defensible.

The first commitment is layered separation between systems of record, the canonical layer, and the consumption surfaces. The alternative, which occurs frequently in legacy SAP landscapes, is a mesh of point-to-point integrations between systems of record and consuming applications, each with its own transformation logic, error handling, and timing assumptions. Such a mesh grows quickly as you add systems. The reference architecture grows linearly, because every new system connects to one canonical layer rather than to every other system. This is not a stylistic preference but a scale property without which AI grounding, governed automation, and audit are not feasible.

The second commitment is that the canonical data spine sits inside SAP BTP rather than in any system of record. The alternative is to treat one of the systems of record—usually SAP S/4HANA—as the de facto canonical layer and to arrange other systems as satellites. This pattern is attractive because it avoids the apparent overhead of a separate platform, but it conflates two functions that need to remain distinct. A system of record holds authoritative state for its own entities. The canonical layer holds reconciled, versioned, lineage-aware representations across all entities for downstream consumption. Asking a system of record to do both weakens its ability to do either well.

The third commitment is that AI services consume the data spine through the same governed APIs as any other consumer. The alternative is to give AI services privileged access on the assumption that AI is just doing analytics and does not need the same controls as users. This alternative produces the most consequential audit findings of any pattern this book discusses. AI output is consumed as decisions; decisions consumed without traceable grounding are decisions that cannot be defended. Section 2.7 develops the identity model that makes this commitment enforceable, and Chapter 14 develops the grounding discipline that makes it useful.

If you find these commitments unsurprising, you can skip ahead to the next chapter. If one or more of them feels novel, we recommend pausing and considering whether the current state of your organization's integration practice actually honors the commitment in question. This exercise can help you identify gaps that the rest of the book is organized to address.

2.2 Authoritative Systems of Record

A system of record is the authoritative source for a defined set of entities. It is the only system who can author those entities, whose audit log is dispositive when those entities are disputed, and whose lifecycle controls determine when those entities become eligible for downstream consumption. Naming the systems of record explicitly, and naming the entities each one owns, is the first act of architectural discipline in any integrated landscape. Programs that skip this step inevitably discover, usually in their first audit cycle, that they have multiple systems claiming authority over the same entity.

SAP Signavio is the system of record for process semantics. It owns the L4 and L5 process model, including activity definitions, roles, control intent, application references, variants, and the publication state that determines which process version is operationally authoritative. Chapter 3 covers the SAP Signavio data model in detail. For the purposes of reference architecture, the salient point is that no other system writes process semantics; downstream systems consume them.

SAP LeanIX is the system of record for enterprise architecture. It owns the application catalog, the interface inventory, the dependency chains between applications, and the lifecycle states that govern when an application is in production, deprecated, or retired. Chapter 5 discusses this in detail. SAP LeanIX does not describe how processes execute; it describes which applications exist and how they relate. The two perspectives are complementary and must be linked, but this connection is implemented in SAP BTP, not in either source system.

SAP S/4HANA is the system of record for finance, controlling, and supply chain master data, including the cost center structure, general ledger accounts, profit centers, materials, and the configuration objects that govern transactional behavior. Master data here means data that is authored once, governed centrally, and read in many places. SAP S/4HANA's master data is referenced from SAP Signavio (in process control intent) and from SAP LeanIX (in capability alignment), but it is not duplicated by either. Reference, not duplication, is the architectural rule.

SAP SuccessFactors is the system of record for human capital master data: positions, employees, organizational units, and the reporting structure. The integration between SAP SuccessFactors and SAP S/4HANA at the position-and-cost-center boundary is one of the most important master data integrations in any enterprise, and Chapter 9 walks through it using a detailed example. For reference architecture, the relevant point is that human master data does not belong in SAP S/4HANA, and finance master data does not belong in SAP SuccessFactors. The boundary is preserved in the canonical layer.

WalkMe is the system of record for adoption telemetry. It captures user-interaction signals at the level of individual SAP Fiori apps, screens, and workflow steps, and it does so in a form that can be linked back to L4 and L5 process activities through the process and SAP Fiori app identifiers maintained in SAP Signavio. WalkMe does not own process definitions, application catalogs, or master data; it owns the empirical record of how users actually interact with the systems that execute the process. Chapter 8 covers the telemetry model in detail.

2.3 Data Ownership and Write Boundaries

Naming the systems of record is the first step. Specifying who is permitted to write what, and who is restricted to reading or deriving, is the second. Figure 2.2 captures this as a matrix. Each row in the matrix names a canonical entity, and each column names a system. Each cell records whether that system holds authoritative write authority, has read access, is permitted to derive a downstream representation, or has no access by design. When read horizontally, the matrix names the system of record for each entity; when read vertically, it specifies the rights each system holds across all entities.

In the following sections, we will introduce you to the three rules that govern write boundaries, along with boundary enforcement. We end with a concrete scenario to work through.

	SAP Signavio	SAP LeanIX	SAP S/4HANA	SAP SuccessFactors	WalkMe	SAP BTP (Data Spine)	SAP Build Process Automation
Process Definition (L4/L5)	Write	Read	–	–	–	Read	Read
Process Version/Approval State	Write	–	–	–	–	Derived	–
Application Catalog	–	Write	–	–	–	Read	Read
Interface/Dependency	–	Write	–	–	–	Read	–
Master Data—Cost Center	–	–	Write	–	–	Read	–
Master Data—Position	–	–	–	Write	–	Read	–
Adoption Telemetry	–	–	–	–	Write	Derived	–
Canonical Data Spine Entities	–	–	–	–	–	Write	–
Workflow Instance State	–	–	–	–	–	Read	Write
AI Suggestion (Proposed)	–	–	–	–	–	Derived	Read

Legend
Write: Authoritative Writer (System of Record)
Derived: Derived Downstream View
Read: Read access
–: No Access By Design

Figure 2.2: Data Ownership and Write Boundaries

2.3.1 The Three Write Boundary Rules

Three rules govern the write boundaries shown in Figure 2.2. The first rule is that every entity has exactly one writer. There is no entity in the architecture for which two systems hold concurrent write authority. Where two systems appear to need to author the same entity, the modeling response is to identify the underlying structural distinction and split the entity, not to allow shared writes. The cost center/position case in Chapter 9 illustrates this: The cost center reference held in an SAP SuccessFactors position is a reference to an SAP S/4HANA-owned entity, not an SAP SuccessFactors-owned copy.

The second rule is that SAP BTP holds write authority only for entities that are intrinsic to the canonical layer itself: the data spine entities, lineage records, validation outcomes, and publication states. SAP BTP never holds write authority for entities that belong to an upstream system of record. This rule prevents the most common pattern of canonical layer drift, in which the integration layer slowly accumulates side-channel writes that diverge from the upstream source.

The third rule is that AI services never hold write authority. AI services may produce candidate artifacts, propose changes, and surface anomalies, but they do so by writing to dedicated suggestion entities that are explicitly typed as proposals rather than as canonical truth. This boundary is enforced at the service layer through the role-scoped tokens described in Section 2.7, and it is reinforced by the AI grounding rules described in Chapter 14.

2.3.2 Why Boundary Enforcement Cannot Be Conventional

Write boundaries enforced by convention will be violated in practice. Documentation drifts, teams change, and well-intentioned developers reach for the convenient solution of writing back through a layer that should be read-only. The architectural answer is to enforce write boundaries at the service layer, using authorization scopes that are issued only to principals with documented write authority for the entity in question. The SAP Authorization and Trust Management service configuration that implements this is described in Section 2.7. The discipline that produces the scope inventory is described in Chapter 19.

2.3.3 Worked Example: Tracing a Cost Center Update Across the Architecture

The boundary discipline described here is easier to internalize when applied to a concrete scenario. Consider this routine event: The finance organization restructures a department, retiring an existing cost center and creating its replacement. The event originates in SAP S/4HANA (the system of record for finance master data) and it must propagate through the integrated landscape without violating any of the write boundaries. Tracing this event end to end illustrates how the matrix in Figure 2.2 governs runtime behavior.

The change begins inside SAP S/4HANA, where a finance master data steward submits the cost center retirement and creation through the appropriate transaction or SAP Fiori app. SAP S/4HANA is the only system in Figure 2.2 that holds write authority for the master data cost center entity. The change is recorded in SAP S/4HANA's master data tables, the change log captures the event, and the SAP S/4HANA-side audit trail records who made the change and when.

From SAP S/4HANA, the change propagates to the canonical data spine on SAP BTP through an ingestion flow, which will be described in detail in Chapter 11. The ingestion flow does not write to an SAP S/4HANA-owned entity; it reads the change event, validates it against the published canonical schema, and writes the corresponding update to the canonical master data reference table on the data spine. Note the entity boundary: The data spine does not duplicate the cost center as if it were authoritative but instead records a reference that points back to the SAP S/4HANA-authoritative version, with lineage information that names the source, the change identifier, and the time of ingestion. This is the derived relationship shown in Figure 2.2.

Downstream of the data spine, three things happen in parallel. First, SAP Signavio's process models that reference this cost center in their control intent are flagged for review. They are not modified, because Signavio is the system of record for process semantics, but they are flagged so that their references can be reviewed against the new structure. Second, any SAP SuccessFactors position whose master data entry still references the retired cost center is flagged for human resources (HR) review using the same mechanism. Third, the canonical layer publishes a new version of its master data snapshot. This version is increment-only, and the previous version is kept so that lineage can be traced.

When users access SAP Build Work Zone the morning after this change, the navigation surface reflects the new cost center structure. When Joule answers questions about cost-center allocation, it grounds its answer in the new published version. When AI agents request grounding bundles, they receive the version they were pinned to—either the new published version, if their pin allows movement, or the old version if their pin is fixed. In all cases, the audit trail records the request, the principal, the version consumed, and the time.

The structural property that makes this workflow safe is that no consumer ever bypasses the data spine, no AI service ever writes back to a system of record, and the only system that has WRITE authority for the cost center at any point in the chain is SAP S/4HANA. The matrix in Figure 2.2 makes these rules visible and enforceable, not just asserted.

2.4 Integration and Extension Responsibilities

The middle band of Figure 2.1, SAP BTP, discharges three distinct responsibilities, and these responsibilities should be kept conceptually separate even when they are physically located on the same platform. The first responsibility is integration: ingesting data from the systems of record, validating it, reconciling it across sources, and persisting it in the canonical data spine. The second responsibility is extension: building execution services and AI services that consume the data spine and deliver value to the consumption surfaces. The third responsibility is governance: enforcing lifecycle, publication, authorization, and audit across both the integration and extension activities. Let's look at each layer in a little more detail:

- **The integration layer**
  The integration layer is built on Cloud Integration, API Management, Event Mesh, and SAP Integration Suite. Its job is to move data from the systems of record into the canonical layer in a controlled, validated, and reconciled fashion. Chapter 11 describes the ingestion flows in detail. The reference architecture imposes three rules on the integration layer that appear in every flow. First, every payload must be validated against a published schema before it is accepted into the canonical layer. Second, every successful ingestion must be reconciled against the source's authoritative count or checksum. Third, every error must be observable, recoverable, and traceable to the originating record.
- **The extension layer**
  The extension layer is built on SAP BTP runtime services (SAP BTP, Cloud Foundry runtime; SAP BTP, Kyma runtime; and SAP HANA Cloud) and the Python-based AI libraries

described in Appendix G. Its job is to expose execution services and AI services that consume the canonical data spine on behalf of users and AI agents. Chapter 12 discusses the execution services, and Chapter 13 covers SAP's AI architecture. The reference architecture imposes one rule on the extension layer that is nonnegotiable: Extension services consume the canonical layer through the same governed APIs as any other consumer, with role-scoped tokens, schema-validated payloads, and audit trails. There is no internal path that bypasses these controls.

- **The governance layer**
  The governance layer is the cross-cutting set of disciplines that apply to both integration and extension. It includes lifecycle management of canonical entities, publication and approval workflows, authorization modeling, audit logging, and the reconciliation processes that detect drift between the data spine and the systems of record. Governance is not a separate platform; it is a set of policies, configurations, and operational practices that run through the integration and extension layers. Chapter 10 covers governance and auditability of the data spine in detail. Chapter 19 describes security, controls, and audit across the integrated landscape.

Integration layers fail in characteristic ways, and the reference architecture is partly designed to make those failures visible rather than to prevent them entirely. Three failure modes are common enough that any operating model should explicitly anticipate them, as follows:

- The first is schema drift. A system of record changes its outbound schema—usually because of an upgrade, a hotfix, or a configuration change made by a different team than the integration owner—and the ingestion flow continues to accept payloads that no longer match the canonical schema. The drift may be silent for weeks, during which the data spine accumulates records that pass validation but mean different things than the schema implies. The architectural defense is strict schema validation at ingestion combined with reconciliation against source counts, so that drift produces either a validation rejection or a reconciliation failure rather than silent corruption.
- The second is partial load drift. An ingestion flow runs successfully but processes only a subset of the source records due to a transient error, a paging bug, or a timing window that closed mid-flow. The data spine ends up with a state that is valid against the schema but incomplete relative to the source. Again, reconciliation against an authoritative source count is the mechanism that catches this. Without reconciliation, partial loads can persist for months until a downstream consumer notices a missing record and traces it back.
- The third is reconciliation drift, which is the failure of the reconciliation mechanism itself. The reconciliation report runs, produces a clean result, and is filed. Months later, an audit discovers that the reconciliation logic was checking the wrong fields, comparing against a stale baseline, or operating against a sampled subset rather than the full source. This is a meta-failure: The mechanism that should detect drift is itself drifting. The architectural

defense is independence between the ingestion logic and the reconciliation logic, treating them as adversarial rather than cooperative components, and periodic verification of the reconciliation logic itself against synthetic test scenarios.

A robust operating model treats each of these failure modes as a probability rather than a possibility. The reference architecture provides the structure within which they can be detected; Chapter 18 describes the operational discipline that ensures they actually are.

2.5 Execution and Consumption Layers

The bottom band of Figure 2.1 contains the consumption surfaces that provide canonical entities to users and AI agents. These are not simply user interfaces but governed surfaces that carry authentication, authorization, and audit obligations of their own. The reference architecture distinguishes four consumption surfaces with distinct roles, as follows:

- **SAP Build Work Zone as the process-aware surface**
  SAP Build Work Zone is the process-aware surface for human users. It consumes the canonical data spine to deliver role-based, process-driven navigation: a user sees the activities that their role is authorized to execute, in the order specified by the current published process version, with links to the appropriate SAP Fiori apps. Chapter 17 describes this in detail. The structural point for the reference architecture is that SAP Build Work Zone is the layer where canonical process semantics meet the user interface, and that layer must consume the canonical data spine through governed APIs rather than through duplicated configuration.
- **Joule as the conversational surface**
  Joule is the conversational surface that allows users to ask questions about processes, applications, and operational state in natural language. Joule consumes the canonical data spine through the same APIs as SAP Build Work Zone, with the additional discipline of grounding: Every Joule response must be grounded in published, validated, role-scoped artifacts. Chapter 14 covers grounding in detail. The reference architecture treats Joule as a privileged but governed consumer, not as an unconstrained query interface.
- **SAP Build Process Automation as the workflow surface**
  SAP Build Process Automation hosts the human-in-the-loop workflows that handle approvals, exception remediation, and AI-assisted recommendations. Chapter 16 covers these patterns. For the reference architecture, the salient points are that SAP Build Process Automation reads the canonical data spine to drive workflow logic, that it writes back only to its own workflow instance state, and that any update to canonical entities produced by an automation flow must transit the appropriate system of record rather than being applied directly to the data spine.
- **AI agents as first-class principals**
  AI agents are first-class principals in the reference architecture. They authenticate, they carry tokens, they are subject to authorization checks at every resource server, and their

activity is logged in the same audit infrastructure as human activity. Section 2.7 develops the identity model that supports this. The architectural point is that an AI agent is not a back channel to the data; it is a named principal with documented rights, the same as a human user.

These four consumption surfaces coexist on top of the canonical data spine, but they do not all consume the same things. Each surface has a characteristic relationship to the data spine entities, and the differences are worth calling out because they appear repeatedly in later chapters.

SAP Build Work Zone consumes the L4 and L5 process model, the application catalog, the role assignments, and the user-to-position mappings. From these, it constructs the process-aware navigation that appears to the user as a contextualized work environment. A finance user logging in sees the activities their position is authorized to execute, in the order specified by the current published process version, with deep links to the SAP Fiori apps that implement those activities. The navigation is rebuilt from the data spine on each session; it is not persisted in the user's browser or in a cache that could drift from the canonical state.

Joule reads the process semantics, the application catalog, and the dependency information needed to answer the user's question, and then constructs a grounding bundle that is sent to the AI inference layer along with the question. The grounding bundle is a snapshot of the relevant spine state at the time of the question, version-pinned and audit-logged. Chapter 14 covers the grounding discipline in detail. The architectural point here is that Joule does not consume the data spine as a whole; it consumes a specific, narrow, role-scoped view of the data spine, constructed on the fly, and persisted only as audit evidence rather than as runtime state.

SAP Build Process Automation's workflows reference the process model to know which activities have automated, which have human approval steps, and which carry AI-recommendation hooks. When a workflow executes, it reads the data spine to determine routing, applies the rules embedded in the process model, and writes back only to its own workflow instance state. The boundary between workflow execution and canonical writes—described earlier in Section 2.3—shows up concretely here: a workflow that completes does not modify an SAP Signavio-owned process model or an SAP S/4HANA-owned master data record. It produces an outcome that, if it requires changes to canonical entities, transits the appropriate system of record through a separate flow.

AI agents are scoped to specific entity types and lifecycle states, version-pinned to particular published versions, and constrained to read-only operations. The grounding citations they produce link back to specific data spine entities at specific versions, providing the traceability that distinguishes a grounded AI suggestion from an ungrounded one. Chapter 14 describes grounding in further detail, Chapter 15 covers process suggestions specifically, and Chapter 19 discusses auditing AI agent activity.

These four consumption patterns share a common discipline: Each surface consumes the data spine through a governed API, with a role-scoped token, and produces audit evidence at the time of consumption. The differences between them are differences of scope and shape, not governance.

It's important to note that these four consumption surfaces (SAP Build Work Zone, Joule, SAP Build Process Automation, and AI agents) are not the only options available on SAP BTP. We chose them because together they cover the four distinct interaction modes that an integrated landscape needs to support without any overlap or gaps.

SAP Build Work Zone covers the navigation interaction mode: The user knows what they want to do and needs to be guided to the right activity. Joule covers the inquiry interaction mode: The user has a question and needs an answer grounded in the canonical state. SAP Build Process Automation covers the workflow interaction mode: A process has been initiated and needs to be routed, approved, or remediated through a sequence of steps. AI agents cover the autonomous interaction mode: A system needs an analytical or predictive output without a human in the immediate loop. These four modes cover every interaction with the canonical state and none of them overlap.

Surfaces that the reference architecture deliberately does not include in this list, despite their availability on SAP BTP, fall into two categories:

- Surfaces that build on top of the four named ones (custom SAP Fiori apps, embedded analytics, custom dashboards) are treated as instances of SAP Build Work Zone or Joule rather than as separate surfaces.
- Surfaces that bypass the canonical layer (direct SAP S/4HANA reporting tools, ad hoc database queries, exports to spreadsheets) are deliberately excluded because they do not honor the architectural discipline that makes the rest of the architecture work. Programs that find themselves needing to surface canonical data through such a bypass should treat the requirement as a signal to extend the canonical layer or one of the four sanctioned surfaces, not as a license to introduce a fifth.

2.6 Security Architecture Principles for Process- and AI-Driven Integration

In the reference architecture, security is treated as a structural concern rather than as a set of bolt-on controls. Five principles govern the design. They appear in every section that follows and in every chapter that develops a specific layer of the architecture. We introduce them here and then elaborate on them in Section 2.7 through Section 2.11:

1. **Principle 1: Trust does not propagate inward by default**
   Each crossing between trust zones must reestablish identity, audience, scope, and version. A token that authenticated a user to the edge does not automatically authorize that user to read the canonical data spine. A service-to-service call that authenticated between two SAP BTP services does not automatically grant that service permission to query AI services. Trust zones are shown in Figure 2.3. Each zone boundary is a control point with explicit obligations.

2. **Principle 2: Identity is the primary security object**
   Authorization decisions in the reference architecture are made against authenticated identities, not against network locations or shared secrets. Every call carries a bearer token. Every resource server validates the token, extracts the principal and the scopes, and authorizes accordingly. Section 2.7 describes the identity topology in detail.
3. **Principle 3: Authorization is role-bound and entity-scoped**
   Authorization scopes are bound to roles, and roles are bound to specific canonical entities and operations. A role that is permitted to read process definitions in finance is not automatically permitted to read process definitions in HR. Section 2.8 covers the role-based exposure of the data spine in detail. The scoping discipline is the architectural mechanism that makes AI grounding workable in Chapter 14.
4. **Principle 4: Audit is part of the design, not an afterthought**
   Every operation that crosses a trust zone, every write to a canonical entity, and every consumption of an AI grounding bundle produces a tamper-evident audit record. Audit trails are not produced by sampling or by retroactive log analysis; they are a structural property of the operation. Section 2.10 describes auditability, traceability, and non-repudiation in detail.
5. **Principle 5: AI services inherit security rather than bypass it**
   AI services in the reference architecture are subject to the same trust zone boundaries, identity propagation, authorization scopes, and audit obligations as any other consumer. There is no AI fast path. Section 2.9 discusses the security boundaries between execution services, AI services, and user interfaces and explains why this principle is the most consequential one in the entire chapter.

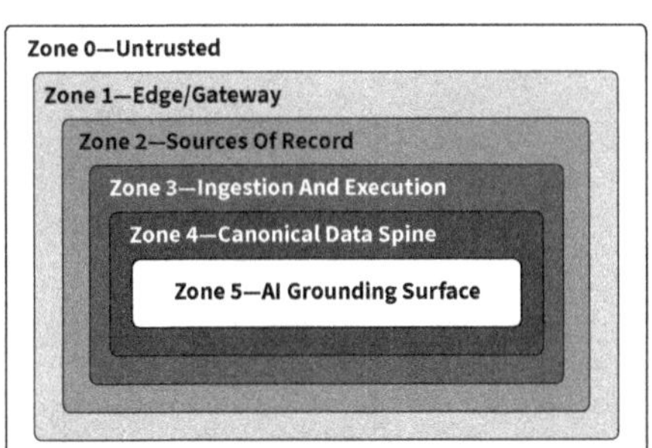

What's In Each Zone

Zone 0	Public internet, end-user devices, third-party Identity Providers (IdP)
Zone 1	API Management, Identity Authentication, BTP Routing, Web Application Firewall (WAF)
Zone 2	SAP Signavio, SAP LeanIX, SAP S/4HANA, SAP SuccessFactors, WalkMe
Zone 3	SAP BTP Integration Flows, Validation, Execution Services
Zone 4	Governed Canonical Entities, Published Versions Only
Zone 5	SAP AI Core, Joule—rEad-Only, Role-Scoped, Version-Pinned

Boundary	Control mechanism	Threat blocked at this boundary	Audit obligation
Z0 → Z1	Transport Layer Security (TLS) termination, IdP token validation	Anonymous traffic, replayed credentials	Edge access log, Identity Authentication log
Z1 → Z2	OAuth 2.0/OpenID Connect (OIDC) with audience scope	Wrong tenant, stale or escalated tokens	Token issuance log, service call log
Z2 → Z3	Service-to-service mutual TLS (mTLS)	Source spoofing, payload injection	Ingestion log with source fingerprint
Z3 → Z4	Write authority check at data spine API	Schema drift, ungoverned writes	Data spine commit log with version lineage
Z4 → Z5	Role scope and version pin on AI request	Drafts to AI, over-privileged AI calls	AI grounding log with bundle hash

Figure 2.3: Security Trust Zones

Figure 2.3 illustrates the trust zone model. It has six concentric zones, each bounded by a control point with explicit obligations. Zone 0 represents the untrusted exterior. Zone 1 is the edge, where Transport Layer Security (TLS) terminates and corporate identity is validated. Zone 2 contains the systems of record. Zone 3 is the SAP BTP ingestion and execution band. Zone 4 is the canonical data spine. Zone 5 is the AI grounding surface. Each zone boundary, shown in the table beneath the diagram, names the control mechanism, the threat blocked at the boundary, and the audit obligation that the crossing produces.

2.7 Identity, Authentication, and Authorization in SAP Signavio, SAP LeanIX, and SAP BTP

Identity is the primary security object in the reference architecture. Every call carries an authenticated principal. Every authorization decision is made against that principal, the scopes attached to it, and the entity being requested. Figure 2.4 shows the identity and authorization topology end to end, from the corporate identity provider down to the resource servers on SAP BTP and beyond. Authorization is enforced at every resource server using the scopes carried in the token. AI agents are first-class principals subject to the same SoD rules as humans.

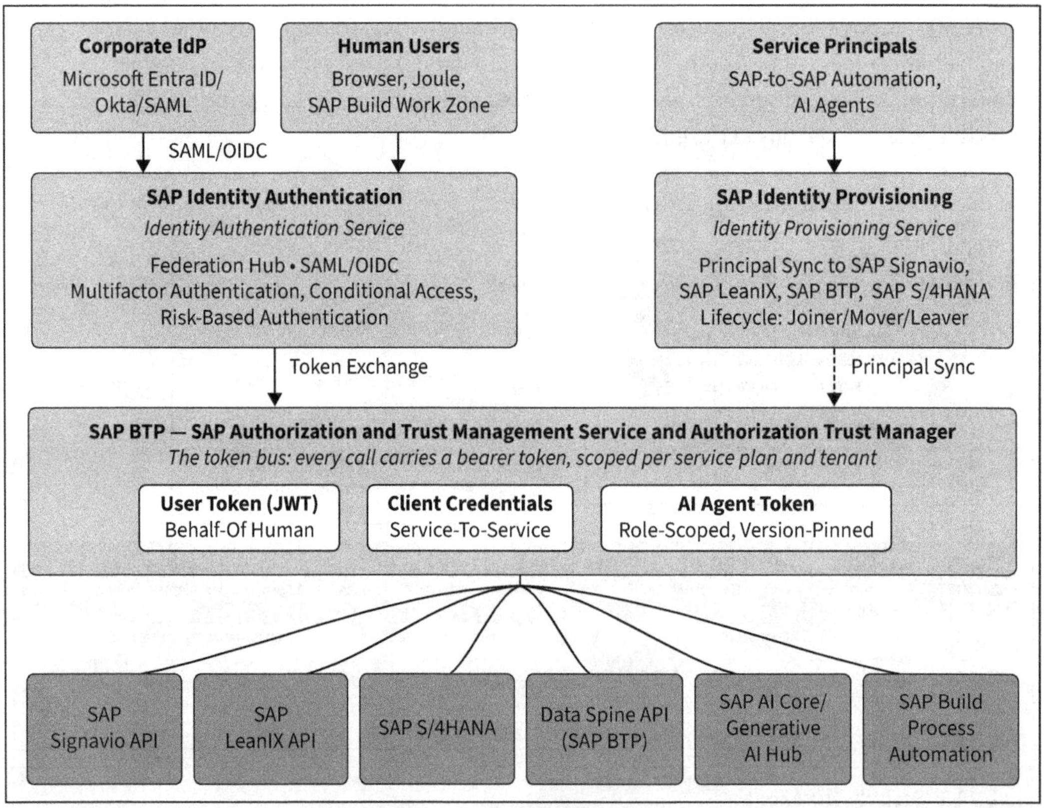

Figure 2.4: Identity, Authentication, and Authorization Topology

Let's look at each major element in turn:

- **The federation hub: Identity Authentication**
  The Identity Authentication service is the federation hub for the integrated landscape. Human users authenticate to Identity Authentication through the corporate identity provider using SAML or OpenID Connect (OIDC). Identity Authentication validates the assertion, applies multifactor authentication and conditional access policies, and issues a session that is trusted by SAP Signavio, SAP LeanIX, and the SAP BTP services. The federation pattern means that the corporate identity provider remains the source of truth for human identity; Identity Authentication is the trust anchor for the SAP-side services.
- **Principal synchronization: Identity Provisioning**
  The Identity Provisioning service is responsible for synchronizing both human users and service principals across the integrated systems. When a user is added to the corporate identity provider, Identity Provisioning provisions that user into SAP Signavio, SAP LeanIX, the SAP BTP subaccounts, and the relevant SAP Fiori catalogs. When the user moves roles or leaves the organization, Identity Provisioning reconciles the change across all subscribed systems. This lifecycle discipline is what prevents the orphan-account problem that audits consistently surface in fragmented SAP landscapes.
- **Token issue: SAP Authorization and Trust Management service**
  On SAP BTP, SAP Authorization and Trust Management service is the OAuth 2.0/OIDC authorization server. It issues the tokens that authenticated principals carry to BTP resource servers. Three token types matter for the reference architecture. User tokens (JSON Web Tokens [JWTs]) are issued on behalf of human users and carry the scopes corresponding to the user's roles. Client credentials tokens are issued for service-to-service automation, with scopes that reflect the service's operational role. AI agent tokens are a special class of client credentials tokens, scoped to the specific grounding bundles and entity types the agent is permitted to read, and version-pinned to ensure the agent reads only published artifacts.
- **Authorization enforcement at the resource server**
  Authorization is enforced at the resource server, not at the client. A consumer of the canonical data spine API presents a token; the data spine API validates the token, extracts the scopes, compares the requested entity and operation against the scopes, and either fulfills or rejects the request. The same enforcement applies for the AI grounding API, the execution service APIs, and the SAP Build Process Automation APIs. The consistency of this pattern across resource servers is what makes audit and compliance tractable. There is no resource server in the reference architecture that performs authorization differently from any other.

The token discipline described here is easier to understand with an example. A typical user token issued by the SAP Authorization and Trust Management service for a finance analyst accessing SAP Build Work Zone would carry, at a minimum, the principal identity from

Identity Authentication (the user's email or stable identifier), the tenant in which the principal is authenticated, the audience for which the token is valid (typically a specific SAP BTP service plan), and the set of scopes that name the operations the principal is authorized to perform. The scopes are the load-bearing element. In a system without scope discipline, the token would carry a role like `finance_user` and the resource servers would interpret that role differently. In the reference architecture, the scopes are explicit operations against named entity types: `spine.read.process.finance.published`, `spine.read.application.finance.published`, `audit.write.spine_consumption`, and so on.

The structural pattern is that scopes name what, not who. They do not encode the user's department, manager, or position. They encode the operations the user is authorized to perform, derived from the user's role in the corporate identity provider plus any additional bindings configured in the SAP Authorization and Trust Management service. This separation is what allows roles to be reorganized without rewriting authorization logic. When a user changes positions, their corporate identity provider (IdP) membership changes; Identity Provisioning propagates the change; the next token they receive carries different scopes; the resource servers behave accordingly. No code is changed.

For service-to-service calls, the pattern is identical, but the principal is a service rather than a human. A scheduled ingestion job that pulls data from SAP S/4HANA carries a client-credentials token whose scopes name only the operations the job needs to perform: read from a specific SAP S/4HANA OData service, write to a specific spine entity, and produce audit records. The token carries the narrow set of privileges this job requires rather than general privileges. If the job is compromised, the blast radius is bounded by the scope set, not by the platform's overall privileges.

For AI agents, the additional discipline is the version pin. An AI agent's token carries scopes such as `spine.read.process.finance.v3.7`—the v3.7 suffix is the version pin. The data spine API parses this and returns only the published artifacts at version 3.7 in the finance domain. If a new version is published while the agent is operating, the agent does not see it until either the agent's token is reissued with a new pin or the operator explicitly updates the pinned version. This produces deterministic AI behavior at the cost of explicit version management, which is a trade-off the reference architecture makes deliberately. Chapter 14 develops the version-pin discipline in detail.

The token is the architectural mechanism by which authorization decisions are made deterministic and auditable. A request without a token is rejected at the resource server. A request with a token has its authorization determined by the scopes the token carries, which are determined by the role the principal holds, which in turn is determined by the corporate IdP plus the SAP Authorization and Trust Management service configuration. Every link in this chain is auditable. This is what makes the architecture defensible in front of an external assessor.

2.8 Data Access Segmentation and Role-Based Exposure of the Data Spine

The canonical data spine on SAP BTP is, by design, a sensitive asset. It contains process semantics that describe how the enterprise operates, architecture semantics that describe what it operates with, master data references that describe whom it operates on behalf of, and adoption telemetry that describes how users actually behave inside the operating model. Exposing this asset in an undifferentiated form would be a security failure and an audit failure. The reference architecture enforces role-based exposure: Each consumer sees only the subset of the data spine that its role permits.

2.8.1 Three Dimensions of Segmentation

Segmentation operates along three dimensions simultaneously. The first dimension is functional domain: A consumer may be scoped to finance, HR, supply chain, a specific line of business, or a defined cross-functional view. The second dimension is entity type: A consumer may be scoped to processes, applications, interfaces, master data references, adoption signals, or AI suggestions. The third dimension is lifecycle state: A consumer may be scoped to published artifacts only, drafts and published artifacts, or a specific named version. The intersection of these three dimensions defines the exact subset of the data spine that a given role can see.

2.8.2 Segmentation in the Token

Segmentation is implemented in the OAuth 2.0 scopes carried by tokens. A scope of the form `spine.read.process.finance.published` communicates the functional domain, the entity type, and the lifecycle state in a single string. The data spine API parses the scope and applies the corresponding filter to the query before returning data. This pattern is uniform across all entity types in the data spine. Chapter 10 describes the scope vocabulary in detail. For the purposes of architecture, segmentation is a data access discipline encoded in tokens, not a presentation layer filter applied after the data has been retrieved.

2.8.3 Segmentation for AI Consumers

AI consumers are subject to the same segmentation discipline as human consumers, with one additional constraint: Their tokens are version-pinned. An AI agent that is scoped to `spine.read.process.finance.v3.7` reads only the published process artifacts at version 3.7 in the finance domain, and it produces grounding citations that include the version pin. This ensures that AI output is reproducible: A question asked today and the same question asked tomorrow will produce identical grounding, regardless of whether new versions have been published in the interim. Reproducibility is a precondition for audit, and version pinning is the architectural mechanism that delivers it.

2.8.4 The Scope Vocabulary in Practice

The scope vocabulary deserves a closer look because its consistency is what makes role-based exposure tractable across hundreds of operations and dozens of entity types. The vocabulary follows a deliberate four-part structure: <service>.<operation>.<entity_type>.<qualifier>, where each part has a closed set of allowed values that the architecture documents and enforces.

The service identifies the SAP BTP service that owns the operation: spine for the canonical data spine, ai for the AI grounding API, audit for audit operations, and bpa for the SAP Build Process Automation engine. The service prefix is significant because it tells the resource server immediately whether the request is even addressed to it; a token carrying only data spine scopes that arrive at the AI grounding API is rejected at the boundary, not at the operation level.

The operation names the verb: read, write, derive, publish, retire, and suggest, among others. This determines whether the operation is a state change or a query, whether it requires an additional confirmation step, and whether it produces a different audit record category. read operations produce read audits, write operations produce mutation audits with before-and-after snapshots, and suggest operations (used by AI services) produce inference audits with grounding citations.

The entity_type names what is being acted upon: process, application, interface, master_data, telemetry, suggestion, grounding_bundle, audit_record, and the canonical layer-specific lineage. Each entity type has its own specialized authorization logic and its own audit schema, so naming the type at the scope level lets the resource server route to the right enforcement path.

The qualifier is what makes the vocabulary expressive enough to handle real organizational segmentation. It can be a functional domain (finance, hr, supply), a lifecycle state (published, draft, deprecated), a version pin (v3.7), or a combination of all three. A scope of the form spine.read.process.finance.published.v3.7 communicates everything a resource server needs to know: read the data spine, fetch process entities, restricted to finance, published only, version 3.7 only. The resource server applies these as filters before returning data, with no code path that can bypass them.

Programs that adopt this vocabulary usually find within the first few weeks that most discussions center on the qualifier set. Defining the qualifier set translates organizational structure into authorization scopes. Get this right and roles map cleanly to scopes; get it wrong and the role catalog becomes a tangle of one-off entries with overlapping privileges. Chapter 19 describes the methodology for designing the qualifier set in detail. The architectural point for this chapter is that the discipline of the vocabulary is what makes segmentation operate at scale, and that scale is what makes AI grounding feasible.

2.9 Security Boundaries Between Execution Services, AI Services, and User Interfaces

Within SAP BTP, three groups of services coexist on the same platform: execution services that read the data spine to support business operations, AI services that consume the data spine to produce suggestions and explanations, and the user interface services that surface both to humans. Co-location does not imply shared trust. The reference architecture maintains explicit security boundaries between the three groups, and every cross-boundary call is subject to the same controls that apply to external calls. Let's look at the groups in more detail:

- **Execution services do not call AI services implicitly**
  An execution service that needs an AI suggestion does not invoke an AI service through an internal back channel. It issues an authenticated, scoped, audited call to the AI service through the same mechanism any external client would use. The AI service, in turn, validates the token, applies the grounding rules, and produces a result that includes the grounding citations. This discipline keeps the execution and AI lifecycles independently auditable and prevents the architectural pattern in which AI output becomes silently embedded in execution logic without explicit governance.
- **AI services do not surface directly to users**
  AI output reaches users through one of two paths: through Joule, which is the conversational surface that consumes AI services according to the rules outlined in Chapter 14, or through SAP Build Work Zone and SAP Build Process Automation, which provide AI suggestions only after they enter the suggestion entity type in the canonical data spine and acquire the appropriate approval status. There is no direct path from a raw AI inference to a user interface; every AI output is mediated by either the Joule grounding discipline or the canonical suggestion lifecycle. This is the architectural answer to the hallucination problem.
- **User interfaces do not aggregate across boundaries**
  User interfaces in the reference architecture—SAP Build Work Zone, Joule, SAP Fiori apps, custom UIs—consume the canonical data spine and the AI services through governed APIs and render the results. They do not aggregate data across security scopes, cache across token boundaries, or construct privileged views that are not available through the underlying APIs. The reason is structural: An aggregation performed in the UI is invisible to the audit infrastructure, which observes the API calls, and an aggregation that is invisible to audit cannot be governed.

2.10 Auditability, Traceability, and Non-Repudiation by Design

Audit obligations in the reference architecture are met by structural design choices rather than retroactive log collection. Every operation that crosses a trust zone, every write to the canonical data spine, every AI grounding call, and every user action that produces a state change generates a structured audit record at the time the operation occurs. The audit records are tamper-evident, time-stamped, and linked to the authenticated principal that performed the operation.

Auditability in the reference architecture has three components. The first is completeness: Every relevant operation produces an audit record, with no operation slipping through uninstrumented code paths. The second is integrity: Audit records cannot be altered after the fact, and the audit log carries cryptographic evidence of its own integrity. The third is queryability: Audit records are stored in a form that supports the operational and regulatory queries the enterprise is likely to face. Chapter 19 describes the audit infrastructure in detail, including the schema, retention policy, and evidence handling procedures.

Traceability is the ability to follow a single operation across multiple systems and trust zones. A user action in SAP Build Work Zone that triggers an execution service call, which in turn calls an AI service, which in turn reads the canonical data spine, must produce a single traceable thread that links all four operations through a correlation identifier. The correlation identifier is propagated in tokens, in HTTP headers, and in audit records. Without it, post-incident reconstruction becomes a manual log-correlation exercise that rarely produces a definitive answer in time to be useful.

Non-repudiation is the property that a principal cannot credibly deny having performed an operation. In the reference architecture, non-repudiation rests on three pillars: strong authentication of the principal at the edge, signed tokens that carry the principal's identity through every call, and tamper-evident audit records that bind the operation to the principal at the time of the call. AI agents are subject to non-repudiation in the same way that humans are. An AI agent that produces a suggestion cannot later be detached from that suggestion; the audit record names the agent, the version, the grounding bundle, and the time.

Auditability is concrete only when the audit records themselves are concrete. The reference architecture defines a uniform audit-record structure that applies to every cross-zone operation, every data spine write, and every AI inference call. The uniformity matters because the audit query patterns the organization will need—incident reconstruction, regulatory reporting, and AI explainability—operate against a single schema rather than against the schema of each individual service.

A typical spine read audit record carries seven core fields. The `principal` field names the authenticated identity that performed the operation in a stable form derived from the corporate identity provider. The `scopes` field captures the OAuth 2.0 scopes carried in the token at the time of the request. The `operation` field names the verb (read, write, suggest, and so on) and the entity type. The `target` field identifies the specific entity instances accessed, including their version pins. The `result` field records whether the operation succeeded, was rejected by authorization, or failed for other reasons. The `correlation_id` field carries the trace identifier that links this audit record to the broader request thread. The `timestamp` field records the time the operation was processed by the resource server, in a tamper-evident form.

For write operations, the record adds two further fields: the before snapshot and the after snapshot of the affected entity. These snapshots are kept as compact diffs rather than full entity copies, both to control storage cost and to make change inspection efficient. For AI inference operations, the record adds a grounding field that names the grounding bundle consumed, the version pinned, and the inference output's content hash so that the inference can be reproduced from its inputs. This last property is what makes AI activity reviewable rather than opaque.

The records are stored on the SAP BTP audit infrastructure, with retention policies that match the organization's regulatory environment. Chapter 19 describes the retention policy, access control on audit records, and queryability. The architectural point for this chapter is that audit records are not log entries that happen to mention what occurred; they are structured artifacts that can be queried, joined, and aggregated to answer specific operational and regulatory questions. The shift from logs to records is what makes the architecture defensible.

2.11 Allocation of Security Responsibilities Across Platforms

The reference architecture spans multiple platforms, and the security responsibilities are allocated explicitly. Each platform has obligations it must discharge for the overall architecture to be secure. Treating the integrated landscape as if security were the responsibility of any single platform produces gaps that adversaries—and auditors—find first. This book uses the following allocation throughout its entirety:

- **SAP Signavio responsibilities**
  SAP Signavio is responsible for authentication of users entering the modeling environment, for role-based authorization of modeling actions (who can edit, who can publish, who can approve), for the integrity of process artifacts under version control, and for the audit log of modeling and publication activity. SAP Signavio is not responsible for downstream consumption controls; once an artifact is published, the data spine and the consuming services are responsible for governing its consumption.
- **SAP LeanIX responsibilities**
  SAP LeanIX is responsible for the authentication and authorization of users entering the architecture management environment; for role-based control over which users can edit applications, interfaces, and lifecycle states; and for the audit log of architecture changes. As with SAP Signavio, SAP LeanIX is not responsible for downstream consumption; it produces authoritative artifacts and the rest of the architecture governs their use.
- **WalkMe responsibilities**
  WalkMe is responsible for the integrity of telemetry capture, the privacy controls that govern which user interaction signals are collected and at what granularity, and the secure transmission of telemetry to the SAP BTP ingestion endpoints. Privacy considerations are particularly consequential here because adoption telemetry can, if misused,

expose individual user behavior in ways that violate employment law or local data protection regulation. Chapter 7 covers the governance discipline that protects against this.

- **SAP BTP responsibilities**
  SAP BTP holds the largest set of security responsibilities. It is responsible for ingestion validation, canonical layer integrity, service layer authorization, OAuth 2.0/OIDC token issuance and validation through the SAP Authorization and Trust Management service, audit logging across every cross-zone operation, the role-based exposure of the data spine described in Section 2.8, and the boundaries between execution, AI, and UI services described in Section 2.9. SAP BTP is the platform on which the reference architecture's security model is most concretely instantiated and where most of the operational security work occurs.
- **Consuming application responsibilities**
  SAP Build Work Zone, Joule, SAP Build Process Automation, and AI agents are responsible for consuming the canonical data spine solely through governed APIs, honoring the scopes carried in their tokens, refusing to aggregate across security boundaries, and producing audit records of their own user-facing activity that complement the audit records produced by the underlying APIs.
- **Joint responsibilities**
  Two responsibilities are jointly held. Identity lifecycle management is shared between the corporate identity provider, Identity Authentication, and Identity Provisioning. Audit consolidation is shared between every platform that produces audit records and the audit infrastructure on SAP BTP that aggregates them into a single queryable trail. Joint responsibilities require explicit operational agreements, which Chapter 18 covers in detail. Without those agreements, the boundaries become easy to dispute, and disputable boundaries are exactly what this allocation is designed to prevent.

An allocation that lives only on a slide is an allocation that will not survive the first incident. The reference architecture treats the allocation in this section as the starting point for a small set of operational artifacts that translate it into runtime behavior. Each artifact names a responsibility owner, a corresponding consumer, and an evidence pattern that demonstrates the responsibility is being discharged.

The first artifact is a service catalog that names every operation each platform exposes, the principal that is authorized to invoke it, and the audit category produced. The catalog is maintained by the platform owner, reviewed quarterly by the security architect, and consumed by the audit team during evidence-gathering exercises. Chapter 19 describes the schema for this catalog and the publication discipline.

The second artifact is a set of joint operational agreements between platforms whose responsibilities interact. The corporate identity provider and Identity Authentication hold such an agreement, covering principal onboarding latency, token revocation propagation, and break-glass procedures. SAP BTP audit infrastructure and the consuming platforms hold an agreement covering audit ingestion frequency, schema versioning, and retention.

These agreements name owners on both sides and specify what evidence each side produces to demonstrate the agreement is being honored.

The third artifact is the regular review cadence that verifies the allocation has not drifted. New consumers appear; old ones retire; platform versions update; regulatory requirements shift. A quarterly review of the allocation against actual operational state is what keeps the documented architecture aligned with the running architecture. Without this cadence, allocation documents become artifacts of the past rather than descriptions of the present, which is the most common failure pattern in any governance regime.

Chapter 18 describes these artifacts in operational detail. For the reference architecture, the relevant point is that the allocation is the source of an explicit, observable, and reviewable operating model rather than an aspirational diagram. Architectures that fail to make this transition tend to fail in audit. Architectures that make it tend to scale across programs without losing coherence.

2.12 Summary

This chapter described the reference integration architecture that the rest of the book configures, integrates, and governs. The architecture is organized as a four-band logical view: systems of record at the top, SAP BTP in the middle (organized internally into ingestion, the canonical data spine, execution services, and AI services), consumption surfaces at the bottom, and an identity and audit rail running across all of them. The chapter named the five systems of record—SAP Signavio, SAP LeanIX, SAP S/4HANA, SAP SuccessFactors, and WalkMe—and specified the entities each one authoritatively writes. It introduced the data ownership and write boundaries matrix as the artifact that records who is permitted to write what, and explained why those boundaries must be enforced at the service layer rather than by convention.

The chapter also established security as a first-class architectural concern. It introduced the trust zone model shown in Figure 2.3, the identity and authorization topology shown in Figure 2.4, and the segmentation discipline that implements role-based exposure of the canonical data spine through OAuth 2.0 scopes. It defined the security boundaries between execution services, AI services, and user interfaces, and specified the allocation of security responsibilities across SAP Signavio, SAP LeanIX, WalkMe, SAP BTP, and the consuming applications. It treated auditability, traceability, and non-repudiation as structural properties of the architecture rather than as retroactive controls.

The chapters that follow develop these architectural commitments into implementation discipline. Chapter 3 and Chapter 4 show you how to configure SAP Signavio for execution and AI readiness. Chapter 5 and Chapter6 help you configure SAP LeanIX for dependency and context data. Chapter 7 and Chapter 8 cover WalkMe configuration for process-aware adoption and telemetry. Chapter 9 and Chapter 10 discuss the SAP BTP canonical data spine in detail, including the table designs, lineage strategy, and governance mechanisms that this chapter only introduced. Chapter 11 and Chapter 12 describe the integration flows and execution services. Chapter 13 through Chapter 15 cover the AI architecture, data preparation, and process suggestions. Chapter 16 and Chapter 17 walk through the SAP Build consumption surfaces. Chapter 18 and Chapter 19 cover operations, security, and audit.

2.12.1 Key Learning Points

- The reference architecture is a four-band logical view: systems of record, SAP BTP organized as ingestion/canonical data spine/execution services/AI services, consumption surfaces, and a cross-cutting identity and audit rail.
- Each band has a different governance regime. Treating the integrated landscape as a single platform collapses governance regimes that must remain distinct in order to scale and audit.
- Five systems of record—SAP Signavio, SAP LeanIX, SAP S/4HANA, SAP SuccessFactors, and WalkMe—hold authoritative writes for non-overlapping sets of entities. Reference, not duplication, is the architectural rule for linking them.
- Every canonical entity has exactly one writer. SAP BTP holds write authority only for entities that are intrinsic to the canonical layer itself. AI services never hold write authority. These boundaries are enforced at the service layer through OAuth 2.0 scopes, not by convention.
- Trust does not propagate inward by default. Each crossing between trust zones must reestablish identity, audience, scope, and version. The trust zone model is the structural answer to defense in depth.
- Identity is the primary security object. Every call carries a bearer token. Every resource server validates the token, extracts scopes, and authorizes accordingly. AI agents are first-class principals, subject to the same authorization discipline as humans.
- Role-based exposure of the canonical data spine operates along three dimensions: functional domain, entity type, and lifecycle state. AI consumers carry version-pinned scopes, which is what makes AI output reproducible and auditable.
- Audit, traceability, and non-repudiation are designed in, not layered on. Every cross-zone operation produces a tamper-evident audit record bound to an authenticated principal and propagated through a correlation identifier.
- Security responsibilities are allocated explicitly across platforms. The boundaries are documented, the joint responsibilities have operational agreements, and the result is an architecture in which no responsibility is deniable.
- The four sanctioned consumption surfaces—SAP Build Work Zone for navigation, Joule for inquiry, SAP Build Process Automation for workflow, and AI agents for autonomous interaction—cover the four distinct interaction modes an integrated landscape needs to support. Surfaces that bypass the canonical layer are deliberately excluded from the architecture and signal a need to extend one of the four rather than introduce a fifth.

- Integration layers fail in characteristic ways: schema drift, partial-load drift, and reconciliation drift. The reference architecture is designed to make these failures visible rather than to prevent them entirely, through schema validation, source reconciliation, and independence between ingestion and reconciliation logic.
- The OAuth 2.0 scope vocabulary in the reference architecture follows a four-part structure (`service`, `operation`, `entity_type`, `qualifier`) that translates organizational structure into authorization decisions. The qualifier set is where the organizational specifics enter the architecture, and getting it right is the act that makes role-based exposure scale.
- Audit records in the reference architecture are structured artifacts, not log entries. Each carries a principal, scope, operation, target, result, correlation identifier, and timestamp. Write operations carry before-and-after snapshots; AI inference operations carry grounding citations. The shift from logs to records is what makes the architecture defensible in audit.

2.12.2 Common Design Pitfalls

- Treating SAP BTP as middleware rather than as an execution fabric with three distinct responsibilities (integration, extension, and governance). The result is collapsed governance regimes and a platform that cannot scale across programs.
- Allowing two systems to claim concurrent write authority for the same entity. This is the most common cause of canonical drift. The architectural response is to identify the underlying structural distinction and split the entity, not to permit shared writes.
- Implementing write boundaries by convention, with documentation rather than service-layer authorization. Documentation drifts. Authorization scopes do not.
- Granting AI services privileged or back-channel access to the canonical layer. AI must consume through the same governed APIs as any other consumer, with role-scoped, version-pinned tokens.
- Aggregating data in user interfaces across security scopes. Aggregations performed in the UI are invisible to the audit infrastructure and ungovernable in practice.
- Treating the audit as a retroactive log-collection exercise. Audit records must be produced as a structural property of every cross-zone operation, in real time, by the operation itself.
- Failing to allocate security responsibilities explicitly across platforms, leaving joint responsibilities disputable. Disputable boundaries are exactly what the allocation is designed to prevent.
- Treating the allocation document as a one-time deliverable rather than an operational artifact that is reviewed quarterly. The running architecture constantly drifts from the documented one, and only regular verification keeps the two aligned.

- Introducing a fifth consumption surface to satisfy a specific team's request for direct access, rather than extending one of the four sanctioned surfaces or extending the canonical layer itself. Each unsanctioned surface becomes a place where governance discipline is harder to enforce in the next audit cycle.

2.12.3 Review Questions

1. Identify the four bands of the reference architecture and describe the governance regime that applies to each. Why must these regimes remain distinct?
2. Choose any canonical entity from Figure 2.2 and walk through the row: Which system writes it, which systems read or derive it, and which systems are forbidden access? Justify the allocation in architectural terms.
3. Explain why write boundaries cannot be enforced by convention. What is the mechanism the reference architecture uses instead, and where in the SAP BTP stack is it implemented?
4. Trace a request from a human user in SAP Build Work Zone to the canonical data spine, naming the trust zones it crosses, the controls applied at each crossing, and the audit records produced.
5. What additional constraint applies to AI agent tokens that does not apply to user tokens, and why is that constraint necessary for AI output to be auditable?
6. Describe the three dimensions along which the canonical data spine is segmented for role-based exposure, and explain how the OAuth 2.0 scope vocabulary encodes the intersection.
7. If the corporate identity provider is the source of truth for human identity, what role does Identity Authentication play, and why is it not redundant?
8. Identify the joint security responsibilities in the allocation specified in Section 2.11, and explain why they require explicit operational agreements.

2.12.4 Student Companion Notes

This chapter is the architectural reference for the entire book. We recommend four exercises for understanding reference architecture in practice.

First, take Figure 2.2 and apply it to your own organization. Replace the row labels with the canonical entities your transformation program actually depends on, and replace the column headers with the systems of record that exist in your landscape. This exercise is most valuable when you discover rows with two writers, columns with no write operations, or cells whose current behavior contradicts what the matrix says they should be. Each contradiction is a place where governance is currently deniable.

Second, sketch the trust zone model from Figure 2.3 for your own landscape. Name the systems that sit in each zone, describe the controls currently in place at each boundary, and identify the boundaries where the controls are weak, missing, or undocumented. The result is a security gap analysis grounded in the reference architecture rather than in a generic framework.

Third, walk through the identity flow from Figure 2.4 for a single concrete use case in your organization—a finance approval, an HR position change, or a procurement event—naming the corporate identity provider, the Identity Authentication configuration, the Identity Provisioning provisioning, and the SAP Authorization and Trust Management service scopes that would be issued. This exercise is most valuable when you identify the place where the current identity flow leaves a gap in audit evidence, because that gap is the one your AI rollout will expose first.

Fourth, take the worked example in Section 2.3—the tracing of a cost-center update across the architecture—and rewrite it for an entity that matters in your own organization. The natural candidates are master data entities that are written by one system, referenced by many others, and currently propagated through ad hoc integrations. As you write the trace, name the system that holds `WRITE` authority, the systems that hold `READ` or `DERIVED` rights, and the consumption surfaces that depend on the entity. The places where you have to add qualifications such as *usually* or *it depends* are where the architectural discipline is weakest in your landscape.

The chapters that follow build on this reference architecture in detail. Chapter 3 opens Part II by establishing the SAP Signavio data model required to support downstream integration and AI.

PART II

Configuring SAP Signavio for Execution and AI Readiness

Chapter 3

SAP Signavio Data Model for Integration

A process model is not only a picture but a contract. Every shape on the canvas is a promise to a downstream system, an auditor, an AI agent, and a human employee. The data behind the shape is what makes the promise enforceable.

This chapter establishes the data model that turns SAP Signavio from a process documentation tool into the system of record for executable process semantics. It is the longest chapter in Part II because every chapter that follows—variants, KPIs, AI grounding, the WalkMe and SAP LeanIX bridges, the BTP integration spine—depends on the discipline established here. We begin with the L4 and L5 object requirements that govern what can be modeled in the first place. We then walk the process and activity identifier standard, the seven mandatory execution attributes that turn a shape into an integration contract, the role and variant taxonomy, the lifecycle and versioning state machine, and finally the boundary discipline, which defines what must never be modeled in SAP Signavio because authority for that content rightfully resides in another system.

There is a temptation in any large transformation to treat the process modeling tool as a place to put everything that has not yet found a home. SAP Signavio becomes the default choice for architects who have not yet identified the right system for application metadata reach, program managers who do not yet have a runtime telemetry tool, and AI teams who have not yet decided where prompt templates should live. The result is a process repository that has become a junk drawer. It looks complete. It is in fact unenforceable. None of the downstream systems can read from it reliably, none of the auditors can rely on it, and none of the AI agents can be safely grounded against it. That's why this chapter outlines the discipline of saying no.

The data model presented here is selective. It is not a list of every SAP Signavio attribute that exists; it is the minimum sufficient set required for a process model to function as the integration contract for an SAP-led transformation. Every attribute we mandate earns its place by being read by another system at runtime, an AI grounding bundle at design time, or an auditor at review time. Attributes that none of those three use have been deliberately excluded. A process model burdened with attributes no one reads will eventually stop being maintained, and a process repository that is not maintained becomes an active source of risk rather than a control.

Where the Process Model Begins

Process modeling does not begin in SAP Signavio. It begins in presales, in the hours of conversation that a digital discovery assessment (DDA) executes between an industry-experienced practitioner and an organization's leadership team. The DDA produces three artifacts that are the source for every later SAP Signavio model: an industry process benchmark report that names the L1 and L2 domains in scope, a quantified value hypothesis that anchors key performance indicators (KPIs) to L3 processes, and a scope proposal process tree that determines which L4 steps will receive modeling effort during delivery. Organizations we have seen succeed treat these artifacts as the basis of the SAP Signavio prepare phase. Organizations who fail treat SAP Signavio as the starting point and spend three months recreating, by interview, what the DDA could have produced in two hours.

We will return to this point in Section 3.5 when we discuss the digital thread that connects DDA to Business Process Model and Notation (BPMN) diagrams to runtime. For now, the operational implication is simple: The data fields described in this chapter are not authored from a blank canvas. Every L4 step we model in SAP Signavio inherits its identifier, its KPI anchor, and its initial role allocation from a DDA artifact, and the modeler's job is to govern, refine, and version the inheritance, not to invent it.

3.1 L4 and L5 Process Object Requirements

In this book, the hierarchy that governs SAP Signavio spans five levels. L1, L2, and L3 are organizational. They describe the shape of the enterprise, the major process domains it operates in, and the named processes within those domains. L4 is the level at which integration begins, because it is the first level where individual steps are visible, attributable to a role, and capable of being mapped to an SAP Fiori app or a transaction code. L5 is the level at which the work instruction lives. It involves the narrative or procedural documentation an end user reads when they are uncertain how to perform a step.

The reason this distinction matters for an integration architecture book is that no integration occurs at L1, L2, or L3. A capability called "Finance" cannot be wired to an SAP Fiori app. A process group called "Record-to-Report" cannot trigger a telemetry event. Only L4 steps can be wired to systems and only L5 instructions can be retrieved by an AI assistant. Time spent modeling L1–L3 in elaborate detail produces a beautiful map with no execution surface; time spent modeling L4 with discipline produces an integration contract that compounds value across every subsequent chapter of this book.

Figure 3.1 shows an example process hierarchy in SAP Signavio. This is an SAP Signavio Process Explorer view for the manufacturing industry, in the finance L1 domain. L4 and L5 are highlighted as the integration depth, meaning the levels at which SAP Signavio carries data that other systems read.

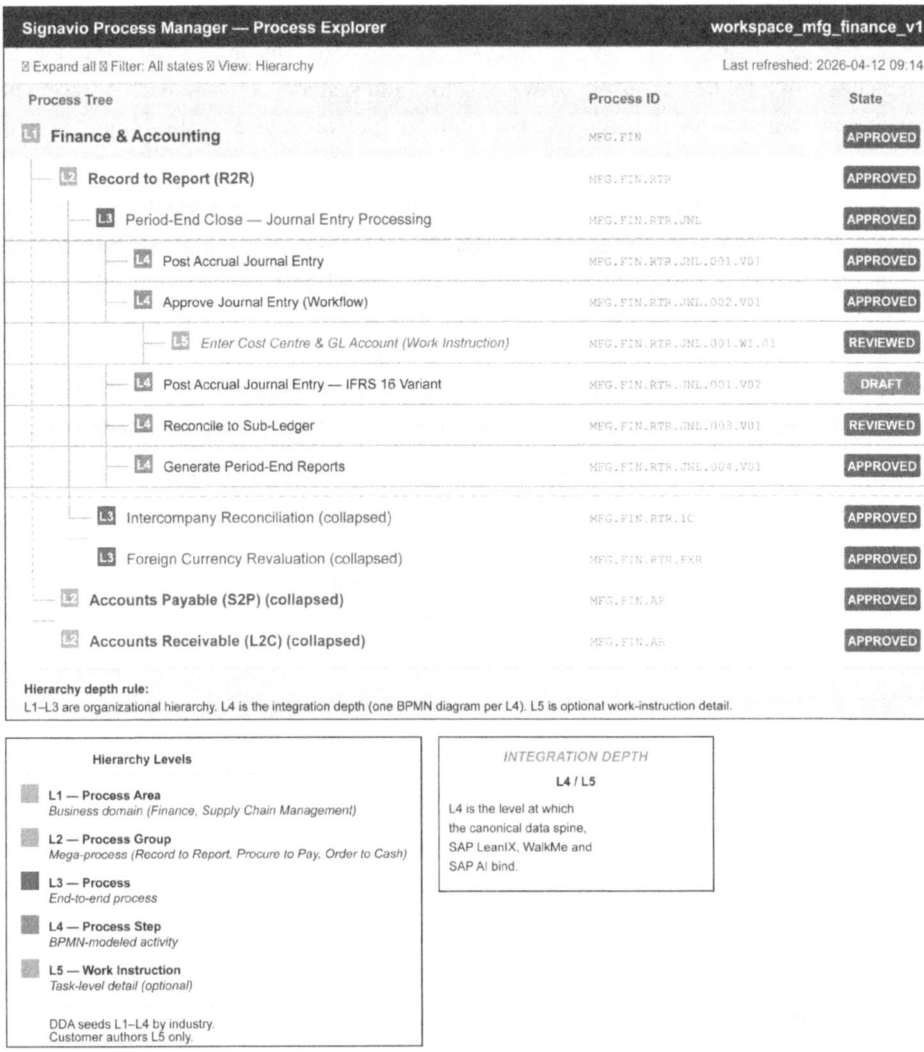

Figure 3.1: Process Hierarchy from L1 Capability to L5 Work Instruction

In the following sections, we'll introduce you to L4 process objects, L5 work instruction objects, and DDA as the source for L4 steps.

3.1.1 L4 Process Step Objects

An L4 step is the smallest unit of work that has a name, an owner, and a system in which it executes. It is the level at which the process model becomes operational. An L4 step in your framework must satisfy four mandatory criteria, and any candidate step that fails any one of them is either not an L4 step or is being modeled at the wrong granularity.

First, an L4 step must be performable by a single role within a single system in a continuous span of attention. "Review the invoice and either approve it or send it back for clarification" is an L4 step because the reviewer holds the task in attention throughout. "Receive the invoice, review it, approve it, post it to the ledger, and notify the vendor" is four or five L4 steps because attention transfers between actors and systems between each phrase. The test is operational rather than linguistic: If a different person, role, or system would pick up the work at any point, the step ends there.

Second, an L4 step must produce or consume a data object that an auditor or an integration would care about. A step that has no input and no output is not a step but rather a heading, a narrative aside, or a step whose data interaction has been omitted from the model. Data objects in this sense include documents (purchase order, invoice, journal entry), decisions (approval, rejection, reclassification), and notifications (email to vendor, alert to controller). The key is that the object has identity outside the step, persists somewhere after the step ends, and could be referenced by a later step or a downstream system.

Third, an L4 step must be triggerable. There must be an upstream event that causes the step to begin, and that event must itself be modeled. The trigger may be a preceding step, a timer, an external message, or a user action, but it must exist on the canvas. Steps without a trigger are most often editorial leakage from a Word document being copied into BPMN, and they create a false impression of completeness. At any step, a reviewer reading the model should be able to ask, "What happened just before this?" and find the answer on the canvas.

Fourth, an L4 step must be referenceable by a stable identifier that survives renaming and refactoring. We will discuss the identifier rules in detail in Section 3.2; the requirement at this stage is that an L4 step is an entity with identity, not a label. When the controller of the record-to-report process renames a step from "Book Journal" to "Post Journal Entry" two years from now, the integration to SAP S/4HANA must not break. Identity carries the integration; the label is for human readers.

3.1.2 L5 Work Instruction Objects

L5 is the level at which work instruction lives. An L5 object is attached to its parent L4 step as a child element and contains the narrative, screenshots, decision logic, or procedural reference that an end user needs in order to perform the step correctly. In an SAP-led transformation, L5 is also the level at which Joule's in-application assistance is grounded, WalkMe tooltips draw their content, and an auditor reads to confirm that a process is performable as designed.

L5 objects in this framework satisfy three criteria:

- First, they are scoped to a single L4 step. An L5 instruction that spans steps is a sign that the parent step has been modeled too coarsely and should be decomposed.
- Second, they are role-scoped. An L5 instruction tells one role how to perform the work, not all the roles involved.
- Finally, they are version-locked to their parent. When the parent L4 step is versioned, the L5 instruction is reviewed and either revised, retired, or carried forward by explicit decision.

L5 instructions that drift out of sync with their parent step are the single largest source of audit findings in process repositories we have reviewed.

An L5 instruction can take many physical forms in SAP Signavio. Some are text-based narrative paragraphs, some are embedded screenshots, some are decision tables, and some are short videos. The decision between forms is editorial and is governed by what its audience needs. The data model treats them all identically: Each one is a child object of the parent L4 step, carries the same metadata fields, and is governed by the same lifecycle.

3.1.3 Digital Discovery Assessment as the Source for L4

We promised to return to the question of where the model begins. This section covers the operational answer for L4 specifically. For every L3 process in scope, a typical DDA produces a candidate list of L4 steps drawn from an SAP industry best-practice repository, annotated with the organization's expected variations and the SAP system that will execute each step. These candidate steps are not the final model but the starting point against which the organization's process experts validate, refine, and deviate from. However, they should still be high quality: industry-correct, system-aware, and premapped to SAP Fiori apps and transaction codes for the corresponding SAP S/4HANA release.

Organizations who use the DDA in this way enter the SAP Signavio prepare phase with 60 to 70 percent of their L4 model already drafted and spend their modeling effort on the 30 to 40 percent that genuinely differs from the industry standard. Organizations who treat SAP Signavio as a blank canvas spend the same effort recreating the 70 percent that already existed in the DDA, lose their value-driven differentiation in the noise of generic process documentation, and arrive at the realize phase with a model that contains everything except the few things that actually differentiate their business. The disciplines outlined in this chapter—immutable identifiers, mandatory attributes, governed variants—are hardest to establish on a blank-canvas model and easiest to establish on a DDA-based one.

There is also a governance argument for DDA that is easy to overlook. When the source is industry standard, every deviation from it is conscious, documented, and traceable to a specific business reason. The variant tree we will discuss in Section 3.4 becomes meaningful because a variant's classification (mandatory, value-driven, or legacy) has a nontrivial baseline to deviate from. When the canvas is blank, everything looks like an intentional design choice and nothing looks like a deviation. The variant tree degrades into a decorative attribute, and the $BPMN(t) = BPMN(DDA) + \Sigma\Delta$ equation we will introduce in Section 3.5.3 reduces to $BPMN(t)$ = whatever the modeler drew on a particular Tuesday afternoon.

3.2 Process and Activity Identifiers

An identifier is a contract with the future. The day a process step receives an identifier in SAP Signavio is the day every other system that will ever reference that step makes an assumption about how the identifier will behave. SAP LeanIX assumes the identifier will not change when the step is renamed. WalkMe assumes the identifier will resolve to the same step three releases from now. Joule grounding bundles assume the identifier uniquely names the step across the entire enterprise, with no collisions between domains. An audit assumes the identifier names a single approved version of a single step, not a family of similar steps that have evolved together. If any of these assumptions is broken by silent renaming, reuse, or duplication, the integration fails silently. There is no compiler error for a broken process identifier. There is only a Joule answer that cites the wrong step, a WalkMe tooltip that slowly migrates into the wrong context, and a regulator who notices, six months too late, that the control evidence does not match the control description.

In the following sections, we will walk you through the standards for SAP Signavio process identifiers, discuss when they become immutable, and show how they are mapped to your system.

3.2.1 The Identifier Standard

We use a six-segment identifier separated by periods, of the form Industry.L1.L2.L3.Step.Variant. Each segment is a fixed-length code drawn from a controlled vocabulary, and the full identifier is read left to right as a path through the process hierarchy. The example used throughout this book, MFG.FIN.RTR.JNL.001.V01, names the first variant of the journal entry posting step within the record-to-report process group within the finance domain within the manufacturing industry. The example is deliberately verbose; in production it functions as a database key.

Figure 3.2 shows the anatomy of an SAP Signavio process identifier. The six segments (industry, L1 domain, L2 group, L3 process, L4 step, and variant) are drawn from controlled vocabularies and read left to right as a path through the hierarchy.

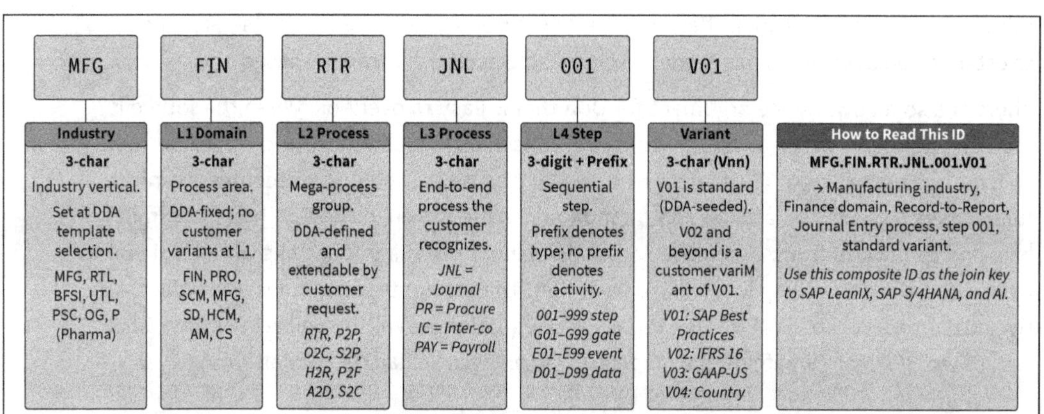

Figure 3.2: Anatomy of a Process Identifier

The industry segment uses a three-letter code drawn from a registered list maintained at the enterprise level: MFG for manufacturing, RTL for retail, BSFI for banking, UTL for utilities, and so on. Most enterprises operate in a single industry and never expose the segment to end users, but it exists because acquisitions, divestitures, and joint ventures do happen, and a process model that has assumed a single industry context will fracture the day a different one is introduced. The cost of carrying the segment from day one is trivial; the cost of retrofitting it later is enormous.

The L1 domain segment uses a three- or four-letter code from a fixed list of eight: FIN, PROC, SD, MFG, SCM, HCM, AM, and CS. These eight correspond to the eight L1 KPI frameworks that anchor every benefit case in the book: finance, procurement, sales and distribution, manufacturing, supply chain, human capital, asset management, and customer service. The list is deliberately closed. Domains outside the eight are either subdomains of one of them (treasury is part of finance, quality is part of manufacturing) or are not in scope for the integration architecture that this book describes.

The L2 group segment uses a three-letter code drawn from each L1 domain's list of process groups. RTR (record-to-report), OTC (order-to-cash), PTP (procure-to-pay), and HTR (hire-to-retire) are the canonical examples. These codes are stable enough that they appear in SAP's marketing material, in the SAP industry best-practice repository that forms the basis of the DDA, and in most enterprise process catalogs that predate SAP S/4HANA. We adopt the canonical codes wherever they exist and deviate only in clearly-named circumstances, which we record in the enterprise process register.

The L3 process segment is a three-letter abbreviation of the named process within its L2 group. Within RTR, JNL is journal posting, REC is reconciliation, ACL is account closing, and REP is reporting. These segments are determined by the organization rather than inherited, because the names of L3 processes vary meaningfully between enterprises in ways that L1 and L2 do not. We commit them in the enterprise process register, version the register itself, and treat the L3 vocabulary as a first-class architectural artifact rather than an editorial convenience.

The step segment is a three-digit zero-padded number starting at 001 and incrementing within an L3 process. The numbers are not semantic. Step 001 is not necessarily the first step in execution order, nor the most important step. It is simply the first step that happened to be modeled in this L3 process. Non-semantic numbering allows the execution order to change. If a regulator demands a new control step, an automation removes a manual step, or a process redesign swaps two steps, a numbering scheme that was meaningful the day it was assigned becomes misleading the day it is broken. Non-semantic numbers do not lie when the process evolves.

The variant segment is V01 for the parent variant and V02, V03, etc., for siblings. Variants are full citizens of the data model and we discuss them in detail in Section 3.4. For identifier purposes, the rule is that V01 is the parent and every later variant inherits the first five segments and differs only in the sixth. This rule is important because it means a search for "MFG.FIN.RTR.JNL.001.*" returns the complete family of variants of the journal entry posting step, which is what every downstream system needs to render the right variant for the right country, channel, or business unit at runtime.

Forbidden Patterns

In addition to the rules on identifier renaming (Section 3.2.2) and reuse (Section 3.5), we'd like to call attention to some additional patterns that should not be used, as they'll break integration:

- **Free-text suffixes**
  Don't use "_v2", "_old", "_FINAL", "_copy". The variant slot Vnn is the only mechanism for derivative versions.
- **Skipping segments**
  All six segments are mandatory. "FIN.RTR.JNL.001" is invalid—it has no industry and no variant. AI grounding refuses such records.
- **Display name drift**
  Display names may change with translation or rebranding, but the ID stays. Treat the ID as the contract; the name is presentation.

3.2.2 Immutability and Renaming

An identifier is immutable as soon as its owning step reaches the **Approved** state. This is the single most important rule in the chapter, and it is the rule most often violated by modelers who do not yet understand that the identifier is a contract. The temptation to rename is constant: A step may be renamed because its label improved, a regulator changed the terminology, or a consultant suggested a clearer name. Yet each rename applied to the identifier severs the integration. The SAP Fiori app that was launched from the old identifier no longer resolves. The WalkMe tooltip that pointed at the old identifier no longer matches. The Joule grounding bundle that contained the old identifier returns nothing.

The discipline is to separate the identifier from the label. The identifier is fixed at the moment of approval and may never change. The label is the human-readable name displayed on the BPMN canvas and can be changed at any time, with the appropriate approval, without affecting the identifier. Reviewers and editors interact with the label; integrations and audits interact with the identifier. The two are kept deliberately distinct. Modelers who insist that the identifier should reflect the current label are welcome to rename steps that are still in **Draft** state; the moment a step is **Approved**, the identifier is locked.

There are two narrowly defined exceptions to identifier immutability, both involving the lifecycle states we will discuss in Section 3.5. Once a step is retired and moved to **Retired** state, a new step with a new identifier can be created in its place. The original identifier is preserved in the historical record but no longer resolves at runtime. Alternatively, a step can be split into multiple successor steps, each with a new identifier, with the original identifier preserved as the ancestor. In both cases, the identifier is not renamed; it is retired, and a new identifier is created. The integration contract is broken in a controlled way, with explicit downstream notification.

3.2.3 Identifier-to-System Mappings

The process identifier is the join key for every cross-system integration we describe in this book. In the SAP Signavio dictionary, each L4 step's identifier is associated with one or more execution system references: an SAP Fiori app ID for the UX that executes the step, a transaction code for the legacy SAP GUI fallback, an SAP LeanIX application reference for the architectural layer the step traverses, and, where applicable, a WalkMe content ID for the runtime guidance. These mappings are data attributes of the L4 step in SAP Signavio; we discuss them in Section 3.3 under mandatory execution attributes. The point at this stage is that the identifier is what each of those references points back to. Without the identifier, the references degrade into copy-pasted strings that drift out of sync with their underlying steps. With the identifier, the references form a star schema with the process step at the center and every other system's metadata as a referenced spoke.

Figure 3.3 shows an SAP Signavio dictionary view. The dictionary is the canonical register of all L4 steps with their identifiers, owners, SAP Fiori app IDs, transaction codes, and SAP LeanIX application references. Other systems join to the dictionary by identifier, never by name, because identifiers are immutable. A retired ID (e.g., **MFG.FIN.AP.PR.999.V01**) is never reused. It remains in the dictionary as audit history. The variant suffix (.V02) preserves parent linkage. Organization-specific deviations are tracked rather than hidden. The **Fiori App ID** and **LeanIX App** ID columns drive the canonical data spine join keys.

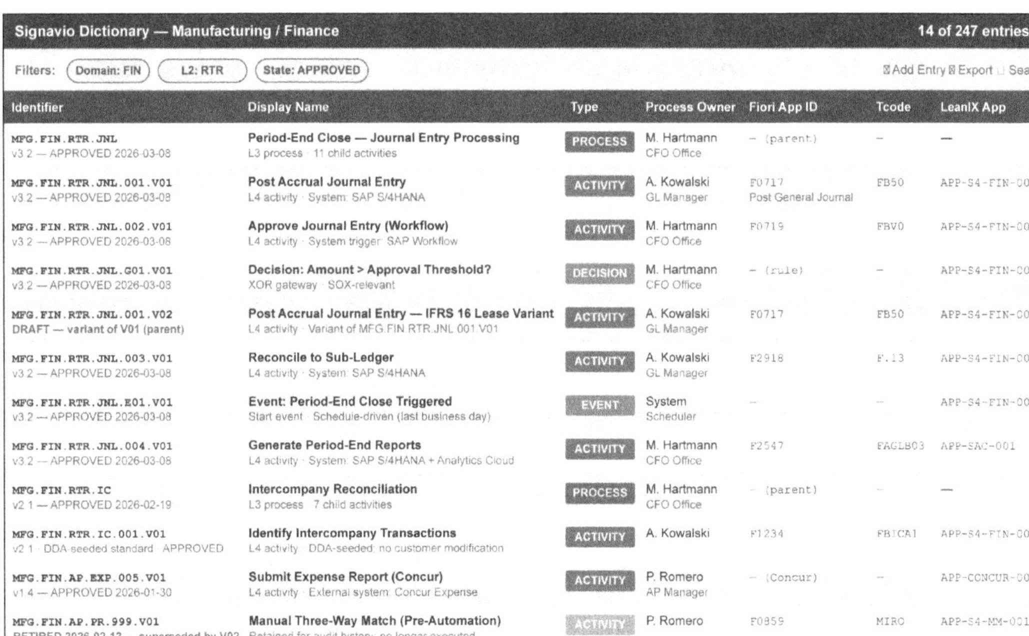
Signavio Dictionary — Manufacturing / Finance — 14 of 247 entries

Filters: Domain: FIN | L2: RTR | State: APPROVED — Add Entry, Export, Searc

Identifier	Display Name	Type	Process Owner	Fiori App ID	Tcode	LeanIX App
MFG.FIN.RTR.JNL v3.2 — APPROVED 2026-03-08	Period-End Close — Journal Entry Processing L3 process · 11 child activities	PROCESS	M. Hartmann CFO Office	– (parent)	–	—
MFG.FIN.RTR.JNL.001.V01 v3.2 — APPROVED 2026-03-08	Post Accrual Journal Entry L4 activity · System: SAP S/4HANA	ACTIVITY	A. Kowalski GL Manager	F0717 Post General Journal	FB50	APP-S4-FIN-001
MFG.FIN.RTR.JNL.002.V01 v3.2 — APPROVED 2026-03-08	Approve Journal Entry (Workflow) L4 activity · System trigger: SAP Workflow	ACTIVITY	M. Hartmann CFO Office	F0719	FBV0	APP-S4-FIN-001
MFG.FIN.RTR.JNL.G01.V01 v3.2 — APPROVED 2026-03-08	Decision: Amount > Approval Threshold? XOR gateway · SOX-relevant	DECISION	M. Hartmann CFO Office	– (rule)	–	APP-S4-FIN-001
MFG.FIN.RTR.JNL.001.V02 DRAFT — variant of V01 (parent)	Post Accrual Journal Entry — IFRS 16 Lease Variant L4 activity · Variant of MFG.FIN.RTR.JNL.001.V01	ACTIVITY	A. Kowalski GL Manager	F0717	FB50	APP-S4-FIN-001
MFG.FIN.RTR.JNL.003.V01 v3.2 — APPROVED 2026-03-08	Reconcile to Sub-Ledger L4 activity · System: SAP S/4HANA	ACTIVITY	A. Kowalski GL Manager	F2918	F.13	APP-S4-FIN-001
MFG.FIN.RTR.JNL.E01.V01 v3.2 — APPROVED 2026-03-08	Event: Period-End Close Triggered Start event · Schedule-driven (last business day)	EVENT	System Scheduler	–	–	APP-S4-FIN-001
MFG.FIN.RTR.JNL.004.V01 v3.2 — APPROVED 2026-03-08	Generate Period-End Reports L4 activity · System: SAP S/4HANA + Analytics Cloud	ACTIVITY	M. Hartmann CFO Office	F2547	FAGLB03	APP-SAC-001
MFG.FIN.RTR.IC v2.1 — APPROVED 2026-02-19	Intercompany Reconciliation L3 process · 7 child activities	PROCESS	M. Hartmann CFO Office	– (parent)	–	—
MFG.FIN.RTR.IC.001.V01 v2.1 · DDA-seeded standard · APPROVED	Identify Intercompany Transactions L4 activity · DDA-seeded: no customer modification	ACTIVITY	A. Kowalski	F1234	FBICA1	APP-S4-FIN-001
MFG.FIN.AP.EXP.005.V01 v1.4 — APPROVED 2026-01-30	Submit Expense Report (Concur) L4 activity · External system: Concur Expense	ACTIVITY	P. Romero AP Manager	– (Concur)	–	APP-CONCUR-001
MFG.FIN.AP.PR.999.V01 RETIRED 2026-02-12 — superseded by V02	Manual Three-Way Match (Pre-Automation) Retained for audit history; no longer executed	ACTIVITY	P. Romero	F0859	MIRO	APP-S4-MM-001

Figure 3.3: SAP Signavio Dictionary View

We show the dictionary view of SAP Signavio rather than a BPMN canvas in Figure 3.3 for a reason. Modelers spend most of their visible time in the BPMN canvas, but the integration value of SAP Signavio is concentrated in the dictionary. The dictionary is the canonical register; the canvas is its presentation. When auditors review a process, they read the dictionary. When AI agents are grounded against the process model, they query the dictionary. When a downstream integration developer asks "Does this step exist and what is its SAP Fiori app ID?" they look in the dictionary. A modeling team that maintains beautiful BPMN canvases but a thin and inconsistent dictionary has built a presentation layer without the system of record beneath it.

We are now returning to a finance lens that recurs throughout the book. In a financial planning and analysis (FP&A) organization, the canonical register is the chart of accounts. Every transaction in every subsidiary system joins to the chart of accounts by a general ledger code. Renaming a general ledger account does not change the general ledger code; the code is what every consolidation, reconciliation, and regulatory report joins on. The SAP Signavio dictionary plays the same role for process semantics that the chart of accounts plays for financial semantics, and the discipline of identifier immutability is the same discipline that any controller will recognize from the day they last had to retire a general ledger account. You do not rename the code; you retire the code and create a new one. The integration depends on it.

3.3 Mandatory Execution Attributes

An identifier locates a step in the universe of process steps. Mandatory execution attributes turn that step into an integration contract. Without the attributes, the identifier names a shape on a canvas and nothing more. With the attributes, the identifier is the anchor of a star schema that joins to every system involved in executing, observing, or improving the step. We require seven mandatory attributes on every L4 step in our framework. The seven are not arbitrary; they are the minimum set required for a downstream system to render the step at runtime, an auditor to evaluate the step at review time, and an AI agent to ground a response against the step at design time. Every attribute on the list earns its place by being read by at least one of those three audiences. Attributes that none of them read have been deliberately excluded.

Figure 3.4 shows how the seven mandatory execution attributes are visible as a side panel attached to a selected L4 step on the SAP Signavio canvas. The attributes are what turn the shape into an integration contract. The activity selected is **MFG.FIN.RTR.JNL.001.V01** (post accrual journal entry).

In the following sections, we'll look at each of the seven mandatory attributes in turn.

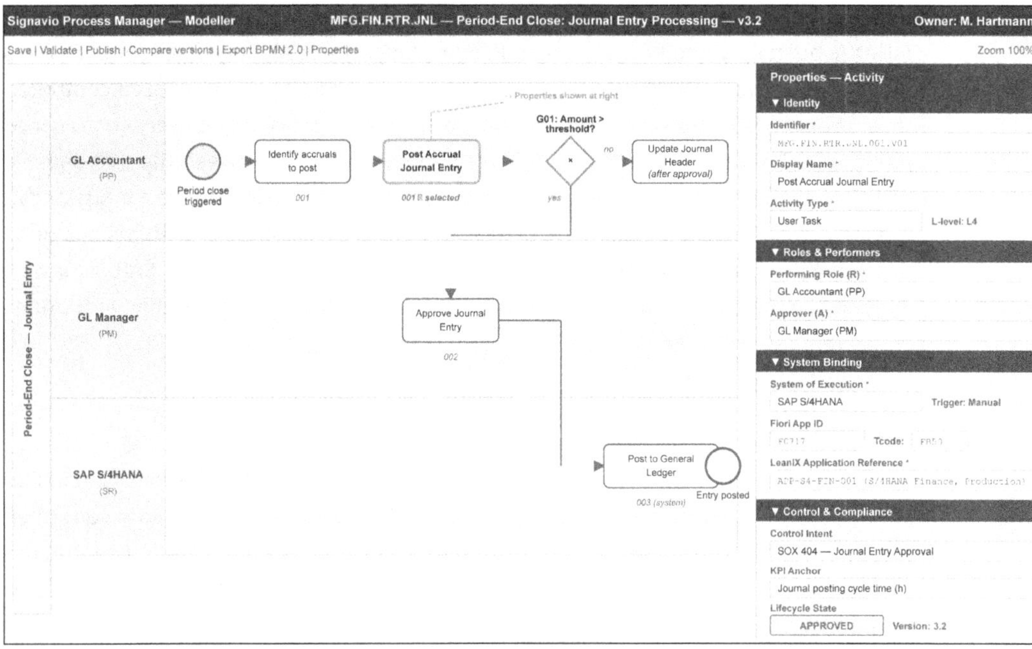

Figure 3.4: BPMN Canvas with Attribute Panel

3.3.1 Step Identifier

The first mandatory attribute is the step identifier in the six-segment form discussed in Section 3.2. The identifier is mandatory because every other attribute is keyed to it. Without an identifier, the step has no identity beyond its current label, and label-based integration is exactly what this chapter is designed to prevent.

3.3.2 Performing Role

The second mandatory attribute is the role that performs the step. Roles are drawn from the controlled taxonomy we will discuss in Section 3.4: process owner, process manager, process expert, process participant, system role, and external actor. Every L4 step is assigned to exactly one role. Steps that appear to be performed by multiple roles are in fact multiple steps that have been collapsed for visual convenience and should be decomposed. The role assignment drives swimlane allocation on the BPMN canvas and is the anchor for runtime authorization checks. The user who appears in the role at runtime is authorized to perform the step, and the audit trail of who performed which step is keyed to the role plus the user identity.

3.3.3 System Trigger

The third mandatory attribute is the system that triggers the step. *Trigger* is used in its precise technical sense here: the system that becomes the active execution context the moment the step begins. For most L4 steps in an SAP-led transformation this will be SAP S/4HANA, but it may also be SAP SuccessFactors, SAP Ariba, SAP Concur, or a non-SAP system invoked through SAP BTP integration. The system trigger is captured as a controlled-list value drawn from the SAP LeanIX application register, ensuring that every step's trigger system is itself a registered architectural element. The lookup between the SAP Signavio attribute and the SAP LeanIX register is performed at attribute-edit time, so a modeler cannot enter a free-text system name that does not exist in the architecture. This is one of the most valuable boundary-discipline mechanisms in the entire data model. SAP Signavio is forced to use the architecture register that SAP LeanIX maintains, rather than authoring a parallel register of its own.

3.3.4 Application Reference

The fourth mandatory attribute is the application reference, which is the specific SAP Fiori app ID, transaction code, or external application URL through which the step is performed. Where SAP Fiori is the runtime, the SAP Fiori app ID (such as F0703 or F2207) is captured as the canonical reference, supplemented where the legacy SAP GUI is still in use by the corresponding transaction code (such as F-02 or FB50). The SAP Fiori app ID is preferred because it survives backend version upgrades that sometimes change transaction codes. For non-SAP applications, an external reference is captured (typically a URL or a named integration endpoint) and the SAP LeanIX register anchors the entry. The application reference is what makes the step launchable from the process model. A user reading the step in WalkMe Discovery or Joule can be deep-linked directly into the application that performs it, with the L5 work instruction provided for guidance.

3.3.5 Control Intent

The fifth mandatory attribute is the control intent of the step. Every step has one of three control intents: it performs work (a productive step), it controls work (a verification, approval, or compliance step), or it observes work (a measurement or notification step). The classification is forced by the attribute schema. Every step must be one of the three, and only one of the three. This classification is mandatory rather than optional because control intent drives audit scope. Auditors review the control steps in detail and the productive steps summarily; without the classification, every step is audited as if it were a control step, which is both expensive and dilutive. The classification also drives separation of duties (SoD) analysis: a single user must not perform both a productive step and the control step that approves it.

3.3.6 Data Objects In and Out

The sixth mandatory attribute is actually a pair: the data objects consumed by the step and the data objects produced by the step. Data objects in this sense are the same named entities (documents, decisions, and notifications) that satisfied the second L4 criteria in

Section 3.1. Each is captured as a controlled-list reference, and the list of valid data objects is itself a maintained register that is shared with the master data and document type catalogs in SAP S/4HANA. By capturing data objects as references rather than free text, you make the process model legible to reconciliation and lineage tooling. A question such as "Where does this invoice number get created and where does it get consumed?" becomes answerable directly from the process model without manual interpretation of step descriptions.

There is a finance-lens caveat worth attaching to this attribute. Data objects in the process model are references; they are not the data itself. A purchase order data object in SAP Signavio is a pointer that says a purchase order is consumed there; the actual purchase order record lives in SAP S/4HANA, and the master data behind it lives in SAP Master Data Governance. A process repository that holds actual data values—such as specific purchase order numbers, journal entries, or customer records—has crossed the boundary into territory we will explicitly forbid in Section 3.6. Process models hold references; transactional systems hold data.

3.3.7 Key Performance Indicator Anchor

The seventh and final mandatory attribute is the KPI anchor, the named KPI to which the step contributes, drawn from the L1 KPI framework that governs the step's domain. A record-to-report journal posting step would anchor to a days-to-close KPI, an order-to-cash credit check step would anchor to a days-sales-outstanding KPI, and a procure-to-pay approval step would anchor to a maverick-spend-reduction KPI. The anchor is mandatory because it is what makes the process model usable for value tracking and continuous improvement. A redesign of the step that does not move the anchored KPI is, by definition, value-neutral, and a step with no anchored KPI cannot be evaluated for value at all. We will return to KPI anchoring in much greater detail in Chapter 9 through Chapter 15; for now, you should refuse to approve any L4 step whose KPI anchor is unfilled on the grounds that an unanchored step cannot be improved.

3.4 Roles, Variants, and Localization

Roles, variants, and localization are three different expressions of the same underlying problem: how to model a process that is performed differently in different contexts without losing the identity of the process itself. A role captures who performs a step. A variant captures how a step is performed differently in different business contexts. Localization captures how a step is performed differently in different countries or jurisdictions. All three deserve first-class status in the data model because they introduce variability that an integration must navigate at runtime, and a process model that has flattened them into a narrative description has handed the runtime a problem it cannot solve.

3.4.1 The Six-Role Taxonomy

We use a six-role taxonomy in this book, applied consistently across every process model in scope. The roles are process owner, process manager, process expert, process participant, system role, and external actor, as shown in Figure 3.5. The taxonomy is closed: no seventh role is permitted, and every L4 step is assigned to exactly one of the six roles. The closure is what makes the taxonomy useful. Open taxonomies that allow modelers to introduce new roles at will produce process models with twenty or thirty different role labels, no two of which mean the same thing across processes, and the swimlane on a BPMN canvas degrades into a label rather than an integration contract.

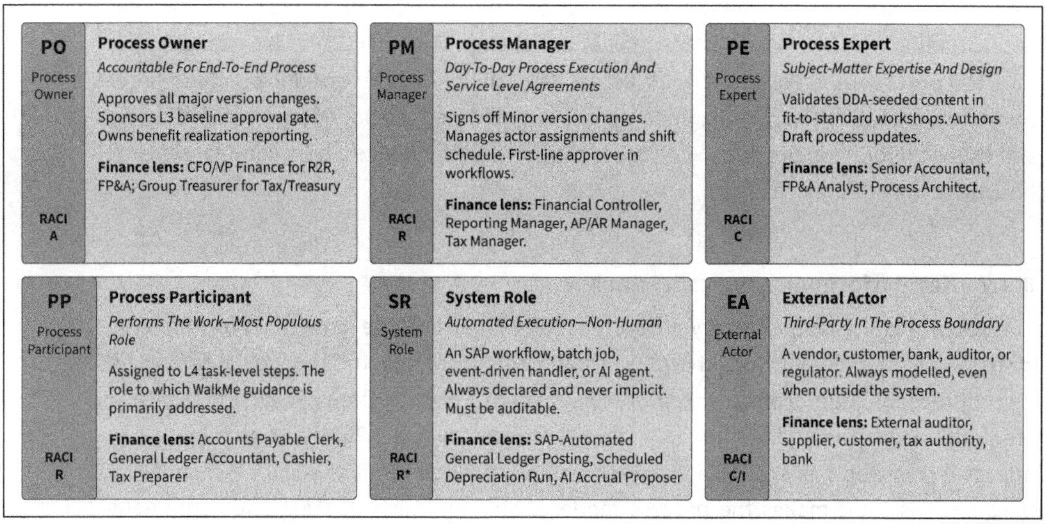

Figure 3.5: The Six-Role Taxonomy

Swimlane Allocation Rules

The swimlane allocation rules are as follows:

- **Rule 1: One swimlane per role.**
  Each lane has exactly one role from the taxonomy in Figure 3.5. For example, there is no "Finance team" lane that mixes the process owner, process manager, and process participant.
- **Rule 2: System and external actors get their own lanes.**
  System and external lanes are mandatory whenever automated steps or third parties exist. Implicit steps that "happen automatically" are forbidden.
- **Rule 3: Lane order conveys responsibility flow.**
  From top to bottom, responsibility flows as follows: process owner/process manager (oversight), process expert (design), process participant (execution), system role (automation), and external actor (external activities). This convention is enforced by the SAP Signavio style guide.

- **Rule 4: Variant pools are siblings, not nested lanes.**
When a process has variants (e.g., country-specific), each variant is a separate BPMN diagram tagged with the same L3 ID and a different .V## suffix. It never takes a new lane.

The process owner is accountable for the process at the L3 level. They are typically a senior business leader such as a controller for record-to-report, a procurement director for procure-to-pay, or a supply chain vice-president for supply planning. The process owner does not perform L4 steps; they approve the L3 process and its variants, sign off lifecycle transitions, and bear accountability for the KPIs anchored to the process. In the data model, the process owner is captured as an attribute of the L3 process node, not the L4 step, because ownership accountability is at L3.

The process manager is responsible for the day-to-day performance of the process. They run the operational rhythm—the daily, weekly, and monthly forums at which the process is reviewed, the metrics that drive the operational scorecard, and the escalation path for process exceptions. In a record-to-report process, the process manager is typically a senior accountant or finance manager. In the data model, the process manager is captured at the L3 process level, alongside the process owner.

The process expert is the deep subject-matter authority for the L4 step. They author the L5 work instruction, validate the step's mandatory attributes, and approve variants and localizations. The process expert role is the busiest role in the modeling team; we sometimes describe it as the role that knows the most about the smallest scope. Process experts are scoped to L3 processes. A person can be the process expert for one or two L3 processes, but not credibly for ten. In the data model, the process expert is the role that owns most L4 steps, recorded as the performing role attribute discussed in Section 3.3.

The process participant is the role that performs the transactional work, meaning the person who actually posts the journal entry, processes the invoice, or releases the goods. Most L4 steps are performed by a process participant rather than a process expert; the expert authored the step and the participant executes it. The participant role is the closest to the SAP runtime, and it is the role most often anchored to the SAP Fiori app ID captured as a mandatory attribute.

The system role is what we use to model automated steps, which are performed by SAP, an integration, or a Joule agent rather than a human user. The system role is an important part of the taxonomy because automated steps deserve the same data discipline as manual steps. They have control intent, they consume and produce data objects, they anchor to KPIs, and they are governed through the same lifecycle as manual steps. The temptation to omit them from the process model on the grounds that "they just happen" is the temptation to leave the most reliable parts of the process undocumented while documenting the parts that depend on human variability. Resist the temptation; system role steps should be modeled with the same rigor as process participant steps.

The external actor is a participant outside the enterprise, such as a customer, a vendor, a regulator. External actors are modeled because their interactions with the process are real and consequential, but the steps they perform are observed rather than authored. You capture what an external actor does and the data they exchange with the process, but you do not model their internal procedures because you do not control them. The boundary between the external actor lane and the rest of the swim lane diagram is the boundary of the enterprise itself, and in security terms, it is also the boundary of the trust zone we discussed in Chapter 2.

3.4.2 Swimlane Discipline

Swimlanes on the BPMN canvas are the visual expression of the role taxonomy. Every step is placed in the swimlane of the role that performs it, with no exceptions. Modelers will occasionally argue that a step is performed jointly by two roles, such as a process participant who submits a journal entry and a process expert reviews it before approval. The argument is correct in the sense that two roles are involved, but it confuses the boundaries of the steps. The submission is one L4 step in the participant lane; the review is another L4 step in the expert lane. Two steps, two lanes, two integration contracts, joined by a sequence flow on the canvas. Collapsing them into one step in a shared lane produces a model that is visually compact and operationally incoherent, because no integration can fire against a shared-lane step; the runtime authorization cannot determine which user is permitted to perform it.

Swimlanes are also the device by which you make role-based responsibility visible at a glance. A finance reviewer scanning the record-to-report process model can see at a glance which steps are performed by process participants (the bulk of the volume), process experts (the review and approval gates), system roles (the automated postings and reconciliations), and external actors (the auditor confirmations and bank responses). The visual layout is the audit trail. A flat diagram with no swimlanes is one in which responsibility is smeared across the canvas, and a finance reviewer cannot tell at a glance whether an approval gate is in the right lane or has migrated, in some quiet refactoring, into a lane it does not belong in.

3.4.3 Variant Classification

A variant captures a substantive difference in how a step is performed across business contexts. The journal entry posting step, for example, may have a manufacturing variant aligned with International Financial Reporting Standards (IFRS), a sales variant aligned with United States Generally Accepted Accounting Principles (US GAAP), and a legacy-batch variant inherited from the previous enterprise resource planning (ERP) system that has not yet been retired. Each variant is a sibling of the parent step in the data model, sharing the first five identifier segments and differing only in the sixth. Each carries its own role, system trigger, application reference, control intent, data objects, and KPI anchor. From the runtime's point of view, the variants are independent steps; from the data model's point of view, they are siblings of a common parent.

Figure 3.6 shows a variant classification tree. The parent variant V01 anchors a family of siblings (mandatory, value-driven, and legacy), each classified by why it exists. The classification governs the lifecycle.

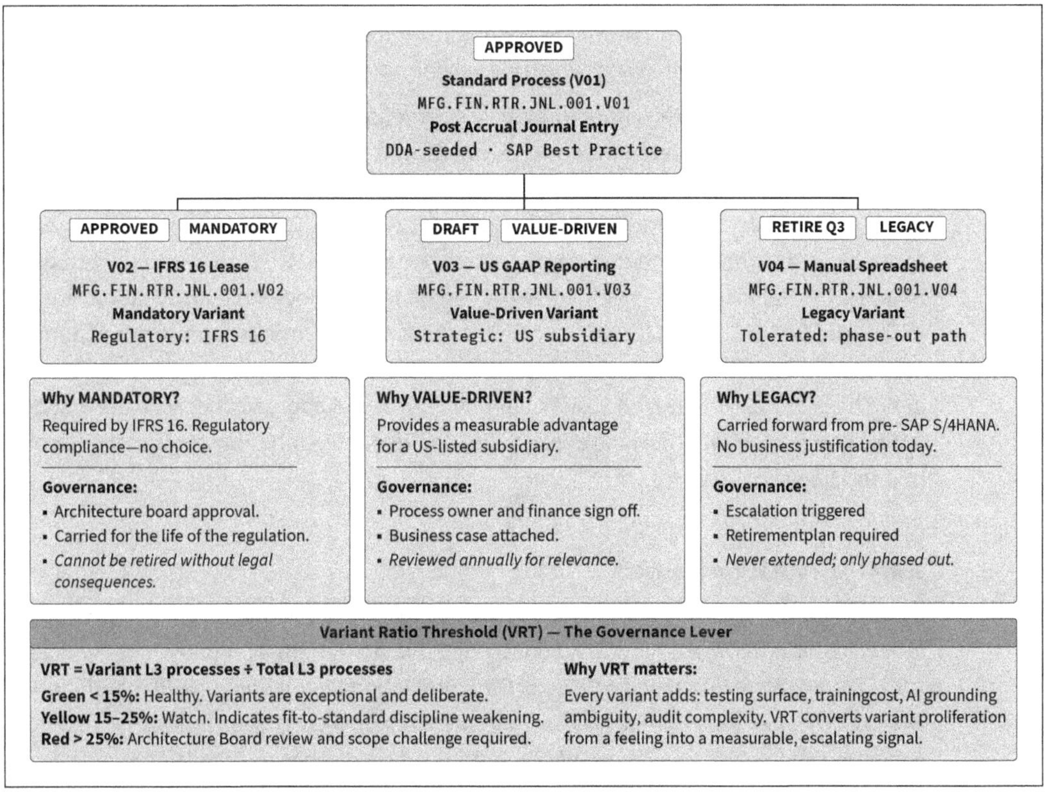

Figure 3.6: Variant Tree for an L4 Step

We classify variants into three categories, and the classification is mandatory. Mandatory variants exist because law, regulation, or external standard requires them: an IFRS variant exists because IFRS reporting is required, a country-specific tax calculation variant exists because the country's revenue authority requires it. Mandatory variants are not optional, are not subject to value-based rationalization, and persist for as long as the underlying regulation persists. Value-driven variants exist because they produce measurably better business outcomes than the parent—for example, a streamlined approval variant for low-value transactions or a machine learning-augmented variant for high-volume reconciliations. Value-driven variants must justify themselves with a quantified KPI improvement, and a value-driven variant whose improvement has not materialized is a candidate for retirement. Legacy variants exist because they were inherited from a previous system or process and have not yet been rationalized. They are tolerated for a defined transition period and are automatically reviewed for retirement at each major release.

The classification governs the lifecycle. Mandatory variants follow the underlying regulation's lifecycle and are reviewed annually. Value-driven variants are reviewed against their anchored KPI on a defined cadence, typically at the end of each fiscal quarter, and are retired when the value case no longer holds. Legacy variants are placed on a retirement schedule from the moment they are classified, and the schedule itself is reported through the variant ratio threshold (VRT), which we will introduce in Section 3.5.

This classification scheme is important for a number of reasons. Organizations running on SAP S/4HANA Cloud Private Edition have variance. It is not eliminable, and the common consultant claim that SAP S/4HANA Cloud Public Edition removes variance is, in our experience, a marketing claim rather than reality. SAP S/4HANA Cloud Public Edition reduces some types of variance and introduces others. What SAP S/4HANA Cloud Private Edition allows is for variance to be governed: every variant is classified, has a justification, and has a lifecycle. SAP S/4HANA Cloud Public Edition tries to suppress variance by configuration restriction; SAP S/4HANA Cloud Private Edition allows variance and demands governance of it. The data model in this chapter is built for the second approach because the first does not survive contact with the kind of regulated, multicountry, multi-business unit enterprise that this book is written for.

3.4.4 Localization Markers

Localization is a special case of variance, and we treat it as a variant rather than as a separate concept in the data model. A country-specific tax variant of a journal entry posting step is a variant; an EU-specific General Data Protection Regulation (GDPR)–handling variant of a customer data update step is a variant; a regional language variant of a customer-communication step is a variant. We do not introduce a separate localization object type because doing so would create a parallel vocabulary for what is operationally the same thing, and a process model with two parallel vocabularies for variance ends up with steps that are localized in one vocabulary and variant in the other, with no clear rule for which applies when. One vocabulary, applied consistently, scales further than two vocabularies that overlap.

What we do introduce is a localization marker on the variant itself—a controlled list attribute that captures the country, region, or language scope of the variant. The marker is what allows the runtime to select the right variant for the right user without the BPMN designer having to encode the logic by hand. A user in Germany executing the journal entry step is offered the IFRS variant; a user in the United States is offered the US GAAP variant; a user in Japan is offered the variant aligned with Japan's Financial Instruments and Exchange Act (JSOX). The selection logic is in the runtime, the marker is in the data model, and the BPMN canvas remains uncluttered with the conditional logic that would otherwise have to render every locale in a single diagram.

3.5 Versioning and Lifecycle Management

A process model that does not include versioning is not true a process model. It is a snapshot, useful at the moment it was taken, and increasingly misleading thereafter. The discipline of versioning is what makes a process repository worth integrating against, because it is what allows downstream systems to ask "What is the approved version of this step right now?" and receive an authoritative answer. Without versioning, every reference to a step is implicitly a reference to whatever the modeler happened to have saved most recently, which may or may not be reviewed, approved, and or the version that the audit examined last quarter. The purpose of this section is to specify the version state machine, the change rules that govern transitions between states, and the digital thread that connects DDA to BPMN to runtime.

3.5.1 The Version State Machine

Every L4 step in our framework lives in one of four states: **Draft, Reviewed, Approved**, or **Retired**. The states form a sequence: a step is created in **Draft**, advances to **Reviewed** when its process expert has validated it, advances to **Approved** when its process owner has signed it off, and eventually moves to **Retired** when its useful life has ended. Transitions between states are explicit acts performed by named users, recorded with timestamps, and audited. There is no implicit promotion: a **Draft** step does not become **Reviewed** because someone forgot to mark it; it remains **Draft** until someone explicitly reviews it. Equally, a step does not regress unilaterally: an **Approved** step cannot be silently edited back to **Draft**. To change an **Approved** step, the modeler must initiate a new version, which begins life in **Draft**, with the previous **Approved** version still resolving in production until the new version itself reaches **Approved**. Figure 3.7 shows the path these states follow. In parallel, major, minor, patch, and variant rules govern when version numbers advance, as shown in Table 3.1.

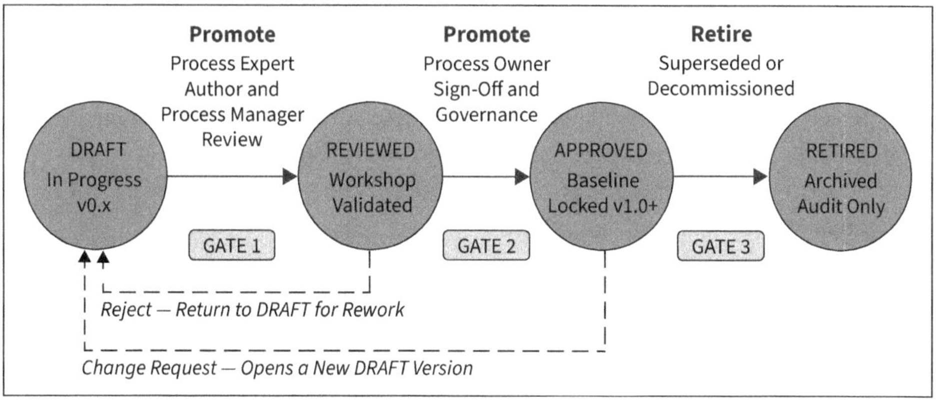

Figure 3.7: The Version State Machine and Change Rules

	Major v1.0 → v2.0	Minor v1.0 → v1.1	Patch v1.0 → v1.0.1	Variant V01 → V02
When	Structural change to flow, scope, or actor model	Textual update, attribute change, actor correction, or non-flow change	Typo fix, metadata correction, link repair, or broken reference	Organization-specific deviation that must coexist with the DDA standard
Approval	Full governance. Process owner and architecture board must re-approve.	Process owner approval is sufficient. Skips architecture board.	No gate; log entry only. Implemented within 48 hours by the SAP Signavio administrator.	Tracked separately. Must reference the parent V01.
Next step	*Returns to* **Draft**.	*May go directly to* **Reviewed**.	*Stays* **Approved**.	*Carries a classification:* **M, V,** *or* **L**.

Table 3.1: Version Number Discipline—Four Change Types

The state machine matters because it is what every downstream system reads. A WalkMe content publication is keyed to **Approved** steps. A Joule grounding bundle is built from **Approved** steps. An auditor reviews **Approved** steps. A regulator who asks "What is your current process for X?" receives an answer drawn from **Approved** steps. By making the state machine explicit, you make the question "Which version is in production?" answerable mechanically, by query, rather than by negotiation between modelers and integrators.

3.5.2 Change Rules: Major, Minor, Patch, and Variant

When an **Approved** step needs to change, the change is classified into one of four rules, and the rule determines how the version number advances and what governance applies. Major changes alter the structure of the step, including its scope, its actor, its trigger, or its core data objects. The version number advances from v1 to v2, and the change requires process owner approval. A major change is the most expensive change to make, because every downstream integration must be revalidated against the new structure, and every L5 work instruction must be reviewed and updated.

Minor changes alter the wording, the L5 instruction, the control phrasing, or other features that do not change the step's structural identity. The version number advances from v1.0 to v1.1, and the change requires process expert approval but not process owner approval. A minor change passes downstream without breaking integrations because the structural attributes are unchanged, but it is recorded in the change log and reviewed at the next process council.

Patch changes are corrections such as a typographical fix, a broken hyperlink in the L5 instruction, or a misclassified KPI anchor that has now been reanchored. The version number advances from v1.0 to v1.0.1, and the change requires only process manager approval. Patches are the most frequent category of change in a mature process repository, and we deliberately keep their governance lightweight to avoid creating bureaucratic resistance to the kind of housekeeping that keeps the model usable.

Variant changes introduce a new sibling variant of an existing step. The variant identifier increments from V01 to V02, V03, and so on, and the new variant inherits the structural properties of its parent. Variant changes require the same governance as the variant's classification implies. A mandatory new variant is approved by the process owner, a value-driven new variant is approved by the process owner with a value case attached, and a new legacy variant is typically not created because new variants should not be inheriting legacy classification.

3.5.3 The Digital Thread: DDA to BPMN to Runtime

We have referred several times to the digital thread that connects DDA, BPMN, and runtime. This section makes that connection explicit. The digital thread is the proposition that a process model is not authored once at the start of a project and then maintained as a separate artifact thereafter. It comes from the DDA, evolves through six lifecycle phases of the implementation program, and continues to evolve in operations after go-live, with every change captured as a versioned delta against the previous state. The full equation is *BPMN(t) = BPMN(DDA) + ΣΔ*, where *BPMN(t)* is the process model at time *t*, *BPMN(DDA)* is produced during pre-sales discovery, and *ΣΔ* is the sum of governed deltas applied since. Figure 3.8 shows this process.

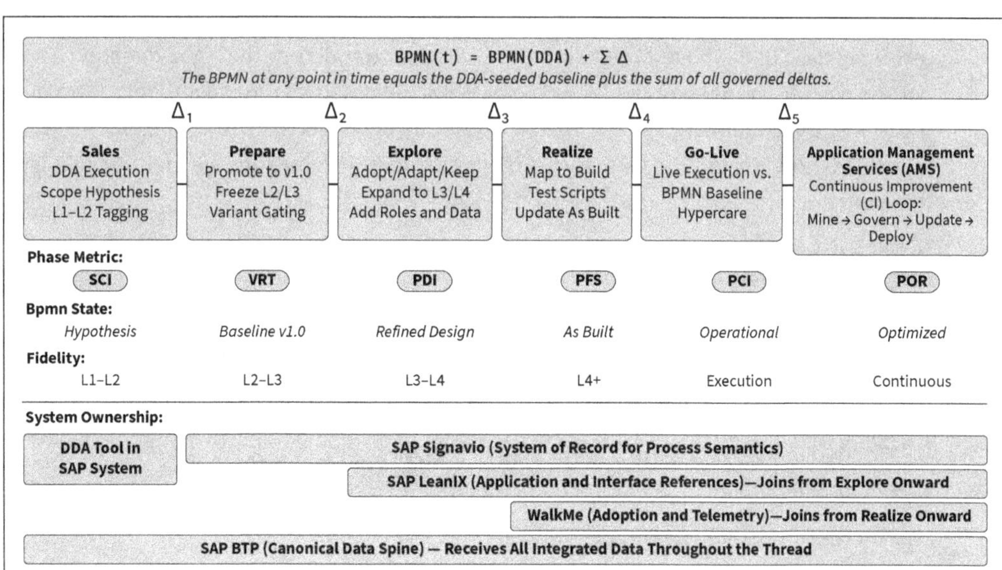

Figure 3.8: The DDA to BPMN to SAP Signavio Digital Thread

Why This Matters for the Data Model

Every attribute defined in the rest of this chapter—identifiers, mandatory execution attributes, role assignments, lifecycle states—must be designed to be *populated by DDA where DDA can* and *augmented through governed deltas thereafter*. This is the design constraint that determines whether the L4 attribute model is workable in practice or only in theory.

The six lifecycle phases are sales, prepare, explore, realize, go-live, and application management services (AMS). These are the same six phases that the SAP Activate methodology uses for its delivery framework. In each phase, a different set of stakeholders interacts with the process model, and a different set of metrics governs its evolution. In sales, the DDA produces the starting point; the metric is the scope confidence index (SCI), which measures the proportion of L4 steps in the proposed scope that are covered by industry-standard practices. In the prepare phase, the modeling team validates and refines these practices against the organization's context; the metric is the VRT, which measures the proportion of candidate steps that have been classified as mandatory, value-driven, or legacy. In explore, the team applies the adopt/adapt/keep framework to each variant; the metric is the process decision index (PDI). In realize, the model is wired to the SAP runtime; the metric is the process-to-functional-spec (PFS) coverage. In go-live, the model becomes the operational reference; the metric is process compliance index (PCI), which measures the proportion of transactional events that occur as the model predicts. In AMS, the model evolves through governed deltas; the metric is the process optimization rate (POR).

These metrics are the answer to the question "How do we govern process model evolution at scale?" Without them, modelers and architects argue qualitatively about whether the model is in a good state, and the arguments do not converge. With them, the argument is mechanical: SCI is above target or below; VRT has trended in the right direction or the wrong one; PCI is meeting the threshold or is not. The metrics turn governance from judgment into measurement, and the digital thread that connects them is what allows the same metric to be evaluated consistently from sales through AMS over the lifespan of an engagement.

The Practitioner's View

There are two views of BPMN practice. The first assumes that BPMN can evolve without governance, variants can appear without classification, L4 steps can be authored from a blank canvas, and the SAP Signavio repository is simply a place to document whatever was implemented. This approach is widely held in the SAP partner community and produces process repositories that look complete but are operationally useless.

The second view—the one that we hold and the one this chapter is built on—is that BPMN evolution is governed by the equation, every change is classified and approved, every variant has a justification and a lifecycle, and the SAP Signavio repository is a system of record rather than a documentation artifact. The two views produce visually indistinguishable BPMN diagrams, but their integration outcomes are radically different. The book is written from the perspective of the second view because the organizations we have seen succeed all follow it.

3.6 What Must Not Be Modeled in SAP Signavio

The discipline of saying yes to what belongs in the process model is matched by the discipline of saying no to what does not. This section outlines the second discipline. It exists because every process repository we have reviewed in our professional life has accumulated content that does not belong in it, that another system would hold better, and that has migrated into the process model because the process model was the first place a modeler looked and the migration was never reversed. The accumulation is incremental and rarely deliberate, and by the time it is noticed, the repository is heavy enough that the cleanup feels expensive. We have come to believe that the cleanup is always less expensive than the continuing operational cost of carrying the content in the wrong system, and that boundary discipline is the cheapest investment a transformation program can make.

Table 3.2 provides a summary of what does and does not belong in SAP Signavio. Six categories belong in SAP Signavio (process semantics, identifiers, attributes, roles, variants, and lifecycle); six categories do not (architecture, configuration, telemetry, master data, AI rules, and external actor models). Each excluded category names the system that holds it instead.

What Belongs in SAP Signavio	What Doesn't Belong in SAP Signavio
Process semantics: This defines what happens, in what order, and by whom, including BPMN flow, swimlanes, gateways, events, and subprocesses. It is the story of how work moves across roles and systems, captured at L4 step granularity and narrated at L5 instruction granularity. This is the irreducible core; no other tool tells this story coherently.	Application architecture and capability maps: This includes application IDs, deployment topologies, technology stacks, integration patterns, and data flows between systems. SAP Signavio references applications by SAP LeanIX ID; it does not catalog them.

Table 3.2: The Signavio Boundary: What Belongs and What Does Not

What Belongs in SAP Signavio	What Doesn't Belong in SAP Signavio
Process and activity identifiers: The identifier follows the naming structure Industry.L1.L2.L3.Step.Variant ID and is immutable once approved. It is the integration key that joins LeanIX architecture, SAP S/4HANA transactions, WalkMe telemetry, and Joule prompts. The dictionary lives here; everywhere else only references it.	SAP configuration, customizing, and code: This includes IMG entries, BAdIs, business transaction events (BTEs), ABAP enhancements, document types, posting keys, and organizational structures. The process model says "Post journal entry"; the configuration says how to do so. The two are joined by an SAP Fiori app ID, not by duplication. This belongs in SAP S/4HANA.
Mandatory execution attributes: This includes the step ID, role, system trigger, SAP Fiori app ID, control intent, data objects in/out, and KPI anchor. This is the seven fields that turn a shape on a canvas into a governed integration contract. In other words, replace references—never the system itself. Each attribute holds a *reference* to the executing system like the SAP LeanIX application ID, the SAP Fiori app ID, the transaction code and not the application record, configuration, or runtime state, which remain in their respective systems of record.	Runtime telemetry and click-stream behavior: This includes user adoption metrics, click paths, time-on-task, error rates, abandonment, and A/B test outcomes. These are observed at runtime; the design is in SAP Signavio and the observation is in WalkMe and SAP Signavio Process Insights.
Role allocation and swimlane discipline: This encompasses the six roles of process owner, process manager, process expert, process participant, system role, and external actor. Every step appears in exactly one lane. Role definitions happen at the chapter level and are not redefined per process. This involves assignments, not policy authoring. Policy is part of identity and access management (IAM) and SoD tools and is referenced by ID.	Master data, reference data, and transaction data: This includes customer records, vendor records, material masters, general ledger accounts, cost centers, document numbers, and posting amounts. Process model defines that the customer is read; it never holds the customer record itself. This belongs in SAP S/4HANA and SAP Master Data Governance.
Variant classification and localization markers: Every variant is classified as mandatory, value-driven, or legacy. Country, regulatory, and channel splits are modeled as variants of the parent, never as parallel processes. VRT is measured here; no VRT means no governance.	AI decision rules, prompts, and Joule skill logic: This includes Joule skill code, prompt templates, model selection, agent orchestration, and threshold rules. SAP Signavio marks that the step is AI-augmented by attribute; the rule engine and prompt live in SAP BTP. There are no prompts inside BPMN annotations.

Table 3.2: The Signavio Boundary: What Belongs and What Does Not (Cont.)

What Belongs in SAP Signavio	What Doesn't Belong in SAP Signavio
Lifecycle state and version history: This involves the state machines from **Draft** to **Reviewed** to **Approved** to **Retired**, along with major/minor/patch/variant versioning, change rationale, and approver identity. It is the audit trail of how the contract evolved and the single source of truth for what is approved right now.	Enterprise architecture models and strategic capability roadmaps: This involves business capability hierarchies, value streams as architecture artifacts, target operating model diagrams, and enterprise transformation roadmaps. SAP Signavio executes the processes; ARIS or SAP LeanIX BTM holds the architecture they sit within.

Table 3.2: The Signavio Boundary: What Belongs and What Does Not (Cont.)

Let's look more closely at what does not belong in SAP Signavio:

- **Application architecture and capability maps**
  The catalog of applications, their deployment topology, the technology stack each runs on, and the integration patterns between them belong in SAP LeanIX or an equivalent enterprise architecture register. SAP Signavio references applications by their SAP LeanIX identifier; it does not catalog them. The temptation to capture application details in SAP Signavio attributes using language like "This step runs in S/4HANA 1909 SPS04" produces a parallel catalog that drifts out of sync with SAP LeanIX the moment the application is upgraded, and the drift is invisible because no one is comparing the two registers. The discipline is to reference, not duplicate. An SAP Signavio attribute names the SAP LeanIX identifier; the version, deployment, and technology stack are found in SAP LeanIX at evaluation time.
- **SAP configuration, customizing, and code**
  IMG entries, document types, posting keys, organizational structures belong in the SAP system, either in the IMG itself or in a configuration documentation tool such as SAP Solution Manager or SAP Cloud ALM. SAP Signavio captures the process intent ("Post a journal entry to the appropriate ledger"); the configuration captures how the system implements the intent ("Document type SA, posting key 40 for debit, posting key 50 for credit"). The two are joined by the SAP Fiori app ID and transaction code captured as mandatory attributes; the configuration itself is not duplicated into SAP Signavio. Modelers will sometimes argue that capturing the configuration in SAP Signavio makes the process model self-contained. It does, briefly, until the configuration changes and the SAP Signavio copy does not. From that moment, the SAP Signavio copy is an active source of misinformation rather than a reference, and the integration that depended on the SAP Signavio copy fails silently.
- **Runtime telemetry and click-stream behavior**
  User adoption metrics, click paths, time-on-task, error rates, and abandonment patterns belong in WalkMe and SAP Signavio Process Insights, not in SAP Signavio. SAP Signavio is the design-time system of record; WalkMe and SAP Signavio Process Insights are the

runtime observation systems. The two interact through the process identifier, but they do not share a database. A modeler who tries to capture adoption metrics in SAP Signavio attributes is conflating design with observation: the design-time view says "This is how the process should be performed," and the observation time view says "This is how the process is being performed in practice." The two views are easiest to compare when each is kept in its proper place and linked by an identifier. Trying to fuse both into a single tool blurs the distinction and results in a tool that does neither job well.

- **Master data, reference data, and transaction data**
  Customer records, vendor records, material masters, general ledger accounts, and cost centers belong in the master data system, typically SAP S/4HANA backed by SAP Master Data Governance. Reference data, such as country codes, currency codes, and tax codes, belongs in the same place. Transaction data, such as specific journal entries, specific purchase orders, and specific shipments, belongs in the SAP transactional system. SAP Signavio holds none of these. It holds the data object references, captured as the data objects attribute, which name the master data and transaction data types in use without holding the values themselves. A process repository that has holds actual master data values has crossed a boundary it will not survive. Master data discipline depends on a single system of record, and an SAP Signavio system that holds duplicate master data is a source of audit findings rather than control evidence.
- **AI decision rules, prompts, and Joule skill logic**
  The prompts that drive Joule skills, the thresholds that govern automated decisions, and the agent orchestration logic that selects which model to invoke when belong in SAP BTP, specifically in the SAP AI Core and generative AI hub services. SAP Signavio marks an L4 step as AI-augmented through a controlled-list attribute ("Step is executed by Joule skill X"), but the prompt that powers the skill, the model used, the grounding bundle that constrains its responses, and the threshold rules that govern when the skill is invoked all live in SAP BTP. The temptation to capture prompt templates in BPMN annotations is high; often, the modeler wants the prompt to be visible in context with the step it powers. However, we believe the exclusion is nonnegotiable. Prompts evolve faster than process models, are versioned differently, are tested differently, and are governed by an AI risk framework that is distinct from process governance. Mixing the two produces an SAP Signavio repository that is neither a clean process model nor a clean prompt library as well as an SAP BTP environment that has drifted out of sync with the SAP Signavio annotations no one is updating in real time.
- **Enterprise architecture and strategic capability roadmaps**
  Business capability hierarchies, value stream diagrams as architectural objects, target operating models, and multi-year transformation roadmaps belong in an enterprise architecture tool such as ARIS, business transformation management, or a similar capability-mapping platform. SAP Signavio executes the processes that operate within the architecture; it does not hold the architecture itself. The distinction matters because architecture and process operate at different time scales and serve different audiences. Architecture changes annually or less; process changes monthly. Architecture is read by

senior leadership and mergers and acquisitions teams; process is read by operating teams and auditors. Holding both in SAP Signavio produces a repository that updates too slowly for process needs and too quickly for architecture needs, and a tool that serves neither audience well.

These six exclusions are the boundary of SAP Signavio in our framework. Every one of them has appeared in process repositories we have inherited, and every one has had to be removed because its presence in the process repository broke the integration trust that SAP Signavio is supposed to provide. Refusing to model them in the first place is the discipline in this chapter with the largest financial payback. We will return to it in the chapters that follow, particularly in Chapter 4 (SAP LeanIX as the architecture register) and Chapter 7 (SAP BTP as the AI rule engine).

3.7 Summary

This chapter established the SAP Signavio data model that the rest of Part II depends on. We began at the L4 step level because integration begins there, and we walked through the four criteria that determine whether a candidate is a real L4 step or has been modeled at the wrong granularity. We treated L5 work instructions as first-class child objects, scoped to a single step and version-locked to their parent. We explained, with operational evidence, that L4 modeling does not begin in Signavio but in pre-sales DDA, and that the $BPMN(t) = BPMN(DDA) + \Sigma\Delta$ equation is the discipline that connects the two.

We specified the six-segment process identifier, Industry.L1.L2.L3.Step.Variant, and the immutability rule that makes it usable as an integration key. We separated label from identifier and treated the label as a presentation concern that may change while the identifier remains fixed. We described the SAP Signavio dictionary as the canonical register of identifiers and the BPMN canvas as its presentation and drew a parallel to the role that a chart of accounts plays in financial integration.

We described the seven mandatory execution attributes on every L4 step: the step identifier itself, the performing role, the system trigger, the application reference, the control intent, the data objects in and out, and the KPI anchor. Each attribute earns its place by being read at runtime, review time, or design time.

We instantiated a closed six-role taxonomy (process owners, process managers, process experts, process participants, system roles, and external actors) and required every L4 step to be assigned to exactly one role, with swimlane discipline on the BPMN canvas as the visual expression of the assignment. We classified variants into mandatory, value-driven, and legacy categories with different lifecycles, and we treated localization as a special case of variance rather than as a separate concept. We discussed how variance is governed in SAP S/4HANA Cloud Private Edition rather than eliminated and that the discipline of governance is what separates the organizations who succeed on private cloud from those who do not.

We specified the four-state version machine (**Draft**, **Reviewed**, **Approved**, and **Retired**) and the four change rules (major, minor, patch, and variant) that determine how versions advance. We connected the version states to the digital thread that runs from DDA through six lifecycle phases to AMS, with phase-gate metrics (SCI, VRT, PDI, PFS, PCI, and POR) that turn governance from judgment into measurement. We listed six categories of content that do not belong in SAP Signavio: application architecture, SAP configuration, runtime telemetry, master data, AI decision rules, and enterprise architecture.

Chapter 4 builds directly on this foundation. This chapter established the process step as the primary object of integration, while Chapter 4 establishes the Joule grounding bundle and the WalkMe content reference as the primary mechanisms by which the process step is consumed at runtime. The discipline of identifier immutability, the seven mandatory attributes, the lifecycle states, and the boundary discipline established here are all prerequisites for what follows. If you've internalized the data model in this chapter, you're ready for everything that comes next; if you haven't, you will have a shaky foundation for later chapters.

3.7.1 Key Learning Points

- Integration occurs only at L4 and L5. L1 through L3 organize the hierarchy but carry no executable semantics. Time spent modeling L1–L3 in elaborate detail produces a beautiful map with no execution surface; time spent on L4 with discipline produces an integration contract.
- L4 modeling begins in presales DDA, not in SAP Signavio. The DDA produces an industry-standard starting place that covers 60 to 70 percent of the eventual L4 model. Organizations who treat Signavio as the starting point spend three months recreating what DDA could have produced in two hours.
- The process identifier is a six-segment, immutable contract: Industry.L1.L2.L3.Step.Variant. Identifiers are fixed the moment a step reaches the **Approved** state and are never renamed. The label may change; the identifier may not.
- The SAP Signavio dictionary is the canonical register; the BPMN canvas is its presentation. Other systems join to the dictionary by identifier, never by name. The relationship is the same one a chart of accounts has with financial reporting.
- Seven mandatory execution attributes turn an L4 step into an integration contract: identifier, performing role, system trigger, application reference, control intent, data objects in and out, and KPI anchor. Each is read by at least one of the three downstream audiences (runtime, audit, and AI grounding).
- The role taxonomy is closed at six roles: process owner, process manager, process expert, process participant, system role, and external actor. Open taxonomies become incoherent swimlanes that cannot anchor runtime authorization.
- Variants are first-class citizens of the data model and are classified as mandatory, value-driven, or legacy. Each classification has a distinct lifecycle. Localization is a special case

of variance, not a separate concept. SAP S/4HANA Cloud Private Edition governs variance; it does not eliminate it.

- The lifecycle state machine follows the sequence **Draft**, **Reviewed**, **Approved**, and **Retired**, with explicit transitions, named approvers, and audit trails. **Approved** is the state that downstream systems read. Changes to **Approved** steps create new versions; they do not silently overwrite.
- $BPMN(t) = BPMN(DDA) + \Sigma\Delta$ is the digital-thread equation. Every governed delta carries a classification (major, minor, patch, and variant) and a phase-gate metric (SCI, VRT, PDI, PFS, PCI, and POR). The metrics turn governance from judgment into measurement.
- The boundary of SAP Signavio is drawn explicitly. There are six categories that do not belong in SAP Signavio: application architecture, SAP configuration, runtime telemetry, master data, AI decision rules, and enterprise architecture. Each has an alternate system that holds it better. The discipline of refusing to model them is the cheapest investment a transformation can make.

3.7.2 Common Design Pitfalls

- Treating identifiers as renameable. This is the single most damaging approach to take. Every rename breaks an integration somewhere downstream, and the failure is silent because there is no compiler error for a broken process identifier.
- Modeling L1–L3 in detail and L4 superficially. This produces a process repository that looks complete to senior leadership but is operationally useless. The integration depth is L4 and below; everything above it is taxonomy.
- Authoring L4 from a blank canvas instead of starting with the DDA. This wastes 60 to 70 percent of the modeling effort recreating industry standard and produces a model with no traceability to the value hypothesis sold during presales.
- Allowing the role taxonomy to grow beyond six roles. Each new role looks reasonable in isolation; collectively, they degrade swimlane discipline and break the runtime authorization model.
- Treating variants as decorative attributes rather than first-class objects with their own classification, lifecycle, and metrics. A process model with unclassified variants has no governance over its variance.
- Modeling configuration, telemetry, master data, or AI prompts in SAP Signavio. This type of content is better held in another system. Their presence in SAP Signavio creates parallel registers that drift out of sync with the systems of record.
- Skipping the version state machine and relying on whatever was saved most recently. The state machine is what makes the question "What is the approved version?" answerable mechanically. Without it, every reference is implicit and every audit becomes a negotiation.

- Treating the BPMN canvas as the system of record and the dictionary as a derived artifact. The relationship is the inverse: the dictionary is canonical; the canvas is its presentation. Modeling teams that have it the wrong way around build beautiful canvases on thin dictionaries.

3.7.3 Review Questions

1. Why is L4 the integration depth and not L3 or L5? What specifically can be wired to systems at L4 that cannot be wired at L3?
2. What are the four mandatory criteria for an L4 step, and what is the operational test that distinguishes one step from two?
3. Why is the process identifier immutable from the **Approved** state onward? What happens to downstream integrations when this rule is violated, and why are the failures typically silent?
4. Name the seven mandatory execution attributes and explain, for each one, which of the three downstream audiences (runtime, audit, or AI grounding) reads it.
5. Why is the role taxonomy deliberately closed at six roles? What goes wrong when modeling teams introduce a seventh, eighth, or ninth role on a per-process basis?
6. Explain the difference between mandatory, value-driven, and legacy variants. Which lifecycle applies to each, and why does the classification matter for the VRT?
7. Explain the equation $BPMN(t) = BPMN(DDA) + \Sigma\Delta$ in your own words. What is $BPMN(DDA)$, what is $\Sigma\Delta$, and why is the equation the discipline that prevents process drift?
8. List the six categories of content that should not be modeled in SAP Signavio, and for each, name the system that should hold it instead. Why is the boundary discipline cheaper than the alternative of carrying the content in the wrong system?

3.7.4 Student Companion Notes

If you are reading this chapter as part of a process governance or enterprise integration course, the practical exercise that consolidates it most efficiently is to take a single L4 step from a familiar process such as a journal posting, invoice approval, or goods receipt and write out the seven mandatory attributes, the role assignment, the variant classification, and the lifecycle state in long-form. Then redo the exercise for a step in the same process that you initially thought was the same step but that operates differently in a different country, business unit, or product line. The second exercise is harder than the first because it forces a decision: Is the difference a variant of the first step, or is it a different step entirely? The decision itself is what teaches the data model.

A second exercise that consolidates the boundary discipline is to take an existing process repository and identify ten attributes or entries that should be removed because they belong in another system. (If you do not have access to a real process repository, the SAP Best Practices repository is a reasonable proxy.) The exercise can be uncomfortable because the offending entries are usually entries that someone took the trouble to create, and the question "What does it cost to leave them where they are?" is harder to answer than the question "What does it cost to remove them?" The discipline of this chapter is to answer the harder question, because the answer is always larger than it appears.

Finally, consider this thought experiment that consolidates the identifier discipline. Imagine that an organization's lead controller approaches the modeling team six months after go-live with a request to rename one approved step from "Book Journal" to "Post Journal Entry" for clarity. Walk through, in writing, exactly what should happen: the label changes, the identifier does not, the integrations do not break, the WalkMe content does not need republishing, the Joule grounding bundle continues to resolve, and the audit history continues to point to the same step. Then walk through the alternate scenario in which the team just renamed it everywhere. The resulting contrast illustrates why identifier discipline matters.

Chapter 4

Extending SAP Signavio for Downstream Consumption

A process repository becomes valuable to the enterprise only when its semantics can leave the modeling tool intact. Extension is not decoration; it is the discipline that turns a process model into an integration contract.

Chapter 3 established what must exist inside SAP Signavio to make process content execution-ready: stable identifiers, mandatory L4/L5 attributes, declared roles and variants, controlled lifecycle states, and a clear boundary between what belongs in SAP Signavio and what does not. That foundation is necessary, but it is not sufficient. Once the modeling discipline is in place, the next obligation is to expose that content to downstream systems in a way that is governed, predictable, and reversible.

This chapter focuses on the extension layer. We explore how SAP Signavio is configured, attributed, and instrumented so that other components of the architecture (SAP LeanIX, the canonical data spine on SAP Business Technology Platform [SAP BTP], SAP Build Process Automation, SAP Build Work Zone, WalkMe, and ultimately SAP AI Services) can consume process semantics without ambiguity. We treat extension as an architectural responsibility rather than a configuration exercise. Specifically, the chapter covers:

- How standard SAP Signavio attributes differ from custom attributes, and how to decide which to use for which purpose
- How external references are modeled so that links to applications, interfaces, configuration objects, and adoption assets remain stable across versions
- How SAP Signavio application programming interfaces (APIs) are prepared for extraction, including authentication, scoping, and pagination patterns suitable for enterprise loads
- When event-driven export is appropriate, when batch export is appropriate, and how the two patterns are reconciled
- How validation and data quality checks are implemented at the boundary between SAP Signavio and the SAP BTP data spine, and how violations are routed back to process owners

The intent is practical. By the end of the chapter, you should be able to specify the full extension contract for a process domain without ambiguity, including the attribute set, reference patterns, API surface, export mode, and validation rules. Each section proceeds from principle to design pattern to a concrete example drawn from the integrated SAP Signavio, SAP LeanIX, and SAP BTP landscape introduced earlier in the book.

4.1 Standard Attributes Versus Custom Attributes

SAP Signavio ships with a rich set of standard attributes. Process names, identifiers, descriptions, owners, status fields, lifecycle markers, and several Business Process Model and Notation (BPMN)–native fields are available immediately. For most modeling work, the standard attribute set is adequate. For execution-grade integration, it is rarely sufficient on its own.

The architectural question is not whether to use custom attributes but how to balance the two. A landscape that relies entirely on standard attributes will not be able to express execution context, configuration linkage, control intent, or AI eligibility. A landscape that relies excessively on custom attributes becomes brittle, hard to govern, and unstable across upgrades. The right balance is established deliberately, documented in a process attribute model, and enforced through SAP Signavio's dictionary.

In the following sections, we will walk through the role of attributes, how and when to use both standard and custom attributes, and how data types and controlled enumerations fit into the picture. We will end with a practical example of custom attributes.

4.1.1 The Role of Attributes in an Integration-Ready Process Model

Attributes serve four distinct purposes in an integration-ready process model. Each purpose corresponds to a different consumer:

- Identification and governance attributes establish what the process is and who owns it. These are consumed by every downstream system and by SAP Signavio.
- Execution attributes describe how the process step is executed in the field. They indicate which application, transaction, or SAP Fiori app is used. These are consumed primarily by SAP Build, the data spine, and WalkMe.
- Control and compliance attributes describe what risks the step mitigates and which controls it embodies. These are consumed by audit, SAP GRC solutions, and SAP Process Control integrations.
- AI eligibility attributes declare whether a step is permitted to be exposed to AI grounding and which constraints apply. These are consumed by SAP AI Core grounding pipelines as well as Joule.

A process step that lacks any one of these layers will work for some purposes and fail for others. A step without execution attributes will appear correct in process documentation but will not produce a usable AI grounding bundle. A step without control attributes will be readable by governance, risk and compliance (GRC) tooling only at a generic level. A step without AI eligibility attributes will either be uniformly excluded from AI consumption or, in less disciplined landscapes, uniformly included—both of which are unsafe defaults.

4.1.2 Standard Attributes: What They Cover and Where They Stop

SAP Signavio's standard attributes cover identification, basic ownership, description, and lifecycle state. They are stable, well-supported, and available through all standard APIs. When a need can be met by a standard attribute, use the standard attribute. They survive platform upgrades without remediation, while custom attributes do not always do so. Table 4.1 summarizes the standard attribute categories that are most relevant to integration and the typical limitation that arises when you rely on them alone.

Standard Attribute Category	What It Provides	Typical Limitation for Integration
Process and activity identifiers	Stable IDs generated by SAP Signavio for processes, models, and BPMN shapes	IDs are unique within the SAP Signavio workspace but do not encode domain or version semantics
Name and description	Human-readable label and documentation text	Free-text fields drift over time and are not safe for machine parsing
Status and lifecycle	The current status: draft, in review, approved, published, or archived	Statuses reflect the modeling lifecycle, not the execution publication state in the data spine
Owner and responsible role	Single-valued ownership reference	These are insufficient for landscapes with separate process, control, and execution owners
BPMN-native fields	Lane, pool, gateway type, or task type	These mechanical attributes do not express business semantics such as control intent or AI eligibility

Table 4.1: Standard SAP Signavio Attribute Categories and Integration Limitations

The pattern is consistent. Standard attributes are reliable and well-typed, but they describe the model as an artifact rather than as an execution contract. To bridge that gap, custom attributes should be introduced deliberately, sparingly, and with a clear lineage to the consuming systems.

4.1.3 A Disciplined Custom Attribute Strategy

Custom attributes in SAP Signavio are powerful because they can encode anything. That is also why they require governance. An ungoverned custom attribute set produces three predictable failure modes: ambiguous semantics (two attributes that mean almost the same thing), volatile naming (attributes renamed casually, breaking downstream consumers), and orphan attributes (defined once for a single use case and never removed).

A disciplined strategy treats the custom attribute set as a contract. The contract is small, named, versioned, and reviewed. Three principles govern its design:

- **Necessity**
  A custom attribute is created only when no standard attribute or external reference can satisfy the requirement. Convenience is not a proper justification.
- **Stability**
  Once a custom attribute is in production use, its name and data type do not change. Renaming a custom attribute is treated as a breaking change to the integration contract and requires explicit migration.
- **Scope**
  A custom attribute is defined at the smallest scope that satisfies its activity, process, lane, or workspace and is not promoted to a higher scope opportunistically.

These principles produce a custom attribute model that is small enough to remember, stable enough to integrate against, and structured enough to govern. Table 4.2 illustrates the recommended custom attributes for an execution-ready L4/L5 activity. The attributes are drawn from the integration contract used by the data spine in subsequent chapters.

Custom Attribute	Type	Scope	Consumer	Purpose
`Activity_Stable_ID`	String	Activity	Data spine, AI	Long-lived ID independent of model revisions
`Execution_Application_ID`	Reference	Activity	Data spine, SAP Build, WalkMe	Stable SAP LeanIX-aligned application reference
`Trigger_Application_ID`	Reference	Activity	Data spine	Application that initiates the step
`Fiori_App_ID`	String	Activity	WalkMe, SAP Build Work Zone	Stable SAP Fiori catalog identifier
`Transaction_or_SSCUI`	String	Activity	Data spine	Backend execution anchor
`Control_Intent`	Enum	Activity	GRC, audit	Preventive, detective, corrective, none

Table 4.2: Recommended Custom Attribute Set for an Execution-Ready Activity

Custom Attribute	Type	Scope	Consumer	Purpose
Variant_Key	String	Activity	Data spine, AI	Variant discriminator (e.g., region, legal entity)
AI_Exposure_Class	Enum	Activity	AI grounding	Permitted, restricted, prohibited
Process_Domain	Enum	Process	Spine, governance	Lead-to-cash, hire-to-retire, record-to-report
Approved_Version_Tag	String	Process	Spine, AI	Tag emitted on publication used by the data spine

Table 4.2: Recommended Custom Attribute Set for an Execution-Ready Activity (Cont.)

Two observations are worth making about this table. First, every custom attribute has a named consumer. Attributes without a consumer are not part of the contract and should not be created. Second, the attributes that look like references (Execution_Application_ID, Trigger_Application_ID, and Fiori_App_ID) are explicitly typed as references rather than as free text. Reference mechanics are addressed in Section 4.2.

4.1.4 Naming Conventions and Dictionary Management

Custom attributes are managed through SAP Signavio's dictionary. The dictionary is the single point at which custom attributes are defined, typed, scoped, and published to modelers. Treating the dictionary as code, rather than as an administrative surface, is the difference between a stable extension contract and a sprawling one. The naming convention used throughout this book follows three rules:

- Use PascalCase with underscores between semantic units. For example, use Execution_Application_ID and not executionApplicationId or exec-app-id.
- Use suffix-driven type signaling: _ID for identifiers, _Tag for short discriminators, _Class or _Intent for enumerated values, and _Path for hierarchical strings.
- Include a domain prefix only when needed. Most attributes are shared across the workspace. Where a domain-specific attribute is genuinely required, the domain is the leading token: HR_Position_Action_Type, FIN_Cost_Center_Subtype.

The dictionary itself is governed through a small number of artifacts: a published attribute catalog, a change log for the catalog, and an enumeration registry for fields whose values are constrained. The catalog and the change log are both exported nightly and persisted alongside the data spine, so that the data spine can validate incoming process exports against the attribute contract that was in force at the time of export.

Operationally, the dictionary is treated as a deliverable with an owner and a release cadence. A named dictionary steward is responsible for changes. Proposed changes flow through a lightweight intake form that captures the requested attribute name, type, scope, intended consumer, expected enumeration values, and justification. The intake form is not bureaucratic theater; it is the first place where the attribute's purpose is articulated clearly enough to evaluate whether a custom attribute is genuinely required or whether the need can be met by a standard attribute or an existing custom attribute.

Approved changes are released on a fixed cadence, typically monthly. A release bundles the new attributes, any deprecations, and any enumeration changes into a versioned dictionary release. The release is announced to modeling teams with a clear effective date and a remediation window for content that is affected by deprecations. Emergency changes outside the cadence are permitted but logged separately, and recurring emergencies are treated as a signal that the cadence itself needs to be revisited.

Three operational artifacts complete the picture. The dictionary release notes that describe what changed and why in language that modelers can act on. The attribute usage report, generated weekly, identifies attributes that have not been populated on any new model in the trailing ninety days; persistent zero-usage attributes are candidates for deprecation. The dictionary diff report, which comes out alongside every release, allows downstream consumers to see exactly which attributes, types, or enumerations changed since the previous release. The diff is consumed by the validation pipeline to detect schema drift between SAP Signavio and the data spine. Figure 4.1 shows the custom attribute dictionary in SAP Signavio.

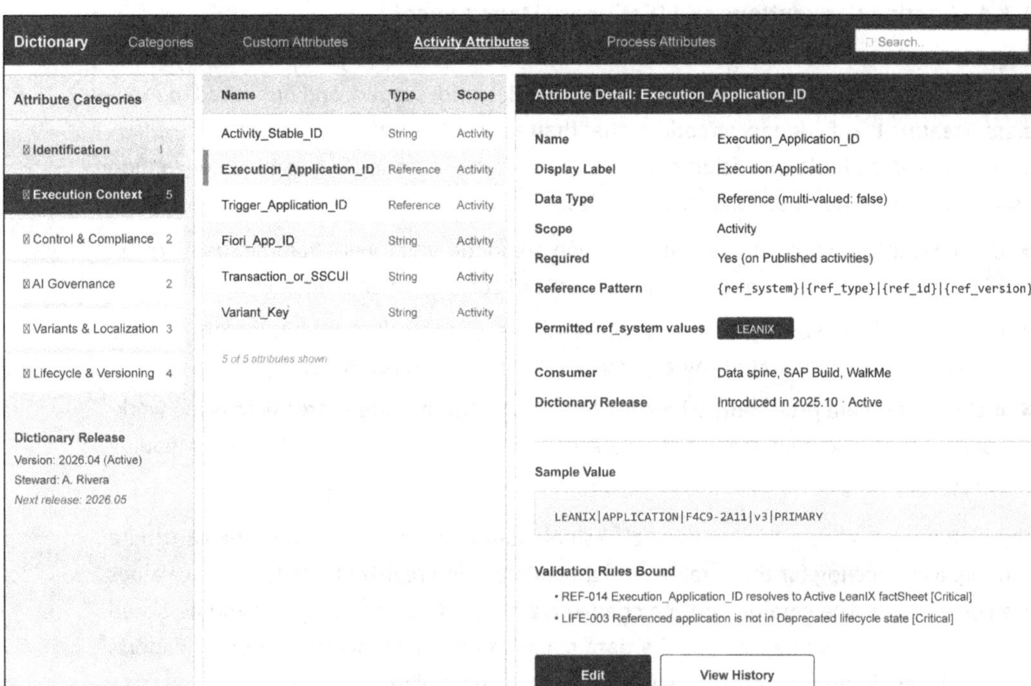

Figure 4.1: Custom Attribute Dictionary in SAP Signavio Process Manager

4.1.5 Common Pitfalls in Attribute Design

The following three pitfalls recur in landscapes that do not treat the attribute model as a contract:

- **Free-text capture of structured information**
  Modelers are tempted to record application names, transaction codes, and SAP Fiori app IDs as free-text fields in the activity description. This is convenient at modeling time but unusable downstream. Free text drifts, is not validated, and forces every consumer to implement its own parsing. The remedy is to extract these values into typed custom attributes from the start.
- **Attribute proliferation**
  Each modeling team adds attributes to solve its immediate need. Within a year, the dictionary contains dozens of attributes with overlapping semantics: `Application`, `App_Name`, `System`, `System_Of_Execution`, and so on. The remedy is a single point of approval for new custom attributes and a quarterly catalog review.
- **Schema-by-attribute**
  Modelers attempt to encode complex structures (lists of dependencies, key-value pairs, and nested references) as concatenated strings in a single attribute. This pattern always degrades into parser fragility downstream. If a field is genuinely structured, it belongs in an external reference (see Section 4.2) or in the data spine, not in an SAP Signavio attribute.

4.1.6 Attribute Scoping: Process, Activity, Lane, and Workspace

Every custom attribute has a scope that determines where in the SAP Signavio model the attribute can be set and which consumers can rely on it being populated. Choosing the smallest correct scope is one of the design choices that has the largest downstream effect.

Four scopes are commonly used (see Table 4.3). Workspace-scope attributes are defined once for the entire SAP Signavio workspace and apply uniformly. Process-scope attributes are set on a process header and apply to the process as a whole. Activity-scope attributes are set on each individual activity. Lane-scope attributes are set on lanes within a model and propagate to the activities that sit in the lane. The four are not interchangeable.

Scope	Where Assigned	Typical Example	Consumer Expectation
Workspace	Once, by the dictionary steward	Enumerations, controlled vocabularies	Always available; never absent
Process	Process header in the model	`Process_Domain`, `Approved_Version_Tag`	Required on every published process

Table 4.3: Custom Attribute Scopes and Consumer Expectations

Scope	Where Assigned	Typical Example	Consumer Expectation
Lane	Lane editor in the model	Default `Execution_Application_ID` for the lane	Inherited unless overridden at activity
Activity	Each L4/L5 activity	`Activity_Stable_ID`, `Trigger_Application_ID`, `Fiori_App_ID`	Required where execution context applies

Table 4.3: Custom Attribute Scopes and Consumer Expectations (Cont.)

Lane-scope attributes are routinely misused. Lanes in BPMN typically represent a role or organizational unit, not a system. Setting `Execution_Application_ID` at lane scope is reasonable when every activity in the lane genuinely executes in the same application; it is unreasonable when the lane represents a role that crosses several systems. The integration team will not detect this kind of misuse from the model alone, which is why lane-scope attributes are reviewed during validation under the cross-domain consistency rules in Section 4.5.

4.1.7 Data Types and Controlled Enumerations

SAP Signavio supports several attribute data types: text, number, date, Boolean, single-select, multi-select, and reference. The integration contract is built almost entirely on the typed forms. Free-text fields are reserved for genuinely human-readable content (such as notes and descriptions); structured data uses typed fields.

Single-select and multi-select fields require a controlled enumeration. The enumeration is defined in the SAP Signavio dictionary, versioned, and exported alongside the attribute values during ingestion. Three principles govern enumeration design:

- **Closed-world**
  Enumerations are closed; values not in the enumeration are not permitted. "Other" is not an enumeration value; it is an indication that the enumeration is incomplete and should be expanded explicitly.
- **Stable codes**
  The persisted value is a short stable code (e.g., `PREVENTIVE`), not the displayed label (e.g., Preventive Control). Labels can change for clarity without breaking integrations; codes do not change.
- Versioned with a deprecation lane: A retired enumeration value is marked **Deprecated** rather than deleted. Existing references continue to resolve, but no new uses are permitted. Deletion happens only after a structural sweep confirms no live references.

Boolean fields are similarly disciplined. Where a flag has more than two meaningful states (e.g., a control may be **Preventive, Detective**, **Corrective**, or **Not Applicable**), the field is modeled as a single-select, not as several Booleans. Three Booleans where one enumeration is correct is a recognized anti-pattern.

4.1.8 Worked Example: Custom Attributes Across Three Process Domains

Custom attribute design is easier to understand with an example. Table 4.4 takes one activity from each of the three process domains commonly modeled in SAP Signavio at L4/L5 depth (lead-to-cash, hire-to-retire, and record-to-report) and shows how the same custom attribute set is populated in each case. The attribute set is identical; the values are domain-specific.

Custom Attribute	Lead-to-Cash: Approve Customer Credit Limit	Hire-to-Retire: Create Position in SAP SuccessFactors Employee Central	Record-to-Report: Post Period-End Adjustment
Activity_Stable_ID	ACT-L2C-CRD-040	ACT-H2R-POS-012	ACT-R2R-PEC-027
Process_Domain	LEAD_TO_CASH	HIRE_TO_RETIRE	RECORD_TO_REPORT
Execution_Application_ID	S4HANA_CORE	SUCCESSFACTORS_EC	S4HANA_FINANCE
Trigger_Application_ID	S4HANA_CORE	S4_MDG	S4HANA_FINANCE
Fiori_App_ID	F1604	F2547	F3093
Transaction_or_SSCUI	FD32	(EC) Position Manage	F.5
Control_Intent	PREVENTIVE	PREVENTIVE	DETECTIVE
Variant_Key	REGION:NA	LEGAL_ENTITY:US01	ENTITY:GROUP
AI_Exposure_Class	RESTRICTED	RESTRICTED	PROHIBITED

Table 4.4: Same Custom Attribute Set Populated Across Three Process Domains

Two observations emerge. First, the attribute schema is invariant across domains; only the values change. This is what makes a single ingestion contract usable across the entire process portfolio. Second, the `AI_Exposure_Class` differs meaningfully across the three. The record-to-report adjustment is marked `PROHIBITED` because the activity touches material financial postings that the organization has chosen to keep entirely out of AI-grounded automation. The hire-to-retire and lead-to-cash activities are restricted, meaning AI may be used in suggestion mode but never in autonomous-execution mode. The classification is per-activity, not per-domain, and is set by the process owner in consultation with the AI governance function.

To summarize, the attribute model is small, named, scoped, consumer-driven, and uniform across domains. Standard attributes are preferred where they suffice. Custom attributes are introduced deliberately and governed as a contract.

4.2 External Reference Patterns

The previous section established that several execution-relevant attributes—application, trigger application, SAP Fiori app ID, configuration anchor—should be modeled as references rather than as free text. This section describes how those references are constructed, why URL-based linking is avoided, and how the references survive lifecycle changes in upstream systems.

The objective of an external reference is simple: A downstream consumer reading the SAP Signavio export should be able to resolve the reference deterministically against another system of record, without inferring meaning from a label, URL, or screen name. Determinism is the operative word. Anything that requires interpretation is not a reference; it is documentation.

4.2.1 Defining an External Reference

An external reference, in the sense used throughout this book, has three properties. It carries a stable identifier from an upstream authoritative system, the type of the referenced object (so that consumers know how to resolve it), and an optional version anchor (when the referenced object is itself versioned). Any names, URLs, or descriptions beyond that are auxiliary.

The minimal reference structure used in SAP Signavio custom attributes is as follows:

```
{ref_system, ref_type, ref_id, ref_version}
```

In SAP Signavio, this structure is represented either as a single delimited string (when only one reference is permitted on the attribute) or as a multi-valued attribute where each value follows the same shape. Following are some examples:

```
LEANIX   | APPLICATION      | F4C9-2A11-... | v3
S4HANA   | TRANSACTION      | KS01
FIORI    | APP_ID           | F1612
SPRO     | SSCUI            | 101005        | 2024.1
```

The four fields are pipe-delimited so that downstream parsers can split deterministically. Whitespace within fields is forbidden. The `ref_system` field is enumerated and governed; it cannot be invented by modelers. The `ref_type` field is enumerated within the chosen `ref_system`.

The choice of pipe as the delimiter is deliberate. Commas are unsafe because identifiers in some upstream systems contain commas. Slashes are unsafe because IMG paths and URI segments use slashes. Tabs are unsafe because some modeling tools strip tab characters silently. Pipes are rare enough in real identifiers that they can serve as a reliable delimiter, and where a pipe does occur in an upstream identifier, the integration team is alerted by a parser error rather than by silent truncation.

Two further conventions reduce ambiguity. Empty fields are written as an empty string between pipes rather than omitted; a four-field reference always has exactly three pipes regardless of whether all fields are populated. Trailing pipes are not used to indicate that the field is reserved for future use; the reference shape is fixed at four fields and is extended only by formal contract change.

4.2.2 Why URL-Based Linking Is Avoided

It is tempting to embed URLs directly into SAP Signavio attributes. You might want to include a link to an SAP LeanIX factsheet, an SAP for Me application page, or an SAP Fiori launchpad tile. URLs feel concrete because you can open them in a browser. However, they are poor integration tools for four reasons:

- URLs encode runtime endpoints, not identity. When the SAP LeanIX URL pattern changes, every embedded URL becomes stale even though the underlying object is unchanged.
- URLs cannot be validated semantically. A consumer can confirm that a URL is reachable but not that it points to the correct object class.
- URLs are environment-bound. A URL that resolves in production may not resolve in test or QA. Process content must remain environment-neutral.
- URLs invite human authoring errors. Copy-pasting a URL is fast at modeling time and unverifiable at integration time.

The remedy is to record only the identity of the referenced object in the attribute. URLs, where they are needed for human navigation, are constructed at consumption time by the consuming system using its current endpoint pattern and the stable reference. This decoupling is what makes process content portable across environments and resilient to upstream URL changes.

4.2.3 Reference Patterns by Target System

Each reference target in the integrated landscape has a recommended pattern. The patterns share the four-field structure but differ in which fields are mandatory and which enumerations apply. Table 4.5 contains reference patterns by target system.

Target	ref_system	ref_type values	ref_id source	ref_version
SAP LeanIX application	LEANIX	APPLICATION, IT_COMPONENT, BUSINESS_CAPABILITY	SAP LeanIX factSheetID	Optional; SAP LeanIX revision tag
SAP LeanIX interface	LEANIX	INTERFACE	SAP LeanIX interface factSheetID	Optional
SAP S/4HANA transaction	S4HANA	TRANSACTION	Transaction code (e.g., Transaction KS01)	Not used
SAP S/4HANA configuration object	SPRO	SSCUI, IMG_NODE	Self-service configuration user interface (SSCUI) ID or IMG path key	Release tag (e.g., 2024.1)
SAP Fiori app	FIORI	APP_ID	SAP Fiori library app ID (e.g., F1612)	Not used
SAP Build automation artifact	BUILD	PROCESS_DEFINITION, AUTOMATION	SAP Build resource ID	Resource version
WalkMe asset	WALKME	WALKTHRU, SHOUTOUT, RESOURCE	WalkMe asset ID	Asset version
Joule skill or grounding bundle	AI	SKILL, BUNDLE	SAP AI Core deployment ID	Bundle hash

Table 4.5: Reference Patterns by Target System

Not every activity requires every reference. A typical L4/L5 activity carries between one and four references: the application in which it executes, the trigger application (if different from the executing application), the SAP Fiori app ID for any UI-driven step, and a configuration anchor for any step whose behavior depends on backend customizing. Adding more references than the activity genuinely needs is a form of attribute proliferation and should be avoided.

4.2.4 Multi-Valued and Qualified References

Some activities legitimately reference more than one object of the same type. An activity that consumes data from two upstream systems may reference two SAP LeanIX interface factsheets. A configuration step that touches several SSCUIs may carry a list of Transaction SPRO references. SAP Signavio supports multi-valued custom attributes for this purpose; each value follows the four-field shape.

Where the relationship between the activity and the reference needs to be qualified—for example, distinguishing between a primary execution application and a secondary read-only system—a qualifier is appended as a fifth field, as follows:

```
LEANIX | APPLICATION | F4C9-2A11 | v3 | PRIMARY
LEANIX | APPLICATION | A77B-1F33 | v2 | READ_ONLY
```

The qualifier vocabulary is small and enumerated. Permitted values include `PRIMARY`, `SECONDARY`, `READ_ONLY`, `FALLBACK`, and `DEPRECATED`. Anything outside the enumeration is rejected by the validation pipeline described in Section 4.5.

4.2.5 Governance of References Across Lifecycles

A reference can become invalid in three ways:

- The referenced object can be retired in its source system.
- The reference type can be reclassified.
- The version anchor can fall behind the referenced object's current version.

Each of these failures is handled differently. When the referenced object is retired, the reference becomes a dangling pointer. The data spine detects this during reconciliation and raises a structural validation finding against the SAP Signavio process. The process owner is notified through the validation feedback channel described in Section 4.5; the activity is not silently rewritten.

When the reference type is reclassified, the reference is technically valid but semantically wrong. Detection is harder and depends on the upstream system signaling the reclassification through its event stream. The discipline introduced in Section 4.4 (event-driven export) addresses this case.

When the version anchor falls behind, the data spine treats the reference as valid but emits a metadata flag indicating that the activity is grounded against a back-version. AI grounding pipelines may choose to exclude such activities from current-state bundles. The version anchor is therefore not merely informational; it directly affects AI eligibility. Figure 4.2 shows how upstream object lifecycle changes affect activity reference in the canonical data spine. Each lifecycle transition triggers a deterministic spine handling rule. No transition is silently absorbed.

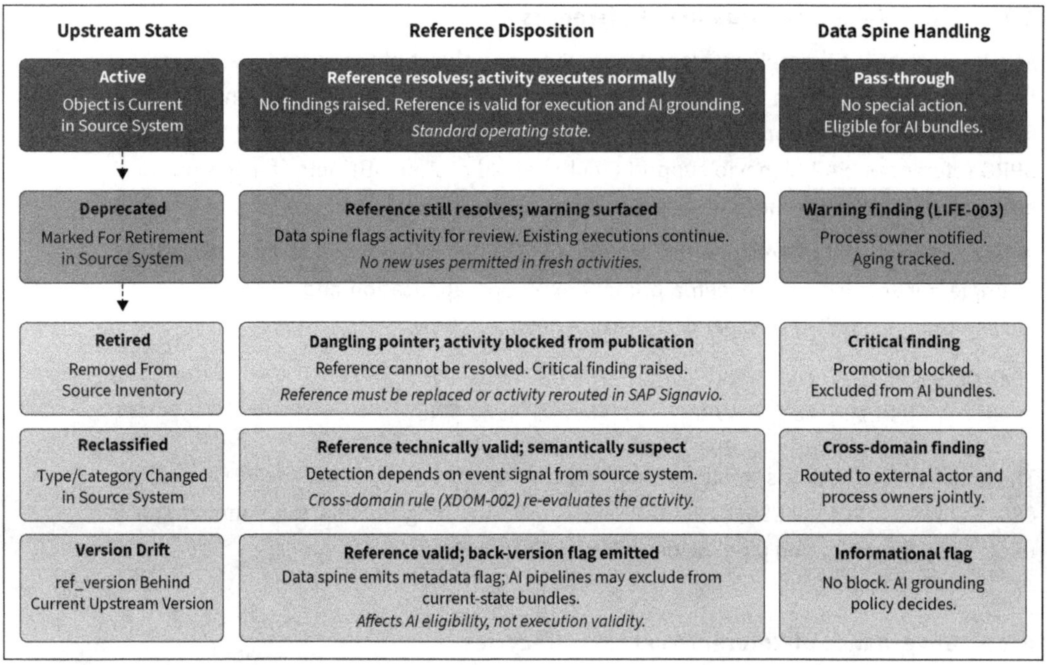

Figure 4.2: Reference Lifecycle States and the Corresponding Spine Handling

The principle behind these patterns is that references are not best-effort hyperlinks. They are typed, governed, validated identity assertions. Their integrity is what makes the canonical data spine trustable for execution and AI grounding.

4.3 Preparing SAP Signavio APIs for Extraction

Section 4.1 and Section 4.2 addressed what should exist inside SAP Signavio. This section addresses how that content leaves SAP Signavio. The mechanism is the SAP Signavio API surface. The objective is to expose process and dictionary content to SAP BTP integration flows in a way that is authenticated, scoped, paginated, and resilient to typical enterprise loads.

We treat the API surface as a contract between the SAP Signavio workspace administrator and the SAP BTP integration team. The contract specifies which endpoints are used, which credentials and scopes are issued, how rate limits are respected, and how schema changes are communicated. A well-defined contract prevents the most common operational failures: credential sprawl, accidental privilege escalation, throttling under peak loads, and silent breakage on SAP Signavio platform updates.

4.3.1 The SAP Signavio API Landscape

SAP Signavio exposes several API surfaces. Not all are appropriate for integration with the canonical data spine. The relevant surfaces, in order of typical use, are as follows:

1. The Process Manager Public API, which exposes models, diagrams, dictionary entries, custom attribute values, and process collections.
2. The Process Intelligence API, which exposes process mining outputs, conformance results, and event-log queries.
3. The Process Governance API, which exposes workflow tasks, approvals, and lifecycle transitions.
4. The Collaboration Hub APIs, which expose comments, ratings, and engagement data.

For the extension contract described in this chapter, the Process Manager Public API is the primary surface. It carries the structured process content and dictionary metadata that the data spine ingests. The Process Intelligence API is consumed only by analytics pipelines, not by the data spine. The Governance API is consumed by the publication-state synchronization flow described in Chapter 11. The Collaboration Hub APIs are not consumed by the data spine.

4.3.2 Authentication, Tokens, and Service Identities

SAP Signavio APIs require authenticated access. Three authentication patterns are available, each appropriate for different contexts:

- **Personal API tokens**
  These are issued to a named user. They are useful for ad hoc extraction during design or for running validation scripts manually. Personal tokens inherit the user's permissions. They are not appropriate for production integration because their lifetime is tied to the user, and their scope is harder to audit.
- **Service account tokens**
  These are issued to a dedicated SAP Signavio service identity. The service identity is created with the minimum permissions required: read access to the relevant workspace, the dictionary, and the published model collection. Write access is not granted. The service token is the standard credential for production extraction flows.
- **OAuth client credentials (where the SAP Signavio tenant supports it)**
  An OAuth client is registered in the SAP Signavio identity layer and used by SAP Integration Suite or SAP BTP destinations. Tokens are short-lived and rotated automatically. This pattern is preferred wherever it is available because it removes long-lived secrets from the integration flow.

Regardless of the chosen pattern, three operational rules apply:

- Credentials are stored only in SAP BTP destinations or the SAP Credential Store, never in code or configuration files.

- Credential rotation is scheduled and rehearsed.
- Each credential carries metadata identifying its purpose, owner, and expiry; orphaned credentials are revoked on a fixed cadence.

The credential rotation procedure deserves emphasis because it is the most common source of avoidable production incidents. A workable rotation pattern has four steps:

1. A successor credential is provisioned alongside the active credential with identical scoping.
2. The integration flow is updated to read the credential reference from a configurable indirection so that the active credential can be changed without redeploying code.
3. The successor credential is activated during a low-traffic window and observed for at least one full extraction cycle.
4. Once the successor is confirmed working, the predecessor is revoked and removed from the credential store. Skipping the parallel-activation step turns rotation from a routine maintenance task into a high-risk cutover.

Two further practices reduce the operational burden. The first is to align credential expiry with a calendar that the integration team controls, avoiding expiries on weekends, on quarter-end days, or during change-freeze windows. The second is to instrument the integration flow to emit a warning when a credential is within thirty days of expiry; reactive rotation in response to expiry alarms is recoverable, while reactive rotation in response to authentication failures is an incident.

4.3.3 Endpoint Inventory for Spine Ingestion

The data spine ingestion flow described in Chapter 11 consumes a defined subset of the Process Manager Public API. The endpoints in scope are listed in Table 4.6.

Endpoint Purpose	Method	Notes
List process collections in the workspace	`GET`	Enumerates publishable scopes
List models in a collection	`GET`	Filtered by published state
Retrieve a model with its diagram and shapes	`GET`	Returns BPMN structure plus attribute values
Retrieve dictionary entries by type	`GET`	Enumerates roles, applications, and reference targets
Retrieve custom attribute definitions	`GET`	Validates the attribute contract version

Table 4.6: Process Manager Public API Endpoints Used by Spine Ingestion

Endpoint Purpose	Method	Notes
Retrieve revision history of a model	`GET`	Used for lineage and version reconciliation
Retrieve a published export bundle	`GET`	Optional, used in batch mode

Table 4.6: Process Manager Public API Endpoints Used by Spine Ingestion (Cont.)

Each endpoint is documented in Appendix E with the exact path, query parameters, response schema, and pagination behavior. The integration flow does not assume the response shape; it parses defensively and validates against a stored schema version. Schema changes upstream are detected as validation failures rather than runtime crashes.

4.3.4 Scoping and Permissioning

A common mistake in SAP Signavio integration is to grant the service identity workspace-administrator privileges in the hopes of making extraction simpler. However, this shortcut is unsafe and unnecessary. The data spine requires read access to a small set of resources and nothing more. Scoping the service identity to that minimum has three benefits: It reduces blast radius if the credential is compromised, it makes audit reviews short and decisive, and it protects production process content from accidental modification by automation. The recommended scoping is:

- Read access to the production process collection only. Draft and personal collections are not visible to the service identity.
- Read access to the dictionary, including custom attribute definitions and enumerations.
- Read access to revision metadata for the in-scope collection.
- No write access to any object. Write operations are reserved for human modelers and governance workflows.

Where the workspace contains both production and non-production collections, two service identities are recommended (one per environment) with no cross-environment privileges. This mirrors the production-discipline pattern applied to other SAP credentials.

4.3.5 Rate Limits, Pagination, and Backoff

The Process Manager Public API enforces rate limits at the workspace level. The exact limits depend on the SAP Signavio tier; published values must be confirmed against the current SAP Signavio documentation for the workspace in question. The integration flow assumes that rate limits apply and is designed to operate well below them.

Three patterns make this practical. First, requests are paginated using the API's native pagination tokens; large collections are processed in pages of bounded size rather than fetched in a single call. Second, requests are batched and sequenced; concurrent requests are capped at a configurable concurrency limit set for each environment. Third, when a request fails because of a transient error or because the API has signaled that the rate limit has been exceeded, the flow waits before retrying. The wait doubles on each successive failure (exponential backoff), and a small random offset (jitter) is added so that multiple flows recovering from the same incident do not retry in lockstep. The flow never retries in a tight loop, which would amplify a transient failure into a sustained outage. A simple ingestion loop, expressed in pseudo-code in Listing 4.1, illustrates the pattern.

```
for page in paginate(collection_endpoint, page_size=50):
    for model_ref in page.items:
        model = with_backoff(get_model, model_ref.id)
        validate(model, schema_version)
        emit(model)
    sleep(per_page_pause)
```

Listing 4.1: Simple Ingestion Loop

The `per_page_pause` and `concurrency_limit` values are environment-tuned. In development and quality assurance (QA), they are small enough to keep extractions short. In production, they are chosen so that a full extraction completes well within the maintenance window and never approaches the workspace rate limit. These parameters are documented in the runbook for the integration flow. Figure 4.3 shows the SAP Signavio API token configuration screen with scoped service identity.

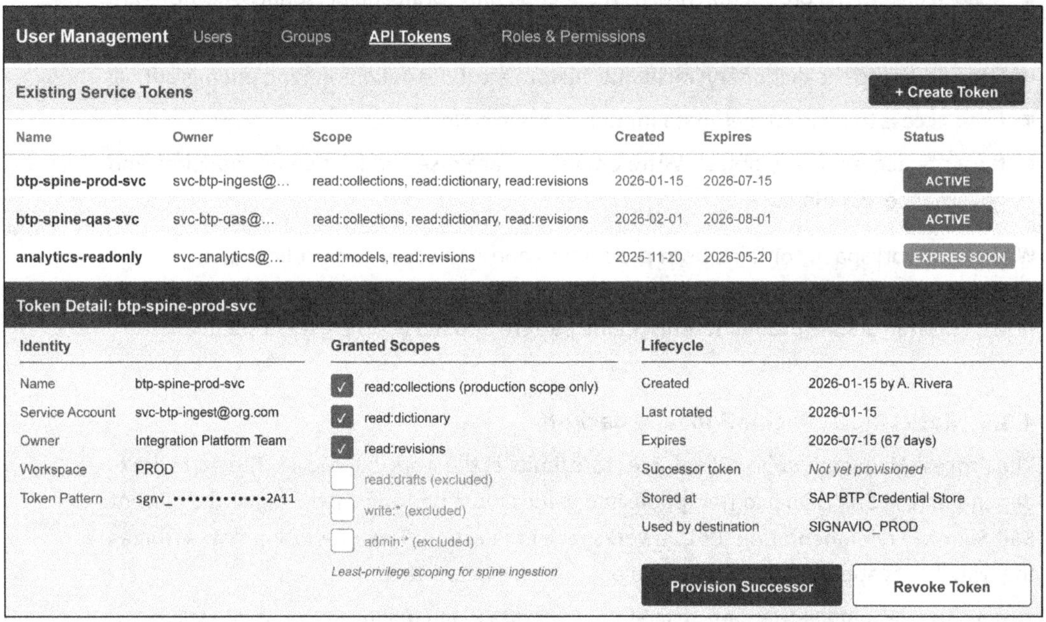

Figure 4.3: SAP Signavio API Token Configuration Screen with Scoped Service Identity

4.3.6 Schema Versioning and Compatibility Discipline

As SAP Signavio evolves, endpoints gain new fields, attribute definitions add enumerations, response shapes occasionally restructure, and so on. The integration flow must accommodate this evolution without producing silent data corruption.

Two practices address this. The first is schema pinning, where the integration flow records the schema version of the API response (alongside each ingested model) and the version of the custom attribute contract that was in force at extraction time. The second is forward-compatible parsing, where unknown fields are preserved but ignored and missing fields produce explicit validation findings rather than null propagation. Together these practices ensure that an upstream change is observable, traceable, and recoverable.

Schema versions are stored in a small registry on SAP BTP. Each registry entry records the API endpoint, the schema digest, the date observed, and a human-readable change note. The registry is itself an audit artifact and is consumed by the validation pipeline described in Section 4.5.

4.3.7 Sample API Calls and Response Handling

A small set of representative API calls illustrates the operational pattern. The examples that follow are simplified for readability; the actual integration flows include error handling, telemetry, and schema validation that are omitted here. The path patterns shown are illustrative. The exact paths must be confirmed against the SAP Signavio API documentation for the workspace tier in use.

The following code is for list models in a published collection:

```
GET /api/v2/workspaces/{ws_id}/collections/{coll_id}/models?status=
published&limit=50
Authorization: Bearer {service_token}
Accept: application/json
```

The response is a paginated envelope containing a list of model summaries and a continuation token. The integration flow does not assume the page size is honored exactly; it consumes whatever is returned and follows the continuation token until exhausted.

The following code retrieves a single model with shapes and attribute values:

```
GET /api/v2/workspaces/{ws_id}/models/{model_id}?include=
shapes,attributes,revisions
Authorization: Bearer {service_token}
```

The response contains the model header, the BPMN shape graph, the attribute values for each shape, and a compact revision summary. The integration flow extracts only the fields named in the schema contract; unknown fields are preserved in a raw payload archive but not used for canonical fields.

An abbreviated response payload, with nonessential fields elided, illustrates the structure that the parser consumes, as shown in Listing 4.2.

```
{
  "modelId": "M-H2R-POS-CREATE",
  "revision": 14,
  "publishedState": "PUBLISHED",
  "processDomain": "HIRE_TO_RETIRE",
  "shapes": [
    {
      "shapeId": "shape-027",
      "type": "Task",
      "name": "Create Position in EC",
      "customAttributes": {
        "Activity_Stable_ID": "ACT-H2R-POS-012",
        "Execution_Application_ID": "LEANIX|APPLICATION|F4C9-2A11|v3|
PRIMARY",
        "Trigger_Application_ID":  "LEANIX|APPLICATION|A12B-7733|v2|
PRIMARY",
        "Fiori_App_ID": "FIORI|APP_ID|F2547",
        "Control_Intent": "PREVENTIVE",
        "AI_Exposure_Class": "RESTRICTED"
      }
    }
  ]
}
```

Listing 4.2: Abbreviated Response Payload

The parser maps each named field into the data spine's canonical schema. Fields outside the contract are not consumed; their presence in the raw payload archive supports later forensic analysis if a defect is discovered after ingestion.

The following code retrieves dictionary entries for a custom attribute:

```
GET /api/v2/workspaces/{ws_id}/dictionary?type=custom_attribute&include=
enumerations
Authorization: Bearer {service_token}
```

Dictionary calls are made first in each ingestion run. The dictionary version retrieved at this point becomes the schema contract for the rest of the run. If the dictionary version differs from the version persisted by the previous successful run, the run records a schema change event in the registry and continues; subsequent validation rules are evaluated against the new version.

4.3.8 Calling SAP Signavio APIs: SAP Integration Suite Versus Direct from SAP BTP

Two architectural choices are available for placing the API calls. Each has legitimate use cases, as follows:

- **SAP Integration Suite**
  Cloud Integration is used as the orchestration layer. The integration flow is modeled as a Cloud Integration integration flow (iFlow), with steps for authentication, paginated retrieval, schema validation, and persistence to the data spine. This pattern is preferred when the same flow needs to combine data from multiple sources (e.g., SAP Signavio plus SAP LeanIX), when the team has Cloud Integration expertise, or when monitoring and alerting are standardized on Cloud Integration's capabilities.
- **Direct from SAP BTP services**
  A custom service running on SAP BTP, Cloud Foundry runtime or SAP BTP, Kyma runtime calls the SAP Signavio API directly. The service uses an SAP BTP destination for the connection profile and SAP Credential Store for the service token. This pattern is preferred when the ingestion logic is closely coupled to the data spine's persistence model, when complex transformation is required that would be awkward in Cloud Integration, or when the integration team prefers code-first development.

Both patterns can coexist. A common topology uses Cloud Integration for the well-bounded extraction flow (list models, retrieve each, persist to staging) and a direct SAP BTP service for the data spine-side processing (validation, normalization, promotion). Whichever choice is made, the connection profile is the same: a destination definition pointing to the SAP Signavio workspace, with the service token referenced from the credential store. Hard-coded URLs and credentials are not permitted in either pattern. Figure 4.4 shows an example of each option for SAP Signavio API consumption; both consume the same scoped service token.

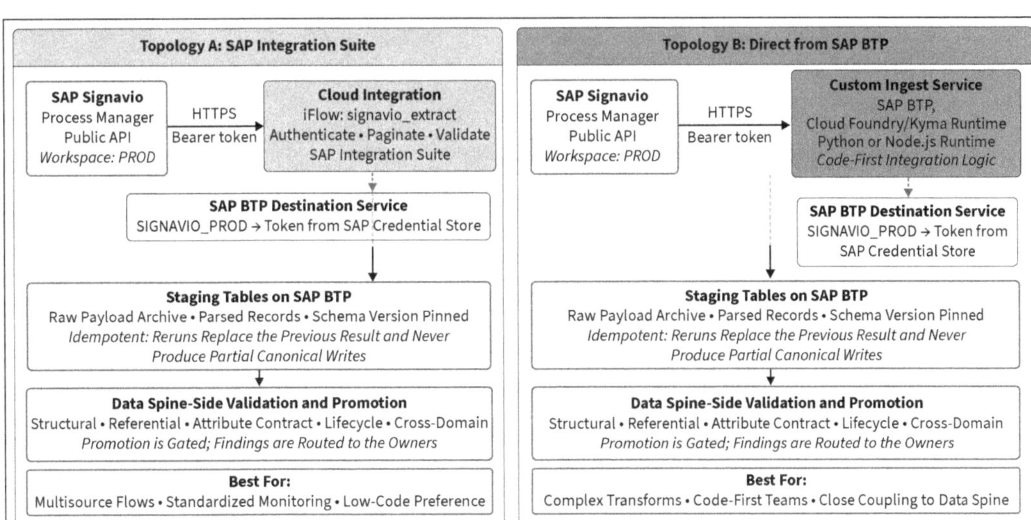

Figure 4.4: Two Valid Topologies for SAP Signavio API Consumption

4.3.9 Operational Telemetry for the Extraction Flow

An extraction flow that runs unattended for months will eventually drift. The workspace will grow, the rate at which models are revised will change, and the API surface will evolve. Telemetry is what makes drift observable before it becomes an incident. Four metrics are tracked by every production extraction run:

- Run duration, broken down by phase (dictionary, list, retrieve, validate, persist).
- Records ingested, broken down by collection and by published-state transition (new, updated, unchanged, deprecated).
- API response codes, broken down by endpoint, with a particular focus on rate-limit responses and timeouts.
- Validation findings emitted, broken down by severity and rule category (see Section 4.5).

Metric thresholds are tuned per environment. A production run that takes twice as long as its baseline is investigated even if it succeeds. A sudden spike in validation findings is investigated even if the integration flow itself reports success. Metrics are persisted in SAP BTP's monitoring stack and reviewed weekly. The objective is to detect quiet degradation, not just outright failure.

4.4 Event Versus Batch Export Considerations

Once the API surface is prepared, the next decision is when to call it. Two patterns dominate enterprise integrations: event-driven export, where SAP Signavio (or its surrounding workflow) signals changes as they occur, and batch export, where the full or incremental content is pulled on a schedule. Each pattern has a legitimate role; neither is universally correct. The decision is made for each integration flow, not by landscape.

This section discusses when each pattern applies, how they are combined, and how their outputs are reconciled in the data spine.

4.4.1 Two Patterns, Two Failure Modes

Batch export is conceptually simple. On a schedule, the integration flow calls the SAP Signavio API, retrieves the in-scope content, and writes it to the data spine. Batch export is reliable, predictable, easy to operate, and easy to audit. Its failure mode is latency: Changes that occurred between scheduled runs are invisible until the next run completes. For a daily batch, the worst-case latency is roughly twenty-four hours.

Event export is conceptually more demanding. The integration flow subscribes to change notifications and reacts to each notification as it arrives. Event export is low-latency and well-suited to use cases that depend on near-real-time alignment. It can result in message loss, out-of-order delivery, and duplicate delivery. None of these are out of the ordinary, and all of them must be handled separately.

The choice between batch and even export is driven by two questions: how quickly downstream systems need to see a change and how tolerant downstream systems are of missing or replayed events. The answers vary by consumer. Table 4.7 walks through recommendations for when to use each pattern.

Consumer	Latency Tolerance	Recommended Pattern
Canonical data spine (publication state)	Hours	Daily batch with intra-day delta refresh
Canonical data spine (model content)	Day	Daily batch
WalkMe authoring (process step changes)	Hours	Daily batch and on-demand pull on publication event
SAP Build Process Automation (workflow definitions)	Days	Weekly batch with manual trigger on major release
AI grounding bundle refresh	Day	Daily batch tied to spine publication
Compliance and audit replay	Variable	Batch with full history retention

Table 4.7: Recommended Export Pattern by Consumer

4.4.2 Batch Export Design

Batch export is the default for the canonical data spine. The reasons are practical: Process content changes on the order of days rather than minutes, the data spine is built around versioned snapshots, and audit replay is a first-class requirement that batch supports naturally. The following three design questions shape the batch pattern:

- **Full versus delta**
  A full export retrieves every published model in the in-scope collection. A delta export retrieves only models whose revision has advanced since the last successful run. Delta export is faster and lighter, but it depends on a reliable watermark. The recommended approach is a hybrid: nightly delta with a weekly full reconciliation. The latter will serve as a structural checkpoint that catches any silent drift.
- **Window selection**
  Batch runs are scheduled in a defined window, typically off-peak local hours for the SAP Signavio workspace. The window is chosen so that the run completes well within it, with margin for transient errors and retries. Batch jobs that overrun their window are treated as incidents.

- **Idempotency**
  A batch run that fails mid-way must be safely rerunnable. The ingestion flow writes to staging tables in the data spine and only promotes to canonical tables after the run is confirmed complete and validation has passed. A rerun replaces the staging content; canonical content is never partially overwritten.

These three properties are the operating discipline for batch export. Chapter 11 presents a more detailed sequence diagram and error-handling flow.

4.4.3 Event-Driven Export Design

Event-driven export is reserved for cases where latency genuinely matters. In the integrated landscape described in this book, the principal candidate is the publication-state event: when a process version transitions from **Approved** to **Published** in SAP Signavio Process Governance, downstream systems benefit from learning quickly so that AI grounding bundles, WalkMe content, and SAP Build automations can be refreshed against the new version.

The event channel is an SAP Integration Suite event subscription that consumes SAP Signavio Process Governance lifecycle events. Each event carries the following:

- The SAP Signavio model identifier and revision
- The lifecycle transition (e.g., **Approved** to **Published**)
- The actor and timestamp
- A correlation ID used to tie the event to its triggering workflow

The event handler does not write to the canonical spine directly. Instead, it triggers a targeted pull through the Process Manager Public API for the affected model and a publication-state update in the data spine. This indirection is deliberate. It keeps the data spine's content path consistent because every model in the data spine arrived through the API ingestion flow and never through an event payload, while still benefitting from the event's low latency for state transitions.

Event handling addresses the three classical failure modes explicitly. Message loss is mitigated by reconciling the event-driven publication state with the daily batch publication state; any divergence is flagged. Out-of-order delivery is mitigated by treating the latest revision in the event as authoritative only after validating against the API response. Duplicate delivery is mitigated by an idempotency key derived from the model ID, revision, and transition; duplicate keys are detected and ignored. Figure 4.5 illustrates event and batch export combined into a single ingestion contract. Events drive low-latency state transitions, batch export carries content, and reconciliation closes the loop

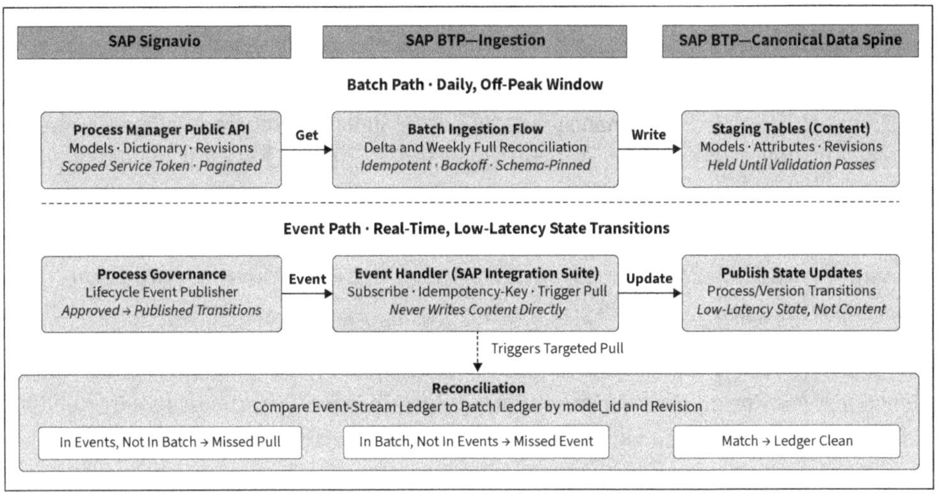

Figure 4.5: Event and Batch Export Combined into a Single Ingestion Contract

4.4.4 Reconciliation Between Event and Batch Exports

Where both patterns are active, reconciliation is mandatory. Reconciliation is not optional; it is the mechanism by which the integration team detects silent drift between SAP Signavio and the data spine. The following reconciliation procedure runs at the end of each batch cycle:

1. Compute the set of model revisions present in the data spine after the batch has completed.
2. Compute the set of model revisions implied by the event stream during the same window.
3. Compare the two sets. Any revision present in events but missing from the batch is a missed pull. Any revision present in the batch but missing from events is a missed event.
4. Both kinds of mismatch are recorded as findings and routed to the integration owner. Recurring mismatches trigger an investigation into the underlying cause (subscription health, API behavior, or workflow timing).

Reconciliation is what allows the integration team to make confident statements about the data spine's currency. Without reconciliation, the question "Is the data spine up to date?" has no defensible answer.

4.4.5 Decision Heuristics: Choosing the Right Pattern

The following heuristics summarize the decision logic (they are intentionally simple, as complexity breeds inconsistency across integration flows):

- If the consumer can tolerate a day of latency, batch export is sufficient. Do not introduce events.

- If the consumer is publication-state sensitive (AI grounding, WalkMe activation, or SAP Build deployment), use event-driven export for the transition signal and batch export for the content.
- If the consumer is conformance-sensitive (SAP Signavio Process Intelligence outputs feeding alerts), event-driven export is appropriate, but the channel is the SAP Signavio Process Intelligence event surface, not the process manager.
- If the consumer is a dashboard that will be read by human employees, batch export with a short refresh interval is almost always preferable to event-driven export. Events make dashboards harder to operate, not easier.

The principle behind these heuristics is that the consumer pays for complexity. An integration team that introduces events without a corresponding consumer requirement has added operational burden without adding value. The default is batch export; events are introduced where the latency requirement is real and the operational cost is justified.

4.4.6 Worked Example: A Reconciliation Cycle

To make reconciliation more tangible, consider a single twenty-four-hour window in a production landscape. The SAP Signavio workspace contains roughly 400 published models. The event subscription is active. The batch export runs at 2 a.m. local time. During the window, the following activity occurs in SAP Signavio:

- Sixteen models are revised and republished by the modeling team during business hours.
- Three models are transitioned from **Deprecated** to **Archived**.
- One new model is published for the first time.
- Two models are reverted to a prior revision after a finding from the previous day's batch.

The event subscription is expected to deliver twenty-two events corresponding to these transitions. The batch run that follows is expected to register the same set of changes through the API. At reconciliation time, the integration team compares two ledgers, shown in Table 4.8.

Source	New	Republications	Reversions	State Transitions
Event stream (during window)	1	16	2	3
Batch (next morning)	1	16	2	3
Difference	0	0	0	0

Table 4.8: A Clean Reconciliation Cycle

The clean case is the goal. In practice, the more interesting case is divergence. Suppose instead that the batch reports fifteen republications and the event stream shows sixteen.

The reconciliation procedure identifies the missing record by joining the two sets on the model identifier and isolating the entry present in events but absent in the batch. Investigation typically reveals one of three causes: the missing model was reverted to a prior revision after the event fired but before the batch ran; the batch encountered a transient error on that model and the run completed without retry; or the model was moved out of the in-scope collection between the event and the batch.

Each cause has a different remediation. A revert is benign and is logged for audit. A transient error indicates a retry policy that needs tightening. A scope change indicates that someone moved a model without notifying the integration team and is being treated as a governance finding rather than a technical one.

The point of this example is that reconciliation produces actionable signals only when both ledgers are populated and compared on the same key. A subscription that captures events but does not log them, or a batch that does not record the set of revisions it processed, leaves the team without the data needed to reason about divergence.

4.4.7 Failure Handling and Operational Runbooks

Both export patterns fail in characteristic ways. A short operational runbook for each pattern reduces resolution time when failures occur. Table 4.9 shows typical batch failures and responses and Table 4.10 shows event-driven failures and responses.

Failure	Detection	Response
API rate-limit response	Phase telemetry shows back-off retries	Reduce concurrency for next run; investigate workspace-side limit
Authentication failure	Run aborts in dictionary phase	Confirm credential rotation; verify credential store entry
Schema mismatch	Validation phase emits schema-version finding	Pin parser to known version; raise change request to update parser
Partial completion within window	Run telemetry shows incomplete pages	Rerun in next window; do not promote staging until complete

Table 4.9: Batch Export Failure Modes and Runbook Responses

Failure	Detection	Response
Subscription disconnect	Heartbeat failure on monitoring channel	Reconnect; reconcile against last seen sequence

Table 4.10: Event-Driven Failure Modes and Runbook Responses

Failure	Detection	Response
Out-of-order delivery	Sequence number gap	Buffer and await reordering; escalate if timeout is exceeded
Duplicate delivery	Idempotency key already seen	Discard duplicate; log for trend
Event without subsequent API resolution	Event handler unable to fetch model	Hold event and retry; alert if window is exceeded

Table 4.10: Event-Driven Failure Modes and Runbook Responses (Cont.)

These runbooks are part of the operational documentation maintained alongside the integration flow. They should be reviewed during incident retrospectives and updated when new failure modes are observed.

4.5 Validation and Data Quality Checks

The extension contract is only as strong as the validation that enforces it. Without validation, the boundary between SAP Signavio and the data spine becomes a one-way pipe that propagates drift, missing attributes, and stale references into the data spine. With validation, the boundary becomes a control surface—a place where defects are detected, named, and routed back to the modeling team for correction.

This section describes how validation is structured at the SAP Signavio export boundary, what categories of checks apply, how violations are handled, and how feedback loops are closed so that quality improves over time rather than degrades.

4.5.1 Where Validation Runs in the Extension Contract

Validation occurs at three points along the export path, described in Table 4.11. Each point catches a different class of defect. Combining them produces a defense-in-depth pattern that prevents both obvious and subtle quality failures.

Validation Point	What It Catches	Outcome on Failure
At export, in the integration flow	Syntactic and structural defects, schema mismatches	Run aborts; finding raised; no data spine write
At spine ingestion, on staging tables	Semantic defects, referential integrity, attribute contract violations	Affected records quarantined; batch reconciliation halted

Table 4.11: Validation Points Along the SAP Signavio Extension Path

Validation Point	What It Catches	Outcome on Failure
At publication, before promotion	Cross-domain consistency: process ↔ architecture ↔ master data	Promotion blocked; process owner and external actor owner both notified

Table 4.11: Validation Points Along the SAP Signavio Extension Path (Cont.)

Each point uses a different rule set. Syntactic and structural rules belong at the export boundary because they can be evaluated without joining to other systems. Semantic and referential rules require the data spine's normalized model. Cross-domain consistency rules require both the data spine and SAP LeanIX, master data, and adoption surfaces, so they run at the publication gate.

4.5.2 Categories of Validation Checks

Validation rules are organized into five categories. Each rule has a category, a severity, an applies-to scope (process, activity, reference, attribute), and a remediation guidance line that is included in the finding payload, as follows:

- **Structural checks**
  Confirm that mandatory attributes are present, identifiers are unique within scope, and the BPMN structure is well-formed. For example, every L4/L5 activity has a populated `Activity_Stable_ID`, every published process has a `Process_Domain`, and no two activities in the same model share a stable ID.
- **Reference integrity checks**
  Confirm that every external reference in a custom attribute resolves against the corresponding upstream system. For example, every `LEANIX|APPLICATION` reference resolves to an active SAP LeanIX factsheet, every `FIORI|APP_ID` reference resolves against the SAP Fiori library, and every `SPRO|SSCUI` reference resolves against the configured release tag.
- **Attribute contract checks**
  Confirm that custom attribute values conform to the declared dictionary. For example, `AI_Exposure_Class` matches one of `{Permitted, Restricted, Prohibited}`, `Control_Intent` matches one of `{Preventive, Detective, Corrective, None}`, and `Variant_Key` matches the registered enumeration for the activity's domain.
- **Lifecycle consistency checks**
  Confirm that the lifecycle state of each artifact is consistent with its dependencies. For example, a published process version does not reference a deprecated application, a published activity does not reference a retired SAP Fiori app, and a published variant does not reference an archived parent.
- **Cross-domain consistency checks**
  Confirm that the relationships between SAP Signavio, SAP LeanIX, master data, and

adoption are coherent. For examples, every `Execution_Application_ID` also has a corresponding SAP LeanIX-to-process mapping, every variant referenced in the data spine has a valid SAP LeanIX deployment scope, and every `Fiori_App_ID` with WalkMe content has matching adoption telemetry channel registration.

The full rule catalog is maintained as a versioned artifact in the data spine, separate from the rules engine that executes it. This separation—rules as data, executor as code—allows the rule set to evolve without redeploying the validation pipeline.

4.5.3 Severity Levels and Outcome Handling

Not all findings are equally serious. A small severity model keeps responses proportional and prevents alert fatigue. The following three severities are sufficient:

- Critical findings block the activity, model, or version from being promoted to a **Published** state in the data spine. The defect must be corrected in SAP Signavio before promotion can resume. Examples include a missing `Activity_Stable_ID`, a dangling LeanIX reference on a published activity, and `AI_Exposure_Class` missing on an activity flagged for AI grounding.
- Warning findings do not block promotion but are shown to the process owner and tracked. Examples include a variant attribute set on an activity that has only one variant in scope and a deprecated reference qualifier being used.
- Informational findings are recorded for trend analysis and quality scoring but do not require immediate action. Examples include long descriptive text in a custom attribute that should be a structured field and an attribute used in only one model across the workspace.

Each finding carries a stable rule ID, the severity, the affected entity, the offending value, and a remediation note. The note is the operative piece. A finding without remediation guidance is a defect in the rule itself, not in the data.

4.5.4 Quarantine Versus Reject: Handling Bad Records

When a record fails validation, the integration flow has two choices. It can reject the record outright, refusing to write it to the data spine, or it can quarantine the record, writing it to a staging area where it is preserved but not promoted. The right choice depends on the failure.

Rejection is appropriate when the record is structurally unparseable—for example, when the schema version of the response does not match any known version. Continuing past such a failure risks corrupting the data spine. Quarantine is appropriate when the record is parseable but fails specific rules. Quarantining preserves the record for the modeling team's review and avoids the operational pain of a hard rejection that loses context.

The quarantine area is a separate set of staging tables in the data spine. Quarantined records are visible to the modeling team through a validation dashboard and are retained for a configurable period (typically thirty days). Once the underlying defect is corrected in SAP Signavio and the record is reexported, it leaves quarantine and proceeds through normal promotion. Figure 4.6 shows the validation finding lifecycle from detection through routing, remediation, and revalidation.

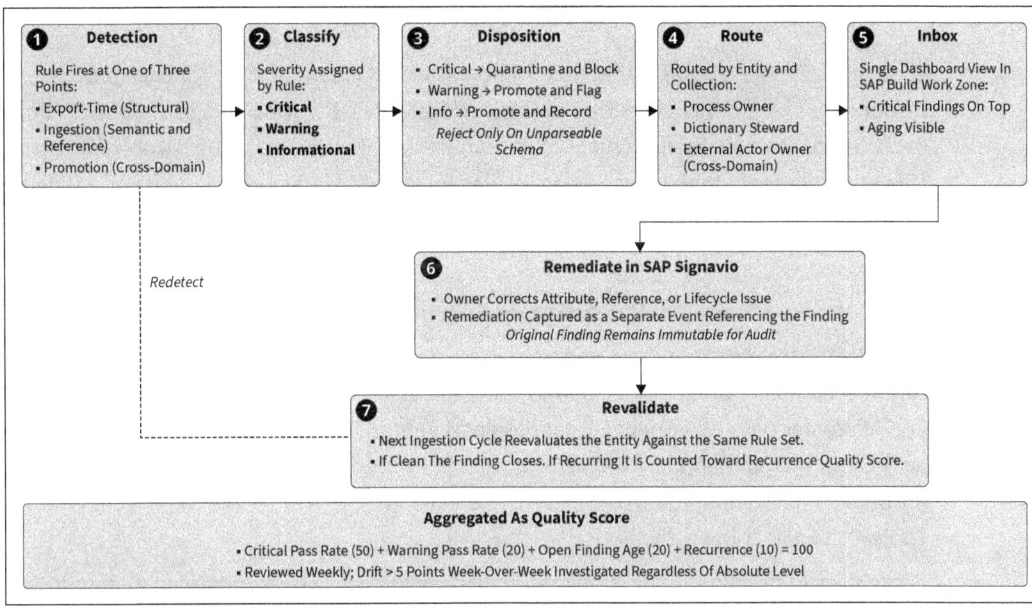

Figure 4.6: Validation Finding Lifecycle from Detection Through Remediation

4.5.5 Closing the Loop with Modeling Teams

Validation findings have value only if they reach the people who can act on them. In a poorly governed landscape, findings accumulate in logs that no one reads, and the same defects recur indefinitely. In a well-governed landscape, findings are routed to named owners with reasonable expectations and visible aging. The feedback loop is implemented in three layers, as follows:

- **Named ownership**
  Every process collection in SAP Signavio has a designated process owner. Every dictionary entry has a designated dictionary steward. Findings are routed automatically by entity type and collection to the corresponding owner. Where ownership is ambiguous, the routing rule fails closed and the finding is escalated to the integration governance lead.
- **A single inbox**
  Owners receive findings through a single dashboard view in SAP Build Work Zone. Critical findings appear at the top with their age. Warning and informational findings are

grouped. The dashboard also exposes the underlying rule and the exact attribute or reference that failed so that the owner does not need to dig through logs.

- **Aging and reporting**
  Findings older than a defined threshold are flagged in a weekly report to the process governance forum. Aging is a signal that the feedback loop is unhealthy; persistent aging indicates either ambiguous ownership, an unactionable rule, or insufficient capacity in the modeling team. Each cause has a different remedy.

Over time, the same dataset that drives findings drives a quality score for each process collection: the share of activities that pass all critical rules, the median age of open findings, and the rate at which findings recur after remediation. The quality score is published alongside the process content. It becomes a leading indicator of integration health and a tangible measure of the maturity of the modeling discipline.

4.5.6 Worked Example: Validating an Execution-Ready Activity

To make these categories more concrete, consider a single L4 activity drawn from the hire-to-retire domain: "Create a position in SAP SuccessFactors Employee Central referencing an SAP Master Data Governance–governed cost center." The activity is modeled in SAP Signavio with the custom attribute set introduced in Section 4.1 and the reference patterns introduced in Section 4.2. Upon export, the validation pipeline runs the checks listed in Table 4.12 against the activity.

Rule ID	Category	Check	Severity
STR-001	Structural	`Activity_Stable_ID` is present and unique	Critical
REF-014	Reference integrity	`Execution_Application_ID` resolves to an active SAP LeanIX application	Critical
REF-022	Reference integrity	`Trigger_Application_ID` resolves to an active SAP LeanIX application	Critical
REF-031	Reference integrity	`Fiori_App_ID` resolves against the SAP Fiori library catalog	Warning
ATTR-007	Attribute contract	`AI_Exposure_Class` matches one of `{Permitted, Restricted, Prohibited}`	Critical
ATTR-011	Attribute contract	`Control_Intent` matches one of `{Preventive, Detective, Corrective, None}`	Warning
LIFE-003	Lifecycle consistency	No referenced applications is deprecated	Critical

Table 4.12: Validation Rules Applied to the Worked Example Activity

Rule ID	Category	Check	Severity
XDOM-002	Cross-domain	SAP Master Data Governance–governed cost center master data reference is reachable from the activity context	Critical
XDOM-005	Cross-domain	The WalkMe asset, if registered, points to a `Fiori_App_ID` consistent with this activity's reference	Warning

Table 4.12: Validation Rules Applied to the Worked Example Activity (Cont.)

Suppose the activity passes seven of the nine rules but fails REF-031 (the `Fiori_App_ID` does not appear in the SAP Fiori library catalog) and XDOM-005 (the WalkMe asset registered against the activity points to a different `Fiori_App_ID`). Both findings are warnings, not critical. The activity is therefore promoted to the data spine, but two findings appear on the process owner's dashboard with aging tracked.

If the activity also failed ATTR-007 (`AI_Exposure_Class` missing), the activity would be quarantined in the data spine's staging tables. AI grounding pipelines would not see the activity until the defect is corrected in SAP Signavio. The execution services described in Chapter 12 would treat the activity as not-yet-published and would surface a clear message to any consumer attempting to use it.

This example illustrates the operating principle of the validation layer. Critical defects are caught and isolated. Warning defects are visible but do not block forward progress. The system never silently consumes broken content; it also never grinds to a halt over minor inconsistencies. That balance is what makes the extension contract safe to operate at enterprise scale.

4.5.7 Quality Scoring as a Leading Indicator

A quality score condenses the validation dataset into a single number per process collection. The score is not a substitute for inspecting individual findings but a leading indicator of whether the modeling discipline is improving or eroding. A simple, defensible scoring formula combines four components, each contributing to a total of 100, as shown in Table 4.13.

Component	Weight	Computation
Critical pass rate	50	Share of activities passing all critical rules
Warning pass rate	20	Share of activities passing all warning rules

Table 4.13: Quality Score Components for a Process Collection

Component	Weight	Computation
Open finding age	20	*20 × (1 – median open age / 30 days)*, capped at 20
Recurrence	10	*10 × (1 – share of remediated findings that recur within 90 days)*

Table 4.13: Quality Score Components for a Process Collection (Cont.)

A score above 90 indicates a healthy collection. A score between 75 and 90 indicates that improvement work is required but the collection is operating safely. A score below 75 indicates that the collection should not be expanded further until the underlying defects are addressed; promoting new content into a low-quality collection accelerates degradation rather than diluting it.

The score is computed weekly and tracked against its trailing four-week baseline. A score that falls by more than five points in a week is investigated, even if the absolute score remains acceptable. Drift is more informative than absolute level; a stable score of 82 is preferable to a score that swings between 95 and 80.

4.5.8 Worked Example: Validation Across a Process Variant

The first worked example focused on a single activity. A more demanding case is a process variant, which models the same logical process with regional or legal-entity differences. Variants exercise the cross-domain consistency rules more aggressively because the relationships between SAP Signavio, SAP LeanIX, and master data may differ across variants.

Consider the hire-to-retire process for a multinational organization with three legal-entity variants: US01, DE01, and IN01. Each variant carries the same activity skeleton but differs in the executing application stack. US01 executes against SAP SuccessFactors Employee Central, DE01 executes against SAP SuccessFactors Employee Central plus a country-specific add-on, and IN01 executes against a partner HR system bridged through SAP BTP.

On export, the validation pipeline runs the rules introduced earlier across all three variants. Variant-specific findings emerge, as shown in Table 4.14.

Variant	Findings	Severity Mix
US01	All checks pass on 47 of 50 activities. Three activities have warnings on REF-031 (SAP Fiori app) and ATTR-011 (`Control_Intent`).	0 critical, 3 warning
DE01	Two activities reference an add-on application whose lifecycle state in SAP LeanIX is sunset. LIFE-003 fires as critical.	2 critical, 5 warning

Table 4.14: Findings Across Three Legal-Entity Variants of the Same Process

Variant	Findings	Severity Mix
IN01	One activity references a partner system via a non-SAP LeanIX reference type. The reference is technically valid but flagged by XDOM-002 because the partner system is not yet onboarded into the SAP LeanIX inventory.	1 critical, 4 warning

Table 4.14: Findings Across Three Legal-Entity Variants of the Same Process (Cont.)

The handling differs by severity. The two critical findings on DE01 block promotion of the variant; until the lifecycle state of the referenced add-on is resolved (either an upgrade path is committed or the activity is rerouted to the standard application), DE01 cannot be published. The critical finding on IN01 also blocks promotion, but the remediation is different: the partner system must be onboarded into SAP LeanIX, after which the activity's reference can be retyped to `LEANIX|APPLICATION`.

Meanwhile, US01 promotes successfully despite three warnings. The warnings are visible to the process owner with aging tracked. The variant-level quality scores reflect the differential health: US01 scores 96, DE01 scores 71, and IN01 scores 79.

This example demonstrates that the validation framework does not treat a process as monolithic. It evaluates each variant on its own merits, blocks only what must be blocked, and produces enough signal for the process owner and the external actor team to coordinate the remediation across the variants.

4.5.9 Auditability of the Validation Layer

Every validation finding is itself an audit artifact. The finding records the rule applied, the entity evaluated, the value found, the severity, the timestamp, the rule version, and the dictionary version against which evaluation occurred. Findings are immutable once written; remediation is captured as a separate event that references the original finding.

This pattern of immutable findings and separately recorded remediations preserves the audit trail across multiple cycles of correction and revalidation. An auditor reviewing a published process can reconstruct exactly which findings were raised, which were remediated, when, and by whom. The reconstruction does not depend on log archaeology; it is a structured query against the data spine's validation tables.

The audit benefit is not abstract. In landscapes regulated under the Sarbanes–Oxley Act (SOX), the General Data Protection Regulation (GDPR), or industry-specific frameworks, the ability to demonstrate that process content was validated before exposure to AI grounding or automation is a tangible compliance asset. The validation layer therefore serves two purposes simultaneously: It improves operational quality, and it produces the evidence required to defend that quality to auditors.

4.6 Summary

This chapter treated SAP Signavio extension as an architectural responsibility rather than a configuration exercise. Process content has integration value only when it can leave the modeling tool with its semantics intact, and preserving those semantics requires a deliberate contract between SAP Signavio and every downstream consumer.

The chapter began with the attribute model. Standard attributes were positioned as the preferred default, with custom attributes introduced sparingly and governed as a contract. A small named, scoped, and consumer-driven custom attribute set was shown to support execution context, control intent, and AI eligibility without introducing the proliferation that destabilizes ungoverned dictionaries.

External references were then formalized as typed identity assertions rather than as URLs or free-text labels. The four-field reference shape—`ref_system`, `ref_type`, `ref_id`, and `ref_version`—was applied to SAP LeanIX applications, SAP S/4HANA transactions, Transaction SPRO configuration objects, SAP Fiori app IDs, SAP Build artifacts, WalkMe assets, and AI grounding bundles. Reference governance addressed retirement, reclassification, and version drift in upstream systems.

API preparation followed. The Process Manager Public API was identified as the primary surface for spine ingestion. Service identities, scoped credentials, paginated and rate-limited extraction patterns, and forward-compatible schema parsing were treated as the operating discipline rather than as optional refinements.

We discussed how batch export differs from event-driven export and assigned each its appropriate role. Batch is the default; events are introduced where latency genuinely matters and operational cost is justified. Reconciliation between the two patterns was identified as the mechanism by which silent drift is detected.

We closed the chapter with a section on validation, where we discussed elevating the boundary between SAP Signavio and the data spine into an explicit control surface. Five rule categories (structural, reference integrity, attribute contract, lifecycle consistency, and cross-domain consistency) were combined with a three-level severity model and a quarantine-versus-reject decision logic. Named owners use feedback loops with aging and quality scoring to improve quality rather than let it decay.

The result is an extension contract that is small enough to operate, strict enough to trust, and visible enough to govern. With the contract established, the next part of the book turns to SAP LeanIX as the corresponding system of record for architecture context.

4.6.1 Key Learning Points

- Standard SAP Signavio attributes describe the model as an artifact; custom attributes turn the model into an integration contract. Both are needed, and the balance should be set deliberately.

- Custom attributes succeed only as a governed contract. Necessity, stability, and minimal scope are the principles; the dictionary is the enforcement mechanism.
- External references must encode identity, not navigation. URLs, free text, and labels are insufficient. The four-field reference shape is the minimum viable structure.
- API preparation is an operating discipline, not a one-time setup. Scoped service identities, rate-aware extraction, and schema-pinned parsing are the baseline.
- Batch is the default export pattern. Events are introduced where the latency requirement is real and the operational cost is justified. Reconciliation between the two is mandatory wherever both are active.
- Validation is a control surface, not a log. Findings have severity, ownership, aging, and a remediation note. Quality scores follow naturally from the same dataset.

4.6.2 Common Design Pitfalls

- Treating custom attributes as a free-form scratchpad. Without a governed dictionary, the attribute set sprawls within months and becomes unstable for downstream consumers. The solution is a single point of approval and a quarterly catalog review.
- Embedding URLs as references. URLs encode endpoints rather than identity. They go stale on environment changes and cannot be validated semantically. The solution is to record only stable identifiers and to construct URLs at consumption time.
- Granting workspace-administrator privileges to integration service identities. This shortcut is unsafe and unnecessary. The solution is read-only scoping to the minimum set of resources required by the data spine.
- Choosing event-driven export by default. Event handling is operationally heavier than batch and only pays for itself where latency genuinely matters. The solution is to default to batch export and introduce events only when a named consumer requirement justifies them.
- Operating without reconciliation. When both event and batch export are in use, the absence of reconciliation guarantees that drift will eventually appear and will not be detected until a downstream incident exposes it. The solution is a routine reconciliation job whose findings are reviewed.
- Treating validation as advisory. Findings that do not block promotion and do not have named owners accumulate without effect. The solution is severity-based outcome handling and routed feedback with aging.

4.6.3 Review Questions

1. Why is a small, named custom attribute contract preferable to a permissive attribute model that allows modelers to add fields freely?
2. What are the three properties of an external reference, and why is each property necessary for downstream consumers?

3. Why are URLs not appropriate as primary references between SAP Signavio activities and applications recorded in SAP LeanIX?
4. What scoping should be applied to an SAP Signavio service identity used by the canonical data spine ingestion flow, and why?
5. Under which conditions is event-driven export preferable to batch export, and which failures must be addressed when events are used?
6. How does reconciliation between event-driven and batch export protect the integrity of the data spine?
7. Explain the difference between rejecting and quarantining a record that fails validation. When is each approach appropriate?
8. How does a validation severity model prevent both alert fatigue and silent failure?
9. What is the role of the quality score, and how does it become a leading indicator of integration health?

4.6.4 Student Companion Notes

If you are a student approaching this chapter, you should focus first on the attribute contract. Take any single L4 activity from a familiar process and attempt to express it using only the standard SAP Signavio attribute set. This exercise quickly reveals where custom attributes are required and why their definitions must be stable.

Second, practice writing external references in the four-field shape for at least three different target systems. The discipline of choosing the correct `ref_type` and the correct `ref_id` source is what separates a robust integration from a fragile one.

Third, read the SAP Signavio Process Manager API reference and identify the endpoints used by the data spine. Trace one endpoint from authentication through pagination to response parsing, paying particular attention to where rate limits and schema changes would manifest.

This chapter completes Part II. The extension contract established here is the basis for everything that follows: how SAP LeanIX is configured for dependency and context data (Part III), how WalkMe content is mapped to process steps (Part IV), how the canonical data spine ingests and normalizes the exported content (Part V), and how the integration flows and execution services consume it (Part VI). The integrity of those downstream layers depends, structurally, on the integrity of the contract defined in this chapter.

PART III

Configuring SAP LeanIX for Dependency and Context Data

Chapter 5

SAP LeanIX Data Model for Integration

An architecture register answers a single question with discipline: what exists, in what state, with what dependencies. The moment it tries to answer a second question, it stops being a register and starts being a competing system of record. The first responsibility of an integration-ready SAP LeanIX model is to know which question it owns.

In Chapter 3 and Chapter 4, we specified what SAP Signavio must carry and how that content can be safely extracted so that process activities can be consumed by anything downstream, including automation, the canonical data spine on SAP Business Technology Platform (SAP BTP), and AI services. Chapter 3 covered the L4 and L5 attributes each activity needs: identifiers, control intent, roles, variants, and lifecycle. Chapter 4 covered the extension contract: which fields are standard, which are custom, how the application programming interfaces (APIs) are shaped for extraction, and how exports are validated. Together, they make SAP Signavio's contents readable and actionable outside of SAP Signavio.

However, process descriptions do not run on their own. An SAP Signavio activity that says "Validate the cost center mapping in SAP Master Data Governance, Financials before posting" cannot be acted on until something names which application SAP Master Data Governance, Financials actually is, which interface carries the mapping, what state that interface is in, and which surfaces are approved to read it. That second body of knowledge is the landscape—applications, interfaces, dependencies, and lifecycle—and SAP LeanIX is where it lives. Part III of this book covers it, and Chapter 5 starts with the data model SAP LeanIX must hold for the rest of the integration to be possible.

SAP Signavio answers the question "What should happen?" and SAP LeanIX answers the question "What exists, and how is it connected?" The two are deliberately distinct, and a great deal of the trouble that appears later during execution, integration, and AI grounding can be traced back to a moment when one tool was asked to answer the other tool's question. The first discipline of this chapter is therefore a boundary one. SAP LeanIX models systems. SAP Signavio models execution. Neither one is allowed to encroach on the other's scope.

With that boundary in place, we will walk through the SAP LeanIX data model required to support execution and AI on top of the canonical data spine on SAP BTP. The objective is practical. By the end of the chapter, you should be able to write down the complete factsheet inventory, attribute set, identifier scheme, lifecycle rules, and boundary disciplines that SAP LeanIX must enforce for SAP BTP integration and SAP AI consumption to be safe. Specifically, this chapter covers:

- The application objects that SAP LeanIX must hold to be a credible integration register, including the identity, ownership, lifecycle, and reference fields that downstream services rely on
- The interface and dependency model required for change-impact analysis and AI grounding to both operate on the same data
- The lifecycle attribute discipline required so that only valid, approved applications and interfaces are exposed to integration and AI consumers
- The stable-identifier scheme that connects SAP LeanIX factsheets to SAP Signavio activities, the canonical data spine on SAP BTP, and SAP S/4HANA objects without semantic drift
- The explicit boundary that prevents semantic overlap with SAP Signavio, including the anti-patterns that put that boundary at risk

As in earlier chapters, each section proceeds from principle to design pattern to a concrete example drawn from the running cost center–position scenario that we will continue to develop throughout the remainder of the book.

A note on scope: This chapter stops at the SAP LeanIX boundary. The next chapter picks up the thread by showing how SAP LeanIX artifacts are mapped to L4 and L5 process steps inside SAP Signavio and how that mapping becomes the substrate for change impact analysis and for AI context preparation. The factsheet model defined here is precisely what that mapping work expects to consume. Figure 5.1 shows where SAP LeanIX sits in the integrated landscape developed across Chapter 1 through Chapter 4, between the design layer (SAP Signavio and WalkMe) and the execution layer (canonical data spine on SAP BTP).

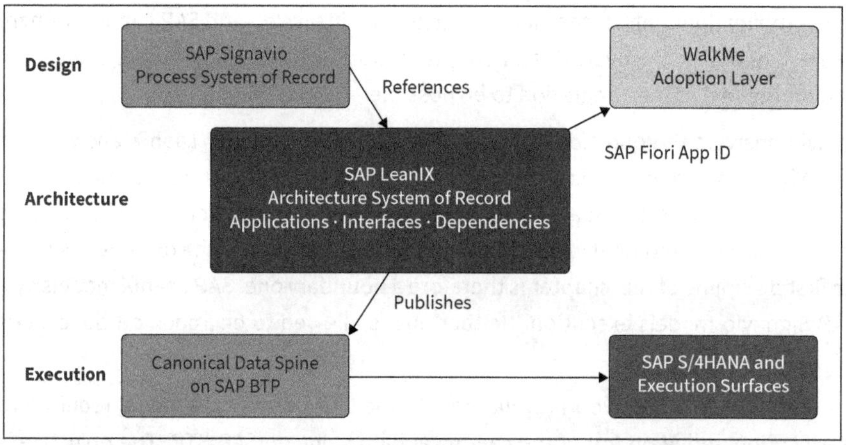

Figure 5.1: SAP LeanIX in the Integrated Landscape, the Architecture System of Record That the Data Spine Reads From

5.1 The Running Scenario: Cost Center to Position

Every section of this chapter ends with a callout that shows how its concept applies to a single running scenario. The scenario was introduced in process terms in Chapter 3, and we will carry it forward into every subsequent chapter. In this section, we will introduce the scenario in SAP LeanIX terms so that each later callout reads against a shared inventory of applications and interfaces.

The scenario describes the propagation of a new cost center from its creation in SAP Master Data Governance through to the assignment of that cost center to a position in SAP SuccessFactors Employee Central. The propagation crosses three applications and three interfaces, and it is governed by three L5 activities recorded in SAP Signavio. The sequence is:

1. `A_COST_CENTER_CREATE` runs in SAP Master Data Governance, Financials. A finance controller creates a new cost center and publishes it. The activity produces the if-mdg-cc-01 interface payload.
2. `A_COST_CENTER_PROPAGATE` runs in SAP S/4HANA Finance. It consumes if-mdg-cc-01, applies controlling-area mapping and validity-period checks, and produces the if-s4-sf-cc-02 interface payload for downstream human resources (HR) consumption.
3. `A_POSITION_COST_CENTER_ASSIGN` runs in SAP SuccessFactors Employee Central. It consumes if-s4-sf-cc-02 and assigns the cost center to one or more positions, then produces if-sf-sac-pos-03 for reporting into SAP Analytics Cloud.

In SAP LeanIX, the three applications and four supporting interfaces are each recorded as a factsheet with a stable `factSheetId`. The `factSheetId`s used throughout this chapter and continued into Chapter 6 are listed in Table 5.1.

factSheetId	displayName	Hosts Which Activity?
fs-mdg-fin-01	SAP MDG-FIN (Europe, the Middle East, and Africa [EMEA])	`A_COST_CENTER_CREATE`
fs-s4hana-fin-emea-01	SAP S/4HANA Finance (EMEA)	`A_COST_CENTER_PROPAGATE`
fs-sf-ec-01	SAP SuccessFactors Employee Central	`A_POSITION_COST_CENTER_ASSIGN`
fs-sac-01	SAP Analytics Cloud	Downstream reader; no activity binding

Table 5.1: Application Factsheets in the Cost Center to Position Scenario

Table 5.1 names the application factsheets that participate in the scenario. Each row corresponds to one SAP LeanIX application factsheet whose mandatory attribute set is the subject of Section 5.2. Table 5.2 names the interface factsheets that carry the propagation.

Three of the four are domain-specific, and if-ad-user-04 is a cross-cutting identity interface that every activity consumes. We will explore the interface model in Section 5.3.

factSheetId	Direction	Payload
if-mdg-cc-01	SAP Master Data Governance, Financials to SAP S/4HANA Finance	Cost center master data (controlling area and validity dates)
if-s4-sf-cc-02	SAP S/4HANA Finance to SAP SuccessFactors Employee Central	Cost center reference (effective date and cost center identifier)
if-sf-sac-pos-03	SAP SuccessFactors Employee Central to SAP Analytics Cloud	Position and cost center binding (for reporting only)
if-ad-user-04	Active Directory to all three apps	User identity and role context (read by every activity)

Table 5.2: Interface Factsheets in the Cost Center to Position Scenario

You should treat the entries in Table 5.1 and Table 5.2 as a fixed cast. Every example in the remainder of the chapter references these `factSheetIds`.

5.2 Application Objects Required for Integration

The application factsheet is the central object in the integration-ready SAP LeanIX model. Every other object in the model—interfaces, IT components, business capabilities, providers, and technology stacks—either belongs to an application, connects two applications, or describes the deployment surface of an application. For that reason, the application model is discussed first, and the rest of the model is layered on top of it.

When an architecture team first stands up an SAP LeanIX tenant, the temptation is to record every system the enterprise can name. Inventories of this kind are easy to justify and impossible to use. An integration-ready application model is not the sum of every system in the landscape. It is the subset of systems whose existence, identity, and lifecycle are required to support execution, integration, and AI on the canonical data spine. Anything outside that subset is admissible in SAP LeanIX as a background record, but only the integration-eligible subset is governed to the standards described in this section.

This section develops the application model in six steps. First, it defines what makes an application integration-eligible: the three conditions that must hold simultaneously before SAP LeanIX records the application as governable. Second, it lays out the mandatory attribute set: the four blocks of identity, lifecycle, references, and governance that every integration-eligible factsheet must carry. Third, it draws the boundary between an application and an IT component, since these two object types are most easily confused. Fourth, it walks through a worked anatomy of a real factsheet, naming each field by name. Fifth, it sets out the inventory discipline that keeps the application register from drifting silently over time. Finally, it supplies the full build specification, including every field and controlled-list value as well as the GraphQL and JSON contracts that downstream consumers will read.

5.2.1 The Integration-Eligible Application

An application is integration eligible when three conditions are simultaneously true. First, at least one L4 or L5 process activity in SAP Signavio references it as an execution location. Second, at least one interface defined on the application is consumed by the canonical data spine, by another integration-eligible application, or by an end-user surface such as SAP Build Work Zone. Third, the application carries a stable identifier that is not expected to change across the lifetime of the applications and interfaces that reference it. If any of the three conditions is missing, the application can still be recorded in SAP LeanIX, but it does not yet belong to the integration-governed subset.

This rule has a finance-control parallel that will be familiar if you have a financial planning and analysis (FP&A) background. Not every cost center in the chart of accounts is eligible for reporting. Some are statistical, some are preproduction placeholders, and some are closed but retained for historical comparability. The same kind of discipline applies to the application register. Existence is not the same as eligibility. An integration register that does not maintain that distinction will be unstable for downstream consumers within a single planning cycle.

5.2.2 The Mandatory Attribute Set

Every integration-eligible application carries a mandatory attribute set. The attributes are grouped into four blocks: identity, lifecycle, references, and governance. Each block has a different consumer pattern in the canonical data spine, and each is enforced at factsheet save time rather than at export time. This is what prevents the architecture register from drifting silently and then producing failures at the boundary.

The identity block answers the question "What is this?" It contains the SAP LeanIX `factSheetId`, the `displayName` used in factsheet views, the IT component reference that identifies the deployable unit, and at least one external identifier that links the SAP LeanIX record back to the system it represents. The external identifier is recorded as a structured field, not as free text. For an SAP application, the external identifier is the SAP system ID and the SAP product code/ For a third-party application, it is the vendor product code and the deployment instance identifier. For a custom-built application, it is the SAP BTP space and the application route. The point is not that one of these is correct and the others are wrong but that each is recorded in a known field with a known parser, so that downstream resolution does not depend on string matching.

The lifecycle block answers the question "What is its state?" The full lifecycle model will be covered in Section 5.4, but every integration-eligible application must carry, at minimum, a lifecycle phase value, a planned end-of-life (EOL) date if known, a flag indicating whether the application is approved for integration, and a separate flag indicating whether the application is approved for AI consumption. The two approval flags are deliberately distinct. An application can be eligible for integration without being eligible for AI. The most common case is an application whose data classification permits machine-to-machine transfer under contractual safeguards but does not permit exposure to a foundation model. The data spine and AI services consult the two flags independently.

The reference block answers the question "What is it connected to?" It records the business capabilities the application supports, the interfaces it provides and consumes, and the process activities that execute against it. The process-activity references are recorded by `Activity_Stable_ID`, not by activity name. The binding mechanism is covered in detail in Chapter 6; what matters here is that the reference block contains the keys that allow the data spine to assemble the change impact view without reading anything beyond SAP LeanIX in the first hop.

The governance block answers the question "Who is responsible?" It records the application owner, the technical owner, the legal-entity and country scope, and the data classification. Every one of these fields has at least one named consumer. The application owner is consulted when an interface change is proposed. The technical owner is consulted when a deployment topology changes. The legal entity and country scope determines which variants in SAP Signavio may reference the application. The data classification gates AI eligibility regardless of the AI approval flag, because classification is enforced upstream of the approval workflow. Figure 5.2 shows the four mandatory blocks as a single anatomical view of an application factsheet, with each block labeled and its representative fields shown.

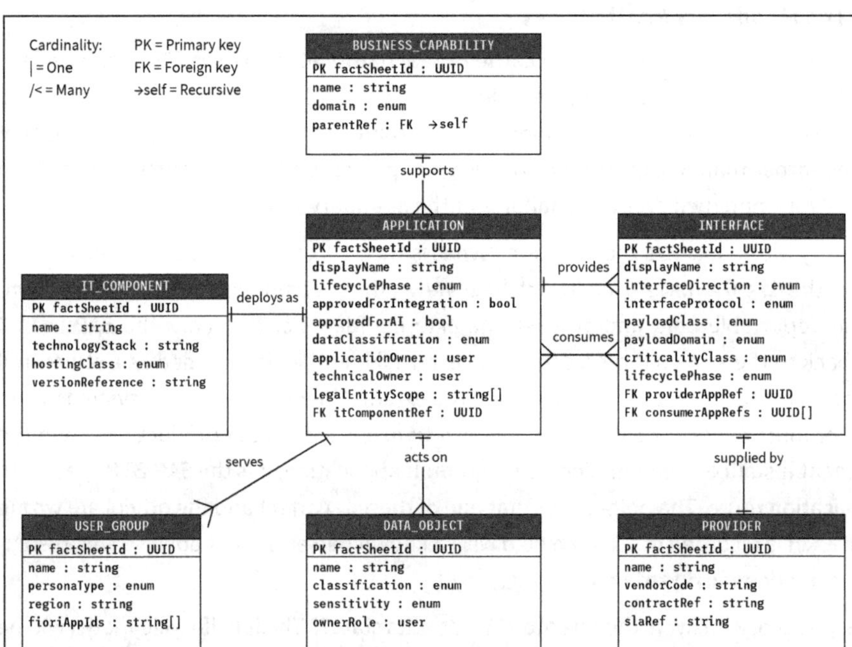

Figure 5.2: The Factsheet Hierarchy Required for Integration; Every Layer Carries a Stable Identifier

5.2.3 Application Versus IT Component

A common source of confusion in early SAP LeanIX deployments is the distinction between the application factsheet and the IT component factsheet. The two objects look alike in the user interface and are sometimes used interchangeably by enthusiastic modelers. The integration model demands that they be kept distinct.

The application is the unit of execution that the business recognizes. SAP S/4HANA Finance, SAP SuccessFactors Employee Central, and SAP Ariba Sourcing are applications. Each carries a single business purpose, a single owner, and a single AI-eligibility decision. The IT component is the unit of deployable software that the operations team recognizes. The SAP S/4HANA core, the SAP NetWeaver kernel, the underlying database, and the operating system are IT components. A single application is supported by several IT components, and a single IT component can support several applications, particularly in shared-platform deployments.

The integration register cares about applications. The operations register cares about IT components. The data spine joins the two only when a change in a shared IT component would propagate to several applications and therefore to several process variants. In that narrow case, the IT component identifier is the join key. In every other case, the IT component is invisible to the data spine. Allowing IT component semantics to leak into the application factsheet, or vice versa, produces a model that operations teams find too high-level and architecture teams find too noisy. Neither audience ends up trusting the register.

5.2.4 Worked Anatomy of an Application Factsheet

To help you understand the attribute discipline, consider a single application factsheet for SAP S/4HANA Finance in a deployment that supports the cost center–position scenario carried through this book (shown in Figure 5.3). The factsheet identity block records the SAP LeanIX `factSheetId` (assigned at creation and never reused), the `displayName` (SAP S/4HANA Finance – EMEA Production), the IT component reference (pointing to the shared SAP S/4HANA core record), and the external identifiers (the SAP S/4HANA system ID, the SAP product code, and the SAP for Me deployment record).

The lifecycle block records that the application is in the active phase, has no planned EOL date, is approved for integration, and is approved for AI consumption at the level of master data references and process metadata but not at the level of transactional payload bodies. The two AI approvals are recorded as separate fields rather than as a single flag, because the data spine consults them independently when assembling the grounding bundle for any given AI request.

The reference block records that the application supports the record-to-report and hire-to-retire business capabilities, provides interfaces if-s4-sf-cc-02 (cost center master data to SAP SuccessFactors) and if-s4-sac-fi-03 (financial postings to SAP Analytics Cloud), and consumes interfaces if-mdg-cc-01 (master data from SAP Master Data Governance) and if-ad-user-04 (user identity from Active Directory). The activity references include `A_COST_CENTER_PROPAGATE`, `A_INVOICE_POST`, and `A_POSITION_BUDGET_CHECK`, each represented by its `Activity_Stable_ID`.

The governance block records that the application owner is the Global Finance function, the technical owner is the SAP S/4HANA platform team, the country scope is the EMEA legal-entity cluster, and the data classification is confidential. The classification value is significant: it gates AI eligibility above the level of the AI approval flag, and any change to it triggers a review of the dependent process variants in SAP Signavio.

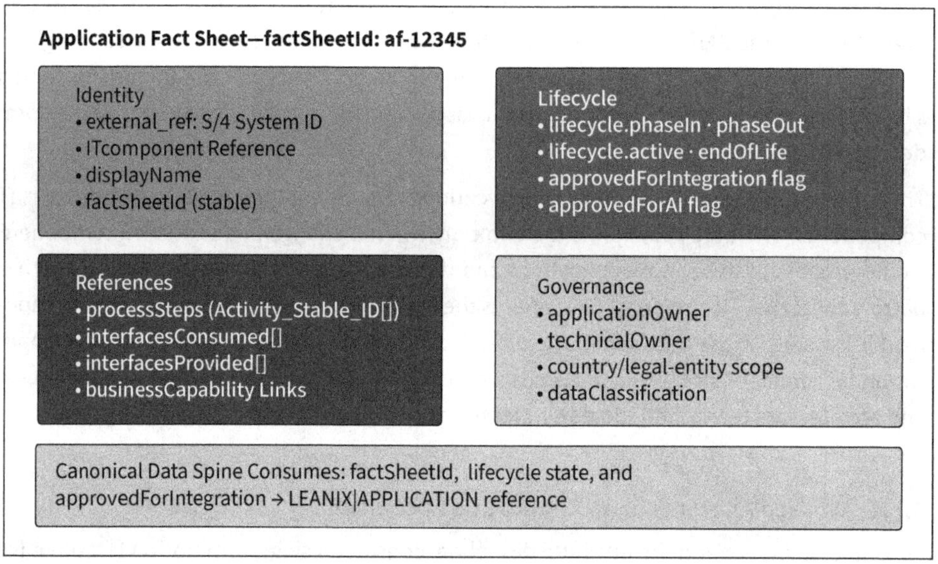

Figure 5.3: Anatomy of an Integration-Eligible Application Factsheet; Four Mandatory Blocks, Each with Named Consumers

5.2.5 The Application Inventory Discipline

A mature integration register is not produced by a single inventory exercise but by a discipline that operates continuously. The discipline has three rules. The first is that no application enters the integration-eligible subset without a named owner and a complete mandatory attribute set. The application can exist as a background record before the attributes are complete, but it cannot be referenced by any spine-published process version until the record is whole. This keeps the integration register from being polluted by half-finished records that will not survive a downstream consumer's validation.

The second rule is that ownership changes are first-class governance events. When the owner of an application changes, the architecture team records the change as a dated event in SAP LeanIX, not as a silent attribute update. The data spine reads the event and rebroadcasts the affected activity bindings to subscribing services. Without this rule, ownership drift accumulates silently and surfaces only during an audit, by which time the chain of custody for the application's integration decisions has been broken.

The third rule is that every integration-eligible application is reviewed at least once per planning cycle, regardless of whether anything appears to have changed. The review is short and consists of a check that owners are still in role, the lifecycle phase still reflects reality, and the interface inventory still matches what the operations team is actually running. Most reviews change nothing. The value of the discipline lies in the small number of reviews that catch a silent change before it becomes a data spine–level inconsistency.

Cost Center to Position Scenario: Application Factsheets

For the scenario, the four mandatory blocks outlined in Section 5.2.2 are populated as follows:

- **Identity**
  fs-mdg-fin-01, fs-s4hana-fin-emea-01, and fs-sf-ec-01 each carry a unique `factSheetId` and a structured `externalId` (SAP system ID [SID] for the SAP applications).
- **Lifecycle**
  All three are in the active phase, with `approvedForIntegration = true`. `approvedForAI = true` on fs-mdg-fin-01 and fs-s4hana-fin-emea-01 with the scope qualifier master-data-reference; `approvedForAI = true` on fs-sf-ec-01 with the qualifier position-reference only (the personnel subject area is excluded).
- **References**
  Each application points to one IT component, to the business capabilities it supports (cost center master data management for the finance apps and position management for the HR app), and to its providing and consuming interfaces.
- **Governance**
  The finance applications are owned by the finance domain owner, fs-sf-ec-01 is owned by the HR domain owner, and `legalEntityScope` is EMEA for all three. This pattern is the per-row reading of the build specification in the next section.

5.2.6 Application Factsheet: Build Specification

We have now described the four mandatory blocks in functional terms. Table 5.3 makes them concrete at the field level. This is the specification an administrator can take to SAP LeanIX and use to configure the application factsheet so that the application becomes consumable by the canonical data spine and by AI services downstream. Every field in the Mandatory column must be present and populated before the application can be considered eligible for integration; the data spine validation rejects factsheets that omit any of them.

Field (SAP LeanIX)	Type	Mandatory	Notes/Controlled Values
`factSheetId`	`string (UUID)`	Yes (auto)	Assigned by SAP LeanIX at fact-sheet creation. Opaque, immutable, and never reused.
`displayName`	`string`	Yes	Human-readable name (e.g., SAP S/4HANA Finance). May change over time; never used as a key.
`alias`	`string[]`	No	Alternate display names used historically; preserved for human search but never for joins.
`description`	`string (markdown)`	Yes	150–500 character functional summary. Read by AI surfaces during grounding.
`itComponentRef`	`factSheetId ref`	Yes	Points to the IT component fact-sheet that names the deploy-able unit.
`externalId.sapSID`	`string`	Conditional	Required if `technologyStack` includes SAP; the three-character system ID (SID; e.g., PRD).
`externalId.vendorSku`	`string`	Conditional	Required for third-party apps; the vendor's product code.
`externalId.btpRoute`	`string`	Conditional	Required for SAP BTP-hosted apps; the Cloud Foundry route or subaccount identifier.
`lifecyclePhase`	`enum`	Yes	Includes plan, phase-in, active, phase-out, and EOL.
`lifecycle.startDate`	`date`	Yes	ISO 8601 date when the application entered its current phase.
`lifecycle.endDate`	`date`	Conditional	Required during phase-out or EOL; planned removal date.
`approvedForIntegration`	`boolean`	Yes	Default false. True only after integration governance board sign-off.

Table 5.3: Application Factsheet—Full Field Specification

Field (SAP LeanIX)	Type	Mandatory	Notes/Controlled Values
approvedForAI	boolean	Yes	Default false. True only after AI eligibility review sign-off. Independent of approvedForIntegration.
freezeWindow	object[]	No	List of {startDate, endDate, reasonCode}. Read by commit gateway to reject writes during the window.
businessCapabilityRef	factSheetId ref[]	Yes	One or more business capability factsheets. Used by the data spine to resolve capability and owner.
providesInterface	factSheetId ref[]	No	Outbound interface factsheets. Empty list is valid.
consumesInterface	factSheetId ref[]	No	Inbound interface factsheets. Empty list is valid.
userGroupRef	factSheetId ref[]	No	Optional persona references for adoption analytics (WalkMe).
dataObjectRef	factSheetId ref[]	No	Optional DataObject factsheets describing material data domains.
fioriAppId	string	Conditional	Required for SAPUI5/SAP Fiori apps. Secondary identifier; never used as primary key (PK).
applicationOwner	user ref	Yes	Business owner. Read by the commit gateway as the approval target for changes.
technicalOwner	user ref	Yes	Technical owner. Receives interface failure alerts.
legalEntityScope	string[]	Yes	ISO 3166-1 alpha-2 country codes where the application operates.
dataClassification	enum	Yes	Includes public, internal, confidential, and restricted.

Table 5.3: Application Factsheet—Full Field Specification (Cont.)

Field (SAP LeanIX)	Type	Mandatory	Notes/Controlled Values
`complianceTags`	`string[]`	No	Controlled list (e.g., Sarbanes–Oxley Act [SOX], General Data Protection Regulation [GDPR], and the Health Insurance Portability and Accountability Act ([HIPAA]). Read by the AI grounding filter.

Table 5.3: Application Factsheet—Full Field Specification (Cont.)

Two patterns in the table are worth flagging. First, the three external ID fields are individually conditional but at least one must be present; the data spine rejects an application that has none. Second, the two approval flags default to false. This is deliberate. The integration-eligible subset of the landscape is built by explicit approval, not by inheritance from everything that exists.

Listing 5.1 demonstrates the canonical extract that the SAP BTP ingestion job runs against the SAP LeanIX GraphQL API to pull the full integration-eligible application population. The filter restricts to `approvedForIntegration = true` and `lifecyclePhase` in (phase-in, active, and phase-out). The result is paged in batches of 200 factsheets.

```
query GetIntegrationEligibleApplications($cursor: String) {
  allFactSheets(
    factSheetType: Application,
    filter: {
      facetFilters: [
        { facetKey: "approvedForIntegration", keys: ["true"] },
        { facetKey: "lifecyclePhase", keys: ["phase-in", "active",
"phase-out"] }
      ]
    },
    first: 200, after: $cursor
  ) {
    pageInfo { hasNextPage endCursor }
    edges {
      node {
        id
        displayName
        description
        ... on Application {
          lifecyclePhase
          approvedForIntegration
          approvedForAI
```

```
          dataClassification
          externalId { sapSID vendorSku btpRoute fioriAppId }
          relApplicationToITComponent { edges { node { id displayName }
} }
          relApplicationToBusinessCapability { edges { node { id
displayName } } }
          relApplicationToInterfaceProvides { edges { node { id
displayName } } }
          relApplicationToInterfaceConsumes { edges { node { id
displayName } } }
          applicationOwner { id email }
          technicalOwner { id email }
          legalEntityScope
          complianceTags
          freezeWindow { startDate endDate reasonCode }
        }
      }
    }
  }
}
```

Listing 5.1: GraphQL Query for the Integration-Eligible Application Set

The GraphQL response is normalized by the SAP BTP ingestion job into the flat, spine-shaped JSON object shown in Listing 5.2. This is the contract that downstream consumers read. Every field in the contract corresponds to one or more fields in the SAP LeanIX factsheet; nothing is invented at the data spine layer, and nothing is renamed.

```
{
  "factSheetId": "7f3e2c01-9d4a-4e7b-bc88-91a2e6d0f5a3",
  "displayName": "SAP S/4HANA Finance",
  "factSheetType": "Application",
  "externalIds": {
    "sapSID": "S4P",
    "fioriAppId": "F0707"
  },
  "lifecycle": {
    "phase": "active",
    "startDate": "2024-04-01",
    "endDate": null
  },
  "governance": {
    "approvedForIntegration": true,
    "approvedForAI": true,
    "dataClassification": "confidential",
```

```
    "complianceTags": ["SOX", "GDPR"],
    "applicationOwner": "finance.lead@example.com",
    "technicalOwner": "basis.lead@example.com",
    "legalEntityScope": ["DE", "US", "IN"]
  },
  "references": {
    "businessCapabilities": ["cap-finance-gl", "cap-finance-ap"],
    "providesInterface": ["if-s4-sf-cc-02"],
    "consumesInterface": ["if-mdg-cc-01"],
    "itComponent": "itc-s4hana-2024"
  },
  "freezeWindows": [
    { "startDate": "2025-12-28", "endDate": "2026-01-05", "reasonCode":
"YEAR_END_CLOSE" }
  ],
  "manifest": {
    "version": "2026-05-20T14:22:00Z",
    "manifestHash": "sha256:a14f...c021"
  }
}
```


Listing 5.2: Data Spine–Shaped JSON for One Integration-Eligible Application

5.3 Interface and Dependency Modeling

If the application factsheet answers "What exists?" the interface factsheet answers "How does everything connect?" For an integration-ready SAP LeanIX deployment, the interface model is not a documentation aid. It is the substrate on which change impact analysis, deprecation handling, and AI grounding all operate. An SAP LeanIX tenant that is rich in applications and thin in interfaces will produce confident answers to questions about systems and useless answers to questions about consequences.

This section develops the interface model in three steps. First, it defines what belongs on an interface factsheet so that the data spine can resolve identity, direction, protocol, and semantic payload. Second, it develops the dependency model in which interfaces and applications are arranged into chains whose criticality cascades. Third, it returns to the running scenario and walks the cost center–position propagation through the interface and dependency layer.

5.3.1 What Belongs on an Interface Factsheet

The interface factsheet records seven things at minimum. Each is recorded as a structured field rather than as descriptive text. The first is identity: the SAP LeanIX `factSheetId` of the interface itself. The second is direction: which application is the provider and which is the consumer. The third is protocol: the technical transport, declared as a controlled-list value (OData v4, REST/JSON, SOAP, event-stream over Kafka, file-based Secure File Transfer Protocol [SFTP], Remote Function Call [RFC], and so on). The fourth is frequency: real-time, near-real-time, scheduled batch with its cadence, or on-demand. The fifth is semantic payload: a named business object or event type drawn from a controlled vocabulary that the architecture and process functions maintain jointly. The sixth is data classification: the same classification scheme used on the application factsheet, with the rule that the interface inherits the more restrictive of the provider and consumer classifications. The seventh is governance: the interface owner, the runbook reference, and the change management contact.

None of the seven fields listed in Table 5.4 is decorative. The provider/consumer direction is consulted by the data spine when computing the blast radius. The protocol is consulted by the integration layer when selecting a connector. The frequency is consulted by the validation layer when deciding whether an event-driven or batch reconciliation pattern applies. The semantic payload is consulted by the AI grounding service when deciding whether a given interface is admissible in a particular request's grounding bundle. The data classification gates AI eligibility. The owner and runbook are consulted at incident time. An interface factsheet that omits any of the seven fields is technically present but operationally invisible.

Field	Type	Why the Data Spine Needs It
`factSheetId`	SAP LeanIX-assigned ID; stable	PK for all interface references in the data spine
`direction`	Controlled list: provider, consumer, bidirectional	Drives blast radius computation on lifecycle changes
`protocol`	Controlled list: OData v4, REST, SOAP, Kafka, RFC, SFTP	Selects the SAP Integration Suite connector at runtime
`frequency`	Controlled list: real-time, near-real-time, batch, on-demand	Selects event-driven or batch reconciliation pattern
`semanticPayload`	Controlled vocabulary: business object or event type	AI grounding eligibility; payload-level governance
`dataClassification`	Public, internal, confidential, or restricted	Inherited from the more restrictive provider or consumer; gates AI
`interfaceOwner`	Named role or person	Change-management routing and incident escalation

Table 5.4: Interface Factsheet—Seven Mandatory Fields and What the Data Spine Reads from Each

Figure 5.4 shows an interface factsheet. It avoids the anti-pattern of modeling the interface as URLs only; there is no payload semantics, no protocol, and no direction. The data spine cannot resolve impact, and AI cannot ground on it.

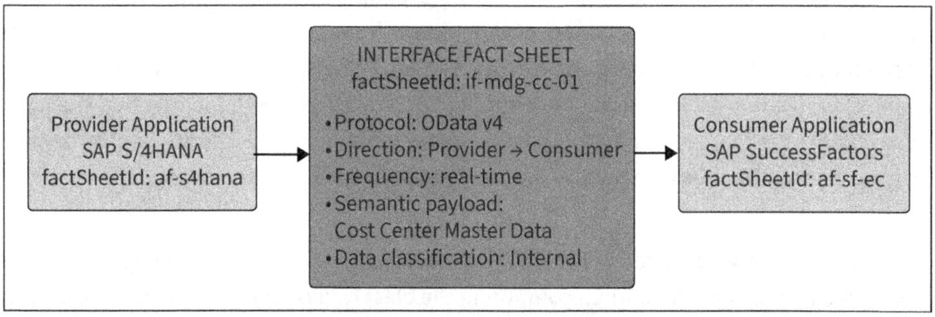

Figure 5.4: Interface Factsheet with Provider, Consumer, and Semantic Payload

A recurring early-stage modeling pattern records each interface as a URL on the application factsheet itself. The pattern looks economical: One click takes the user to the running endpoint and the architecture team feels that the model is alive. The pattern is structurally wrong, and it surfaces as failure in three places.

First, the URL encodes a runtime endpoint, not an identity. The endpoint migrates every time the underlying application changes its hostname, port, or environment path. Each migration breaks every downstream reference, including references the architecture team did not know existed. The right discipline, which we established in Chapter 4 for SAP Signavio references, is to reference by stable identifier and resolve to a URL only at consumption time.

Second, the URL does not carry direction. A consumer that reaches a provider URL is indistinguishable, at the level of the URL itself, from a provider that exposes a callback to the consumer. The data spine cannot compute blast radius without direction, and AI cannot ground on an interface whose direction is implicit.

Third, the URL is silent about the semantic payload. Two URLs that look identical may carry payloads with very different data classification profiles. The same endpoint may serve cost center master data under one route and salary data under another. Conflating them by recording the interface as a URL eliminates the classification distinction that downstream governance depends on.

The URL is a useful field on the interface factsheet, but only as an environment-specific runtime reference, never as the interface identity. The architecture register holds the interface identity; the destination service on SAP BTP holds the URL.

5.3.2 Dependency Modeling and Chains

Once applications and interfaces are recorded with discipline, the dependency model emerges almost mechanically. A dependency is a directional statement that one application requires another to perform a process activity. Dependencies are recorded on the application factsheet as references to other factsheets, with an explicit interface reference where one exists. A dependency without an interface reference is admissible only in the case of indirect dependencies that surface through shared IT components.

Dependencies form *chains*, ordered sequences of applications connected by interfaces that support a single process activity or a single process variant. A typical hire-to-retire activity may run through an identity provider, a master data governance system, an enterprise resource planning system, a human capital management system, and an analytics surface. Each step in the chain depends on the step before it. A change at any step propagates downstream as long as the chain is connected.

The integration register records two properties of every chain that the data spine consumes. The first is the criticality of each node, recorded as a value in a controlled vocabulary (standard, high, or mission-critical). The criticality of a node is not the criticality of the chain; the criticality of the chain is computed downstream as the maximum criticality of any of its nodes, weighted by the variants that depend on the chain. The second is the recovery tier of each node, recorded separately because criticality and recovery tier are not the same property. A mission-critical node may have a recovery tier in hours, while a high criticality node may have a recovery tier in minutes because the volume of dependent activities is larger. Figure 5.5 shows a single interface dependency chain—the five-link sequence from cost center master data in SAP Master Data Governance to position assignment in SAP SuccessFactors Employee Central.

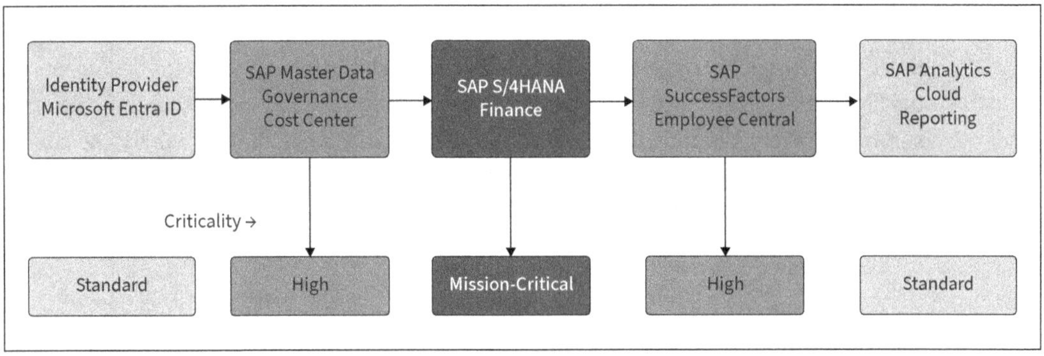

Figure 5.5: A Dependency Chain with Criticality Propagating Along the Chain; The Data Spine Reads the Chain to Compute the Blast Radius

5.3.3 Worked Example: Cost Center to Position

The cost center–position scenario, which will return in later chapters, begins with the creation of a new cost center in SAP Master Data Governance and ends with the assignment of a new position to that cost center in SAP SuccessFactors Employee Central. Between the two events, several interfaces and several applications are involved, and any one of them can fail in a way that produces a confident but wrong recommendation if the dependency chain is not modeled.

The interface inventory for the scenario is short but disciplined. The first interface, if-mdg-cc-01, carries cost center master data from SAP Master Data Governance to SAP S/4HANA Finance. Its direction is provider-to-consumer, its protocol is OData v4, its frequency is real-time, its semantic payload is cost center master data, and its classification is confidential. The second interface, if-s4-sf-cc-02, propagates the same cost center reference from SAP S/4HANA Finance to SAP SuccessFactors Employee Central. Its direction is provider-to-consumer, its protocol is REST/JSON, its frequency is near-real-time, its semantic payload is cost center reference (not the full master data record), and its classification is internal. The third interface, if-sf-sac-pos-03, carries the resulting position and cost center binding from SAP SuccessFactors to SAP Analytics Cloud for reporting. Its direction is provider-to-consumer, its protocol is OData v4, its frequency is scheduled batch with a daily cadence, its semantic payload is position cost center binding, and its classification is internal.

The dependency chain composes these three interfaces with the four applications they connect. The chain is recorded with criticality values. SAP Master Data Governance is mission-critical because every downstream record originates here. SAP S/4HANA Finance is mission-critical because the financial postings depend on its cost center hierarchy. SAP SuccessFactors Employee Central has high criticality because position changes can be deferred for short periods without business impact. SAP Analytics Cloud has standard criticality because reporting can tolerate a day's delay without harm. The chain as a whole is mission critical because it inherits the maximum criticality of any node.

With the chain recorded, several downstream consequences follow automatically. A deprecation of any one interface raises a deprecation notice on the chain, which the data spine routes to every process activity that depends on the chain. A change in the data classification on any interface triggers a reevaluation of AI eligibility for every variant that consumes the chain. An incident on SAP Master Data Governance opens an event that the data spine surfaces to every dependent activity within seconds, allowing in-flight work to be paused before incorrect master data flows through. Without the recorded chain, none of these consequences is automated; the chain has to be reconstructed from memory every time it is needed.

5.3.4 Hub Interfaces and the Limits of Direct Modeling

Real SAP landscapes contain interfaces that do not fit a clean provider/consumer pattern. The most common case is a hub interface, in which an integration layer such as SAP Integration Suite mediates between several providers and several consumers. The architecture register has to be honest about hubs because pretending they are point-to-point produces a model that is locally precise and globally misleading.

A hub interface is recorded as two interface factsheets, not one. The first records the provider-to-hub direction; the second records the hub-to-consumer direction. The integration suite itself is recorded as a first-class application factsheet, with its own owner, lifecycle, classification, and AI-eligibility evaluation. This is more work than recording a single virtual interface, but it produces a model that is truthful about who owns what. When SAP Integration Suite is upgraded, the change propagates correctly to both halves of the hub interface, and the dependent process activities see the change at the appropriate scope.

A small number of hub interfaces, particularly those that mediate event streams, are recorded as a single hub interface factsheet with a list-valued provider field and a list-valued consumer field. This compression is admissible only when the hub applies the same protocol, frequency, semantic payload, and classification to every provider and consumer in the list. The moment any one of those properties varies, the compression must be unwound into separate factsheets. The architecture register is precise about this restriction because the data spine will silently produce wrong blast-radius computations if the restriction is violated.

Cost Center to Position Scenario: Interface Dependency Chain

The dependency chain for the scenario reads as a single causal path through four interface factsheets:

- if-mdg-cc-01 has fs-mdg-fin-01 as its provider and fs-s4hana-fin-emea-01 as its sole consumer. Its `payloadClass` is `Master`, its `payloadDomain` is `CostCenter`, and its `criticalityClass` is `Critical` because every downstream cost center reference depends on it.
- if-s4-sf-cc-02 has fs-s4hana-fin-emea-01 as its provider and fs-sf-ec-01 as its consumer. Its `payloadClass` is `Reference`, which carries the cost center identifier and an effective-date pair, not the master-data record itself.
- if-sf-sac-pos-03 has fs-sf-ec-01 as its provider and fs-sac-01 as its consumer. Its `criticalityClass` is `Medium` (a reporting interface, not an operational one).
- if-ad-user-04 is consumed by all three applications.

When the dependency graph is asked “What depends on if-mdg-cc-01?” the data spine reads the provider and consumer edges and returns the chain from if-mdg-cc-01 to if-s4-sf-cc-02 to if-sf-sac-pos-03 with the intermediate applications resolved by reference.

5.3.5 Interface Factsheet: Build Specification

Interfaces are first-class factsheets in SAP LeanIX, not attributes hanging off applications. The field list in Table 5.5 is what every interface factsheet must carry to be reachable from the data spine. The discipline described there is what makes the dependency graph navigable; an interface with a missing provider or no payload class cannot be assembled into a chain at extraction time.

Field (SAP LeanIX)	Type	Mandatory	Notes/Controlled Values
factSheetId	string (UUID)	Yes (auto)	Assigned at creation. Same opacity rules as for applications.
displayName	string	Yes	Human-readable name. Recommended pattern: if-[provider-short]-[consumer-short]-[payload] (e.g. if-s4-sf-cc-02).
description	string	Yes	Functional summary of what the interface does.
interface-Direction	enum	Yes	Includes inbound, outbound, and bidirectional (relative to the provider).
interfaceProtocol	enum	Yes	Includes REST, OData, SOAP, RFC, IDoc, Kafka, file, and Java Database Connectivity (JDBC).
interfacePattern	enum	Yes	Can be synchronous, asynchronous, batch, event-driven, or streaming.
payloadClass	enum	Yes	Includes master, transactional, reference, configuration, and telemetry.
payloadDomain	enum	Yes	Controlled list (e.g. cost center, position, general ledger account, customer, or order).
payloadFormat	enum	Yes	Includes JSON, XML, CSV, IDOC, EDI, Avro, and Protobuf.
providerAppRef	factSheetId ref	Yes	The single application factsheet that publishes the interface.
consumerAppRef	factSheetId ref[]	Yes	One or more application factsheets that consume the interface.
middlewareRef	factSheetId ref	Conditional	Required when traffic flows through SAP Integration Suite, Cloud Integration, or other middleware.

Table 5.5: Interface Factsheet—Full Field Specification

Field (SAP LeanIX)	Type	Mandatory	Notes/Controlled Values
`technicalEndpoint`	`string`	No	Reference URL or topic name. Not used for matching; metadata only.
`dataVolume.daily`	`integer`	No	Average daily record count; consumed by capacity planning.
`latencyClass`	`enum`	Yes	Includes real time (under 1s), near real time (under 60s), hourly, daily, and weekly.
`lifecyclePhase`	`enum`	Yes	Includes plan, phasein, active, phaseout, and EOL (independent of provider/consumer phase).
`criticalityClass`	`enum`	Yes	Includes critical, high, medium, and low. Cascades into the dependency graph.
`securityClassification`	`enum`	Yes	Includes public, internal, confidential, and restricted (independent of payload).
`slaRef`	`string`	No	Reference to operational service-level agreement (SLA) document; metadata only.
`lastValidated`	`datetime`	Yes	Set by the interface owner at each governance review.

Table 5.5: Interface Factsheet—Full Field Specification (Cont.)

Note that `providerAppRef` is a single reference, not a list. An interface has exactly one provider. If two applications publish what looks like the same interface, they are in fact two interfaces with two `factSheetIds` and two sets of consumers. The temptation to share an interface across providers is the usual source of payload-semantics drift that the data spine cannot reconcile. Listing 5.3 shows a GraphQL query that extracts active interfaces with provider and consumers.

```
query GetActiveInterfaces($cursor: String) {
  allFactSheets(
    factSheetType: Interface,
    filter: {
      facetFilters: [
```

```
        { facetKey: "lifecyclePhase", keys: ["phase-in", "active"] }
      ]
    },
    first: 200, after: $cursor
  ) {
    pageInfo { hasNextPage endCursor }
    edges {
      node {
        id
        displayName
        description
        ... on Interface {
          interfaceDirection
          interfaceProtocol
          interfacePattern
          payloadClass
          payloadDomain
          payloadFormat
          latencyClass
          criticalityClass
          securityClassification
          lifecyclePhase
          lastValidated
          relInterfaceToProviderApplication { edges { node { id
displayName } } }
          relInterfaceToConsumerApplication { edges { node { id
displayName } } }
          relInterfaceToITComponent { edges { node { id displayName } } }
}
        }
      }
    }
  }
}
```

Listing 5.3: GraphQL Query for Active Interfaces with Provider and Consumer Edges

Listing 5.4 shows a normalized JSON payload sent to the data spine. The GraphQL response from Listing 5.3 is normalized by the SAP BTP ingestion job into the flat JSON object shown in Listing 5.4. Three groups of fields appear: a topology group (`providerAppRef` and `consumerAppRefs`) that names the apps the interface connects, a payload group (`class`, `domain`, and `format`) that names what the interface carries, and a runtime group (`protocol`, `pattern`, `direction`, and `latencyClass`) that names how it carries it. The shape mirrors the GraphQL response field for field so that the ingestion job is a near-direct mapping rather than a derivation, and the `factSheetId` in the topology block joins the interface back to the application payload from Listing 5.2 by the same key.

```
{
  "factSheetId": "4c8a91e2-5b3d-44e9-a812-c0f7be2d8a14",
  "displayName": "if-mdg-cc-01",
  "factSheetType": "Interface",
  "topology": {
    "providerAppRef": "7f3e2c01-9d4a-4e7b-bc88-91a2e6d0f5a3",
    "consumerAppRefs": [
      "a210ff8b-d4e6-4f00-b9d3-77c2a05f1428",
      "f9b1c0a7-2e58-4cb5-9d70-3a8e6c5d0b91"
    ],
    "middlewareRef": "itc-btp-is-prd"
  },
  "payload": {
    "class": "master",
    "domain": "CostCenter",
    "format": "JSON"
  },
  "runtime": {
    "protocol": "REST",
    "pattern": "Asynchronous",
    "direction": "Outbound",
    "latencyClass": "NearRealTime"
  },
  "governance": {
    "criticalityClass": "Critical",
    "securityClassification": "confidential",
    "lifecyclePhase": "active",
    "lastValidated": "2026-04-15T09:30:00Z"
  },
  "manifest": {
    "version": "2026-05-20T14:22:00Z",
    "manifestHash": "sha256:b27c...e405"
  }
}
```

Listing 5.4: Data Spine–Shaped JSON for One Active Interface

5.4 Lifecycle and Change-Related Attributes

Lifecycle is the attribute that turns a static architecture inventory into an executable register. An application that exists is not the same as an application that is operational, and an application that is operational is not the same as one that is approved for integration or for AI consumption. The lifecycle attribute discipline makes those distinctions explicit and enforces them at the point of publication to the canonical data spine.

The discipline has a finance control counterpart that is worth naming. A new general ledger account exists in the chart of accounts long before it is posted to, and it continues to exist long after it is closed to new postings. The accounting register distinguishes between the existence of an account, its eligibility for posting, its eligibility for reporting, and its eligibility for consolidation. The same kind of distinction applies to the application register, and the cost of conflating any two of these states is comparable: silent errors that surface at a downstream reconciliation rather than at the point of failure.

This section describes the lifecycle attribute discipline in six steps. First, it lays out the five-phase model (plan, phase-in, active, phase-out, and EOL) and the consumer visibility rules that follow from each phase. Second, it explains the two approval flags, `approvedForIntegration` and `approvedForAI`, and why they must be kept independent. Third, it shows how lifecycle transitions become governance events on the data spine rather than silent attribute updates. Fourth, it extends the lifecycle attribute set to include version references, planned changes, and freeze windows that the commit gateway reads at write time. Fifth, it consolidates the rules into a phase-to-consumer visibility matrix that the data spine uses to decide which factsheets to expose to which surface. Finally, it supplies the full build specification for every lifecycle and change-related attribute.

5.4.1 The Five-Phase Lifecycle Model

The lifecycle model recognized by the integration register has five named phases. The phases are recorded as values in a controlled list on every factsheet (both applications and interfaces). The data spine consults the phase before exposing the factsheet to any consumer. The phases are as follows:

1. **Plan**
   The application is approved in principle and a factsheet has been created, but no production deployment exists yet. Plan phase factsheets are not visible to the data spine, but they are visible to architecture-planning users.
2. **Phase-in**
   The application is being deployed and is reachable, but its integration is incomplete. The data spine surfaces the factsheet to integration developers and impact analysis services in read-only mode. AI services do not see the factsheet in this phase, and SAP Signavio activities are not permitted to reference it as an execution location for a published process variant.
3. **Active**
   The application is fully operational and approved for production integration. The data spine exposes the factsheet to all consumers subject to the `approvedForIntegration`

and `approvedForAI` flags. This is the only phase in which a published process variant may reference the application as an execution location.

4. **Phase-out**
   The application is being decommissioned, but it is still operational. The data spine exposes a deprecation notice on every reference to the factsheet and requires that any new process variants that would create new references first be reviewed by the architecture and process functions jointly.

5. **EOL**
   The application is no longer operational. The data spine removes the factsheet from active consumption and retains it in the archive layer for historical references. Process variants that referenced the factsheet are flagged as obsolete until the references are removed or redirected.

5.4.2 The Two Approval Flags

In addition to the lifecycle phase, every integration-eligible application carries two approval flags that operate at the same level. The first, `approvedForIntegration`, records whether the application is permitted to participate in machine-to-machine data exchange through the canonical data spine. The second, `approvedForAI`, records whether the application's data is permitted to flow into AI grounding bundles, AI training corpora, or AI inference contexts.

The two flags are not redundant. An application can be approved for integration and denied for AI—for instance, because its data classification permits machine-to-machine exchange under a contractual safeguard but does not permit exposure to a foundation model. The reverse is less common but admissible: An application may be approved for AI consumption (its data is being used to ground a research model) while not yet being approved for production integration. The flags exist to keep these cases separable rather than collapsing them into a single permission.

Each approval flag carries three pieces of metadata: the approval date, the approving authority, and the scope of the approval. The scope is recorded as a controlled list value that names the variants, country scopes, or data classifications under which the approval is valid. An application may, for example, be approved for AI consumption for global variants but denied for variants scoped to specific countries whose privacy regimes prohibit it. The data spine evaluates the scope at the point of every grounding request rather than relying on a single global flag.

5.4.3 Lifecycle Transitions as Governance Events

Lifecycle phase changes are not silent attribute updates. Every phase transition is recorded as a dated event with an approving authority, an effective date, and a reason code. The data spine subscribes to these events and rebroadcasts them to dependent services, including the SAP Signavio governance API for process-variant reconciliation.

Three transitions deserve specific discipline. The transition from plan to phase-in is the moment the application becomes visible to the data spine. Before this transition, the application can be referenced only by planning artifacts; after it, the application can be referenced by integration developers and impact analysis services. The transition has to be made explicit because the visibility cost and benefit trade-off changes at this point: The data spine begins to spend effort tracking the application, and downstream services start to depend on its presence.

The transition from phase-in to active is the moment the application becomes executable for published process variants. This transition must be preceded by a completeness check: identity attributes filled, lifecycle dates valid, approvals granted with named authorities, interface inventory complete, and dependency chain recorded. The data spine refuses the transition if any of the completeness checks fails, and the refusal is recorded with the specific check that failed so that the architecture team can remediate.

The transition from active to phase-out is the moment the application is marked for decommissioning. The data spine triggers a deprecation notice on every dependent process activity and every dependent interface. The notice does not block continued operation; it requires the affected process owners to acknowledge it and to record their remediation plan. The deprecation notice has a configurable aging window. Notices older than the window without acknowledgment are escalated to the application owner and the chief architect.

5.4.4 Beyond Phases: Other Change-Related Attributes

Phases are the dominant lifecycle attribute, but they are not the only one. Three additional change-related attributes are required on every integration-eligible factsheet.

The first is the version reference. The application factsheet records the major release version of the underlying software (for example, SAP S/4HANA 2023, SAP SuccessFactors 1H/2025, or SAP BTP, Cloud Foundry runtime 4.21). The version reference is consulted by AI services when assembling a grounding bundle, because two versions of the same application can have different attribute sets and different interface behaviors. A reasoning service that grounds against the wrong version produces fluent but stale recommendations, which are difficult to detect at the recipient and expensive to unwind.

The second is the planned change reference. When a known change is scheduled (a version upgrade, an interface deprecation, a hosting migration), the factsheet records the change as a structured event with a planned date, a description, and a reference to the change management record. The data spine reads planned change events and surfaces them in impact views, allowing the architecture team to anticipate the propagation rather than react to it. Planned changes that arrive without prior recording on the factsheet are treated as audit events; the data spine does not refuse to process them, but it records the absence of prerecording as a governance finding.

The third is the freeze window reference. Many enterprise SAP landscapes operate under change-freeze windows tied to the financial calendar (period close, fiscal year close, or regulatory reporting cycles). The application factsheet records the freeze windows that apply to it, expressed as recurring date ranges with a controlled-list reason code. The data spine consults the freeze windows when validating whether a proposed change can take effect, and it refuses publication of process variants that depend on changes scheduled inside a freeze window unless an exception has been approved. This discipline is one that you will recognize if you have a finance-control background, as it is a direct counterpart of the posting period control. The calendar gates the transaction, not the other way around.

5.4.5 Phase-to-Consumer Visibility Matrix

Table 5.6 summarizes the visibility of each phase to each major consumer. The matrix is enforced by the data spine itself, not by application code in any individual consumer. A consumer that needs to override the visibility rule must request an exception through the data spine's governance interface; bypassing the matrix at the application layer is not permitted.

Phase	Visible to Integration Developers?	Visible to AI Services?
Plan	No	No
Phase-in	Yes (read-only)	No
Active	Yes	Yes (subject to `approvedForAI` scope)
Phase-out	Yes (with deprecation notice)	Yes (with deprecation notice)
EOL	No (archive only)	No (archive only)

Table 5.6: Phase-to-Consumer Visibility Matrix—What Each Lifecycle Phase Exposes to Integration and AI Consumers

Cost Center to Position Scenario: Lifecycle and Freeze Windows

All three scenario applications are in active phase at the time of writing. fs-mdg-fin-01 carries a recurring `freezeWindow` tied to year-end and quarter-end (`YEAR_END_CLOSE` and `QUARTER_CLOSE` reason codes) because cost center master data must not change while the books are being closed. The commit gateway reads the freeze window before allowing if-mdg-cc-01 to publish a new cost center. A publish attempt during the window is refused with the reason code attached. fs-s4hana-fin-emea-01 has a `plannedChange` entry of `changeType = Upgrade` with `plannedDate` set to the next fiscal-year boundary. The change reference (CR-2027-0142) is consulted by the AI grounding service when assembling a bundle that includes this application, so that recommendations are qualified by the upcoming change. fs-sf-ec-01 has no planned change and no freeze window. None of the three applications is in phase-out, so the data spine exposes all activity bindings without deprecation notices.

5.4.6 Lifecycle Attribute Set: Build Specification

Lifecycle attributes live on both application and interface factsheets, and the data spine reads them at every lookup. Table 5.7 documents the full set, including the controlled list values that the data spine validates against. An out-of-list value is rejected at ingestion; this prevents accidental introduction of new phases or new approval states that the gateway and AI surfaces would not know how to interpret.

Field	Type	Applies To	Controlled Values/Notes
`lifecyclePhase`	`enum`	App, interface	Includes plan, phase-in, active, phase-out, and EOL.
`lifecycle.startDate`	`date`	App, interface	ISO 8601. Date of entry into current phase.
`lifecycle.endDate`	`date`	App, interface	ISO 8601. Required when in phase-out or EOL.
`lifecycle.plannedNextPhase`	`enum`	App, interface	Optional. Same domain as `lifecyclePhase`. Forecast only.
`lifecycle.plannedNextDate`	`date`	App, interface	Required if `plannedNextPhase` is set.
`approvedForIntegration`	`boolean`	App	Default false. Must be approved by the integration governance board.
`approvedForIntegrationDate`	`date`	App	Date of approval. Required when `approvedForIntegration` is true.
`approvedForIntegrationBy`	`user ref`	App	User who recorded the approval.
`approvedForAI`	`boolean`	App	Default false. Must be approved by AI eligibility review.
`approvedForAIDate`	`date`	App	Date of approval. Required when `approvedForAI` is true.
`approvedForAIBy`	`user ref`	App	User who recorded the approval.

Table 5.7: Lifecycle and Change-Related Attributes—Full Field Specification

Field	Type	Applies To	Controlled Values/Notes
approvedForAIScope	string[]	App	Optional scope qualifier (e.g., [read-only, metadata-only, summary-only]).
versionReference	string	App, interface	Major release identifier (e.g., SAP S/4HANA 2023 FPS02 or OData v4).
plannedChange	object[]	App, interface	List of {plannedDate, changeType, changeRef, description}. changeType includes upgrade, deprecate, replatform, and hosting.
freezeWindow	object[]	App	List of {startDate, endDate, reasonCode}. reasonCode includes YEAR_END_CLOSE, QUARTER_CLOSE, RELEASE_FREEZE, and COMPLIANCE_AUDIT.
deprecationNotice	object	App, interface	List of {noticeDate, effectiveDate, replacementFactSheetId, agingDays}. Sent to all consumers when phase moves to phase-out.
lastReviewed	datetime	App, interface	Set by the owner at each governance review. Stale values trigger an aging alert.

Table 5.7: Lifecycle and Change-Related Attributes—Full Field Specification (Cont.)

Listing 5.5 shows the lifecycle slice of a data spine payload for one application that has entered the phase-out phase. The payload carries the current lifecycle state, the planned next phase and date, the two approval flags with their approval-event metadata, a list of planned changes (one deprecate entry pointing at the replacement factSheetId), two active freeze windows aligned to the financial calendar, and the deprecation notice that the data spine sends to every consumer of the factsheet. Each block in the payload corresponds directly to fields specified in Table 5.7; the listing is the runtime view of what those fields look like once populated.

```
{
  "factSheetId": "7f3e2c01-9d4a-4e7b-bc88-91a2e6d0f5a3",
  "lifecycle": {
    "phase": "phase-out",
    "startDate": "2025-10-01",
    "endDate": "2026-12-31",
    "plannedNextPhase": "eol",
```

```
    "plannedNextDate": "2026-12-31"
  },
  "approvals": {
    "approvedForIntegration": true,
    "approvedForIntegrationDate": "2024-04-01",
    "approvedForIntegrationBy": "igb.chair@example.com",
    "approvedForAI": false,
    "approvedForAIScope": []
  },
  "plannedChanges": [
    {
      "plannedDate": "2026-08-01",
      "changeType": "Deprecate",
      "changeRef": "CR-2026-0741",
      "description": "Replaced by SAP S/4HANA Public Cloud (factSheetId
4f73...)"
    }
  ],
  "freezeWindows": [
    { "startDate": "2025-12-28", "endDate": "2026-01-05", "reasonCode":
"YEAR_END_CLOSE" },
    { "startDate": "2026-03-29", "endDate": "2026-04-03", "reasonCode":
"QUARTER_CLOSE" }
  ],
  "deprecationNotice": {
    "noticeDate": "2025-10-01",
    "effectiveDate": "2026-12-31",
    "replacementFactSheetId": "4f73abc0-1234-5678-9abc-def012345678",
    "agingDays": 90
  }
}
```

Listing 5.5: Lifecycle Slice for an Application in Phase-Out

5.5 Stable Identifiers and Keys

Identifiers are the silent infrastructure of an integration register. When they are correct, no one notices. When they are wrong, every downstream system fails in a slightly different way, and the cause is hard to find. In this section, we describe the identifier discipline in seven steps. First, we explain why the factSheetId is the PK and what its opaque, immutable, and unique properties mean in practice. Second, we show how external identifiers (SIDs, vendor codes, SAP BTP routes, and Active Directory IDs) are recorded as resolvable references rather than as PKs. Third, it treats identifier changes (merge, split, and deprecation) as governance events rather than as routine attribute updates. Fourth, it explains how SAP Signavio activities bind to SAP LeanIX through stable references that the data spine resolves at lookup time. Fifth, it covers the SAP Fiori app ID as a secondary identifier used by the WalkMe adoption layer. Sixth, it supplies the full build specification for the identifier set across all participating systems. Finally, the cross-chapter field-name glossary maps every name introduced in this chapter to the canonical data spine names used in later parts of the book.

The general principle, which we established in Chapter 3 for SAP Signavio and reinforced in Chapter 4 for the extension contract, is that identity is owned by exactly one system. Every other system references that identity but does not mint new identifiers for the same object. For applications and interfaces, the system of record is SAP LeanIX. SAP Signavio references SAP LeanIX identifiers in its dictionary. The canonical data spine references SAP LeanIX identifiers in its application and interface keys. WalkMe references SAP LeanIX identifiers through the SAP Fiori app ID linkage, which is addressed in Chapter 7 and Chapter 8. None of these consumers is permitted to invent a new identifier scheme for an application or interface that already exists in SAP LeanIX.

5.5.1 The factSheetId as a Primary Key

Every SAP LeanIX factsheet carries a `factSheetId`, assigned at creation by the SAP LeanIX platform and stable for the lifetime of the factsheet. The `factSheetId` is the PK by which integration-eligible applications and interfaces are referenced throughout the rest of the integrated landscape. Three properties of the `factSheetId` are exploited explicitly by the data spine.

First, the `factSheetId` is opaque. It carries no semantic information about the application it identifies. This opacity is a feature, not a defect: It prevents downstream consumers from inferring meaning from the identifier itself and from building logic that depends on the identifier's structure. An identifier that is interpretable invites the very semantic drift the architecture register is meant to prevent.

Second, the `factSheetId` is immutable. Once assigned, it never changes. Renaming the application, moving it between business units, reclassifying its data, or upgrading its underlying technology do not change it. The only path by which an application's identifier changes is the creation of a new factsheet, an explicit merge or split operation, or a deprecation followed by a fresh registration. Each of these is a governance event with explicit approval.

Third, the `factSheetId` is universally unique within the SAP LeanIX tenant. Cross-tenant federation, where applicable, is handled by namespacing the `factSheetId` with the tenant prefix at the data spine boundary. Cross-tenant federation refers to architecture register setups where multiple SAP LeanIX workspaces (typically one per region or one per acquired entity) are synchronized into a single canonical view. The data spine maintains a federation registry that resolves prefixed identifiers to their owning tenant. This is more elaborate than most enterprises need on day one, but adding it later is significantly more expensive than establishing it from the start.

5.5.2 External Identifiers as Resolvable References

The `factSheetId` is necessary but not sufficient. The integration register also records external identifiers that resolve back to the systems the factsheet represents. The external identifiers are recorded in structured fields with controlled naming, so that automated resolution does not depend on string-pattern matching.

For SAP applications, the external identifier discipline is straightforward. The application factsheet records the SAP SID (the three-character SID used by the SAP NetWeaver and SAP BTP runtimes), the SAP product code (drawn from the SAP product catalog), and the SAP for Me deployment record. Each of these is recorded in a separate field. The data spine resolves an SAP LeanIX `factSheetId` to an SAP runtime by consulting the SID; it resolves an SAP runtime event back to an SAP LeanIX `factSheetId` by consulting the same field in reverse. Neither direction requires text pattern matching.

For third-party applications, the discipline is slightly more elaborate because the identifier conventions vary by vendor. The factsheet records the vendor name (drawn from a controlled list), the vendor product code, and the deployment instance identifier. The deployment instance is significant: A single vendor product may be deployed in several instances across the enterprise, and the integration register has to distinguish between them. Treating two instances of the same vendor product as a single record is one of the most common modeling errors and is among the most expensive to unwind.

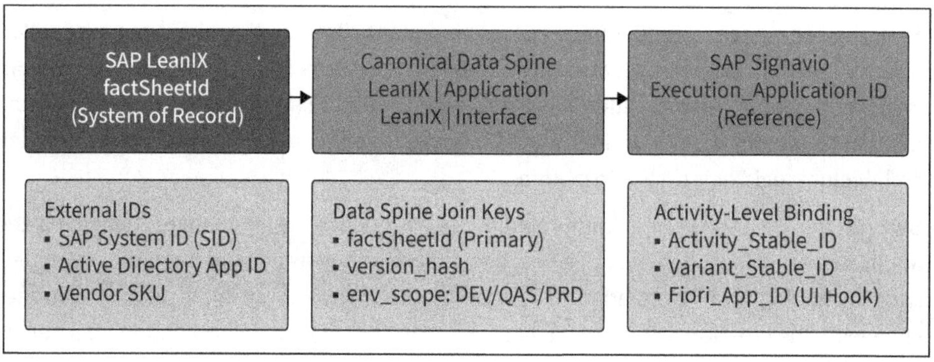

Figure 5.6: Stable Identifier Model Linking SAP LeanIX, the Canonical Data Spine, and SAP Signavio Without Semantic Drift

For custom-built applications hosted on SAP BTP, the factsheet records the subaccount, the space, the application route, and the destination service binding. The identifier set is denser because the deployment surface is denser, but the principle is the same: Each field carries a known resolver and a known consumer. Figure 5.6 shows how the stable identifier model spans SAP LeanIX, the canonical data spine, and SAP Signavio, with each system carrying its own identifier scheme and the `factSheetId` functioning as the system-of-record key.

5.5.3 Identifier Changes as Governance Events

The fact that identifiers do not change is enforced at the SAP LeanIX boundary. The platform rejects routine updates to the `factSheetId`; the only operations that produce a new `factSheetId` are explicit governance events with named approvers. There are three of these events.

The first is the merge. Two factsheets representing the same underlying application are merged into one, the surviving `factSheetId` becomes the canonical reference, and the retired `factSheetId` is recorded in a deprecation table that the data spine consults when resolving historical references. The merge is permitted only when both factsheets are in the same lifecycle phase and have compatible classifications, owners, and reference inventories. The merge produces a structured `AUDIT_EVENT` that flows into the audit chain, and the data spine publishes the merge as a deprecation notice to every dependent consumer.

The second is the split. A single factsheet that was incorrectly modeled as one application is split into two or more factsheets, each with its own `factSheetId`. The split is permitted only when the architecture team can demonstrate that the original factsheet conflated two distinct business purposes, two distinct owners, or two distinct classifications. The split produces a structured `AUDIT_EVENT`, and the data spine routes the change to every dependent consumer with a remediation deadline.

The third is the deprecation and reregistration. An application is retired and a new application is registered to take its place. The two events are linked, the new `factSheetId` is recorded as the successor of the old one, and the data spine resolves historical references to the new identifier transparently for a configurable transition window. After the window expires, historical references continue to resolve, but a deprecation finding is opened on each one.

None of these three events is automated. Each requires named approvers from both the architecture and process functions. This is not done for bureaucratic reasons but because an identifier change is one of the few operations in the integration landscape that can break audit trails if it is performed casually. The data spine's `AUDIT_EVENT` chain depends on the immutability of identifiers; the moment that property is violated, the chain must be rebuilt with explicit traversal logic that crosses the merge, split, or deprecation event.

5.5.4 Binding Identifiers to Activities

On the SAP Signavio side, the binding from a process activity to an SAP LeanIX application is recorded as a structured reference using the four-field shape established in Chapter 4: `ref_system`, `ref_type`, `ref_id`, and `ref_label`. For SAP LeanIX bindings, `ref_system` is

LEANIX, ref_type is APPLICATION or INTERFACE, ref_id is the SAP LeanIX factSheetId, and ref_label is the displayName at the moment of binding (recorded as a courtesy for human readers but never relied upon for resolution).

The Activity_Stable_ID owns the binding. It is the field on the SAP Signavio activity that holds the LEANIX|APPLICATION reference. The pipe character (|) in LEANIX|APPLICATION (and LEANIX|INTERFACE) is the SAP Signavio reference separator introduced in Chapter 4. When the data spine assembles the EXECUTION_CONTEXT for an activity at runtime, it resolves the Activity_Stable_ID, reads the LEANIX|APPLICATION reference, and looks up the current state of the SAP LeanIX factsheet. The lookup uses the factSheetId directly. The displayName, the URL, the lifecycle phase, the classification, and the approval flags are all read at lookup time, not cached on the SAP Signavio side.

The same shape applies to interface references. An activity that depends on an interface records a LEANIX|INTERFACE reference whose ref_id is the interface factSheetId. The data spine resolves the reference to the current interface state at lookup time. A variant that depends on two interfaces records two references; an activity that consumes a hub interface records two references (one for the provider-to-hub direction and one for the hub-to-consumer direction), consistent with the discipline we established in Section 5.3.

5.5.5 SAP Fiori App ID as a Secondary Identifier

For SAP applications that present an SAP Fiori user interface, the application factsheet records the SAP Fiori app ID alongside the SID and product code. The SAP Fiori app ID is a five-character code (for example, F0703 or F2207) that uniquely identifies an SAP Fiori app in the SAP Fiori apps reference library. The SAP Fiori app ID is the binding key by which WalkMe attaches adoption content and telemetry to a specific user interface, and it is the binding key by which SAP Signavio activities reference the user-facing surface in which they execute.

The SAP Fiori app ID is a secondary identifier in the sense that it does not replace the factSheetId. A single SAP LeanIX application factsheet may carry several SAP Fiori app IDs, one for each SAP Fiori app that presents a user-facing surface of the application. The factsheet records the SAP Fiori app IDs as a list-valued field, with each entry carrying a controlled list tag that names the role for which the surface is intended. The data spine and WalkMe consume the list together: The data spine uses the factSheetId for application-level resolution, and WalkMe uses the SAP Fiori app ID for surface-level adoption content.

The discipline that the SAP Fiori app ID is secondary, not primary, is important. A common modeling error is to treat the SAP Fiori app ID as the application identifier itself, particularly when the architecture team's familiarity with SAP Fiori is stronger than its familiarity with SAP LeanIX. The error manifests as soon as a single SAP Fiori app is split into two, or two SAP Fiori apps are merged into one, without any change at the SAP LeanIX application level. The data spine sees an SAP Fiori-level reorganization as a series of changes; if the references depend on the SAP Fiori app ID alone, every reference must be manually redirected.

If the references depend on the `factSheetId` with the SAP Fiori app ID as a secondary identifier, the data spine handles the reorganization transparently.

Cost Center to Position Scenario: Identifier Resolution

For the scenario, the data spine resolves four distinct identifier systems back to the LeanIX `factSheetId`. The SAP Signavio side records `A_COST_CENTER_CREATE` with a `LEANIX|APPLICATION` reference (`ref_id` fs-mdg-fin-01) and a `LEANIX|INTERFACE` reference (`ref_id` if-mdg-cc-01) with role tag `PRODUCES`. The SAP side reports `sapSID = M1P` for fs-mdg-fin-01, `sapSID = S4P` for fs-s4hana-fin-emea-01, and an SAP SuccessFactors tenant ID for fs-sf-ec-01. The SAP Fiori side carries an SAP Fiori app ID on fs-s4hana-fin-emea-01 (the Cost Center Maintenance app) that lets the WalkMe layer bind in-app guidance to the activity. The data spine resolution table holds one row per external identifier. For example, the query "Which SAP LeanIX application is sapSID = M1P?" returns fs-mdg-fin-01 in a single lookup, regardless of whether the identifier system is asking from SAP S/4HANA, Active Directory, or the data spine itself. This is the discipline that lets you record the activity binding without inventing a new identifier scheme, as we will describe in Chapter 6.

5.5.6 Identifier Set: Build Specification

Identifiers are the substrate of every join the data spine performs. Table 5.8 documents the full identifier set carried by application and interface factsheets, the system that owns each identifier, and the way the data spine resolves it. The discipline is that SAP LeanIX owns the `factSheetId`; every other identifier is recorded as a reference to a system that owns it.

Identifier	Owner	Mandatory	Resolution/Notes
`factSheetId`	SAP LeanIX	Yes	PK. Opaque universally unique identifier (UUID). Used by the data spine and SAP Signavio as the join key.
`externalId.sapSID`	SAP system	Conditional	Three-character SID. Required if `technologyStack` includes SAP. Resolved by the data spine to the SAP LeanIX `factSheetId` at lookup time.
`externalId.vendorSku`	Vendor	Conditional	Vendor product code. Required for third-party apps. Data spine treats it as a non-unique secondary key.

Table 5.8: Identifier Set—Full Specification Across SAP LeanIX, SAP BTP, SAP Fiori, and SAP Signavio

Identifier	Owner	Mandatory	Resolution/Notes
`externalId.btpRoute`	SAP BTP	Conditional	Cloud Foundry route or sub-account identifier. Required for SAP BTP–hosted apps.
`externalId.activeDirectoryId`	Active Directory	No	Active Directory application identifier. Optional; used by identity-aware grounding.
`fioriAppId`	SAP Fiori	Conditional	Five-character SAP Fiori app ID (e.g., F0707). Required for SAPUI5 apps. Secondary identifier; used by WalkMe binding.
`itComponentRef`	SAP LeanIX	Yes	`factSheetId` of the IT component. Resolves to the deployable unit, not the logical application.
`activityStableRef`	SAP Signavio	No (read-only on SAP LeanIX side)	List of `Activity_Stable_IDs` that bind activities to this application. Set by SAP Signavio; surfaced on SAP LeanIX via the data spine join.
`variantStableRef`	SAP Signavio	No (read-only on SAP LeanIX side)	List of `Variant_Stable_IDs` applicable to activities running against this application.
`mergeSourceRef`	SAP LeanIX	Conditional	When this factsheet survives a merge, it lists the prior `factSheetIds` it absorbed. Required if the factsheet was created by a merge governance event.
`splitParentRef`	SAP LeanIX	Conditional	When this factsheet is created by a split, it stores the `factSheetId` of the original parent. Required if the factsheet was created by a split governance event.
`deprecationReplacementRef`	SAP LeanIX	Conditional	When this factsheet is being phased out, it lists the `factSheetId` that replaces it. Required when `lifecyclePhase` is phase-out or EOL and a successor exists.

Table 5.8: Identifier Set—Full Specification Across SAP LeanIX, SAP BTP, SAP Fiori, and SAP Signavio (Cont.)

On the data spine side, identifiers are stored in a resolution table that lets any system look up the SAP LeanIX `factSheetId` from whatever identifier it knows. The JSON in Listing 5.6 shows the identifier part of the application payload and the corresponding entries in the data spine resolution table.

```
// Identifier slice of the application payload
{
  "factSheetId": "7f3e2c01-9d4a-4e7b-bc88-91a2e6d0f5a3",
  "identifiers": {
    "primary": { "system": "LeanIX", "value": "7f3e2c01-9d4a-4e7b-bc88-
91a2e6d0f5a3" },
    "external": [
      { "system": "SAP",   "key": "sapSID",   "value": "S4P"  },
      { "system": "Fiori", "key": "fioriAppId","value": "F0707"},
      { "system": "AD",    "key": "adAppId",  "value": "app-s4fin-prd" }
    ],
    "history": {
      "mergeSources": ["1a7e0000-...", "8c920000-..."],
      "deprecationReplacement": null
    }
  }
}

// Spine resolution table entries (one row per external identifier)
[
  { "system":"SAP",   "key":"sapSID",
    "value":"S4P",          "factSheetId":"7f3e2c01-..." },
  { "system":"Fiori", "key":"fioriAppId",
    "value":"F0707",        "factSheetId":"7f3e2c01-..." },
  { "system":"AD",    "key":"adAppId",
    "value":"app-s4fin-prd","factSheetId":"7f3e2c01-..." },
  { "system":"LeanIX","key":"merge",
    "value":"1a7e0000-...","factSheetId":"7f3e2c01-..." },
  { "system":"LeanIX","key":"merge",
    "value":"8c920000-...","factSheetId":"7f3e2c01-..." }
]
```

Listing 5.6: Identifier Slice and the Data Spine Resolution Table Entries It Produces

5.5.7 Cross-Chapter Field Name Glossary

The SAP LeanIX field names introduced in this chapter are the upstream side of a longer pipeline. The canonical data spine on SAP BTP (covered in Part V) uses a different naming convention, and the AI preparation layer (covered in Part VII) adds a further layer of eligibility metadata. To keep our vocabulary consistent across chapters, this section lists the end-to-end mapping for every field name you will encounter in later chapters.

Two conventions are worth stating before we look at the tables. First, SAP LeanIX field names follow camelCase (factSheetId, lifecyclePhase); spine field names follow UPPER_SNAKE_CASE (APPLICATION_ID, LIFECYCLE_STATE). The change of case is intentional: When a name appears in camelCase in a later chapter, it is being read from the SAP LeanIX side; when it appears in UPPER_SNAKE_CASE, it is being read from the data spine. Second, the SAP LeanIX lifecycle has five values (plan, phase-in, active, phase-out, and EOL) but the data spine consolidates these into four (PLAN, ACTIVE, PHASE_OUT, and RETIRED). Phase-in is mapped to ACTIVE on the data spine because, from the data spine consumer's perspective, an application reachable for read is operationally active even when its deployment is still being staged.

Table 5.9 shows the application-side field mapping.

SAP LeanIX (This Chapter)	Canonical Data Spine Name	AI/Eligibility (Chapter 14)	Notes
factSheetId	APPLICATION_ID	Row identity in BUNDLE_MANIFEST	PK. Stable across all three layers.
displayName	APPLICATION_NAME	(Not exposed)	Trimmed, whitespace-normalized on the data spine.
lifecyclePhase	LIFECYCLE_STATE	Feeds ELIGIBILITY_POLICY filter	Plan/phase-in/active/phase-out/EOL maps to PLAN/ACTIVE/ACTIVE/PHASE_OUT/RETIRED.
approvedForIntegration	(Governance attribute)	(Not directly mapped)	Gate on whether spine accepts the row at all.
approvedForAI	(Governance attribute)	ELIGIBILITY_POLICY.ai_eligible_flag input	Necessary but not sufficient for AI eligibility; ELIGIBILITY_POLICY adds jurisdiction and sensitivity checks.

Table 5.9: Application-Side Field Mapping—SAP LeanIX to Canonical Data Spine to AI Eligibility

SAP LeanIX (This Chapter)	Canonical Data Spine Name	AI/Eligibility (Chapter 14)	Notes
dataClassification	(Carried)	ELIGIBILITY_POLICY.sensitivity_class input	Includes public, internal, confidential, and restricted.
applicationOwner	BUSINESS_OWNER	(Role context input)	Resolved against IDENTITY on the data spine.
technicalOwner	IT_OWNER	(Role context input)	Same resolution as BUSINESS_OWNER.
itComponentRef	IT_COMPONENT_ID	(Not exposed)	Foreign key (FK) into IT_COMPONENT on the data spine.
businessCapabilityRef	CAPABILITY_ID list	(Grounding bundle scope)	Many-to-many via map table on spine.
legalEntityScope	JURISDICTION list	ELIGIBILITY_POLICY.jurisdiction	ISO 3166-1 alpha-2 codes.
complianceTags	COMPLIANCE_TAG list	ELIGIBILITY_POLICY filter input	SOX/GDPR/HIPAA–controlled list.
freezeWindow	FREEZE_WINDOW table	(Read by commit gateway)	Spine-stored; consulted at write time.

Table 5.9: Application-Side Field Mapping—SAP LeanIX to Canonical Data Spine to AI Eligibility (Cont.)

Table 5.10 shows the interface-side field mapping.

SAP LeanIX (This Chapter)	Canonical Data Spine Name	AI/Eligibility (Chapter 14)	Notes
factSheetId	INTERFACE_ID	Row identity in BUNDLE_MANIFEST	PK.
displayName	(Carried)	(Not exposed)	Trimmed, whitespace-normalized on the spine.

Table 5.10: Interface-Side Field Mapping—SAP LeanIX to Canonical Data Spine to AI Eligibility

SAP LeanIX (This Chapter)	Canonical Data Spine Name	AI/Eligibility (Chapter 14)	Notes
providerAppRef	SRC_APP_ID	(Not exposed)	FK into APPLICATION on spine.
consumerAppRefs	TGT_APP_ID list	(Not exposed)	Many-to-many via map table.
interfaceProtocol	PROTOCOL	(Not exposed)	Canonical enumeration: REST, OData, SOAP, IDoc, event, and file.
criticalityClass	CRITICALITY	ELIGIBILITY_POLICY input	LOW/MEDIUM/HIGH/MISSION_CRITICAL on the data spine.
payloadClass	PAYLOAD_CLASS	(Grounding scope)	Includes master, transactional, reference, configuration, and telemetry.
payloadDomain	PAYLOAD_DOMAIN	(Grounding scope)	Controlled list (cost center, position, general ledger account, etc.).
lifecyclePhase	LIFECYCLE_STATE	Feeds ELIGIBILITY_POLICY filter	Same enumeration and mapping as APPLICATION.
securityClassification	(Carried)	ELIGIBILITY_POLICY input	Inherits the more restrictive provider or consumer classification.

Table 5.10: Interface-Side Field Mapping—SAP LeanIX to Canonical Data Spine to AI Eligibility (Cont.)

Table 5.11 shows the fields involved with dependency, capability, and data object mapping.

SAP LeanIX	Canonical Data Spine Name	Notes	Where Consumed
(Consumer/provider edges)	DEPENDENCY (USES/EXTENDS/ REPLACES)	Edge kind preserved as enumeration on the data spine.	Ingestion (Chapter 11), impact services (Chapter 12)
businessCapability	CAPABILITY	Code becomes CAPABILITY_ID; parent becomes PARENT_ID; domain preserved.	Transformation (Chapter 11), navigation (Chapter 17)
dataObject.classification	DATA_OBJECT_REF. CLASSIFICATION	Public, internal, confidential, or restricted preserved verbatim.	Eligibility filter (Chapter 14)
dataObject.sensitivity	DATA_OBJECT_REF. SENSITIVITY	Canonical enumeration aligned to data privacy policy.	Eligibility filter (Chapter 14)

Table 5.11: Activity-Binding Fields—SAP Signavio-Side References That Point Back to SAP LeanIX

Table 5.12 shows the activity-binding fields (SAP Signavio references that point back to SAP LeanIX).

SAP Signavio Field	Value Range	Cardinality	Notes
LEANIX|APPLICATION ref_id	factSheetId of an application	Exactly one per activity	Read by spine to resolve the execution location of the activity.
LEANIX|INTERFACE ref_id	factSheetId of an interface	Zero or more per activity	Each binding carries a role tag: CONSUMES, PRODUCES, or CONSUMES_AND_PRODUCES.
Activity_Stable_ID	SAP Signavio-minted opaque ID	PK on the activity	Stable across process versions; used by the data spine as the activity join key.

Table 5.12: SAP Signavio Activity Binding Fields

SAP Signavio Field	Value Range	Cardinality	Notes
Variant_Stable_ID	SAP Signavio-minted opaque ID	One per variant	Filters activities by region or business unit at grounding time.

Table 5.12: SAP Signavio Activity Binding Fields (Cont.)

You should now be able to start with a field name and trace it through the tables to its source and destination. Our intent is not for you to memorize the mapping but to state it once, in one place, so that we can refer to fields by their canonical name without ambiguity throughout the rest of the book.

5.6 Avoiding Overlap with SAP Signavio

So far in this chapter, we have described what SAP LeanIX must contain to be an integration register: applications, interfaces, dependencies, lifecycle attributes, and identifiers. Our argument has been almost entirely inward-looking, focused on the shape of SAP LeanIX itself. In this section, we reverse that perspective. We look at the boundary between SAP LeanIX and SAP Signavio and ask what should not appear in SAP LeanIX because it belongs in SAP Signavio and what should not appear in SAP Signavio, because it belongs in SAP LeanIX. The two repositories share a working surface in the data spine, but they do not share content. The discipline of keeping them separate keeps the data spine coherent.

The refrain we established in Chapter 1 and reinforced through Chapter 3 and Chapter 4 applies with particular force here: The boundary is the absence of a direct connection. There is no link from an SAP Signavio activity to an SAP LeanIX application that exists outside the data spine. There is no link from an SAP LeanIX interface to an SAP Signavio process that exists outside the data spine. The two systems do not call each other directly. They publish into the data spine, and the data spine joins their content using stable identifiers. The absence of the direct connection is what makes the boundary auditable. If a direct connection existed, the join would be hidden inside one of the products, and the audit trail of who said what about which object would degrade. The data spine is the audit-bearing surface, and it remains audit-bearing only as long as the boundary holds.

SAP LeanIX is the system inventory and contains the following:

- What applications exist
- How they connect (interfaces)
- The lifecycle state of each one
- Who owns them
- Their technology stack

SAP Signavio has the process semantics, as follows:

- What activities happen
- Who performs them (roles)
- Which variants apply (country, business unit)
- Tontrol intent of each step
- Which application a step runs in

5.6.1 The Division of Responsibility

SAP LeanIX answers questions about the landscape: What runs, where does it run, and how is it connected? SAP Signavio answers a different set of questions: What work happens, in what order, by whom, under what controls, and with what variants? These types of questions are not synonyms. They are answered using different vocabularies, by different governance bodies, on different cadences, and for different audiences. The shape of SAP LeanIX is the shape of the answer to the first set of questions. The shape of SAP Signavio is the shape of the answer to the second.

From an FP&A perspective, the division is the same one that operates in finance between the chart of accounts and the closing checklist. The chart of accounts answers what exists: which general ledger accounts, which cost centers, which profit centers, which company codes, and how they relate. The closing checklist answers what gets done: which steps run, in what order, by which controller, with which sign-offs, and against which controls. The chart of accounts and the closing checklist are not the same thing. They refer to each other through stable references—the checklist step mentions a general ledger account number; the general ledger account number does not embed a checklist step—but they do not duplicate each other's content. The SAP LeanIX and SAP Signavio relationship is the same as it as at the level of applications and activities.

Table 5.13 summarizes this division. Each row names a topic and identifies the system of record for it. The third column records the join that surfaces in the data spine, which is the mechanism by which a consumer that needs both sides reads them together.

Topic	System of Record	Data Spine Join
Application identity and lifecycle	SAP LeanIX (`factSheetId`)	Activity references `LEANIX|APPLICATION`
Interface identity and topology	SAP LeanIX (`factSheetId`)	Activity references `LEANIX|INTERFACE`
Process identifier and hierarchy	SAP Signavio (`Process_Stable_ID`)	Resolved by data spine to publishing version

Table 5.13: System of Record and Topic Division

Topic	System of Record	Data Spine Join
Activity identifier and semantics	SAP Signavio (Activity_Stable_ID)	Resolved by data spine to publishing version
Variant definition and applicability	SAP Signavio (Variant_Stable_ID)	Resolved by data spine; variant scope filters lookups
Role-to-activity assignment	SAP Signavio (role registry)	Data spine applies role_scope_intersect at access time
Control statement (intent, frequency, owner)	SAP Signavio (control attributes)	Joined to application via activity reference
Phase-in and phase-out dates	SAP LeanIX (lifecycle attributes)	Filters activity reachability
Approval for integration/ approval for AI	SAP LeanIX (governance flags)	Read by commit gateway before write
Audit trail of suggestion and verdict	AUDIT_EVENT chain on the data spine	Not in either source repository

Table 5.13: System of Record and Topic Division (Cont.)

The table is short because the division itself is short. There are only a few topics, and each one has a clear home. The discipline is not in extending the list of topics; it is in honoring the existing ones. The pressure to extend usually comes from a particular kind of meeting: a workshop in which the SAP LeanIX modeler and the SAP Signavio modeler discover that they both want to record the same fact, and a senior stakeholder, weary of the debate, suggests that both teams should just keep it. The compromise feels diplomatic in the room but is corrosive in the data spine. The fact must have exactly one system of record. The other system, if it needs the fact, must read it from the data spine.

5.6.2 Anti-Patterns and Their Costs

The easiest way to understand a boundary is to look at what goes wrong when it is breached. The anti-patterns listed in the following sections are common, partly because each one seems reasonable when it is proposed. Each is paired with the cost it imposes on the data spine and the consumers downstream.

Anti-Pattern 1: Modeling Process Flows Inside SAP LeanIX

The pattern: An SAP LeanIX modeler adds a process steps list to an application factsheet, recording the order in which users typically operate the application. The intent is to give architects a quick view of what the application does without leaving SAP LeanIX.

The cost: The list duplicates SAP Signavio at lower fidelity. It carries no identifiers, no variants, no role bindings, and no controls. It will drift because nothing forces it to track SAP Signavio publishing events. When AI is grounded in the data spine and reads from both sources, it will encounter two answers to the same question and have no principled way to reconcile them. Worse, the list in SAP LeanIX will be used for impact analysis (because it is right there next to the application), and the impact analysis will be silently wrong whenever the SAP Signavio activities have changed and the SAP LeanIX list has not.

The remedy: Remove the list. If architects need a quick view of what an application does, the data spine surfaces the activities that reference the application as `LEANIX|APPLICATION`, and the architect reads the activities in the surface that has them at full fidelity.

Anti-Pattern 2: Application Names as Free Text in SAP Signavio

The pattern: An SAP Signavio modeler records the application a step runs against as a text attribute on the activity—for example, "SAP Concur" or "SAP S/4HANA Finance." The intent is to make the activity readable on its own, without forcing the reader to dereference a reference.

The cost: The text is unjoinable. The data spine cannot resolve it to an SAP LeanIX factsheet because the text does not carry the `factSheetId`. When the application is renamed in SAP LeanIX (be it a marketing change, an acquisition, or a productization), the text in SAP Signavio does not update. Impact analysis from the SAP LeanIX side cannot find the SAP Signavio activities because there is no key to match. AI grounded on the data spine sees an orphan: an activity that names an application but does not point to one.

The remedy: Every activity that runs against an application records a `LEANIX|APPLICATION` reference whose `ref_id` is the `factSheetId`. The free-text display name is read by the data spine from the SAP LeanIX factsheet at lookup time. The SAP Signavio activity never repeats the name.

Anti-Pattern 3: Payload Semantics Duplicated on Both Sides

The pattern: An SAP LeanIX interface factsheet records what the interface carries (for example, cost center master data), and an SAP Signavio activity that consumes the interface also records what the interface carries. The two descriptions are written by different authors at different times and do not match exactly.

The cost: Drift occurs. The SAP LeanIX description will be updated when the interface specification changes, because the interface team owns the SAP LeanIX factsheet. The SAP Signavio description will be updated when the activity is refactored, which is on a different cadence. The two descriptions diverge slowly, and a consumer that reads both has to decide which one to trust. The data spine cannot adjudicate because it does not own either text.

The remedy: The SAP LeanIX interface factsheet owns the description of what the interface carries. The SAP Signavio activity records that it consumes the interface (`LEANIX|INTERFACE` reference) and describes the work the activity does with the payload it receives. The activity does not redescribe the payload.

Anti-Pattern 4: Controls Modeled in SAP LeanIX

The pattern: A control (for example, "All journal entries above $50,000 require dual approval") is recorded as an attribute on the SAP LeanIX application factsheet for the application in which the control operates. The intent is to make controls visible to architects evaluating the application.

The cost: Controls are activity-level constructs. They have an owner, a frequency, a control intent, a population, and a sample. They belong on the SAP Signavio activity (or activities) that they govern, where the role, the control owner, and the control intent are also recorded. Modeling them on the SAP LeanIX application puts them at the wrong level of granularity—an application typically supports many activities, and a control governs only some of them—and detaches them from the role and variant context that makes them executable.

The remedy: The control is an SAP Signavio activity attribute. The SAP LeanIX application factsheet does not record controls. If the control is so broad that it applies to every activity that runs against the application, it is likely an information security control on the application rather than a business process control, and it belongs in a separate control catalog (a topic not covered in this chapter).

Anti-Pattern 5: Capability Owners as Application Attributes in SAP Signavio

The pattern: An SAP Signavio modeler records the business capability and the capability owner against an activity, instead of letting the data spine resolve the capability and owner from the SAP LeanIX application factsheet.

The cost: the capability and owner are SAP LeanIX-owned attributes. When the organization rebrands or restructures, the SAP LeanIX factsheet is updated; the SAP Signavio attribute is not. The data spine sees two answers and has to pick one. The governance team has to chase the SAP Signavio updates, which is not the team's job.

The remedy: The activity records only the `LEANIX|APPLICATION` reference. The data spine resolves the capability and owner from SAP LeanIX at lookup time.

5.6.3 Worked Example: Boundary: Cost Center to Position Propagation

The cost center to position scenario, introduced in earlier chapters and carried forward through this chapter, is the clearest example of how the boundary behaves in practice. In this scenario, a new cost center is created in SAP Master Data Governance, Financials. It must propagate as a valid cost center assignment in the position object in SAP SuccessFactors Employee Central. The propagation runs through an enrichment activity and a write-back activity, both of which are governed by SAP Signavio process versions and both of which run against applications and interfaces registered in SAP LeanIX. The AI proposes a mapping, and a human approves it. The commit gateway is the only path through which the write reaches SAP SuccessFactors Employee Central.

Looking at the scenario from the boundary perspective, the question is: Which facts belong in which system, and how does the data spine knit them together? Figure 5.7 shows the propagation as a sequence of stages, with the system of record annotated for each step.

At stage one, the SAP Master Data Governance, Financials event publishes a new cost center. The cost center itself is master data, not landscape data, and its system of record is SAP Master Data Governance, Financials. SAP LeanIX does not record cost center identifiers as application attributes, and SAP Signavio does not record cost center identifiers as activity attributes. Both reference SAP Master Data Governance, Financials through the data spine.

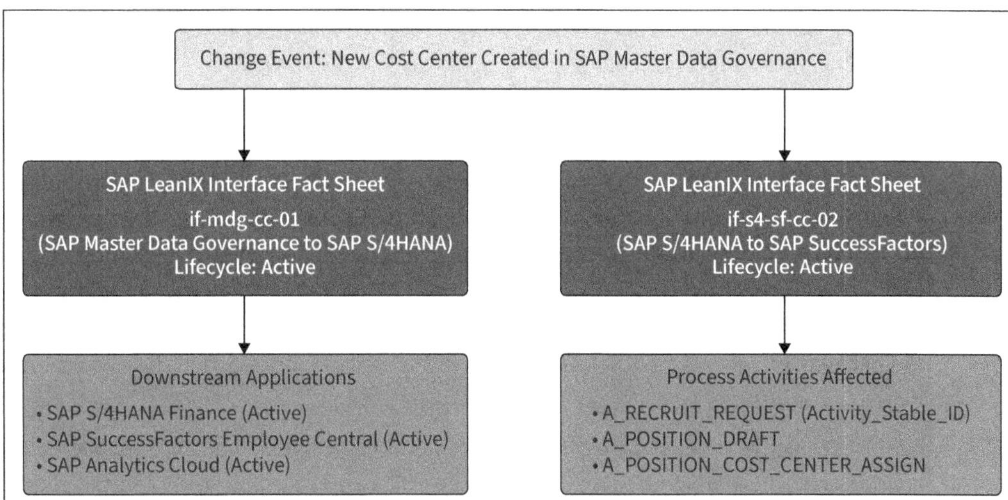

Figure 5.7: Cost Center to Position Propagation Across the Boundary, Where Each Stage Names Its System of Record and the Data Spine Performs the Join

At stage two, the data spine receives the SAP Master Data Governance, Financials event and looks up which activities are eligible to act on cost center events. The eligibility is determined by SAP Signavio: An activity is eligible if its publishing version is approved and its control intent includes cost center propagation. The data spine reads the eligible activities from SAP Signavio.

At stage three, each eligible activity has, in its SAP Signavio definition, a `LEANIX|INTERFACE` reference to the interface that carries the propagation and a `LEANIX|APPLICATION` reference to the application in which it executes. The data spine resolves both references against SAP LeanIX. From SAP LeanIX, the data spine reads the interface's technology and direction, the application's lifecycle phase, and the application's approvedForAI flag. The activity is reachable only if the application is in an active or phase-in state and approvedForAI is true.

At stage four, AI proposes the mapping from the cost center to the validity date range and the eligible position population. The proposal is a SUGGESTION with a grounding bundle whose manifest_hash records exactly which SAP LeanIX and SAP Signavio versions were read. The grounding bundle is published through the data spine; the AI does not call SAP LeanIX or SAP Signavio directly, and it does not write to either.

At stage five, a human reviews the SUGGESTION and issues a VALIDATION_VERDICT of ACCEPT, MODIFY, or REJECT. The verdict is recorded on the data spine, with the role_scope_intersect of the reviewer narrowing the population to those positions the reviewer is authorized to act on. The commit gateway uses the verdict to decide whether to release the write.

At stage six, if the verdict is ACCEPT or MODIFY, the commit gateway calls the interface registered against the activity. The call is a single, governed write into SAP SuccessFactors Employee Central. The interface itself is the SAP LeanIX-registered interface, the activity is the SAP Signavio-registered activity, the verdict is the data spine AUDIT_EVENT, and the grounding bundle is retrievable by manifest_hash for audit reconstruction at any later point.

What this scenario shows is that the boundary, when properly maintained, is not an obstacle to execution. It is the structure that lets execution happen at scale. Every fact has exactly one home, every reference is resolved at lookup time, and every change in either system is visible to the other through the data spine, without either system having to know about the other directly. The AI participates by proposing, the human participates by deciding, and the gateway is the only path to the write. The boundary is, once again, the absence of a direction connection.

For a controller who has run a month-end close, this structure is familiar. The general ledger does not know which closing tasks the controller will run tomorrow, and the closing checklist does not know which general ledger accounts will receive postings. The link between them is the posting itself: a journal entry that references both the general ledger account and the closing task. The posting is the audit event. The general ledger and the checklist are the systems of record. The link is resolved by reference, not by duplication. SAP LeanIX and SAP Signavio, joined through the data spine, behave the same way at the level of applications and activities.

Cost Center to Position Scenario: Boundary in Action

In this example scenario, the boundary between SAP LeanIX and SAP Signavio is visible in concrete terms. SAP LeanIX records that fs-mdg-fin-01 exists, that it is in active phase, that it provides if-mdg-cc-01, and that it is owned by SAP Master Data Governance, Financials. SAP LeanIX does not record the activity A_COST_CENTER_CREATE that runs in the application; that record lives in SAP Signavio with its own `Activity_Stable_ID`. SAP Signavio records the activity, its control intent (dual-control review for cost centers above a threshold), the variants applicable by country and business unit, and the `LEANIX|APPLICATION` reference that names fs-mdg-fin-01 by its `factSheetId`. SAP Signavio does not record the lifecycle of fs-mdg-fin-01, the freeze window, or the `approvedForAI` flag; those facts live in SAP LeanIX. When the data spine assembles the grounding bundle for an AI surface reasoning about a cost center change, it reads fs-mdg-fin-01 from SAP LeanIX (lifecycle, flags, and ownership) and `A_COST_CENTER_CREATE` from SAP Signavio (activity, control, and variant), joining the two by `factSheetId`. Neither system calls the other; the data spine is the only point at which the two records meet. This is the discipline that we will keep returning to throughout this book: the boundary is the absence of a direct connection.

5.7 Summary

In this chapter, we built the SAP LeanIX data model that the rest of the book depends on, and we have done so under a single discipline: Every attribute, every reference, and every lifecycle state has been justified by what the data spine, the commit gateway, and the AI surface will need to do with it. The model is not a generic application register. It is the integration-bearing surface of the landscape, structured so that the boundary between SAP LeanIX and SAP Signavio is the absence of a direct connection and so that every write that AI proposes can be grounded, approved, and committed through a single governed path.

Five threads ran through the chapter. The first was the definition of the integration-eligible application: The application factsheet carries identity, lifecycle, references, and governance, four mandatory blocks that together make the application addressable, reachable, locatable, and safe to write to. The second thread was interface and dependency modeling: Every interface is a first-class factsheet with its own `factSheetId`, technology, direction, provider, consumers, and payload class, and dependency chains are read along the consumer-provider edges rather than encoded as ad hoc attributes on either endpoint. The third thread was lifecycle: The five-phase model (plan, phase-in, active, phase-out, and EOL) governs visibility on the data spine, with the `approvedForIntegration` and `approvedForAI` flags as separate, explicit governance events. The fourth thread was identifiers: The `factSheetId` is the PK, opaque and immutable, external identifiers (SID, vendor codes, and SAP BTP routes) are recorded as references rather than PKs, and SAP Signavio activities bind to SAP LeanIX through `LEANIX|APPLICATION` and `LEANIX|INTERFACE` references that the data spine resolves at lookup time. The fifth thread was the boundary: SAP LeanIX owns landscape, SAP Signavio owns activity, the data spine joins them through stable references, and the AI commit gateway is the only path through which a proposal can become a write.

The next chapter, Chapter 6, picks up the cost center to position scenario at the level of mapping: how SAP LeanIX applications and interfaces are bound to specific L4 and L5 activities in SAP Signavio, how that mapping is maintained as both sides change, and how it underwrites the impact analysis and AI grounding work that the later chapters will build on. The model we covered in this chapter is the foundation. Chapter 6 shows how the foundation carries weight.

A five-stage data preparation flow prepares SAP LeanIX content for publication into the data spine. Extract reads the factsheets and their references through the GraphQL API. Validate checks the mandatory attribute set, the identifier integrity, and the lifecycle consistency. Stage holds the validated payload in a versioned area against which spine consumers can be regressed. Enrich resolves the external-identifier references (SIDs, SAP BTP routes, and SAP Fiori app IDs) into the shapes the data spine expects. Publish releases the bundle to the data spine with a `manifest_hash` that consumers will use to ground their reads. The flow is governed by the same approval discipline as every other write into the data spine; a publish event is itself a governance event, recorded as an `AUDIT_EVENT` and resolvable by hash for any later audit reconstruction.

5.7.1 Key Learning Points

- An application is integration-eligible only when its factsheet carries all four mandatory blocks: identity, lifecycle, references, and governance. Missing any one block disqualifies the application from being a write target for the commit gateway, regardless of how widely the application is used.
- The `factSheetId` is the PK. It is opaque, immutable, and unique. Display names, SIDs, and vendor product codes are references, not keys. Joining anything to LeanIX by display name is an anti-pattern that breaks on the first rename.
- Interfaces are first-class factsheets. An interface has its own `factSheetId`, lifecycle, provider, consumer references, and payload class. Modeling an interface as an attribute on an application loses the consumer side of the relationship and breaks impact analysis.
- Lifecycle phase governs visibility. The five-phase model (plan, phase-in, active, phase-out, and EOL) is read by the data spine; activities that reference an application or interface in plan or EOL are invisible to the consumer surface.
- The two approval flags are separate and orthogonal. `approvedForIntegration` authorizes spine reads and synchronous writes; `approvedForAI` authorizes grounded reads by AI surfaces. Conflating them is the most common source of silent AI access to data the AI was never approved to ground on.
- SAP Signavio binds to SAP LeanIX through `LEANIX|APPLICATION` and `LEANIX|INTERFACE` references on the activity. The references carry `factSheetIds`; the data spine resolves the rest. Free-text application names on activities are unjoinable and degrade over time.

- The boundary between SAP LeanIX and SAP Signavio is the absence of a direct connection. The two systems publish independently into the data spine and never call each other directly. The audit trail of who said what about which object is preserved only as long as the boundary holds.
- The commit gateway is the only path through which AI-proposed changes become writes. AI proposes, a human decides, and the gateway, governed by `approvedForIntegration` and `approvedForAI`, executes. Bypassing the gateway breaks every audit promise that the rest of the architecture depends on.

5.7.2 Common Design Pitfalls

- Recording lifecycle phases but not filtering them. The application register correctly maintains the lifecycle attribute, but the filter is applied at modeling time rather than at lookup time, and phase-out applications continue to surface in AI grounding bundles even though they are formally on the way out. The solution is to enforce the lifecycle filter in the data spine's read path. That way, phase-out applications remain visible to impact analysis but are excluded from AI grounding by default.
- Conflating `approvedForIntegration` with `approvedForAI`. The two flags are treated as a single integration toggle, and a security-sensitive application ends up grounded by AI surfaces because the AI flag was assumed to inherit from the integration flag. The solution is to keep the two flags independent, with `approvedForAI` defaulting to false until an explicit AI grounding approval has been recorded as a distinct governance event.
- Using three factsheets for one application. An application is modeled from import sources without identifier reconciliation, producing three near-duplicate factsheets with slightly different display names and three different technical IDs. The solution is to reconcile to a single `factSheetId`, record the prior `factSheetIds` as merge-source references on the survivor, and record the merge as a governance event with an effective date so that downstream consumers can follow the consolidation in their own audit trails.
- Recording interface consumers as text rather than relationships. An interface factsheet lists its consumers in the description field rather than as consumer relationship edges, and the data spine reads an empty consumer list because text is not part of the relational graph. The solution is to record every consumer as a first-class relationship from the consumer application factsheet to the interface factsheet; the relationship, not the text, is what the data spine consults during impact analysis.
- Treating the freeze window as a notification rather than a gate. The freeze window attribute is correctly recorded on the application factsheet, but the commit gateway does not read it at write time, and a write proceeds during a closed window because nothing structurally prevents it. The solution is to make the freeze window a gateway-enforced constraint: The gateway reads the window at write time and refuses the write if the window is open, with the reason code attached to the refusal record.

- Binding an activity to the SAP Fiori app ID instead of the `factSheetId`. An SAP Signavio activity references the SAP Fiori app ID directly, and the binding breaks the moment the SAP Fiori app ID is reassigned during an SAP Fiori catalog reorganization. The solution is to reference the `factSheetId` as the primary identifier on every SAP Signavio activity; the SAP Fiori app ID is a secondary identifier on the application factsheet, which can change without breaking the activity bindings.
- Recording deprecation events in SAP LeanIX but not propagating them to cached data spine reads. A consumer continues to receive an application that has been formally deprecated in SAP LeanIX because the data spine read was cached against the prior SAP LeanIX state and nothing invalidated the cache. The solution is to resolve data spine reads against the current SAP LeanIX state at lookup time and to bound bundle caching by `manifest_hash` so a new hash invalidates the prior cache automatically.

5.7.3 Review Questions

1. An architect proposes that the application display name should be the PK in SAP LeanIX with the reasoning that everyone knows applications by name. What are the three problems that result from this design choice?
2. An SAP LeanIX workspace records 600 applications. Of these, 410 carry an SID, 180 carry a vendor product code without an SID, and 10 are custom-built and carry neither. What is the discipline that allows all 600 to remain eligible for integration without forcing artificial identifiers on the 10 custom-built applications?
3. An interface factsheet records its provider as Application A and its consumers as Applications B, C, and D. Application D is moved to phase-out with an EOL date six months from now. What state does the data spine show for the interface itself, and what state does it show for the interface-to-D dependency? Justify the difference.
4. The SAP Signavio activity "Validate cost center mapping" carries a `LEANIX|APPLICATION` reference to `factSheetId` 87FE1A and a `LEANIX|INTERFACE` reference to `factSheetId` 9C2BB7. The SAP LeanIX application 87FE1A is moved to phase-out. What does the data spine return when a consumer reads the activity? What does the AI surface see when it grounds on the activity?
5. A new control intent is to be modeled: "Cost center master data changes above a defined threshold require dual approval." In which system—SAP LeanIX or SAP Signavio—does this control belong, and why? What references does the chosen system carry to the other system?
6. The architecture team wants to add a list of process steps directly to the SAP LeanIX application factsheet so that architects can see the activities it supports without leaving SAP LeanIX. Why is this proposal an anti-pattern, and what is the structural alternative that meets the underlying need?

7. A vendor consolidation merges two applications, formerly identified by `factSheetIds` 1A7E and 8C92, into a single survivor application identified by a new `factSheetId`, 4F73. What three things must happen in SAP LeanIX, and what three things must happen on the data spine, for the consolidation to be ready for audit?
8. Distinguish `approvedForIntegration` from `approvedForAI`. Give one example of an application for which the first is true and the second is false, and explain why the asymmetry is principled rather than an oversight.

5.7.4 Student Companion Notes

If you are a student working through this chapter, you should focus first on the factsheet contract. Take any application in your own landscape and attempt to populate its four mandatory blocks (identity, lifecycle, references, and governance) using only the field definitions in Section 5.2.2. The exercise quickly reveals which attributes are missing in your current SAP LeanIX practice and where the integration-eligible boundary is silently being crossed. The applications that cannot be completely populated are not yet ready for integration, regardless of how operationally important they are.

Second, practice writing the identifier set for three different application types: an SAP-native application with an SID, a third-party application with a vendor product code, and a custom-built application that has neither. The discipline of choosing the right PK and the right external identifiers is what separates a robust integration register from one that drifts each time an application is renamed or rehosted. The custom-built case is the most instructive because it forces a decision about what an internally minted identifier looks like and how it is recorded as a reference, not as a PK.

Third, take an interface in your own landscape that has more than one consumer and trace its propagation through the data spine. Identify the provider, the consumer list, the payload class, and the lifecycle phase. Then walk through what would happen if the provider moved to phase-out: what the data spine would expose to integration developers, AI surfaces, and impact analysis. The contrast between the three views, each of which sees a different part of the same factsheet, is the lifecycle discipline this chapter teaches.

This chapter opens Part III, which covers the architecture layer of the integrated landscape. The SAP LeanIX data model defined here is the foundation for everything that follows: how SAP LeanIX is mapped to process semantics in SAP Signavio (Chapter 6), how the architecture register is published to the canonical data spine (Part V), and how AI services consume the resulting view (Part VII). The integrity of those downstream layers depends, structurally, on the integrity of the factsheet contract defined in this chapter.

Chapter 6
Mapping SAP LeanIX to Process Semantics

A register that lists applications is a catalog. A register whose entries are bound to the activities that run against them is an execution surface. Mapping is the act that turns one into the other, and the discipline of mapping is the discipline that keeps the join honest after both sides have changed.

Chapter 5 specified the SAP LeanIX data model that the rest of the book depends on: the four mandatory attribute blocks on every integration-eligible application, the interface factsheet with its provider, consumer, direction, and semantic payload, the five-phase lifecycle with its two approval flags, and the stable identifier discipline that holds the model together. The chapter stopped at the SAP LeanIX boundary, on the understanding that landscape data, on its own, does not execute. An application that exists in SAP LeanIX, with every attribute populated correctly, still does nothing until an SAP Signavio activity names it as the place where work happens.

This chapter performs the binding. It takes the SAP LeanIX factsheets defined in Chapter 5 and connects them to the L4 and L5 process activities defined in Chapter 3. The binding is not a freehand annotation. It is a structured, key-bearing reference that the canonical data spine resolves at lookup time. When the binding is correct, an AI suggestion that touches a cost center knows which application will receive the write, which interface will carry it, which role is authorized to approve it, and which lifecycle phase the receiving application is in. When the binding is wrong, every layer above it inherits the error, including the `SUGGESTION` that AI publishes and the `VALIDATION_VERDICT` the human records against it.

The mapping discipline we use in this chapter is not entirely new. It is the continuation of the boundary commitment established in Chapter 1 and reinforced through Chapter 3 and Chapter 5: process semantics and landscape semantics live in different repositories, on different governance cadences, owned by different functions, and the join between them lives on the data spine. The refrain is the same one carried forward from the earlier chapters: The boundary is the absence of a direct connection. There is no direct integration from an SAP Signavio activity to an SAP LeanIX factsheet. There is no direct integration from an SAP LeanIX interface to an SAP Signavio process. Both publish into the data spine, and the data spine joins them using the stable references this chapter defines.

If you are a financial planning and analysis (FP&A) practitioner, this discipline might be familiar. The general ledger does not know which closing tasks will run tomorrow. The closing checklist does not know which general ledger accounts will receive postings. The two artifacts are joined by the posting itself: a journal entry that references both. The posting is the audit event. The ledger and the checklist are the systems of record. The link is resolved by reference, not by duplication. SAP LeanIX and SAP Signavio, joined through the data spine, behave the same way at the level of applications and activities. This chapter is about the references that make the posting possible.

Specifically, this chapter covers:

- How SAP LeanIX applications are bound to L4 and L5 process activities through structured references on the SAP Signavio activity and how the data spine resolves those references at lookup time without forcing either side to know about the other directly
- How interfaces, which are first-class factsheets in SAP LeanIX, are bound to the activities that consume them and how that binding underwrites both change impact analysis and AI grounding
- How business capabilities and domains, modeled in SAP LeanIX as the organizing overlay on the application register, are projected onto the process taxonomy so that an end-to-end process can be read against the capability map it exercises
- How the bound model supports four families of change impact use cases (application phase-out, interface deprecation, role reorganization, and control-intent change) without any of them requiring reconstruction from memory
- How the mapped SAP LeanIX data is prepared for ingestion into the SAP Business Technology Platform (SAP BTP) canonical data spine, including the five-stage handshake (extract, validate, stage, enrich, and publish) that the data spine consumers depend on.

Each section moves from principle to pattern to a worked example drawn from the cost center to position scenario that we have carried through the previous chapters. The mapping is the moment where the two sides of the integration first behave like a single system and the moment where, if the discipline is breached, they begin to diverge silently.

Before we begin, a note on scope: This chapter stops at the mapping itself. The execution flows that consume the mapping, the suggestion-and-verdict pattern that AI participates in, the commit gateway that performs the write, and the audit trail that records the whole sequence are picked up in later chapters of the book. What this chapter covers is the joinable surface: the keys, the cardinalities, the lifecycle intersections, and the validation gates that make the join trustworthy. Figure 6.1 shows the mapping surface. SAP Signavio activities and SAP LeanIX factsheets publish into the data spine; the data spine performs the join through `LEANIX|APPLICATION` and `LEANIX|INTERFACE` references. The boundary is the absence of a direct connection.

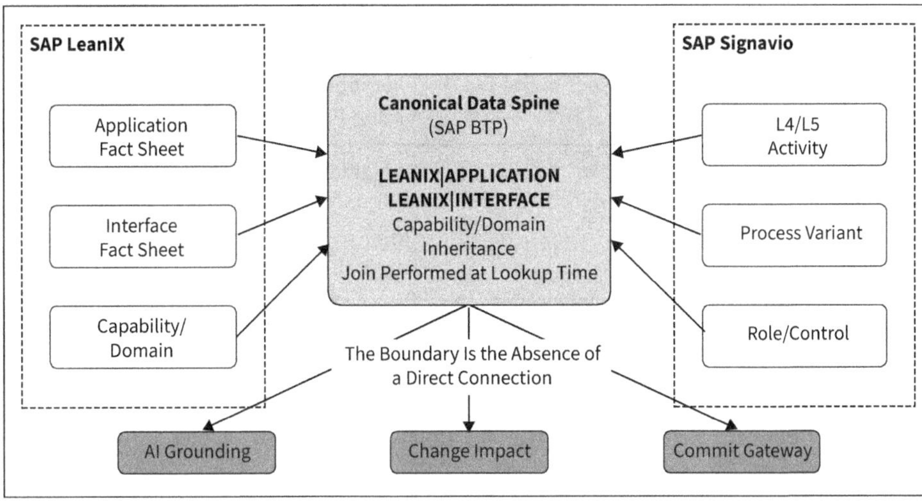

Figure 6.1: The Mapping Surface

6.1 Linking Applications to L4/L5 Activities

The SAP Signavio activity is where work is described, and the SAP LeanIX application is where the work executes. The two facts must be bound to each other for any downstream actions to be possible. Without a binding, an activity is a promise about behavior with no addressable target, and an application is an addressable target with no description of what it is supposed to do. The binding is the structural fact that turns the description and the target into a single executable statement.

This section explains the binding in three steps. First, it defines the structured reference that carries the binding on the SAP Signavio side. Second, it states the cardinality rules that govern how many bindings any one activity may hold and how many activities a single application may receive. Third, it returns to the running cost center to position scenario and walks through a concrete binding example. The binding pattern is intentionally narrow: There is one shape for the binding, one mechanism for resolution, and one place where the resolution is recorded.

6.1.1 The LEANIX|APPLICATION Reference

Every L4 or L5 activity that runs against an application records exactly one `LEANIX|APPLICATION` reference. The reference is the structured field shape we established in Chapter 4 for downstream consumption: four fields, each carrying a known purpose. Table 6.1 specifies the field contract. The shape is deliberately minimal, as the same four-field shape that carries every external reference in SAP Signavio, with the `ref_system` and `ref_type` fields naming the consumer of the reference.

Field	Type/Source	Cardinality	Purpose
`ref_system`	Controlled list; fixed value `LEANIX`	1 of 1	Identifies the target system class. Drives spine routing of the resolution query.
`ref_type`	Controlled list; fixed value `APPLICATION`	1 of 1	Identifies the target factsheet type within SAP LeanIX. Required because the same `ref_system` also carries `INTERFACE` references.
`ref_id`	SAP LeanIX `factSheetId` (opaque, immutable)	1 of 1	The only join key. Resolves to the current factsheet at lookup time.
`ref_label`	`displayName` at binding time (free text)	0 of 1	Courtesy label for human readers and audit aids. Never used for resolution; may go stale without consequence.

Table 6.1: LEANIX|APPLICATION Reference—Field Specification

The reference lives on the SAP Signavio activity, not on the SAP LeanIX factsheet. This direction is not symmetric. SAP LeanIX does not record the list of activities that bind to it on the application factsheet. The reverse view ("Which activities run against this application?") is reconstructed by the data spine at lookup time by querying the published SAP Signavio variants for all activities whose `LEANIX|APPLICATION` `ref_id` matches the `factSheetId`. The data spine maintains an index for this lookup, but the index is a derived structure; the SAP Signavio activity remains the source of truth.

The direction is important because it preserves the boundary. SAP LeanIX owns the landscape; SAP Signavio owns the activity. If SAP LeanIX recorded the activity list on the application factsheet, the architecture team would inherit responsibility for keeping that list in sync with the process team's publishing decisions, and the responsibility would arrive without the authority to act on it. The asymmetric binding (activity points at application or an application being queried by the data spine) keeps each function in its own register and lets the data spine compute the join. Figure 6.2 expresses the binding as an entity-relationship diagram with three entities: the SAP Signavio activity, the structured reference, and the SAP LeanIX application factsheet, joined by cardinality.

These three entities are joined by cardinality 1:1 on the activity side and N:1 from reference to factsheet. `ref_id` is the only join key; every factsheet attribute is resolved at lookup time.

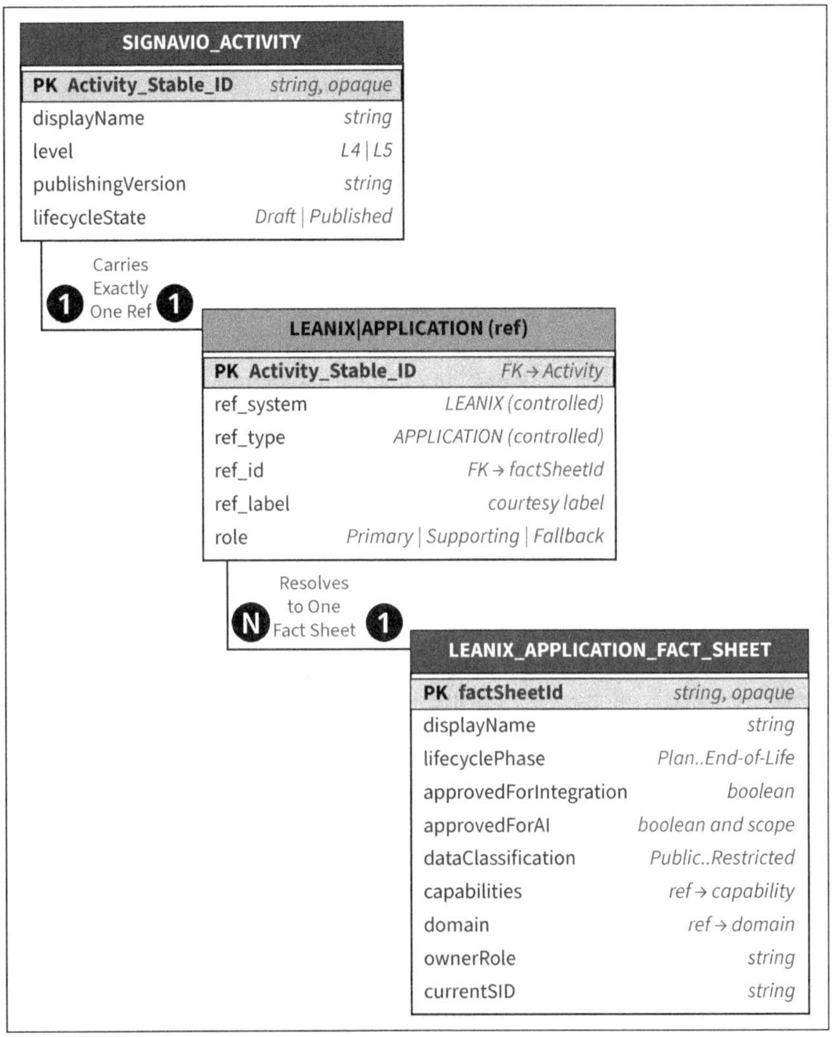

Figure 6.2: Entity-Relationship Diagram of the LEANIX|APPLICATION Binding

6.1.2 Cardinality and the Single-Execution Location Rule

An L4 or L5 activity records exactly one `LEANIX|APPLICATION` reference. We state this rule in the singular form for a reason: An activity that appears to run against two applications is, in almost every case, two activities that should have been decomposed at the L5 level. The temptation to record a list of applications on a single activity arises most often when the process modeler has captured a higher-level step (the L3 or L4 view) and is reluctant to decompose further. The discipline we established in Chapter 3 was that the L5 view names the execution location uniquely. The binding pattern we use here is the same rule from the other direction: If a single binding is not possible, the activity is at the wrong level. Table 6.2 summarizes the cardinality rules that govern how many bindings an activity may carry on each side of the relationship.

Side	Cardinality	Rule
Activity → Application	1:1	Each L4 or L5 activity records exactly one `LEANIX|APPLICATION` reference. Two-application activities indicate an L5 activity that has not been decomposed.
Activity → Application (multi-role)	1:(1+N)	Narrow exception. An activity may carry up to one primary binding plus optional supporting and fallback bindings, each tagged with a controlled list role attribute. The data spine consumes the primary binding for execution context assembly. Supporting bindings carry ancillary applications that participate without being the execution location. Fallback bindings are used for resilience and incident handling. Aligns with `ACTIVITY_APPLICATION_MAP.Relationship_Type` as described in Chapter 9.
Application → Activity	1:N	Open. A single LeanIX application receives as many activity bindings as the process register publishes against it. Constant-time lookup in `factSheetId`.
`Activity_Stable_ID` ownership of binding	1:1	The `Activity_Stable_ID` is the immutable owner of the reference. Renaming or retitling does not change the binding; only split or merge governance events do.

Table 6.2: Cardinality Rules for LEANIX|APPLICATION Bindings

The single-execution-location rule has principled exceptions. When an activity is performed by a human against an application, and the human may at any time substitute a fallback application for the same business purpose (a typical case is an enterprise reporting activity that the user can run in SAP Analytics Cloud or in a downloaded Excel file during an outage), the activity records two bindings: one primary and one fallback, each tagged with a controlled-list role attribute. A third value, supporting, is admissible when an activity engages an ancillary application (an authentication or audit-logging service) that is materially involved in the execution but is not the place where the business work happens. The three role values (primary, supporting, and fallback) are the same ones that surface on the data spine as the `Relationship_Type` attribute of the `ACTIVITY_APPLICATION_MAP` table we introduced in Chapter 9. The data spine consumes the primary binding for execution context assembly, the supporting binding for completeness in change impact computation, and the fallback binding for resilience and incident handling. The multi-role case is rare in well-modeled L5 activities, but when fallback or ancillary paths exist, they must be captured to prevent audit gaps during an outage.

6.1.3 Activity_Stable_ID as the Owner of the Binding

The binding is recorded on the SAP Signavio activity, and the activity is identified by its Activity_Stable_ID. The Activity_Stable_ID is the primary key (PK) established in Chapter 3 for activities: opaque, immutable, and unique across the SAP Signavio tenant. The LEANIX|APPLICATION reference is recorded as a structured attribute on the activity, keyed by the Activity_Stable_ID, and preserved across every refactoring operation that does not change the activity's identity.

The discipline that activities are identified by Activity_Stable_ID and not by display name has two consequences for the binding. The first is that activity renaming, retitling, or relocalization does not change the binding. The cost center mapping activity can be renamed in an SAP Signavio publishing event without any disturbance to the LEANIX|APPLICATION reference; the data spine continues to resolve the activity to the same factSheetId. The second is that activity splitting and merging (governance events recorded in SAP Signavio with their own audit trail) are the only operations that change the binding. A split produces two new Activity_Stable_IDs, each of which must be bound explicitly; a merge produces one surviving Activity_Stable_ID with the binding carried forward from the canonical predecessor and the alternative binding recorded as a governance event.

6.1.4 Resolution at Lookup Time, Not at Binding Time

The binding records the factSheetId. It does not cache the application's displayName, its lifecycle phase, its data classification, its approval flags, or any of the other attributes recorded on the SAP LeanIX factsheet. Those attributes are read by the data spine at lookup time, against the current state of the SAP LeanIX factsheet. This discipline is prevents staleness. An attribute that is cached at binding time goes stale the moment SAP LeanIX updates the factsheet, and the staleness is silent. An attribute read at lookup time is current by design; if the application has moved into phase-out since the binding was recorded, the data spine sees the new state the first time the binding is resolved.

Lookup-time resolution has a finance-control parallel that will be familiar to controllers. A journal entry posted against general ledger account 4321 records the account number, not the account's description, the current owner, or the current period's status. The description, owner, and status are read from the chart of accounts at the moment the entry is queried. If the account is closed before the next reporting cycle, queries against the entry reflect the closure without any back-rewrite of the entry. The same architecture applies here: The binding records the identifier, and the attributes are read by reference.

Lookup-time resolution underwrites three downstream behaviors that the rest of the book depends on. First, lifecycle changes in SAP LeanIX propagate to activity-level views without any change on the SAP Signavio side. If the application is moved from active to phase-out, the data spine surfaces the deprecation marker on every dependent activity automatically.

Second, approval flag changes propagate to AI grounding without any change on the SAP Signavio side. If `approvedForAI` is revoked, the data spine excludes the activity from the grounding bundle on the next assembly cycle. Third, ownership changes in SAP LeanIX surface as governance events without any change on the SAP Signavio side. The data spine routes the change to every dependent activity owner, and the owners decide whether the activity needs to be refactored or rebound.

6.1.5 Worked Example: Cost Center to Position Application Bindings

The cost center to position scenario, which we introduced in Chapter 3 and developed through Chapter 4 and Chapter 5, binds three L5 activities to three SAP LeanIX applications. Table 6.3 records the three bindings concretely, with the `Activity_Stable_ID`, the application `factSheetId`, and the resolved attributes the data spine reads at lookup time. This example is the same scenario carried forward from the prior chapters; what we add in this section is the explicit binding shape that turns the scenario into an integrated execution surface.

Activity_Stable_ID	Application factSheetId	displayName (Resolved)	lifecyclePhase	approvedForAI
`A_COST_CENTER_CREATE`	fs-mdg-fin-01	SAP MDG-FIN (Europe, the Middle East, and Africa [EMEA])	Active	True (master-data reference)
`A_COST_CENTER_PROPAGATE`	fs-s4hana-fin-emea-01	SAP S/4HANA Finance (EMEA)	Active	True (master-data reference)
`A_POSITION_COST_CENTER_ASSIGN`	fs-sf-ec-01	SAP SuccessFactors Employee Central	Active	True (position reference only)

Table 6.3: Cost Center to Position Scenario—LEANIX|APPLICATION Bindings as Resolved by the Data Spine

Each binding is a single `LEANIX|APPLICATION` reference. None of the three activities lists more than one application. The propagation from cost center creation to position assignment crosses three applications, but the crossing is performed by three separate activities each bound to a single execution location, not by one activity bound to three locations. The decomposition at L5 is what makes the binding unambiguous.

At data spine lookup time, an AI surface that reasons about the scenario assembles its grounding bundle by resolving each binding in turn. For `A_COST_CENTER_CREATE`, the data spine reads fs-mdg-fin-01 and finds that it is in the active phase, approvedForIntegration is true, approvedForAI is true at the master data reference level, and the data classification is confidential. The activity is included in the grounding bundle with those attributes. For A_COST_CENTER_PROPAGATE, the data spine reads fs-s4hana-fin-emea-01 and finds the same pattern. For A_POSITION_COST_CENTER_ASSIGN, the data spine reads fs-sf-ec-01 and finds that approvedForAI is restricted to position-reference data and excludes compensation data; the AI surface receives the activity with that scope qualification.

The chain of three bindings, resolved independently against three factsheets, is what allows the AI to propose the propagation as a SUGGESTION with a complete grounding bundle. The manifest_hash that the SUGGESTION carries records exactly which factSheetId versions and which Activity_Stable_ID publishing versions were read. If any of the three applications later transitions phase, the SUGGESTION is invalidated by the change in the underlying state, and any audit reconstruction will be able to recover precisely what was read at the moment the SUGGESTION was published. This is the same cost center to position chain we described in process terms in Chapter 3 and described in landscape terms in Chapter 5; the bindings in Table 6.3 are what joins the two descriptions into a single execution surface.

6.2 Interface-to-Process Relationships

Section 6.1 bound activities to the applications in which they execute. That binding answers the question "Where does this activity run?" It does not answer the second question every multisystem process asks, which is "Which connections carry the work between the applications?" An activity that posts a journal entry into SAP S/4HANA Finance does not run in isolation; it depends on at least one interface that delivered the cost center master data the entry references, and it produces at least one downstream interface that propagates the entry to consumers. The interface bindings recorded in this section are how those dependencies are made explicit on the data spine.

Interfaces are first class factsheets in SAP LeanIX, as we established in Chapter 5, Section 5.3, with their own factSheetId, direction, protocol, frequency, semantic payload, data classification, and governance fields. The binding from an SAP Signavio activity to an SAP LeanIX interface follows the same shape as the application binding (the four-field structured reference) but is read by different downstream consumers and carries different semantic weight. Where the application binding answers *where*, the interface binding answers *through what*.

This section develops the interface binding in six steps. The first names the LEANIX|INTERFACE reference and specifies its six fields, including the role tag and payload-shape expectation that distinguish it from the simpler application binding. The second unpacks the semantics of the role tag (CONSUMES, PRODUCES, and the rare CONSUMES_AND_PRODUCES) and shows why each value is consumed differently by change impact analysis and AI grounding. The third addresses hub interfaces and the two-sided binding discipline they require. The fourth defines the activity-level data flow statement that emerges, by composition, from the application and interface bindings together. The fifth returns to the cost center to position scenario and traces all six interface bindings concretely through Table 6.6. The closing section states what the interface binding deliberately does not carry and why the omissions are principled rather than oversights.

The interface binding differs structurally from the application binding in three ways you should keep in mind from the start. It is multivalued where the application binding is single-valued, because an activity may legitimately consume from several interfaces and produce to several others. It is direction-bearing through the role tag, which records whether the activity sits upstream or downstream of the interface, because the two cases have different downstream consumers and different failure modes. Finally, it is contract-bearing through the payload shape expectation, which records what the activity expects to read from or write to the interface payload, because the contract is what allows the data spine's validation layer to detect silent payload drift before it becomes silent integration breakage.

Read together, the application binding from Section 6.1 and the interface bindings described here produce a single, structured statement of how an activity executes: This is the application in which the work happens, these are the interfaces through which the work crosses application boundaries, these are the role tags that record direction, and this is the contract each binding carries. That statement is what the architecture in every subsequent chapter reads when it reasons about the activity, and it is what makes the bindings load-bearing rather than decorative.

6.2.1 The LEANIX|INTERFACE Reference

Every L4 or L5 activity that consumes or produces an interface records a `LEANIX|INTERFACE` reference. The shape is identical to the application reference: the same four fields with the same purposes. Two additional fields on the binding side carry the activity-specific contract with the interface: a controlled list role attribute and a payload shape expectation. Table 6.4 specifies all six fields.

Field	Type/Source	Cardinality	Purpose
`ref_system`	Controlled list; fixed value `LEANIX`	1 of 1	Identifies target system class.
`ref_type`	Controlled list; fixed value `INTERFACE`	1 of 1	Distinguishes from `APPLICATION` references on the same activity.

Table 6.4: LEANIX|INTERFACE Reference—Field Specification with Role Tag and Payload Shape Expectation

Field	Type/Source	Cardinality	Purpose
`ref_id`	SAP LeanIX `factSheetId` of the interface (opaque)	1 of 1	Join key. Resolves to the current interface factsheet at lookup time.
`ref_label`	`displayName` at binding time (free text)	0 of 1	Courtesy label for human readers; not used for resolution.
`role`	Controlled list: `CONSUMES`, `PRODUCES`, and `CONSUMES_AND_PRODUCES`	1 of 1	Names the activity's relationship to the interface. Drives change impact partition and AI grounding scope.
`payloadShape`	Schema fragment identifier (qualified name)	1 of 1	The activity's declared contract with the interface payload. Validated at publication against the SAP LeanIX semantic payload.

Table 6.4: LEANIX|INTERFACE Reference—Field Specification with Role Tag and Payload Shape Expectation (Cont.)

Unlike the application binding, the interface binding is multivalued. An activity that consumes two interfaces records two `LEANIX|INTERFACE` references; an activity that consumes an interface and produces an interface records two references with different role tags. The decision to allow multivalued interface bindings, where application bindings are single-valued, reflects a real asymmetry. An activity has exactly one execution location: the application in which the work happens. An activity may depend on many connections to do its work: master data flowing in from one interface, configuration values from another, and a write-back interface that delivers the result. The binding cardinality follows the underlying structure rather than imposing a uniform rule that does not match it.

6.2.2 Role Tag Semantics

The role tag is a small attribute that carries a large amount of semantic content. Table 6.5 summarizes the three values, the downstream consumers that read each, and the question each value answers in a change impact or AI grounding context.

Role Tag	Meaning	Read By (Change Impact)	Read By (AI Grounding)
CONSUMES	Activity reads from the interface; the interface is a precondition.	Upstream dependency query: "If this interface is deprecated, which activities lose a precondition?"	Included when the AI surface reasons about preconditions; excluded when reasoning about downstream impact.
PRODUCES	Activity writes to the interface; the interface is a postcondition.	Downstream dependency query: "If this activity is removed, which interfaces lose a producer?"	Included when the AI surface reasons about downstream impact; excluded when reasoning about preconditions.
CONSUMES_AND_PRODUCES	Rare. Activity is an in-place transformation that reads from and writes to the same interface.	Included in both upstream and downstream queries; flagged for closer review.	Included in both directions; usage is rare and warrants review of whether the model is collapsing two activities.

Table 6.5: Role Tag Values—Semantics and Downstream Consumers

Each role tag has different downstream consumers. CONSUMES bindings are read by the change impact analysis surface to compute upstream dependencies; PRODUCES bindings are read by the same surface to compute downstream dependencies. The two computations are different and use different role-tag filters; conflating them by recording an ambiguous binding ("The interface is involved somehow") makes both computations unreliable.

The role-tag discipline also gates AI grounding. An AI surface that reasons about an activity's preconditions reads the CONSUMES interfaces only; the PRODUCES interfaces are not preconditions and are explicitly excluded from the grounding bundle in that context. An AI surface that reasons about an activity's downstream impact reads the PRODUCES interfaces only. The distinction is what allows the AI to answer two different questions about the same activity without conflating them.

6.2.3 Hub Interfaces and Two-Sided Bindings

Hub interfaces, which we established in Chapter 5, Section 5.3 as two interface factsheets (provider-to-hub and hub-to-consumer) rather than one virtual interface, require a corresponding two-sided binding on the activity. An activity that depends on a hub-mediated propagation records two LEANIX|INTERFACE references: one to the provider-to-hub interface (with the role tag CONSUMES if the activity is downstream of the hub) and one to the hub-to-consumer interface (with the role tag PRODUCES if the activity is upstream of the hub). The activity does not record a binding to the hub application itself; the hub is the SAP LeanIX interface's registered route, not the activity's execution location.

The two-sided binding looks heavier than recording the interface as a single reference, and on a first encounter, this discipline often seems unnecessary. The cost of compressing the two sides into one is small at modeling time and large at impact time. When SAP Integration Suite (the hub) is upgraded, the upgrade propagates correctly to both interface factsheets and therefore to both sides of every activity binding. When the hub is replaced (a less frequent but more consequential event), the replacement is recorded as a deprecation of the two interface factsheets and a reregistration of the new pair; every dependent activity sees the change through the data spine without any modification on the SAP Signavio side. Figure 6.3 shows interface bindings on an activity. CONSUMES role tags carry preconditions; PRODUCES role tags carry postconditions. Hub interfaces appear as two distinct factsheet bindings, one for each side of the hub.

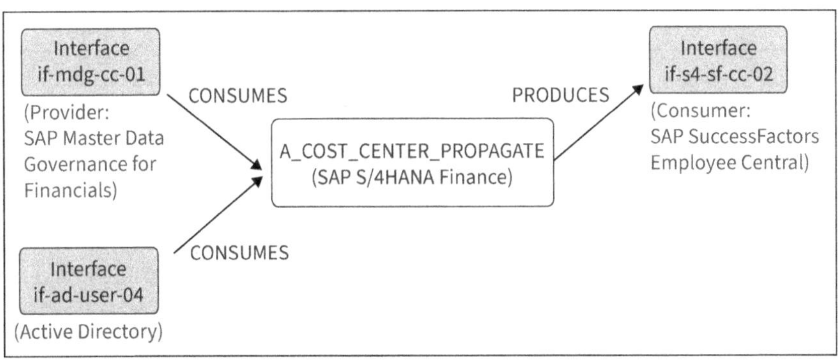

Figure 6.3: Interface Bindings on an Activity

6.2.4 The Activity-Level Data Flow Statement

The combined LEANIX|APPLICATION and LEANIX|INTERFACE bindings on a single activity produce, by design, a complete data-flow statement: This activity runs in this application, reads from these interfaces, and writes to these interfaces. The statement is not authored as a free-text narrative; it is constructed from the structured bindings the activity carries. A consumer that reads the activity through the data spine sees the statement at full fidelity, with current attributes resolved from SAP LeanIX. The same consumer reading the activity directly from SAP Signavio would see the binding identifiers but would not be able to resolve them; the resolution is the data spine's contribution.

The activity-level data flow statement is the unit of analysis that we will build on in every subsequent chapter. The change impact service reads it to compute the blast radius. The AI grounding service reads it to assemble the SUGGESTION's context. The commit gateway reads it to determine which interface receives the write when a VALIDATION_VERDICT of ACCEPT is recorded. The audit event chain stores the statement at the moment of each suggestion and verdict so that the conditions under which any past decision was made can be reconstructed exactly.

6.2.5 Worked Example: Cost Center to Position Interface Bindings

To continue our cost center to position scenario, consider that each of the three activities introduced in the prior example carries interface bindings in addition to its application binding. Table 6.6 records all interface bindings across the three activities. Read as a chain, Table 6.6 traces the full propagation from the cost center creation event in SAP Master Data Governance, Financials to the position and cost center binding read in SAP Analytics Cloud.

Activity_Stable_ID	Interface factSheetId	Role Tag	payloadShape
A_COST_CENTER_CREATE	if-mdg-cc-01	PRODUCES	CostCenter.MasterDataOut.v3
A_COST_CENTER_CREATE	if-ad-user-04	CONSUMES	Identity.Authenticated.v1
A_COST_CENTER_PROPAGATE	if-mdg-cc-01	CONSUMES	CostCenter.MasterDataIn.v3
A_COST_CENTER_PROPAGATE	if-s4-sf-cc-02	PRODUCES	CostCenter.Replicated.v2
A_POSITION_COST_CENTER_ASSIGN	if-s4-sf-cc-02	CONSUMES	CostCenter.Replicated.v2
A_POSITION_COST_CENTER_ASSIGN	if-sf-sac-pos-03	PRODUCES	Position.WithCostCenter.v1

Table 6.6: Cost Center to Position Scenario—LEANIX|INTERFACE Bindings with Role Tag and Payload Shape Expectation

The table makes three structural facts explicit. First, the propagation is a chain of CONSUMES and PRODUCES bindings: A_COST_CENTER_CREATE PRODUCES if-mdg-cc-01 and A_COST_CENTER_PROPAGATE CONSUMES it, A_COST_CENTER_PROPAGATE PRODUCES if-s4-sf-cc-02 and A_POSITION_COST_CENTER_ASSIGN CONSUMES it, and so on through if-sf-sac-pos-03. Second, the payload shape expectations match on both sides of each interface (CostCenter.MasterDataOut.v3 produced, CostCenter.MasterDataIn.v3 consumed; CostCenter.Replicated.v2 on both sides of the SAP S/4HANA-to-SAP SuccessFactors interface). Third, the only non-master-data dependency is A_COST_CENTER_CREATE's CONSUMES of Active Directory for authentication; the row is included to show that non-business interfaces are bound through the same shape.

Figure 6.4 shows the forward and reverse traversal of the cost center to position chain. The same bindings serve both directions; the role tags in Table 6.6 determine the walk order.

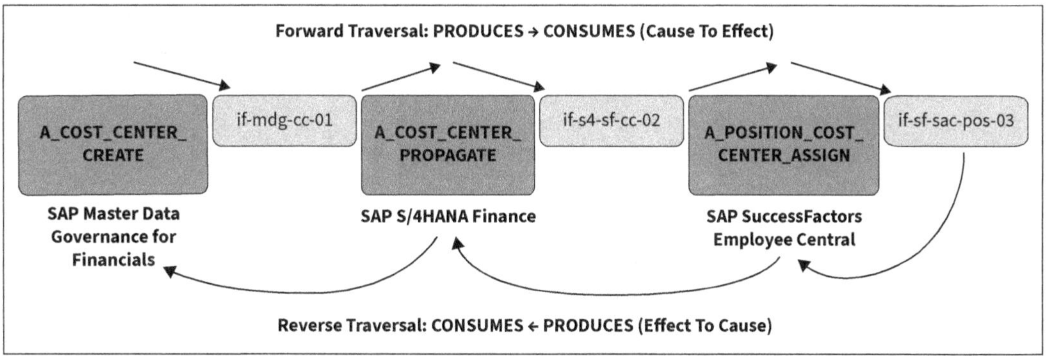

Figure 6.4: Forward and Reverse Traversal of the Cost Center to Position Chain

At spine lookup time, the chain is read in two directions independently. When read forward (cause to effect), the data spine starts from the SAP Master Data Governance, Financials cost center event, finds the activities that `CONSUME` the produced interface, and walks downstream activity by activity. When read in reverse (effect to cause), the data spine starts from the SAP Analytics Cloud reporting query, finds the activity that `PRODUCES` the interface SAP Analytics Cloud reads, and walks upstream activity by activity. Both directions return the same chain of activities and interfaces because the bindings are symmetric statements of the same underlying relationship. The two directions answer different questions, and each is the natural surface for a different downstream consumer.

6.2.6 What the Interface Binding Does Not Carry

The interface binding records the `factSheetId`, the role tag, the payload shape expectation, and the `ref_label`. It does not record the protocol, the frequency, the semantic payload itself, the data classification, or any other attribute that lives on the SAP LeanIX interface factsheet. Those attributes are the SAP LeanIX side of the join; they are read at lookup time, not cached at binding time.

The same temptation that appeared on the application side appears again here: to copy the interface's protocol or classification onto the SAP Signavio activity, on the grounds that the activity reader will then see everything in one place. This temptation is the anti-pattern we called out in Chapter 5, Section 5.6: payload semantics duplicated on both sides. The moment the protocol or classification is recorded in both places, the two copies will drift, and the data spine will be unable to adjudicate. The discipline is the same on the application side: The activity records the identifier and the role tag; the data spine reads the rest from the SAP LeanIX factsheet.

The payload shape expectation is the one exception, and it is recorded on the activity side deliberately. Two activities consuming the same interface may expect different subsets of the payload. The interface is the shared transport; the payload shape expectation is the activity's contract with that transport. The data spine's validation layer enforces the expectation at lookup time by comparing the activity's declared shape against the interface's semantic payload, which is how a control intent change on the activity surfaces as a structured exception when the change has unintentionally modified the produced payload.

6.3 Capability and Domain Alignment

Section 6.1 and Section 6.2 bound individual activities to individual applications and interfaces. The bindings are point-to-point: one `Activity_Stable_ID` to one `factSheetId` on the application side, one `Activity_Stable_ID` to one or more `factSheetIds` on the interface side. These bindings are correct, but they do not yet support the kind of cross-process questions every transformation program eventually asks: "Which activities support order-to-cash?" "Which applications carry the record-to-report capability?" "Which domains are affected by this regulatory change?" These questions are not answered by a single binding but by an overlay that organizes the bindings into a coherent capability map.

This section establishes that overlay. The capability and domain structures live in SAP LeanIX, where they are owned by the architecture function; the projection onto activities lives in SAP Signavio, where each activity inherits the capabilities and domains of the application it executes in. The data spine reads the projection at lookup time, like every other SAP LeanIX-owned attribute, and assembles cross-process views from the inherited values rather than from a separate capability-to-activity binding.

6.3.1 The Capability Map in SAP LeanIX

A business capability, in the SAP LeanIX vocabulary, is what the enterprise does, expressed independently of how it does it. Examples include record-to-report, hire-to-retire, order-to-cash, source-to-pay, plan-to-manufacture, and issue-to-resolution. Capabilities are organized into a hierarchy that the architecture function maintains: a top-level capability (finance) decomposes into mid-level capabilities (record-to-report, controlling, and treasury), each of which decomposes into specific capabilities (cost-center master data management, period close, and liquidity forecasting). The hierarchy is the organizing skeleton of the SAP LeanIX tenant's capability factsheets.

Every SAP LeanIX application factsheet records the capabilities it supports. The recording is not exhaustive (an enterprise resource planning system supports dozens of capabilities, not all of which are operationally meaningful) but selective: Each application records the capabilities for which it is the canonical execution surface. SAP S/4HANA Finance, in the running scenario, records cost center master data management, period close, and several others. SAP SuccessFactors Employee Central records position management, compensation, and workforce planning. SAP Analytics Cloud records financial reporting and operational analytics. The recording is one of the four mandatory references we established in Chapter 5, Section 5.2.

Because capabilities are recorded on the application factsheet, they are read by the data spine through the same lookup mechanism that resolves every other application attribute.

An activity that binds to SAP S/4HANA Finance through its LEANIX|APPLICATION reference inherits, by design, the capabilities recorded on the SAP S/4HANA Finance factsheet. The data spine does not require a separate binding from the activity to the capability; the inheritance is automatic, and it is current by design because it is read at lookup time.

6.3.2 The Domain Layer Above the Capability Map

Capabilities organize what the enterprise does. Domains organize who owns it. A domain, in the SAP LeanIX vocabulary, is a governance container: a named business area with an accountable owner, a defined scope, and a set of capabilities that fall under its remit. Finance, human resources (HR), supply chain, customer engagement, and risk are typical domains are all factsheets in SAP LeanIX, each with its own factSheetId, ownership attributes, and capability references.

The domain layer matters for two reasons. The first is operational: Change impact analysis, which is the subject of the next section, is almost always scoped by domain. A cost center change in finance has an impact radius that is bounded by finance and its adjacent domains; a user-interface change in customer engagement does not propagate into finance unless an explicit cross-domain dependency exists. The domain boundary is the natural scope of an impact analysis query.

The second reason is governance. Approval routing, change boards, and the data spine's commit-gateway authorization rules are all scoped by domain. A SUGGESTION that AI publishes against a finance-owned cost center routes its VALIDATION_VERDICT through finance's domain approval chain, not through HR's. The domain is what makes the routing automatic; without it, every approval would have to be routed by application, which is too granular to be operationally useful.

6.3.3 Capability and Domain Inheritance Through the Data Spine

An L4 or L5 activity in SAP Signavio inherits its capabilities and its domain from the SAP LeanIX application it is bound to. The inheritance is not recorded on the activity; it is computed by the data spine when the activity is read. This is the same architectural choice we made in Section 6.1 for application attributes: Cache nothing, resolve everything at lookup time, and let the SAP LeanIX side own the definition.

The inheritance produces a useful side effect. When the architecture function refines the capability map (a common occurrence when transformation programs revisit their domain structure), the refinement propagates to every dependent activity through the next spine read. The process team does not need to be notified; the change is visible in the activity's capability list the moment it is queried. The asymmetry, again, preserves the boundary: Capability is SAP LeanIX's responsibility, activity is SAP Signavio's, and the data spine is where the join is computed.

Figure 6.5 shows the cross-process view of the capability and domain projection. Capabilities live on the SAP LeanIX application factsheet and activities inherit them by composition through the `LEANIX|APPLICATION` binding. The data spine assembles the cross-process view at lookup time.

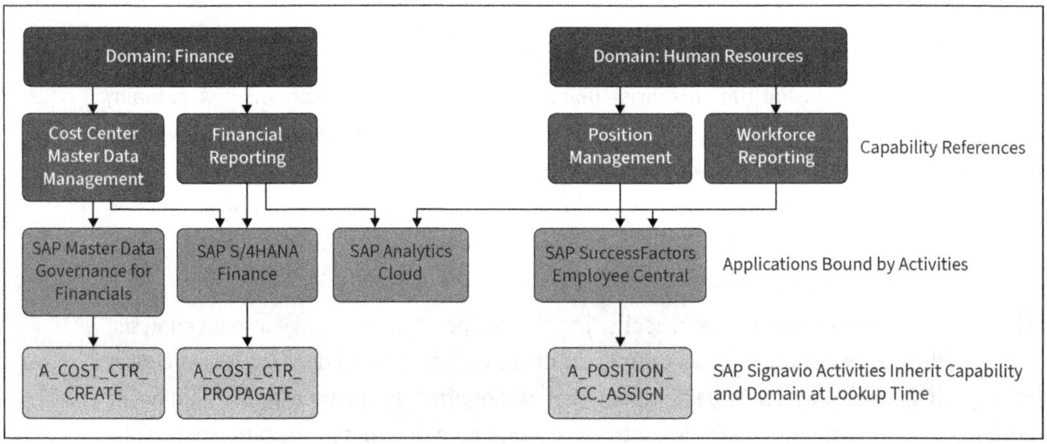

Figure 6.5: Capability and Domain Projection

6.3.4 End-to-End Process Views from Inherited Capabilities

The combination of activity bindings and capability inheritance produces a structural surface that supports end-to-end process views without any new binding. An end-to-end process—say, hire-to-retire—is defined in SAP Signavio as a sequence of L4 and L5 activities. Each activity carries its `LEANIX|APPLICATION` binding, and through the binding inherits a set of capabilities and a domain. The data spine reads the activities in sequence, resolves the inherited values for each, and produces a view that lists every application, every capability, and every domain the end-to-end process touches.

This view is useful for two distinct audiences. For the architecture function, it answers the question "Which applications are needed to execute this end-to-end process?" with an answer that is precise enough to support deprecation planning. If an application that supports a single capability will be retired, the architecture team can identify which end-to-end processes depend on the capability so they can be remediated before retirement. For the process function, the view answers the question "Which domains are affected by changes to this end-to-end process?" with an answer that determines who needs to be at the change review meeting.

Neither view requires anything beyond the bindings we established in Section 6.1 and Section 6.2 and the inheritance described in this section. Both views are computed by the data spine at query time, against current SAP LeanIX and SAP Signavio state. Both views are stable across the cosmetic changes that would otherwise force remapping work: renaming an application, restructuring a capability hierarchy, or reassigning a domain owner. The structure is what the data spine reads, what stays stable, and what supports the cross-process questions that we build on throughout the rest of the book.

6.3.5 Worked Example: Cost Center to Position Capability and Domain Inheritance

When read with a capability and domain lens, the cost center to position scenario crosses two domains and four capabilities. Table 6.7 records the inheritance for each of the three activities: the bound application, the inherited capabilities (from the application factsheet), and the inherited domain. No row records a direct activity-to-capability binding; every value is resolved by the data spine from the `LEANIX|APPLICATION` reference shown in Table 6.3.

Activity_Stable_ID	Bound Application	Inherited Capabilities	Inherited Domain
`A_COST_CENTER_CREATE`	fs-mdg-fin-01 (SAP Master Data Governance, Financials)	Master data management, cost center master data management	Finance
`A_COST_CENTER_PROPAGATE`	fs-s4hana-fin-emea-01 (SAP S/4HANA Finance)	Cost center master data management, period close	Finance
`A_POSITION_COST_CENTER_ASSIGN`	fs-sf-ec-01 (SAP SuccessFactors Employee Central)	Position management, workforce planning	HR

Table 6.7: Cost Center to Position Scenario—Inherited Capabilities and Domain

Read as an end-to-end view, the scenario crosses finance and HR, exercises four capabilities, and touches three execution applications (with a fourth, SAP Analytics Cloud, reached through the `PRODUCES` interface from `A_POSITION_COST_CENTER_ASSIGN`). The cross-domain intersection matters for governance: a change to the scenario requires approval from both finance and HR domain owners. The four-capability exercise matters for impact analysis. Any capability rescope (for example, a finance reorganization that moves cost center master data management out of master data management and into controlling) propagates through the bindings without any remapping. The next read against Table 6.7 simply returns the new capability label.

The contrast with the prior chapters is worth stating explicitly. Chapter 3 described the cost center to position scenario in process terms (the activity sequence, the L5 decomposition, the control intents). Chapter 5 described it in landscape terms (the application factsheets, the interface factsheets, the lifecycle phases, the approval flags). Neither chapter could answer “Which domains does the scenario cross?” or “Which capabilities does it exercise?”

in a structured way. The bindings in this chapter—Table 6.3, Table 6.6, and Table 6.7 in particular—are what make the cross-cutting answer possible.

6.4 Change Impact Use Cases

The bindings we described in Section 6.1 through Section 6.3 are not interesting in themselves but because of what they support. The single most consequential thing that the bound model supports is change impact analysis: the ability to answer, when something changes on the SAP LeanIX side or the SAP Signavio side, the question "What else has to change with it?" The question is asked dozens of times a year in any active transformation program, and the cost of answering it wrong is either silent breakage downstream or months of remediation work after the fact.

This section describes four families of change impact use cases that the bound model handles structurally. Table 6.8 lays out the four families against five common dimensions: which side originates the change, which binding the change is read through, which spine computation answers the impact query, what gate the data spine applies, and what the typical remediation path looks like. The remainder of the section describes each family in turn and then closes with a composite change example against the cost center to position scenario, because real transformation programs almost always encounter the composite case rather than the four families in isolation.

Use Case	Originating Side	Binding Read Through	Data Spine Computation	Remediation Path
Application phase-out	SAP LeanIX (`lifecyclePhase`)	`LEANIX|APPLICATION` reverse index	List activities whose `ref_id` matches; surface deprecation marker	Refactor binding, redirect via translation table, or retire activity
Interface deprecation	SAP LeanIX (interface `lifecyclePhase`)	`LEANIX|INTERFACE` bindings partitioned by role tag	Two queries: `CONSUMES` (lost preconditions) and `PRODUCES` (lost postconditions)	Rebind to successor interface; `CONSUMES` failures need fallback and `PRODUCES` failures need new producer or postcondition removal

Table 6.8: Change Impact Use Case Matrix—the Four Families That the Bound Model Handles Structurally

Use Case	Originating Side	Binding Read Through	Data Spine Computation	Remediation Path
Role reorganization	SAP Signavio (role registry)	Activity role attributes joined to SAP LeanIX governance block	`role_scope_intersect` recomputed at lookup time across affected activities	Verdict authority routes automatically to successor role; no rebinding required
Control-intent change	SAP Signavio (activity control)	`LEANIX|INTERFACE` `payloadShape` on `PRODUCES` bindings	Compare declared `payloadShape` against SAP LeanIX interface semantic payload	Update SAP LeanIX interface specification, back out the control change, or split into two activities

Table 6.8: Change Impact Use Case Matrix—the Four Families That the Bound Model Handles Structurally (Cont.)

A note on perspective: From the standpoint of a controller running month-end close, change impact analysis is the structural equivalent of running a what-if scenario before committing a journal posting. It asks what the system will look like after the change is implemented. The discipline of running the analysis before the change, and acting on what it returns, is what separates a clean transformation program from one that is constantly remediating its own decisions.

This section covers four families of change impact use cases in turn. The first use case addresses application phase-out, the case in which an SAP LeanIX-side lifecycle transition propagates to every dependent activity through the reverse index. The second use case addresses interface deprecation, where the role tag is what partitions the affected activities into precondition loss and postcondition loss populations. The third use case addresses role reorganization, an SAP Signavio-side change whose effects surface through the data spine's `role_scope_intersect` mechanism without modifying any SAP LeanIX factsheets. Finally, the fourth use case addresses the control-intent change, the subtle case in which an SAP Signavio-side modification propagates into a `PRODUCES` interface and back into SAP LeanIX through the payload shape expectation.

6.4.1 Use Case 1: Application Phase-Out

The first use case is the one that motivates most architecture registers in the first place: an application is to be retired, and the architecture function needs to know which activities will be affected. The question, posed as an SAP LeanIX-side change, is "If an application factSheetId X transitions from active to phase-out, which SAP Signavio activities lose their execution location?"

The query is answered structurally by reading every SAP Signavio activity whose LEANIX|APPLICATION ref_id equals X in every published process version. The result is a list of Activity_Stable_IDs, each with its publishing version and its dependent process variants. The data spine performs the read using the derived index we established in Section 6.1; no traversal of either SAP LeanIX or SAP Signavio is required at query time. The result is current by design because it is computed against current state.

The phase-out transition does not, by itself, force any of the affected activities to be modified. What it does is open a deprecation notice on each one, with the application's planned end-of-life (EOL) date as the deadline. The process owners of each activity receive the notice and decide whether to refactor the activity to bind to a different application, redirect the activity to a successor application through the data spine's deprecation translation table (as established in Chapter 5, Section 5.5 for the merge case and applied here for the phase-out case), or retire the activity along with the application.

The phase-out transition does, however, immediately affect AI grounding. An application in phase-out is still visible to integration consumers with a deprecation marker, per the phase-to-consumer visibility matrix we established in Chapter 5, Section 5.4. It remains visible to AI services with the same marker, but the marker is consulted by the grounding service when assembling the bundle. A SUGGESTION that depends on an activity bound to a phase-out application carries the marker forward into the VALIDATION_VERDICT review, where the human reviewer sees that the suggestion involves a deprecated execution location and can factor the deprecation into the verdict. Figure 6.6 shows the application phase-out. The SAP LeanIX lifecycle transition surfaces on every bound activity through the data spine's derived index; the deprecation marker propagates to AI grounding and to human review.

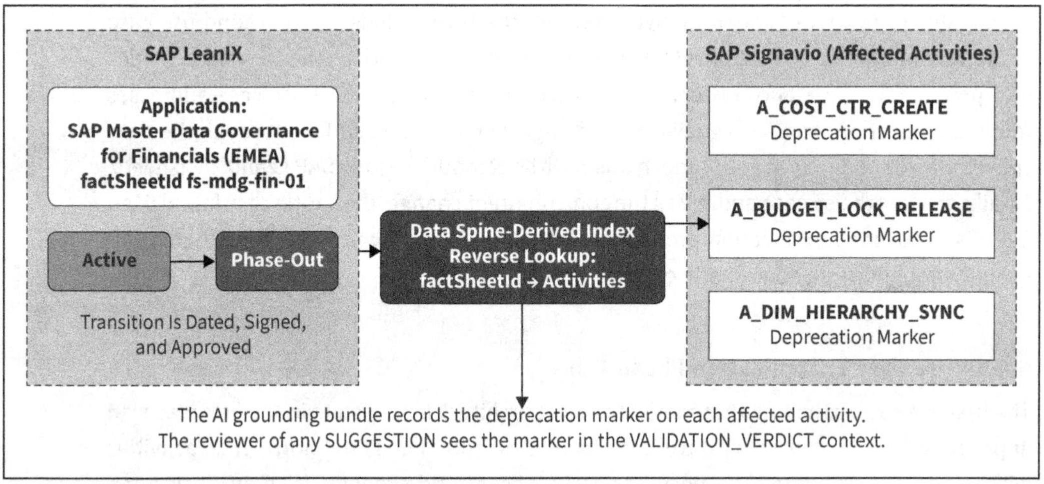

Figure 6.6: Application Phase-Out Impact Propagation

6.4.2 Use Case 2: Interface Deprecation

The second use case is interface deprecation. An interface will be retired (typically as part of a hosting migration, a protocol upgrade, or a vendor consolidation), and the architecture function needs to know which activities depend on it. The question, posed as an SAP LeanIX-side change, is "If interface `factSheetId` Y transitions from active to phase-out, which SAP Signavio activities lose a precondition or a postcondition?"

The query is answered by reading every SAP Signavio activity whose `LEANIX|INTERFACE` `ref_id` equals Y, with the result partitioned by role tag. Activities with the `CONSUMES` role tag lose a precondition: They will fail at execution time unless the interface is replaced. Activities with the `PRODUCES` role tag lose a postcondition: They will execute successfully but produce nothing downstream. The partition matters because the two failure modes have different remediation paths. A `CONSUMES` failure requires either a successor interface or a fallback data source; a `PRODUCES` failure requires either a successor interface or an explicit decision that the postcondition is no longer needed.

The data spine surfaces the deprecation on every dependent activity as a structured event with the planned EOL date and the role tag. Process owners receive the event and decide on a remediation path, which is recorded as a planned change against the activity. The planned change is itself a structured event that the data spine's validation layer consults when subsequent publishes are attempted: A republish of the activity after the deprecation date, against the deprecated interface, is rejected at publication time.

6.4.3 Use Case 3: Role Reorganization

The third use case is structurally different from the first two. Application phase-out and interface deprecation are SAP LeanIX-side changes that propagate to SAP Signavio activities through bindings. Role reorganization is an SAP Signavio-side change—a reassignment of responsibility from one role to another—that propagates to execution context through the data spine's `role_scope_intersect` mechanism. The change does not modify any SAP LeanIX factsheets, but it changes which humans are authorized to record `VALIDATION_VERDICT`s against `SUGGESTION`s that involve activities bound to SAP LeanIX applications.

The question, posed as an SAP Signavio-side change, is "If role R is reorganized into roles R1 and R2, which activities are reachable to each successor role, through which SAP LeanIX-bound applications?" The query is answered by reading the activities that previously named R as a responsible role, mapping each activity's responsibility to the appropriate successor (R1 or R2 depending on the reorganization rules), and computing the `role_scope_intersect` with the SAP LeanIX-side approval flags on each bound application.

The `role_scope_intersect`, which we established in Chapter 3 and carried forward through the data spine's execution model, is the intersection of what the role is authorized to do (from the SAP Signavio role registry) and what the SAP LeanIX-bound application authorizes the role to do (from the application's governance block). The intersection is computed at lookup time, and a role reorganization changes the intersection without changing the underlying authorization grants. The reorganization is an SAP Signavio event, but its effects are visible in every execution context that involves an SAP LeanIX-bound application.

6.4.4 Use Case 4: Control Intent Change

The fourth use case is the most subtle. A control intent change is a modification to the control statement carried on an SAP Signavio activity: the dollar threshold above which dual approval is required, the frequency at which a reconciliation is performed, or the scope of the population that the control samples. The change does not modify the activity's execution location (the `LEANIX|APPLICATION` binding is unchanged) or its data flow (the `LEANIX|INTERFACE` bindings are unchanged). It modifies the behavior the activity exhibits when it runs.

The question, posed as an SAP Signavio-side change, is "If the control intent on activity A is changed, which SAP LeanIX-bound applications and interfaces are affected in their integration behavior?" In most cases, the answer is none. A control-intent change is internal to the activity, and the bound applications and interfaces are unaware of it. But there are cases—particularly when the control affects the payload of a `PRODUCES` interface (such as a new validation, exclusion, or threshold field)—where the change propagates into the interface specification and therefore back into SAP LeanIX.

This use case is structural because it is the most common source of silent integration breakage. A control-intent change that modifies a `PRODUCES` interface's payload, without a corresponding update to the SAP LeanIX interface factsheet, produces a binding whose `payloadShape` no longer matches the interface specification. The data spine's validation layer detects the mismatch at publication time and refuses the republish, but the detection requires that both sides be modeled. If the control change is recorded only as free text on the SAP Signavio activity (the anti-pattern of modeling controls in reverse in SAP LeanIX that we called out in Chapter 5, Section 5.6), the data spine cannot adjudicate, and the mismatch surfaces only at execution time, when it is much more expensive to remediate. Figure 6.7 shows the bidirectional impact analysis surface. SAP LeanIX-side changes propagate to SAP Signavio activities; SAP Signavio-side changes propagate to SAP LeanIX-bound applications and interfaces. The data spine performs both walks against the same set of bindings.

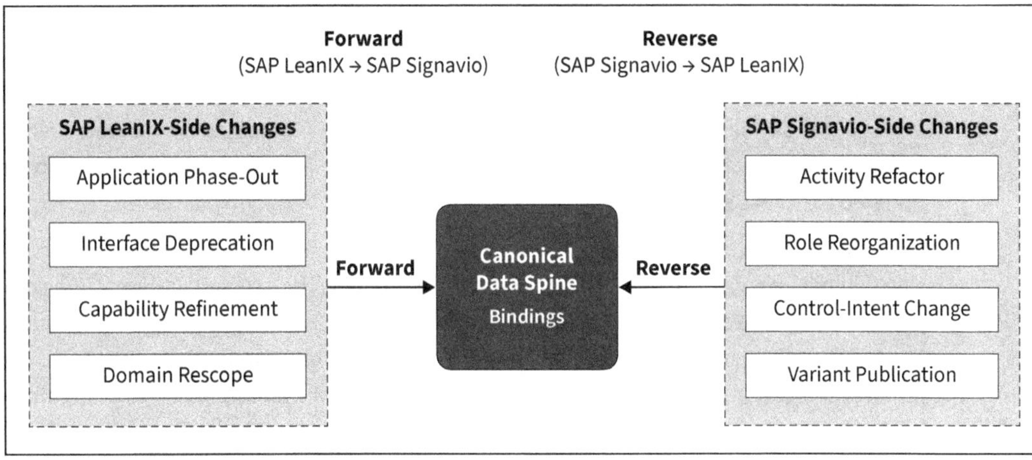

Figure 6.7: The Bidirectional Impact Analysis Surface

6.4.5 Worked Example: Cost Center to Position Composite Change Impact

In real transformation programs, the four use cases we outlined in the previous sections almost always arrive together, not in isolation. Consider a finance landscape consolidation against the cost center to position scenario we used in our previous examples. The consolidation involves four concurrent changes:

- In the first use case (SAP LeanIX), the legacy SAP Master Data Governance, Financials application fs-mdg-fin-01 transitions to phase-out and is replaced by a new instance fs-mdg-fin-02 on SAP BTP. Planned EOL is six months out.
- In the second use case (SAP LeanIX), the if-mdg-cc-01 interface that carries master data out of SAP Master Data Governance, Financials is deprecated and replaced by if-mdg-cc-02 with the same CostCenter.MasterDataOut.v3 payload shape.
- In the third use case (SAP Signavio), the finance role `FIN_CostCenter_Steward` is reorganized into `FIN_MDM_Steward` and `FIN_CostCenter_Reviewer`. Responsibility for `A_COST_CENTER_CREATE` moves to `FIN_MDM_Steward` and responsibility for `A_COST_CENTER_PROPAGATE` moves to `FIN_CostCenter_Reviewer`.
- In the fourth use case (SAP Signavio), a new control on `A_COST_CENTER_PROPAGATE` introduces a regional-restriction field on the produced payload, advancing the `payloadShape` from CostCenter.Replicated.v2 to .v3.

When read against the cost center to position scenario, the composite change produces a single, structured impact view. `A_COST_CENTER_CREATE` inherits the application phase-out marker from fs-mdg-fin-01 (the first use case) and a rebinding planned change to fs-mdg-fin-02 once it is validated. The if-mdg-cc-01 `PRODUCES` binding on `A_COST_CENTER_CREATE` carries the interface deprecation marker (the second use case) and a planned rebinding to if-mdg-cc-02. The role reorganization (the third use case) splits the verdict authority across two roles, recomputed by the data spine through `role_scope_intersect` at the next lookup. The payload-shape advance (the fourth use case) is detected by the data spine's validation layer, because if-s4-sf-cc-02 (the SAP LeanIX interface factsheet) has been updated to .v3 in parallel by the architecture function, the validation passes.

Each change is recorded as a structured event. The data spine surfaces all four events in a single impact view, scoped by the finance and HR domains the scenario crosses. The composite view reviewer (a process owner, architecture lead, or controller leading the close cycle that the scenario supports) sees the four changes as one event rather than four. The remediation can be planned holistically: Refactor the bindings, approve the new control intent, accept the role reorganization, and schedule the cutover. The bindings are the foundation, the data spine is the surface, and the discipline of recording every change as a structured event is what makes the surface trustworthy.

6.5 Data Preparation for SAP BTP Ingestion

The previous section developed the mapping between SAP LeanIX factsheets and SAP Signavio activities. The mapping is correct, the bindings are joinable, and the impact views are computable. However, none of this is available to the canonical data spine on SAP BTP yet. The mapping must be extracted from the two source repositories, validated against the discipline defined in this chapter and the previous chapters, staged for consumption, enriched with the resolved attributes the data spine expects, and published as a versioned bundle that downstream consumers can read.

This section describes that five-stage handshake from the SAP LeanIX side. We specified the SAP Signavio side, with which it composes, in Chapter 4. The two sides are deliberately symmetric. Both are five-stage pipelines, publish through the data spine's `manifest_hash` mechanism, record their publications as `AUDIT_EVENT`s, and are governed by the same approval discipline. Table 6.9 summarizes the five stages in a single view. The five stages (extract, validate, stage, enrich, and publish) are developed in turn through the remainder of this section, each with its own gating discipline and audit event signature. We close the section by returning to our running example with a concrete bundle that demonstrates every stage.

Stage	Inputs	Outputs/Artifacts	Gating	AUDIT_EVENT Recorded
1. Extract	SAP LeanIX GraphQL application programming interface (API) and last cursor watermark	Raw payload of applications, interfaces, capabilities, and domains	Incremental by default; full extract scheduled or on demand	`EXTRACT_RUN` with cursor delta
2. Validate	Extract payload and Chapter 5 discipline	Either advancement of payload or structured exception	Six validation classes (identity, attributes, lifecycle, references, flags, and classification)	`VALIDATION_PASS` or `VALIDATION_FAIL` per class
3. Stage	Validated payload	Versioned holding pen and regression-suite fixtures	Regression must pass against prior published bundle	`STAGE_WRITTEN` with generation number
4. Enrich	Stage payload, SAP BTP destination service, and WalkMe registry	Enriched payload (system ID [SID] to URL, SAP Fiori to WalkMe, and capability expansion)	Read-only against source; never writes back	`ENRICH_COMPLETED`
5. Publish	Enriched payload	`bundle_id` and `manifest_hash`; consumer notification	Approval by named domain data steward	`BUNDLE_PUBLISHED` with `manifest_hash`

Table 6.9: SAP LeanIX Five-Stage Data Preparation Pipeline—Stages, Gates, and Audit Events

6.5.1 Stage 1: Extract

The extract stage reads SAP LeanIX through its GraphQL API. The API is the canonical read surface against which every spine-side extractor operates; direct database access to SAP LeanIX is not permitted, and alternative APIs (REST endpoints intended for specific UI flows, for example) are admissible only when the GraphQL API does not support the required query shape. The discipline is intentional: GraphQL is what SAP LeanIX guarantees as a stable contract, and writing extractors against any other surface couples them to internal structure that may change without notice.

Three classes of queries run in the extract stage. The first is the application inventory: every application factsheet in an integration-eligible lifecycle phase (phase-in, active, or phase-out), with its four mandatory attribute blocks and its capability and domain references. The second is the interface inventory: every interface factsheet in the same lifecycle phases, with its provider, consumer, protocol, frequency, semantic payload, classification, and governance attributes. The third is the dependency inventory: the provider-consumer relations among interfaces, the application-to-interface relations on both sides of each interface, and the chain criticality and recovery-tier attributes recorded on each node.

Each query is run with a stable cursor that records the last factsheet revision read on the previous extraction cycle. Incremental extraction is the default; full extraction is admissible at scheduled intervals (typically weekly) and on demand when a reconciliation event triggers it. The cursor-based extraction has a finance-control parallel, the same kind of incremental discipline that an enterprise data warehouse uses to load posting deltas from the general ledger rather than rereading the entire ledger nightly. The semantic is identical: Process what changed, against a known watermark, with the option to reread fully under controlled conditions.

6.5.2 Stage 2: Validate

The validate stage applies the disciplines from Chapter 5 and this chapter to the extracted payload. The validations are not a courtesy; they are gating checks that determine whether the payload may proceed to subsequent stages. A payload that fails any validation is held in a quarantine area, an exception event is opened, and the stage is not advanced until the underlying issue is remediated in SAP LeanIX. The data spine does not attempt to repair invalid data in flight.

Six classes of validations run in order. The first checks identity integrity: Every `factSheetId` is non-null, opaque (meaning there is no embedded semantic meaning), and unique within the tenant. The second checks the mandatory attribute set: Every integration-eligible application carries its four mandatory blocks completely populated, and every integration eligible interface carries its seven mandatory fields. The third checks lifecycle consistency: Phase values are drawn from the five-phase controlled list, and transitions between phases are recorded as dated events with approving authorities. The fourth checks reference integrity: Every `LEANIX|APPLICATION` and `LEANIX|INTERFACE` reference recorded in the staged SAP Signavio payload resolves to a `factSheetId` that the SAP LeanIX side has also published. The fifth checks approval flag consistency: The two approval flags are present, each carries its three-piece metadata, and the scope fields are drawn from controlled lists. The sixth checks classification inheritance on interfaces: The recorded classification matches the stricter of the provider and consumer classifications.

Each validation produces either a pass (meaning the payload advances) or a structured exception with the specific check that failed, the factsheet identifier that failed it, and a remediation hint. The exceptions are routed to the SAP LeanIX modeler responsible for the affected factsheet, not to a generic operations queue, because remediation is almost always a modeling change rather than a pipeline change. The discipline that exceptions go to the modeler is what keeps the SAP LeanIX register from quietly degrading.

6.5.3 Stage 3: Stage

The stage area is a versioned holding pen for validated payloads. Every payload that passes validation is written to the stage area with a generation number, a timestamp, and the cursor watermark that produced it. Stage data is visible to integration developers, regression test fixtures, and the data spine's reconciliation service, but it is not visible to production consumers. This distinction matters: The stage area exists to support testing and regression against a known payload before that payload is published and to support reconciliation queries that compare current state to a recent past state without disturbing production.

Two important behaviors are bound to the stage area. The first is the regression discipline. Before any payload is published, a regression suite runs against it in the stage area, comparing the behavior of the data spine's key APIs (process lookup, change impact queries, and AI grounding bundle assembly) against the same APIs run against the prior published payload. The regression is automatic, and a failure blocks publication. The second is the historical retention. Stage payloads are retained for a configurable window (typically thirty days) so that audit reconstruction has access to recent intermediate states even when the publication cadence is high.

6.5.4 Stage 4: Enrich

The enrich stage resolves external identifiers and inserts the resolved values into shapes the data spine expects. SAP LeanIX records SAP system IDs, vendor product codes, SAP BTP routes, and SAP Fiori app IDs as structured fields on the application factsheet (see Chapter 5, Section 5.5). The data spine's production consumers expect these identifiers to be present in a canonical form. Enrichment performs the transformations that make the shapes match.

Three enrichment patterns predominate. The first is the SID-to-URL resolution, where a recorded SAP system ID is enriched with the currently bound URL endpoint, drawn from the SAP BTP destination service. The URL is not a primary identifier (as we noted in Chapter 5, Section 5.3), but it is needed by execution services at runtime, and enrichment is the stage that supplies it. The second is the SAP Fiori app ID-to-WalkMe binding, where a recorded

SAP Fiori App ID is enriched with the WalkMe content reference, drawn from the WalkMe registry that Chapter 7 will develop. The third is the capability-and-domain expansion, where capability and domain references recorded as `factSheetIds` are enriched with their `displayNames`, owners, and parent references, so that the data spine's capability view can be assembled without further lookups.

Enrichment is read-only against SAP LeanIX. None of the enrichment operations writes back to SAP LeanIX, and the enriched shapes live exclusively on the data spine. This is the same discipline we applied on the SAP Signavio side in Chapter 4: Enrichment supplies the data spine's consumer surface without disturbing the source repository. The boundary is preserved because the enrichment is downstream of the source, not upstream of it.

6.5.5 Stage 5: Publish

The publish stage releases the validated, staged, and enriched payload as a versioned bundle. The bundle is assigned a `bundle_id`, a `manifest_hash` that uniquely identifies the contents at the byte level, and a publication timestamp. The publication is recorded as an `AUDIT_EVENT` in the data spine's audit chain, with the publishing authority, the validation pass results, and the regression suite outcome attached.

The `manifest_hash` is the load-bearing identifier downstream of publication. Every AI `SUGGESTION` that grounds against the published payload records the `manifest_hash` in its grounding bundle. Every `VALIDATION_VERDICT` recorded against a `SUGGESTION` carries the `manifest_hash` forward into its audit record. Every change impact view computed against the payload records the `manifest_hash` so that the view can be reproduced exactly at a later time. The `manifest_hash` is what allows the data spine to be honest about which data was read at which moment; it is the foundational audit record that the rest of the book's reasoning depends on.

Publication is the only stage that surfaces the payload to production consumers. Stage and enrich populate the data, but publish is what makes the data readable by the integration services, the AI grounding service, and the commit gateway. The transition from stage to publish is therefore a governance event in its own right. It is approved by a named authority (the data steward for the affected domain), recorded with the regression-suite outcome and the validation results, and reversible only by publishing a subsequent bundle that supersedes it. The data spine does not support unpublishing in place; the only path to undo a publication is to publish a corrected successor, which carries its own `AUDIT_EVENT`.

Figure 6.8 shows the published bundle. The `bundle_id` and `manifest_hash` identify the bytes. The manifest carries the source watermarks, the validation outcomes, and the `AUDIT_EVENT` linkage. Every downstream consumer records the `manifest_hash` with its read.

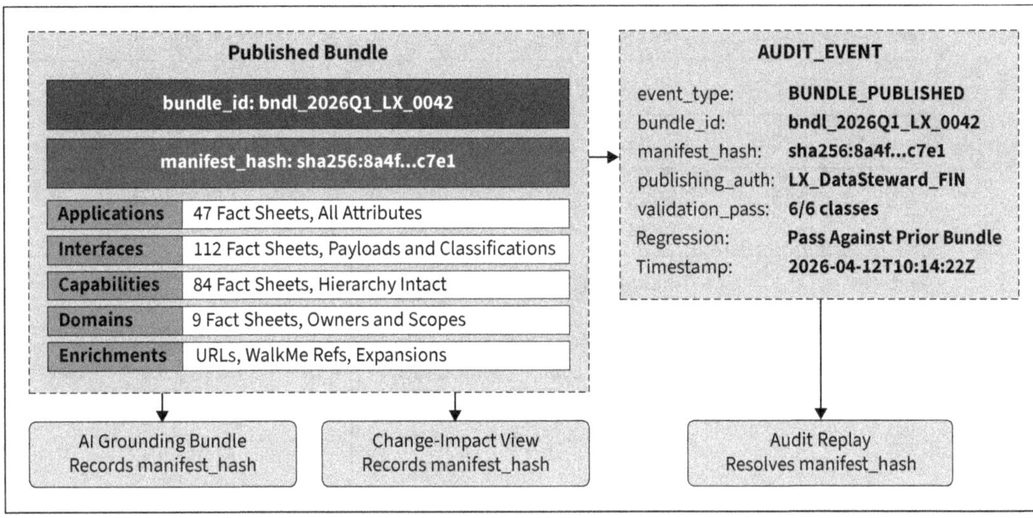

Figure 6.8: The Published Bundle

6.5.6 Composition with the SAP Signavio Pipeline

The SAP LeanIX five-stage pipeline and the SAP Signavio five-stage pipeline established in Chapter 4 run independently, on their own cadences, and produce their own bundles. The data spine joins the two bundles at lookup time using the `LEANIX|APPLICATION` and `LEANIX|INTERFACE` references the SAP Signavio side carries; neither pipeline waits for the other, and neither pipeline's `manifest_hash` depends on the others.

This independence has two practical consequences. The first is that an SAP LeanIX-side change that does not affect the join surface (a capability hierarchy refinement, for example) publishes without any SAP Signavio activity. The data spine's consumers see the change the first time they query the inherited capability list, and the change is visible without any SAP Signavio republication. The second is that an SAP Signavio-side change that does not affect the binding (a control intent change that does not modify the `PRODUCES` interface payload, for instance) publishes without any SAP LeanIX activity. The data spine's consumers see the change the first time they query the activity, and the change is visible without any SAP LeanIX republication.

6.5.7 Worked Example: Cost Center to Position Bundle Publication Record

The cost center to position scenario, when traced through the SAP LeanIX five-stage pipeline, passes through every stage. Table 6.10 records the concrete bundle that is produced for the scenario's SAP LeanIX-side footprint: four application factsheets, three interface factsheets, and the associated capability and domain context. The `bundle_id` and `manifest_hash` we use are illustrative; in a live system, they are assigned at publication time and carried forward into every downstream consumer that reads against the bundle.

Bundle Artifact	Content	Source Watermark
bundle_id	bndl_2026Q1_LX_0042	Assigned at publish
manifest_hash	sha256:8a4f...c7e1 (illustrative)	Computed over bundle bytes
Applications (4)	fs-mdg-fin-01, fs-s4hana-fin-emea-01, fs-sf-ec-01, fs-sac-01	GraphQL cursor: 2026-04-12T08:33Z
Interfaces (3)	if-mdg-cc-01, if-s4-sf-cc-02, if-sf-sac-pos-03	GraphQL cursor: 2026-04-12T08:33Z
Capabilities (4)	Master data management, cost center master data management, position management, and financial reporting	Inherited via application references
Domains (2)	Finance and HR	Inherited via application references
Enrichments	SIDs resolved to SAP BTP URLs, SAP Fiori app IDs bound to WalkMe content, and capability hierarchy expanded	Read against destination service
Validation	6 of 6 classes passed	Stage 2 results attached
Regression	Pass against prior bundle bndl_2026Q1_LX_0041	Stage 3 fixture comparison
Publishing authority	FIN_DataSteward (cross-signed by HR_DataSteward for cross-domain scope)	Recorded in AUDIT_EVENT

Table 6.10: Cost Center to Position Scenario—SAP LeanIX Bundle Publication Record (Stage 5 Output)

The next time an AI surface assembles a SUGGESTION for the scenario, the grounding bundle records manifest_hash sha256:8a4f...c7e1. The same SUGGESTION twelve months later, replayed for audit reconstruction, reads the preserved manifest_hash and replays the exact same SAP LeanIX state that was available at the moment of the original suggestion. The discipline of the five-stage pipeline is what makes the replay honest. The bundle does not change after publication; new bundles supersede old bundles, and old bundles remain available for reconstruction. The audit trail is preserved by design.

When read together, the examples in this chapter—Table 6.3, Table 6.6, Table 6.7, and Table 6.10—trace the cost center to position scenario from individual activity bindings to interface bindings with role tags to capability and domain inheritance to the published bundle that the data spine's consumers actually read. The scenario began in Chapter 3 as a process description and was a landscape register in Chapter 5. This chapter is the binding that makes the two descriptions compatible, and the bundle in Table 6.10 is the artifact that downstream services consume.

6.6 Cross-Chapter Vocabulary Continuity

Chapter 6 introduces vocabulary that is consumed and refined by every subsequent chapter in the book. The names this chapter uses are deliberately the SAP LeanIX-side names—camelCase identifiers carried on the factsheet (`factSheetId`, `displayName`, `lifecyclePhase`) and the structured reference shapes carried on the SAP Signavio side (`LEANIX|APPLICATION`, `LEANIX|INTERFACE`). These names cross the boundary onto the canonical data spine during the ingestion pipeline described in Section 6.5, where they are transformed into the `UPPER_SNAKE_CASE` entity and column names that govern the data spine's tables. The transformation is not a renaming for cosmetic reasons; it is the moment when SAP LeanIX-owned vocabulary leaves SAP LeanIX and becomes data spine-owned vocabulary, with the data spine's naming conventions and stricter referential integrity.

Table 6.11 lists the principal mappings. Each row names a concept introduced in this chapter, the data spine entity or column that consumes it after ingestion, and the chapter where the consuming entity is described in detail. This chapter defines the SAP LeanIX side of each mapping, and each referenced chapter defines its own data spine side. Table 6.11 connects the two.

SAP LeanIX Side	Data Spine Entity or Column	Developed in Full
`factSheetId` (application)	`APPLICATION.LeanIX_App_ID` (external bridge); `APPLICATION.App_ID`(spine PK)	Chapter 9
`displayName` (application)	`APPLICATION.Application_Name`	Chapter 9
`lifecyclePhase` (plan through EOL)	`APPLICATION.Lifecycle_State` (plan, active, phase-out, and retired; phase-in folds into active; EOL becomes retired)	Chapter 9
`approvedForIntegration`/ `approvedForAI` flags	`ELIGIBILITY_POLICY` (`ai_eligible_flag`, `integration_eligible_flag`)	Chapter 14
`dataClassification`	`ELIGIBILITY_POLICY` scope inputs	Chapter 14
Capabilities, domain (inherited via binding)	`APPLICATION.Category` and capability factsheet references; consumed by change impact queries	Chapter 12

Table 6.11: Cross-Chapter Vocabulary Continuity

SAP LeanIX Side	Data Spine Entity or Column	Developed in Full
ownerRole on the application factsheet	APPLICATION business and IT owner attributes; consumed by role-scope resolution	Chapter 14
factSheetI (interface)	INTERFACE.Interface_ID	Chapter 9
Provider/consumer on interface factsheet	INTERFACE.Provider_App_ID/Consumer_App_ID (separate foreign keys [FKs] to APPLICATION.App_ID)	Chapter 9
Interface protocol	INTERFACE.Interface_Type (API, event, File, IDoc, RFC, OData)	Chapter 9
Interface service-level agreement (SLA)/frequency	INTERFACE.SLA_Level	Chapter 9
LEANIX|APPLICATION reference on activity	ACTIVITY_APPLICATION_MAP (Activity_ID, App_ID, and Relationship_Type—primary, supporting, and fallback)	Chapter 9
LEANIX|INTERFACE reference on activity	ACTIVITY_INTERFACE_MAP (Activity_ID and Interface_ID)	Chapter 9
CONSUMES/PRODUCES role tag	Derived at ingestion: PRODUCES when the activity's App_ID matches Provider_App_ID and CONSUMES when it matches Consumer_App_ID	Chapter 9
payloadShape expectation on activity binding	Validated against INTERFACE semantic payload at ingestion; surfaces as a VALIDATION_RESULT on mismatch	Chapter 11, Chapter 9
Trigger → target → downstream chain	EXECUTION_CONTEXT (Trigger_App_ID, Target_App_ID, and Downstream_App_ID; three separate FKs to APPLICATION)	Chapter 9
bundle_id, manifest_hash (see Section 6.5)	BUNDLE_MANIFEST.bundle_id and BUNDLE_MANIFEST.manifest_hash	Chapter 14
Cost center master data (running scenario)	MASTER_DATA_REFERENCE and MASTER_DATA_ATTRIBUTE, derived at ingestion: PRODUCES when the activity's App_ID matches Provider_App_ID and CONSUMES when it matches Consumer_App_ID ACTIVITY_MASTERDATA_REF	Chapter 9

Table 6.11: Cross-Chapter Vocabulary Continuity (Cont.)

SAP LeanIX Side	Data Spine Entity or Column	Developed in Full
AI SUGGESTION grounded on bound activity	SUGGESTION (suggestion_id, bundle_id FK, inference_event_id FK, proposed_state, and risk_class)	Chapter 15
VALIDATION_VERDICT recorded against SUGGESTION	VALIDATION_VERDICT (verdict_id, suggestion_id FK, and verdict_value: ACCEPT, MODIFY, REJECT, and validator_principal)	Chapter 15
AUDIT_EVENT chain (publication, suggestion, verdict)	AUDIT_EVENT (first-class data spine table)	Chapter 9
role_scope_intersect on verdict authority	Computed at verdict time from role attributes joined to APPLICATION governance block	Chapter 14

Table 6.11: Cross-Chapter Vocabulary Continuity (Cont.)

Three observations follow from the mapping. First, the data spine's APPLICATION and INTERFACE entities are referenced rather than replicated. The LeanIX_App_ID column on APPLICATION is the external bridge; the data spine does not become a second SAP LeanIX. In Chapter 9, we will apply the same discipline to MASTER_DATA_REFERENCE a controlled pointer into the source system rather than a copy of the source's master data records. The boundary of referencing, not duplicating, holds at every layer of the model.

Second, the lifecycle vocabulary collapses on crossing the boundary. The five-phase SAP LeanIX model (plan, phase-in, active, phase-out, and EOL) becomes a four-state spine model (plan, active, phase-out, and retired). This is intentional, because phase-in and active are equivalent from the data spine's execution eligibility perspective, and EOL and retired are the same terminal state under different labels. The transformation is recorded in the extract stage, and it is the same transformation we will document in Chapter 9 on the data spine side.

Third, the AI vocabulary we will introduce in Chapter 13–Chapter 15 (SUGGESTION, VALIDATION_VERDICT, GROUNDING_BUNDLE, INFERENCE_EVENT, BUNDLE_MANIFEST, CITATION_RECORD, ELIGIBILITY_POLICY, and HUMAN_VALIDATION_QUEUE) operates entirely on the data spine side of the boundary. It consumes the bindings described in this chapter but does not modify them. AI may propose, but it never publishes a new factsheet, and it never modifies an Activity_Stable_ID's LEANIX|APPLICATION reference. The discipline we established in Chapter 1—AI may propose, a human must decide, and only a governed handler may write—is what keeps the boundary intact through the AI-mediated workflows we cover in subsequent chapters.

Looking forward, the next part of the book picks up the bound model we described in Chapter 5 and Chapter 6 and shows how WalkMe attaches to it as a process-aware adoption layer. The SAP Fiori app IDs that SAP LeanIX records as secondary identifiers on the application factsheet are the binding key by which WalkMe content is mapped to specific user interfaces. The activity bindings described in this chapter are what lets adoption telemetry flow back into the data spine as a structured signal that downstream AI services can ground on. The mapping discipline established here is the foundation that every subsequent chapter builds on.

6.7 Summary

In this chapter, we performed the binding between SAP LeanIX and SAP Signavio that the rest of the book depends on. The bindings we developed are not new identifiers, new repositories, or new tools; they are structured references on existing activities that name existing factsheets, with the data spine performing the join at lookup time. The discipline is light at modeling time and load-bearing at every downstream stage.

Five threads ran through the chapter. The first thread was the `LEANIX|APPLICATION` reference: Every L4 or L5 activity binds to exactly one application through a four-field structured reference whose `ref_id` carries the `factSheetId`. The reference lives on the SAP Signavio side, and the data spine resolves the SAP LeanIX attributes at lookup time without caching anything that could go stale. The single-execution-location rule keeps the binding unambiguous; the multirole exception (primary, supporting, and fallback) is rare and explicitly tagged, and aligns with the `Relationship_Type` values that the data spine's `ACTIVITY_APPLICATION_MAP` table will use in Chapter 9.

The second thread was the `LEANIX|INTERFACE` reference. Activities carry one binding per interface they consume or produce, each tagged with a role attribute (`CONSUMES`, `PRODUCES`, or the rare `CONSUMES_AND_PRODUCES`). Hub interfaces appear as two distinct factsheet bindings on the activity side, matching the two-factsheet modeling of hubs that we established in Chapter 5. The role tag is what makes the binding interpretable for change impact analysis and AI grounding in opposite directions; it is also what allows the data spine to populate `ACTIVITY_INTERFACE_MAP` rows with a meaningful directionality at ingestion.

The third thread was capability and domain inheritance. Capabilities and domains are SAP LeanIX-owned attributes recorded on the application factsheet. SAP Signavio activities inherit them by composition through the `LEANIX|APPLICATION` binding; the data spine reads the inherited values at lookup time and assembles cross-process views from them. The inheritance is what allows end-to-end process views to be computed without any new binding work, and it is what keeps the views current as the architecture function refines the capability map.

The fourth thread was change impact analysis. The bindings developed in the first three sections support four families of impact use cases structurally: application phase-out, interface deprecation, role reorganization, and control-intent change. Each family is answered by a structured query against the bindings, with the data spine performing the read and surfacing the result with the appropriate deprecation markers, lifecycle constraints, and approval requirements. Composite changes—which are what real transformation programs almost always encounter—are answered by composing the individual queries into a single impact view.

The fifth thread was data preparation. The SAP LeanIX five-stage pipeline (extract, validate, stage, enrich, and publish) is the symmetric counterpart to the SAP Signavio pipeline we described in Chapter 4. The two pipelines run independently and publish independently; the data spine joins their outputs at lookup time. The `manifest_hash` recorded at publication is the load-bearing identifier that downstream AI grounding bundles and audit reconstruction depend on. The published bundle described in Section 6.5 becomes, on the consumer side, the `BUNDLE_MANIFEST` entity introduced in Chapter 14, and the `manifest_hash` it carries becomes the cite key in `CITATION_RECORD` rows that join every AI suggestion to the evidence that grounded it.

This chapter's refrain, carried forward from Chapter 1 and reinforced in Chapter 3 and Chapter 5, holds here without modification: the boundary between SAP LeanIX and SAP Signavio is the absence of a direct connection. There is no direct integration between the two systems; both publish into the data spine, and the data spine joins them through stable references. The absence of a connection is what makes the join auditable. If a direct connection existed, the join would be hidden inside one of the products, and the audit trail of who said what about which object would degrade. The data spine is the audit-bearing surface, and it remains audit-bearing only as long as the boundary holds.

6.7.1 Key Learning Points

- Activities bind to applications through a single `LEANIX|APPLICATION` reference whose `ref_id` carries the `factSheetId`. The reference lives on the SAP Signavio side; SAP LeanIX does not maintain a reverse list. The data spine computes the reverse view through its derived index.
- Activities bind to interfaces through one or more `LEANIX|INTERFACE` references, each tagged with a `CONSUMES` or `PRODUCES` role attribute. The role tag is consulted differently by change impact analysis and by AI grounding; conflating the two degrades both computations.
- The single-execution-location rule keeps the application binding unambiguous. An activity that appears to run against two applications is, in almost every case, two activities awaiting L5 decomposition. The fallback exception is rare and explicitly tagged.
- Lookup-time resolution is the architectural choice that keeps the binding current. The activity records the `factSheetId` only; the data spine reads the lifecycle phase, the approval flags, the data classification, and every other attribute from SAP LeanIX at the moment of the query.

- Capability and domain are SAP LeanIX-owned attributes inherited by SAP Signavio activities through the application binding. The inheritance is computed by the data spine; no separate capability-to-activity binding exists. Refining the capability map in SAP LeanIX propagates to every activity on the next read.
- Change impact analysis is supported structurally across four use cases: application phase-out, interface deprecation, role reorganization, and control-intent change. Composite changes compose the individual queries into a single impact view scoped by domain.
- Data preparation runs through the same five-stage pipeline that Chapter 4 defined for SAP Signavio: extract, validate, stage, enrich, and publish. The pipelines run independently; the data spine joins their outputs at lookup time. The `manifest_hash` recorded at publish is the load-bearing identifier for every downstream consumer.
- The boundary between SAP LeanIX and SAP Signavio remains the absence of a direct connection. The mapping does not create a direct connection; it creates a joinable surface on the data spine that both repositories publish into independently. The audit-bearing nature of the data spine depends on the boundary holding.

6.7.2 Common Design Pitfalls

- Binding an activity to a list of applications instead of a single one. An L5 activity is recorded with a `LEANIX|APPLICATION` reference that names two or three applications, on the grounds that the work touches all of them. The activity is in fact at the L4 level and conceals two or more L5 activities that should each have their own execution location, but the modeler stopped one level of decomposition too soon. The remedy is to decompose at L5, bind each child activity to its single execution location, and preserve the L4 view as an organizational container without an execution binding.
- Not including a deprecation marker in the AI grounding because the binding was cached. An AI grounding bundle includes an activity whose bound application has just been moved to phase-out, and the grounding service is unaware of the transition because the lifecycle phase was cached at binding time rather than resolved at lookup time. The remedy is to ensure every spine read resolves the SAP LeanIX factsheet at lookup time and treats the caching of variable attributes as a defect to be corrected; the `manifest_hash` bounds bundle staleness and invalidates the cache as soon as the underlying state changes.
- Binding a change impact query to stale or missing domain references. A change impact query scoped by domain returns activities that visibly do not belong to that domain when the capability or domain references on the bound application factsheet were stale or missing entirely at extraction time. The remedy is to treat missing capability and domain references as gating exceptions during the validate stage rather than as soft warnings, so an application without a completed references block is held in quarantine until the modeling gap is closed in SAP LeanIX.

- Having your republish rejected because a `LEANIX|INTERFACE ref_id` no longer resolves. An SAP Signavio activity is republished and the validation layer refuses the publication with a message that an interface `ref_id` does not resolve. The underlying SAP LeanIX interface was deprecated in a prior cycle, but the dependent SAP Signavio activity was not rebound and the deprecation date has passed. The remedy is to handle interface deprecation as a structured event that requires explicit acknowledgment on every dependent activity before the deprecation date, rather than allowing deprecations to advance with open dependent bindings.
- Binding an activity to an application that records no capabilities. The data spine resolves an activity through its `LEANIX|APPLICATION` binding but the inherited capability list comes back empty, and downstream cross-process views show the activity as belonging to no capability. The application factsheet was created without completing the references block, which means the application is not integration-eligible per the discipline we established in Chapter 5. The remedy is to either complete the references block in SAP LeanIX or remove the binding entirely. An activity bound to an integration-ineligible application is an audit gap waiting to surface.
- Compressing a hub interface into a single binding. A hub-mediated interface is recorded on the consuming activity as a single `LEANIX|INTERFACE` binding, and an upgrade to SAP Integration Suite breaks both halves of the binding silently because the data spine reads the hub as one factsheet when it is structurally two. The remedy is to record two bindings on every activity that depends on a hub-mediated interface (one for the provider-to-hub direction and one for the hub-to-consumer direction), following the two-factsheet hub modeling discipline established in Chapter 5, Section 5.3.
- Allowing control-intent change to create a silent payload mismatch on a `PRODUCES` interface. A control-intent change on an activity modifies the payload of a `PRODUCES` interface without a corresponding update to the SAP LeanIX interface specification, and the mismatch surfaces only at runtime when downstream consumers fail to validate the new shape. The remedy is to enforce the payload-shape expectation on the activity binding at publish time, validating it against the interface's semantic payload and treating every mismatch as a gating exception that requires either an interface-side or activity-side update before the publication can advance.

6.7.3 Review Questions

1. An activity is bound to one application through a single `LEANIX|APPLICATION` reference. A modeler proposes that activities should be allowed to bind to two or three applications when the work spans several systems. What are the structural reasons this proposal is rejected, and what is the principled alternative for the cases where it appears justified?
2. Interface bindings are multivalued, with each binding tagged `CONSUMES` or `PRODUCES`. Why is the role tag essential rather than decorative? Give an example of a change impact query that would produce the wrong answer if the role tag were absent.

3. An SAP LeanIX modeler refines the capability map by splitting cost center master data management into two capabilities: cost center master data maintenance and cost center hierarchy governance. The modeler updates the application factsheets to record the new capability references. What, if anything, must the SAP Signavio modelers do in response?
4. An application is being phased out. The data spine's change impact query returns twelve activities bound to the application. Three of the activities are marked for retirement; the other nine are in active use. What are the structural actions required before the application can complete its phase-out, and what is the role of the deprecation translation table in supporting them?
5. Distinguish the SAP LeanIX five-stage data preparation pipeline (extract, validate, stage, enrich, and publish) from the SAP Signavio five-stage pipeline. Where do the two pipelines converge, and where do they remain independent? What is the role of the `manifest_hash` in each one?
6. The cost center to position scenario crosses the finance and HR domains. A `SUGGESTION` is generated for the scenario and a `VALIDATION_VERDICT` is required. Which domain owner records the verdict, and what determines the routing? What happens if the two domains disagree?
7. A control-intent change on activity A modifies the payload of the `PRODUCES` interface binding from if-s4-sf-cc-02 to a new shape with an additional threshold field. The SAP LeanIX interface factsheet is not updated correspondingly. What does the data spine's validation layer do at the next SAP Signavio publication, and what is the path to remediation?
8. This chapter's refrain is that the boundary between SAP LeanIX and SAP Signavio is the absence of a direct connection. Articulate, in your own words, what is meant by this refrain and why the auditability of the integrated platform depends on it. Give one concrete scenario in which violating the refrain would degrade the audit trail.

6.7.4 Student Companion Notes

If you're approaching this chapter as a student, you should focus on the binding shape itself. Take any L5 activity in your own SAP Signavio workspace and verify that it carries exactly one `LEANIX|APPLICATION` reference whose `ref_id` resolves to a current SAP LeanIX factsheet, plus one `LEANIX|INTERFACE` reference per interface it consumes or produces, each tagged with a role attribute drawn from the controlled list. The exercise quickly reveals which activities are at the L4 level masquerading as L5, which bindings reference stale factsheets, and which activities are missing role tags. The activities that fail any of these checks are not yet ready for spine ingestion, regardless of how well their process content is modeled.

Second, take the cost center to position scenario and trace it through Table 6.3, Table 6.6, Table 6.7, and Table 6.10 in sequence, recording at each step which side of the boundary owns the fact being read. The first three tables show the bindings as they exist on the SAP

Signavio activity, and the fourth shows the bundle that the data spine publishes after ingestion. Tracing the scenario end to end will help you build intuition for what the data spine actually reads when an AI surface assembles a grounding bundle or a change impact query computes blast radius. You can also try articulating (without looking up the answer) why the data spine resolves the bindings at lookup time rather than caching the attributes and what specifically goes wrong, in audit terms, if any layer caches them.

Third, take an interface in your own landscape that has more than one consumer activity and apply the four change impact use cases from Section 6.4 to it in turn: What happens if the interface is deprecated? What happens if one of the consumer applications enters phase-out? What happens if the role responsible for the consumer activity is reorganized? What happens if a control intent on the producer activity changes the payload? The four answers help you test role-tag partitioning, cross-domain approval routing, and payload-shape validation discipline in a single artifact. This exercise is the most useful when your chosen interface crosses two domains, because cross-domain bindings are where most early-stage mistakes first appear.

This chapter closes Part III, which covered the architecture layer of the integrated landscape through Chapter 5 and Chapter 6. The bindings we described are what enable the adoption layer in Part IV (WalkMe content, SAP Fiori-level telemetry, and the user-facing surface of the bound model), publication to the canonical data spine in Part V, and the AI services that ground on the bound model in Part VII . The integrity of every downstream layer depends, structurally, on the integrity of the bindings from this chapter. The next chapter picks up the same bound model and shows how user-facing adoption content attaches to it as a process-aware layer.

PART IV

WalkMe for Process-Aware Adoption and Telemetry

Chapter 7
Designing WalkMe for Process-Aware Adoption

Adoption is where a process either earns its existence or quietly stops existing. A user in the middle of an SAP Fiori screen, with a deadline behind them and a half-finished task in front of them, has no patience for documentation. They will follow whatever is rendered next to the field, or they will improvise. WalkMe exists so that what is rendered next to the field is the same thing that was modeled, governed, and approved upstream, and that each use produces evidence that can be examined.

Chapter 3 and Chapter 4 specified what SAP Signavio must contain and how that content can be safely extracted. Chapter 5 and Chapter 6 discussed the SAP LeanIX object model and the structured references that bind SAP LeanIX applications and interfaces to SAP Signavio activities through the canonical data spine. Now, in Part IV, we turn to the third instrumented tool in the integrated landscape: WalkMe.

WalkMe has a different role from SAP Signavio and SAP LeanIX, and getting that role right is the central focus of this chapter. SAP Signavio is the authoritative source of process semantics, and SAP LeanIX is the authoritative source of architecture and landscape semantics; both are systems of record. WalkMe is neither. WalkMe is the in-application adoption layer that renders guidance, captures user interaction telemetry, and reports both back to the data spine. WalkMe consumes the stable identifiers that SAP Signavio and SAP LeanIX produce rather than authoring them. The distinction matters because it determines the direction of every integration in this chapter and, by extension, the direction of every AI interaction with WalkMe in Chapter 13 through Chapter 15. WalkMe is a target for upstream truth and a target for AI agents and never a source. Specifically, this chapter covers:

- How WalkMe is positioned in the process-driven architecture as a target layer that renders guidance bound to `Process_ID`, `Activity_ID`, `factSheetId`, and `fiori_app_id`, and never as a system of record that writes canonical truth.
- How WalkMe content is mapped to L4 and L5 process steps through a single binding entity on the data spine (`WALKME_ASSET_BINDING`) that resolves at user launch time and obeys `role_scope` and `variant_key` constraints carried from upstream.
- How WalkMe aligns with SAP Signavio on the process side and SAP Build Work Zone on the launch side, so that the user experiences a single navigable surface even when three governance regimes sit behind it.

- How in-app guidance is governed end-to-end: The lifecycle states a WalkMe asset passes through, the separation of duties (SoD) rules that gate each transition, the audit evidence each transition emits, and the change-coordination mechanics that keep WalkMe content from drifting when SAP Signavio or SAP LeanIX changes.
- Why WalkMe is a target for AI agents and what that means operationally for the commit gateway, the `SUGGESTION` and `VALIDATION_VERDICT` records, and the boundary between proposing a guidance asset and writing it into the WalkMe tenant.

The chapter introduces one knowledge graph and two new canonical entities. The knowledge graph organizes the WalkMe domain into typed nodes (assets, processes, activities, applications, SAP Fiori apps, role scopes, telemetry channels) and typed edges between them. The two new canonical entities are `WALKME_ASSET`, which carries the identity and lifecycle of a guidance object, and `WALKME_ASSET_BINDING`, which carries the structured reference from that asset to the upstream SAP Signavio and SAP LeanIX identifiers it depends on. These entities take their place in Appendix D alongside the data spine tables we'll introduce in Chapter 9. Chapter 8 covers the telemetry side of the same architecture, showing how the events emitted by WalkMe assets become a governed input back into the data spine and, downstream, into AI grounding and process improvement.

From a financial planning and analysis (FP&A) practitioner's point of view, the design problem this chapter solves will be familiar. Anyone who has tried to standardize a month-end close has run into the same gap: The controller's binder describes what should happen, the system configuration determines what can happen, and the analyst sitting in front of the transaction code at 8 p.m. on the third business day determines what actually happens. WalkMe is the discipline that closes the third gap without pretending the first two are the same. The SAP Signavio activity says what should happen. The SAP LeanIX-resolved application says where it can happen. The WalkMe overlay says, at the moment of execution, how it will happen. None of the three overrules the others. The binding is what makes them coherent.

7.1 Positioning WalkMe in a Process-Driven Architecture

In this book, we position WalkMe as a target layer in the reference architecture that renders guidance, captures adoption signals, and emits telemetry. It is not a process repository, an architecture register, or a master data system. This positioning is consequential because every subsequent integration decision in this chapter—and every AI interaction with WalkMe in Chapter 13 through Chapter 15—follows from it.

Figure 7.1 illustrates this positioning. Three bands sit above the adoption layer: the systems of record that hold write authority, the canonical data spine that joins their content on stable identifiers, and the adoption and consumption layer where WalkMe and SAP Build Work Zone render the joined view to end users. WalkMe and SAP Build Work Zone are siblings on the same band; they consume the data spine through governed application programming interfaces (APIs) and they emit telemetry back to it. Neither writes canonical state. The telemetry path back to the data spine, shown as a dashed line, will be discussed in detail in Chapter 8.

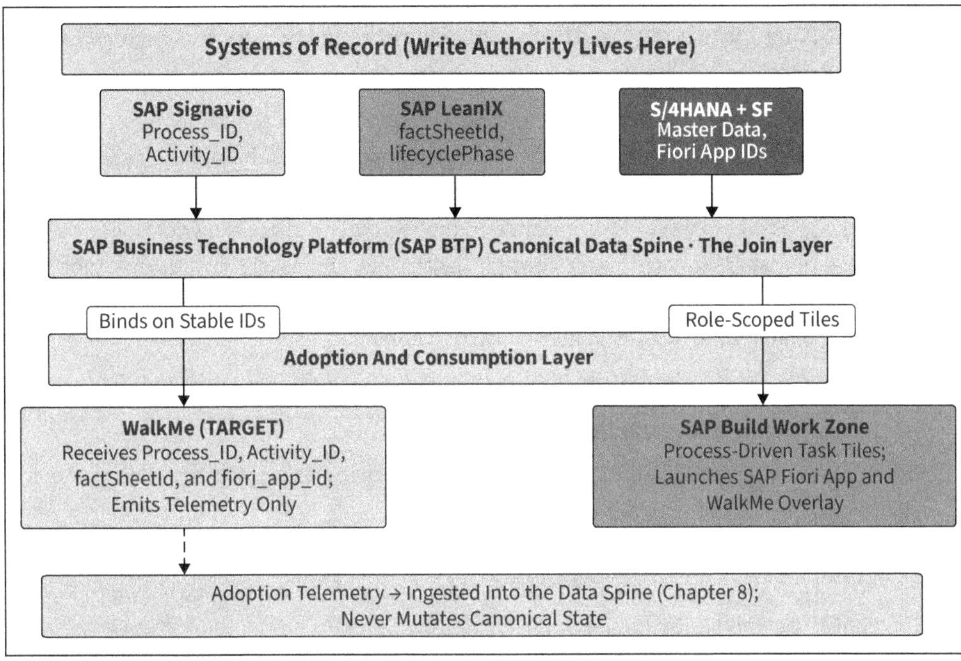

Figure 7.1: Positioning WalkMe in the Process-Driven Architecture

In the following sections, we'll discuss why referring to WalkMe as a *target* layer is important, the role of WalkMe as a knowledge graph, and where WalkMe sits in the reference architecture.

7.1.1 Why the Target Distinction Matters

If you're new to this architecture, you may be tempted to treat WalkMe as a low-friction place to store work instruction content that does not have a natural home elsewhere. The logic is appealing. WalkMe's authoring tools are accessible, its rendering is user-facing, and its versioning is straightforward. The cost of that shortcut, however, is durable. Once any piece of canonical truth lives only in WalkMe, three downstream guarantees of this architecture break.

First, the audit trail breaks. The `AUDIT_EVENT` chain established later in Chapter 9 records every state change on the data spine and every commit through the governed gateway. Content authored only in WalkMe never enters that chain. An auditor reconstructing what an end user was told months ago cannot reconstruct it from the data spine alone; they must subpoena the WalkMe tenant separately. Two systems of truth means no system of truth in reality.

Second, AI grounding breaks. The grounding bundle defined in Chapter 14 contains Signavio activities, SAP LeanIX factsheets, master data references, and execution context. Content that exists only in WalkMe is not in the bundle, which means an AI suggestion grounded on that bundle is unaware of the work instruction the user actually follows. The model proposes one thing, the user is guided to do something else, and the proposal is judged against the wrong reality.

Third, change coordination breaks. When the SAP Signavio process owner publishes a new version of an activity, the impact analysis service described in Chapter 12 walks the data spine to find every binding affected. A WalkMe asset that is bound only to itself and not to a `Process_ID` and `Activity_ID` is invisible to that walk. The process owner does not know they have rendered a downstream content asset stale; the asset's owner does not know that upstream truth has moved. Both become aware of the mismatch only when user friction exposes it, which is the most expensive form of feedback.

7.1.2 WalkMe as a Knowledge Graph

The easiest way to help you understand WalkMe's place in this architecture is to model it as a knowledge graph, as shown in Figure 7.2.

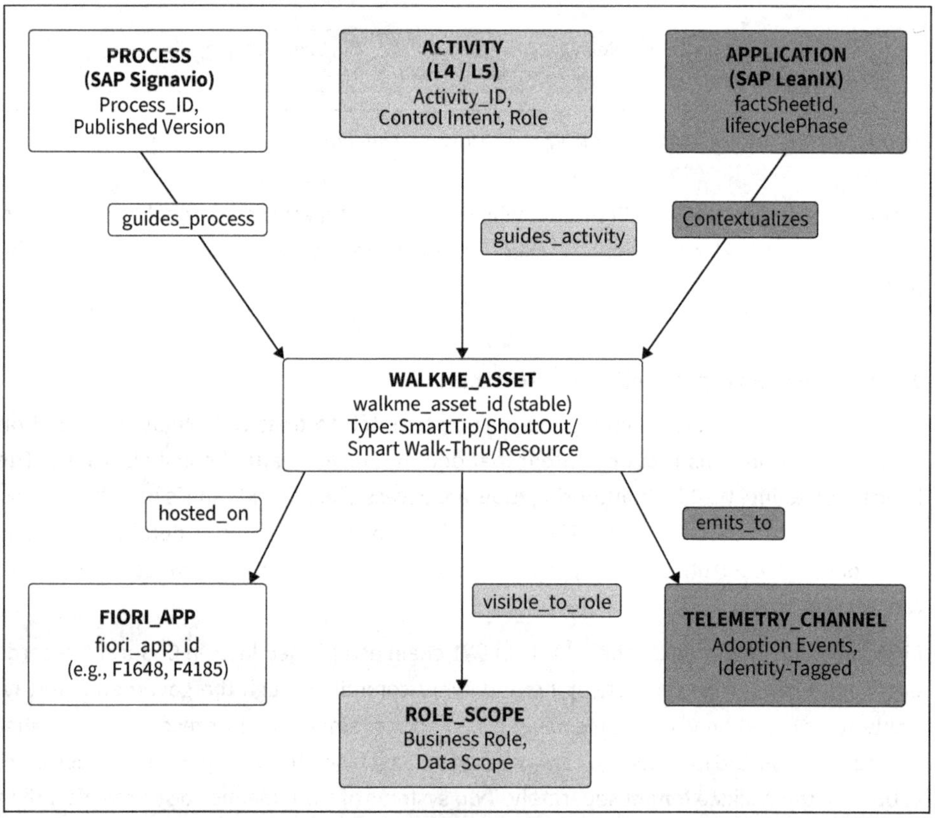

Figure 7.2: The WalkMe Knowledge Graph

WALKME_ASSET sits in the center. Five typed edges connect it to upstream and downstream nodes: guides_process binds the asset to an SAP Signavio Process_ID, guides_activity binds it to a specific L4 or L5 Activity_ID, contextualizes binds it to one or more SAP LeanIX applications through factSheetId, hosted_on binds it to an SAP Fiori app identifier, visible_to_role binds it to a role_scope, and emits_to binds it to a telemetry channel that flows back to the data spine. Every asset is a node bound to four upstream identifiers (process, activity, application, and SAP Fiori app) and two scoping nodes (role and telemetry channel). Edges are typed; the graph itself is the integration contract.

Treating WalkMe as a knowledge graph has three practical consequences. The first is that change becomes a graph traversal rather than a documentation exercise. When an SAP Fiori app is replaced by a successor, the binding service walks all WALKME_ASSET nodes connected through hosted_on edges to the retired identifier and produces an impacted-content report. The second is that role scoping becomes a query rather than a manual list. An auditor asking which guidance assets a finance master data steward in the Europe, Middle East, and Africa (EMEA) variant sees today is answered by intersecting role_scope and variant_key across the graph. The third is that AI grounding becomes typed. When the data preparation pipeline in Chapter 14 assembles a grounding bundle for a process suggestion, it can include the relevant WalkMe assets by traversing guides_activity from the Activity_ID, rather than by full-text search across the WalkMe tenant. Typed retrieval beats textual retrieval for grounding stability, and typed retrieval is what a knowledge graph offers.

The knowledge graph is implemented in the data spine, not in WalkMe itself. WalkMe stores the assets; the data spine stores the typed identifiers, the edges, and the lifecycle state. The two are joined through one canonical entity, WALKME_ASSET_BINDING, which Section 7.2 specifies in detail. The author of a WalkMe asset names the upstream identifiers; the binding is then created on the data spine and audited there. If the author does not name an upstream Process_ID, the asset is rejected at the data spine boundary. There is no orphan-content path.

7.1.3 WalkMe in the Reference Architecture's Trust Zones

Chapter 2 introduced the trust zone model that organizes security obligations across the integrated landscape. WalkMe sits in the user-facing adoption zone, alongside the SAP Fiori shell and SAP Build Work Zone. The trust boundary between that zone and the canonical spine is enforced through OAuth 2.0 scopes issued by the SAP Authorization and Trust Management service. A WalkMe render request reaching the data spine carries a token with the user's authenticated identity, the scopes corresponding to their assigned roles, and the variant context of the current session. The data spine resolves WALKME_ASSET_BINDING records that satisfy both the role_scope and the variant_key constraints. Assets that do not satisfy the constraints are simply not returned. The user sees no error, no empty placeholder, and no hint that other content exists. The experience is that of a filtered view, not a redacted document.

The trust-zone discipline has a quiet but important implication for AI: A WalkMe asset proposed by an AI agent in Chapter 13 enters the same authorization model as any asset proposed by a human author. It does not bypass role_scope at render time or approval at commit time. The boundary between proposing and writing is identical regardless of who or what authored the proposal. This is the principle the rest of this chapter operationalizes.

7.2 Mapping WalkMe Content to L4/L5 Process Steps

Section 7.1 positioned WalkMe as a target layer and established the knowledge graph view that organizes its objects. This section specifies the binding mechanism that makes the knowledge graph real on the data spine. The binding is the act of attaching a WalkMe asset to the L4 or L5 process step it guides, the application it runs inside, the SAP Fiori app that hosts it, and the role and variant scope that determines who sees it. The binding lives on the data spine, not in WalkMe; the data spine is what permits the join across the three governance regimes of SAP Signavio, SAP LeanIX, and WalkMe.

7.2.1 Stable Identifiers as the Binding Mechanism

Every binding in this architecture is built from stable identifiers, not from labels, URLs, or screen captures. The same discipline Chapter 4 applied to SAP Signavio external references and Chapter 6 applied to SAP LeanIX mappings now applies to WalkMe content. Four identifiers are mandatory on every WALKME_ASSET_BINDING record:

- Process_ID: The SAP Signavio identifier of the L3 or L4 parent process that contextualizes the asset. It is inherited from Chapter 3's identifier discipline; the format and validation are unchanged.
- Activity_ID: The SAP Signavio identifier of the L4 or L5 step the asset directly guides. An asset that cannot name an Activity_ID is rejected at the data spine boundary; guidance without a step is undisciplined documentation.
- factSheetId: The LeanIX identifier of the application in which the activity executes. It is inherited from Chapter 5's stable identifier model. Several factSheetId entries may be referenced on one binding when the activity spans a trigger and target application as defined in Chapter 9's coverage of EXECUTION_CONTEXT.
- fiori_app_id: The SAP Fiori library identifier of the screen the user is on when the asset is rendered. It is carried into WalkMe as the hosted_on edge in the knowledge graph and used at render time to scope which assets become eligible.

Two additional fields scope the binding without changing its identity. The role_scope field is the set of business role tags from Chapter 3 that determine which authenticated users can see the asset at render time. The variant_key field is the controlled-list enumeration from Chapter 3's variant model that restricts the asset to a specific geographic, regulatory, or business unit variant of the activity. Together, the four mandatory identifiers and two scoping fields qualify the binding without entangling it in the asset's content. The content is in WalkMe; the binding is on the data spine.

7.2.2 The WALKME_ASSET_BINDING Entity and the Data Model

Figure 7.3 shows the full data model. At the top sit the four upstream reference entities the binding depends on: `PROCESS_VERSION` (from SAP Signavio, keyed on `Process_ID` and `Activity_ID`, the primary keys [PKs] for this entity), `APPLICATION` (from SAP LeanIX, keyed on `factSheetId`), `FIORI_APP` (the SAP Fiori library, keyed on `fiori_app_id`), and `ROLE_SCOPE` (the data spine's role catalog). In the center sits `WALKME_ASSET_BINDING`, the join entity, which carries foreign keys (FKs) to each of the four upstream entities plus the two scoping fields and a `binding_state` attribute. Below the binding sits `WALKME_ASSET`, the entity that owns the asset body and content version. One asset can be referenced by many bindings, meaning the relationship is 1:N. For example, the same Smart Walk-Thru can be bound to different `Activity_ID`s with different `role_scopes`. The bottom row carries the AI and audit entities: `SUGGESTION`, `VALIDATION_VERDICT`, `AUDIT_EVENT`, and `TELEMETRY_EVENT`. The binding entity is the single point of truth that holds every routing decision; the asset, the AI flow, and the audit chain all reference it.

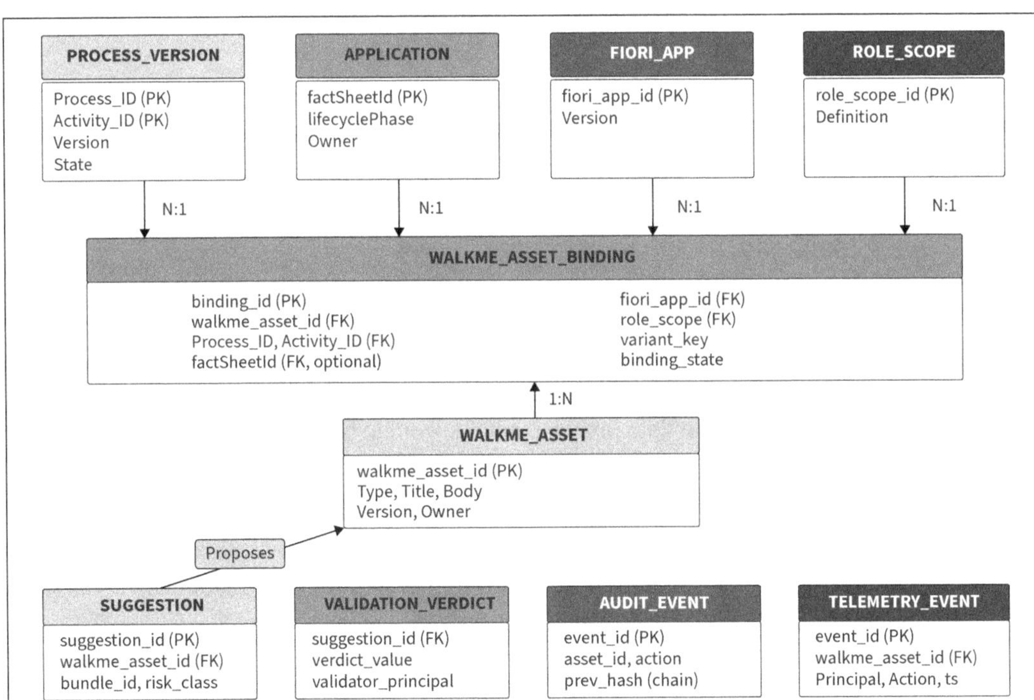

Figure 7.3: WalkMe Data Model

Neither the SAP Signavio activity nor the WalkMe asset references the other directly; both reference the binding, and the data spine resolves the join at lookup time. This is the same boundary refrain carried forward from Chapter 1, Chapter 3, Chapter 5, and Chapter 6: the boundary is the lack of a direct connection. There is no point-to-point integration from SAP Signavio to WalkMe, and there is no point-to-point integration from WalkMe to SAP LeanIX. Both publish their content into the data spine, and the data spine joins them through the binding.

The binding_state attribute is essential to the model. It carries one of four values (DRAFT, PUBLISHED, DEPRECATED, or RETIRED) and is the attribute that determines whether the binding is visible to the render path. A DRAFT binding is visible only to its author and to approvers. A PUBLISHED binding is visible to all users whose role_scope and variant_key satisfy the binding constraints. A DEPRECATED binding is visible only to users who saw it previously, with a banner indicating supersession. A RETIRED binding is not visible to render at all but is retained indefinitely for audit. Section 7.4 discusses this lifecycle in detail.

7.2.3 Worked Example: Cost Center to Position with WalkMe Overlay

Our running scenario carries forward four activities across two systems of record. WalkMe binds one asset to each activity. The following IDs are the canonical references each asset will use; they appear unchanged in every subsequent chapter:

- A_COST_CENTER_CREATE: SAP Master Data Governance, Financials, SAP Fiori app ID F1648, factSheetId fs-mdg-fin-01
- A_COST_CENTER_PROPAGATE: Controlling in SAP S/4HANA, SAP Fiori app ID F0763, factSheetId fs-s4hana-fin-emea-01
- A_POSITION_DRAFT: SAP SuccessFactors Employee Central, SAP Fiori app ID F2647, factSheetId fs-sf-ec-01
- A_POSITION_COST_CENTER_ASSIGN: SAP SuccessFactors Employee Central, SAP Fiori app ID F0846, factSheetId fs-sf-ec-01 and fs-s4hana-fin-emea-01

Process_ID p-r2r-cc-001 covers the cost center side while Process_ID p-htr-pos-001 covers the position side. The two processes meet at A_POSITION_COST_CENTER_ASSIGN, which is the activity that links human resources (HR) to finance through the canonical cost center identifier.

Figure 7.4 carries forward the scenario from Chapter 6 and adds the WalkMe overlay this chapter defines. Each of the four activities has its own L4 process model in SAP Signavio, its own factSheetId in SAP LeanIX, and its own SAP Fiori screen in production. Telemetry from each asset is grouped by Process_ID and will appear in Chapter 8.

Each activity in the scenario receives a WalkMe asset, and each asset is bound on the data spine. The cost center create activity is overlaid by asset wm-cc-create-001, a Smart Walk-Thru that steps the finance master data steward through the seven fields SAP Master Data Governance, Financials requires for an EMEA cost center, including the validation against the controlling area and the company code consistency check. The binding for this asset carries Process_ID p-r2r-cc-001, Activity_ID A_COST_CENTER_CREATE, factSheetId fs-mdg-fin-01, fiori_app_id F1648, role_scope {FIN_MD_Steward}, and variant_key EMEA.

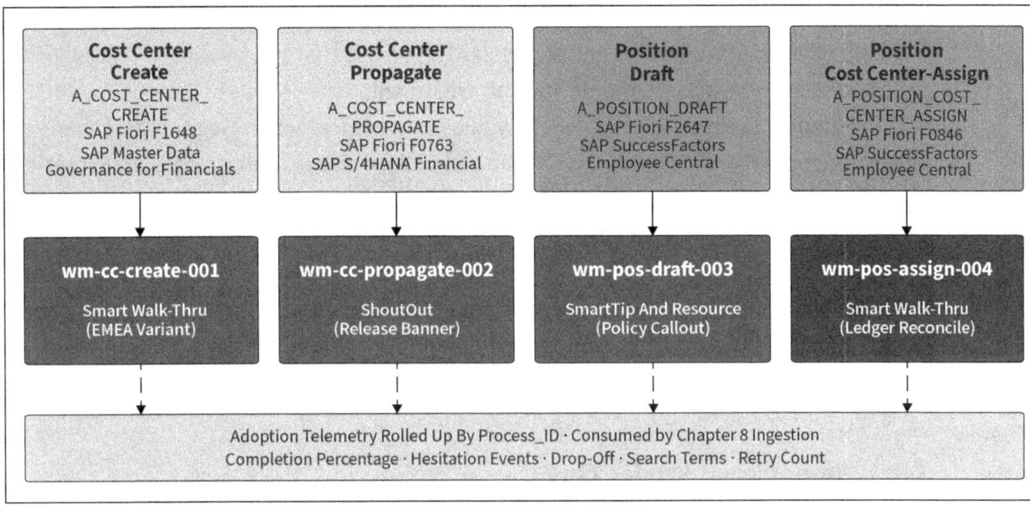

Figure 7.4: Running Scenario Continued—WalkMe Overlay on the Cost Center to Position Flow

The cost center propagate activity is overlaid by asset wm-cc-propagate-002, a ShoutOut that informs the SAP S/4HANA controller of newly created cost centers waiting for activation in their controlling area. The binding carries the same Process_ID, Activity_ID A_COST_CENTER_PROPAGATE, factSheetId fs-s4hana-fin-emea-01, fiori_app_id F0763, role_scope {CO_Controller}, and variant_key EMEA. The asset is intentionally brief, rather than a detailed guided walk-through, because the activity is largely automated and the user's role is approval and not data entry.

The position draft activity is overlaid by asset wm-pos-draft-003, a SmartTip that calls out the organizational unit policy on direct reports and the cost center selection rule, plus a Resource link to the HR organizational design policy. The binding carries Process_ID p-htr-pos-001 (the parent hire-to-retire process), Activity_ID A_POSITION_DRAFT, factSheetId fs-sf-ec-01, fiori_app_id F2647, role_scope {HR_OrgDesigner, HR_BusinessPartner}, and variant_key GLOBAL. Note that this is the first asset whose role_scope crosses functions: The HR business partner role is granted visibility for collaboration, even though only the organizational designer can edit the draft.

The position cost center assign activity is overlaid by asset wm-pos-assign-004, the most consequential Smart Walk-Thru in the sequence. This is the activity where the two domains meet—an HR position is assigned to a finance-governed cost center—and the WalkMe overlay enforces the cross-domain validation. The asset walks the user through the cost center lookup against the active master-data version, presents the controller's name and finance approval state from MASTER_DATA_REFERENCE, and blocks submission if the cost center's lifecycle phase is anything other than active. The binding carries Process_ID p-htr-pos-001, Activity_ID A_POSITION_COST_CENTER_ASSIGN, two factSheetId values (fs-sf-ec-01 and fs-s4hana-fin-emea-01, because the activity spans HR and finance), fiori_app_id F0846, role_scope {HR_OrgDesigner}, and variant_key GLOBAL.

Two observations follow from this example. The first is that the binding model scales to cross-domain activities without special treatment. The A_POSITION_COST_CENTER_ASSIGN binding holds two factSheetId values because the activity legitimately spans two applications; the binding entity accepts an ordered list rather than a single value precisely so this case is handled without schema gymnastics. The second is that the role_scope and variant_key fields are what permit the same WalkMe tenant to render legitimately different content to different users on the same screen. The cost center create asset is rendered in Finland one way and in Germany another way only because variant_key is consulted at render time. The two variants are different bindings to different assets, both of which are governed, both of which emit telemetry, both of which appear in the audit chain.

7.2.4 Resolution at Render Time

When a user opens SAP Fiori app F0846 in SAP SuccessFactors Employee Central, the SAP Build Work Zone shell (described further in Section 7.3 and in detail in Chapter 17) resolves the Process_ID associated with the user's current task tile and presents it to the process-aware execution service, as we will cover in Chapter 12. The service queries the data spine for WALKME_ASSET_BINDING records where Process_ID, fiori_app_id, role_scope, variant_key, and binding_state simultaneously match the request. The matching asset_ids are returned to WalkMe through a server-to-server token exchange, and WalkMe renders only those assets. The user never sees a query, a token, or an unmatched asset. Guidance appears automatically and disappears when it no longer applies.

The resolution path has an important property: It is symmetric with the AI grounding path defined in Chapter 14. When the grounding service assembles a bundle for an AI suggestion about an activity, it walks the same WALKME_ASSET_BINDING table on the same identifiers and includes the matching assets in the bundle. AI sees what the user sees, scoped by the same role and variant constraints. This is the precondition for AI to be trusted as a draft author of WalkMe content in Chapter 13: The model is grounded on the same surface the user is rendered on. If AI proposes a guidance asset that contradicts what is currently rendered, the contradiction is visible at grounding time, not render time, and the contradiction can be reviewed in the validation funnel before any commit reaches the WalkMe tenant.

7.3 Aligning WalkMe with SAP Signavio and SAP Build Work Zone

SAP Signavio, WalkMe, and SAP Build Work Zone may converge on the user's screen, but they sit in different governance regimes and serve different constituencies. SAP Signavio is governed by the process owner community and answers the question "What should happen?" SAP Build Work Zone is governed by IT operations and answers the question "How does the user reach the screen where it happens?" WalkMe is governed by the WalkMe content manager and the same process owner community jointly, and it answers the question "How is the work performed once the user is there?" As you will see in the following sections, alignment among the three is a structural commitment expressed on the data spine.

Figure 7.5 shows the three-way alignment as a single binding layer. Each tool publishes its responsibilities into the data spine using the same stable identifiers. SAP Signavio publishes `Process_ID`, `Activity_ID`, variant, role, and control intent. SAP Build Work Zone publishes the role-based tile assignments and the SAP Fiori app launches associated with them. WalkMe publishes the `WALKME_ASSET_BINDING` records that scope guidance to those same identifiers. None of the three writes into the others; all three write into the data spine, and the data spine renders a coherent picture to anyone who queries it. Each tool publishes into a shared binding layer on the data spine using the same stable identifiers. The user experiences a single navigable surface.

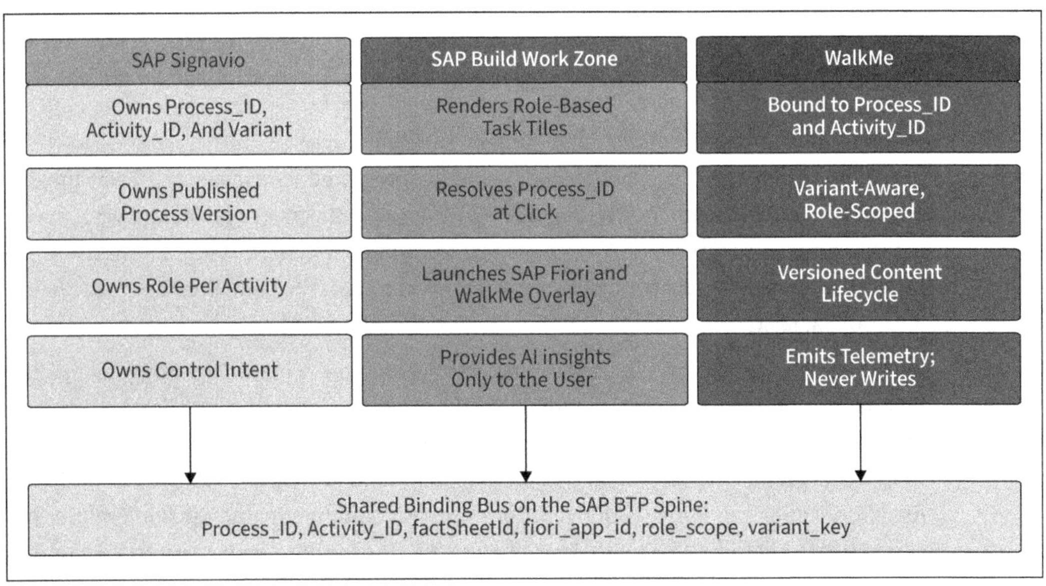

Figure 7.5: Three-Way Alignment of SAP Signavio, WalkMe, and SAP Build Work Zone

7.3.1 SAP Signavio: The Authoritative Process Reference

The SAP Signavio responsibility in this alignment is unchanged from Chapter 3. Process owners maintain the L4 and L5 process models, including the mandatory execution attributes the data spine consumes: the `Activity_Stable_ID`, the `Performing_Role`, the `System_Trigger`, the `App_Reference`, the `Control_Intent` classification, the variant enumeration, and the lifecycle state. WalkMe is a downstream consumer of these attributes; it does not request changes to them. When a WalkMe content manager discovers that an asset cannot be cleanly bound because the upstream Activity_Stable_ID is missing or because the variant enumeration does not include a needed value, the remediation path runs through SAP Signavio governance, not around it. The asset stays in **Draft** status until the upstream attribute is published. This is the same discipline Chapter 6 imposed on SAP LeanIX-side bindings, and it has the same effect: It forces the modeling debt to be settled at the source rather than papered over downstream.

One SAP Signavio responsibility is specific to the WalkMe alignment. Process owners must classify each activity for adoption-content eligibility using a controlled list attribute named Guidance_Required. Three values are permitted. The value Required means the activity is sufficiently complex or risk-sensitive that a WalkMe asset is mandated; absence of a binding is reported as a defect by the validation layer in Chapter 4. The value Optional means a WalkMe asset is permitted and welcomed but is not required. The value Prohibited means a WalkMe asset is forbidden, typically because the activity is internal to system processing and has no human-facing screen or the activity is so sensitive that out-of-system explanation could itself create a control risk. The attribute is part of the SAP Signavio extension contract established in Chapter 4 and is consumed by the WalkMe authoring discipline in Section 7.4.

7.3.2 SAP Build Work Zone: The Launch Surface

SAP Build Work Zone is the launch surface in the integrated landscape. It is what the user opens, what presents them with the task tiles their role makes available, and what carries them into the underlying SAP Fiori app where the work happens. Chapter 17 discusses SAP Build Work Zone in detail; the relevant part in this section is the contract it offers to WalkMe and SAP Signavio.

When a user clicks a task tile in SAP Build Work Zone, three identifiers flow downstream before the SAP Fiori app finishes loading. The Process_ID identifies the parent process the tile represents; this is the same identifier SAP Signavio publishes. The fiori_app_id identifies the screen being launched; this is what SAP Build Work Zone resolves from the tile manifest. The role_scope is derived from the user's authenticated session and the role assignments held in SAP Cloud Identity Services. These three identifiers reach the data spine through a process-aware execution service call, which returns the matching WALKME_ASSET_BINDING records as Section 7.2 described. The user's experience is that the SAP Fiori app opens, and within a second or two the WalkMe overlay appears. The overlay is not loaded by the user, selected from a menu, or searched for; it is rendered because the binding resolved.

The SAP Build Work Zone responsibility in this alignment is to deliver the Process_ID and role_scope context to the data spine in a form the data spine can match. The tile manifest in SAP Build Work Zone carries the Process_ID and the fiori_app_id as first class properties; these are populated at tile-creation time from the same SAP Signavio extract that Chapter 4 specified. A tile whose Process_ID is missing or whose fiori_app_id does not resolve against the SAP Fiori library is flagged by the validation layer and is not promoted to

Published in SAP Build Work Zone. As with SAP Signavio, the discipline runs at the boundary, not after the user encounters a broken tile.

7.3.3 WalkMe: The Guidance Surface

WalkMe's responsibility in the alignment is to render the matching assets and to emit telemetry for each render and each user interaction within the asset. The rendering is governed at three boundaries: the `role_scope` must be satisfied, the `variant_key` must match, and the `binding_state` must be **Published**. These three checks happen on the data spine, not in WalkMe; the WalkMe tenant receives a scoped list of `asset_ids` and renders only those. A WalkMe administrator who logs into the WalkMe console can see all assets in all binding states, but the rendering path itself honors only **Published** bindings whose scopes match the requesting session. This is a deliberate separation of editorial and runtime concerns.

Telemetry from WalkMe is emitted as a stream of events tagged with `walkme_asset_id`, `Process_ID`, `Activity_ID`, `fiori_app_id`, user principal (pseudonymized at the boundary, as Chapter 8 will specify), `session_id`, and `event_type`. The `event_type` values include `asset_rendered`, `step_started`, `step_completed`, `step_skipped`, `hesitation_detected`, `drop_off`, and `survey_response`. The events flow into the data spine through a single ingestion endpoint that Chapter 8 covers in detail. From the perspective of this chapter, the relevant fact is that the same identifiers that determine what is rendered also determine what is recorded. Adoption telemetry joins back to SAP Signavio activity definitions and SAP LeanIX application contexts through the same keys. The full closure of design, delivery, telemetry, and improvement—the loop introduced in Chapter 1—is operationalized through this single key discipline.

7.3.4 Joint Operating Cadence

Three governance regimes meet at the binding. They do not merge, but they cooperate on a cadence that mirrors the lifecycle of the underlying activity. That cadence is the practical answer to the question every transformation program eventually asks: who decides when an activity, an application, or an SAP Fiori app changes, and who acts on the decision.

When a process owner in SAP Signavio publishes a new version of an activity, the data spine's impact analysis service, which we'll introduce in Chapter 12, walks the `WALKME_ASSET_BINDING` table and produces a list of affected bindings. Each affected binding has its `binding_state` set to in review and its owning WalkMe content manager is notified. The manager has a defined window (typically two weeks for non-control activities and five business days for control activities) to either revise the underlying asset, retire it, or request an extension. If no action is taken in the window, the binding is auto-deprecated; the asset continues to render with a deprecation banner to existing users, but no new users see it. The intent of auto-deprecation is not punitive; it prevents stale guidance from becoming a permanent fixture due to inaction.

When an IT operations team in SAP LeanIX retires an application (meaning the lifecycle phase transitions from active to phase-out), the same impact analysis service walks the binding table and identifies `WALKME_ASSET_BINDING` records that reference the retired `factSheetId`. These bindings are placed in review automatically and their owning manager is notified. The manager must either reauthor the asset against the successor application or retire the asset. The discipline is identical to the SAP Signavio path; the trigger is different, but the operating model is the same.

When a WalkMe content manager initiates a content change on their own initiative—such as a clarification, typo fix, or regulatory update—the change is authored in draft, submitted for approval, as Section 7.4 will describe, and promoted to **Published** without disturbing SAP Signavio or SAP LeanIX. The cadence here is one-way: WalkMe-initiated changes do not propagate upstream. If the change reveals that the underlying activity has shifted in a way that warrants an SAP Signavio update, the manager opens a change request through SAP Signavio governance. This is the same discipline a controller would apply if they discovered during the monthly close that a journal-entry convention had drifted from the documented policy. The close would be finished using the current policy, the policy gap would be documented, and the policy would be updated through the correct governance channel. WalkMe content drift never silently rewrites the SAP Signavio process.

7.4 Governance of In-App Guidance and Change

The previous sections specified what WalkMe is in the architecture, how its assets are bound to upstream identifiers, and how three governance regimes coordinate at the binding boundary. This section operationalizes the governance itself: the lifecycle a WalkMe asset passes through, the approvals that gate each transition, the audit evidence each transition emits, and the discipline that determines how AI agents propose new assets without bypassing the same controls a human author would face.

7.4.1 The Lifecycle of a WalkMe Asset

Figure 7.6 shows the lifecycle of a WalkMe asset. It passes through five binding states: **Draft**, **In Review**, **Published**, **Deprecated**, and **Retired**. Each transition is co-signed by two distinct roles and gated by an explicit approval. Additionally, each transition emits an `AUDIT_EVENT` to the data spine's hash-chained audit ledger that will be established in Chapter 9. The binding states are not editorial conveniences; they are the runtime filter that determines whether an asset reaches a user's screen.

A draft asset is one the author is still building. It is visible only to the author and to any reviewers explicitly named. It is excluded from all rendering paths, regardless of role or variant. The **Draft** state is also the only state in which the asset body itself can be edited; subsequent states allow metadata changes (`binding_state`, lifecycle attributes, retirement dates) but not body changes. An asset that needs significant rework after publication must be cloned to a new `walkme_asset_id`, taken through the full lifecycle, and then have its predecessor deprecated in a paired transition. This is a deliberate forcing function: It makes content history easy to reconstruct because every body of guidance has its own stable identifier and its own audit trail.

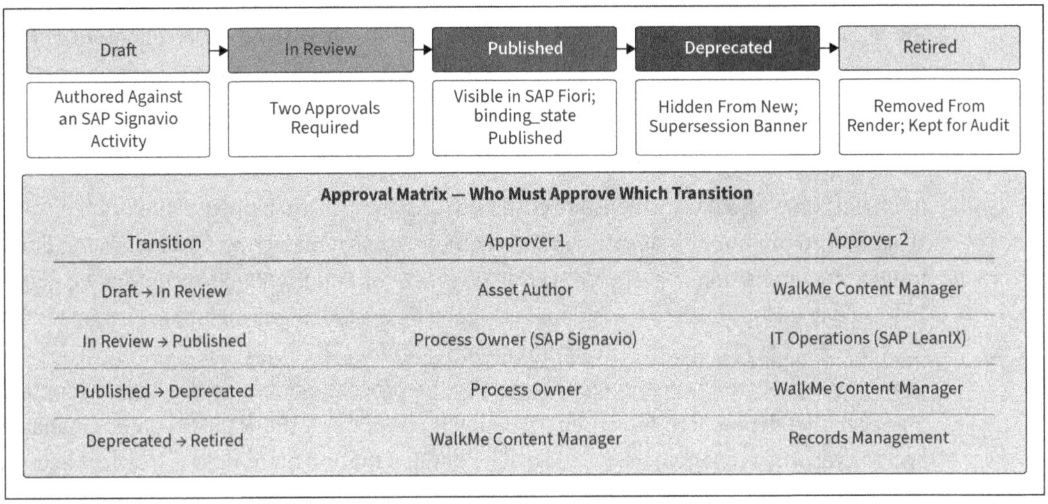

Figure 7.6: WalkMe Content Lifecycle and Approval Workflow

The **In Review** state is the approval gate. The asset has been submitted by its author and is awaiting two co-signing approvals as we'll discuss throughout this section. Rendering is still suppressed; the only change from **Draft** is that editing is locked and the asset is visible to the approval queue. The **In Review** state has a configurable timeout, defaulting to fifteen business days; an asset that has not received both approvals by the timeout is automatically returned to **Draft** and the author is notified. This prevents indefinite parking of assets in the approval queue, which is a common failure in adoption-content programs that lack timeout discipline.

The **Published** state is the only state in which the asset's binding is visible to the render path. The transition into **Published** is the event at which the asset becomes user-facing. The transition is recorded in the audit chain with the principals of both approvers, the timestamp, the binding constraints in effect, and a hash of the asset body. From this moment forward, every render of the asset emits telemetry that joins back to the audit record on `Process_ID`, `Activity_ID`, and `walkme_asset_id`. The asset is in production.

The **Deprecated** state is for assets that have been superseded but must continue to render to users who have started, but not completed, a session that included them. A deprecated asset renders with a banner indicating the supersession and a link to the successor asset. Deprecated bindings are visible only to users whose session predates the deprecation; new sessions see only published bindings. The **Deprecated** state has its own timeout, defaulting to thirty days, after which the asset is automatically retired. This window is calibrated to absorb users on extended leave or seasonal absences without forcing a permanent presence for stale content.

The **Retired** state is terminal. A retired asset does not render and is not visible to any user. The asset record itself is retained indefinitely in the WalkMe tenant and the binding record is retained indefinitely on the data spine. Retention is required because audit reconstruction of a past user session may need to display the guidance that was active at the time of the session. Audit reconstruction is the only path through which retired content is rendered, and the rendering happens in an audit console rather than in the user-facing path.

7.4.2 Approval and Separation of Duties

Approvals on WalkMe transitions are cosigned. Cosigning is not a process preference; it is a structural commitment that satisfies SoD requirements. In the permission model, the draft-to-in-review transition is cosigned by the asset author and the WalkMe content manager. The in-review-to-published transition is cosigned by the SAP Signavio process owner for the bound `Activity_ID` and the IT operations approver for the bound `factSheetId`. The published-to-deprecated transition is cosigned by the process owner and the WalkMe content manager. The deprecated-to-retired transition is cosigned by the WalkMe content manager and the records management function, who verifies that no audit hold is in effect against the asset. In short, the authoring, approval, and render functions are three independently managed permission sets, as follows:

- **Authoring**
  The WalkMe content author role can edit draft and in-review assets but cannot publish.
- **Approval**
  The process owner and IT operations can approve published transitions; the cosigning is captured in the data spine.
- **Render**
  The end user only sees assets where the `role_scope` corresponds to user roles.

Three SoD rules are enforced at the data spine and are not configurable by tenant administrators. The first rule is that an asset cannot be approved by its author. An approval attempt by the author's principal is rejected at the gateway and logged as a denied `AUDIT_EVENT`. The second rule is that the in-review-to-published transition requires one process owner approver and one IT operations approver. Two process owner approvers, or two IT operations approvers, do not satisfy the cosigning requirement. This is the same `role_scope_intersect` logic Chapter 14 applies to AI validation verdicts: The two principals must come from intersecting but distinct role scopes. The third rule is that render scope confers no edit or approval rights. A user who sees an asset rendered does not gain the ability to modify or approve it; visibility and authority are independently provisioned. These three rules together prevent the failure modes that audit findings on adoption programs most commonly cite.

7.4.3 WalkMe as a Target for AI Agents

WalkMe is a target for AI agents, not a source. AI agents read WalkMe assets and the data spine's `WALKME_ASSET_BINDING` entries during grounding. They can propose new or modified WalkMe assets in the form of `SUGGESTION` records but they cannot write to the WalkMe tenant directly. Every write to WalkMe goes through the commit gateway on the data spine, and every write requires a `VALIDATION_VERDICT` from a human reviewer before the gateway accepts it. There is no direct integration from the SAP AI Core, the generative AI hub, or Joule to the WalkMe tenant.

Figure 7.7 illustrates the data flow. The AI layer in the top band (SAP AI Core and the generative AI hub) observes telemetry patterns from the WalkMe channel (i.e., friction events on a specific `Activity_ID`, repeated drop-off at a specific step, or search terms that cluster around a topic), grounds against the corresponding SAP Signavio activity and SAP LeanIX application, and proposes a guidance asset. The proposal is emitted as a `SUGGESTION` record on the data spine, carrying `suggestion_id`, `bundle_id` (the grounding bundle we will cover in Chapter 14), `inference_event_id`, `proposed_state` (the asset body, scoping, and binding constraints), and `risk_class`.

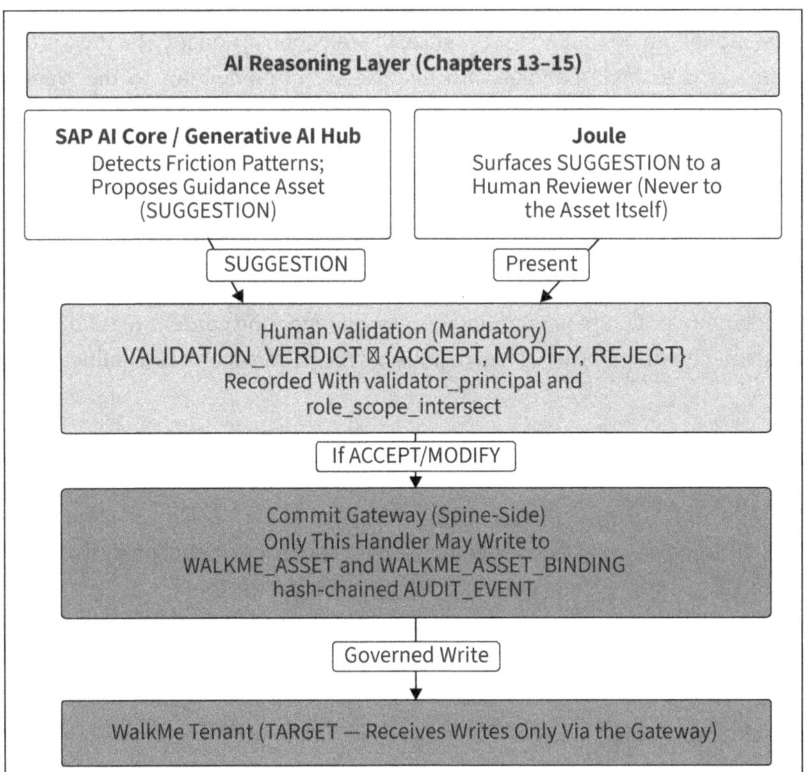

Figure 7.7: WalkMe as a Target for AI Agents

Joule, the interaction layer described in Chapter 13, provides the SUGGESTION to a human reviewer with appropriate role_scope_intersect, typically a process owner or WalkMe content manager. The reviewer records a VALIDATION_VERDICT of ACCEPT, MODIFY, or REJECT. An ACCEPT or MODIFY verdict carries the proposal (possibly with human edits) to the commit gateway on the data spine. The commit gateway is the only handler permitted to write to WALKME_ASSET and WALKME_ASSET_BINDING. The gateway verifies the verdict, checks the role_scope_intersect, emits an AUDIT_EVENT with the prev_hash chain, and forwards the write to the WalkMe tenant through the tenant's management API. The asset enters the **Draft** state and reenters the lifecycle described in the previous section. AI-authored content is no different from human-authored content from this point forward.

Three properties of this flow are worth emphasizing. The first is that AI never skips human validation, regardless of confidence score. A SUGGESTION with a confidence score of 0.99 still requires a VALIDATION_VERDICT before the gateway accepts it. Chapter 1's rule that AI must not be the decider is enforced as a structural property of the gateway, not as a policy that an administrator could relax. The second is that REJECT is a recorded outcome with equal standing to ACCEPT. A rejected SUGGESTION emits an AUDIT_EVENT, contributes to the AI performance signals that will be monitored in Chapter 18, and is preserved for future grounding so the model can be trained to recognize the rejection rationale. The third is that the commit gateway is a single handler with a single set of credentials to the WalkMe tenant. WalkMe tenant administrators cannot create alternative write paths to satisfy an urgent request from the AI layer; the boundary is mechanical, not procedural.

7.4.4 Change Coordination with SAP Signavio and SAP LeanIX

Change coordination across the three governance regimes is the discipline that prevents the most common point of failure in adoption programs: rendering content that no longer matches the system it describes. Three change patterns recur, and each has a defined operating response.

The first pattern is an SAP Signavio-initiated change to an Activity_ID's content. The process owner has revised the L4 or L5 model perhaps the step sequence changed, a new control was added, or a variant was introduced. When the new PROCESS_VERSION transitions to APPROVED in SAP Signavio and is published to the data spine, the impact analysis service walks WALKME_ASSET_BINDING and sets binding_state to **In Review** for every affected record. The owning WalkMe content manager is notified with the diff between the old and new process versions and a calculated impact summary indicating which steps of the activity changed. The manager has a defined window to reauthor the asset against the new version. The asset under the old binding continues to render until the new version's binding is **Published**, at which point the old binding transitions to deprecated. There is never a window in which no guidance renders for an activity that has Guidance_Required = Required; the overlap is intentional.

The second pattern is an SAP LeanIX-initiated change to an application's lifecycle phase. IT operations has determined that an application is moving from active to phase-out or that a successor application is taking over a capability. The same impact analysis service walks the WALKME_ASSET_BINDING table on factSheetId and produces a remediation list. The remediation discipline mirrors pattern one: Bindings affected by the application transition are placed **In Review**, the owning manager is notified, and the asset is either reauthored against the successor application's SAP Fiori app (with a new fiori_app_id on the binding) or retired. The two-week window for non-control activities and five-day window for control activities applies.

The third pattern is an SAP Fiori app retirement or successor relationship in SAP Fiori launchpad. The SAP Fiori library publishes successor relationships when one app is replaced by another; the binding service consumes these relationships and auto-suggests successor fiori_app_id values on affected bindings. The manager confirms or rejects the suggestion, and the binding is updated accordingly. This is the only pattern in which a non-WalkMe initiator can change a binding's fiori_app_id, and the change is always proposed, never enacted, without manager confirmation.

7.4.5 Audit Evidence and Retention

Every WalkMe transition, authoring action, approval, render, and telemetry event produces an AUDIT_EVENT on the data spine. The audit chain is the same one Chapter 6 established: Each event carries a prev_hash that links to the preceding event for the same asset, producing a tamper-evident record. The audit console we cover in Chapter 19 supports four query patterns relevant to WalkMe content: reconstruct the guidance a specific user saw at a specific time, reconstruct the guidance an activity displayed in a specific binding state, list the approval history of a specific walkme_asset_id, and list the AI-authored content currently rendering. These four queries are the practical instruments through which auditors, process owners, and operational teams query the WalkMe content catalog.

The retention policy follows the data spine's general rule: Audit records are retained indefinitely, asset bodies are retained for the regulatory horizon applicable to the bound activity (typically seven years for finance activities and five years for HR activities), and telemetry events are retained for the period defined in the data-protection regime applicable to the user population. The retention discipline is enforced at the data spine through metadata tags on each binding; it is not the WalkMe administrator's responsibility to remember when an asset is eligible for purge. Records management approves the **Deprecated**-to-**Retired** transition only after the retention metadata confirms eligibility.

7.5 Summary

This chapter positioned WalkMe as a process-aware adoption layer inside the architecture we have been constructing chapter by chapter. We first discussed how WalkMe is a target, not a source. It consumes Process_ID from SAP Signavio and factSheetId from SAP LeanIX; it does not author either. Its assets reference these identifiers but never replace them. When an activity is renamed or retired, an application is decommissioned or replaced, a role's scope changes, or a variant is deprecated, the change flows through the data spine and reaches WalkMe through the binding service. WalkMe never originates that change, and the discipline of treating it as a target is the discipline that prevents the adoption layer from drifting into its own version of the process model.

Adoption guidance, like any other artifact that reaches a user's screen, must obey a governed lifecycle. A WalkMe asset moves through **Draft**, **In Review**, **Published**, **Deprecated**, and **Retired** states, and each transition is cosigned by two distinct roles. The runtime selector that decides which asset reaches a given user is the same expression that governs every other rendering decision in this architecture: the intersection of the user's role scope, the user's variant key, and the asset's published binding state. The selector is deterministic, auditable, and identical across the SAP Signavio process tile, the WalkMe overlay, the SAP Build Work Zone group, and the SAP Fiori app. A user who is not authorized to see a step does not see the WalkMe content for that step either.

WalkMe is where AI proposes new adoption content, and it is also where AI consumes operational context to populate that content. The knowledge graph introduced in Section 7.1 is the structure through which both directions of traffic flow: nodes for process versions, activities, applications, SAP Fiori apps, roles, variants, and WalkMe assets as well as edges for the bindings between them. AI agents read the graph to decide which assets are relevant, write `SUGGESTION` rows that propose new assets or modifications, and never write directly to a binding. Every AI write to WalkMe is a proposal; every proposal is gated by the same manager and process owner approvals that govern human authoring; every approval emits an `AUDIT_EVENT` to the hash-chained audit ledger Chapter 9 will establish and Chapter 19 will operationalize.

We also discussed the boundary between WalkMe and the rest of the architecture is not a wire. There is no API between WalkMe and SAP Signavio that creates an activity, no API between WalkMe and SAP LeanIX that updates a `factSheetID`, and no API between WalkMe and the data spine that bypasses the commit gateway. The boundary is the absence of those wires. Adoption is consumed where the user is, but the adoption layer is not a write path into the systems that describe the work. This separation is what makes the architecture explainable to an auditor and stable across years of change. It is also what allows AI agents to operate against WalkMe without the risk of corrupting the process model or the architecture register.

Chapter 8 will pick up the telemetry that WalkMe emits and follow it into the data spine and onward into the AI services that consume adoption signals. The telemetry schema, the friction detection patterns, and the feedback loop that turns adoption events into process improvement candidates are the subjects of that chapter. The design discipline established in this chapter, in particular the discipline that WalkMe is a target rather than a source, sets the conditions under which telemetry can be trusted as evidence rather than treated as ambient noise.

7.5.1 Key Learning Points

- WalkMe is positioned as a target for AI agents and a consumer of the canonical spine. It is not a system of record for process semantics or architecture state. Its assets reference upstream identifiers; they never originate them.
- Every WalkMe asset is bound through a `WALKME_ASSET_BINDING` that names a `walkme_asset_id`, a `Process_ID`, an `Activity_ID`, and optionally a `factSheetId`, `Fiori_app_id`, `role_scope`, `variant_key`, and `binding_state`. The binding is the unit of governance, not the asset body.
- The knowledge graph is the structure that makes WalkMe content addressable by AI. Nodes carry the canonical identifiers from SAP Signavio, SAP LeanIX, and the data spine; edges carry the binding semantics. AI agents traverse the graph to propose; they do not shortcut around it.
- Three-way alignment between WalkMe, SAP Signavio, and SAP Build Work Zone is achieved through identifier convergence, not through copy-paste. The same `Process_ID`, `role_scope`, and `variant_key` drive the SAP Signavio process tile, the SAP Build Work Zone group, and the WalkMe overlay; rendering divergence between the three is a defect, not a degree of freedom.
- The runtime selector for WalkMe content is the intersection of `role_scope`, `variant_key`, and `binding_state`. A user who is not authorized to perform an activity does not see the adoption content for that activity. This is the same `role_scope_intersect` rule that Chapter 14 and Chapter 19 apply to every rendering decision.
- The lifecycle of a WalkMe asset is **Draft**, **In Review**, **Published**, **Deprecated**, and **Retired**. Each transition is cosigned by two distinct roles, each transition emits an `AUDIT_EVENT` to the data spine's hash-chained audit ledger, and only the **Published** state reaches a user's screen. The lifecycle is enforced at the binding, not negotiated by the author.
- AI-authored WalkMe content is created through the `SUGGESTION` and `VALIDATION_VERDICT` entities defined in Chapter 9. AI never writes directly to a binding. The commit gateway is the only path from a proposal to a published asset, and the gateway applies the same human approvals regardless of whether the proposer was an AI agent or a content author.
- The boundary between WalkMe and the systems of record is the absence of a wire. WalkMe does not write to SAP Signavio, SAP LeanIX, or the data spine outside the commit gateway. The architecture is explainable because those writes do not exist, not because they are policed.

7.5.2 Common Design Pitfalls

- Treating WalkMe as a process repository by copying step descriptions into WalkMe rather than referencing the `Process_ID`. This causes the two definitions drift. WalkMe assets reference the activity; they do not redefine it. The binding service refuses to publish an asset whose body asserts a step description that contradicts the SAP Signavio activity name.

- Scoping a Smart Walk-Thru to an SAP Fiori screen rather than to an `Activity_ID`. The asset will become stranded when the screen is replaced. Every asset binds to an `Activity_ID` first; `fiori_app_id` is a rendering attribute, not the binding's primary identifier.
- Publishing an asset without a `variant_key`. This causes the `variant_key` to appear to all users and creates confusion in regions or business units for which the guidance does not apply. Every binding carries an explicit `variant_key`, even when the key is the literal value of `ALL`; the absence of a variant key is treated as a defect, not a default.
- Giving an AI agent write credentials to WalkMe. This allows it to bypass the `SUGGESTION` discipline and publish content that has not been reviewed. AI agents should be issued credentials that only permit writes to the `SUGGESTION` table; the binding service should refuse to accept writes from any identity that is not a process owner or content manager.
- Having the WalkMe administrator decide when adoption events are purged. This causes retention to diverge from the regulatory horizon for the underlying activity. Retention metadata lives on the binding and is set by the process owner and enforced by the data spine. The WalkMe administrator inherits the policy.
- Treating a WalkMe Smart Walk-Thru as evidence that a control was performed, when in fact the user may have dismissed the overlay. Telemetry records what was rendered and what was acknowledged, but control evidence comes from the underlying transaction in the system of record, not from the WalkMe overlay.
- Allowing an author to edit a published asset in place to fix a typo without first establishing a draft cycle to audit. Every edit to a published asset creates a new draft, which advances through review and back to published; the prior published version is moved to deprecated. The audit trail is unbroken because the lifecycle is unbroken.

7.5.3 Review Questions

1. Explain in your own words why WalkMe is positioned as a target for AI agents rather than as a source. What would change in the architecture if WalkMe were treated as a source of activity definitions, and which of those changes would be undesirable?
2. Describe the relationship between a `walkme_asset_id`, a `Process_ID`, and a `factSheetId`. Which of these is the binding's primary identifier, which two are references, and what discipline does the architecture impose when one of the references is updated upstream?
3. Sketch the knowledge graph for a single L4 activity such as `A_POSITION_DRAFT`. Identify the nodes, the edges, and the cardinality of each edge. Explain how an AI agent would traverse the graph to identify candidate WalkMe assets to suggest for a related activity.

4. Trace the rendering decision for a Smart Walk-Thru bound to `A_POSITION_DRAFT`, `variant_key EMEA`, `role_scope HR_Manager_EMEA`, `binding_state` **Published**, for a user whose role scope is `HR_Manager_NAM`. Explain why the overlay does not render, and name the rule that produces this outcome.
5. Describe the lifecycle of a WalkMe asset from **Draft** to **Retired**. For each transition, name the two roles that co-sign, the `AUDIT_EVENT` field set that the transition emits, and a practical reason the transition might fail validation.
6. Walk through the path from an AI-authored proposal to a published WalkMe asset. Name every entity the proposal touches (`SUGGESTION`, `VALIDATION_VERDICT`, `INFERENCE_EVENT`, the commit gateway, `WALKME_ASSET`, and `WALKME_ASSET_BINDING`), and explain at which point in the path the proposal would be rejected if the cited `Process_ID` were deprecated.
7. Explain why telemetry retention is set on the binding rather than on the WalkMe asset itself, and identify a scenario in which the same asset body would be subject to different retention rules in two different bindings.
8. The chapter argues that the boundary between WalkMe and the systems of record is the absence of a direct connection. Identify three connections that organizations are tempted to build, and for each, explain what auditability or operational risk is introduced if it is built.

7.5.4 Student Companion Notes

If you are studying this chapter as part of a degree program or a structured certification track, you will find that WalkMe content is best learned by building it. A useful exercise is to take a single L4 activity, model a Smart Walk-Thru against it, and walk through the binding lifecycle from `DRAFT` to `PUBLISHED`, recording the `AUDIT_EVENT` entries that each transition would emit. It is best to read the WalkMe editor documentation in parallel with this chapter because the in-product terminology (segments, rules, and deployments) is more concrete than the abstract binding language we use. The mapping between the two vocabularies is covered in Appendix C and is worth looking through at least once.

You should pay particular attention to the three-way alignment outlined in Section 7.3. In production landscapes, the most common adoption defect is not a broken WalkMe asset but a WalkMe asset that contradicts the SAP Signavio activity or the SAP Build Work Zone group. The defect is invisible until a user complains, and the diagnosis is usually obscured by the fact that each of the three layers is owned by a different team. Building the binding discipline before content authoring begins is much less expensive than reconciling drift after a portfolio of assets is already in production. If you can take away only one principle from this chapter, it should be this one: Identifier convergence is the substitute for constant reconciliation, and the data spine is what makes convergence enforceable.

If you are preparing for the SAP Certified Application Associate exams in process intelligence or transformation management, you should know that the certification content does not yet cover the binding model described in this chapter because the binding model is a design pattern and not a product feature. The exams expect familiarity with WalkMe authoring, telemetry capture, and basic governance, which are the topics of Appendix C. The material in this chapter sits above the certification floor and is intended for architects, lead consultants, and transformation owners who must defend a design rather than execute against one.

Chapter 8
Using WalkMe Telemetry to Improve Process and Integration Outcomes

A modeled process tells you what should happen. A configured system tells you what can happen. Adoption telemetry tells you what did happen. Of the three, only adoption telemetry produces evidence that earns improvement; the first two are commitments waiting for the third to vindicate or refute them.

In Chapter 7, we positioned WalkMe as a target layer in the integrated landscape. We established that WalkMe content is bound to the stable identifiers that SAP Signavio and SAP LeanIX produce, that `WALKME_ASSET` and `WALKME_ASSET_BINDING` are canonical entities on the data spine, and that the lifecycle of a guidance asset is governed by the same separation of duties (SoD) discipline that governs process versions upstream. The chapter ended at the moment a WalkMe overlay renders on a user's SAP Fiori screen. It did not describe what happens after the user touches the screen.

This chapter focuses on that second half. Every step the user takes inside a WalkMe-guided interaction produces a signal. Most signals are quiet: A step completes on the first attempt, the dwell time sits comfortably inside the cohort median, and the user moves to the next task. Some signals are louder: The user opens a help search, retypes a field three times, abandons the activity mid-flow, or backs out of one application and into another. Every one of those signals is evidence, and that evidence becomes useful only when it is captured with provenance and bound to the same identifiers as the rest of the data spine. Adoption insight without provenance is at best a hunch and at worst a managerial liability.

Specifically, this chapter explains:

- How user interaction signals are captured at the WalkMe overlay, what an event payload contains, and how each event is tagged with the identity context required for downstream join: `user_id`, `role_scope`, `variant_key`, `Process_ID`, `Activity_ID`, `fiori_app_id`, and `walkme_asset_id`
- How the WalkMe analytics event stream is ingested into the SAP Business Technology Platform (SAP BTP) data spine through a five-stage pipeline (emit, stream, validate, resolve, and persist) with idempotent writes, a dead-letter queue, and an `AUDIT_EVENT` row per stage
- How three derived facts are computed from raw events (`FRICTION_INDICATOR`, `DEVIATION_OBSERVATION`, and `TRAINING_GAP`) and why each derivation is a query, not a judgment

- How adoption insights flow into two downstream destinations: the SAP Signavio process-improvement queue and the grounding bundles consumed by SAP AI Core for SUGGESTION generation
- Why none of these elements ever writes canonical state, and why the absence of a direct connection from WalkMe back to the data spine is the architectural commitment that lets the controller, the process owner, and the auditor read the same evidence with the same trust

We introduce two new canonical entities in this chapter, which also appear in Appendix D and the data spine tables in Chapter 9: TELEMETRY_EVENT, which is the append-only event record, and ADOPTION_FACT, which is the period-aggregated rollup. Three derived entities (FRICTION_INDICATOR, DEVIATION_OBSERVATION, and TRAINING_GAP) are materialized views over TELEMETRY_EVENT joined to the upstream identifiers. The knowledge graph we introduced in Chapter 7 acquires its first set of telemetry-derived edges, shown later in Figure 8.7.

The running scenario continues. A new position in SAP SuccessFactors Employee Central must be linked to a finance-governed cost center in SAP S/4HANA. The four activities (A_COST_CENTER_CREATE, A_COST_CENTER_PROPAGATE, A_POSITION_DRAFT, and A_POSITION_CC_ASSIGN, each guided by its own WalkMe overlay since Chapter 7) now appear with their observed signals attached. Section 8.4 closes by walking through the telemetry view of the same four activities and showing how the evidence produces a SUGGESTION that, after a VALIDATION_VERDICT of MODIFY, results in the commit gateway writing a new WalkMe binding. This scenario crosses the boundary the right way: Telemetry feeds a human-validated suggestion and the human-validated suggestion modifies the WalkMe binding. The binding never modifies the underlying process or architecture model.

From a financial planning and analysis (FP&A) and controlling perspective, the design problem we solve in this chapter has the shape of variance investigation. A controller reading a variance does not look at the variance alone; they look at the path that produced it. The variance is the evidence, and the path is the explanation. WalkMe telemetry is the tool that records the path. Without it, a process owner sitting in a monthly operating review is left explaining a number with anecdote. With it, the same review names the activity that drove the number, the role that walked the activity, the cohort that walked it that way, and the WalkMe asset that was (or should have been) in front of the user at the moment the variance was created. In this sense, this chapter gives you the controller's instrumentation for the integrated execution layer.

8.1 Capturing User Interaction Signals

Telemetry capture is the act of recording what a user did, with enough context for someone else to make sense of it later. The WalkMe overlay is the point at which this recording happens, and the WalkMe analytics event stream is the channel through which the recording reaches the data spine. This section specifies four signal channels, the identity tagging that every signal carries, the single governed gateway that admits signals into the data spine, and the rule that no signal is admitted without a binding to an upstream identifier already on the data spine. Figure 8.1 shows adoption telemetry capture, from the four signal channels emitted at the user's SAP Fiori screen, through the ingestion gateway, to the three data spine entities that record the event with full identity context.

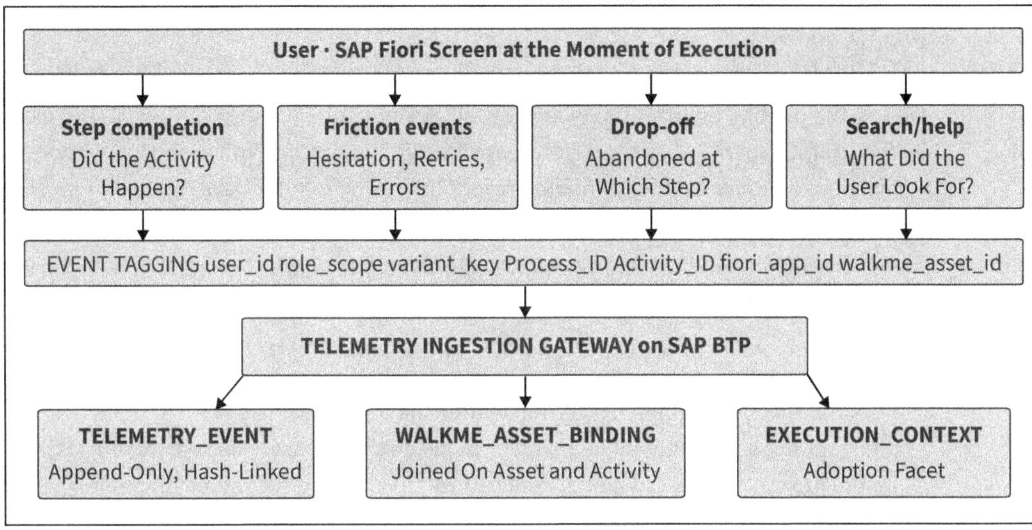

Figure 8.1: Adoption Telemetry Capture

8.1.1 The Four Signal Channels

Every user interaction signal that this chapter treats as canonical belongs to one of four channels. The channels are exhaustive: An event that does not belong to one of them is not admitted to the data spine. Restricting the catalog to four channels is what keeps the schema stable and what lets a controller read an indicator without rederiving the meaning of every event class. Let's look at each of the four in turn:

- **Step completion**
  A completion event fires when the activity referenced by a `WALKME_ASSET_BINDING` reaches a state the overlay recognizes as complete: a save, a submit, a confirm-and-continue, or a navigation away from the screen after the field state is valid. The event payload carries the `binding_id`, the `Activity_ID` that the binding resolves to at fire time, and the elapsed time from the first rendered step of the asset to the completion. Completion is a positive signal. The majority of events on a healthy adoption loop are completions, and the absence of completion is what the drop-off channel measures.
- **Friction events**
  A friction event fires when the overlay detects a pattern that, in the controller's view, looks like hesitation. The patterns are specified in Section 8.3.1 and include step replay (the user reenters the same field three or more times), long pause (the user dwells on a

single step beyond two standard deviations of the cohort median), help storm (five or more help searches in a single session), validation loop (the field-level validator rejects the input twice in succession), and lost in navigation (the user opens more than three tiles before landing on the activity). Each pattern fires its own event with its pattern_type in the payload. A single user interaction can fire several friction events, and that is expected; the friction stream is designed to over-record rather than under-record, and the aggregation that turns events into indicators is the place where the noise is filtered.

- **Drop-off**
  A drop-off event fires when an activity that was started is not completed within the window the binding defines. The window is an attribute of WALKME_ASSET_BINDING introduced in Chapter 7. For short-lived activities, it is measured in seconds, for approval activities in hours, and for periodic reconciliations in calendar days. The event payload records the last step the user reached before abandoning, which is what makes the drop-off stream actionable rather than merely diagnostic. A drop-off without a last-reached step would only tell you that something went wrong; the last-reached step tells you where to look.
- **Search and help**
  A search event fires when the user activates the in-overlay help search, types a search string, or opens an external help link from inside the overlay. The search string itself is recorded with the user identifier reduced to role_scope_id rather than user_id, so that individual search behavior cannot be queried but cohort-level demand for help on a given activity can. The reduction is a privacy-by-design choice that this book treats as nonnegotiable; Chapter 19 expands on the rule.

8.1.2 Identity Tagging at the Edge

Every event leaving the WalkMe overlay carries an identity envelope that the ingestion pipeline relies on for join, validation, and authorization. The envelope contains the following seven fields, none of which are optional or derived after the fact:

- user_id: The authenticated user identifier issued by the Identity Authentication tenant. This is the only place in the envelope where the user is individually identifiable, and Chapter 19 explains the access controls that surround it.
- role_scope_id: The role scope the user is acting under, resolved at session start from the Identity Authentication role profile. A user with multiple role scopes emits events under whichever scope was active when the overlay rendered.
- variant_key: The process variant key resolved from the user's organizational unit and the activity's variant axis. The same Activity_ID can have several variants; the variant_key is what tells the data spine which one the user actually executed.
- Process_ID: The stable SAP Signavio identifier of the L4 process the activity belongs to.
- Activity_ID: The stable SAP Signavio identifier of the L5 activity the WalkMe asset was bound to.

- fiori_app_id: The SAP Fiori application identifier the activity is executed in. For activities outside SAP Fiori, the field carries the canonical application reference resolved from the SAP LeanIX factSheetId.
- walkme_asset_id: The identifier of the specific WalkMe asset that rendered. Together with the activity and variant identifiers, this resolves to a unique WALKME_ASSET_BINDING row.

The envelope is constructed at the overlay, not at the gateway. That is what gives the event its provenance: The gateway can refuse an event whose envelope does not resolve to a published WALKME_ASSET_BINDING, but it cannot manufacture missing fields. An event with a missing field lands in the dead-letter queue, which Section 8.2.3 describes.

8.1.3 The Single Governed Gateway

Every event reaches the data spine through one gateway. The gateway is an SAP BTP, Cloud Foundry environment application running in the same SAP BTP subaccount as the data spine's persistence services. It exposes a single ingestion endpoint, validates the envelope against the current WALKME_ASSET_BINDING table, resolves the binding to its Process_ID and Activity_ID at the version state effective at the event_timestamp, and appends one row to TELEMETRY_EVENT. The gateway is the place where the contract between the WalkMe analytics event stream and the canonical data spine is enforced. The WalkMe tenant does not write to the data spine; the gateway does, on the WalkMe tenant's behalf, only when the envelope passes the contract.

The single-gateway pattern is the same one we used for SAP Signavio in Chapter 4 and SAP LeanIX in Chapter 6. This uniformity is deliberate. A controller asked how an adoption indicator was constructed should be able to trace the answer through the same shape of evidence chain, regardless of which upstream system the evidence originated in: gateway logs first, AUDIT_EVENT rows second, and canonical table rows third. The shape of the answer does not change because the originating system changes; the controller's audit reconstruction is the same query with different join keys.

What WalkMe Telemetry Is Not

WalkMe telemetry is not a productivity surveillance instrument. It does not measure how fast individual users complete activities, and it does not produce per-user performance scores. The cohort reduction at search-event capture and the privacy-preserving access controls described in Chapter 19 are what enforce this boundary. The instrument is calibrated to identify where guidance content needs more depth, where the published process version diverges from the executed path, and where the WalkMe asset bound to an activity is the wrong asset for the role that is actually using it. None of those questions is a question about an individual; all of them are about the design of the process and the guidance.

8.2 Linking Adoption Data to the SAP BTP Data Spine

Capturing a signal is necessary but not sufficient. The signal becomes useful only when it lands on the data spine joined to the identifiers that the rest of the architecture already uses. Section 8.1 specified what is captured at the overlay. This section specifies the data model on the data spine that receives the capture, the five-stage pipeline that gets the event from the WalkMe analytics event stream to that data model, and the discipline that keeps the pipeline idempotent, replayable, and free of gaps in the audit chain.

8.2.1 The Telemetry Data Model

The data spine acquires one new canonical entity for the event itself and four derived entities for the facts computed on top of it. Figure 8.2 illustrates them as an entity-relationship diagram, with the upstream identifiers on the top row, `TELEMETRY_EVENT` in the center, and the four derived facts on the bottom row. As you will see in Section 8.2.2 and Section 8.2.3, `TELEMETRY_EVENT` is the append-only canonical record. `FRICTION_INDICATOR`, `DEVIATION_OBSERVATION`, `TRAINING_GAP`, and `ADOPTION_FACT` are derived materialized facts. Every row in every table carries its `bundle_id` and `manifest_hash`. Figure 8.2 is the reference for the rest of the chapter and for Appendix D.

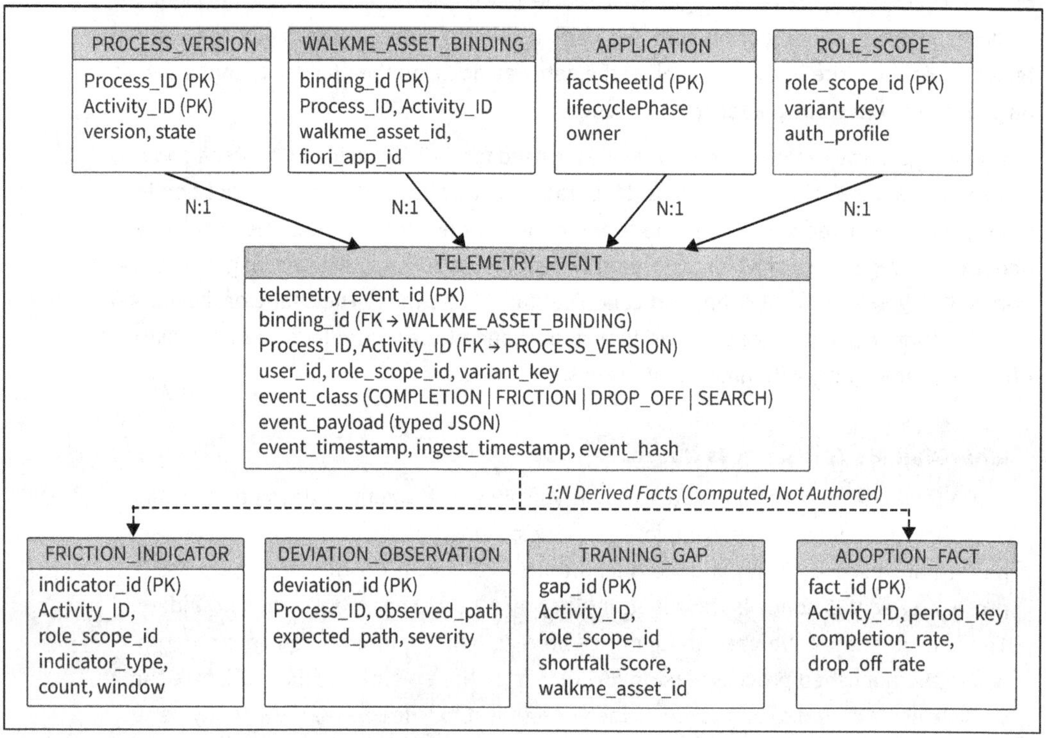

Figure 8.2: Telemetry Data Model

TELEMETRY_EVENT

TELEMETRY_EVENT is the append-only fact table that holds one row per signal admitted by the gateway. The primary key (PK) is telemetry_event_id, a universally unique identifier (UUID) minted at the gateway. The foreign keys (FKs) are binding_id (resolved at gateway time against WALKME_ASSET_BINDING), Process_ID and Activity_ID (resolved against the PROCESS_VERSION effective at event_timestamp), and role_scope_id (resolved against the user's session). The non-key attributes are user_id (subject to the role-scoped access controls in Chapter 19), variant_key, event_class with the enumerated values COMPLETION, FRICTION, DROP_OFF, and SEARCH, event_payload as typed JSON, event_timestamp from the client clock, ingest_timestamp from the gateway clock, and event_hash computed as a SHA-256 of the prior event_hash concatenated with the canonical serialization of the current row. The event_hash is what chains the table into an audit trail; a controller verifying the chain recomputes the hashes and compares them against the stored values.

Two design choices are especially important. First, TELEMETRY_EVENT carries both an event_timestamp and an ingest_timestamp because the two are demonstrably different. The client clock can be delayed by network latency, paused by a screen lock, or skewed by a misconfigured workstation. A query that needs to order events as the user experienced them uses event_timestamp; a query that needs to reconstruct the gateway's view of arrival uses ingest_timestamp. Second, the event_payload is typed JSON, not free-form JSON. The schema for each event_class is registered on the data spine and versioned; the gateway refuses a payload whose shape does not match the registered schema for the event_class, event_class_version pair. The schema discipline is what prevents the telemetry table from becoming a heap that nobody can query, which is the second of the two ways adoption instrumentation usually fails in practice.

FRICTION_INDICATOR, DEVIATION_OBSERVATION, TRAINING_GAP

The three derived entities are governed materialized views. Each one is the output of a query whose definition is registered in the data spine alongside its result table. A controller asked how the indicator was computed reads the registered query; another team can re-derive the indicator by running the same query against the same TELEMETRY_EVENT rows joined to the same PROCESS_VERSION snapshot. Section 8.3 will specify the queries; this section specifies the shape of the result rows.

FRICTION_INDICATOR rows carry an indicator_id, the Activity_ID and role_scope_id the indicator was raised against, the indicator_type (drawn from the five friction patterns we will specify in Section 8.3), the count of events that contributed to the indicator, the time window over which the events were counted, a severity_score in the range 0 to 1, and the bundle_id of the grounding bundle that produced the indicator. DEVIATION_OBSERVATION rows carry a deviation_id, the Process_ID and observed_path (an ordered list of Activity_IDs), the expected_path drawn from the relevant PROCESS_VERSION, a severity classification (LOW, MEDIUM, or HIGH), the cohort size from which the deviation was observed, and the same bundle_id and manifest_hash that govern every spine row produced by an inference. TRAINING_GAP rows carry a gap_id, the Activity_ID and role_scope_id the gap was raised against, a shortfall_score in the range 0 to 1, the walkme_asset_id of the asset whose binding is the proximate candidate for revision, and the bundle_id of the grounding bundle that closed over the row.

ADOPTION_FACT

ADOPTION_FACT is the period-aggregated rollup. One row exists per Activity_ID, period_key, where period_key is the calendar-bound aggregation key the data spine assigns at period close. This is typically the International Organization for Standardization (ISO) week, the calendar month, or the fiscal month, depending on the consumer. The row holds the completion_rate, the drop_off_rate, the median dwell time, the count of friction events normalized by completions, and a small number of derived comparison fields. ADOPTION_FACT is the table that an FP&A or process-owner dashboard reads when the question is "How is adoption trending?" rather than "Where did this specific number come from?"

Worked Example: Cost Center to Position Event Shapes

On a Wednesday afternoon, a human resources (HR) business partner in Europe, the Middle East, and Africa (EMEA) opens A_POSITION_DRAFT in SAP SuccessFactors Employee Central. The WalkMe overlay binds the activity to walkme_asset_id wma-pos-draft-002 and renders the position-template selector. The HR business partner types a position title, opens the help search twice (search_terms recorded at cohort granularity), retypes the cost center reference field three times (the overlay records a friction event of type step replay), backs out of the screen, opens A_COST_CENTER_PROPAGATE in SAP S/4HANA Finance, then returns and completes A_POSITION_DRAFT ninety seconds later. Six TELEMETRY_EVENT rows are written: one COMPLETION, one DROP_OFF (the first abandonment of A_POSITION_DRAFT), two SEARCH events, one FRICTION event, and one second COMPLETION when the HR business partner finally submits. Every row carries the same binding_id, Process_ID, Activity_ID, role_scope_id (HR_Business_Partner), and variant_key (EMEA_standard) but a different event_timestamp and event_hash. Two derived rows follow at the next aggregation window: one FRICTION_INDICATOR for step replay on A_POSITION_DRAFT in the HR_Business_Partner cohort and one DEVIATION_OBSERVATION for the backward jump into A_COST_CENTER_PROPAGATE in cohort EMEA.

8.2.2 The Five-Stage Ingestion Pipeline

The pipeline that moves an event from the WalkMe analytics event stream to TELEMETRY_EVENT has five stages, and each stage has exactly one responsibility. The single-responsibility rule is what makes the pipeline replayable: Replay restarts from the stage at which the failure was recorded, with the prior stages' AUDIT_EVENT rows already in place. Figure 8.3 shows the pipeline end to end. Each stage appends one AUDIT_EVENT row; failed events are quarantined in the dead-letter queue, never silently dropped.

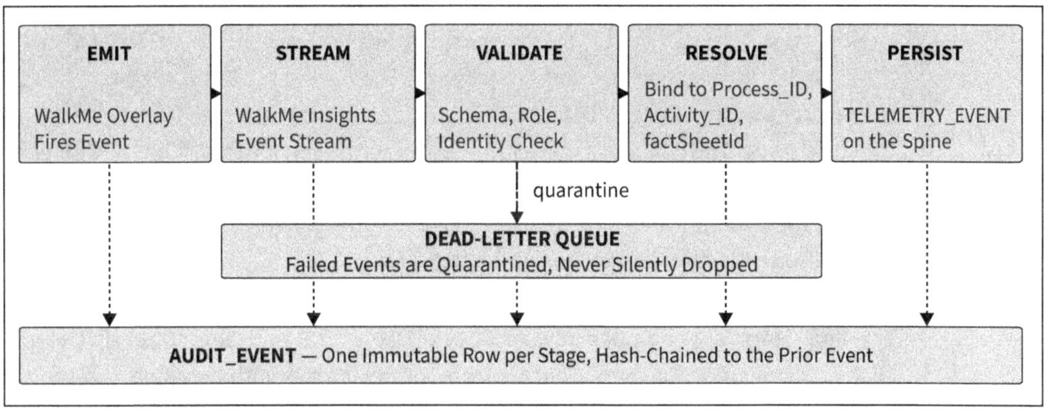

Figure 8.3: The Five-Stage Ingestion Pipeline

Let's look at each stage in order:

1. **Emit**
   The WalkMe overlay constructs the event envelope and fires the event into the WalkMe analytics event stream. Construction at the overlay is what gives the event its provenance; the gateway cannot manufacture envelope fields after the fact. An envelope with a missing field never leaves the overlay. The overlay drops it and writes a client-side diagnostic that the WalkMe tenant administrator can review. The first AUDIT_EVENT row of the pipeline is written here, against the actor identity of the WalkMe tenant service account.

2. **Stream**
   The WalkMe analytics event stream delivers the event to the SAP BTP gateway's stream consumer. The consumer is an Event Mesh subscription configured for at-least-once delivery, which means duplicates are possible. Duplicates are handled at the next stage by the gateway's idempotency check, which uses `telemetry_event_id` as the deduplication key. A duplicate produces a second AUDIT_EVENT row instead of a canonical row so that the controller's audit reconstruction can show exactly how many delivery attempts an event experienced.

3. **Validate**
   The gateway validates the envelope against three contracts in sequence: the `event_payload` against the registered schema for `event_class`, `event_class_version`; the

envelope fields against the current WALKME_ASSET_BINDING and PROCESS_VERSION tables for existence and lifecycle phase; and the user's identity against the Identity Authentication session that the envelope claims. A validation failure at any contract sends the event to the dead-letter queue with a typed failure code. The validation stage does not write TELEMETRY_EVENT; it writes only an AUDIT_EVENT row and, on failure, a dead-letter row.

4. **Resolve**
   Resolution turns the envelope into a fully canonical row. The binding_id is looked up by walkme_asset_id, Activity_ID, variant_key against the PUBLISHED row of WALKME_ASSET_BINDING effective at event_timestamp. The Process_ID and Activity_ID are resolved against the PUBLISHED PROCESS_VERSION effective at event_timestamp, not the current one. The difference matters for replay across version transitions and for accurate reconstruction of past adoption events. The application's factSheetId is resolved through the binding. The role_scope_id is resolved against ROLE_SCOPE. A resolution failure also sends the event to the dead-letter queue.

5. **Persist**
   The final stage appends one row to TELEMETRY_EVENT and one AUDIT_EVENT row chained to the prior. The persistence is idempotent: If a row with the same telemetry_event_id already exists, the stage writes no second canonical row and emits an AUDIT_EVENT with the disposition DUPLICATE_SUPPRESSED. The table is partitioned by ingest_timestamp at calendar-month granularity for query economy at month-end-close time. The indexes that matter for the indicator queries we use in Section 8.3 on Activity_ID, role_scope_id, and Process_ID, event_timestamp, and binding_id are created here.

8.2.3 The Dead-Letter Queue and Replay

Events that fail validation or resolution land in a dead-letter queue rather than being silently dropped. The queue is itself a data spine table with the typed failure code, the original envelope, and the gateway timestamps; the table is read by an operations console that lets a steward triage failed events into one of three resolutions. If the event was malformed at source, the row is closed and the WalkMe asset is flagged for content review. If the event was malformed because an upstream version transition had not yet propagated, the row is retried after the next ingestion of WALKME_ASSET_BINDING or PROCESS_VERSION. If the event was malformed because a controlled change deliberately invalidated the prior binding, the row is closed with a reference to the AUDIT_EVENT row that records the change.

This triage discipline is what keeps the dead-letter queue from growing into the same heap-with-no-query that we warned against in earlier chapters. A dead-letter queue without a triage discipline is a heap with a slightly more dignified name; a dead-letter queue with one is a control surface.

8.3 Detecting Friction, Deviations, and Training Gaps

Once TELEMETRY_EVENT carries the captured signal, the question becomes what to make of it. This section specifies the three detection queries that turn raw events into governed indicators. Each query is registered in the data spine, versioned, and run on a schedule that produces FRICTION_INDICATOR, DEVIATION_OBSERVATION, and TRAINING_GAP rows. The detection logic is deliberately transparent: A controller asked what a friction indicator of type step replay actually means reads the registered query and re-derives the indicator. The indicator is a row, not a feeling.

8.3.1 Friction Detection

Friction is the gap between the path the user intended and the path the screen permitted. Five patterns are recognized as canonical friction. Together, they cover the friction surfaces you have seen across multiple integrated landscapes throughout this book. Figure 8.4 shows the five patterns each flowing into a single FRICTION_INDICATOR row, with the indicator's downstream consumers labeled. Each indicator is consumed by the SAP Signavio improvement queue, the AI grounding bundles, and the training content planner.

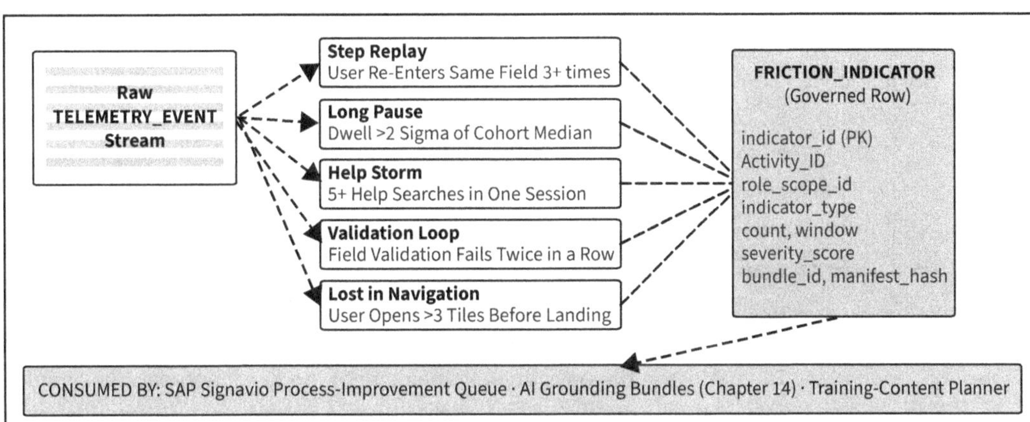

Figure 8.4: Friction Detection

The five patterns are as follows:

- **Step replay**
  Step replay activates when a user enters the same field three or more times within a single binding session. The query joins TELEMETRY_EVENT to WALKME_ASSET_BINDING and counts FRICTION events with pattern_type = step_replay grouped by Activity_ID, role_scope_id, binding_session_id. A count of three or more in a session over a fourteen-day rolling window, in at least fifteen percent of the cohort, raises one FRICTION_INDICATOR row of type STEP_REPLAY. The threshold is a parameter on the registered query, not a constant hidden in the code; changing the threshold creates a new version of the query and a new run.

- **Long pause**
  Long pause activates when a user dwells on a single step for longer than two standard deviations above the cohort median. The cohort is defined as users with the same `role_scope_id` executing the same `Activity_ID` and `variant_key` within the trailing thirty days. The query only computes the median and standard deviation from completed events. `DROP_OFF` events are excluded from the dwell-time baseline because they would skew the cohort median upward and mask the very dwell anomalies the indicator is trying to identify.
- **Help storm**
  Help storm activates when a user issues five or more `SEARCH` events within a single session. The five-search threshold is a heuristic the operations team has found, in practice, to separate exploratory searching from genuine confusion. The threshold is configurable per `WALKME_ASSET_BINDING` because some activities (audit reviewers reading control narratives, for example) naturally have higher search rates than transactional activities. The indicator raised is `FRICTION_INDICATOR` with `indicator_type = HELP_STORM` and a `severity_score` derived from the search event count divided by the cohort baseline.
- **Validation loop**
  Validation loop activates when the same field validator rejects the user's input twice in succession within ninety seconds. The ninety-second window is what distinguishes a validation loop (the user is confused about what the field expects) from a user correcting an obvious typo. Validation loop is the most actionable of the five patterns because the field that rejected the input is named in the `event_payload`. The indicator row carries the field reference, and the SAP Signavio improvement queue receives a remediation candidate that names the field, the role scope, and the activity.
- **Lost in navigation**
  Lost in navigation activates when a user opens more than three tiles in SAP Build Work Zone or the SAP Fiori launchpad before landing on the activity the WalkMe asset was bound to. The pattern is the navigational analogue of the dwell anomaly: It identifies users who knew which activity they needed but could not find it. The remediation is rarely a change to the WalkMe asset itself but a change to the SAP Build Work Zone tile configuration, which the indicator routes to the work-zone owner rather than the asset owner.

8.3.2 Deviation Detection

Deviation is the gap between the path the `PROCESS_VERSION` publishes and the one the cohort actually walked. The query is structurally different from the friction queries. It does not look for a pattern within a single user's session; it reconstructs the sequence of activities a cohort traversed and compares the sequence to the published path. Figure 8.5 illustrates this for the running scenario. The governed path is reconstructed from `PROCESS_VERSION` pv-ccpos-2.1.0; the observed path is reconstructed from the EMEA controller cohorts' TELEMETRY_EVENT rows. A backward jump from position draft to cost center propagate raises a DEVIATION_OBSERVATION.

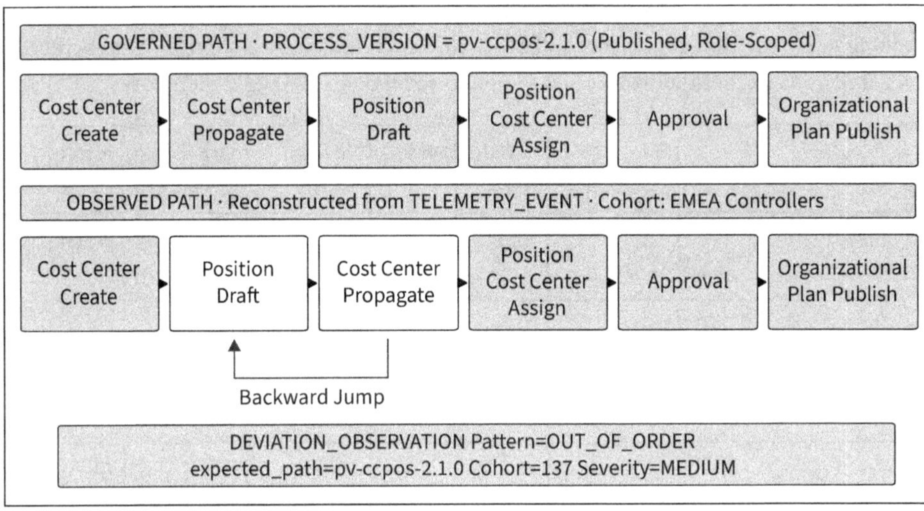

Figure 8.5: Deviation Detection

Three deviation patterns are canonical. OUT_OF_ORDER activates when the observed sequence visits activities in an order that the PROCESS_VERSION graph does not permit; the running scenario shows a backward jump from A_POSITION_DRAFT to A_COST_CENTER_PROPAGATE that is structurally a backward edge in the published process and is therefore out of order by definition. SKIPPED_STEP activates when the observed sequence omits an activity the PROCESS_VERSION requires for the variant the user walked; this is the deviation pattern that controllers typically care about most because skipped steps are where controls are bypassed. UNEXPECTED_INCLUSION activates when the observed sequence visits an activity that does not belong to the variant the user is supposed to be walking; this is the pattern that surfaces variant misassignment, which is a governance issue rather than a user issue.

Severity is assigned by a small table registered alongside the deviation query. OUT_OF_ORDER deviations default to MEDIUM severity. SKIPPED_STEP deviations default to HIGH severity if the skipped activity is flagged as a control step in the SAP Signavio L5 model and to MEDIUM otherwise. UNEXPECTED_INCLUSION defaults to LOW severity except when the included activity has approvedForAI = false in SAP LeanIX, in which case the severity is raised to HIGH. The defaults are overridable per PROCESS_VERSION; the override is itself a versioned, reviewable row, not a configuration entry buried in a script.

8.3.3 Training Gap Detection

A training gap is a structural shortfall between what a role scope is expected to accomplish on an activity and what its cohort actually accomplishes. The shortfall is computed from the same TELEMETRY_EVENT table that feeds friction and deviation detection, but the aggregation is different. Training gaps are derived across role scopes, not within a session or a cohort path. Table 8.1 shows the matrix view that the operations team uses to scan for training gaps. Ok indicates a shortfall of <0.20, watch indicates a shortfall between 0.20–0.45, and gap indicates shortfall of >0.45 results in an open TRAINING_GAP row.

	FIN_ Controller	HR_Business_ Partner	Position_ Owner	Org_ Designer	Audit_ Reviewer
A_COST_CENTER_ CREATE	Ok	Gap	Ok	Ok	Watch
A_COST_CENTER_ PROPAGATE	Watch	Watch	Ok	Ok	Ok
A_POSITION_DRAFT	Ok	Ok	Gap	Ok	Watch
A_POSITION_CC_ ASSIGN	Ok	Watch	Watch	Ok	Ok
A_ORG_PLAN_PUBLISH	Ok	Ok	Ok	Gap	Ok
A_VARIANT_REVIEW	Ok	Watch	Ok	Ok	Gap

Table 8.1: Training-Gap Heat Map

The shortfall_score is a number in the range 0 to 1, computed as a weighted combination of three normalized inputs: the completion_rate of the cohort divided by the global completion_rate for the activity, the average count of friction events per completion (higher is worse), and the average drop_off_rate (higher is worse). The weights are an attribute of the registered query, defaulting to equal weights, and are overridable per Activity_ID, role_scope_id where empirical experience suggests one input is more predictive than the others. A shortfall_score above 0.45 raises one TRAINING_GAP row. A score in the range 0.20 to 0.45 raises a watch state that is visible to the operations team but does not flow downstream until the next aggregation cycle confirms the trend. A score below 0.20 is recorded but does not raise an indicator.

Training gaps name the walkme_asset_id that is the proximate candidate for revision, but they do not propose the revision. Proposing the revision is the work of the AI surface, as we covered in Chapter 15, with the gap indicator carried into the grounding bundle as evidence. The separation matters. The gap is the observation, the proposal is the inference, the verdict is the decision, and the commit is the gateway. Each is a separate row in a separate table, and the four-step shape is what gives the controller a four-row audit trail to read.

Worked Example: Cost Center to Position Detection Outputs

The same EMEA cohort that produced the six TELEMETRY_EVENT rows in Section 8.2 produces, at the next aggregation window, three derived rows. A FRICTION_INDICATOR of type STEP_REPLAY is raised against A_POSITION_DRAFT, HR_Business_Partner with a count of 12 over the trailing week and a severity_score of 0.34; the indicator names cost_center_reference as the field driving the step replay. A DEVIATION_OBSERVATION is raised with the pattern = OUT_OF_ORDER, expected_path drawn from pv-ccpos-2.1.0, observed_path showing the backward jump, severity = MEDIUM, and cohort_size = 137. A TRAINING_GAP is raised against A_POSITION_CC_ASSIGN, HR_Business_Partner with shortfall_score = 0.51, naming wma-pos-ccassign-001 as the candidate WalkMe asset for revision. None of the three rows writes anything to SAP Signavio, SAP LeanIX, the WalkMe tenant, or any transaction system. They write only to themselves and to the AUDIT_EVENT chain. What happens next is the subject of Section 8.4.

8.4 Feeding Adoption Insights into Downstream AI Services

The derived facts we covered in Section 8.3 are only as valuable as the places they feed. This section specifies those places, and there are exactly two: the SAP Signavio process-improvement queue and the grounding bundles that SAP AI Core consumes for SUGGESTION generation. Both destinations consume the same rows; neither has a write path back to the data spine that bypasses the commit gateway or the VALIDATION_VERDICT discipline we established in Chapter 15. This section closes with a knowledge graph view of the chapter as a whole and a worked example from our running scenario that ties the whole loop together.

8.4.1 WalkMe as a Target for AI Agents

Chapter 7 stated that WalkMe is a target for AI agents, not a source. This chapter is where that statement becomes mechanically concrete. The AI surface reads adoption indicators from the data spine through a grounding bundle; it does not write anything to the WalkMe tenant directly. Every change to a WALKME_ASSET or WALKME_ASSET_BINDING that is initiated by an AI inference flows through the same path as any other AI-influenced change to a canonical entity: an INFERENCE_EVENT row, a SUGGESTION row with target_entity = WALKME_ASSET_BINDING and pattern_family = harmonization or control_gap, a HUMAN_VALIDATION_QUEUE entry, a VALIDATION_VERDICT row with verdict_value = ACCEPT or MODIFY, and a commit gateway invocation that performs the write to the WalkMe tenant on behalf of the verdict. This asymmetry—telemetry flows from WalkMe into the data spine but writes to WalkMe flow only through the commit gateway—is an example of the boundary being the absence of a direct connection. The boundary is real because there is no second connection to break it.

8.4.2 The Two Downstream Destinations

Figure 8.6 shows closed-loop feedback. Adoption facts flow into governance feedback (SAP Signavio improvement queue) and AI grounding feedback (grounding bundles); both converge at the commit gateway and write one AUDIT_EVENT row per decision.

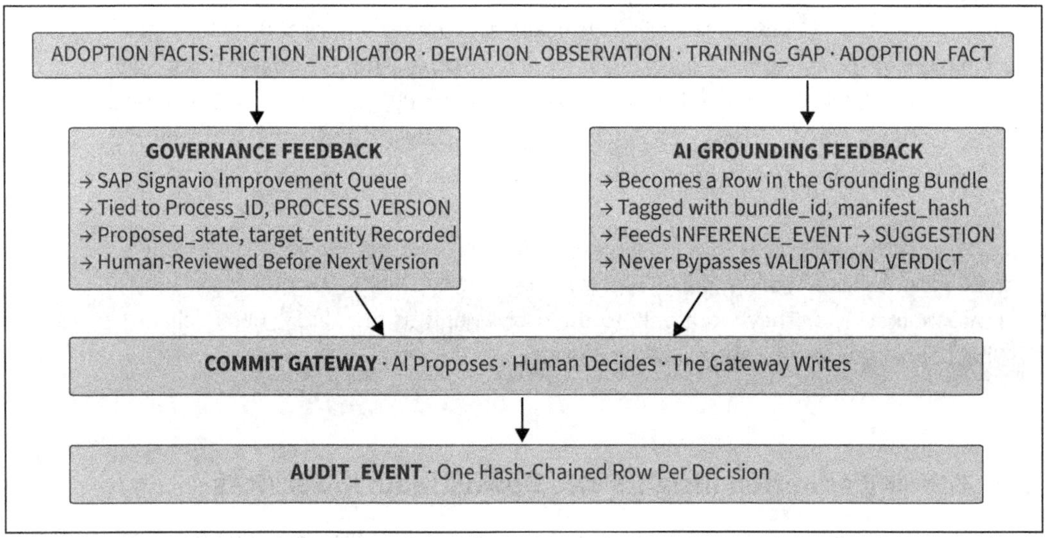

Figure 8.6: Closed-Loop Feedback

The SAP Signavio process improvement queue is the slower, more deliberate of the two destinations. A DEVIATION_OBSERVATION of severity HIGH on a control step does not produce an immediate process change; it produces a candidate improvement that the process owner reviews, accepts or refines, and folds into the next PROCESS_VERSION. The queue carries the same evidence the AI grounding bundle carries (the indicator row, its bundle_id, and the AUDIT_EVENT chain back to the originating TELEMETRY_EVENT rows) but the consumption surface is SAP Signavio Process Manager, not the SAP AI Core inference path. The two paths are deliberately separate because the lifecycle is deliberately different. Process version changes occur on a quarterly cadence in most enterprises; AI-proposed WalkMe binding revisions can occur weekly. The same evidence informs both, but the cadences diverge.

Governance feedback is what closes the loop back to the PROCESS_VERSION that we have been treating as published truth. In this architecture, a controller who reads a deviation observation and wonders whether the process model is wrong has a specific destination for that question: It goes into the improvement queue as a structured candidate. The candidate carries the deviation_id, the Process_ID, the Activity_IDs involved, the cohort statistics, and the proposed alteration written in a form the process owner can evaluate. Whether the candidate is accepted is a governance decision; the system presents the evidence, and the human chooses the response.

The grounding bundle path is the faster of the two. Each indicator row produced in Section 8.3 becomes a candidate input to the next grounding bundle assembled by the AI preparation layer of Chapter 14. The bundle carries the indicator row alongside the canonical context (the relevant PROCESS_VERSION rows, the WALKME_ASSET_BINDING rows, the APPLICATION rows from SAP LeanIX, and the role_scope and variant_key context) and presents the bundle to SAP AI Core for inference. The model emits one or more SUGGESTION rows, each with a target_entity, proposed_state, pattern_family, risk_class, and confidence-derived prioritization. The SUGGESTION enters the HUMAN_VALIDATION_QUEUE we introduced in Chapter 15, where a human validator records a VALIDATION_VERDICT with verdict_value drawn from ACCEPT, MODIFY, or REJECT. Only on ACCEPT or MODIFY does the commit gateway invoke a write.

The pattern_family that adoption indicators most commonly produce is harmonization, where the indicator names a place where two cohorts walking the same activity show divergent paths and the suggestion proposes a WalkMe asset revision that would close the divergence. Control_gap is the second-most common, where a SKIPPED_STEP deviation against a control activity raises a suggestion to strengthen the WalkMe asset's enforcement of the step, which in this architecture means adding a required acknowledgment or a guided override narrative rather than blocking the user. Blocking is the process owner's decision via SAP Signavio governance; strengthening guidance is the WalkMe surface's decision via the suggestion path.

8.4.3 Adoption Signals in the Knowledge Graph

Chapter 7 introduced the knowledge graph that organizes the WalkMe domain into typed nodes and typed edges. The graph only had structural edges: process decomposes_to activity, activity guided_by WalkMe asset, asset renders_in SAP Fiori app, and SAP Fiori app delivered_by application. In this chapter, we add adoption edges, which are edges that did not exist before telemetry started flowing and carry telemetry-derived evidence. Figure 8.7 shows the augmented graph. Solid edges are structural (as described in Chapter 5–Chapter 7); dashed edges are adoption-derived. Every adoption edge carries its bundle_id, and the graph never speaks without provenance.

Adoption edges never introduce new nodes. The graph is a closed vocabulary, and telemetry can only add edges between nodes that already exist. There are five canonical adoption edges: a TELEMETRY_EVENT node observed_at a WalkMe asset, a TELEMETRY_EVENT node exercised against an activity, a cohort executed_by an activity, a role scoped_for an asset, and an event aggregated_from a cohort. Each edge carries the bundle_id of the inference that materialized it, allowing the AI surface that traverses the graph during grounding to trace any edge back to its evidence. The graph is what gives the model a structured, reproducible context to reason against, instead of a flat retrieval over loose document fragments. This discipline is what separates AI grounding in this architecture from the retrieval-augmented patterns that treat enterprise content as a heap of pages.

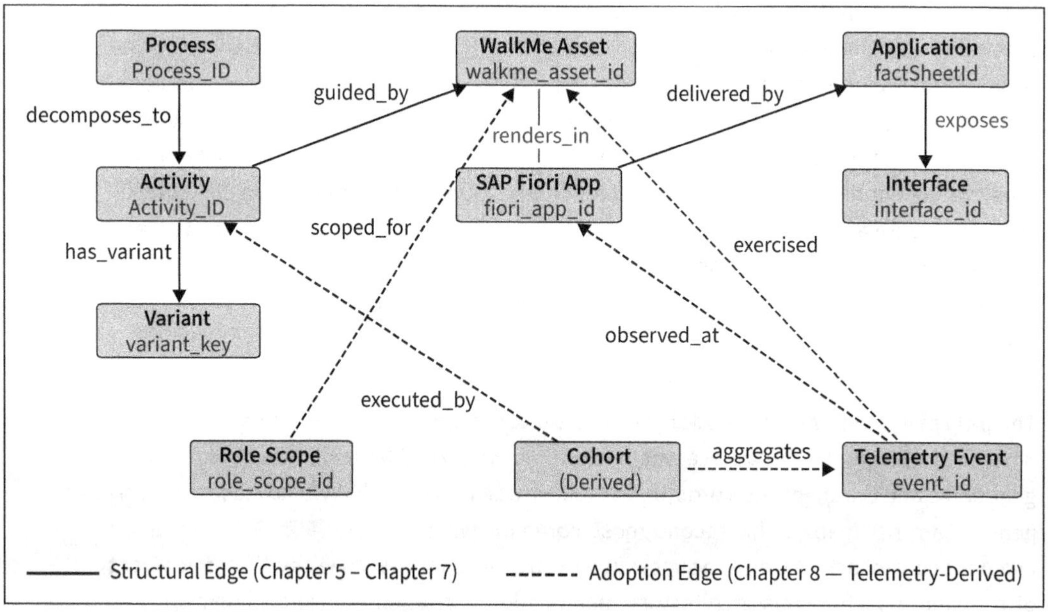

Figure 8.7: The Knowledge Graph with Adoption Edges

8.4.4 The End-to-End Loop

Figure 8.8 places the running scenario in a single view so that you can read the loop end to end. The four activities are the same ones we bound to four WalkMe assets in Chapter 7.

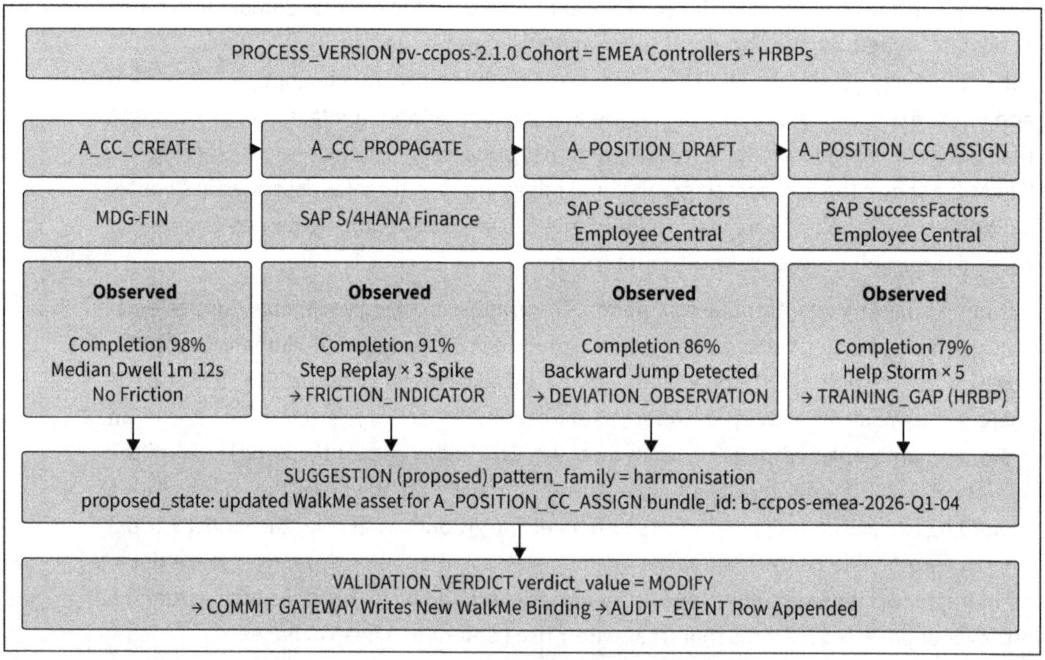

Figure 8.8: Cost Center to Position: The Telemetry View of the Running Scenario

The observed row underneath each activity shows the telemetry-derived state: A_CC_CREATE shows 98 percent completion with no friction, A_CC_PROPAGATE shows 91 percent completion with a step replay friction indicator raised, A_POSITION_DRAFT shows 86 percent completion with a deviation observation raised against the backward jump, and A_POSITION_CC_ASSIGN shows 79 percent completion with a training gap raised against the HR_Business_Partner cohort.

The SUGGESTION row at the bottom of the figure shows the pattern_family = harmonization suggestion that the AI surface produces from the four indicators: a revised WalkMe asset for A_POSITION_COST_CENTER_ASSIGN that surfaces the cost center reference earlier in the flow, reduces the step replay incidence by removing the round-trip back to A_COST_CENTER_PROPAGATE, and includes a deeper explanation of the cost center to position mapping for the HR business partner cohort. The VALIDATION_VERDICT row records verdict_value = MODIFY, with the modification narrowing the scope of the revised asset to the EMEA variant only. The commit gateway then writes the modified WALKME_ASSET_BINDING and appends one AUDIT_EVENT row that chains back through the SUGGESTION, the INFERENCE_EVENT, the grounding bundle, and the originating indicators to the TELEMETRY_EVENT rows.

The audit reconstruction the controller can perform from any row in this chain is a single query per hop: from AUDIT_EVENT to VALIDATION_VERDICTto SUGGESTION to INFERENCE_EVENTto grounding bundle to indicator to TELEMETRY_EVENT. Six hops, six rows, one walk. This is the controller's one-query audit reconstruction that we have been promising since Chapter 1, and this is where the telemetry side of the reconstruction acquires its specific rows.

Figure 8.9 illustrates the loop's six stages. The circle at the center of the diagram is our architectural refrain: The boundary is the absence of a direct connection.

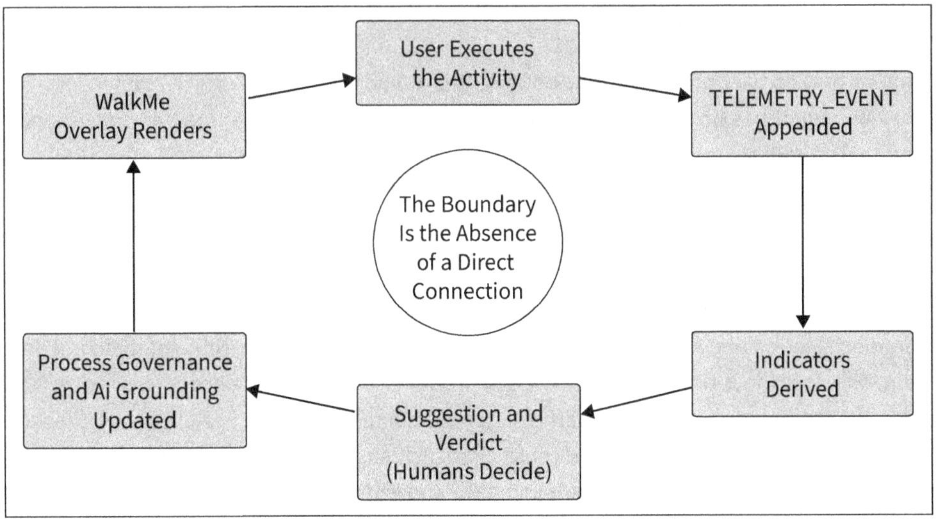

Figure 8.9: The Adoption Loop, End to End

The loop closes through governance and AI, not through a hidden write from the WalkMe tenant to the canonical data spine. If a future operator looks at the architecture and asks "Why is there not a faster path from telemetry to a fix?" the answer is the center of the circle: The faster path exists, and it passes through a human at the verdict stage. That is the feature, not the friction. A change that bypasses the verdict is one that bypasses the controller, and this architecture does not make that exchange.

8.5 Summary

This chapter carried the WalkMe story from positioning to instrumentation. Chapter 7 established that WalkMe is the target layer for upstream truth and the target layer for AI agents; in this chapter, we showed how the target layer produces evidence in return. Four signal channels emit at the overlay. The events flow through a single governed gateway into one append-only spine table. Four derived facts are computed from the events by registered, versioned queries, and the facts feed only two downstream destinations: the SAP Signavio improvement queue and the AI grounding bundles. None of the steps writes canonical state without passing through the commit gateway, and the commit gateway requires a `VALIDATION_VERDICT` row with `verdict_value` of `ACCEPT` or `MODIFY` before it will write. The shape of that loop is the same shape we specified for every AI-influenced change in the architecture in Chapter 15. What we added in this chapter is the evidence end of the loop.

Three convictions carry forward into the chapters that follow. The first is that adoption telemetry is a row, not a feeling. Every indicator is the output of a registered query, and another reviewer can re-derive it. A controller who does not trust an indicator runs the query; a process owner who disagrees with an indicator's severity proposes a new weighting through a versioned override. The instrument is transparent, reproducible, and contestable.

The second conviction is that the boundary is the absence of a direct connection. The WalkMe tenant has no write path back to the data spine; the data spine has no write path back to the WalkMe tenant except through the commit gateway under a `VALIDATION_VERDICT`. The asymmetry is what lets the controller, the process owner, and the auditor read the same evidence with the same trust. Every chapter that follows assumes this asymmetry and would fail at the architectural level if it were relaxed.

The third conviction is that telemetry produces evidence, not verdicts. The friction indicator, the deviation observation, and the training gap are all evidence. None of them is a decision. The decision is the `SUGGESTION` the AI produces from the evidence, the `VALIDATION_VERDICT` the human writes against the suggestion, and the `AUDIT_EVENT` row that records both. Telemetry does not decide what to change; it supplies the rows that humans and AI use to decide what to change next. If you have an FP&A background, you should recognize this distinction from monthly operating reviews: A variance is evidence, an explanation is a hypothesis, an accepted explanation is a decision, and the decision goes into next month's plan as a commitment. Adoption telemetry does the same work for the integrated execution layer.

Chapter 9 returns to the canonical data spine and the entities this chapter and previous chapters have introduced. The data spine is now responsible for governing process semantics from SAP Signavio, architecture context from SAP LeanIX, and adoption evidence from WalkMe. Chapter 10 through Chapter 12 specifies how the data spine is implemented on SAP BTP, how data is ingested into it, and how execution services consume it. Chapter 13 through Chapter 15 specifies how AI consumes the data spine and how the human-in-the-loop discipline closes the loop. Chapter 16 and Chapter 17 specify how SAP Build consumes the data spine for workflow and user experience. The TELEMETRY_EVENT, FRICTION_INDICATOR, DEVIATION_OBSERVATION, TRAINING_GAP, and ADOPTION_FACT entities introduced in this chapter appear in each of the later chapters as a consumed input, never as a side-channel write.

8.5.1 Key Learning Points

- Adoption telemetry is captured at the WalkMe overlay and tagged at the edge with seven identity fields: user_id, role_scope_id, variant_key, Process_ID, Activity_ID, fiori_app_id, and walkme_asset_id. An event without a complete envelope never reaches the data spine.
- Four signal channels are canonical and exhaustive: step completion, friction events, drop-off, and search-and-help. An event that does not belong to one of the four is not admitted to the data spine.
- Two new spine entities were introduced here and will be expanded upon in Appendix D: TELEMETRY_EVENT (append-only) and ADOPTION_FACT (period-aggregated). Three derived materialized views (FRICTION_INDICATOR, DEVIATION_OBSERVATION,and TRAINING_GAP) carry their bundle_id and manifest_hash for provenance.
- The five-stage ingestion pipeline (emit, stream, validate, resolve, and persist) appends one AUDIT_EVENT row per stage. Failed events land in a dead-letter queue with a typed failure code; they are never silently dropped.
- Derived facts feed exactly two downstream destinations: the SAP Signavio process-improvement queue (slower, with governance cadence) and the grounding bundles consumed by SAP AI Core (faster, with weekly cadence). Both converge at the commit gateway.
- The boundary is the absence of a direct connection. WalkMe has no write path back to the data spine; the data spine writes to the WalkMe tenant only through the commit gateway under a VALIDATION_VERDICT of ACCEPT or MODIFY. This asymmetry is the architectural commitment.

8.5.2 Common Design Pitfalls

- Letting WalkMe write back to the data spine. Once a WalkMe tenant is granted a write path, the integrity discipline of the commit gateway collapses. Every other chapter in this book assumes the asymmetry; relaxing it breaks the controller's audit reconstruction at the WalkMe boundary.
- Capturing free-form event payloads. A TELEMETRY_EVENT table whose event_payload is untyped becomes a heap with no queryable structure. The schema discipline at the gateway is what keeps the table answerable to month-end queries.
- Treating indicators as judgments. An indicator is evidence, not a verdict. Forwarding an indicator to remediation without the SUGGESTION-VALIDATION_VERDICT-commit gateway chain is what creates the silent-write paths the architecture is designed to prevent.
- Resolving against the current PROCESS_VERSION instead of the version that is effective at event_timestamp. The current version shortcut breaks replay and corrupts the audit reconstruction across version transitions. The resolve stage must use as-of event timestamp lookups.
- Mixing user identity into cohort analytics. Search-event captures user_id only to reduce it to role_scope_id by design; downstream consumers that add details would enable identity recreation, creating a surveillance tool the architecture deliberately does not provide.
- Letting the dead-letter queue grow without triage. A dead-letter queue without an operations console and a triage discipline is a heap with a slightly more dignified name. The triage workflow is part of the architecture, not an operational afterthought.

8.5.3 Review Questions

1. Why is event identity tagging performed at the WalkMe overlay rather than at the ingestion gateway, and what would be lost if the gateway tagged events instead?
2. A friction indicator is raised against an activity. What is the difference between the indicator, the suggestion it feeds, the verdict that gates the suggestion, and the audit event that records the verdict, and which of the four is the decision?
3. The deviation detection query reconstructs the observed path from TELEMETRY_EVENT and compares it to the path published by PROCESS_VERSION. Why does the query resolve against the PROCESS_VERSION effective at the event_timestamp rather than the current PROCESS_VERSION? What replay scenario would the current version shortcut break?
4. The shortfall_score that raises a TRAINING_GAP combines three normalized inputs with weights that default to equal but are overridable per Activity_ID, role_scope_id. What governance discipline must surround a weight override so that the indicator remains reproducible by an independent reviewer?
5. WalkMe is described in this chapter as both a target for upstream data and a target for AI agents. Explain the asymmetry in concrete terms: Which writes flow into the WalkMe

tenant, which reads flow out of it, and which row in which table enforces the boundary in each direction?

6. A controller reading the AUDIT_EVENT row that records a MODIFY verdict on a WalkMe binding suggestion wants to reconstruct the full evidence chain. List the six rows the controller traverses, in order, naming the table each row belongs to.

8.5.4 Student Companion Notes

If you are a student working through this chapter, you should focus first on the envelope and the binding. Take any WalkMe asset in your own tenant that is bound to an SAP Fiori activity, and verify that the analytics event stream from that asset carries the seven envelope fields specified in Section 8.1 (user_id, role_scope_id, variant_key, Process_ID, Activity_ID, fiori_app_id, and walkme_asset_id), each resolvable to an upstream identifier already on the data spine. This exercise quickly reveals which assets are emitting envelopes whose fields cannot be resolved (the dead-letter queue is the visible symptom), which bindings are pointing at deprecated PROCESS_VERSION rows, and which assets are running without a role_scope_id at all. The assets that fail any of these checks are not yet ready for adoption indicators to be raised against them, regardless of how well the overlay itself is designed.

Second, take the cost center to position scenario and trace the six TELEMETRY_EVENT rows recorded in the worked example from Section 8.2 forward through the five-stage ingestion pipeline, the three derived facts from Section 8.3, the SUGGESTION and VALIDATION_VERDICT from Section 8.4, and the AUDIT_EVENT chain that records the commit. The discipline of tracing the scenario row-by-row is what builds intuition for what a controller actually reads when reconstructing an evidence chain six weeks after the fact: six hops, six rows, and one walk. A useful variation of this exercise is to articulate, without looking up the answer, why the resolve stage uses an as-of event timestamp lookup rather than the current PROCESS_VERSION, and what specifically breaks, in audit terms, if any stage takes the current-state shortcut.

Third, take a friction pattern in your own landscape that you suspect is significant but for which you have only anecdotal evidence, and write the pattern out as one of the five canonical queries specified in Section 8.3. The query needs the threshold, the window, the cohort definition, and the registered schema reference. Once the query exists as a registered artifact, run it against a week of TELEMETRY_EVENT rows and compare the FRICTION_INDICATOR rows it raises against the anecdote. This exercise is the most useful when your anecdote turns out to be partially right: The pattern is real, but the cohort or the threshold the anecdote assumed is not where the indicator actually activates. That is the moment the architecture earns its place. The anecdote becomes a hypothesis, the query becomes the test, and the indicator becomes the evidence.

This chapter closes the part of the book devoted to the adoption layer of the integrated landscape. The telemetry, indicators, and grounding-feedback paths developed here are what allow the canonical data spine to present adoption evidence to FP&A dashboards in Chapter 9, the AI consumption layer to ground suggestions on observed cohort behavior rather than authored content alone in Chapter 14 and Chapter 15, and the governance and audit sections in Chapter 18 and Chapter 19 to reconstruct any AI-influenced WalkMe change through one six-row query. The integrity of every downstream layer depends, structurally, on the integrity of the events captured at the WalkMe overlay. The next chapter takes the canonical data spine as a whole and shows how Chapter 5 through Chapter 8 form a single entity-relationship model that the rest of the book reads from.

PART V

The Data Spine on SAP BTP

Chapter 9

Designing the Canonical Data Spine

Enterprise architecture is not merely descriptive but prescriptive. Integration is not about moving data but preserving meaning. A governed semantic backbone turns isolated systems into a unified execution platform where process truth endures and automation can be trusted.

The canonical data spine is the governed semantic integration layer implemented on SAP Business Technology Platform (SAP BTP). It harmonizes SAP Signavio process semantics, SAP LeanIX architecture context, WalkMe adoption telemetry, execution context modeling, configuration dependencies, and SAP AI consumption into a single, stable integration framework.

It is not a replication repository, and it does not replace upstream systems of record. It is a normalization and governance layer designed to stabilize identity, versioning, system ownership, and cross-system relationships so that execution services and AI-driven configuration generation operate on consistent, auditable context.

This chapter is the structural core of the book. It moves from conceptual necessity to implementation discipline and explains:

- Why point-to-point integrations fail structurally
- The canonical entity model
- Stable identifier and version design
- Execution context modeling
- Validation and publication workflows
- A detailed SAP Master Data Governance/SAP SuccessFactors Employee Central use case
- Transaction SPRO configuration object mapping
- An SAP S/4HANA configuration scenario
- Application programming interface (API) contracts
- Transport and governance alignment

9.1 Why Point-to-Point Integrations Fail

Most integration landscapes begin pragmatically. Process data is extracted from SAP Signavio. Application and interface metadata is extracted from SAP LeanIX. SAP Integration Suite connects them. Downstream systems consume the combined dataset.

Initially, the architecture appears stable, but over time, it fails for structural reasons. The earliest stages of an integration program often rely on direct, point-to-point synchronization. Process models exported from SAP Signavio are joined to application inventories from SAP LeanIX, and adoption signals from WalkMe are stitched into integration flows. At first glance, this feels productive—the tooling appears aligned, and basic connectivity is established quickly. However, this approach is fragile because each system evolves independently:

- Process semantics in SAP Signavio adapt as users refine operational intent.
- Architecture semantics in SAP LeanIX change as applications are upgraded, retired, or rehosted.
- Adoption telemetry and enablement content in WalkMe evolve as user interface (UI) changes, user behaviors shift, and task flows are optimized.
- Integration mappings or middleware transformations often live outside any governed layer.

Over time, independent semantic drift emerges. Without a canonical semantic layer, disparate identifiers, inconsistent object versions, and mismatched enablement links degrade predictability. For example:

- WalkMe helps users complete an activity that was renamed in SAP Signavio but never remapped.
- SAP LeanIX retires an interface that is still referenced in a process variant.
- A process export contains multiple draft versions and the integration layer cannot reliably determine which one is authoritative.
- AI bundles trained on ungoverned combinations of process, architecture, and telemetry produce inconsistent recommendations because the underlying semantics are ambiguous.

The result is not just mapping errors but a slow erosion of trust in the integrated landscape. AI consumption magnifies these weaknesses because AI tolerates incomplete data but not ambiguous identity, mixed versions, or unclear ownership.

If draft processes leak into AI grounding bundles, or if adoption signal contexts are mismatched, suggestions become untraceable. If an application is deprecated in SAP LeanIX but still mapped to activities, AI may recommend defunct solutions from the core architectural elements, as follows:

- **Independent semantic evolution**
  Process models evolve independently of architecture lifecycles. Application dependencies evolve independently of process structure. Adoption artifacts evolve independently of both. When reconciliation logic lives inside middleware transformations rather than a canonical semantic layer, semantic drift becomes inevitable. The same activity may be represented differently across tools. Identifiers may change, versions may overlap, and ownership boundaries blur.

- **Version misalignment**
  SAP Signavio supports versioning. SAP LeanIX tracks lifecycle state. Middleware rarely models lineage explicitly. Without a canonical structure separating identity from version state, version ambiguity propagates downstream. AI systems are particularly sensitive to version ambiguity.
- **Ownership diffusion**
  When integration mappings are embedded in pipelines or scripts, no system owns the integrated view. When inconsistencies arise, teams debug symptoms rather than fix structural issues.
- **AI amplification**
  AI requires:
  - Stable identifiers
  - Approved version state
  - Explicit execution context
  - Clear ownership boundaries

Without these guarantees, AI suggestions lose explainability and trust. The canonical data spine exists to prevent this decay.

9.2 Canonical Data Spine Entities and Relationships

The canonical data spine solves these issues by externalizing identity, relationships, and governance into a layer that all tools (SAP Signavio, SAP LeanIX, WalkMe, execution services, and AI) can reference consistently. It normalizes semantics and ensures that walk-forward integration, telemetry, and change governance are all anchored to a stable, auditable backbone rather than ad hoc mappings.

The canonical data spine exists to harmonize multiple systems of record without collapsing them into one another. To understand its structure, it is important to think in terms of domains rather than tables. The canonical data spine organizes integration into six logical domains. Each domain represents a distinct type of enterprise truth:

- Process semantics
- Architecture semantics
- Execution context
- Master data
- Adoption and telemetry
- Governance and publication

Each of these domains originates in a different system of record and is governed differently, and each domain maps to a system of record upstream and a publication boundary downstream. The data spine does not redefine these truths. It stabilizes the relationships between them. Before examining the individual tables of each domain, we will first clarify how the domains relate conceptually.

9.2.1 Domain Model Overview and Architectural Boundaries

Table 9.1 summarizes the six domains and their roles in the integration architecture.

Domain	Primary Purpose	Upstream System of Record	Consumed By	Why It Exists in the Data Spine
Process semantics	Stable process identity, version, L4/L5 activities	SAP Signavio	SAP BTP services, SAP AI	To preserve execution intent independent of tooling
Architecture semantics	Applications, interfaces, lifecycle state	SAP LeanIX	SAP BTP services, SAP AI	To anchor process steps to real systems
Execution context	Trigger/target ownership per activity	Derived in data spine	Execution services, SAP AI	To eliminate cross-system ambiguity
Master data	Financial and organizational reference objects	SAP S/4HANA, SAP SuccessFactors	Execution services	To ensure process steps reference valid enterprise data
Adoption and telemetry	WalkMe assets and interaction signals	WalkMe	Operational improvement, SAP AI	To connect modeled intent with actual user behavior
Governance and publication	Validation results, approval state, audit trail	SAP BTP workflow	All consumers	To enforce publication discipline before consumption

Table 9.1: Domain Model Overview

The data spine does not duplicate upstream schemas. It normalizes the minimum necessary semantics required to support cross-system execution and AI consumption. The following separation is deliberate:

- SAP Signavio defines what should happen.
- SAP LeanIX defines where it can happen.
- Master data defines what enterprise objects are valid.
- WalkMe reveals how users actually interact.

- Execution context clarifies who owns which step.
- Governance ensures nothing becomes executable without approval.

Without this separation, integration logic becomes either overcentralized or uncontrolled. Everything collapses into a single opaque layer, or ownership fragments across tools with no structural alignment.

9.2.2 From Domains to Canonical Entities

Once domain are defined and established, they are translated into canonical entities. The data spine does not mirror entire upstream schemas. It extracts only what is required for execution and AI grounding. Table 9.2 lists the core canonical entities .

Canonical Entity	Purpose	Domain
PROCESS	Stable process identity	Process semantics
PROCESS_VERSION	Version and approval state	Process semantics
ACTIVITY	Executable L4/L5 step	Process semantics
APPLICATION	Executable system	Architecture semantics
INTERFACE	Dependency between systems	Architecture semantics
EXECUTION_CONTEXT	Trigger/target system per activity	Execution context
WALKME_ASSET	Enablement artifact mapped to activity	Adoption
TELEMETRY_EVENT	Runtime interaction signal	Adoption
PUBLISH_STATE	Governance boundary	Governance

Table 9.2: Canonical Entity with Its Purpose at Each Domain

There is a clear progression in how these entities function:

1. Process defines what should happen.
2. Version defines which definition is valid.
3. Activity defines how that intent is executed step by step.
4. Application defines which system supports the step.
5. Execution context defines ownership and direction.
6. Master data defines the enterprise constraints that the step must respect.
7. Adoption reveals real-world behavior.
8. Governance defines when the model becomes authoritative.

This layered translation is what gives the canonical data spine structural strength.

9.2.3 Execution Context as the Structural Pivot

Execution context is the most common source of architectural confusion. In many landscapes, a process diagram shows a sequence of activities without explicitly stating which system triggers the activity, which system executes it, and which system consumes its output.

In a single-system environment, this ambiguity may be tolerable. In integrated landscapes (such as SAP Master Data Governance, SAP S/4HANA, SAP S/4HANA Cloud Private Edition, and SAP SuccessFactors), it is not. Consider a simplified example:

- A position is planned in SAP Analytics Cloud during headcount planning.
- A cost center request originates in SAP Master Data Governance.
- The cost center is activated in controlling in SAP S/4HANA.
- The cost center is replicated downstream.
- A position is created in SAP SuccessFactors referencing that cost center.

Even this simple environment spans multiple ownership boundaries. In real-world programs, additional non-SAP systems may participate, making the need for explicit ownership even stronger. These responsibilities are not interchangeable. The `EXECUTION_CONTEXT` table contains the following three fields, formalizing this ownership:

- `Trigger_Application_ID`
- `Target_Application_ID`
- `Context_Role`

By modeling this explicitly, the data spine removes inference from the integration layer. AI systems no longer guess which system owns a configuration step, and execution services no longer assume where changes should occur. Governance workflows can route approvals accurately. Without execution context, ambiguity becomes embedded in configuration and automation.

9.2.4 Logical Consolidation of the Canonical Model

At this point, you can think of the canonical data spine not as a collection of tables but as a structured relationship model, as follows:

- Process identity sits at the core.
- Version state branches from identity.
- Activities anchor execution intent.
- Applications anchor execution capability.
- Execution context binds intent to capability.
- Adoption artifacts bind modeled intent to real user behavior.
- Governance binds everything to publication discipline.

The relationships between the following elements are deliberate:

- A PROCESS owns multiple PROCESS_VERSION records.
- Each PROCESS_VERSION contains multiple ACTIVITY records.
- Each ACTIVITY maps to one or more APPLICATION entities.
- Each ACTIVITY defines an EXECUTION_CONTEXT that clarifies which application triggers and which application executes the step.
- WALKME_ASSET and TELEMETRY_EVENT records attach to ACTIVITY and APPLICATION entities, reflecting runtime interaction.
- PUBLISH_STATE records gate the exposure of PROCESS_VERSION, ACTIVITY, and related entities to execution services and AI.

Nothing in this structure is incidental. Every relationship exists to prevent ambiguity:

- Identity is separated from version.
- Execution ownership is separated from architecture inventory.
- Adoption signals are separated from process semantics.
- Governance is separated from ingestion.

When visualized as an entity relationship diagram, the model reveals a clear center of gravity: PROCESS leads to PROCESS_VERSION which leads to ACTIVITY. From ACTIVITY, relationships radiate outward:

- To APPLICATION (via mapping tables)
- To EXECUTION_CONTEXT (trigger/target clarity)
- To MASTER_DATA entities (cost center, position, and cost element)
- To WALKME_ASSET (enablement linkage)
- To TELEMETRY_EVENT (runtime signals)
- To CONFIG_OBJECT and master data references (configuration impact)

Governance entities (VALIDATION_RESULT and PUBLISH_STATE) overlay the entire model without redefining identity. They enforce lifecycle discipline without distorting structure.

This separation between structural relationships and governance overlays is what makes the canonical data spine resilient. It allows upstream systems to evolve independently while preserving a stable integration backbone.

Figure 9.1 illustrates these relationships and shows how identity, execution context, and governance interconnect across domains. A full entity-relationship diagram with all keys and constraints is available in Appendix D, which you can download at *https://www.sap-press. com/6352*.

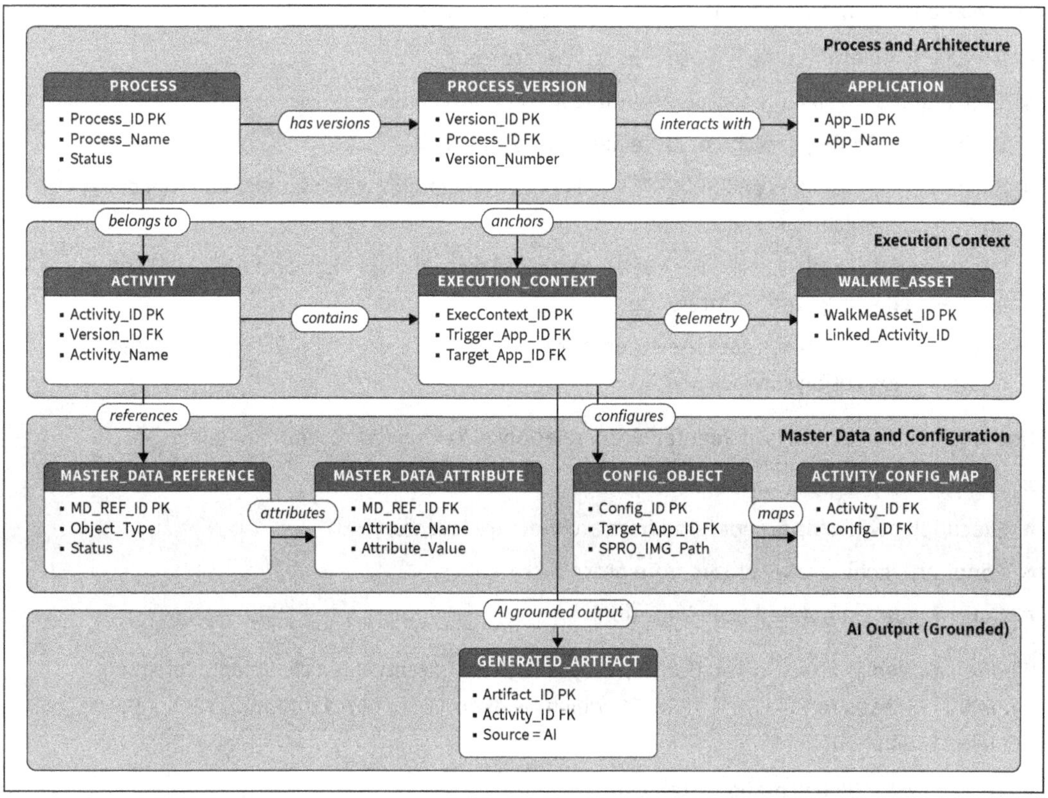

Figure 9.1: High-Level Entity-Relationship Diagram for Integration for Process with AI

9.3 Process, Activity, Application, and Interface Keys

With the domain model established in Section 9.2, the next step is to express that structure as a concrete schema. The objective here is not to mirror SAP Signavio, SAP LeanIX, or WalkMe. The data spine is intentionally smaller than any of those systems. It captures only what is required to keep identity stable, make execution ownership explicit, enforce publication discipline, and provide reliable context to execution services and SAP AI. The following four design decisions shape everything that follows:

- Identity is stable; version is state.
- Version must remain separate from identity.
- Relationships must be explicit.
- Systems of record remain authoritative; the data spine governs integration and publication.

The schema is therefore organized into structural identity tables, mapping tables, and governance overlays.

This section introduces nine structural building blocks of the data spine in dependency order. It begins with stable identity and version separation for processes and activities (Section 9.3.1), and then anchors the architecture side with the application and interface keys carried over from SAP LeanIX (Section 9.3.2). It connects the two through activity-to-architecture mapping tables (Section 9.3.3) and pins each connection to a single point of system ownership through the execution context (Section 9.3.4). Master data referencing (Section 9.3.5) and adoption telemetry (Section 9.3.6) attach to the structural backbone without becoming systems of record on the data spine. Configuration objects and activity-to-configuration mappings (Section 9.3.7) record the customizing dependencies that downstream AI consumers must respect. The governance overlay covering validation and publication (Section 9.3.8) is the gate every entity passes before consumption. Finally, the AI artifact tables (Section 9.3.9) separate AI-generated proposals from canonical truth, which closes the schema. Every use case that follows, including the one in Section 9.4, reads against this set of entities in exactly this order.

9.3.1 Stable Identity and Version: Process and Activity

Process semantics originate in SAP Signavio, but the data spine must preserve identity independent of naming and version changes. A process may be revised many times; it must still be referable as the same process. Listing 9.1 shows the data definition language (DDL) definition for the canonical `PROCESS` table, the structural identity anchor for every process represented on the data spine.

```
CREATE TABLE PROCESS (
  Process_ID          NVARCHAR(50) PRIMARY KEY,
  Process_Name        NVARCHAR(255) NOT NULL,
  Domain              NVARCHAR(100),
  Status              NVARCHAR(30) CHECK (Status IN
('Active','Deprecated','Archived')),
  Created_At          TIMESTAMP DEFAULT CURRENT_TIMESTAMP,
  Updated_At          TIMESTAMP
);
```

Listing 9.1: Create Table Process

Version state is separate from identity. Only approved, published versions can be exposed to AI and execution services. Listing 9.2 shows the DDL definition for `PROCESS_VERSION`, where version state, approval status, and lineage are recorded independently of the stable process identity defined.

```
CREATE TABLE PROCESS_VERSION (
  Version_ID          NVARCHAR(50) PRIMARY KEY,
  Process_ID          NVARCHAR(50) NOT NULL,
  Version_Number      INTEGER NOT NULL,
```

```
  Effective_From     DATE,
  Effective_To       DATE,
  Approval_Status    NVARCHAR(30) CHECK (Approval_Status IN
('Draft','Validated','Approved','Published','Deprecated','Archived')),
  Parent_Version_ID  NVARCHAR(50),
  Change_Description NVARCHAR(1000),
  Created_At         TIMESTAMP DEFAULT CURRENT_TIMESTAMP,
  Updated_At         TIMESTAMP,
  FOREIGN KEY (Process_ID) REFERENCES PROCESS(Process_ID)
);
```

Listing 9.2: Create Table Process Version

Activities are anchored to a version but maintain stable identity across revisions. Sequence can change without breaking downstream links. Listing 9.3 shows the DDL definition for `ACTIVITY`, which anchors each L4 or L5 step to a specific process version while preserving stable activity identity across revisions.

```
CREATE TABLE ACTIVITY (
  Activity_ID       NVARCHAR(50) PRIMARY KEY,
  Process_ID        NVARCHAR(50) NOT NULL,
  Version_ID        NVARCHAR(50) NOT NULL,
  Activity_Name     NVARCHAR(255) NOT NULL,
  Sequence_No       INTEGER,
  Activity_Type     NVARCHAR(50),
  Responsible_Role  NVARCHAR(100),
  Created_At        TIMESTAMP DEFAULT CURRENT_TIMESTAMP,
  Updated_At        TIMESTAMP,
  FOREIGN KEY (Process_ID) REFERENCES PROCESS(Process_ID),
  FOREIGN KEY (Version_ID) REFERENCES PROCESS_VERSION(Version_ID)
);
```

Listing 9.3: Create Table Activity

9.3.2 Architecture Keys and Identity: Applications and Interfaces

SAP LeanIX remains the system of record for application and interface portfolios. The data spine normalizes what is required for integration and impact analysis while retaining SAP LeanIX identifiers for traceability. Listing 9.4 shows the DDL definition for `APPLICATION`, which records each executable system on the data spine while retaining its SAP LeanIX `factSheetId` for traceability back to the architecture layer.

```
CREATE TABLE APPLICATION (
  App_ID            NVARCHAR(50) PRIMARY KEY,
  LeanIX_App_ID     NVARCHAR(50) UNIQUE,
```

```
  Application_Name  NVARCHAR(255) NOT NULL,
  Category          NVARCHAR(50),      -- S4, DATASPHERE (extendable)
  Lifecycle_State   NVARCHAR(50),
  Criticality_Level NVARCHAR(20),
  Created_At        TIMESTAMP DEFAULT CURRENT_TIMESTAMP,
  Updated_At        TIMESTAMP
);
```

Listing 9.4: Create Table Application

Interfaces represent dependency edges. They are essential for change impact and for execution services that need to route or validate integration paths. Listing 9.5 shows the DDL definition for `INTERFACE`, which records each integration dependency edge between a provider application and a consumer application, retaining the interface type and SLA level for change-impact analysis.

```
CREATE TABLE INTERFACE (
  Interface_ID      NVARCHAR(50) PRIMARY KEY,
  Provider_App_ID   NVARCHAR(50) NOT NULL,
  Consumer_App_ID   NVARCHAR(50) NOT NULL,
  Interface_Type    NVARCHAR(50),
  SLA_Level         NVARCHAR(30),
  Created_At        TIMESTAMP DEFAULT CURRENT_TIMESTAMP,
  Updated_At        TIMESTAMP,
  FOREIGN KEY (Provider_App_ID) REFERENCES APPLICATION(App_ID),
  FOREIGN KEY (Consumer_App_ID) REFERENCES APPLICATION(App_ID)
);
```

Listing 9.5: Create Table Interface

9.3.3 Mapping Tables: Process-to-Architecture and System Involvement

Point-to-point integration decays largely because the detail of where the step happens is only recorded in informal documentation. The data spine makes this explicit through mapping tables. Listing 9.6 shows the DDL definition for `ACTIVITY_APPLICATION_MAP`, the mapping table that records which application executes each activity and makes the system where the step happens a structural fact, rather than keeping it in informal documentation.

```
CREATE TABLE ACTIVITY_APPLICATION_MAP (
  Activity_ID       NVARCHAR(50) NOT NULL,
  App_ID            NVARCHAR(50) NOT NULL,
  Relationship_Type NVARCHAR(30) CHECK (Relationship_Type IN
('ExecutesIn','Triggers','Consumes','Produces')),
  Created_At        TIMESTAMP DEFAULT CURRENT_TIMESTAMP,
  PRIMARY KEY (Activity_ID, App_ID, Relationship_Type),
```

```
  FOREIGN KEY (Activity_ID) REFERENCES ACTIVITY(Activity_ID),
  FOREIGN KEY (App_ID) REFERENCES APPLICATION(App_ID)
);
```

Listing 9.6: Create Table Activity Application Map

Listing 9.7 shows the DDL definition for `ACTIVITY_INTERFACE_MAP`, the optional but recommended mapping table that records which interfaces an activity consumes or produces. The role tag from Chapter 6 is carried forward here into the canonical schema.

```
CREATE TABLE ACTIVITY_INTERFACE_MAP (
  Activity_ID   NVARCHAR(50) NOT NULL,
  Interface_ID  NVARCHAR(50) NOT NULL,
  PRIMARY KEY (Activity_ID, Interface_ID),
  FOREIGN KEY (Activity_ID) REFERENCES ACTIVITY(Activity_ID),
  FOREIGN KEY (Interface_ID) REFERENCES INTERFACE(Interface_ID)
);
```

Listing 9.7: Create Table Activity Interface Map

These mappings allow execution services to answer simple questions reliably:

- Which systems execute this step?
- Which interfaces are involved?
- Which consumers are impacted if an interface changes?

9.3.4 Execution Context: Ownership Clarity in Multisystem Processes

Execution context is the structural pivot of the model. It prevents ambiguity in cross-system steps by explicitly recording trigger and target application ownership. Listing 9.8 shows the DDL definition for `EXECUTION_CONTEXT`, the structural pivot that records the trigger application, target application, and context role for every activity, which prevents ambiguity in cross-system steps.

```
CREATE TABLE EXECUTION_CONTEXT (
  ExecContext_ID     NVARCHAR(50) PRIMARY KEY,
  Activity_ID        NVARCHAR(50) NOT NULL,
  Trigger_App_ID     NVARCHAR(50) NOT NULL,
  Target_App_ID      NVARCHAR(50) NOT NULL,
  Context_Role       NVARCHAR(30) CHECK (Context_Role IN
('Initiator','Executor','Producer','Consumer')),
  Downstream_App_ID NVARCHAR(50),
  Created_At         TIMESTAMP DEFAULT CURRENT_TIMESTAMP,
  Updated_At         TIMESTAMP,
  FOREIGN KEY (Activity_ID) REFERENCES ACTIVITY(Activity_ID),
  FOREIGN KEY (Trigger_App_ID) REFERENCES APPLICATION(App_ID),
```

```
  FOREIGN KEY (Target_App_ID) REFERENCES APPLICATION(App_ID),
  FOREIGN KEY (Downstream_App_ID) REFERENCES APPLICATION(App_ID)
);
```

Listing 9.8: Create Table Execution Context

This single table eliminates a surprising number of integration and governance errors by forcing teams to encode system ownership explicitly.

9.3.5 Master Data Referencing (Not Replication)

The data spine should support configuration and AI grounding, but it cannot become a second system of record for master data. Therefore, master data is modeled as references with optional attributes for constraint validation. Listing 9.9 shows the DDL definition for MASTER_DATA_REFERENCE, which records a reference to a master data object held in the upstream system of record without replicating its content into the data spine.

```
CREATE TABLE MASTER_DATA_REFERENCE (
  MD_REF_ID        NVARCHAR(50) PRIMARY KEY,
  OBJECT_TYPE      NVARCHAR(50) NOT NULL,   -- COST_CENTER, POSITION, etc.
  SOURCE_SYSTEM    NVARCHAR(50) NOT NULL,   -- S4, DATASPHERE, etc.
  SOURCE_OBJECT_ID NVARCHAR(50) NOT NULL,
  VALID_FROM       DATE,
  VALID_TO         DATE,
  STATUS           NVARCHAR(30),
  Created_At       TIMESTAMP DEFAULT CURRENT_TIMESTAMP
);
```

Listing 9.9: Create Table Master Data Reference

Listing 9.10 shows the DDL definition for MASTER_DATA_ATTRIBUTE, the extensible side table that holds the minimum attribute set required for validation rules and AI grounding, with provenance and confidence recorded per attribute.

```
CREATE TABLE MASTER_DATA_ATTRIBUTE (
  MD_REF_ID        NVARCHAR(50) NOT NULL,
  ATTRIBUTE_NAME   NVARCHAR(100) NOT NULL,
  ATTRIBUTE_VALUE  NVARCHAR(255),
  PRIMARY KEY (MD_REF_ID, ATTRIBUTE_NAME),
  FOREIGN KEY (MD_REF_ID) REFERENCES MASTER_DATA_REFERENCE(MD_REF_ID)
);
```

Listing 9.10: Create Table Master Data Attributes

Listing 9.11 shows the DDL definition for ACTIVITY_MASTERDATA_REF, the optional linkage table that records which master data references an activity creates, references, or approves, using the Usage_Type controlled list.

```
CREATE TABLE ACTIVITY_MASTERDATA_REF (
  Activity_ID  NVARCHAR(50) NOT NULL,
  MD_REF_ID    NVARCHAR(50) NOT NULL,
  Usage_Type   NVARCHAR(30),    -- Requires, Creates, Updates, References
  PRIMARY KEY (Activity_ID, MD_REF_ID),
  FOREIGN KEY (Activity_ID) REFERENCES ACTIVITY(Activity_ID),
  FOREIGN KEY (MD_REF_ID) REFERENCES MASTER_DATA_REFERENCE(MD_REF_ID)
);
```

Listing 9.11: Create Table Activity Master Data Reference

This approach supports additional systems (SAP Ariba, SAP Concur, etc.) by extending OBJECT_TYPE and SOURCE_SYSTEM without schema redesign.

9.3.6 Adoption and Telemetry: WalkMe

WalkMe is not documentation. In this architecture, it is the adoption layer that ties modeled intent to real user behavior. The data spine captures only what is required to map enablement and telemetry to canonical process steps. Listing 9.12 shows the DDL definition for WALKME_ASSET, which maps each WalkMe enablement artifact to the canonical activity it supports, turning adoption content into a process-aware layer rather than freestanding documentation.

```
CREATE TABLE WALKME_ASSET (
  WalkMeAsset_ID      NVARCHAR(50) PRIMARY KEY,
  Asset_Type          NVARCHAR(50),
  Target_App_ID       NVARCHAR(50) NOT NULL,
  Linked_Activity_ID  NVARCHAR(50) NOT NULL,
  Version             NVARCHAR(30),
  Created_At          TIMESTAMP DEFAULT CURRENT_TIMESTAMP,
  FOREIGN KEY (Target_App_ID) REFERENCES APPLICATION(App_ID),
  FOREIGN KEY (Linked_Activity_ID) REFERENCES ACTIVITY(Activity_ID)
);
```

Listing 9.12: Create Table WalkMe Asset

Listing 9.13 shows the DDL definition for TELEMETRY_EVENT, which captures runtime user interaction signals against canonical activities. This is the raw material for adoption analysis, change impact feedback, and the AI improvement loop, as we will cover in Section 9.4.

```
CREATE TABLE TELEMETRY_EVENT (
  Event_ID          NVARCHAR(50) PRIMARY KEY,
  Activity_ID       NVARCHAR(50) NOT NULL,
  App_ID            NVARCHAR(50) NOT NULL,
  Event_Type        NVARCHAR(50),
  Event_Timestamp   TIMESTAMP,
  User_Role         NVARCHAR(100),
  FOREIGN KEY (Activity_ID) REFERENCES ACTIVITY(Activity_ID),
  FOREIGN KEY (App_ID) REFERENCES APPLICATION(App_ID)
);
```

Listing 9.13: Create Table Telemetry Event

WalkMe signals can be used as an input to AI suggestions, but they should never overwrite canonical truth.

9.3.7 Configuration Objects and Activity Mapping

To support AI-driven configuration specification generation, configuration dependencies must be modeled explicitly. Listing 9.14 shows the DDL definition for `CONFIG_OBJECT`, which records each configuration node (typically a Transaction SPRO or IMG menu path) that an activity depends on for correct execution in the target system.

```
CREATE TABLE CONFIG_OBJECT (
  Config_ID           NVARCHAR(50) PRIMARY KEY,
  IMG_Path            NVARCHAR(500),
  SSCUI_ID            NVARCHAR(50),
  Config_Object_Name  NVARCHAR(255),
  Target_App_ID       NVARCHAR(50),
  Config_Type         NVARCHAR(30),
  Created_At          TIMESTAMP DEFAULT CURRENT_TIMESTAMP,
  FOREIGN KEY (Target_App_ID) REFERENCES APPLICATION(App_ID)
);
```

Listing 9.14: Create Table Config Objects

Listing 9.15 shows the DDL definition for `ACTIVITY_CONFIG_MAP`, the mapping table that links activities to the configuration objects whose state they consume, which allows AI configuration-specification generators to ground proposals in attested customizing nodes rather than inferred ones.

```
CREATE TABLE ACTIVITY_CONFIG_MAP (
  Activity_ID       NVARCHAR(50) NOT NULL,
  Config_ID         NVARCHAR(50) NOT NULL,
```

```
  Required_Flag    TINYINT DEFAULT 0 NOT NULL,
  Validation_Rule  NVARCHAR(255),
  PRIMARY KEY (Activity_ID, Config_ID),
  FOREIGN KEY (Activity_ID) REFERENCES ACTIVITY(Activity_ID),
  FOREIGN KEY (Config_ID) REFERENCES CONFIG_OBJECT(Config_ID)
);
```

Listing 9.15: Create Table Activity Config Map

This is what allows an AI suggestion to reference a precise IMG path instead of a generic recommendation.

9.3.8 Governance Overlay: Validation and Publication

Validation and publication enforce discipline. They prevent draft leakage into execution and AI. Listing 9.16 shows the DDL definition for `VALIDATION_RULE`, which records each rule the data spine applies before allowing a structural entity to transition from validated to published.

```
CREATE TABLE VALIDATION_RULE (
  Rule_ID          NVARCHAR(50) PRIMARY KEY,
  Rule_Category    NVARCHAR(50),
  Applies_To       NVARCHAR(50),
  Severity         NVARCHAR(20),
  Rule_Description NVARCHAR(500),
  Is_Active        TINYINT DEFAULT 1 NOT NULL,
  Created_At       TIMESTAMP DEFAULT CURRENT_TIMESTAMP
);
```

Listing 9.16: Create Table Validation Rules

Listing 9.17 shows the DDL definition for `VALIDATION_RESULT`, which records the outcome of every validation rule applied to every canonical entity. This provides the audit trail every transition between governance states depends on.

```
CREATE TABLE VALIDATION_RESULT (
  Validation_ID    NVARCHAR(50) PRIMARY KEY,
  Batch_ID         NVARCHAR(50),
  Rule_ID          NVARCHAR(50) NOT NULL,
  Entity_Type      NVARCHAR(50),
  Entity_ID        NVARCHAR(50),
  Severity         NVARCHAR(20),
  Result           NVARCHAR(10),
```

```
  Notes             NVARCHAR(500),
  Validated_At      TIMESTAMP DEFAULT CURRENT_TIMESTAMP,
  FOREIGN KEY (Rule_ID) REFERENCES VALIDATION_RULE(Rule_ID)
);
```

Listing 9.17: Create Table Validation Results

Listing 9.18 shows the DDL definition for PUBLISH_STATE, the governance overlay that gates exposure of canonical entities to execution services and AI consumers. Only published entities are visible to them.

```
CREATE TABLE PUBLISH_STATE (
  Entity_ID     NVARCHAR(50) NOT NULL,
  Entity_Type   NVARCHAR(50) NOT NULL,
  State         NVARCHAR(30) CHECK (State IN
('Draft','Validated','Approved','Published','Deprecated','Archived')),
  Approved_By   NVARCHAR(50),
  Approved_At   TIMESTAMP,
  PRIMARY KEY (Entity_ID, Entity_Type)
);
```

Listing 9.18: Create Table Publish State

9.3.9 AI Artifacts: Separated from Canonical Truth

AI outputs are stored as governed artifacts, not canonical records. Listing 9.19 shows the DDL definition for GENERATED_ARTIFACT, where AI-produced configuration specifications, process suggestions, and workflow definitions are recorded as typed proposals. They are distinct from canonical truth and gated by human review before any structural promotion.

```
CREATE TABLE GENERATED_ARTIFACT (
  Artifact_ID         NVARCHAR(50) PRIMARY KEY,
  Artifact_Type       NVARCHAR(50),
  Linked_Activity_ID NVARCHAR(50),
  Target_Application NVARCHAR(50),
  Publish_State       NVARCHAR(30),
  Generated_At        TIMESTAMP DEFAULT CURRENT_TIMESTAMP,
  Approved_By         NVARCHAR(50),
  FOREIGN KEY (Linked_Activity_ID) REFERENCES ACTIVITY(Activity_ID),
  FOREIGN KEY (Target_Application) REFERENCES APPLICATION(App_ID)
);
```

Listing 9.19: Create Table Generated Artifacts

At this point, the canonical data spine is fully expressed as a structural model. Identity, versioning, execution context, master data referencing, configuration dependencies, governance overlays, and AI artifact segregation have all been formalized. The data spine is no longer conceptual; it becomes a governed integration backbone capable of supporting execution services, configuration traceability, and AI grounding without ambiguity.

What remains is to demonstrate how these structures operate together in practice. The following section introduces a simplified dataset derived from the SAP Master Data Governance on SAP S/4HANA and SAP SuccessFactors position creation scenario, illustrating how the model anchors real cross-system behavior while preserving system of record boundaries.

9.4 Use Case: Creation of a New Position

Let's explore how those structures operate together in a real enterprise scenario aligned with SAP Best Practices. Our use case will follow the creation of a new position when cost centers are governed by finance.

The remainder of this chapter applies every structural element introduced in Section 9.3 to a single concrete scenario: the creation of a position in SAP SuccessFactors Employee Central against a cost center governed in SAP Master Data Governance, Financials on SAP S/4HANA. The section opens with the sample data set (Section 9.4.1) and the SAP Best Practice components it aligns with (Section 9.4.2). It then traces the cross-system flow end to end (Section 9.4.3) and isolates execution context as the mechanism that makes the flow unambiguous (Section 9.4.4). Master data referencing (Section 9.4.5) and configuration dependencies (Section 9.4.6) attach to the flow without replicating any system of record. Validation and publication discipline (Section 9.4.7) gate the flow before any execution service or AI consumer sees it, and the AI-governed suggestion flow (Section 9.4.8) closes the loop by feeding telemetry from runtime back into the design surface. The architectural outcome (Section 9.4.9) names what the data spine has accomplished in this scenario that would otherwise require informal coordination across four systems.

9.4.1 Sample Data Set

The sample data set in Figure 9.2 shows how master data references work without replicating SAP S/4HANA or SAP SuccessFactors schemas. Two governed master data references on the canonical data spine: a cost center and a position. The data spine stores a reference and lifecycle, not a copy of the source record.

RECORD 1: MD_CC_1000 · COST_CENTER

Identity and Source

MD_REF_ID	OBJECT_TYPE	SOURCE_SYSTEM	SOURCE_TENANT	SOURCE_OBJECT_ID	EXTERNAL_KEY	VALID_FROM	VALID_TO
MD_CC_1000	COST_CENTER	SAP S/4HANA	S4CLNT100	1000	S4-KOSTL	2026-01-01	9999-12-31

Canonical State and Governance

SOURCE_STATUS	CANONICAL_STATUS	LEGAL_SCOPE	ORG_SCOPE	MDG_CHANGE_REQ	WORKFLOW_STATUS	REPLICATION	LAST_REPLICATED
Active	Published	US01	A100/PC100	CR_4500123	Approved	Confirmed	2026-01-15 10:22

RECORD 2: MD_POS_5001 · POSITION

Identity and Source

MD_REF_ID	OBJECT_TYPE	SOURCE_SYSTEM	SOURCE_TENANT	SOURCE_OBJECT_ID	EXTERNAL_KEY	VALID_FROM	VALID_TO
MD_POS_5001	POSITION	SAP SuccessFactors	SF-PROD	5001	SF-Position	2026-01-15	9999-12-31

Canonical State and Governance

SOURCE_STATUS	CANONICAL_STATUS	LEGAL_SCOPE	ORG_SCOPE	MDG_CHANGE_REQ	WORKFLOW_STATUS	REPLICATION	LAST_REPLICATED
Active	Published	US_LE	HR_ADMIN	–	–	N/A	–

Figure 9.2: Sample Master Data Reference with Mandatory Fields

Table 9.3 shows the `MASTER_DATA_ATTRIBUTE` sample dataset. It contains the mandatory governance fields per attribute, with provenance and confidence scoring. Confidence below 1.00 marks values that are inferred or AI-classified and should never be treated an authoritative source. The first four entries are `COST_CENTER` attributes, with the final two being `POSITION` attributes.

MD_REF_ID	ATTRIBUTE_NAME	VALUE	SOURCE	CONFIDENCE	EFFECTIVE_FROM	EFFECTIVE_TO	CLASSIFICATION	SENSITIVITY
MD_CC_1000	COMPANY_CODE	US01	S/4MDG	1.00	2026-01-01	9999-12-31	Internal	Financial
MD_CC_1000	CONTROLLING_AREA	A100	S/4MDG	1.00	2026-01-01	9999-12-31	Internal	Financial
MD_CC_1000	PROFIT_CENTER	PC100	S/4MDG	0.98	2026-01-01	9999-12-31	Internal	Financial
MD_CC_1000	COSTCENTER_CATEGORY	HR	MDG	0.95	2026-01-01	9999-12-31	Internal	Financial
MD_POS_5001	LEGAL_ENTITY	US_LE	SF	1.00	2026-01-15	9999-12-31	Internal	HR
MD_POS_5001	DEPARTMENT	HR_ADMIN	SF	0.97	2026-01-15	9999-12-31	Internal	HR

Table 9.3: Sample Master Attributes Data with Mandatory Fields

Table 9.4 shows the `ACTIVITY_MASTERDATA_REF` sample mapping. Activities reference master data without owning it; usage type expresses the relationship. A single activity can reference master data it does not own, and ownership stays with the source domain.

ACTIVITY_ID	MD_REF_ID	USAGE_TYPE	REQUIRED	DOMAIN
ACT_APPROVE_COSTCENTER	MD_CC_1000	Creates/approves	Yes	Finance
ACT_CREATE_POSITION	MD_CC_1000	References	Yes	HR
ACT_CREATE_POSITION	MD_POS_5001	Creates	Yes	HR

Table 9.4: Sample Activity Data Reference with Mandatory Fields

9.4.2 Business Context and SAP Best Practice Alignment

This scenario aligns with several SAP Best Practice components:

- Cost center accounting in SAP S/4HANA (controlling)
- SAP Master Data Governance, Financials
- SAP SuccessFactors Employee Central's position management

In most enterprise landscapes:

- Finance governs cost centers in SAP S/4HANA, frequently through SAP Master Data Governance workflows.
- HR governs positions in SAP SuccessFactors Employee Central.
- Positions cannot be created unless a valid, approved cost center exists and is available for reference.

Individually, these processes are well defined within each application. The complexity arises at the boundary between them. The canonical data spine does not redefine SAP Best Practices. It governs the semantic and execution alignment across them.

9.4.3 End-to-End Cross-System Flow

The creation of a new position when cost centers are governed by finance can be decomposed into L4/L5 steps. For each step, we will provide the system of record:

1. Submit cost center request: SAP Master Data Governance on SAP S/4HANA
2. Approve cost center: SAP Master Data Governance on SAP S/4HANA
3. Activate cost center: Controlling in SAP S/4HANA
4. Publish cost center to integration: SAP S/4HANA to SAP BTP

5. Validate cost center publication: Canonical data spine
6. Create position: SAP SuccessFactors Employee Central
7. Confirm referential integrity: SAP SuccessFactors and canonical data spine

Each of these steps maps to canonical entities and execution context definitions.

9.4.4 Execution Context as Structural Control

Execution context removes ambiguity in system ownership. Table 9.5 shows a representative execution of context mapping.

Activity	Trigger_App_ID	Target_App_ID	Context_Role
Submit cost center request	S4_MDG	S4_MDG	Initiator
Approve cost center	S4_MDG	S4_MDG	Executor
Activate cost center	S4_MDG	S4_CORE	Executor
Publish cost center	S4_CORE	BTP_INTEGRATION	Producer
Create position	BTP_INTEGRATION	SUCCESSFACTORS_EC	Executor
Validate position	SUCCESSFACTORS_EC	SUCCESSFACTORS_EC	Consumer

Table 9.5: Execution Context Mapping for the Cost Center to Position Flow

This explicitly models ownership boundaries and eliminates inference from integration flows or AI consumers. Figure 9.3 shows a cross-system process governed by the canonical data spine on SAP BTP. Each system retains ownership; the data spine governs the cross-system semantic transition.

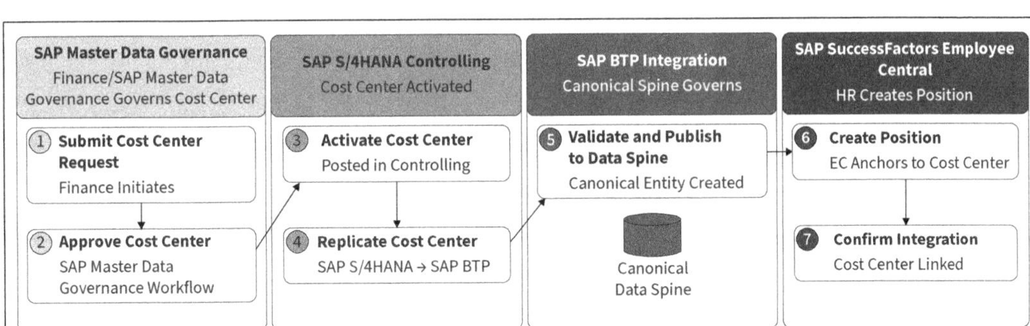

Figure 9.3: Cost Center to Position Integration Flow

9.4.5 Master Data Referencing in the Data Spine

The data spine does not replicate SAP S/4HANA CSKS tables or the SAP SuccessFactors position schema. Instead, it stores the following:

- A master data reference (OBJECT_TYPE, SOURCE_SYSTEM, or SOURCE_OBJECT_ID)
- Status and lifecycle state
- Minimal attribute set required for validation

Examples of this include:

- MASTER_DATA_REFERENCE
- MD_CC_1000 (COST_CENTER, S4, 1000, or Approved)
- MD_POS_5001 (POSITION, SUCCESSFACTORS, 5001, or Active)
- MASTER_DATA_ATTRIBUTE
- COMPANY_CODE = US01
- CONTROLLING_AREA = A100
- COSTCENTER_CATEGORY = HR
- COSTCENTER_ID (for Position) = 1000

This provides sufficient grounding for validation rules, configuration mapping, and AI inference without duplicating system of record responsibilities.

9.4.6 Configuration Dependencies

Cost center configuration may affect whether a position can be created. For example, you may have the following dependencies:

- Cost center category must be defined.
- Standard hierarchy must exist.
- Controlling area must be consistent.
- Certain validation rules may restrict usage.

These dependencies are captured through CONFIG_OBJECT and ACTIVITY_CONFIG_MAP. Table 9.6 shows an example of the configuration objects that the activate cost center activity depends on, with the Transaction SPRO/IMG path that defines each

Activity	Config_ID	IMG Path
Activate cost center	CFG_CC_CATEGORY	Transaction SPRO • **Controlling** • **Cost Center Accounting** • **Master Data** • **Cost Centers** • **Define Cost Center Categories**
Activate cost center	CFG_CC_HIER	Transaction SPRO • **Controlling** • **Cost Center Accounting** • **Master Data** • **Standard Hierarchy** • **Maintain Standard Hierarchy**

Table 9.6: Configuration Object Dependencies for the Activate Cost Center Activity

This linkage allows the data spine to trace from process to activity to configuration object to target system.

9.4.7 Validation and Publication Discipline

Before the position can be created, the following must be true:

- Cost center must be approved in SAP Master Data Governance.
- Cost center must be activated in SAP S/4HANA.
- Cost center must be published in the data spine.
- Execution context must be defined.
- No draft process version may be consumed.

`VALIDATION_RULE` and `PUBLISH_STATE` enforce these gates. Only published entities are visible to execution services, SAP Build workflows, and SAP AI Core grounding bundles. This ensures that AI and automation operate only on approved semantic truth.

9.4.8 AI-Governed Suggestion Flow

When friction or validation anomalies occur (e.g., repeated failure during position creation), it creates a feedback loop–based artifact creation process, as shown in Figure 9.4.

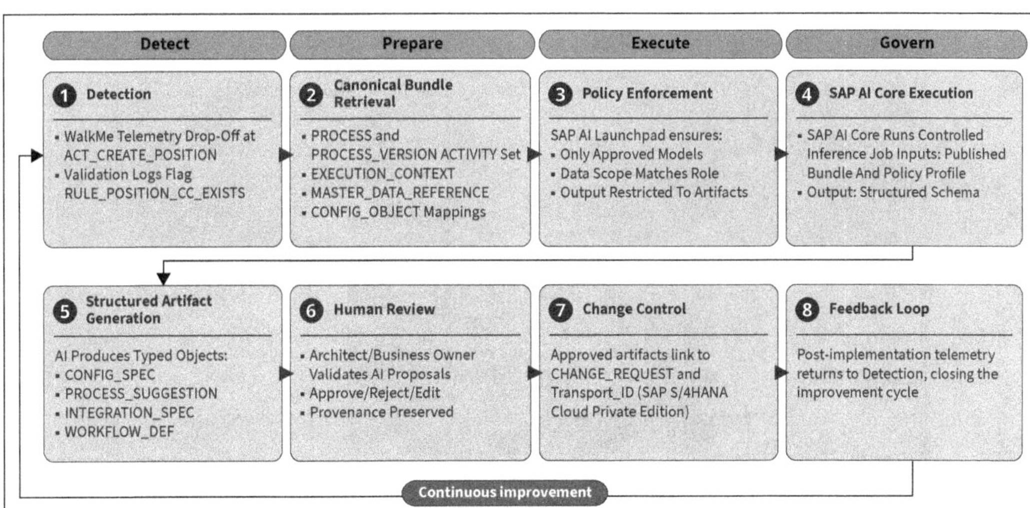

Figure 9.4: AI Governed Suggestion Flow

The full JSON schemas for each AI artifact payload are provided in Appendix E, and the full canonical OpenAPI contracts, together with the complete entity-relationship diagram, are provided in Appendix D. You can download the appendices at *https://www.sap-press.com/6352/*.

The runtime execution of AI core jobs, model lifecycle, launchpad configuration, and deployment pipelines are covered in Chapter 10.

9.4.9 Architectural Outcome

This use case demonstrates the purpose of the canonical data spine, as follows:

- Preserve system of record boundaries.
- Stabilize identity and version.
- Make execution ownership explicit.
- Enforce publication before consumption.
- Enable configuration traceability.
- Ground AI suggestions in structured semantics.

The data spine transforms what would otherwise be an informal cross-system workflow into a governed execution platform. Figure 9.5 shows the canonical data spine. AI does not produce free-form output. Every artifact is typed, traceable, and bound to a canonical entity. AI only generates these artifact types; everything else stays within the data spine's authority.

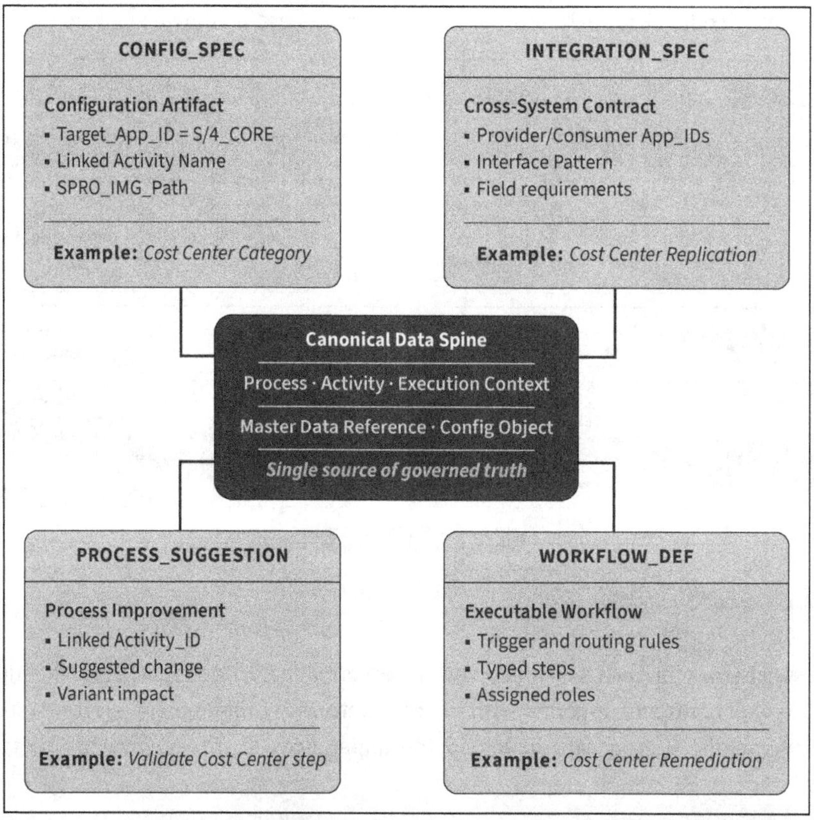

Figure 9.5: High-Level Data Model for Integration

9.5 Versioning and Lineage Strategy

The canonical data spine is designed to evolve without destabilizing execution. Versioning and lineage are the mechanisms that allow controlled change while preserving semantic continuity. Versioning in this architecture is not a documentation practice but a structural guarantee that execution services consume only approved semantic states, configuration mappings remain traceable, and AI inferences remain reproducible over time.

This section shows how version identity, lineage, temporal validity, and dependency tracking operate together.

9.5.1 Structural Identity and Version Separation

The data spine distinguishes between structural identity and structural realization. A `PROCESS` represents a persistent business capability. A `PROCESS_VERSION` represents a specific structural definition of that capability at a specific point in time. This separation ensures that identifiers remain stable even as definitions evolve. This distinction is foundational. Consider a process for position creation. Over time, the organization may add a validation gate, change sequencing, introduce a regional branch, or modify configuration dependencies.

If the identity of the process changes with each revision, downstream relationships collapse. Instead, the data spine ensures that `Process_ID` remains stable while `Version_ID` reflects structural evolution. Listing 9.20 shows the example DDL pattern that preserves `Process_ID` as a stable identifier across every version while letting `Version_ID` carry the structural evolution, which is the foundation for every change impact query that follows.

```
PROCESS_ID: P_HR_POSITION
VERSION_ID: V1.0  (Initial)
  VERSION_ID: V1.1  (Added cost center validation)
  VERSION_ID: V1.2  (Integration refinement)
```

Listing 9.20: Example: Process ID

The `PROCESS_ID` does not change. The `PROCESS_VERSION` defines the structure. The structural relationships of `ACTIVITY`, `EXECUTION_CONTEXT`, and `CONFIG_OBJECT` mappings are always bound to a specific `Version_ID`, never directly to `Process_ID`. This distinction is critical for downstream mappings (`ACTIVITY_APPLICATION_MAP`), execution context resolution, configuration dependencies, and artifact traceability. Without this separation, a structural revision would cascade identifier changes across integration layers.

9.5.2 Version Graph and Lineage Modeling

Versioning is not linear in real enterprise systems. It is graph-based. Each `PROCESS_VERSION` can reference a `Parent_Version_ID`. This produces a directed acyclic graph (DAG) rather than a flat revision list. Following is an example of lineage:

```
V1.0
 └── V1.1
      ├── V1.2
      └── V1.2_EU
```

In this model, V1.2 extends V1.1 globally, V1.2_EU branches for regulatory requirements, and both share structural ancestry with V1.1. This lineage graph enables controlled branching for regional variants, inheritance of stable structure, explicit divergence points, and impact analysis across branches.

Lineage is not inferred from timestamps. It is explicitly modeled. The lineage graph supports dependency recalculation, artifact traceability, AI inference reconstruction, and regulatory audit replay.

9.5.3 Temporal Validity and Runtime Resolution

Versioning is both structural and temporal. `PROCESS_VERSION` contains `Effective_From` and `Effective_To`. This allows runtime resolution, such as the one shown in Listing 9.21.

```
SELECT Version_ID
FROM PROCESS_VERSION
WHERE Process_ID = 'P_HR_POSITION'
AND Effective_From <= :executionDate
AND (Effective_To IS NULL OR Effective_To >= :executionDate)
AND Approval_Status = 'Published';
```

Listing 9.21: Resolution Query Example

This capability ensures historical replay, compliance reporting, and AI artifact reconstruction. AI should not implicitly ground itself on the latest version; it must ground on what was published and effective at a specific time.

9.5.4 Publication as an Execution Structural Boundary

A version may exist in multiple states, as follows:

- Draft
- Validated
- Approved
- Published
- Deprecated
- Archived

Only published versions can:

- Serve as execution services version
- Feed AI grounding bundles
- Cross reference configuration impact resolution

- Feed execution services
- Populate AI grounding bundles
- Serve as reference for configuration mapping

Draft or validated versions may exist for modeling purposes, but they are explicitly excluded from runtime exposure. Draft versions can coexist, but they are intentionally invisible to runtime APIs. This boundary is enforced at the data access layer of SAP BTP services. Publication is not advisory; it is enforced.

9.5.5 Dependency Revalidation on Structural Change

Version transitions must account for cross-entity dependencies. When a new version is created, the data spine evaluates the following:

- Are all referenced applications active?
- Are all execution contexts defined?
- Are configuration mappings valid?
- Are master data reference types supported?
- Do validation rules remain consistent?

Consider this scenario: Version V1.3 introduces a new activity, `ACT_VALIDATE_COSTCENTER_CATEGORY`. This activity references `APPLICATION = S4_CORE` and `CONFIG_OBJECT = CFG_CC_CATEGORY`.

Before V1.3 can transition to published, the data spine verifies the following four conditions:

- Every application referenced by an activity in V1.3 exists on the data spine and is itself in a published state.
- Every configuration object referenced through `ACTIVITY_CONFIG_MAP` exists on the data spine and is attested by a named role.
- Every activity in V1.3 has a complete `EXECUTION_CONTEXT` row (meaning `Trigger_Application_ID`, `Target_Application_ID`, and `Context_Role` are all populated).
- No `VALIDATION_RULE` in a draft state applies to the entities included in V1.3. Only when all four conditions pass does the transition to published proceed.

The versioning discipline must account for structural dependencies. A new process version does not exist in isolation. It is linked to:

- Activities
- Application mappings
- Configuration objects
- Master data references
- Validation rules

The data spine tracks relationships between:

- PROCESS_VERSION
- ACTIVITY
- APPLICATION
- CONFIG_OBJECT
- MASTER_DATA_REFERENCE
- VALIDATION_RULE

When a version transitions to published, the data spine recalculates dependencies. Suppose Version 1.3 introduces a new activity, ACT_VALIDATE_CC_CATEGORY. This activity references APPLICATION = S4_CORE and CONFIG_OBJECT = CFG_CC_CATEGORY.

Before Version 1.3 can be published, the data spine must verify:

- The referenced application exists and is not deprecated.
- The configuration object mapping exists.
- No orphaned master data references exist.
- Execution context is defined.

A simplified revalidation logic may resemble the code shown in Listing 9.22.

```
SELECT COUNT(*)
FROM ACTIVITY_CONFIG_MAP ACM
LEFT JOIN CONFIG_OBJECT CO ON ACM.Config_ID = CO.Config_ID
WHERE CO.Config_ID IS NULL;
```

Listing 9.22: Spine Verification Logic

If any configuration mapping is missing, publication is blocked. This ensures that structural evolution does not invalidate implementation coherence.

9.5.6 AI Grounding and Bundle Immutability

AI reproducibility depends on version immutability. Every AI inference must reference the following:

- Process_ID
- Version_ID
- Bundle extraction timestamp
- Policy profile
- Hash of structural snapshot

Listing 9.23 shows an example of an AI-generated artifact's metadata as it is recorded in the GENERATED_ARTIFACT record that anchors the artifact to a canonical version, names the model that produced it, records the policy profile under which it ran, and preserves the bundle manifest hash used to ground the inference.

```
{
  "processId": "P_HR_POSITION",
  "versionId": "V1.3",
  "bundleTimestamp": "2026-03-14T10:22:31Z",
  "policyProfile": "EA_GOVERNANCE_V1",
  "bundleHash": "a93f88e2..."
}
```

Listing 9.23: Example Artifact Metadata

The `bundleHash` is computed from the following:

- `PROCESS_VERSION`
- `ACTIVITY` set
- `EXECUTION_CONTEXT` mappings
- `CONFIG_OBJECT` links
- `VALIDATION_RULE` set

If V1.4 is later introduced, artifacts grounded on V1.3 remain reproducible. This immutability ensures AI traceability, legal defensibility, and governance transparency. Without version discipline, AI suggestions lose historical context.

9.5.7 Managing Variant and Regulatory Branches

Global enterprises require controlled divergence. A regional branch may add regulatory validation, modify execution context, and introduce additional configuration dependency.

Rather than duplicating the entire process, the data spine supports branching from a stable parent version. Branching allows shared structure retention, differential validation rules, and region-specific configuration overlays.

AI must ground on the correct branch explicitly. Branch ambiguity is prohibited.

9.5.8 Versioning as an Operational Control System

Versioning is not a passive record of change. It is an active control system. It governs:

- Which structure integration services execute
- Which configuration mappings are authoritative
- Which validation rules apply
- Which AI artifacts are valid

Without disciplined versioning and lineage execution services drift, configuration traceability breaks, AI reproducibility collapses, and governance becomes reactive. With disciplined versioning, change is controlled, structural evolution is explicit, AI remains explainable, and audit remains defensible. The canonical data spine therefore becomes a controlled evolutionary model rather than a static repository.

9.6 Ownership and Lifecycle Rules

If the canonical data spine defines structure and versioning defines change discipline, ownership and lifecycle rules define authority. Without explicit ownership boundaries and controlled state transitions, even a well-modeled data spine will eventually degrade under operational pressure.

Ownership in the data spine is not conceptual. It is enforced through access control, state transitions, immutability rules, and dependency revalidation. This section formalizes who owns which entities, who may transition lifecycle states, and how governance boundaries are technically enforced across process, architecture, configuration, and AI artifacts.

This section describes who owns which entities, who may transition lifecycle states, and how governance boundaries are enforced across process, architecture, configuration, and AI artifacts.

9.6.1 Entity Ownership Boundaries

The data spine does not invent ownership; it reflects enterprise reality. Process semantics originate in SAP Signavio and are owned by business process stakeholders. Architecture semantics originate in SAP LeanIX and are owned by enterprise architecture. Master data objects remain owned by their source systems—finance for cost centers in SAP S/4HANA and HR for positions in SAP SuccessFactors. The data spine governs how these domains align, but it does not assume domain authority.

However, once domain entities participate in cross-system execution, their participation becomes governed by the data spine. Execution context, configuration mappings, validation rules, and artifact governance are owned within the data spine because they exist to coordinate across domains rather than redefine them. Two layers of authority therefore coexist:

- Domain authority governs content inside the system of record. The domain system owns truth.
- Data spine authority governs cross-domain participation and runtime eligibility. The data spine owns coherence.

9.6.2 Identity and Mutability Boundaries

The integrity of the data spine depends on a clear distinction between immutable anchors and mutable state, as you can see in the following box:

Mutability Boundaries

```
+-----------------------------+
|          IMMUTABLE          |
+-----------------------------+
| Process_ID                  |
| Activity_ID                 |
| Version_ID (once Published) |
| Config_ID                   |
| MD_REF_ID                   |
| Artifact_ID                 |
+-----------------------------+

+-----------------------------+
|           MUTABLE           |
+-----------------------------+
| Activity_Name               |
| Sequence_No                 |
| Approval_Status             |
| Validation_Notes            |
| Artifact_Status (pre-pub)   |
+-----------------------------+
```

Once an entity transitions to published, its structural anchors cannot change. If semantic meaning evolves, a new version must be created. This immutability ensures the following:

- Historical traceability
- Reproducible AI grounding
- Audit defensibility
- Configuration alignment stability

Identity anchors must never change once published. If structural meaning changes, a new version must be created. Variable attributes support refinement prior to publication, but they cannot alter historical traceability.

This distinction protects lineage, AI reproducibility, and audit stability. Variable attributes may be refined prior to publication, but they cannot retroactively alter a published structural state. Immutability is not restrictive; it is protective.

9.6.3 Lifecycle State Model

Each canonical entity progresses through controlled lifecycle states. These states are not symbolic; they determine consumption eligibility.

A process version, for example, transitions through draft, validated, approved, and published states. Only the published state exposes the version to execution services and AI grounding bundles. Once published, structural modification is no longer permitted. A new version must be created to introduce change.

Artifacts follow a similar but distinct lifecycle. An AI-generated CONFIG_SPEC may be in a draft, proposed, approved, implemented, or retired state. These states govern whether it is actionable, not whether it exists. Lifecycle modeling enforces temporal discipline. Without it, semantic drift is inevitable.

9.6.4 Validation and Publication Authority

Validation and publication are not administrative steps inserted at the end of a modeling exercise. They are architectural control gates that determine whether semantic content is safe for execution.

Validation exists to ensure structural integrity before exposure. It confirms that execution context is defined and unambiguous, master data references are valid and aligned to their systems of record, configuration dependencies are explicitly mapped, and no draft versions of processes or artifacts are unintentionally exposed to runtime services. In effect, validation enforces coherence across domains before a process definition becomes consumable.

Publication, however, is a separate act. Validation determines correctness; publication determines authority. Only designated governance roles may transition a canonical entity into the published state. That transition typically requires coordinated review by the process owner, the architecture authority, and the governance function. Publication is therefore the moment at which semantic content becomes executable and eligible for AI grounding. Listing 9.24 shows the service-layer enforcement query that prevents invalid version transitions. The database refuses any state change that does not satisfy the verification conditions, so the publication gate cannot be bypassed in application code.

```
UPDATE PROCESS_VERSION
SET Approval_Status = 'Published',
    Approved_By = :user,
    Approved_At = CURRENT_TIMESTAMP
WHERE Version_ID = :versionId
AND Approval_Status = 'Approved';
```

Listing 9.24: Service-Layer Enforcement Must Prevent Invalid Transitions

No API endpoint, integration service, or AI runtime is permitted to bypass this boundary. The separation between validated structure and published authority ensures that execution remains deliberate rather than accidental.

9.6.5 Cross-Domain Change Discipline

In integrated enterprise environments, change rarely remains confined to a single domain. A modification to a financial master record may ripple into workforce processes. An application lifecycle change in SAP LeanIX may invalidate execution mappings. A configuration update in SAP S/4HANA may require new validation constraints before downstream systems can consume updated semantics.

The canonical data spine treats these interactions as structural dependencies rather than incidental side effects. When a dependent entity transitions state—whether a cost center is deprecated, an interface is retired, or a configuration parameter is altered—the data spine enforces revalidation across affected entities. This prevents silent misalignment and ensures that cross-domain dependencies remain explicit.

Change discipline, therefore, is not reactive. It is encoded into the lifecycle model. The data spine does not wait for failure; it anticipates dependency impact through structured relationships and revalidation triggers. The query in Listing 9.25 identifies published versions that must be revalidated due to architectural change.

```
SELECT PV.Version_ID
FROM PROCESS_VERSION PV
JOIN ACTIVITY A ON PV.Version_ID = A.Version_ID
JOIN ACTIVITY_APPLICATION_MAP AM ON A.Activity_ID = AM.Activity_ID
JOIN APPLICATION APP ON AM.App_ID = APP.App_ID
WHERE APP.Lifecycle_State = 'Deprecated'
AND PV.Approval_Status = 'Published';
```

Listing 9.25: Query to Identify Published Versions

Change discipline is encoded into the structural model. The data spine anticipates dependency impact rather than reacting to runtime failure.

9.6.6 Separation of Duties and Access Control

Ownership boundaries are meaningful only when they are technically enforced. The data spine's governance model must align with enterprise separation of duties (SoD) principles.

Process modelers define semantics but cannot approve publication. Integration architects manage execution context but cannot modify process definitions. AI services generate artifacts but cannot alter canonical entities. Governance authorities authorize publication but do not modify upstream master data. These separations are implemented through role-based access control within SAP BTP, typically using Identity Authentication or SAP Authorization and Trust Management service role collections and service-layer authorization. Lifecycle transitions are bound to role entitlements rather than convention. Publication only transitions via governance roles

This ensures that governance remains technical rather than ceremonial. Authority is encoded into the platform, not assumed. Lifecycle transitions are bound to entitlements rather than convention. Governance becomes structural when authority is encoded in the platform itself.

9.6.7 AI Consumption Boundaries

AI operates as a structured consumer of published semantic bundles. It does not own canonical entities and does not participate in lifecycle transitions beyond artifact generation. Within this architecture, AI may analyze published process definitions, evaluate execution context, correlate telemetry, and generate structured artifacts such as configuration specifications or workflow definitions. It may propose refinements to process structure or suggest integration adjustments.

It cannot publish new versions, modify canonical records, execute configuration changes, or override validation failures. These constraints are not optional safeguards; they are foundational design principles. AI augments governance, but it does not replace it.

By restricting AI to structured artifact generation and enforcing publication discipline, the data spine preserves trust while still enabling automation.

9.6.8 Governance as Structural Integrity

Ownership and lifecycle rules are not external policies layered onto the data model. They are embedded within the structure of the data spine itself. Without explicit ownership boundaries, versioning becomes ambiguous. Without lifecycle enforcement, validation becomes discretionary. Without publication discipline, AI grounding becomes unsafe. Without structural traceability, configuration changes lose accountability.

When ownership and lifecycle rules are encoded directly into canonical entities and their state transitions, governance becomes part of the architecture rather than an afterthought. The canonical data spine then functions not merely as a semantic alignment layer but as a controlled execution platform capable of supporting explainable AI and disciplined change management. Structural integrity, in this context, is not theoretical but operational.

9.7 Summary

This chapter defined the canonical data spine as a governed semantic backbone rather than a data aggregation layer. It established why point-to-point integrations degrade over time and why stable identifiers, explicit execution context, and disciplined lifecycle control are prerequisites for reliable execution.

The domain model clarified that enterprise truth is distributed: process semantics in SAP Signavio, architecture semantics in SAP LeanIX, master data in source systems such as SAP S/4HANA and SAP SuccessFactors, and adoption telemetry in WalkMe. The data spine does not replace these systems; it stabilizes their relationships and governs publication boundaries.

The schema formalized structural identity, separated version from state, and elevated execution context as a first-class construct. We referenced master data rather than replicated it. We explicitly linked configuration objects to process activities, enabling traceable alignment between business intent and technical implementation.

The applied use case demonstrated how finance-governed cost centers and HR-governed positions can operate coherently across systems when anchored in canonical identity and lifecycle discipline. Ownership and lifecycle rules completed the architecture by defining authority boundaries, validation gates, and AI consumption constraints.

The result is an execution-grade semantic model capable of supporting SAP BTP services and SAP AI without compromising governance, auditability, or system of record integrity.

9.7.1 Key Learning Points

- The canonical data spine exists to preserve meaning across systems. Its primary contribution is not integration connectivity but semantic stability.
- Identity must remain stable even as versions evolve. Structural identifiers are long-lived anchors; version states are controlled transitions. Confusing the two leads to drift.
- Execution context must be explicit. Multisystem processes require formal modeling of trigger and target ownership. Without it, ambiguity migrates into configuration and AI suggestion logic.
- Master data should be referenced, not replicated. The data spine stores only what is necessary for validation and grounding, preserving system of record authority.
- Validation and publication are architectural gates. Only published entities are eligible for execution services and AI grounding. Draft leakage undermines trust.
- AI augments governance but does not replace it. Structured artifact generation preserves human accountability while enabling automation.

9.7.2 Common Design Pitfalls

- Treating the data spine as a replication layer. Copying full upstream schemas into SAP BTP increases complexity and erodes ownership clarity. The data spine should normalize, not duplicate.

- Embedding semantic mappings exclusively in middleware flows. When reconciliation logic lives only inside integration scripts, governance becomes opaque, and drift is inevitable.
- Underestimating execution context modeling. Assuming that the system referenced in a process diagram is the same as the system that executes a step leads to configuration misalignment in cross-application landscapes.
- Allowing draft or partially validated entities to be consumed by execution services or AI models. This erodes trust quickly. Publication discipline must be enforced technically, not procedurally.
- Granting AI direct mutation authority over canonical entities. This introduces governance risk. AI should generate structured proposals, not execute change.

9.7.3 Review Questions

1. Why is separating structural identity from version state essential for long-running transformation programs?
2. How does execution context modeling prevent ambiguity in multisystem processes?
3. In the cost center–position scenario, what risks emerge if master data is replicated rather than referenced?
4. Why must validation precede publication in an AI-ready integration architecture?
5. How do configuration object mappings strengthen traceability between business intent and technical implementation?
6. What structural risks arise if AI is allowed to consume draft process versions?
7. How do ownership boundaries preserve auditability and lifecycle discipline?

9.7.4 Student Companion Notes

If you are studying this chapter as part of an SAP University Alliance program or enterprise architecture curriculum, you should focus on understanding the structural principles rather than memorizing table definitions.

First, practice diagramming a simple cross-system process and explicitly labeling trigger and target systems. Second, trace how a process version transitions from draft to published and how that affects eligibility for execution and AI consumption. Third, model a simple configuration dependency and explain how it links back to process intent.

This chapter provides the structural foundation for the rest of the book. Subsequent chapters will demonstrate how that structure is implemented and operated on SAP BTP.

Chapter 10

Implementing the Data Spine on SAP BTP

Architecture is not built by drawing diagrams; it is built by writing the rows, columns, and constraints that make those diagrams true. The data spine becomes real the moment its tables can be queried, its application programming interfaces (APIs) can be called, and its audit trail can be replayed.

Chapter 9 defined the canonical data spine in conceptual and logical terms. It explained why point-to-point integrations fail at scale, identified the canonical entities that must be governed centrally, set out the keys that bind those entities together, and established the versioning, ownership, and lifecycle rules that keep the data spine coherent over time. That was the design. This chapter is the build.

This chapter turns the design into a working implementation on SAP Business Technology Platform (SAP BTP). It explains which SAP BTP services are used to persist the canonical model, how each canonical entity is mapped onto a concrete table or document store, how schemas are organized so that domain ownership is enforced at the database layer rather than only by convention, how the data spine is extended without polluting its canonical core, how APIs expose ingestion and access surfaces with idempotency and replay safety, and how every operation is captured in an audit trail strong enough to serve as legal evidence.

The implementation discipline matters as much as the design. A well-designed data spine that is implemented carelessly will drift within months. Custom columns will accumulate. Foreign keys (FKs) will be dropped temporarily to unblock a load. Audit logs will grow into a heap that nobody can query. AI services will reach into raw tables instead of through governed bundles. This chapter is therefore prescriptive about implementation patterns. The choices we outline are the ones that have proven defensible in practice, and the alternatives we warn against are the ones that erode the data spine over time.

This chapter explains the following:

- Which SAP BTP persistence services map to which canonical responsibilities and why a single store cannot serve them all
- How the canonical schema is realized as SAP HANA Cloud tables, indexes, partitions, and SAP HANA deployment infrastructure (HDI) containers, including SQL data definition language (DDL) for the core entities
- How extension fields and tenant-specific attributes are added through governed side tables rather than by widening canonical tables

- How the data spine exposes three distinct API surfaces (ingestion, access, and governance), each with its own contract, idempotency model, and security regime
- How auditability is engineered into the platform through hash-chained audit events, lineage edges, and tamper-evident retention so that the data spine itself is its own evidence chain
- How AI services consume the data spine through grounding bundles that are read-only, role-scoped, and version-pinned, in keeping with the AI boundaries established in Chapter 1

This chapter is written for anyone who will sit in front of SAP HANA Cloud, SAP BTP cockpit, SAP Cloud Application Programming Model, and SAP Build to deliver these capabilities. Each section moves from principle to pattern to concrete artifact: tables, columns, constraints, endpoints, and event payloads.

10.1 Persistence Options on SAP BTP

The data spine is multi-store by design. A single persistence technology cannot satisfy the simultaneous demands of relational integrity for canonical entities, schemaless flexibility for telemetry payloads, vector retrieval for AI grounding, append-only durability for audit trails, and event-stream semantics for change propagation. Forcing all of these into one store either degrades one capability or compromises governance. The implementation therefore distributes the data spine across several SAP BTP services, with SAP HANA Cloud at the center and other stores attached around it for capabilities that SAP HANA Cloud is not the right choice for.

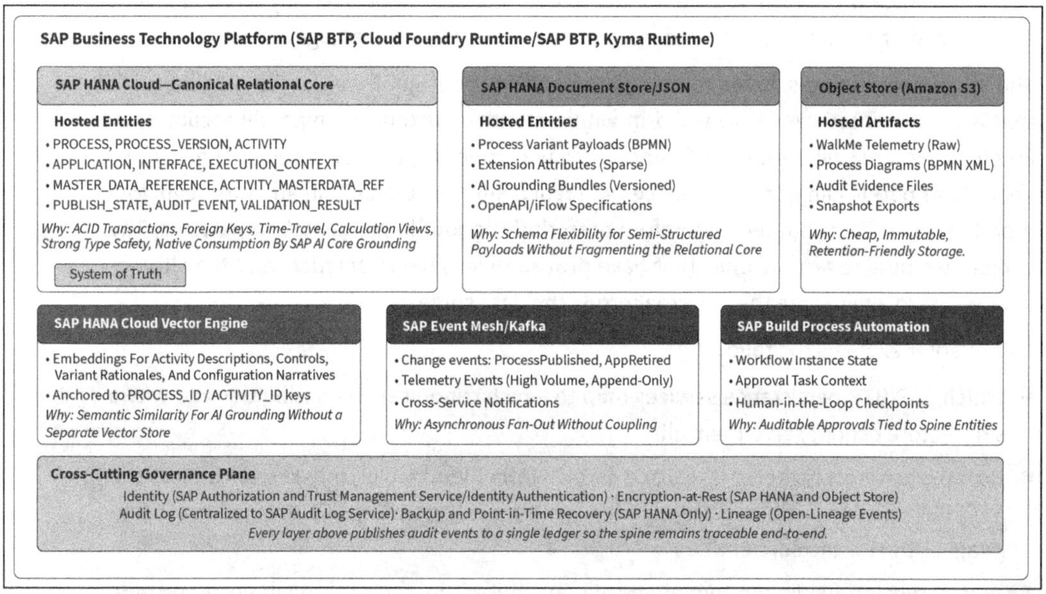

Figure 10.1: SAP BTP Persistence Layers for the Canonical Data Spine

This section catalogs the persistence services (see Figure 10.1 for an overview) used by the data spine, explains which canonical responsibilities each one carries, and sets out the rules that prevent the multi-store architecture from devolving into a mess. The rules are simple, and they are enforced: Each canonical entity has exactly one primary store, secondary stores hold only derived or projected forms, cross-store joins are performed only through governed APIs, and every store is bound to the same governance and audit regime.

10.1.1 SAP HANA Cloud as the Canonical Relational Core

SAP HANA Cloud is the system of record for the canonical entities defined in Chapter 9: `PROCESS`, `PROCESS_VERSION`, `ACTIVITY`, `APPLICATION`, `INTERFACE`, `EXECUTION_CONTEXT`, `MASTER_DATA_REFERENCE`, and the tables that bind them. These entities are relational by nature. They have stable keys, well-defined cardinalities, and integrity constraints that must be enforced atomically. They are also queried in ways that benefit from a real query optimizer: change-impact lookups, lineage traversals, role-scoped reads, and grounding-bundle assembly all involve multi-table joins with predicate pushdown.

We chose SAP HANA Cloud not because it is the only relational option on SAP BTP, but because it provides three properties the data spine depends on. It supports schema isolation through HDI containers, which lets each domain own its tables and grant access explicitly. It offers built-in time-travel through system-versioned tables, which simplifies the lineage and reproducibility requirements identified in Chapter 9. It also supports vector embeddings natively through the vector engine, which keeps the AI grounding layer colocated with the canonical data instead of requiring a separate retrieval store.

SAP HANA Cloud holds the canonical state. It does not hold raw telemetry, large binary payloads, or transient workflow state. Those belong in other stores.

10.1.2 Object Store for Raw Telemetry and Archival Payloads

WalkMe telemetry events, SAP Signavio export payloads, and SAP LeanIX webhook payloads arrive in volumes that would be wasteful to keep in SAP HANA Cloud at full fidelity. The raw form of these payloads is also rarely queried; what matters operationally is the parsed, normalized projection that ends up in SAP HANA. The Object Store service on SAP BTP, backed by an Amazon Simple Storage Service (S3)-compatible bucket, is the right place for the raw form. Each ingestion event writes its raw payload to the bucket under a key derived from the event identifier, and SAP HANA Cloud holds only the parsed projection plus a pointer back to the raw object.

This separation has three benefits. It keeps SAP HANA Cloud focused on query able canonical state, which protects performance and cost. It preserves the unmodified source payload for forensic replay, which is essential when a downstream defect is traced back to an ingestion bug. It also allows retention policies to differ: SAP HANA Cloud retains canonical state indefinitely, while raw payloads can be tiered to cold storage after a defined window without losing the canonical record.

10.1.3 Document Store for Schemaless and Variant Payloads

Some integration data does not fit comfortably into a fixed relational schema. Validation results that include arbitrary nested structures, AI grounding bundle manifests that vary in shape across model versions, and transient enrichment payloads exchanged between execution services are all examples. These can be persisted in SAP HANA Cloud as JSON columns, but when their shape changes frequently or when they are written far more than they are read in structured form, the SAP HANA Cloud document store provides a more direct fit. In this chapter, we use it for noncanonical, nonauthoritative payloads where structural flexibility matters more than referential integrity.

The discipline is that the document store never holds primary canonical state. Anything stored there is either a derived projection of canonical data or a transient artifact. Loss of the document store contents must be recoverable from canonical sources within a defined recovery window.

10.1.4 SAP HANA Cloud Vector Engine for AI Grounding Embeddings

In Chapter 1, we established that AI services consume the data spine only through governed grounding bundles, and in Chapter 9, we made grounding a first-class canonical concern. The implementation requires vector storage for embeddings of process descriptions, activity steps, control intents, and other text spans that AI services retrieve by semantic similarity. The SAP HANA Cloud vector engine provides this capability inside the same database that holds the canonical entities, which simplifies governance considerably. Embeddings can be joined to their source rows in a single query, access controls inherit from the same role model, and the audit trail covers retrieval as a first class operation.

Colocating embeddings with canonical data also makes versioning straightforward. Every embedding row carries the version of the source entity it was derived from, the model identifier that produced it, and the timestamp of generation. When the source row is updated, the embedding is regenerated under a new version, and the old embedding is retained until its retention window closes. This keeps grounding bundles reproducible. A bundle assembled last month can be replayed today against the same embeddings the AI service originally received.

10.1.5 Event Mesh for Change Propagation

The data spine is not a pull-only repository. Downstream services such as execution APIs, AI grounding services, and monitoring dashboards need to know when canonical state changes. Event Mesh is the change-propagation layer. Every committed write to a canonical entity emits a domain event onto a topic named after the entity. Subscribers receive a CloudEvents-formatted message with the entity identifier, version, change type, and a pointer to the canonical row. They do not receive the full row payload; they re-fetch it through the Access API to ensure they always read the latest committed state under their own role context.

This pattern decouples write throughput on the data spine from read latency on its consumers, and it ensures that no consumer becomes the unintended system of record for canonical state. Events are advertisements that something has changed; they are not authoritative carriers of the change.

10.1.6 SAP Build Process Automation for Workflow States

Some operations on the data spine are not single transactions. Approving a process version, retiring an interface, or accepting an AI suggestion is a workflow that spans multiple human and system steps and may take days to complete. SAP Build Process Automation holds the workflow state for these operations. The canonical record of the eventual outcome lives in SAP HANA Cloud for example, the new approved version row in `PROCESS_VERSION`—but the in-flight state of the approval workflow lives in SAP Build Process Automation until it terminates.

This separation prevents the data spine from accumulating partial, in-flight, or speculative state. Either an approval has happened (and SAP HANA reflects it) or it has not (and SAP Build Process Automation is still tracking it). There is no third state where the data spine holds an unapproved version that might or might not be resolved.

10.1.7 Choosing the Right Store for Each Canonical Concern

The mapping from canonical entity to primary store is not an architectural opinion that varies by team. It is a design rule. The decision matrix in Table 10.1 makes that rule explicit: Each canonical entity has one primary store, and any secondary store carries only a derived or projected representation. When teams later propose to move data between stores or to introduce a new store, that proposal is reviewed against the matrix. If it would violate the rule, it is either redesigned or escalated for architectural change.

Canonical Entity	Primary Store	Secondary Store	Volume	Change Frequency	Rationale	Volume Profile
`PROCESS`, `PROCESS_VERSION` *Process semantics domain*	SAP HANA Cloud (relational)	JSON store (Business Process Model and Notation [BPMN] payload)	Low	Quarterly	Strong identity, FK integrity, time-travel	Stable/governed (low write rate)
`ACTIVITY` *L4/L5 executable steps*	SAP HANA Cloud (relational)	Vector store (embeddings)	Medium	Monthly	Joins to `APP`, `EXEC_CTX`; AI grounding	Stable/governed (low write rate)

Table 10.1: Persistence Decision Matrix—Canonical Entity to Store Mapping

Canonical Entity	Primary Store	Secondary Store	Volume	Change Frequency	Rationale	Volume Profile
APPLICATION, INTERFACE *Architecture semantics*	SAP HANA Cloud (relational)	–	Low	Monthly	Lifecycle queries; impact analysis joins	Stable/ governed (low write rate)
EXECUTION_CONTEXT *Trigger/target ownership*	SAP HANA Cloud (relational)	–	Medium	Monthly	Pivotal join; consumed by execution svc	Stable/ governed (low write rate)
MASTER_DATA_REFERENCE *References, never replicas*	SAP HANA Cloud (relational)	–	Medium	Daily	FK to SAP S/4HANA and SAP SuccessFactors keys only	Append-only (high volume, immutable)
TELEMETRY_EVENT *WalkMe interaction signals*	Object Store (raw)	SAP HANA aggregated tables	Very high	Continuous	Append-only, retention-tiered	Stable/ governed (low write rate)
WALKME_ASSET *Adoption artifacts*	SAP HANA Cloud (relational)	–	Low	Weekly	Joined to ACTIVITY for friction analysis	Append-only (high volume, immutable)
AI_GROUNDING_BUNDLE *Curated AI input package*	SAP HANA Cloud and JSON	Object Store (snapshots)	Medium	On publish	Reproducibility: snapshot kept immutably	AI-derived (regenerated from canonical)
AUDIT_EVENT *Append-only ledger*	SAP HANA Cloud (relational)	Object Store (long-term)	High	Per action	No modification; tiered to Amazon S3 after 90 days	Append-only (high volume, immutable)

Table 10.1: Persistence Decision Matrix—Canonical Entity to Store Mapping (Cont.)

The matrix is also the basis for capacity planning. SAP HANA Cloud sizing is driven by the canonical entity counts and their growth rates. Object Store sizing is driven by raw payload volumes and retention windows. The SAP HANA Cloud vector engine is sized by the number of embedded text spans times the embedding dimensionality. Treating these dimensions independently, rather than rolling them into a single SAP BTP cost line, makes operational forecasting tractable.

Implementation Note

Every persistence service used by the data spine is provisioned in the same SAP BTP global account and subaccount as the canonical APIs. Cross-account or cross-region persistence introduces network latency, billing complexity, and audit fragmentation that are not justified by the modest deployment-isolation benefit. If multi-region deployment is required for regulatory reasons, each region runs the entire spine as a self-contained instance, with replication handled at the canonical record level rather than at the storage layer.

10.2 Table and Schema Design

This section covers the working DDL for the canonical core. It defines the SAP HANA Cloud tables that realize the entities introduced in Chapter 9, the keys and constraints that enforce their integrity, the indexes that make them queryable at execution latency, and the HDI container layout that organizes them by domain ownership. The intent is that an implementation team can take these definitions, adapt them to local naming conventions, and deploy them through the SAP Cloud Application Programming Model without rederiving the schema from first principles.

The schema is presented in narrative order in Figure 10.2: first the entities that anchor process and architecture identity, then the entities that bind them to execution context, then the entities that carry adoption and AI consumption metadata, and finally the cross-cutting governance entities. Where a table is introduced, its primary key (PK), FKs, and most consequential constraints are shown; secondary attributes are summarized in prose to keep the listings readable. The complete attribute model is reproduced in Appendix D.

In the following sections, we will review the table structure needed for creating the canonical schema which supports the implementation data spine from the ERD. We also provide SAP HANA SQL DDL needed to create the table with its key attribute. Note that the table structure is minimum viable set of attributes needed for automating generation of the process and does not limit you from expanding the attributes to your need for your implementation.

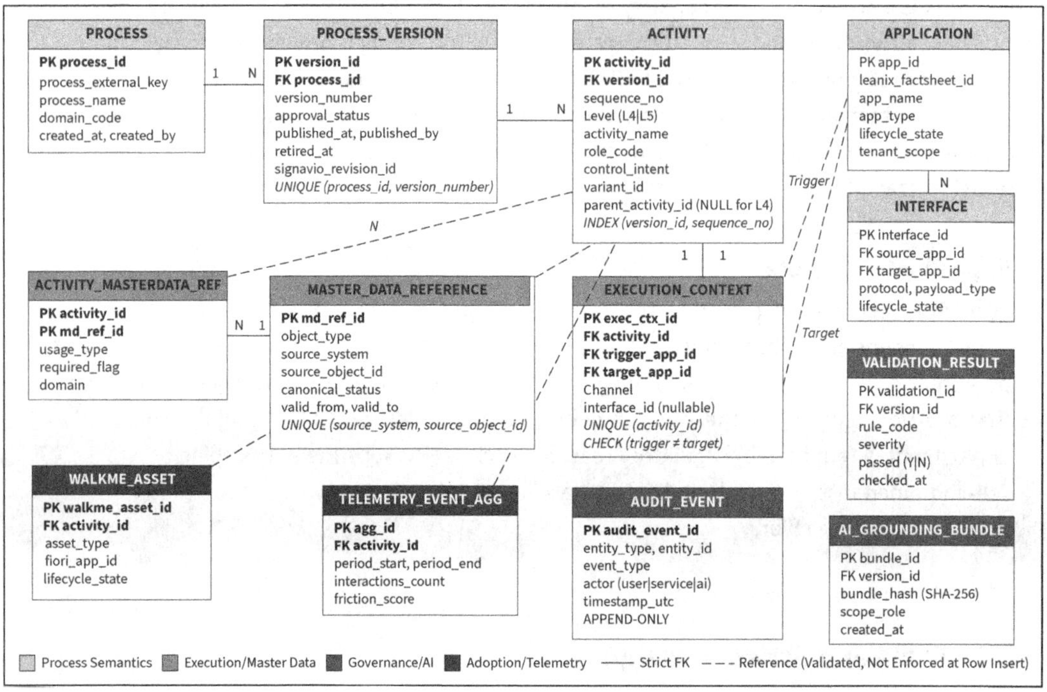

Figure 10.2: Canonical Schema—Entity-Relationship Diagram (SAP HANA Cloud)

10.2.1 Process and Activity Tables

PROCESS is the data spine's representation of an SAP Signavio process at L4. It carries the stable identifier issued by SAP Signavio, the canonical name, the owning domain, and a status that distinguishes draft, published, and retired states. PROCESS_VERSION carries the version-specific attributes that change as the process evolves: version number, effective dates, change reason, and the hash of the canonical content at that version. ACTIVITY belongs to a specific PROCESS_VERSION and represents an L5 step. Its identifier is stable across versions where the activity persists and is reissued only when an activity is added, removed, or semantically reshaped. Listing 10.1 defines the PROCESS and PROCESS_VERSION tables on the process_semantics HDI container.

```
CREATE TABLE process_semantics.PROCESS (
  process_id           NVARCHAR(36)  NOT NULL,
  signavio_process_id  NVARCHAR(64)  NOT NULL,
  canonical_name       NVARCHAR(256) NOT NULL,
  owning_domain        NVARCHAR(64)  NOT NULL,
  status               NVARCHAR(16)  NOT NULL,
  created_at           TIMESTAMP     NOT NULL,
  created_by           NVARCHAR(128) NOT NULL,
  CONSTRAINT pk_process PRIMARY KEY (process_id),
```

```
  CONSTRAINT uq_process_signavio UNIQUE (signavio_process_id),
  CONSTRAINT ck_process_status CHECK (status IN ('DRAFT','PUBLISHED',
'RETIRED'))
);

CREATE TABLE process_semantics.PROCESS_VERSION (
  process_version_id  NVARCHAR(36)  NOT NULL,
  process_id          NVARCHAR(36)  NOT NULL,
  version_number      INTEGER       NOT NULL,
  effective_from      TIMESTAMP     NOT NULL,
  effective_to        TIMESTAMP,
  change_reason       NVARCHAR(512),
  content_hash        NVARCHAR(64)  NOT NULL,
  approved_at         TIMESTAMP,
  approved_by         NVARCHAR(128),
  CONSTRAINT pk_process_version PRIMARY KEY (process_version_id),
  CONSTRAINT fk_pv_process FOREIGN KEY (process_id)
    REFERENCES process_semantics.PROCESS(process_id),
  CONSTRAINT uq_pv_version UNIQUE (process_id, version_number)
);
```

Listing 10.1: Defining the PROCESS and PROCESS_VERSION Tables

The CONTENT_HASH column on PROCESS_VERSION is important. It is computed as the SHA-256 hash of the canonical serialization of the version's content—the activities, their attributes, their relationships, and any references they carry. Two PROCESS_VERSION rows with the same CONTENT_HASH represent the same process content, even if they were ingested at different times. This makes deduplication trivial during reconciliation, and it provides the integrity anchor that the audit trail depends on. A change to any activity inside a version produces a different hash, which means a tampered or corrupted version can be detected without comparing every column. Listing 10.2 defines the ACTIVITY table, scoped to a specific PROCESS_VERSION through an FK.

```
CREATE TABLE process_semantics.ACTIVITY (
  activity_id          NVARCHAR(36)  NOT NULL,
  process_version_id   NVARCHAR(36)  NOT NULL,
  signavio_activity_id NVARCHAR(64)  NOT NULL,
  sequence_number      INTEGER       NOT NULL,
  canonical_name       NVARCHAR(256) NOT NULL,
  control_intent       NVARCHAR(64),
  role_responsible     NVARCHAR(128),
  CONSTRAINT pk_activity PRIMARY KEY (activity_id),
  CONSTRAINT fk_activity_pv FOREIGN KEY (process_version_id)
    REFERENCES process_semantics.PROCESS_VERSION(process_version_id),
```

```
  CONSTRAINT uq_activity_in_version UNIQUE (process_version_id,
signavio_activity_id)
);
```

Listing 10.2: Defining the ACTIVITY Table

ACTIVITY belongs to a specific PROCESS_VERSION rather than to a PROCESS directly. This is intentional: When a process is reversioned, its activities are reissued with their new sequence and any attribute changes, while the previous version's activities remain unchanged in their own version row. Lookups that need the current activities for a process join PROCESS to its latest published PROCESS_VERSION and then to ACTIVITY; lookups that need the activities as they were on a specific date select the PROCESS_VERSION row that was effective on that date and join from there.

10.2.2 Architecture Tables

APPLICATION and INTERFACE are the data spine's representations of SAP LeanIX objects. APPLICATION holds the stable SAP LeanIX factsheet identifier, the canonical name, the lifecycle state, the owning domain, and a few attributes that are referenced often enough by execution services to justify caching them on the data spine rather than fetching them from SAP LeanIX on every call. INTERFACE represents an integration interface (typically an API, an event topic, or a file feed) between two applications, with directionality and transport class. Listing 10.3 defines the APPLICATION and INTERFACE tables on the arch_semantics HDI container.

```
CREATE TABLE arch_semantics.APPLICATION (
  application_id        NVARCHAR(36)  NOT NULL,
  leanix_factsheet_id   NVARCHAR(64)  NOT NULL,
  canonical_name        NVARCHAR(256) NOT NULL,
  lifecycle_state       NVARCHAR(16)  NOT NULL,
  owning_domain         NVARCHAR(64)  NOT NULL,
  hosting_class         NVARCHAR(32),
  CONSTRAINT pk_application PRIMARY KEY (application_id),
  CONSTRAINT uq_application_leanix UNIQUE (leanix_factsheet_id),
  CONSTRAINT ck_application_lifecycle CHECK (
    lifecycle_state IN ('PLAN','ACTIVE','PHASE_OUT','RETIRED'))
);

CREATE TABLE arch_semantics.INTERFACE (
  interface_id          NVARCHAR(36)  NOT NULL,
  leanix_interface_id   NVARCHAR(64)  NOT NULL,
  source_application    NVARCHAR(36)  NOT NULL,
  target_application    NVARCHAR(36)  NOT NULL,
  transport_class       NVARCHAR(32)  NOT NULL,
  canonical_name        NVARCHAR(256) NOT NULL,
```

```
  lifecycle_state      NVARCHAR(16)  NOT NULL,
  CONSTRAINT pk_interface PRIMARY KEY (interface_id),
  CONSTRAINT uq_interface_leanix UNIQUE (leanix_interface_id),
  CONSTRAINT fk_interface_source FOREIGN KEY (source_application)
    REFERENCES arch_semantics.APPLICATION(application_id),
  CONSTRAINT fk_interface_target FOREIGN KEY (target_application)
    REFERENCES arch_semantics.APPLICATION(application_id),
  CONSTRAINT ck_interface_distinct CHECK (source_application <>
                                           target_application)
);
```

Listing 10.3: Defining the APPLICATION and INTERFACE Tables

The CHECK constraint on INTERFACE that source and target applications must differ is the first place the schema captures a rule that is otherwise easily violated. Self-referencing interfaces (an application calling itself) are not interfaces in the architectural sense; they are internal calls, and modeling them as interfaces inflates dependency graphs in misleading ways. Enforcing the distinction at the database level prevents loose ingestion paths from introducing them by accident.

10.2.3 Execution Context: Where Process Meets Architecture

EXECUTION_CONTEXT is the table that binds an activity to the application that performs it and, where applicable, to the interface that carries the work to a subsequent application. It is the most queried table in the data spine; almost every change-impact analysis, AI grounding bundle, and audit reconstruction starts here. Its design therefore prioritizes lookup performance and integrity. Listing 10.4 defines the EXECUTION_CONTEXT table together with the four indexes that keep change-impact and grounding lookups fast at production scale.

```
CREATE TABLE execution_md.EXECUTION_CONTEXT (
  execution_context_id NVARCHAR(36)  NOT NULL,
  activity_id          NVARCHAR(36)  NOT NULL,
  trigger_application  NVARCHAR(36)  NOT NULL,
  target_application   NVARCHAR(36)  NOT NULL,
  interface_id         NVARCHAR(36),
  execution_pattern    NVARCHAR(32)  NOT NULL,
  effective_from       TIMESTAMP     NOT NULL,
  effective_to         TIMESTAMP,
  CONSTRAINT pk_exec_context PRIMARY KEY (execution_context_id),
  CONSTRAINT fk_ec_activity FOREIGN KEY (activity_id)
    REFERENCES process_semantics.ACTIVITY(activity_id),
  CONSTRAINT fk_ec_trigger FOREIGN KEY (trigger_application)
    REFERENCES arch_semantics.APPLICATION(application_id),
  CONSTRAINT fk_ec_target FOREIGN KEY (target_application)
    REFERENCES arch_semantics.APPLICATION(application_id),
```

```
  CONSTRAINT fk_ec_interface FOREIGN KEY (interface_id)
    REFERENCES arch_semantics.INTERFACE(interface_id),
  CONSTRAINT ck_ec_pattern CHECK (
    execution_pattern IN ('SYNC_API','ASYNC_EVENT','BATCH','MANUAL_
HANDOFF')),
  CONSTRAINT ck_ec_distinct_apps CHECK (trigger_application <> target_
application)
);

CREATE INDEX ix_ec_activity ON execution_md.EXECUTION_CONTEXT(activity_
id);
CREATE INDEX ix_ec_trigger ON execution_md.EXECUTION_CONTEXT(trigger_
application);
CREATE INDEX ix_ec_target  ON execution_md.EXECUTION_CONTEXT(target_
application);
CREATE INDEX ix_ec_interface ON execution_md.EXECUTION_
CONTEXT(interface_id);
```

Listing 10.4: Defining the EXECUTION_CONTEXT Table and Its Indexes

Four indexes are nonnegotiable on this table. Change-impact services need the `activity_id` index to find every execution context affected by a process change. Architecture-driven impact analysis needs the trigger and target application indexes to find every activity that touches an application scheduled for retirement. AI grounding services need the interface index to assemble interface-level reasoning bundles. Without these indexes, the lookup latency at production scale rises into seconds rather than milliseconds, and the data spine becomes the bottleneck the architecture was meant to remove.

The `CK_EC_DISTINCT_APPS` check captures the same principle as the equivalent constraint on `INTERFACE`. An execution context where trigger and target are the same application is not meaningful; it is an internal call that should not be modeled here. Enforcing it at the database prevents silent corruption.

10.2.4 Master Data and Adoption References

Activities frequently depend on master data. A posting activity requires a chart-of-accounts lookup, a procurement activity requires a vendor master, and a credit-check activity requires a customer master. These dependencies are tracked through `MASTER_DATA_REFERENCE`, a catalog of master data sources known to the data spine, and `ACTIVITY_MASTERDATA_REF`, the link table that records which activities depend on which master data sources. The link table is necessary for change-impact analysis. When a master data domain is reorganized, every dependent activity must be identifiable in a single query. Listing 10.5 defines the master data reference catalog and the link table that records which activities depend on which master data sources.

```
CREATE TABLE execution_md.MASTER_DATA_REFERENCE (
  md_reference_id      NVARCHAR(36)  NOT NULL,
  domain               NVARCHAR(64)  NOT NULL,
  canonical_name       NVARCHAR(256) NOT NULL,
  source_application   NVARCHAR(36)  NOT NULL,
  CONSTRAINT pk_md_reference PRIMARY KEY (md_reference_id),
  CONSTRAINT fk_md_source FOREIGN KEY (source_application)
    REFERENCES arch_semantics.APPLICATION(application_id),
  CONSTRAINT uq_md_name_in_domain UNIQUE (domain, canonical_name)
);

CREATE TABLE execution_md.ACTIVITY_MASTERDATA_REF (
  activity_id          NVARCHAR(36)  NOT NULL,
  md_reference_id      NVARCHAR(36)  NOT NULL,
  usage_class          NVARCHAR(32)  NOT NULL,
  CONSTRAINT pk_act_md PRIMARY KEY (activity_id, md_reference_id),
  CONSTRAINT fk_amd_activity FOREIGN KEY (activity_id)
    REFERENCES process_semantics.ACTIVITY(activity_id),
  CONSTRAINT fk_amd_md FOREIGN KEY (md_reference_id)
    REFERENCES execution_md.MASTER_DATA_REFERENCE(md_reference_id),
  CONSTRAINT ck_amd_usage CHECK (usage_class IN ('READ','VALIDATE',
'POST'))
);
```

Listing 10.5: Defining the Master Data Reference Tables

WALKME_ASSET and TELEMETRY_EVENT_AGG carry the adoption-layer mapping introduced in Chapter 8. WALKME_ASSET records each WalkMe Smart Walk-Thru, SmartTip, or launcher tied to an activity, with its mapped SAP Fiori app identifier, target audience role, and lifecycle state. TELEMETRY_EVENT_AGG holds aggregated daily counts of friction events, completions, and abandonments per activity per role, computed from raw telemetry that lives in the Object Store. Aggregating at a daily grain gives execution dashboards a usable read path while keeping SAP HANA Cloud volumes manageable; the raw event stream remains available in the Object Store for forensic analysis at finer time grains.

10.2.5 Audit and AI Consumption Tables

AUDIT_EVENT and AI_GROUNDING_BUNDLE close out the canonical core. AUDIT_EVENT is described in detail in Section 10.5; for the schema design discussion in this section, what matters is its append-only nature, its hash-chained integrity, and its role as the data spine's evidence layer. AI_GROUNDING_BUNDLE records every grounding bundle assembled for an AI service: its identifier, the role context under which it was assembled, the entities and versions it included, the assembly hash, and the timestamp at which it was delivered. The bundle itself (the actual content shipped to the AI model) lives in the Object Store; the table holds the manifest. Listing 10.6 defines the `AUDIT_EVENT` table, including the hash-chain columns that make the audit trail tamper-evident.

```
CREATE TABLE governance.AUDIT_EVENT (
  audit_event_id       NVARCHAR(36)  NOT NULL,
  occurred_at          TIMESTAMP     NOT NULL,
  actor_principal      NVARCHAR(256) NOT NULL,
  action               NVARCHAR(64)  NOT NULL,
  entity_type          NVARCHAR(64)  NOT NULL,
  entity_id            NVARCHAR(36)  NOT NULL,
  before_hash          NVARCHAR(64),
  after_hash           NVARCHAR(64),
  payload_hash         NVARCHAR(64)  NOT NULL,
  prev_event_hash      NVARCHAR(64),
  CONSTRAINT pk_audit_event PRIMARY KEY (audit_event_id)
);

CREATE INDEX ix_audit_entity ON governance.AUDIT_EVENT(entity_type,
entity_id, occurred_at);
CREATE INDEX ix_audit_actor  ON governance.AUDIT_EVENT(actor_principal,
occurred_at);
```

Listing 10.6: Defining the AUDIT_EVENT Table with Hash-Chain Columns

The HDI container layout that organizes these tables is shown in Figure 10.3. Each domain (process semantics, architecture semantics, execution metadata, adoption, governance, and AI consumption) owns its own container. Cross-container access is granted explicitly; a service that needs to read both `PROCESS` and `APPLICATION` must hold synonyms for both containers, and the grants are reviewed during code promotion. This makes domain ownership visible at the database layer rather than relying on team conventions.

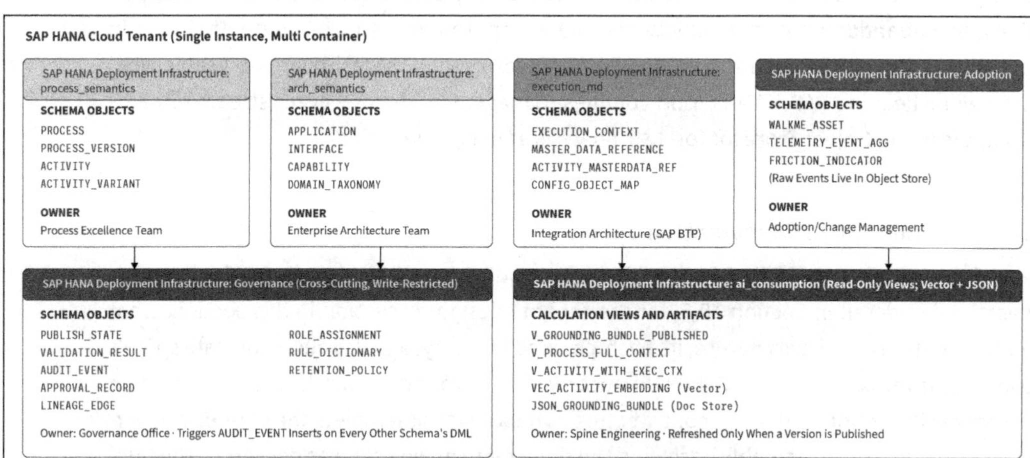

Figure 10.3: HDI Container Layout—Schema Isolation on SAP HANA Cloud

The container model also enables independent deployment. The process semantics container and the architecture semantics container can be updated through different change windows because they correspond to different ingestion sources and different governance owners. A schema change in the architecture container—for example, adding a new lifecycle state value—does not require redeploying the process container, and the audit trail records both events independently.

Implementation Note

SAP HANA Cloud system-versioned tables are enabled on `PROCESS`, `PROCESS_VERSION`, `ACTIVITY`, `APPLICATION`, `INTERFACE`, and `EXECUTION_CONTEXT`. This provides built-in time-travel queries (`SELECT … AS OF SYSTEM TIME`) without application-level versioning logic. The `AS-OF` query path is what the lineage and reproducibility capabilities in Chapter 9 rely on for historical replay; without it, every consumer would have to reimplement temporal logic.

10.3 Data Extension Patterns

No canonical schema survives contact with reality unchanged. Tenants will need attributes the canonical model does not anticipate. Industries will require fields that are essential to one customer and meaningless to another. AI services will introduce derived attributes, such as embeddings, similarity scores, and classification tags, that should not pollute the canonical core. The question is not whether the data spine is extended, but how. The wrong answer is to widen canonical tables with custom columns whenever a need arises; the resulting schema becomes unrecognizable within two release cycles, breaks every consuming integration, and turns governance into archaeology. The right answer is a disciplined extension pattern that keeps the canonical core stable while permitting controlled, audited additions through side tables.

This section describes the extension model, distinguishes the cases where each variant applies, and shows how AI-derived attributes are handled without compromising the read-only contract that AI services operate under. Figure 10.4 shows extension field patterns. New attributes are introduced as side tables, not as core columns, so tenants extend without polluting the canonical schema. Key rules that keep the core stable include:

- Core column additions require a schema-board approval and a version increment of the canonical model.
- Tenant- or project-specific attributes always go in extension tables, never in the core. Extension definitions are versioned with their attributes.

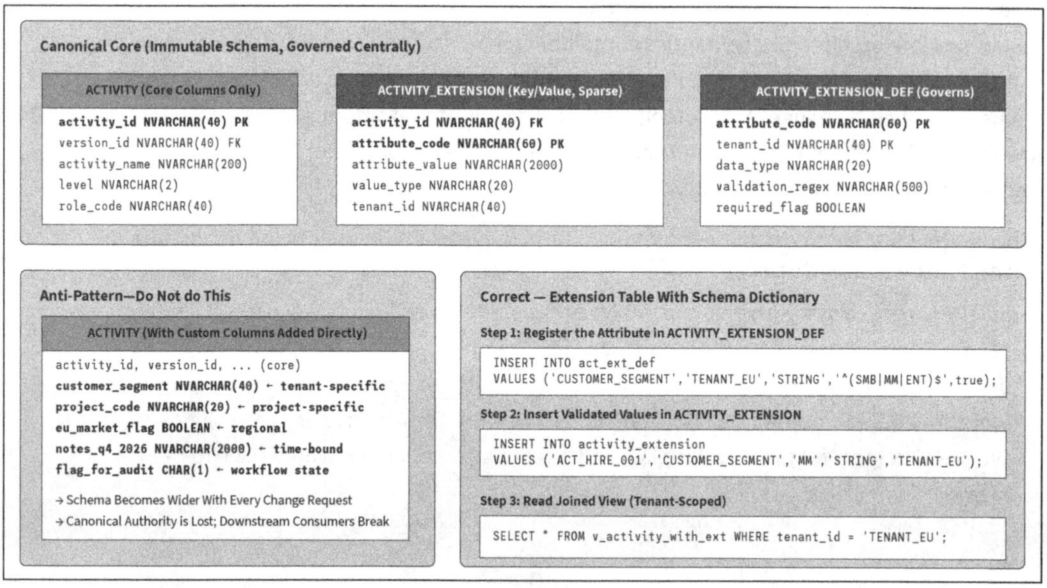

Figure 10.4: Extension Field Pattern—Keeping the Canonical Core Stable

10.3.1 Why the Canonical Core Stays Closed

The canonical tables defined in Section 10.2 form a closed contract. Their column lists are part of the data spine's published interface; consumers such as execution services, AI grounding services, monitoring dashboards, and audit tooling depend on those columns existing with their declared types and meanings. A change to that contract is a breaking change. It must go through the same versioning, review, and migration discipline as a public API change. In practice, this means new columns are added to canonical tables only when a new attribute is genuinely canonical: It is meaningful for every tenant, it is part of the integration contract with upstream sources, and it would be derivable today only by manual, one-off methods.

Most extension needs do not meet that bar. A tenant-specific compliance code, an industry-specific control category, or a project-specific tagging dimension are real needs, but they are not canonical. They belong in extension tables, where they can evolve under their own governance without disturbing the data spine's contract.

10.3.2 The Extension Side-Table Pattern

The basic extension pattern is a side table with the canonical entity's identifier as an FK, a small dictionary that defines which extension attributes are recognized for that entity, and a key-value pair that holds the actual value. The dictionary serves two purposes. It documents the extension surface so that a consumer can discover what extensions exist without scanning rows, and it enforces typing so that an extension defined as `INTEGER` cannot silently accept a string. Listing 10.7 defines the extension dictionary and the side table that carries the actual extension values, leaving the canonical `ACTIVITY` table untouched.

```
CREATE TABLE process_semantics.ACTIVITY_EXTENSION_DEF (
  extension_def_id     NVARCHAR(36)  NOT NULL,
  extension_key        NVARCHAR(64)  NOT NULL,
  data_type            NVARCHAR(16)  NOT NULL,
  description          NVARCHAR(512),
  owning_tenant        NVARCHAR(64)  NOT NULL,
  is_required          BOOLEAN       NOT NULL DEFAULT FALSE,
  CONSTRAINT pk_act_ext_def PRIMARY KEY (extension_def_id),
  CONSTRAINT uq_act_ext_key UNIQUE (extension_key, owning_tenant),
  CONSTRAINT ck_act_ext_type CHECK (
    data_type IN ('STRING','INTEGER','DECIMAL','BOOLEAN','TIMESTAMP',
'JSON'))
);

CREATE TABLE process_semantics.ACTIVITY_EXTENSION (
  activity_extension_id NVARCHAR(36) NOT NULL,
  activity_id           NVARCHAR(36) NOT NULL,
  extension_def_id      NVARCHAR(36) NOT NULL,
  value_string          NVARCHAR(2048),
  value_integer         BIGINT,
  value_decimal         DECIMAL(28,8),
  value_boolean         BOOLEAN,
  value_timestamp       TIMESTAMP,
  value_json            NCLOB,
  CONSTRAINT pk_act_ext PRIMARY KEY (activity_extension_id),
  CONSTRAINT fk_ae_activity FOREIGN KEY (activity_id)
    REFERENCES process_semantics.ACTIVITY(activity_id),
  CONSTRAINT fk_ae_def FOREIGN KEY (extension_def_id)
    REFERENCES process_semantics.ACTIVITY_EXTENSION_DEF(extension_def_id),
  CONSTRAINT uq_ae_one_per_def UNIQUE (activity_id, extension_def_id)
);
```

Listing 10.7: Defining the Activity Extension Tables

This pattern keeps ACTIVITY itself untouched. Tenants add definitions to ACTIVITY_EXTENSION_DEF, and values flow into ACTIVITY_EXTENSION. The OWNING_TENANT column on the definition makes extensions tenant-scoped, meaning a definition created by tenant A is not visible to tenant B unless it is explicitly promoted to a shared scope through a governed process. The UQ_AE_ONE_PER_DEF constraint enforces that each activity carries at most one value per defined extension, preventing the table from accumulating duplicate rows when an upstream system retries.

The same pattern is applied to the other canonical entities that need extensions: PROCESS, APPLICATION, and INTERFACE each have their own EXTENSION_DEF and EXTENSION tables, with the same structure and the same constraints. The repetition is intentional. A unified any-entity extension table would force consumers to filter by entity type and would weaken the FK integrity that the per-entity tables enforce. This mild redundancy is a small cost for the considerable governance benefit.

10.3.3 The Anti-Pattern That Destroys the Data Spine

The extension pattern exists because the alternative is destructive. The alternative is what teams reach for under deadline pressure: Add a custom column to `ACTIVITY` because a tenant needs a tag right now, name it `CUSTOM_FIELD_1` because we will rename it later, and move on. Within six months, the canonical table has eight `CUSTOM_FIELD_N` columns, three of which are used by one tenant, two by another, and three by no one. The semantics are documented on a wiki page that nobody updates. Consuming integrations either ignore the columns and silently lose data, or they hardcode tenant-specific behavior into the canonical contract. The data spine is no longer a spine but a junkyard.

The discipline that prevents this is procedural as much as technical. Every proposed change to a canonical table goes through a schema review. The review answers three questions: Is the attribute meaningful for every tenant, is it derivable only by canonical knowledge, and does any existing extension definition already capture it? If the answers are not all yes, the change is rejected and the team is redirected to the extension pattern. The review is short, with most cases decided in minutes, but it is mandatory.

10.3.4 Industry Variants and Vertical Extensions

Some extensions are not tenant-specific but industry-specific. A regulated industry tenant may require a control reference that is meaningless to a general-industry tenant, and a public sector tenant may require a fund or appropriation tag. These are still extensions, not canonical fields, but they are shared across multiple tenants in the same vertical. The pattern accommodates this through a shared extension scope. An extension definition with `OWNING_TENANT` set to a vertical identifier rather than a single tenant becomes visible to every tenant assigned to that vertical.

Shared extension promotion is itself a governed operation. A tenant proposes promotion of one of its private extensions to a shared scope, the proposal is reviewed by the vertical's owning team, and, if accepted, the extension definition is reassigned. Existing values on tenants outside the original tenant are not autopopulated. They are added through normal ingestion, and the audit trail records both the promotion event and the subsequent value writes.

10.3.5 AI-Derived Extensions

AI services produce derived attributes that look like canonical data but should not be persisted as canonical data. An embedding generated from an activity's description, a similarity score between two activities, and a classification tag inferred by a model are all useful artifacts, but each one is also tied to a specific model, run, and bundle. Persisting them as canonical attributes would conflate model output with authoritative state and would make the canonical contract dependent on whichever model happened to produce the value last.

AI-derived attributes are therefore stored in their own table, `AI_DERIVED_ATTRIBUTE`, with explicit references to the model and run that produced them and an explicit version pin. The table is read-only to consumers other than the AI services that produced the rows. Updates are not in place. A new run produces new rows, and old rows are retained for the lineage window. Listing 10.8 defines the `AI_DERIVED_ATTRIBUTE` table, which holds model-derived values with explicit model and run pinning.

```
CREATE TABLE ai_consumption.AI_DERIVED_ATTRIBUTE (
  derived_attr_id       NVARCHAR(36)  NOT NULL,
  entity_type           NVARCHAR(64)  NOT NULL,
  entity_id             NVARCHAR(36)  NOT NULL,
  attribute_name        NVARCHAR(64)  NOT NULL,
  value_text            NCLOB,
  value_decimal         DECIMAL(28,8),
  model_id              NVARCHAR(128) NOT NULL,
  model_version         NVARCHAR(32)  NOT NULL,
  run_id                NVARCHAR(36)  NOT NULL,
  generated_at          TIMESTAMP     NOT NULL,
  bundle_hash           NVARCHAR(64)  NOT NULL,
  CONSTRAINT pk_ai_derived PRIMARY KEY (derived_attr_id)
);

CREATE INDEX ix_ai_derived_entity
  ON ai_consumption.AI_DERIVED_ATTRIBUTE(entity_type, entity_id,
attribute_name, generated_at);
```

Listing 10.8: Defining the AI_DERIVED_ATTRIBUTE Table

Vector embeddings are a special case of AI-derived attribute. They live in a dedicated table that uses SAP HANA Cloud's `REAL_VECTOR` type, with the same model and version pinning as other derived attributes. The embedding table is the read source for grounding bundle assembly; AI services do not query it directly under their consumer roles, only through the grounding API.

AI Augments, Governance Authorizes

AI-derived attributes are visible to consumers, but they never gate execution decisions. A workflow may surface an AI-suggested classification, but the decision to act on it is made by a human under the governance role. The data spine records the human decision as the canonical truth, and the AI suggestion as the supporting evidence. This is the boundary that Chapter 1 set, made structural in the schema.

10.3.6 Extension Discoverability

An extension model is only useful if consumers can discover what extensions exist. The data spine exposes a discovery endpoint on the Access API that returns, for each canonical entity type, the list of extension definitions visible to the calling role: the keys, data types, descriptions, and scope (tenant, vertical, or shared). Consuming services typically call this endpoint at startup and cache the result; they re-fetch when the cached version timestamp is older than a tunable threshold.

This discoverability mechanism replaces the temptation to introspect the schema directly. Consumers should not run information-schema queries against the canonical tables, as doing so couples them to physical column layouts that may change. The discovery endpoint is the contract, and the underlying definition tables are implementation.

10.4 API Design for Ingestion and Access

The data spine exposes three API surfaces, each with its own purpose, its own contract, and its own security regime. The Ingestion API accepts canonical writes from upstream sources. The Access API serves canonical reads to downstream consumers. The Governance API performs state-transition operations (approving a version, retiring an interface, or accepting a suggestion) that change canonical state through workflow rather than through direct write. Conflating these surfaces into a single API is a frequent design mistake. It creates security regimes that are too permissive on one axis and too restrictive on another, and it makes audit reasoning harder than it needs to be. Figure 10.5 shows the API layer architecture.

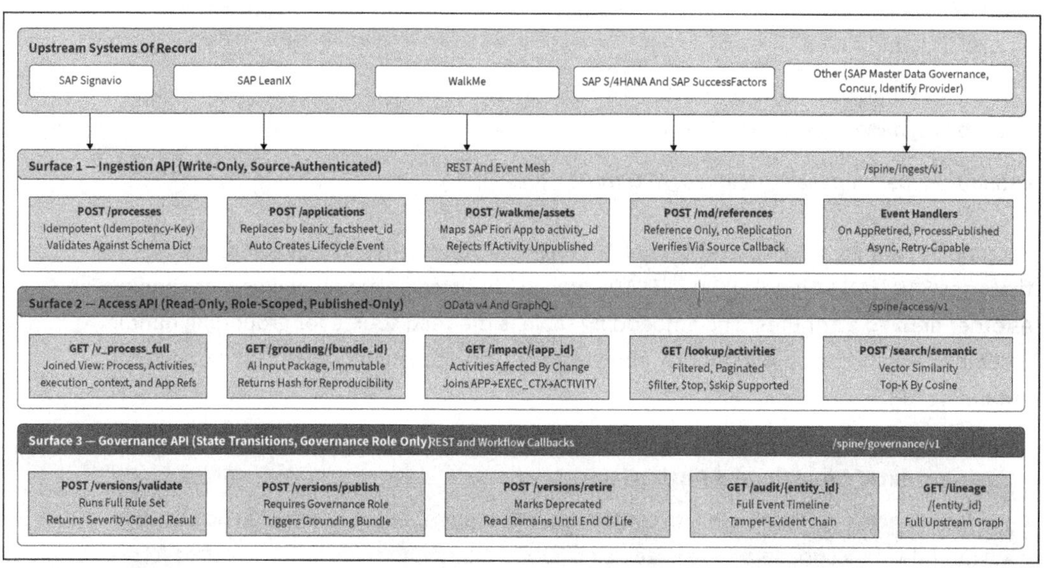

Figure 10.5: API Layer Architecture—Ingestion, Access, and Governance Surfaces

The sections that follow take each of these surfaces in turn. Section 10.4.1 specifies the Ingestion API and its idempotency model. Section 10.4.2 covers the Access API, including

OData query patterns and role-scoped filtering. Section 10.4.3 introduces the Governance API and its workflow-mediated state transitions. Section 10.4.4 details the AI grounding endpoints layered on top of the Access API. Section 10.4.5 explains how Event Mesh advertises canonical changes to downstream consumers. Section 10.4.6 describes the versioning discipline that governs all three surfaces. Section 10.4.7 closes by mapping every previous choice to SAP API Policy v.4.2026a.

10.4.1 The Ingestion API: Idempotent Writes from Upstream Sources

The Ingestion API accepts data from SAP Signavio, SAP LeanIX, and WalkMe either directly or through Cloud Integration integration flows (iFlows) that perform the upstream extraction and translate it into the canonical payload shape. Its contract is narrow. Each endpoint accepts a single canonical entity type, validates the payload against the canonical schema, computes the content hash, and either inserts a new row or updates an existing row depending on whether the hash matches what is already on the data spine. The most consequential design property is idempotency. The same payload posted twice produces the same canonical state, with one audit event recording the first write and a second event recording the no-operation replay.

Idempotency is achieved through an idempotency key supplied by the caller. The key is the SHA-256 hash of the natural identifier and the content payload. If the data spine has already processed an event with that key, it returns the original response without reapplying the write. This makes retries safe: An iFlow that times out on the network can replay the same request and the data spine remains consistent. It also makes replay safe: A backfill that reemits historical events does not corrupt current state. Listing 10.9 shows a complete ingestion call, including the idempotency key header and both the success and idempotent-replay responses.

```
POST /spine/ingest/v1/processes
Headers:
  Idempotency-Key: 7c8a3e1f...   # SHA-256 of natural id + content
  Authorization: Bearer <oauth-token>
  X-Source-System: SIGNAVIO
  Content-Type: application/json

Body:
{
  "signavio_process_id": "sgv-prc-1042",
  "canonical_name":     "Order to Cash - Standard",
  "owning_domain":      "OTC",
  "version_number":     7,
  "effective_from":     "2025-11-15T00:00:00Z",
  "change_reason":      "Q4 controls update",
  "activities": [ ... ]
```

```
}

Response 201 Created:
{
  "process_id":         "prc-9f3a-4c11-...",
  "process_version_id": "pv-1a4e-7c22-...",
  "content_hash":       "9f0c1a3e...",
  "audit_event_id":     "ae-5b2c-8d09-..."
}

Response 200 OK (idempotent replay):
{
  "process_id":         "prc-9f3a-4c11-...",
  "process_version_id": "pv-1a4e-7c22-...",
  "content_hash":       "9f0c1a3e...",
  "replay": true
}
```

Listing 10.9: Posting a Process to the Ingestion API with an Idempotency Key

The error envelope is consistent across all Ingestion API endpoints. A validation failure returns HTTP 422 with a structured list of field-level errors and the canonical schema version that the payload was validated against. An idempotency-key collision with a different payload returns HTTP 409. A rate-limit response returns HTTP 429 with a *Retry-After* header. None of these error paths leave partial state on the data spine; the write is atomic, and either it commits in full or no canonical change occurs.

Figure 10.6 makes the replay path explicit. The iFlow writes the outbound payload to its own outbox before posting to the Ingestion API; if the post fails or the response is lost, the outbox driver retries with the same idempotency key. The data spine recognizes the replay, returns the original response, and the iFlow marks the outbox row complete. This pattern eliminates the dual-write problem that plagues simpler integrations. At no point is the upstream system's belief about its own send state inconsistent with the data spine's record of having received it. Idempotency keys must outlive the operation they protect. Default retention is set to 30 days and archived to Object Store thereafter. Validation runs before any write. Audit and idempotency log are inserted in the same transaction as the entity write so they cannot drift.

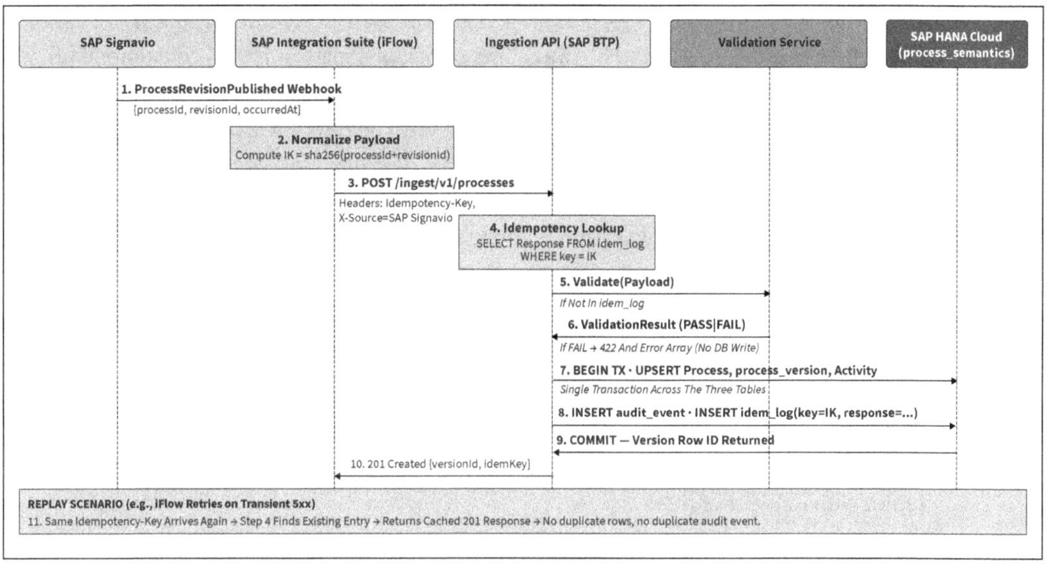

Figure 10.6: Ingestion Flow—SAP Signavio to SAP BTP with Idempotent Replay

10.4.2 The Access API: Read with Role and Version Context

The Access API serves canonical reads. It is implemented as an OData v4 service generated from SAP Cloud Application Programming Model, which gives consumers query capabilities (projection, filtering, sorting, and navigation) without requiring per-query custom endpoints. OData v4 is the right level of abstraction for this surface because the canonical entities are already shaped as a navigable graph; consumers traverse from `PROCESS` to `PROCESS_VERSION` to `ACTIVITY` to `EXECUTION_CONTEXT` through standard OData expansion rather than through bespoke endpoints. Listing 10.10 shows two representative OData v4 queries against the Access API.

```
GET /spine/access/v1/Processes('prc-9f3a-4c11-...')?
       $expand=versions($filter=status eq 'PUBLISHED';
           $expand=activities($expand=executionContexts(
               $expand=triggerApplication,targetApplication,interface)))

GET /spine/access/v1/Activities?
       $filter=processVersion/process/owningDomain eq 'OTC'
           and contains(canonicalName,'Credit Check')
       &$expand=masterDataReferences($expand=mdReference)
       &$top=50
```

Listing 10.10: Querying the Access API with OData v4

Two properties of the Access API are not visible in the URL syntax but matter at runtime. First, every response is filtered by the calling principal's role context. A consumer with read access to the order-to-cash domain sees only order-to-cash processes; the same query under a different role returns a different result set. The filtering is applied at the database layer through the same HDI grants that constrain who can read which container, plus row-level filters on `owning_domain` that are evaluated as predicates on every query. Second, every response is implicitly version-pinned: a query without an `AS-OF` clause returns the current canonical state, but the response includes a version timestamp that the consumer can use to fetch the same state again later. This makes lineage trivially reproducible.

Pagination is mandatory. No Access API endpoint returns an unbounded result set. The OData `$top` parameter has a server-enforced ceiling, and consumers requesting larger windows must paginate through `$skiptoken`. This protects the data spine from accidental denial-of-service by well-meaning consumers who issue *give me everything* queries against tables with millions of rows.

10.4.3 The Governance API: State Transitions Through Workflow

Some operations on the data spine are not direct writes. Approving a process version, retiring an interface, accepting an AI suggestion, and granting a new role to a principal are all state transitions that require authorization, and authorization in turn requires a governance workflow. The Governance API is the entry point to those workflows. Unlike the Ingestion API, it does not write canonical state directly; it creates a workflow instance in SAP Build Process Automation and returns its identifier. The workflow proceeds through human and system steps, and the canonical state change is committed only when the workflow terminates successfully. Listing 10.11 shows a Governance API call that starts a process version publish workflow and the `202 Accepted` response that returns the workflow handle.

```
POST /spine/governance/v1/processVersions/{pvid}/approve
Headers:
  Authorization: Bearer <oauth-token>
  X-Justification: "Q4 controls update; reviewed by OTC governance board"

Body:
{
  "approval_type":     "PROCESS_VERSION_PUBLISH",
  "approvers":         ["jane.doe@example.com","john.smith@example.com"],
  "evidence_refs":     ["s3://spine-evidence/2025/q4/otc-board-minutes.pdf"]
}

Response 202 Accepted:
{
```

```
  "workflow_instance_id": "wf-3c7a-...",
  "expected_resolution": "2025-11-22T17:00:00Z",
  "audit_event_id":       "ae-7e1d-..."
}
```

Listing 10.11: Requesting Process-Version Approval Through the Governance API

The 202 Accepted response is deliberate. The Governance API does not return the result of the approval; it returns the identifier of the workflow that will eventually produce that result. Consumers poll the workflow status through SAP Build Process Automation's API or subscribe to its terminal event on Event Mesh. When the workflow terminates with approval, the data spine writes the corresponding canonical change (a new PROCESS_VERSION row marked PUBLISHED, for example) and emits an audit event linking the canonical write to the workflow instance and to every approver action within it.

This decoupling is what makes the governance audit trail defensible. Every state-transition write on the data spine is preceded by a workflow whose evidence (approver identities, justification, supporting documents, or timing) is captured in SAP Build Process Automation and referenced by audit event identifier. A regulator asking "Who approved this version of the order-to-cash process, and what evidence did they review?" is answered by replaying the workflow, not by reconstructing what happened from scattered logs.

10.4.4 AI Grounding Endpoints

The Access API includes a small set of endpoints dedicated to AI grounding. These endpoints assemble grounding bundles—collections of canonical entities, their versions, and any supporting context—and return them in the shape that AI services consume. The grounding endpoints are read-only; they cannot write canonical state, and they enforce the same role-based filtering as other Access API endpoints. What distinguishes them is the bundle assembly logic: A grounding request specifies the entry-point entity (typically a PROCESS or an ACTIVITY) and the depth and breadth of context required, and the endpoint walks the canonical graph to produce a coherent, version-pinned bundle. Figure 10.7 shows the AI grounding API flow.

The following hard rules are enforced by the grounding service:

- Only PUBLISHED versions enter a bundle.
- Bundles are immutable once hashed.
- No modification API is exposed to AI—it has read-only access.
- Role scope is enforced server-side.
- Every step 6 and 11 audit row is hash-chained, meaning auditors can replay any inference and verify the inputs were exactly what they appear to be.

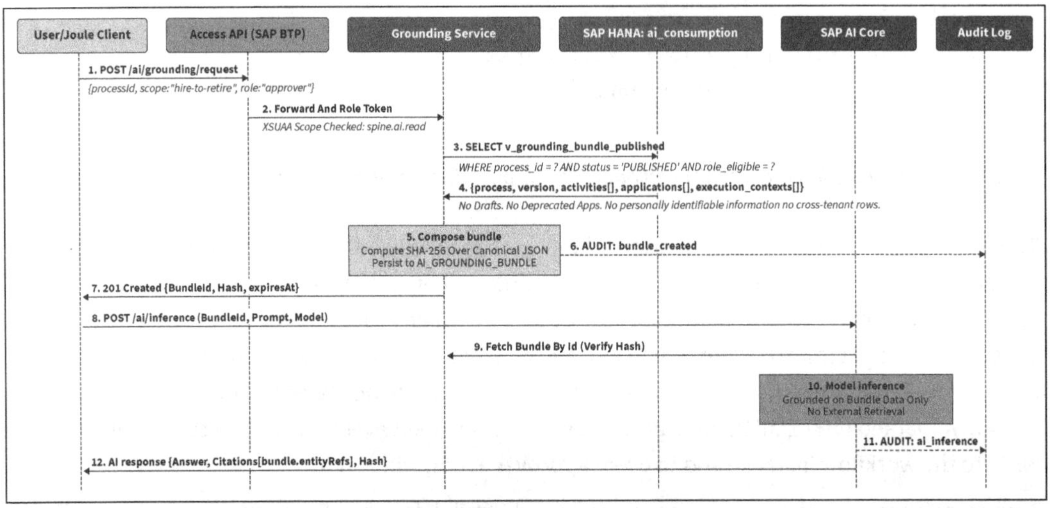

Figure 10.7: AI Grounding API Flow—From Spine to SAP AI Core, Safely

Every assembled bundle is hashed, recorded in `AI_GROUNDING_BUNDLE`, and then delivered. The hash is what makes a bundle reproducible. The same entry-point and depth specification at the same canonical version produces the same bundle, with the same hash, regardless of when it is requested. AI services that record the bundle hash alongside their output gain provenance that survives both model versions and time. When a model produces a suggestion, the suggestion is traceable to the exact canonical context the model saw, down to the row hashes.

> **Read-Only Is Structural, Not Contractual**
>
> The AI consumer role on the data spine has no `INSERT`, `UPDATE`, or `DELETE` privileges on any canonical table. The grounding endpoints assemble bundles by issuing read-only queries through a service principal whose grants are limited to `SELECT` on the relevant containers. There is no path by which an AI service can write canonical state, even if a misconfigured client believed it could. This is governance as schema permission, not runtime check.

10.4.5 Event Mesh Integration

Every successful write through the Ingestion API or commit through the Governance API emits a domain event onto Event Mesh. The event format follows CloudEvents, with a small canonical envelope: event identifier, event type, source URI, time, and a data payload that carries the entity type, entity identifier, version, change type, and a back-reference to the audit event identifier. Subscribers do not receive the full canonical row in the event; they receive the change advertisement and re-fetch the row through the Access API under their own role context. Listing 10.12 shows the CloudEvents payload emitted to Event Mesh when a process version is published.


```
{
  "specversion": "1.0",
  "type":         "com.example.spine.process_version.published.v1",
  "source":       "https://spine.example.com/ingest",
  "id":           "evt-1c4a-7e22-...",
  "time":         "2025-11-15T14:32:08Z",
  "datacontenttype": "application/json",
  "data": {
    "entity_type":      "PROCESS_VERSION",
    "entity_id":        "pv-1a4e-7c22-...",
    "version_number":   7,
    "change_type":      "PUBLISH",
    "audit_event_id":   "ae-5b2c-8d09-..."
  }
}
```


Listing 10.12: Emitting a Process-Version Publication Event to Event Mesh

This event payload is small by design. It carries enough information for subscribers to decide whether they care about the change and to fetch the row if they do; it does not carry the row itself, which would create a parallel state-distribution channel that is harder to govern than the canonical Access API. The `audit_event_id` link is the join point. A subscriber that records the event can later correlate its own actions with the canonical audit trail by audit event identifier, which is essential for cross-system reconstruction.

10.4.6 Versioning the API Surface

The three APIs are versioned in the URL path (*ingest/v1*, *access/v1*, *governance/v1*). A breaking change introduces v2 alongside v1; both run in parallel for a published deprecation window before v1 is decommissioned. Backwards-compatible additions such as new optional fields or endpoints do not bump the version. The discipline that distinguishes a breaking change from a compatible addition is the same as for canonical schema changes: It is reviewed, recorded in a release note, and announced to consumers ahead of the release.

The OpenAPI specification of every API surface is published on the data spine itself, at a stable URL under each version path. Consumers fetch the specification at deployment time and use it to generate clients; the specification is therefore part of the data spine's contract, not a side artifact maintained in a separate repository.

10.4.7 Alignment with SAP API Policy v.4.2026a

The three-surface API design described in this chapter aligns by construction with SAP API Policy v.4.2026a. The policy, published as part of SAP's tightening of AI and agent access patterns, makes three points that bear directly on how the data spine is implemented and operated. None of the three are afterthoughts in the architecture; each one corresponds to a structural choice already made in the preceding sections, and the alignment is made explicit here so that the chapter's design can be cited as policy-conforming during architectural review.

First, the policy distinguishes published APIs (those listed in the SAP Business Accelerator Hub or explicitly identified in product documentation) from internal, private, or SAP-reserved APIs that customer and third-party applications should not consume. The data spine consumes only published APIs: SAP Signavio's documented public integration endpoints, SAP LeanIX's published REST and webhook surfaces, WalkMe's official telemetry export, and SAP AI Core's documented inference and embedding APIs. There is no path in the architecture in this chapter, or in any iFlow that feeds it, where the data spine reaches into an undocumented, internal, or reserved SAP API. Where a published API does not yet exist for a needed integration, the requirement is escalated to SAP through the standard product-feedback channel rather than worked around through scraping, screen capture, undocumented endpoint discovery, or impersonation.

Second, and most consequential for the AI design in this chapter, is the fact that Section 2.2.2 of SAP API Policy v.4.2026a prohibits API usage by AI systems that plan, select, or execute sequences of API calls, except through SAP-endorsed architectures, data services, or service-specific pathways. The grounding architecture established here is precisely such an endorsed pathway. AI services do not plan or execute SAP API call sequences. They do not autonomously decide which SAP Signavio process, SAP LeanIX factsheet, or WalkMe asset to fetch next, and they do not orchestrate calls to SAP APIs based on intermediate model output. Instead, AI services receive preassembled grounding bundles produced by the data spine's own Access API under a deterministic, role-scoped, version-pinned assembly path. The bundle is the unit of AI consumption. The model sees only what the data spine assembled for it; it does not call SAP APIs at all. This is what Section 2.2.2 of SAP API Policy v.4.2026a prescribes for AI access patterns, and it is the reason the AI consumer role on the data spine carries no write scope, no broker scope, and no direct SAP API call scope of any kind.

The same constraint applies to Model Context Protocol (MCP)-style adapters and agent ecosystems. The data spine does not expose SAP-published APIs as agent-callable tools through an MCP server or any equivalent indirection. An agent that needs SAP context calls the data spine's Access API for a grounding bundle. The data spine assembles the bundle by reading from its own canonical store, and the SAP API surface is not exposed to the agent at all. This preserves SAP's monitoring and rate-limit posture. Every call SAP sees is a deterministic, governed call from the data spine's ingestion path, not an unbounded fan-out from an autonomous agent.

Third, the policy prohibits bulk extraction, scraping, harvesting, and systematic large-scale replication outside endorsed data services. The data spine's ingestion is event-driven and idempotent. It accepts canonical writes one entity-version at a time through documented iFlows that subscribe to upstream change events on published APIs. There is no bulk-pull job that scrapes SAP Signavio, SAP LeanIX, or WalkMe at scale, no nightly extract that mirrors the SAP catalog, and no mass replication of SAP system data into the data spine. Where bulk reconciliation is required (for example, after a connectivity outage) it runs through the same idempotent ingestion endpoints, one event at a time, with the same audit, rate-limit, and back-pressure treatment as live traffic. The traffic shape SAP sees from the data spine is the same shape the policy was written to permit.

The policy also reserves the right to monitor usage and take enforcement actions—throttling, suspension, or termination—for noncompliance, and it explicitly prohibits bypassing API controls through proxies, custom gateways, impersonation, or similar mechanisms. The data spine does not introduce any such bypass. Every call to an SAP-published API is made under the consuming organization's own credentials and rate quotas. No shared service principal masquerades for multiple tenants, no cached SAP token is replayed beyond its issued lifetime, and no third-party gateway sits between the data spine and SAP that obscures the call origin. The data spine is a consumer of SAP-published APIs, not a wrapper that hides them from SAP's monitoring.

Why the Policy and the Architecture Converge

The practical reason for this alignment is straightforward. The policy reflects what production-scale operation has already taught: A single prompt to an autonomous AI agent can fan out into thousands of SAP API calls whose pattern, intent, and authority are not traceable to a controlled architecture. Transactional APIs were not designed for that traffic shape, and neither audit trails nor rate limits hold up against it. The grounding architecture replaces the unbounded, agentic pattern with a bounded, governed one: AI sees what governance approved, when governance approved it, in the form governance assembled it. SAP API Policy v.4.2026a and the architecture in this chapter are saying the same thing in different languages.

10.5 Governance and Auditability

Governance is not a layer added on top of the data spine after it is built; it is a property of how the data spine is built. The audit trail is a database-resident sequence of tamper-evident events. Lineage is a query path through canonical and audit tables, not a feature of a separate metadata catalog. Separation of duties (SoD) is a set of role grants that the database itself enforces. Retention is a deliberate tiering across SAP HANA Cloud and Object Store, with cryptographic guarantees that records cannot be retroactively altered. This section describes how each of these properties is engineered into the implementation, with the underlying premise that an auditor or regulator should be able to reconstruct any change to the data spine from the data spine's own data, without external cross-references.

The user-facing question this section answers is the one a control owner or finance leader will eventually ask: When the regulator arrives, can we show who did what, when, on whose authority, with what supporting evidence, in a way that cannot be plausibly disputed? The answer requires the audit trail to be complete, the lineage to be traversable, and the integrity of both to be cryptographically verifiable.

Figure 10.8 shows the audit log and lineage. Every write produces an audit row, and every audit row links to its predecessor—the chain is the evidence. In Figure 10.8, each arrow is a `LINEAGE_EDGE` row. Auditors trace any AI output back to its bundle, the bundle back to the published version, and the version back to its validation result and the original SAP Signavio revision, all without leaving SAP HANA. Hash chaining (`prev_event_hash`) makes silent insertion or deletion of audit rows mathematically detectable.

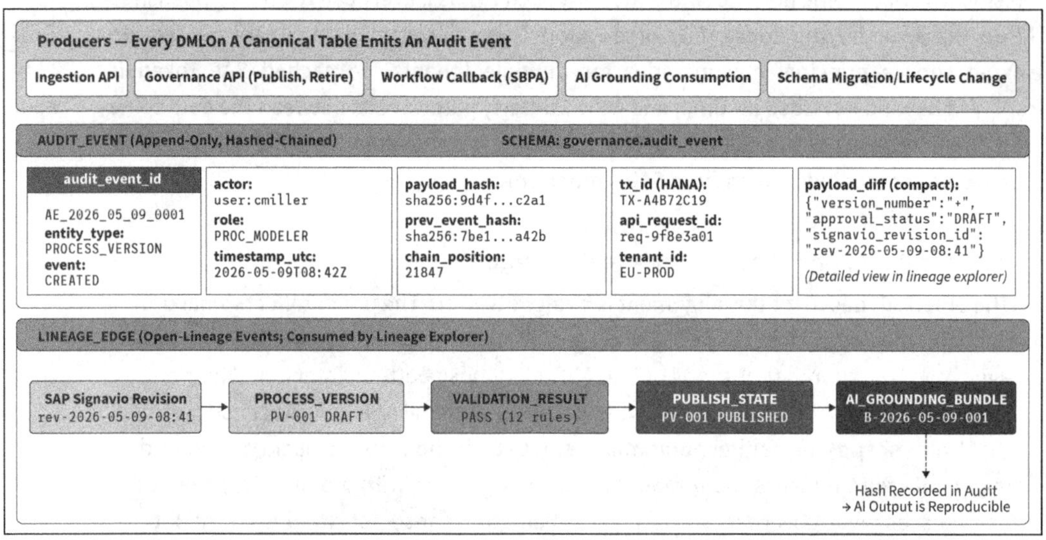

Figure 10.8: Audit Log and Lineage—Tamper-Evident Evidence Chain

10.5.1 Hash-Chained Audit Events

`AUDIT_EVENT`, introduced briefly in Section 10.2, is the data spine's evidence layer. Every canonical write through the Ingestion API, the Governance API, or internal jobs produces an audit event row before the canonical row is committed. The two writes are inside the same database transaction; either both commit or neither does. There is no path by which a canonical change happens without a corresponding audit event.

The audit event row carries the actor principal, the timestamp, the action, the entity type and identifier, the hash of the entity state before the change, the hash of the entity state after the change, the hash of the action payload, and, critically, the hash of the previous audit event. Each row is cryptographically linked to its predecessor through the prev_event_hash column. The chain forms a Merkle-style sequence: Any retroactive alteration to a past row breaks the chain at every subsequent row, which is detectable by recomputing hashes from the genesis event forward. Listing 10.13 shows the hash-chain verification query that the nightly integrity job runs against AUDIT_EVENT.

```
-- Hash-chain verification (run periodically by integrity job)
WITH ordered AS (
  SELECT audit_event_id, occurred_at, payload_hash, prev_event_hash,
         LAG(payload_hash) OVER (ORDER BY occurred_at, audit_event_id) AS
computed_prev
  FROM governance.AUDIT_EVENT
)
SELECT audit_event_id, occurred_at
FROM ordered
WHERE (  (prev_event_hash IS NULL     AND computed_prev IS NOT NULL)
      OR (prev_event_hash IS NOT NULL AND computed_prev IS NULL)
      OR  prev_event_hash <> computed_prev )
  AND occurred_at > (SELECT MAX(genesis_at)
                     FROM governance.AUDIT_GENESIS);
  AND occurred_at > (SELECT MAX(genesis_at) FROM governance.AUDIT_GENESIS);
```

Listing 10.13: Verifying the Audit Hash Chain

The integrity job runs the chain-verification query nightly and at every schema change. Any row where the recorded prev_event_hash does not match the computed predecessor hash is reported as a chain break. Chain breaks are alerted, not silently ignored. The alert notifies the on-call governance engineer, an investigation begins, and the chain is restored either by demonstrating the break is benign (for example, a planned schema migration captured in AUDIT_GENESIS) or by treating it as an integrity incident that must be reconstructed and documented.

The audit table itself is append-only at the database level. The HDI grants for the governance container give every service a SELECT and INSERT privilege on AUDIT_EVENT; nobody—not even the database administrator under normal operations—has UPDATE or DELETE privileges. Privileged operations that need to alter the table (for example, a forensic redaction required by a court order) require a separate, named service principal whose grants are issued explicitly for the duration of the operation and whose every action is audited in a meta-log.

10.5.2 Lineage as a First-Class Query Path

Lineage answers a different question than audit. Audit answers "What changed and who changed it?" Lineage answers "What does this canonical row depend on, and what depends on it?" The two are related, as every lineage edge corresponds to a canonical relationship that was itself created by an audited write, but they are queried differently and serve different consumers. Lineage is what change-impact services traverse; audit is what compliance reviewers replay.

Lineage on the data spine is implemented through a `LINEAGE_EDGE` table that captures every canonical-to-canonical relationship beyond direct FKs. FK relationships are themselves lineage edges, though some lineage relationships do not appear as FKs but as an AI grounding bundle's dependence on a set of canonical rows, an extension definition's inheritance from a vertical-shared template, or an audit event's dependence on the workflow instance that authorized it. `LINEAGE_EDGE` captures these explicitly, in a shape compatible with OpenLineage so that the same edge graph can be projected into broader lineage tooling. Listing 10.14 defines the `LINEAGE_EDGE` table together with the bidirectional indexes that support both forward and backward lineage traversal.

```
CREATE TABLE governance.LINEAGE_EDGE (
  edge_id               NVARCHAR(36)  NOT NULL,
  source_entity_type    NVARCHAR(64)  NOT NULL,
  source_entity_id      NVARCHAR(36)  NOT NULL,
  target_entity_type    NVARCHAR(64)  NOT NULL,
  target_entity_id      NVARCHAR(36)  NOT NULL,
  edge_type             NVARCHAR(32)  NOT NULL,
  established_at        TIMESTAMP     NOT NULL,
  established_by_audit NVARCHAR(36)  NOT NULL,
  CONSTRAINT pk_lineage_edge PRIMARY KEY (edge_id),
  CONSTRAINT fk_lineage_audit FOREIGN KEY (established_by_audit)
    REFERENCES governance.AUDIT_EVENT(audit_event_id),
  CONSTRAINT ck_lineage_type CHECK (edge_type IN (
    'COMPOSITION','VERSION_DERIVES_FROM','GROUNDING_USES',
    'EXTENSION_INHERITS','APPROVAL_AUTHORIZES','REPLACES'))
);

CREATE INDEX ix_lineage_source ON governance.LINEAGE_EDGE
(source_entity_type, source_entity_id);
CREATE INDEX ix_lineage_target ON governance.LINEAGE_EDGE
(target_entity_type, target_entity_id);
```

Listing 10.14: Defining the LINEAGE_EDGE Table and Its Bidirectional Indexes

Both directional indexes are essential. Forward lineage queries ("What does this row depend on?") traverse from the source. Backward lineage queries ("What depends on this row?") traverse from the target. Change-impact services rely on backward lineage; reproducibility services rely on forward lineage. Without both indexes, one direction of traversal degrades to a table scan and the corresponding capability becomes unusable at production scale.

10.5.3 Separation of Duties Through SAP Authorization and Trust Management Service Scopes

SoD on the data spine is enforced through the SAP Authorization and Trust Management service, the SAP BTP authorization service that manages OAuth scopes and role collections. Three principles drive the role design: Scopes are narrow, role collections are explicit, and the AI scope is read-only by structural definition rather than by runtime check.

Narrow scopes mean that each scope grants exactly one capability (read process semantics, write architecture semantics, or approve process versions) and not a bundle of capabilities. A role collection then aggregates the scopes appropriate for a job function. For example, an order-to-cash process owner's collection includes read process semantics, propose process changes, and approve order-to-cash process versions, but not write architecture semantics or grant new roles. A change to a scope does not require redefining the collection, and a change to a job function does not require redefining the scopes. This separation makes role evolution tractable.

The AI scope is the most consequential separation. AI services consume the data spine through a service principal whose role collection contains only read scopes, such as read process semantics, read architecture semantics, read execution metadata, and read AI grounding bundles. There is no write scope of any kind in the AI service principal's collection, which means the database itself rejects any attempt by the AI service to write canonical state. This is what AI not being allowed to author looks like at the implementation layer: not a runtime check that could be misconfigured, but a permission grant that does not exist. Listing 10.15 shows the relevant excerpt of *xs-security.json*, including the read-only AIConsumer role.

```
# XSUAA scope and role collection definitions (xs-security.json excerpt)
{
  "scopes": [
    {"name": "$XSAPPNAME.read.process",       "description": "Read
process semantics"},
    {"name": "$XSAPPNAME.read.arch",          "description": "Read
architecture semantics"},
    {"name": "$XSAPPNAME.read.execution",     "description": "Read
execution metadata"},
    {"name": "$XSAPPNAME.read.grounding",     "description": "Read AI
grounding bundles"},
    {"name": "$XSAPPNAME.write.process",      "description": "Ingest
process semantics"},
    {"name": "$XSAPPNAME.write.arch",         "description": "Ingest
architecture semantics"},
    {"name": "$XSAPPNAME.approve.processv",   "description": "Approve
process version"},
    {"name": "$XSAPPNAME.audit.read",         "description": "Read audit
and lineage"},
```

```
    {"name": "$XSAPPNAME.governance.admin",  "description": "Governance
admin operations"}
  ],
  "role-templates": [
    {"name": "AIConsumer",
     "scope-references": [
        "$XSAPPNAME.read.process",
        "$XSAPPNAME.read.arch",
        "$XSAPPNAME.read.execution",
        "$XSAPPNAME.read.grounding"
     ]},
    {"name": "ProcessOwner",
     "scope-references": [
        "$XSAPPNAME.read.process",
        "$XSAPPNAME.read.arch",
        "$XSAPPNAME.read.execution",
        "$XSAPPNAME.approve.processv"
     ]}
  ]
}
```

Listing 10.15: Defining SAP Authorization and Trust Management Service Scopes and Role Collections

The AIConsumer template is intentionally limited to read scopes and the grounding read scope. There is no write scope, no approval scope, and no governance admin scope. The role is structurally incapable of writing canonical state, regardless of what the AI service code believes it can do. The same reasoning applies to other consumer roles: A monitoring service has audit read scope but not approval, and a process owner has approval within their domain but not governance admin. Each role is a minimum-privilege grant for a specific job function.

10.5.4 Retention Tiering and Cryptographic Anchoring

Different canonical concerns have different retention requirements. Audit events must be retained for the regulatory window applicable to the controlled processes, typically seven years for financial controls and ten or more for some regulated industries. Canonical entity rows must be retained for the lifetime of the entity plus a defined post-retirement window, so that historical reconstruction remains possible. Raw telemetry payloads carry less regulatory weight and can be tiered to colder storage after a shorter window. AI grounding bundle manifests must be retained for as long as any AI-derived suggestion they fed into is referenced.

Retention is implemented as a tiering policy, not a deletion policy. Records older than the in-database retention window are migrated to Object Store under a write-once-read-many configuration; the canonical row is replaced by a tombstone that points to the archived object. Queries that need historical state read the tombstone, fetch the archived object, and reconstruct the row. This keeps SAP HANA Cloud focused on the active retention window while preserving the long tail in cheaper storage with stronger immutability guarantees.

The integrity of the long tail is anchored cryptographically. At the close of every retention period (typically monthly), the audit chain's most recent hash is published to a tamper-evident anchor, either an internal root-of-trust ledger or an external timestamp authority. The anchor allows a future auditor to verify that the audit chain through that point in time is the same chain that existed at the time of anchoring. Combined with the chain's internal hash linkage, this makes retroactive alteration of historical audit events practically infeasible.

10.5.5 Compliance Evidence Chain

The combination of audit events, lineage edges, role-bounded actions, and cryptographic anchoring produces a compliance evidence chain that does not depend on the cooperation of the audited system to be verified. A regulator examining a specific control (for example, the requirement that no order-to-cash process change goes live without governance approval) issues three queries against the data spine.

The first query returns every `PROCESS_VERSION` row for the order-to-cash process where status was set to `PUBLISHED` in the relevant period, joined to the audit events that recorded those transitions. The second query, run from each of those audit events, walks the lineage edges of type `APPROVAL_AUTHORIZES` backward to the workflow instance and from there to the approver actions and supporting evidence captured during workflow execution. The third query verifies that the audit chain segments covering those events have not been altered, by comparing recomputed hashes against the published cryptographic anchors. The three queries together constitute the evidence: Every published version had a corresponding workflow, every workflow had identified approvers, every approval had supporting evidence, and the entire audit record is integrity-protected.

Auditability

This is the discipline that makes the data spine defensible. Auditability is not a side capability that can be retrofitted; it is a property that emerges from the same writes that produce canonical state. Every write earns its audit event, every relationship earns its lineage edge, and every privilege earns its scope. The audit trail is the system, and the system is its own audit trail.

10.6 Summary

This chapter has walked through translating the canonical data spine from design into a working implementation on SAP BTP. The implementation distributes the data spine across the persistence services that match each canonical concern—SAP HANA Cloud for relational canonical state, Object Store for raw payloads and archived records, document store for schemaless variants, the SAP HANA Cloud vector engine for AI grounding embeddings, Event Mesh for change propagation, and SAP Build Process Automation for workflow state—under a clear rule that each canonical entity has exactly one primary store. The rule is the architecture; the services are how it is realized.

The canonical schema is realized in SAP HANA Cloud as a set of tables organized by domain into HDI containers, with FKs that enforce relational integrity, check constraints that enforce semantic rules at the database layer, and indexes that make the data spine queryable at execution latency. Process and architecture identity are anchored in `PROCESS`, `PROCESS_VERSION`, `ACTIVITY`, `APPLICATION`, and `INTERFACE`. Execution context binds them through `EXECUTION_CONTEXT` and the master data reference tables. Adoption signals flow through `WALKME_ASSET` and `TELEMETRY_EVENT_AGG`. AI consumption is mediated by `AI_GROUNDING_BUNDLE` and `AI_DERIVED_ATTRIBUTE`. Governance and audit are embodied in `AUDIT_EVENT` and `LINEAGE_EDGE`.

Extensions are added through a disciplined side-table pattern that keeps the canonical core stable while permitting tenant- and vertical-specific attributes to evolve under their own governance. The pattern's importance is not theoretical, because the alternative (adding custom columns to canonical tables) destroys the data spine within months. AI-derived attributes are kept structurally distinct from canonical attributes, persisted with explicit model and run pinning so that every AI output is traceable to the exact canonical context that produced it.

Three API surfaces expose the data spine without conflating their purposes. The Ingestion API accepts idempotent writes from upstream sources, with an idempotency key that makes retries and replays safe by construction. The Access API serves OData v4 reads under role-scoped, version-pinned filtering, with grounding endpoints that assemble reproducible bundles for AI consumption. The Governance API initiates workflow-mediated state transitions through SAP Build Process Automation, separating the authorization decision from the canonical write so that every approval has supporting evidence captured in workflow.

Auditability and governance are engineered into the platform rather than layered on top of it. Every canonical write produces a hash-chained audit event in the same database transaction, making the audit trail tamper-evident at the database level. Lineage edges capture canonical-to-canonical relationships beyond direct FKs, in a shape compatible with OpenLineage. SoD is enforced through SAP Authorization and Trust Management service scopes and role collections that are minimum-privilege by construction. The AI consumer role is structurally read-only because its role collection contains no write scope. Retention is tiered across SAP HANA Cloud and Object Store with cryptographic anchoring of the audit chain at retention boundaries, producing an evidence chain whose integrity does not depend on the audited system's cooperation.

Table 10.2 traces every canonical design decision we covered in Chapter 9 to its concrete implementation construct in this chapter. The mapping is not one-to-one; some design decisions are realized through a combination of schema, API, and governance constructs, while a few implementation constructs (such as idempotency keys and the SAP HANA Cloud vector engine integration) extend the design rather than merely realizing it. What Table 10.2 makes visible is the tightness of the trace: There is no design element from Chapter 9 that lacks an implementation correspondent here, and there is no implementation construct here that exists only because the platform allows it. The implementation is the design, made executable.

Chapter 9 Design Decision	Chapter 10 Implementation	SAP BTP Construct	Section Reference
Separate identity from version state `PROCESS_ID` *versus* `PROCESS_VERSION`	Two physical tables joined by FK, with `UNIQUE` (`process_id`, `version_no`)	SAP HANA Cloud (`process_semantics`) Time-travel on; FK enforced	Section 10.2
Execution context as structural pivot *Trigger/target/channel*	`EXECUTION_CONTEXT` table with FKs to `APPLICATION` (trigger is not equal to target)	`execution_md container` Check constraint enforces inequality	Section 10.2, Section 10.4
Master data referenced, not replicated *Authority stays at source*	`MASTER_DATA_REFERENCE` holds keys, not values; verified via callback	`execution_md.master_data_ref` Daily reconciliation job	Section 10.4
Validation precedes publication *Architectural gate, not procedure*	Validation service called before any data manipulation language (DML); results stored, audited	SAP Cloud Application Programming Model and `RULE_DICTIONARY` Validation API endpoint	Section 10.5
Only published entities reach AI *No draft leakage*	`v_grounding_bundle_published` view filters with status `PUBLISHED`	`ai_consumption` HDI container Read-only; refresh on publish	Section 10.4, Section 10.5

Table 10.2: From Design to Implementation—Chapter 9 Decisions Realized in This Chapter

Chapter 9 Design Decision	Chapter 10 Implementation	SAP BTP Construct	Section Reference
Cross-tenant participation, not collapse *Tenant extends without polluting core*	Side tables (`ACTIVITY_EXTENSION`) and schema dictionary, tenant-scoped	`EXTENSION` and `EXTENSION_DEF` Tenant ID propagated through	Section 10.3
Audit and lineage as first-class *Evidence chain by design*	`AUDIT_EVENT` (hash-chained) and `LINEAGE_EDGE` (open-lineage)	Governance container; audit log service sink Tiered to Object Store at 90 days	Section 10.5
AI augments governance *No mutation by AI*	Grounding API is read-only; AI proposals enter via SAP Build Process Automation workflows	AI scope is `spine.ai.read` only, never `spine.governance.write`	Section 10.4, Section 10.5

Table 10.2: From Design to Implementation—Chapter 9 Decisions Realized in This Chapter (Cont.)

10.6.1 Key Learning Points

- The data spine is multi-store by design. SAP HANA Cloud is the canonical relational core. Object Store carries raw payloads and archived state. The SAP HANA Cloud vector engine carries AI embeddings. Event Mesh carries change advertisements. SAP Build Process Automation carries workflow state. Each canonical entity has exactly one primary store; secondary stores hold derived or projected forms only.
- Canonical tables are a closed contract. New columns are added only when the attribute is canonical for every tenant; everything else goes into governed extension side tables with their own definition dictionaries and integrity constraints.
- FKs, check constraints, and HDI container grants are governance, not just database hygiene. They turn principles like *trigger application differs from target* and *domain ownership is exclusive* into rules the database itself enforces.
- There are three API surfaces and three governance regimes. The Ingestion API accepts idempotent canonical writes, the Access API serves role-scoped, version-pinned reads, and the Governance API initiates workflow-mediated state transitions. Conflating them produces security regimes that are simultaneously too permissive and too restrictive.
- Idempotency is a structural property, not a runtime hope. The idempotency key derived from natural identifier and content hash makes every replay safe by construction, eliminating the dual-write problem at the integration boundary.

- AI services consume the data spine through grounding bundles assembled by read-only endpoints, recorded with content hashes, and traceable through the `AI_GROUNDING_BUNDLE` and `AI_DERIVED_ATTRIBUTE` tables. The AI consumer role is structurally write-incapable; this is governance as schema permission, not as runtime check.
- Audit events are hash-chained, append-only, and committed in the same transaction as the canonical writes they record. Lineage edges capture relationships beyond direct FKs. Together they form an evidence chain whose integrity is verifiable without trusting the audited system.
- SoD is realized through SAP Authorization and Trust Management service scopes and role collections. Each role is a minimum-privilege grant for a specific job function. The AI consumer role contains no write scope of any kind, by structural definition.
- Retention is tiered across SAP HANA Cloud and Object Store with cryptographic anchoring at retention boundaries. The canonical record is preserved indefinitely, the active query window is bounded, and the long tail is immutable.
- Domain systems remain authoritative for their own data; the data spine never replaces them. It harmonizes identity, version, and relationships across domains so that execution and AI services have a single, consistent context to read from. AI augments governance but does not replace authority.
- Compliance with SAP API Policy v.4.2026a is built into the architecture, not bolted on after the fact. The data spine consumes only published APIs from the SAP Business Accelerator Hub; AI services receive preassembled grounding bundles rather than orchestrating their own SAP API calls (Section 2.2.2 of SAP API Policy v.4.2026a). MCP-style adapters do not expose SAP APIs as agent-callable tools; and bulk extraction is eliminated by event-driven idempotent ingestion. The traffic shape SAP sees is the shape the policy was written to permit.

10.6.2 Common Design Pitfalls

- Adding custom columns to canonical tables under deadline pressure. Within two release cycles, the canonical contract is unrecognizable, and every consumer must hardcode tenant-specific behavior.
- Treating the audit table as an operational log. Operational logs are written with a best-effort approach, often outside transactions, and are frequently truncated. The audit table is canonical state; it must be written transactionally with the change it records.
- Allowing AI services to write into canonical tables just for derived attributes. Once the AI consumer role has any write privilege, the structural argument for AI as a read-only consumer collapses, and the boundary is one runtime misconfiguration away from being crossed.

- Using a single API surface for ingestion, access, and governance. The combined security regime ends up either too open (because ingestion needs broader writes) or too restrictive (because access needs broader reads), and neither problem is reconcilable without decomposition.
- Skipping the idempotency key on ingestion endpoints based on the reasoning that the upstream iFlow only sends each event once. Networks and outboxes guarantee no such thing; the data spine is the only place idempotency can be enforced safely, and skipping it produces silent duplication that is detected only when reconciliation fails months later.
- Storing AI grounding bundles inside the AI consumer's own database. The bundle then lives outside the data spine's audit boundary, and the model output's provenance is no longer traceable to the canonical context it was derived from.
- Treating retention as a deletion policy. Canonical state should be archived to immutable storage, never deleted; deletion destroys the historical reconstruction capability and breaks the lineage chain at the deletion boundary.
- Letting role collections accumulate scopes just in case. Minimum privilege erodes one optional grant at a time; once a role contains both read and approve scopes for the same domain, SoD exists only on paper.
- Allowing extension definitions to proliferate without governance. An ungoverned extension surface is no better than custom columns; the data spine ends up with hundreds of extensions whose semantics nobody tracks, defeating the purpose of the side-table pattern.
- Believing that lineage can be reconstructed later from audit events alone. Some lineage relationships (such as grounding dependencies, extension inheritance, and approval authorization) are not directly implied by canonical writes; they must be recorded as explicit lineage edges at the time the relationship is established.
- Letting an AI agent or MCP-style tooling layer call SAP APIs directly to assemble its own context. This violates SAP API Policy v.4.2026a Section 2.2.2—which prohibits AI systems from planning, selecting, or executing sequences of API calls outside SAP-endorsed architectures—and produces traffic patterns that exceed audit, rate-limit, and governance capacity. AI services consume the data spine's grounding bundles; they do not plan or execute SAP API call sequences.
- Reaching for non-published APIs when a documented endpoint does not yet exist for a needed integration. Internal, private, or SAP-reserved APIs are off-limits regardless of how convenient they appear; the policy-conforming response is to escalate the integration gap to SAP through the standard product feedback channel and to defer the integration until a published API is available, not to work around the gap with scraping, screen capture, or undocumented endpoints.

10.6.3 Review Questions

1. Why was SAP HANA Cloud chosen as the canonical relational core rather than another relational database, and which of its capabilities does the data spine actually depend on?
2. Under what conditions is it acceptable to store data in the document store rather than in SAP HANA Cloud, and what discipline prevents the document store from drifting into a parallel canonical layer?
3. Why does `PROCESS_VERSION` carry a `CONTENT_HASH` column, and how is it used in reconciliation, audit, and replay?
4. Explain the role of the `CK_EC_DISTINCT_APPS` check constraint on `EXECUTION_CONTEXT`. What kind of upstream defect would it catch, and why does the schema enforce it rather than the ingestion service?
5. Compare the extension side-table pattern with the alternative of adding custom columns to canonical tables. What specifically goes wrong in the alternative, and over what timescale?
6. Why are AI-derived attributes stored separately from canonical attributes, and what specifically makes the AI consumer role structurally write-incapable?
7. What is the idempotency key on the Ingestion API, how is it computed, and what failure modes does it eliminate?
8. Why does the Governance API return `202 Accepted` rather than directly committing the requested state change, and how does this affect the audit trail?
9. Describe the hash chain on `AUDIT_EVENT` and explain how a retroactive alteration to a past row would be detected. What additional anchoring is needed beyond the in-table chain?
10. How does role-scoped filtering on the Access API interact with HDI container grants? Why are both layers necessary rather than one?
11. How does the architecture in this chapter satisfy SAP API Policy v.4.2026a, particularly Section 2.2.2's prohibition on AI systems planning, selecting, or executing sequences of API calls? Identify the three structural choices in the chapter that produce this alignment and explain why each is not substitutable.

10.6.4 Student Companion Notes

If you are working through this chapter alongside an actual SAP BTP environment, you can use the following companion notes to identify the artifacts that should accompany each section's content:

- **Section 10.1**
  Provision an SAP BTP subaccount with SAP HANA Cloud, Object Store, Event Mesh, and the SAP Build Process Automation service. Verify each service's namespace and trust configuration before proceeding to schema work.

- **Section 10.2**
  Implement the canonical schema using the SAP Cloud Application Programming model. Generate the HDI deployer, deploy the containers, and verify with a few `SELECT` statements that the FKs and check constraints reject malformed inserts as expected.
- **Section 10.3**
  Add an extension definition for a tenant-specific attribute through the discovery endpoint and post a value, then re-fetch the activity through the Access API and confirm the extension appears under the activity's extensions navigation property.
- **Section 10.4**
  Build a small ingestion test harness that posts the same canonical payload twice with the same idempotency key, then a third time with a modified payload but the same key. Verify that the data spine returns `200 OK` with `replay: true` on the second call and `409 Conflict` on the third.
- **Section 10.5**
  Run the audit chain verification query against a populated `AUDIT_EVENT` table, confirm a clean result, then deliberately alter a past row in a non-production environment and confirm that the verification query reports the chain break at exactly the altered row's successor.

The supplementary materials accompanying this book include a working SAP Cloud Application Programming Model project that realizes the schema and APIs described in this chapter, a Postman collection covering the Ingestion, Access, and Governance API surfaces, and reference scripts for chain verification, lineage traversal, and grounding-bundle replay. These are intended for study and adaptation rather than direct production use; the canonical model in the supplements has been simplified for pedagogical clarity, and a real implementation will refine identifiers, indexes, and partitioning to match the host landscape's volume and access patterns.

The next chapter, Chapter 11, picks up from the Ingestion API surface defined here and works through the upstream side: the iFlows that extract from SAP Signavio and SAP LeanIX, the synchronization and reconciliation patterns that keep the data spine consistent with its sources, along with the error handling and retry logic that make the integration safe at production scale. Chapter 11 builds the journey to this chapter's schema and APIs.

PART VI

Integration Flows and Execution Services on SAP BTP

Chapter 11
Integration Flows from SAP Signavio and SAP LeanIX

Middleware moves data; ingestion preserves meaning. The boundary that decides whether an enterprise has a single version of process truth or many is not drawn at the network layer but in the discipline that governs every payload as it crosses from a system of record into the canonical data spine.

Chapter 9 and Chapter 10 introduced and implemented the canonical data spine on SAP Business Technology Platform (SAP BTP). The data spine is the structural memory of the enterprise; it preserves identity, version state, and ownership across the systems that describe how the enterprise is supposed to operate. Chapter 9 defined the entities, identifiers, and lifecycle rules; Chapter 10 turned that design into SAP HANA Cloud schemas, three governed application programming interface (API) surfaces (ingestion, access, and governance), the AI grounding endpoints, and the hash-chained `AUDIT_EVENT` and lineage layer that make the data spine its own evidence chain. This chapter establishes the path toward Chapter 10's schema and APIs.

We explore how data is ingested from SAP Signavio and SAP LeanIX into the canonical data spine on SAP BTP. We cover ingestion flow design, event versus batch trade-offs, idempotency, schema and semantic validation, drift detection, error classification and retry, dead-letter routing, reconciliation, and the validation log that anchors auditability. We also specify how AI is allowed (and not allowed) to participate in these pipelines, consistent with the boundaries set in Chapter 1: It can detect patterns, score anomalies, and draft remediation suggestions, but it should not publish, modify, or auto-approve canonical artifacts.

The chapter is written for practitioners who must operate these flows in production. It assumes the structural decisions of Chapter 9 (canonical entities, identifiers, version state, and lifecycle status) and the implementation choices of Chapter 10 (persistence, schema, API surface) are in place. Where appropriate, it looks ahead to Chapter 12 (execution services that consume the data spine), Chapter 14 (data preparation rules for SAP AI), and Chapter 18 (operational monitoring of the integrated platform).

Three principles run through the chapter and should be carried into every implementation decision. Identity is preserved across every hop. Idempotency is the foundation of reliability. Every retry must produce the same canonical state. Validation is a precondition for consumption. No payload should reach an execution service or AI grounding bundle until the validation log records a definitive verdict against it. This chapter explains:

- How SAP Signavio process and activity data is ingested into the data spine
- How SAP LeanIX application, interface, capability, and dependency data is ingested into the data spine
- How event-driven and batch ingestion patterns are designed and reconciled
- How synchronization, drift detection, and reconciliation cycles are governed
- How errors are classified, retried, quarantined, or routed to dead-letter queues
- How idempotency and lineage are enforced end-to-end
- How a four-tier validation pipeline gates canonical promotion
- How AI participates in the pipeline as a pattern detector and explainer rather than an authoritative actor

11.1 SAP Signavio to SAP BTP Ingestion Flows

SAP Signavio is the system of record for process semantics. It owns the L4 and L5 process and activity definitions, the variants, the role assignments, the control intent, and the application references that describe how the enterprise is supposed to execute its work. The ingestion of SAP Signavio data into the canonical data spine is the moment at which process truth becomes available to the rest of the integrated landscape: SAP LeanIX-aware execution services, SAP AI grounding bundles, SAP Build Process Automation workflows, and SAP Build Work Zone navigation.

The ingestion flow is therefore not a connector. It is a governance boundary expressed in pipeline form. It enforces that only published process versions enter the data spine, that mandatory attributes are present, that identifiers are stable across revisions, that cross-references resolve, and that any deviation between the source and the data spine is detected, classified, and either auto-resolved within tolerance or escalated to a human authority.

This section defines the reference architecture for SAP Signavio ingestion, the patterns by which it is implemented (event-driven and batch), the end-to-end flow with all stages and decision points, the payload-to-data spine mapping that drives normalization, and the table designs that record idempotency and lineage.

11.1.1 Reference Architecture for SAP Signavio Ingestion

The reference architecture for SAP Signavio ingestion is shown in Figure 11.1. It places SAP Signavio and SAP LeanIX as the upstream systems of record, illustrates the ingestion channels through which their data crosses into SAP BTP, depicts the five governed stages of the SAP BTP integration pipeline, and shows the canonical data spine and its authorized consumers. Figure 11.1 deliberately shows both sources together because the pipelines they drive share the same five stages, the same idempotency discipline, the same validation gating, and the same publication boundary.

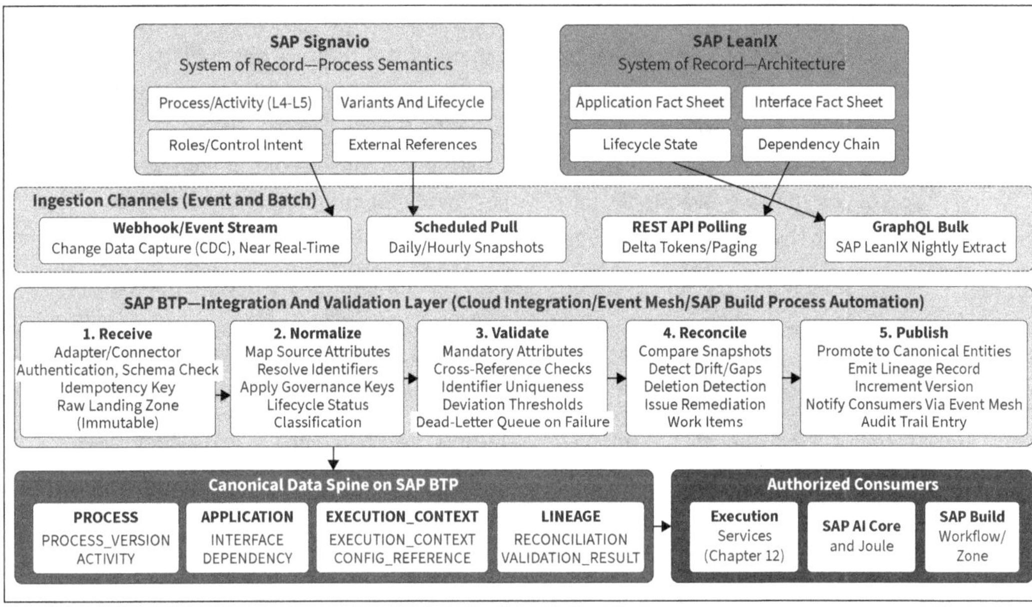

Figure 11.1: Reference Ingestion Architecture (SAP Signavio and SAP LeanIX into the SAP BTP Data Spine)

The ingestion channel—webhook, scheduled pull, REST polling, or GraphQL bulk—is a transport choice; it does not change the validation discipline applied downstream. The five integration stages on SAP BTP (receive, normalize, validate, reconcile, publish) are mandatory; no payload may skip a stage. The canonical data spine is reached only after the publish stage, and only authorized consumers read from it. Execution services (Chapter 12), SAP AI Core (Chapter 13), and SAP Build (Chapter 16 and Chapter 17) consume the data spine through governed APIs, not by querying upstream systems directly.

The architectural separation between SAP Signavio (process semantics) and SAP LeanIX (architecture and applications) is preserved in the pipeline. SAP Signavio payloads carry process and activity identifiers; they reference applications by external key. SAP LeanIX payloads carry application and interface identifiers; they do not own process semantics. The cross-references between them are resolved during the validation stage and persisted as foreign keys inside the data spine, as defined in Chapter 9.

11.1.2 Event-Driven and Batch Ingestion Patterns

Two patterns coexist in production. Event-driven ingestion reacts to changes as they occur in SAP Signavio, typically via webhook delivery from the SAP Signavio webhook framework or via near-real-time polling of the SAP Signavio change feed. Batch ingestion consumes a full or incremental export on a fixed schedule, typically nightly. Each pattern has different latency, throughput, and reconciliation properties, and a mature platform uses both.

The two patterns must be reconciled, not chosen between. Webhooks are useful for low-latency propagation of approved process changes, but they are not sufficient on their own because they can be missed, delayed, or duplicated under network failure. Batch snapshots are useful for completeness, but they are not sufficient on their own because they introduce a large refresh window during which the data spine may be out of step with the source. Combining the two—webhooks for currency, batch for completeness—produces a pipeline that is both responsive and self-healing.

Table 11.1 summarizes the trade-offs and the recommended use of each pattern.

Property	Event-Driven (Webhook/Stream)	Batch (Scheduled Pull/Bulk Export)	Recommended Use
Latency	Seconds to minutes	Hours (next scheduled run)	Use both
Completeness guarantee	None (events can be lost)	Strong (every record is observed)	Batch as truth, event as currency
Failure mode	Lost or duplicate events	Missed window, late detection	Reconcile both daily
Idempotency requirement	Mandatory (correlation ID and payload hash)	Mandatory (snapshot ID and entity hash)	Same discipline applies to both
Operational visibility	High event rate, low per-event detail	Lower frequency, full state visibility	Combine for monitoring
Best fit	Process publication, activity changes	Daily reconciliation, drift detection	See Section 11.3

Table 11.1: Comparison of Event-Driven and Batch Ingestion Patterns

The architectural decision is to treat the batch path as the source of truth for completeness and the event path as the source of truth for currency. The reconciliation cycle described in Section 11.3 compares the two daily and resolves any divergence through the governance pathway. Neither path is permitted to silently overwrite the data spine without passing the same validation pipeline.

Why Both Patterns Are Required

Programs that select only one pattern routinely discover the other was needed. Event-only programs accumulate small, undetected gaps over weeks; batch-only programs fail to propagate urgent process changes (regulatory updates, control adjustments) within their operating window. The cost of running both is small as the same five validation stages serve both patterns, and the resilience benefit is large. Treat the choice as additive, not exclusive.

11.1.3 End-to-End SAP Signavio Ingestion Flow

Figure 11.2 presents the full ingestion flow as a swimlane diagram across SAP Signavio, the Cloud Integration boundary, the SAP BTP ingestion pipeline, and the canonical data spine. The diagram traces both the happy path (publish to consumers) and the error path (dead-letter queue and remediation), and shows where the daily reconciliation cycle joins the pipeline.

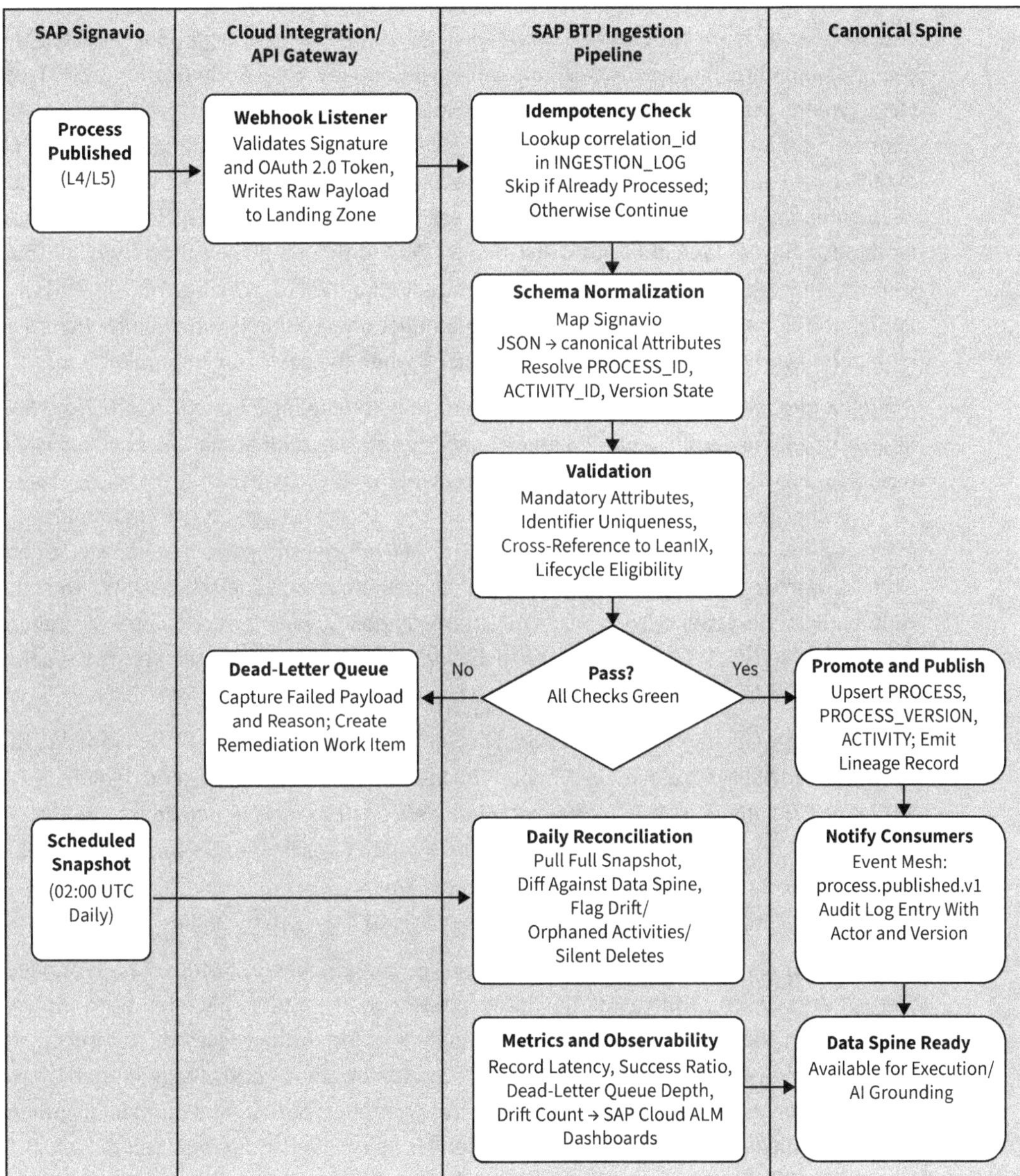

Figure 11.2: SAP Signavio-to-SAP BTP Ingestion Flow

The flow begins inside SAP Signavio with the publication of a process model. Publication is a governance act in SAP Signavio: It transitions a draft into an approved, versioned artifact eligible for downstream consumption. Only published artifacts trigger the webhook; drafts are not propagated. This is the first boundary in the pipeline and the most important. If draft state were allowed to flow, the data spine would acquire artifacts that consumers should not see, and the publication discipline described in Chapter 9 would be compromised.

The first stage is reception. The webhook listener on Cloud Integration validates the request signature, verifies the OAuth2 token issued by SAP Authorization and Trust Management service, canonicalizes the payload, and forwards it to the Ingestion API defined in Chapter 10, Section 10.4.1. The Ingestion API is the single boundary at which an external payload becomes a candidate canonical artifact: It computes the idempotency key, writes the immutable raw record, and acknowledges receipt. The Cloud Integration integration flow (iFlow) does not write to SAP HANA Cloud directly; routing every payload through the API surface preserves the contract Chapter 10 specifies: a 200 OK with `replay: true` on a duplicate payload, a 409 Conflict on a hash mismatch under the same idempotency key, and a 202 Accepted handed off when the payload enters validation.

The second stage is idempotency lookup, performed by the Ingestion API itself. Every SAP Signavio payload carries an SAP Signavio-generated change identifier and is also hashed (SHA-256 over the canonicalized JSON). The combined key (`correlation_id`, `payload_hash`) is checked against the `INGESTION_LOG` table that the Ingestion API maintains. If a matching row exists with status `SUCCESS`, the API returns the cached acknowledgment without reprocessing. This is the property that makes the pipeline self-healing. No retry, whether from the producer or from an internal retry policy, will produce a duplicate canonical state. The `INGESTION_LOG` and the Ingestion API's idempotency contract are the same artifact viewed from inside the platform and from outside it.

The third stage is schema normalization. The SAP Signavio JSON structure is mapped to the canonical attributes defined in Chapter 9. Identifiers are resolved, and `PROCESS_ID`, `ACTIVITY_ID`, and `VARIANT_ID` are extracted. Version numbers are parsed into `VERSION_NO` and `VERSION_STATUS`. Timestamps are converted to Coordinated Universal Time (UTC). Enumerations are mapped to canonical values. This stage is deterministic and does not require human review; if it cannot resolve, the payload moves to the failure path.

The fourth stage is validation. The four-tier validation pipeline of Section 11.5 is applied. Tier 1 checks structural integrity. Tier 2 checks referential integrity: that referenced applications exist in the data spine, variant parentage is valid, and predecessor and successor activities resolve. Tier 3 checks semantic consistency: that control-tagged steps have appropriate roles, variants carry only the deltas from their parent, and domain alignment with SAP LeanIX capabilities is consistent. Tier 4 is AI-assisted anomaly detection, which is advisory and does not gate publication.

The fifth stage is the decision branch. If all blocking tiers pass, the payload moves to promotion. The data spine is updated transactionally (UPSERT into PROCESS, PROCESS_VERSION, ACTIVITY, PROCESS_VARIANT, and EXECUTION_CONTEXT as required), a lineage edge is written linking the canonical update to the source payload hash and ingestion identifier, an entry is appended to the hash-chained AUDIT_EVENT table defined in Chapter 10, Section 10.5.1, and a process.published.v1 event is emitted on Event Mesh. If any blocking tier fails, the payload is routed to the dead-letter queue, a remediation work item is created in SAP Build Process Automation, and the source team is notified. In both branches the AUDIT_EVENT chain advances—either to record the canonical write or to record the gating decision—so that no payload reaches a terminal state without a tamper-evident audit row.

The reconciliation cycle, described in detail in Section 11.3, joins the pipeline at the end of each day. It pulls a complete snapshot from SAP Signavio through the same authentication path, diffs it against the data spine, and either confirms consistency or flags drift through the same validation pipeline. The flow is deliberately symmetric. Events and snapshots converge in the same validation gate, ensuring that no path bypasses the discipline.

11.1.4 Payload Mapping into the Canonical Data Spine

The mapping from an SAP Signavio payload to canonical entities is the most visible expression of the data spine's minimalism. Many fields available in SAP Signavio do not appear in the data spine because they are not required for execution or AI consumption; the data spine carries only what is needed to keep identity stable to make execution context explicit, and support auditability. Figure 11.3 illustrates the mapping for a representative procure-to-pay process and one of its activities.

The mapping is governed by three rules that recur throughout this chapter. First, identity attributes carry forward unchanged. The SAP Signavio process identifier becomes PROCESS_ID, the SAP Signavio activity identifier becomes ACTIVITY_ID, and the SAP Signavio variant identifier becomes VARIANT_ID. These attributes are immutable. They never change between revisions, and they are the keys that downstream services and AI grounding bundles depend on.

Second, version state is added by the data spine, not inherited from the source. SAP Signavio knows whether a process is published; the data spine records that fact as VERSION_STATUS, attaches an EFFECTIVE_FROM date, and increments VERSION_NO. The combination of stable identity and explicit version state is what allows the data spine to support multiple concurrent published versions of the same process when transition periods require it.

Third, cross-system references are resolved against the data spine, not stored as opaque text. When SAP Signavio references an application as text ("SAP S/4HANA Production"), the normalization stage resolves that text against the APPLICATION table, which is populated from SAP LeanIX, and stores the resolved APPLICATION_ID as a foreign key. If the resolution fails, the payload is quarantined; it does not enter the data spine with an ambiguous reference. This is the architectural answer to the process models without execution ownership problem identified in Chapter 1.

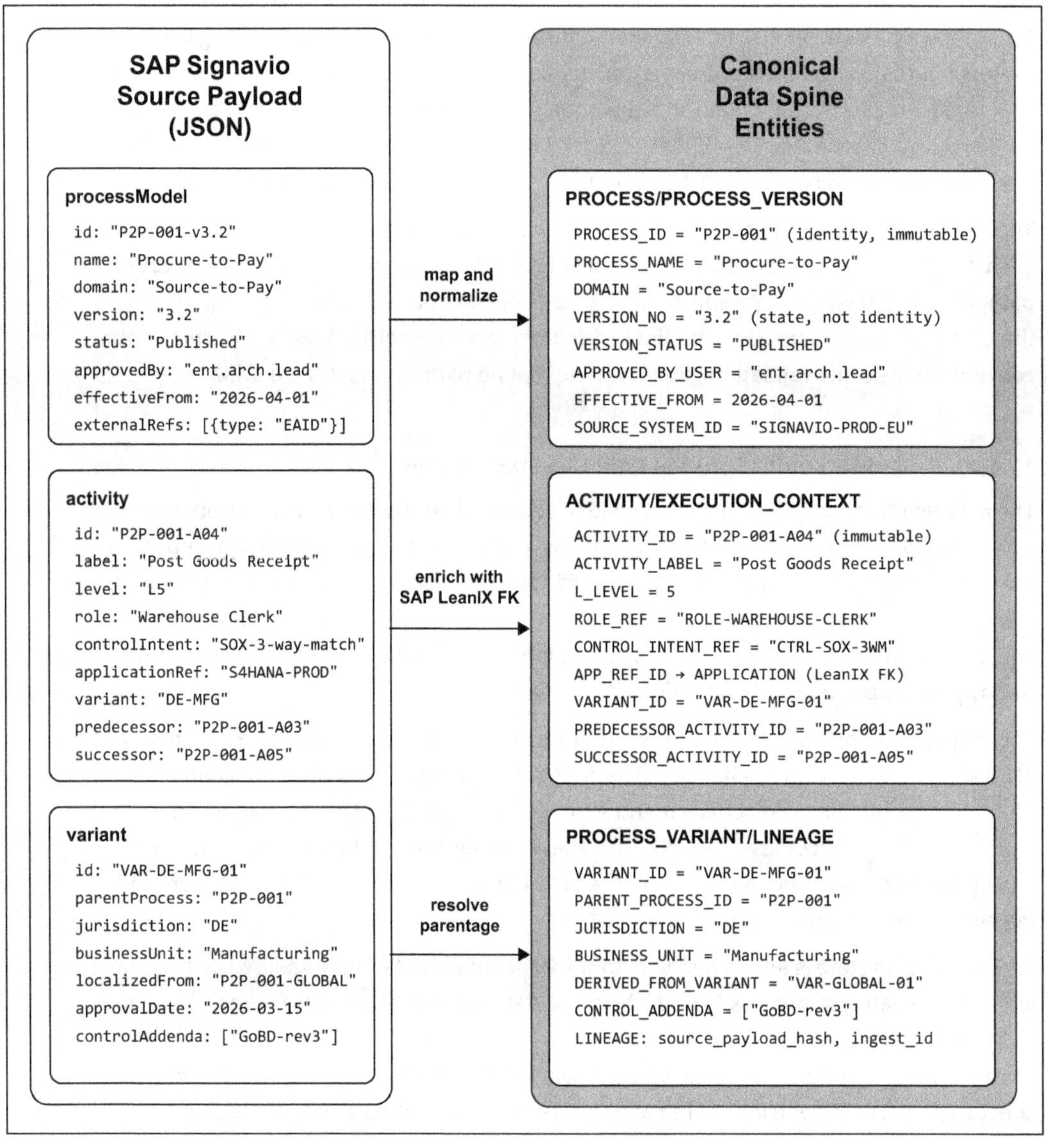

Figure 11.3: SAP Signavio Payload Mapping to Canonical Data Spine Entities

Table 11.2 documents the canonical mapping for the most frequently used SAP Signavio attributes. The mapping is deterministic: For each SAP Signavio attribute on the left, there is exactly one canonical destination, with a defined transformation.

SAP Signavio Source Attribute	Canonical Entity	Canonical Attribute	Transformation Rule
processModel.id	PROCESS	PROCESS_ID	Verbatim; identity, immutable
processModel.name	PROCESS	PROCESS_NAME	Trim whitespace; max 255 characters
processModel.domain	PROCESS	DOMAIN	Mapped to canonical domain enumeration
processModel.version	PROCESS_VERSION	VERSION_NO	Parsed as a semantic-versioning-style value of N.M (major.minor)
processModel.status	PROCESS_VERSION	VERSION_STATUS	Mapped (Published to PUBLISHED); others rejected
processModel.approvedBy	PROCESS_VERSION	APPROVED_BY_USER	Resolved against IDENTITY table
processModel.effectiveFrom	PROCESS_VERSION	EFFECTIVE_FROM	ISO-8601 to UTC timestamp
activity.id	ACTIVITY	ACTIVITY_ID	Verbatim, identity-preserving, immutable
activity.label	ACTIVITY	ACTIVITY_LABEL	Trim; localization handled separately
activity.level	ACTIVITY	L_LEVEL	Integer 4 or 5; reject otherwise
activity.role	ACTIVITY	ROLE_REF	Resolved against ROLE table
activity.controlIntent	ACTIVITY	CONTROL_INTENT_REF	Resolved against CONTROL_REF
activity.applicationRef	EXECUTION_CONTEXT	TARGET_APPLICATION_ID	Resolved against SAP LeanIX-sourced APPLICATION; trigger and target apps recorded distinctly per Chapter 10, Section 10.2.3
variant.id	PROCESS_VARIANT	VARIANT_ID	Verbatim, identity-preserving, immutable
variant.parentProcess	PROCESS_VARIANT	PARENT_PROCESS_ID	Foreign key into PROCESS
variant.jurisdiction	PROCESS_VARIANT	JURISDICTION	ISO-3166 country code

Table 11.2: SAP Signavio-to-Canonical Data Spine Attribute Mapping

Notice that several common SAP Signavio attributes are deliberately omitted from the data spine. Free-text descriptions, modeler comments, diagram coordinates, and revision-specific metadata are not loaded. They are valuable inside SAP Signavio but they are not stable enough or governed enough to support execution. The data spine's minimalism is not designed to be cost-saving; it is the structural reason it can be authoritative without becoming a duplicate of SAP Signavio.

The idempotency discipline that holds the entire pipeline together is implemented as a small, indexed log table on SAP HANA Cloud. The table is consulted at the second stage of every ingestion, including reconciliation. Its schema is intentionally simple.

As shown in Listing 11.1, the composite primary key (CORRELATION_ID, PAYLOAD_HASH) is the technical foundation of idempotency. A retry that carries the same correlation identifier and the same payload finds the row, reads its STATUS, and either returns the cached response (if SUCCESS) or resumes from the recorded checkpoint (if PROCESSING). A retry that carries a new correlation identifier is treated as a new ingestion. A retry that carries the same correlation identifier but a different payload hash is rejected as suspicious and routed to the dead-letter queue.

```
CREATE TABLE INGESTION_LOG (
  CORRELATION_ID     NVARCHAR(64)  NOT NULL,
  PAYLOAD_HASH       NVARCHAR(64)  NOT NULL,
  SOURCE_SYSTEM      NVARCHAR(50)  NOT NULL,
  ENTITY_TYPE        NVARCHAR(50)  NOT NULL,
  ENTITY_ID          NVARCHAR(100),
  STATUS             NVARCHAR(20)  CHECK (STATUS IN ('RECEIVED',
'PROCESSING','SUCCESS','FAILED','QUARANTINED')),
  ATTEMPT_COUNT      INTEGER       DEFAULT 0,
  CACHED_RESPONSE    NCLOB,
  ERROR_CLASS        NVARCHAR(30),
  ERROR_DETAIL       NCLOB,
  RECEIVED_AT        TIMESTAMP     DEFAULT CURRENT_TIMESTAMP,
  COMPLETED_AT       TIMESTAMP,
  PRIMARY KEY (CORRELATION_ID, PAYLOAD_HASH)
);
CREATE INDEX IDX_INGEST_SRC_ENT ON INGESTION_LOG (SOURCE_SYSTEM,
ENTITY_TYPE, ENTITY_ID);
CREATE INDEX IDX_INGEST_STATUS  ON INGESTION_LOG (STATUS, RECEIVED_AT);
```

Listing 11.1: INGESTION_LOG Table—the Durable Record of Every Payload Received, Its Outcome, and the Cached Response Returned on Duplicate Retries

AI Insight: Idempotency Is What Makes AI Audits Possible

When SAP AI agents act on canonical data (e.g., drafting configuration changes, proposing process improvements, or summarizing impact analyses), every action they take is auditable to the same standard as a human user. That auditability depends on the LINEAGE record written at the publish stage, which in turn depends on the INGESTION_LOG. Without idempotency, retries produce duplicate canonical state and fragmentary lineage; with it, every AI-readable artifact has exactly one provenance trail. The discipline that looks like operational hygiene is, in fact, the substrate that defends generative and agentic AI during audit.

11.1.5 Adapter Implementation Notes

The SAP Signavio adapter is implemented on Cloud Integration as a set of iFlows deployed in a single integration package. The package is the unit of versioning and transport; it carries the iFlows themselves, the message mappings, the value mapping tables, and the security material identifiers. Treating the package as the deployable unit ensures that changes to the adapter are auditable and reversible through the same change management process that governs any other transportable artifact in SAP.

Three iFlows compose the adapter. The first is the webhook receiver, which exposes the HTTPS endpoint registered in SAP Signavio. The endpoint is protected by client certificate authentication or by an inbound OAuth 2.0 client secret rotated every ninety days. The second iFlow is the scheduled snapshot puller, which executes on the daily reconciliation window and walks the SAP Signavio export API with cursor-based paging. The third iFlow is the dead-letter replay handler, which is triggered manually from SAP Build Process Automation and reissues a payload through the standard receive path so that idempotency is preserved on retry.

All three iFlows write into the same INGESTION_LOG and emit the same metric set. From a downstream perspective, there is a single entry point to the pipeline regardless of how the payload arrived. This unification is deliberate. It prevents the operational team from having to reason about three different observability surfaces, and it prevents subtle differences in retry semantics between paths from creating divergent canonical state.

11.1.6 Authentication and Credential Rotation

Authentication between SAP Signavio and SAP BTP uses OAuth 2.0 client credentials issued and managed by SAP Authorization and Trust Management service on the SAP BTP side, paired with an SAP Signavio API client identity. The token has a short lifetime (typically one hour) and is refreshed automatically by the iFlow runtime. The client secret is rotated on a fixed schedule (ninety days), and the rotation procedure is encoded as an SAP Build Process Automation workflow so that the rotation, the credential update in the secure store, the iFlow restart, and the verification probe are performed as a single audited transaction.

Two failure points related to authentication recur in production, and both are visible only when the controls are correctly instrumented. The first is a stale token that survives past the rotation date because the iFlow caches it. The receiver returns 401 errors for the cached interval until the cache is invalidated. The second is a partially-rotated credential where the secret is updated in one environment but not another, which produces intermittent 401s that look like transient errors and consume the retry budget. Both failures are caught by the observability metric `auth_failure_ratio`, which alerts at 1 percent over a rolling fifteen-minute window.

11.2 SAP LeanIX to SAP BTP Ingestion Flows

SAP LeanIX is the system of record for enterprise architecture. It owns the application catalog, the interface inventory, the dependency graph, the capability taxonomy, and the lifecycle states of those objects. The ingestion of SAP LeanIX data into the canonical data spine is the moment at which architecture context becomes available to processes, execution services, and AI grounding bundles. Without it, the data spine cannot resolve the application references that SAP Signavio activities carry, reason about change impact, or tell an AI agent which applications are eligible to host an activity in a given jurisdiction.

SAP LeanIX ingestion follows the same five-stage architecture as SAP Signavio ingestion. The differences are in the source semantics and the channels: SAP LeanIX exposes its data primarily through a GraphQL API and nightly bulk exports, and the change events come from the SAP LeanIX webhook framework. The discipline—idempotency, normalization, validation, reconciliation, and publication—is identical.

In the following sections, we'll walk through SAP LeanIX's role as the architecture system of record and the importance of dependency-first ingestion order, as well as capability and domain alignment, webhook semantics, and data classification.

11.2.1 SAP LeanIX as the Architecture System of Record

Figure 11.4 shows the SAP LeanIX ingestion flow with the principal SAP LeanIX factsheet types on the left, the SAP BTP pipeline stages in the middle, and the canonical entities on the right. Five factsheet types matter most for this chapter: application, interface, business capability, IT component (the technical substrate, including version and support lifecycle), and data object. Each maps to one or more canonical entities, as defined in Chapter 9.

The SAP LeanIX pipeline is dependency-first. Applications must be ingested before the interfaces that reference them. Interfaces must be ingested before the dependency edges that connect them. Capabilities must exist before the application-to-capability mappings can be resolved. The pipeline enforces this order through a topological pre-pass over the bulk export. It sorts incoming factsheets so that producers precede consumers, and it queues any orphan reference for the next batch run rather than failing the entire pipeline.

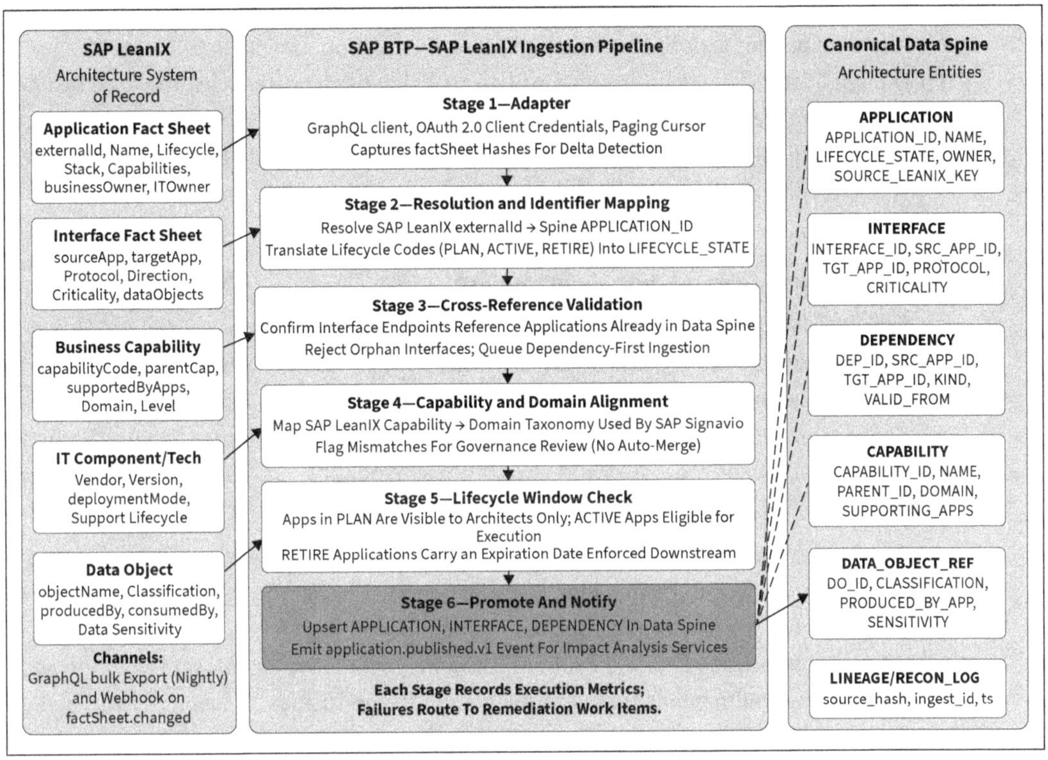

Figure 11.4: SAP LeanIX-to-SAP BTP Ingestion Flow

11.2.2 Dependency-First Ingestion Order

The order in which SAP LeanIX objects are loaded matters. Objects that reference unknown targets cannot be validated, including interfaces that reference unknown applications, dependency edges that reference unknown interfaces, and capability mappings that reference unknown applications. The pipeline respects this hierarchy by processing factsheet types in the following order: application first, then IT component (so application-to-technology references resolve), then business capability (so domain alignment can be checked), then interface (so endpoints resolve), then dependency (so the graph is closed), and finally data object (so producer/consumer references resolve).

This order is not a performance optimization; it is required for model integrity. A pipeline that loads in arbitrary order will produce a snapshot in which some references are temporarily orphaned, and the validation log will be polluted with false negatives. The dependency-first order ensures that every cross-reference resolves on the first attempt, and that any orphan is a real orphan, not a sequencing artifact.

Lifecycle gating is applied at the same stage. SAP LeanIX factsheets carry a lifecycle phase (`PLAN`, `PHASE_IN`, `ACTIVE`, `PHASE_OUT`, and `END_OF_LIFE`). Only `ACTIVE` applications are eligible to host process steps; PLAN applications are visible to architects for design conversations but are not exposed to execution services or AI grounding. PHASE_OUT and END_OF_LIFE applications carry an expiration date that downstream services use to refuse new bindings. The lifecycle gate is enforced in the data spine, not by convention. The consumer-facing APIs described in Chapter 12 read VISIBILITY_SCOPE flags that are set during the SAP LeanIX ingestion publish stage.

11.2.3 Capability and Domain Alignment

Capability data from SAP LeanIX intersects with the domain taxonomy used in SAP Signavio. A purchase-to-pay process in SAP Signavio carries a domain attribute (e.g., source-to-pay); a procurement application in SAP LeanIX is mapped to capabilities (e.g., supplier management, purchase order processing, and invoice verification). Aligning these two views is what makes process-driven impact analysis possible. The pipeline does not auto-merge them; it flags mismatches as governance work items.

When an application supports a capability that does not align with any domain that SAP Signavio currently uses, the pipeline flags the gap as a CAPABILITY_TAXONOMY_GAP. When a capability has no supporting application, the pipeline flags it as a CAPABILITY_UNSUPPORTED. These are not failure conditions; they are governance signals routed to the architecture council. AI may flag the pattern with a message such as "Capability X has no supporting application in three regions," but the architecture council decides whether to retire the capability, onboard an application, or accept the gap.

Table 11.3 documents the canonical mapping for the most frequently used SAP LeanIX attributes.

SAP LeanIX Source Attribute	Canonical Entity	Canonical Attribute	Transformation Rule
application.externalId	APPLICATION	APPLICATION_ID	Verbatim, identity-preserving, immutable
application.name	APPLICATION	APPLICATION_NAME	Trim whitespace; max 255 characters
application.lifecycle	APPLICATION	LIFECYCLE_STATE	Mapped to canonical enumeration (PLAN, ACTIVE, RETIRE...)
application.businessOwner	APPLICATION	BUSINESS_OWNER	Resolved against IDENTITY
application.itOwner	APPLICATION	IT_OWNER	Resolved against IDENTITY

Table 11.3: SAP LeanIX-to-Canonical Data Spine Attribute Mapping

SAP LeanIX Source Attribute	Canonical Entity	Canonical Attribute	Transformation Rule
`interface.externalId`	`INTERFACE`	`INTERFACE_ID`	Verbatim, identity-preserving, immutable
`interface.sourceApp`	`INTERFACE`	`SRC_APP_ID`	Foreign key (FK) into `APPLICATION`
`interface.targetApp`	`INTERFACE`	`TGT_APP_ID`	FK into `APPLICATION`
`interface.protocol`	`INTERFACE`	`PROTOCOL`	Canonical enumeration (REST, OData, SOAP, IDoc, event)
`interface.criticality`	`INTERFACE`	`CRITICALITY`	Canonical enumeration (`LOW`, `MEDIUM`, `HIGH`, `MISSION-CRITICAL`)
`dependency.kind`	`DEPENDENCY`	`KIND`	Mapped to canonical (`USES`, `EXTENDS`, `REPLACES`)
`capability.code`	`CAPABILITY`	`CAPABILITY_ID`	Verbatim; identity-preserving
`capability.parent`	`CAPABILITY`	`PARENT_ID`	FK into `CAPABILITY`
`capability.domain`	`CAPABILITY`	`DOMAIN`	Aligned to SAP Signavio domain taxonomy
`dataObject.classification`	`DATA_OBJECT_REF`	`CLASSIFICATION`	Canonical enumeration (`Public`, `Internal`, `Confidential`, `Restricted`)
`dataObject.sensitivity`	`DATA_OBJECT_REF`	`SENSITIVITY`	Canonical enumeration aligned to data privacy policy

Table 11.3: SAP LeanIX-to-Canonical Data Spine Attribute Mapping (Cont.)

As with the SAP Signavio mapping, several useful SAP LeanIX attributes are not loaded into the data spine. Free-text descriptions, design rationale, and diagram positions remain in SAP LeanIX. The data spine carries only what is needed to support execution and AI consumption, leaving SAP LeanIX as the authoritative source for everything else. Listing 11.2 shows a representative GraphQL query used by the SAP LeanIX adapter to page through application factsheets. Cursor-based paging makes the snapshot replayable; the cursor is stored in the `INGESTION_LOG` and resumed on retry.

```
query allApplications($cursor: String) {
  allFactSheets(
    factSheetType: "Application",
    after: $cursor,
    first: 200
  ) {
    edges { node { id name lifecycle externalId
      ... on Application { businessOwner itOwner technicalSuitability }
      tags { name } } }
    pageInfo { hasNextPage endCursor }
  }
}
```

Listing 11.2: SAP LeanIX GraphQL Fragment for Application Pull

Why Architecture Lifecycle Gates Matter for AI

The most expensive AI failure point in an SAP transformation program is an agent that proposes a process step for an application that has been retired. The proposal may be fluent and confident, but it is operationally wrong. The lifecycle gate enforced during SAP LeanIX ingestion is the architectural defense against this type of failure. By the time AI sees the data spine, `RETIRED` applications are not in the consumer-facing view; `PLAN` applications are visible only with explicit scope. The agent cannot recommend what it cannot see, and the data spine ensures it sees only what is currently authoritative.

11.2.4 SAP LeanIX Webhook Semantics and Bulk Export Sizing

The SAP LeanIX webhook delivers an event when a factsheet is created, updated, archived, or unarchived. The event payload is a thin reference—the factsheet identifier, the event type, and a timestamp—instead of the full factsheet. The receiver must therefore call back into the SAP LeanIX GraphQL API to retrieve the full state at the moment of processing. This callback design has two consequences. First, the data spine reflects the state of the factsheet at the moment of the callback, not the moment of the event. If two events arrive close together, the second event's callback supersedes the first; the data spine never sees the intermediate state. Second, the callback is the operation that authenticates against SAP LeanIX, so any authentication failure surfaces here, not at the webhook receipt.

Bulk export sizing matters at scale. An SAP LeanIX environment with 5,000 applications, 20,000 interfaces, and 1,000 capabilities produces a complete export in the range of 100–250 MB depending on the depth of the included relations. The pipeline streams the export through a server-side cursor rather than loading it into memory, which allows the daily reconciliation to run within a one-hour window even on the largest environments. The cursor checkpoint is committed to the `WATERMARK` table after every page, so a mid-export failure resumes from the last checkpoint rather than restarting the entire pull.

Two operational patterns recur across customer environments. The first is the *survey week* pattern: When SAP LeanIX surveys are collected from application owners, the bulk export grows substantially for the duration of the survey because draft and historical states are temporarily larger. The pipeline absorbs this without configuration change because it pages over the export rather than loading it. The second is the *model cleanup* pattern: When an enterprise architect retires several hundred factsheets in a single week, the reconciliation cycle sees a large `REMOVED` bucket. The bulk-drift threshold described in Section 11.3 catches this and pauses the domain so that the architecture council can confirm the retirements before the data spine absorbs them.

11.2.5 Data Classification and Sensitive Attribute Handling

SAP LeanIX captures data classification (public, internal, confidential, or restricted) and sensitivity (privacy categories aligned to the enterprise data privacy policy) on data object factsheets. These attributes flow into the data spine as `DATA_OBJECT_REF` entries and are referenced by interfaces and applications. The classification is a control attribute. It determines which downstream consumers can include the data object in their context, which AI grounding bundles can carry it, and which roles in SAP Build can see it in user interfaces.

The ingestion pipeline applies two safeguards specifically for sensitive data. First, the tier 3 semantic validation checks that any application referencing a restricted data object also has an authoritative IT owner and a documented control owner; an application that processes restricted data without those references is quarantined. Second, the publish stage emits a separate event topic (`data.sensitive.classified.v1`) when a restricted attribute changes; downstream services that maintain access-control caches subscribe to this topic to invalidate caches promptly. The intent is that classification changes propagate within minutes rather than days, and that no downstream consumer can rely on a stale classification when deciding whether to surface a data object.

11.3 Synchronization and Reconciliation

Synchronization is the steady-state property of the integrated platform: source systems and the canonical data spine agree on the state of every governed artifact. Reconciliation is the discipline that detects, classifies, and resolves divergence when synchronization drifts. The two are inseparable. A platform that synchronizes without reconciling will drift silently; a platform that reconciles without disciplined synchronization will perform expensive, repetitive work.

The synchronization design in this book is event-driven for currency and snapshot-driven for completeness, as introduced in Section 11.1. The reconciliation design that this section introduces is a daily cycle that compares the snapshot to the data spine, classifies any drift as material or immaterial, routes immaterial drift to automatic update and material drift to governance review, and feeds patterns of recurring drift back to the AI services, as we'll describe in Chapter 15.

11.3.1 States and Transitions

Every canonical artifact in the data spine moves through a defined set of states. Figure 11.5 shows the states and transitions an artifact may undergo between its receipt at the SAP BTP boundary and its eventual archival, with every transition logged as an immutable lineage event.

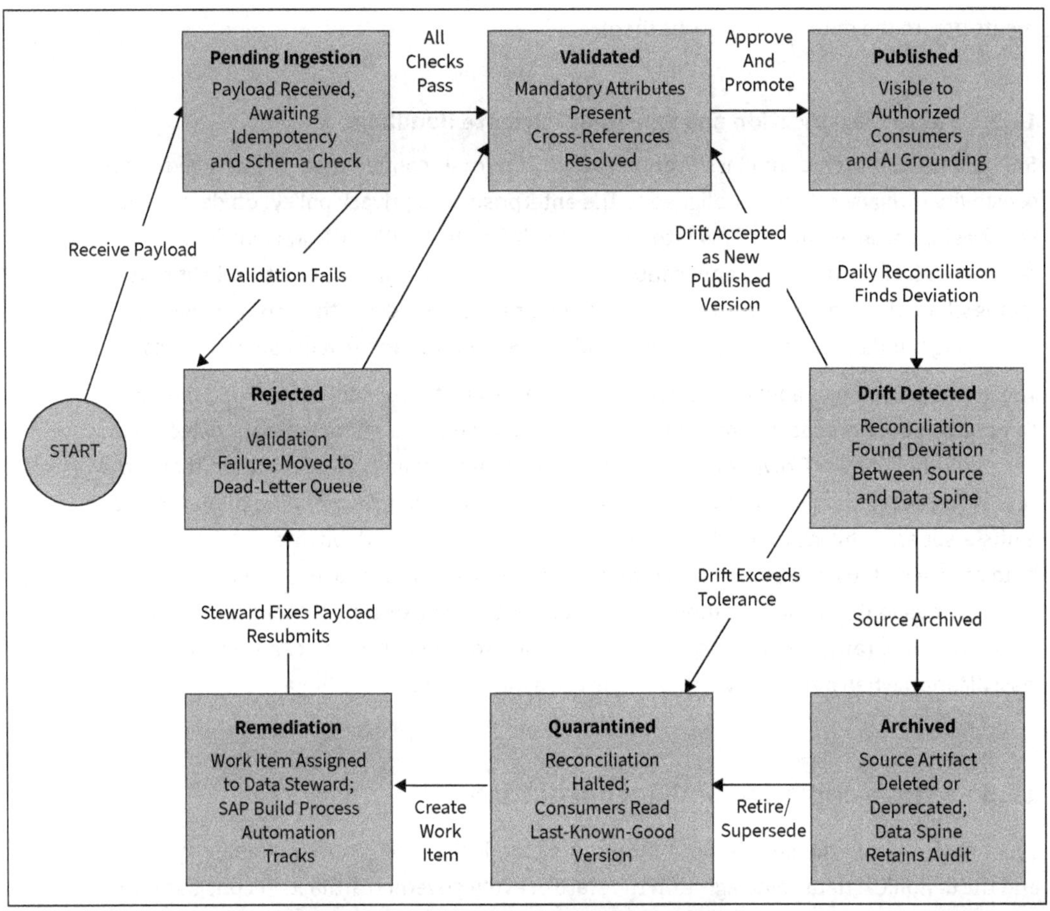

Figure 11.5: Synchronization and Reconciliation State Diagram

Figure 11.5 shows six states:

- Pending ingestion is the initial state, where the payload has been received but has not yet passed idempotency and schema checks.

- Validated is the state after tiers 1 through 3 of the validation pipeline have passed. Published is the consumer-visible state.
- Drift detected is the state into which an artifact transitions when reconciliation finds a deviation between the source and the data spine.
- Quarantined is the state for artifacts whose drift exceeds tolerance and require governance review before any further action; consumers fall back to the last-known-good version while quarantine holds.
- Rejected is the state for payloads that fail the validation pipeline outright.
- Archived is the terminal state for artifacts whose source has been deleted or deprecated; the data spine retains the audit history.

Two architectural decisions are encoded in this diagram. First, the data spine never silently mutates. Every state transition is the result of an explicit event—ingestion, validation, reconciliation, governance approval, or archival—and every transition is logged. Second, the data spine never loses history. An artifact moving from published to drift detected does not overwrite the published version; the published version remains available to consumers under its current `VERSION_NO` until a new version is approved. The drift is recorded as a candidate state, not as an immediate replacement.

11.3.2 Daily Reconciliation Cycle

Figure 11.6 shows the daily reconciliation cycle that operationalizes the state diagram. The cycle runs in ten phases and drives both SAP Signavio and SAP LeanIX in the same job. It produces three artifacts: the updated canonical data spine, the drift report consumed by the dashboards described in Chapter 18, and the AI pattern signal consumed by Chapter 15.

The cycle begins at 02:00 UTC with a snapshot pull from each source system. The pull uses the same authentication path as the event-driven channel, and the resulting snapshot is written to an immutable, dated table (`snapshot_yyyy_mm_dd`) on SAP HANA Cloud. The dated table is preserved for thirty days; it is the artifact a forensic investigation refers back to when reconciling a historical incident.

The diff phase is hash-based. Each canonical entity row in the live spine carries a `SOURCE_HASH` attribute computed at ingestion. The reconciliation job rehashes every snapshot row using the same canonical algorithm and compares the hashes. The output is a four-bucket classification per entity: `ADDED` (in the snapshot, not the data spine), `CHANGED` (hash differs), `REMOVED` (in the data spine, not the snapshot), and `UNCHANGED`. Only the first three are processed further; the `UNCHANGED` rows are counted for telemetry and ignored.

The classification phase distinguishes material from immaterial drift. Immaterial drift includes label changes that do not affect identity or referential structure (for example, a process model whose SAP Signavio label is corrected for spelling). Material drift includes any change to identity attributes, role assignments, control intent, lifecycle state, or cross-system foreign keys. The threshold between the two is governed; it is up to the pipeline to decide. The default policy is conservative: Anything that affects audit, control, or execution is material.

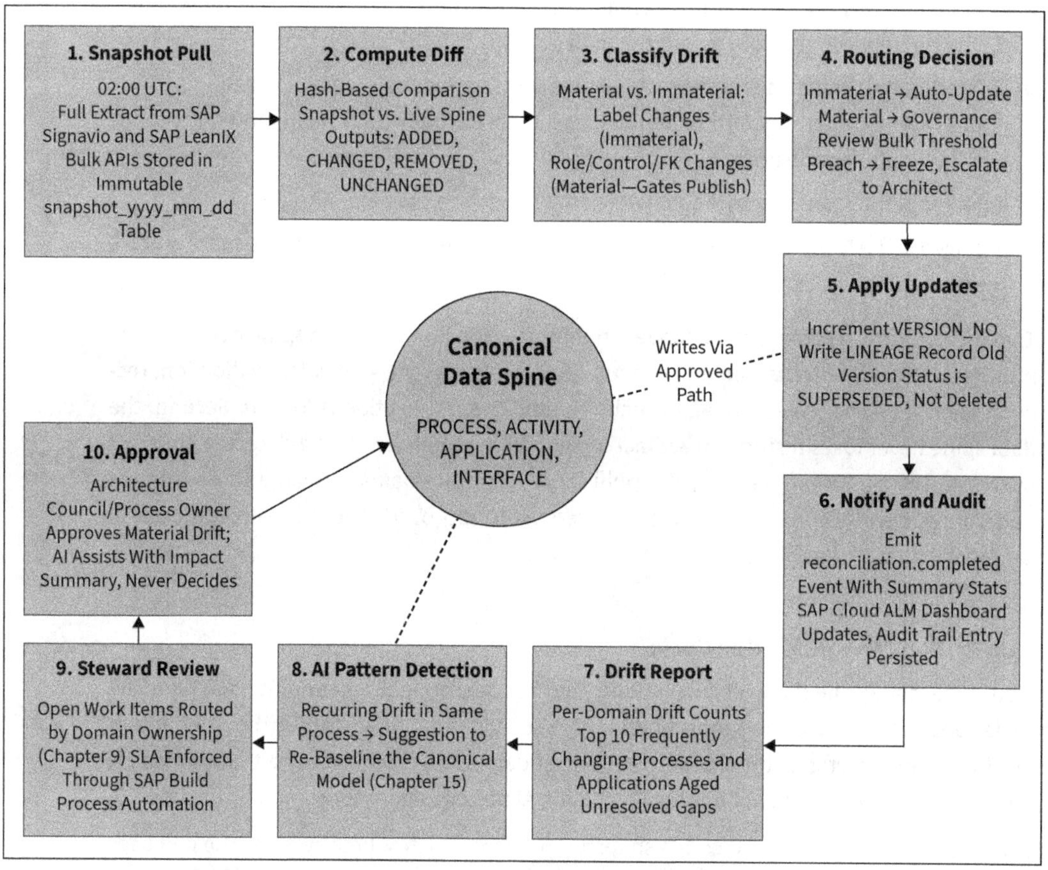

Figure 11.6: Daily Reconciliation Cycle and Drift Detection

The routing phase applies different policies to material and immaterial drift. Immaterial drift is auto-applied to the data spine. The version number is incremented, the `LINEAGE` row is written, and a `process.published.v1` (or `application.published.v1`) event is emitted. Material drift opens a governance work item in SAP Build Process Automation, routed to the data steward and the relevant authority by domain. The artifact remains in its previous published state until the work item is resolved; consumers continue to read the prior version. AI assists with the work item by drafting an impact summary and a recommended resolution, but it does not approve it.

Bulk drift is treated as a special case. When the snapshot diff exceeds a configurable threshold (for example, more than 5 percent of artifacts in a domain show `CHANGED` status), the pipeline does not auto-apply any change, even immaterial ones, and instead freezes ingestion for that domain and pages on-call. The reasoning is empirical. In production, bulk drift is almost always a symptom of an upstream incident (e.g., an SAP Signavio environment refresh, an SAP LeanIX bulk import, or an authentication change) and applying changes during such an event corrupts the data spine. The architectural reflex is to stop and investigate, not to absorb changes.

11.3.3 Drift Classification: Material Versus Immaterial

Table 11.4 summarizes the canonical drift classification used across both SAP Signavio and SAP LeanIX reconciliation. The classification is the rule applied to every CHANGED row from the diff.

Drift Type	Examples	Auto-Apply Allowed?	Routing
Cosmetic	Label spelling, description text, modeler comment	Yes	Auto-apply, increment patch version
Identity	PROCESS_ID change, ACTIVITY_ID change	No	Quarantine; suspect upstream corruption; on-call
Lifecycle	Application moves from PLAN to ACTIVE	Yes (with policy)	Auto-apply, emit lifecycle event
Lifecycle (retirement)	Application moves to PHASE_OUT or END_OF_LIFE	No	Governance review; downstream impact analysis
Control intent	Activity gains or loses Sarbanes-Oxley Act (SOX) tagging	No	Governance review by control owner
Role assignment	Activity's ROLE_REF changes	No	Governance review by separation of duties (SoD) owner
Cross-reference	Activity's APPLICATION_ID changes	No	Governance review by architecture lead
Bulk	>5 percent of domain changed in one cycle	No	Halt domain; on-call; investigate upstream
Silent deletion	Source artifact missing from snapshot	No	Quarantine; requires an explicit retire decision

Table 11.4: Drift Classification and Routing Policy for Daily Reconciliation

Two rows of this table deserve special attention. The treatment of identity drift—a PROCESS_ID or ACTIVITY_ID changing between revisions—is to suspect upstream corruption, not to apply the change. The architectural premise of Chapter 9 is that identity is immutable; an apparent identity change in the source therefore signals that something has gone wrong in the source, not that the data spine should rewrite its keys. The treatment of silent deletion (an artifact present in the data spine but missing from the snapshot) is similar. Rather than infer retirement, the pipeline quarantines and demands an explicit retire decision through the governance pathway. Both rules exist because automatic application of the contrary policy would corrupt the data spine in ways that are invisible to consumers and expensive to recover from.

AI as Pattern Detector for Reconciliation

Reconciliation produces a stream of drift events. Many of those events are routine; some of them, when looked at together, are signals. SAP AI services are configured with read-only access to the reconciliation log and the data spine and can detect patterns that humans typically miss—a particular activity that drifts every Tuesday because of a recurring batch job, a specific application whose lifecycle attribute oscillates because two SAP LeanIX surveys disagree, or a domain whose drift count grows month over month because the process modelers and the architecture team are out of step. AI surfaces these patterns with explanations and confidence scores; the architecture council decides whether to act. The boundary is the same one defined in Chapter 1: AI augments governance, but it does not replace authority.

11.3.4 Snapshot Retention and Forensic Replay

Daily snapshots are retained for thirty days. The retention window is a balance between forensic depth and storage cost. Thirty days is sufficient to cover the typical incident cycle (detection, investigation, remediation, and post-mortem) and is long enough to allow regulators to request a point-in-time view that predates any contested change. Snapshots older than thirty days are archived to cold storage, where they remain accessible but are no longer kept on hot SAP HANA Cloud tables.

Forensic replay is the operation of reconstructing the canonical state as it existed on a specific date. You select the snapshot for that date, materialize it through the same five-stage pipeline with reconciliation suppressed so that no current-day comparison contaminates the result, and then inspect the output. Replays write to a distinct schema (RECON_REPLAY_yyyy_mm_dd) so that the live spine is never disturbed. The capability is used in practice for two scenarios: investigating an audit query ("What did the data spine show on the day this configuration was approved?") and validating a hypothesis about when a particular drift event was introduced.

11.3.5 Watermarks and Hot Restart Semantics

Both event-driven and batch ingestion maintain watermarks, which are monotonic markers indicating how much of the source has been consumed. The webhook listener tracks the highest SAP Signavio sequence number it has acknowledged. The bulk puller tracks the highest SAP LeanIX cursor it has paged through. Watermarks are stored in a small `WATERMARK` table on SAP HANA Cloud and are committed in the same transaction as the canonical update. This helps ensure that a successful ingestion never advances the watermark without a corresponding canonical row and a failed ingestion never advances the watermark at all.

Hot restart is the property that an iFlow restart, node failure, or controlled deployment does not lose data and does not produce duplicates. The watermark and the idempotency log together provide this property. On restart, the runtime reads the watermark and resumes consumption from there. Any payloads that were in flight and reached the receiver but had not yet committed appear again on the next pass; idempotency lookup detects them and either skips them (if `SUCCESS`) or resumes them (if `PROCESSING`). The pipeline therefore tolerates restart without operator intervention—a property that becomes essential at scale, where restarts are routine.

11.3.6 Operational Cadence and On-Call Posture

Reconciliation is a continuous practice, not a quarterly project. The operational cadence layered on top of the daily cycle has three time horizons. At the daily horizon, the on-call engineer reviews the reconciliation summary at handover, confirms that any drift work items are routed to the correct owner, and acknowledges any severity 2 alerts. At the weekly horizon, the integration lead reviews drift trends per domain, identifies recurring patterns, and tunes the bulk-drift threshold for any domain whose noise floor has shifted. At the monthly horizon, the architecture council reviews aggregate health (success ratios, dead-letter queue residency, and validation tier failure rates by source) and decides on platform-level investments, such as adding rules to the registry or adjusting source-system service level agreements (SLAs).

On-call posture is calibrated by severity. Severity 1 events (authorization failures, bulk-drift halts, or identity-drift quarantines) page the on-call engineer immediately and require acknowledgment within fifteen minutes. Severity 2 events (dead-letter queue entries past their retry budget or success-ratio drops below 99 percent for a single source) page within the working day and require resolution within twenty-four hours. Severity 3 events (governance work items or advisory Tier 4 findings) are routed by ticket and tracked against domain-specific SLAs. This is what allows a platform that ingests thousands of payloads a day to be operated by a small, focused integration team rather than become a perpetual war room.

11.4 Error Handling and Retries

Production ingestion pipelines fail. Networks blip, authentication tokens expire, schemas evolve, and foreign keys go stale. The architectural question is not whether failures occur but whether the pipeline classifies them correctly, retries the right ones with the right discipline, quarantines the rest, and produces an audit trail that supports both operational recovery and regulatory defense. This section defines the error-handling architecture for both SAP Signavio and SAP LeanIX ingestion.

11.4.1 Error Classification

Figure 11.7 shows the error-handling and retry state machine. It distinguishes four error classes (transient, schema, semantic, and authorization(and assigns a different strategy to each. The classification is performed at the point of failure; the failing stage records the class along with the error detail in `INGESTION_LOG`. The classification, not the stage of failure, drives the retry decision. Every retry preserves the original `correlation_id` and every failure is logged with classification, attempt count, and final disposition.

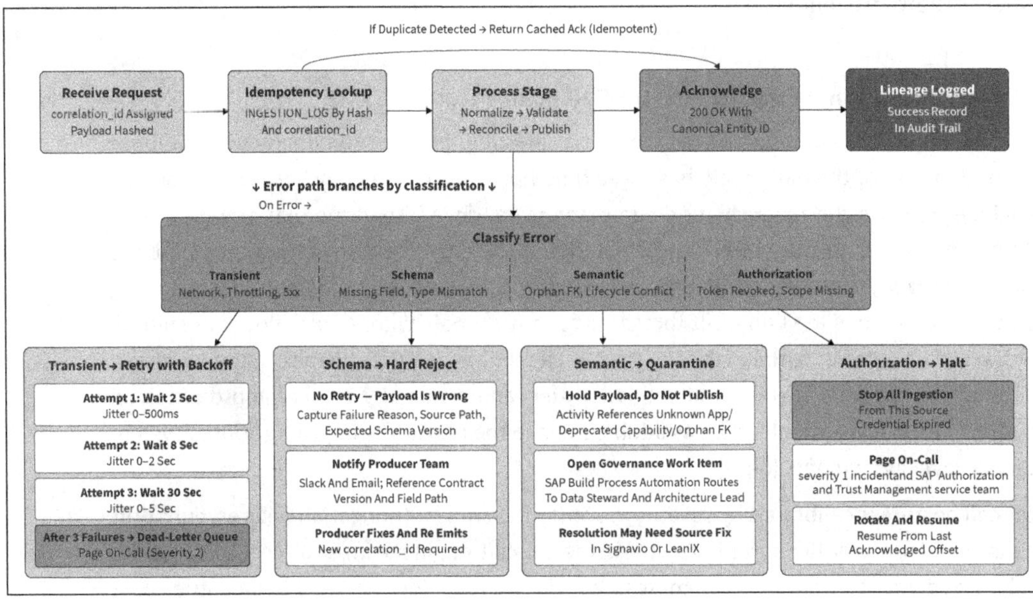

Figure 11.7: Error Handling, Idempotency, and Retry State Machine

The four error classes have very different operational profiles. Transient errors (network timeouts, throttling, and transient 5xx responses) are expected to resolve on their own; the correct response is exponential backoff with jitter. Schema errors (a missing field, a wrong type, or an unknown enumeration) cannot resolve on retry because the payload itself is wrong. The correct response is a hard reject and a notification to the producer. Semantic errors (an activity that references an unknown application, a variant whose parent process does not exist) cannot resolve on retry but may resolve through governance if the missing application is added to SAP LeanIX and the parent process is published in SAP Signavio. The correct response is quarantine and a governance work item. Authorization errors (an expired token, a missing scope, a revoked client) cannot be retried at all; the correct response is to halt all ingestion from the affected source and page on-call.

11.4.2 Idempotency Discipline

Every retry must produce the same canonical state. This is the property that makes the pipeline self-healing, and it is enforced by the INGESTION_LOG schema introduced in Section 11.1. The implementation pattern is a check-then-act under a unique constraint: The listener attempts to insert a row with status RECEIVED and the composite primary key (CORRELATION_ID, PAYLOAD_HASH). If the insert succeeds, processing continues; if the insert fails because the row already exists, the listener reads the existing row and acts according to its STATUS. Listing 11.3 shows the idempotent receipt logic. The unique constraint is the source of truth; the application code does not race against itself or its peers.

```
def receive(payload, correlation_id):
    payload_hash = sha256(canonicalize_json(payload))
    try:
        insert_ingestion_log(correlation_id, payload_hash,
                             status='RECEIVED')
    except UniqueViolation:
        existing = read_ingestion_log(correlation_id, payload_hash)
        if existing.status == 'SUCCESS':
            return existing.cached_response       # idempotent replay
        if existing.status == 'PROCESSING':
            return ack_in_progress(existing)      # retry-safe
        if existing.status == 'FAILED':
            return reject_duplicate_failure()     # do not re-attempt
    process_pipeline(payload, correlation_id, payload_hash)
```

Listing 11.3: Idempotency Lookup Pattern

Two anti-patterns are common in production, and both should be avoided. The first is generating the correlation identifier on the receiving side; this destroys idempotency because every retry from the producer arrives with a fresh identifier. The correct pattern is to require the producer to supply a stable correlation identifier and to reject any payload that does not carry one. The second is checking for duplicates by querying the canonical entity table directly ("Does a row for this PROCESS_ID with this VERSION already exist?"). This is unreliable because the check cannot distinguish between a duplicate retry and a legitimate republication. The INGESTION_LOG is the authoritative record that answers the question "Have we already processed this payload?" The canonical entity tables answer a different question.

11.4.3 Retry, Backoff, and Dead-Letter Routing

Transient retries follow exponential backoff with jitter. The default policy is three attempts at intervals of approximately two, eight, and thirty seconds, each with a random jitter window to avoid thundering-herd retries when a downstream system recovers. After three failed attempts, the payload is moved to a `dead_letter_queue` with status `data_letter_queue`, an operational alert is raised at severity 2, and the payload becomes available for manual replay through the SAP Build Process Automation operator interface.

The dead-letter queue is a service-level instrument, not a parking lot. Aged dead-letter queue entries trigger escalation. An entry that has been in the dead-letter queue for more than twenty-four hours is escalated to severity 1 regardless of its original classification, on the basis that an unresolved dead-letter queue entry represents an artifact that the data spine and the source disagree about, which is precisely the condition reconciliation is meant to prevent.

Table 11.5 summarizes the per-class retry strategy and the operational treatment of failures.

Error Class	Retry?	Backoff Strategy	Routing on Final Failure	Severity
Transient (5xx, throttle, timeout)	Yes, up to 3 attempts	Exponential with jitter (2 seconds, 8 seconds, 30 seconds)	Dead-letter queue	Severity 2; escalated to severity 1 after 24 hours
Schema (missing field, type)	No	N/A	Producer notification and dead-letter queue	Severity 3
Semantic (orphan FK, lifecycle)	No (auto)	Governance work item	Quarantine and work item	Severity 3; SLA-tracked
Authorization (token, scope)	No	Halt source ingestion	On-call paged immediately	Severity 1
Bulk (threshold breach)	No	Domain freeze	Architecture council	Severity 1
Identity drift suspected	No	Quarantine	Architecture council	Severity 2

Table 11.5: Per-Class Retry and Routing Strategy

11.4.4 Observability Tied to Failures

Every failure produces three observable artifacts: an `INGESTION_LOG` row with class and detail, a metric increment on the per-source-per-class counter, and (where the policy requires) an alert routed through SAP Cloud ALM. The metrics are the basis for the operational dashboards described in Chapter 18. The alerts are the basis for incident response. The `INGESTION_LOG` rows are the basis for forensic analysis and audit response.

Two summary metrics deserve continuous attention. The first is the success ratio per source, computed as the rolling fraction of payloads with terminal status `SUCCESS` over a 24-hour window. A drop below 99 percent for either SAP Signavio or SAP LeanIX is a leading indicator of an upstream change that the integration team has not yet absorbed. The second is the dead-letter queue depth, which denotes the number of payloads currently in dead-letter queue status. A growing dead-letter queue depth indicates that retries are not draining and that operational attention is required to clear them before they age into severity 1.

11.4.5 Service-Level Objectives and Alerting Thresholds

The success ratio and dead-letter queue depth are operational metrics, but they are not the service-level objectives (SLOs). The SLOs that govern the platform operate at a higher level: freshness bounds the time between an SAP Signavio publication and the corresponding data spine update; completeness bounds the maximum allowable drift detected by the daily reconciliation; and audit-readiness bounds the maximum time for which any artifact can sit without a definitive validation verdict. The default targets are typically thirty minutes for freshness, 1 percent for completeness, and ten minutes for audit-readiness, although enterprises adjust these based on regulatory and operational context.

Alerting thresholds are derived from the SLOs through a burn-rate calculation. A short-window alert fires when the recent error rate would exhaust the SLO budget within the day; a long-window alert fires when the rolling error rate would exhaust the monthly budget. The two-window approach catches both fast-developing incidents and slow-burning regressions. The pattern is not unique to this platform; it is the same SLO discipline applied to billing systems and customer-facing services, applied here to the integration layer that supports them.

AI Insight: Why Idempotency Is the Foundation

When the AI services described in Chapter 13 through Chapter 15 reason over the canonical data spine, they are reasoning over an artifact whose every change is recorded with a single, immutable lineage entry. This is what makes their outputs auditable. If retries produced duplicate canonical state, the lineage would be ambiguous. Agents could reference one of two valid versions of the same artifact, and the audit trail would have to disambiguate. Idempotency removes that ambiguity at the source. It is not a performance optimization but the architectural property that lets generative and agentic AI be defended in audit. Treat idempotency violations the way you treat data corruption: as a severity 1 incident.

11.4.6 Poison Messages and Circuit Breakers

A poison message is a payload that consistently fails the same way on every retry, typically because it triggers a deterministic exception in a downstream component. Without protection, the system can allow a poison message to consume the retry budget and starve other traffic. The pipeline applies two protections. First, the per-class retry counter in INGESTION_LOG ensures that no payload exceeds its allowed retries. Second, a circuit breaker at the integration runtime opens for the affected class when the failure ratio exceeds a threshold (typically 10 percent over a five-minute window), which causes new arrivals of the same class to be quarantined immediately rather than entering the retry loop.

The circuit breaker is itself observable. When it opens, an alert is raised, the affected payloads are accumulated in a holding queue, and an operator decides whether to drain them through the standard remediation path or to isolate them for separate analysis. The breaker closes automatically once the failure ratio falls below half the threshold for a recovery window. The architectural intent is to limit blast radius: a single misbehaving source or a single problematic class of payload should never bring the entire ingestion platform down.

11.5 Data Consistency Validation

The four-tier validation pipeline is the boundary between received data and canonical truth. It is the gate that decides whether a payload is allowed to update the data spine, whether it must be quarantined for governance review, or whether it must be rejected outright. This section specifies the validation tiers, the rule catalog, and the validation log that anchors auditability. It also walks through rule lifecycle and versioning, validation discipline, and ends with a practical example to demonstrate the concepts covered.

11.5.1 Four-Tier Validation Model

Figure 11.8 shows the validation pipeline. Four tiers are applied in sequence to every payload: structural, referential, semantic, and AI-assisted. The first three tiers are deterministic and gating. If any of them fails, the payload does not proceed to publication. The fourth tier, AI-assisted, is advisory. It produces signals and confidence scores that route artifacts to additional review but does not block publication on its own.

Tier 1 is structural. It checks mandatory attributes (every PROCESS payload carries PROCESS_ID, NAME, DOMAIN, VERSION; every APPLICATION payload carries APPLICATION_ID, NAME, LIFECYCLE_STATE; and so forth), identifier syntax (regex patterns enforced per entity type), data types and formats (timestamps, enums, lengths), and primary-key uniqueness. The checks are fast and require no cross-entity context. A tier 1 failure indicates that the payload is malformed; it cannot be repaired by the pipeline.

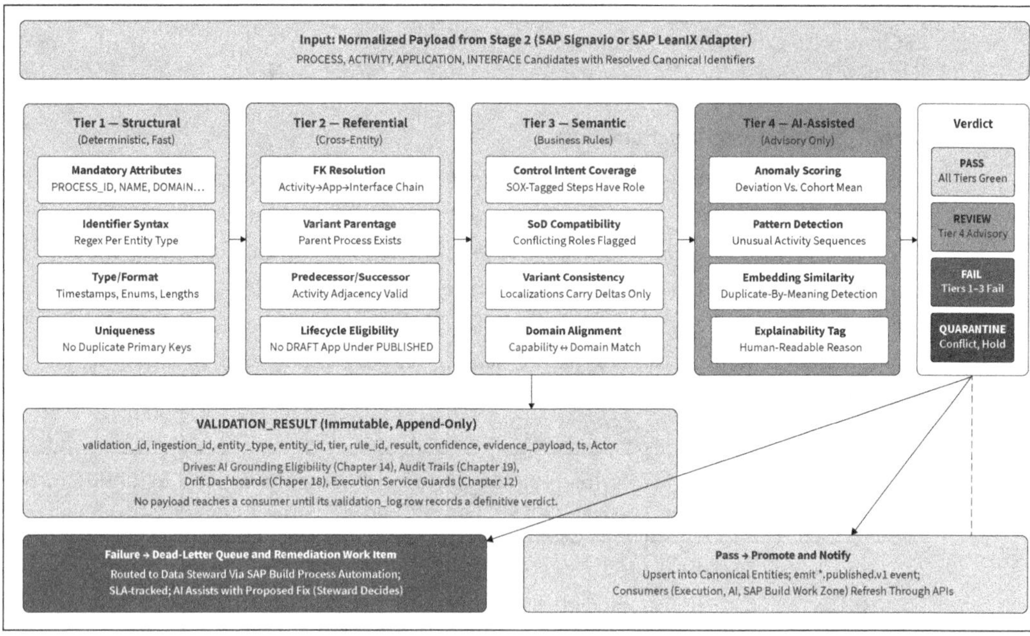

Figure 11.8: Data Consistency Validation Pipeline

Tier 2 is referential. It checks that every foreign key resolves, every activity's `APPLICATION_ID` exists in `APPLICATION`, every interface's `SRC_APP_ID` and `TGT_APP_ID` exist in `APPLICATION`, every variant's `PARENT_PROCESS_ID` exists in `PROCESS`, and every predecessor and successor activity exists. It also checks that lifecycle state is consistent across the dependency graph (a `PUBLISHED` process may not reference a `DRAFT` activity). Tier 2 failures usually indicate a sequencing problem in ingestion or a real semantic gap, and they are quarantined for review rather than rejected.

Tier 3 is semantic. It checks business rules that depend on multiple attributes to ensure that activities tagged with control intent carry an appropriate role, that two activities in the same process do not conflict on SoD, that variant deltas are consistent with their parent process (for example, a DE-MFG variant carries only the deltas required for German manufacturing, not a duplicated copy of the global model), and that capability and domain alignment is consistent between SAP Signavio and SAP LeanIX. Tier 3 failures are typically governance issues and route to a work item with the appropriate authority.

Tier 4 is AI-assisted and advisory. It applies four kinds of signal: anomaly scoring (does this payload deviate substantially from the cohort mean for its entity type?), pattern detection (does this payload introduce an activity sequence that has never been seen before?), embedding similarity (is this artifact semantically a duplicate of one already in the data spine, even though its identifier is new?), and explainability tagging (the AI service writes a human-readable note describing why it produced its score). Tier 4 outputs do not gate publication. They are persisted in the validation log so that human reviewers can use them when triaging tier 2 or tier 3 failures, and so that the process suggestion services described in Chapter 15 can reference them.

11.5.2 Validation Rules Catalog

Table 11.6 catalogs representative rules across the four tiers. The catalog is illustrative, not exhaustive; the operational platform typically carries several hundred rules across both SAP Signavio and SAP LeanIX ingestion. What matters for the architecture is that every rule has a rule identifier, a tier assignment, a verdict policy (gating or advisory), and an evidence schema that is recorded in `VALIDATION_RESULT`.

Tier	Rule	What It Checks	Verdict on Failure	Example Rule ID
1	Mandatory `PROCESS_ID`	Payload includes populated `PROCESS_ID`	`FAIL` (gating)	STR-PROC-001
1	`PROCESS_ID` syntax	Matches ^[A-Z0-9-]{3,64}$	`FAIL` (gating)	STR-PROC-002
1	Timestamp format	ISO-8601 with timezone	`FAIL` (gating)	STR-FMT-014
2	Application FK resolves	Activity's `APPLICATION_ID` in `APPLICATION`	`QUARANTINE`	REF-ACT-021
2	Variant parentage	`PARENT_PROCESS_ID` in `PROCESS`	`QUARANTINE`	REF-VAR-007
2	Lifecycle consistency	`PUBLISHED` process (only for `ACTIVE` apps)	`QUARANTINE`	REF-LIF-031
3	Control role coverage	SOX-tagged activity has role assigned	`GOVERNANCE`	SEM-CTRL-005
3	SoD compatibility	No conflicting roles on adjacent steps	`GOVERNANCE`	SEM-SOD-018
3	Variant delta minimality	Variant carries only delta from parent	`GOVERNANCE`	SEM-VAR-011
3	Domain alignment	Process domain ⇔ capability domain	`GOVERNANCE`	SEM-DOM-002
4	Anomaly score	Deviation from cohort >3 sigma	`REVIEW` (advisory)	AI-ANOM-001
4	Embedding duplicate	Cosine similarity >0.92 vs. existing	`REVIEW` (advisory)	AI-DUP-007

Table 11.6: Representative Validation Rules Across the Four Tiers

Tier	Rule	What It Checks	Verdict on Failure	Example Rule ID
4	Pattern novelty	Unseen activity sequence	REVIEW (advisory)	AI-PAT-014
4	Explainability tag	Human-readable reason persisted	INFO (always)	AI-EXPL-000

Table 11.6: Representative Validation Rules Across the Four Tiers (Cont.)

11.5.3 Validation Log as the Audit Spine

The VALIDATION_RESULT table is the audit spine of the validation discipline. Every rule evaluation, regardless of outcome, writes a row that is also chained into the AUDIT_EVENT hash sequence defined in Chapter 10, Section 10.5.1. The two structures are read together: VALIDATION_RESULT carries the structured evidence per rule per artifact and AUDIT_EVENT carries the tamper-evident proof that the row has not been altered after the fact. The schema is intentionally evidence-rich because the same rows are read by operational dashboards (Chapter 18), audit and compliance (Chapter 19), and the AI grounding endpoints defined in Chapter 10, Section 10.4.4. A payload that has not produced a definitive verdict in VALIDATION_RESULT is, by policy, not eligible to update the data spine nor to appear in any AI grounding bundle returned by those endpoints. Listing 11.4 shows the VALIDATION_RESULT table—the immutable, append-only record of every rule evaluation, used by execution services, audit trails, and AI grounding gates.

```
CREATE TABLE VALIDATION_RESULT (
   VALIDATION_ID      NVARCHAR(64) NOT NULL,
   INGESTION_ID       NVARCHAR(64) NOT NULL,
   PAYLOAD_HASH       NVARCHAR(64) NOT NULL,
   ENTITY_TYPE        NVARCHAR(50) NOT NULL,
   ENTITY_ID          NVARCHAR(100) NOT NULL,
   TIER               INTEGER       CHECK (TIER BETWEEN 1 AND 4),
   RULE_ID            NVARCHAR(50) NOT NULL,
   RESULT             NVARCHAR(20) CHECK (RESULT IN ('PASS','FAIL',
     'QUARANTINE','GOVERNANCE','REVIEW','INFO')),
   CONFIDENCE         DECIMAL(5,4),  -- only for Tier 4
   EVIDENCE_PAYLOAD   NCLOB,          -- JSON evidence
   EVALUATED_AT       TIMESTAMP     DEFAULT CURRENT_TIMESTAMP,
   ACTOR              NVARCHAR(100), -- service or AI model identifier
   PRIMARY KEY (VALIDATION_ID),
   FOREIGN KEY (INGESTION_ID, PAYLOAD_HASH)
     REFERENCES INGESTION_LOG (CORRELATION_ID, PAYLOAD_HASH)
);
CREATE INDEX IDX_VL_ENTITY  ON VALIDATION_RESULT (ENTITY_TYPE, ENTITY_ID);
```

```
CREATE INDEX IDX_VL_TIER    ON VALIDATION_RESULT (TIER, RESULT);
CREATE INDEX IDX_VL_INGEST  ON VALIDATION_RESULT (INGESTION_ID);
```

Listing 11.4: Create Table Validation Result

Three downstream uses of the validation log shape its design. First, the execution services described in Chapter 12 read it before exposing any artifact through their APIs; an artifact without a clean tier 1–3 verdict cannot be served. Second, the AI grounding rules outlined in Chapter 14 use the log to determine which artifacts are eligible to be assembled into a grounding bundle; only artifacts whose latest validation row is PASS or whose advisory tier 4 results have been reviewed and accepted are eligible. Third, the audit framework covered in Chapter 19 uses the log as the primary evidence of control effectiveness. Auditors can sample the log and trace any canonical artifact back to the ingestion that produced it and the rules that approved it.

AI Insight: Validation as the AI Grounding Gate

Chapter 1 argued that AI amplifies whatever coherence or fragmentation already exists in an enterprise. The validation log is the operational mechanism by which that argument becomes concrete. The AI grounding endpoints described in Chapter 10, Section 10.4.4 return only the artifacts whose latest VALIDATION_RESULT row is PASS or whose advisory tier 4 signals have been reviewed and accepted. Anything else is excluded by the endpoint contract, not by convention. Joule agents and Joule Studio agents that consume those bundles therefore inherit the validation discipline of the data spine. This is also the structural basis on which the platform satisfies Section 2.2.2 of SAP API Policy v.4.2026a—the prohibition against AI systems planning, selecting, or executing sequences of API calls outside SAP-endorsed architectures. Agents do not assemble their own context from the data spine's APIs but instead receive validated bundles. The validation log is the substrate that lets agentic AI be safely deployed in regulated SAP environments.

11.5.4 Rule Lifecycle and Versioning

Validation rules are governed artifacts. A new rule does not appear in the pipeline by file edit; it is registered in a RULE_REGISTRY table with a unique RULE_ID, a tier assignment, an effective-from timestamp, an owner identity, and a change-control reference. The pipeline reads the registry on startup and on a scheduled refresh. Adding a rule, deactivating a rule, or changing a rule's severity is therefore a tracked, reviewable operation, not a code commit.

Two operational implications of this discipline matter for production. First, when a new rule is added, the pipeline runs the rule against historical data in shadow mode for a defined period before the rule is activated. Shadow runs write to VALIDATION_RESULT with RESULT='SHADOW' so that the operations team can assess the rule's false-positive rate before it begins to gate publication. Second, when a rule is deactivated, the deactivation timestamp is stored on the registry row. Queries against VALIDATION_RESULT can therefore correctly attribute historical verdicts to the rule version that was in force at the time, which is essential for audit defense.

11.5.5 Validation Discipline and Financial Reporting Trust

A validation pipeline is not only an integration concern but a precondition for trust in any reporting that depends on the integrated estate. Cost control, planning, profitability analysis, and management reporting in SAP S/4HANA increasingly draw on the same canonical data spine that feeds AI grounding bundles. When a controller drills from a profit center variance back to the underlying activities, the activities they see are the ones the integration pipeline published. When a financial planning and analysis (FP&A) analyst pivots a forecast against capability ownership, the capability hierarchy they see is the one SAP LeanIX ingestion populated. If those artifacts are silently inconsistent with the source, the report is wrong even when every formula is correct.

Three practical implications follow. First, the validation log should be reviewed by the FP&A and controlling teams during quarter-end close, not only by the integration team during incident triage, because an artifact in QUARANTINE that participates in a controlling object will produce a reportable gap. Second, the bulk-drift halt described in Section 11.3 is a financial-grade control, not just an operational safeguard. It prevents the data spine from absorbing changes whose audit basis has not been established. Third, the lineage table is the FP&A trace from a reported number back to the canonical artifact and the source payload that produced it, a capability that traditional reporting platforms have struggled to provide and that this architecture provides as a side effect of doing ingestion correctly.

The implication for transformation programs is that the validation pipeline is a deliverable owned jointly by the integration architect, the data steward, and the controller. Treating it as solely a technical concern produces a platform that is technically sound but financially unverified. Treating it as a joint deliverable produces a platform that supports both AI grounding and audited financial reporting from the same logs, with the same discipline.

11.5.6 Worked Example: A Single Activity Through the Pipeline

To better understand this discipline, consider this example of a single activity payload moving through the pipeline. The payload is an activity called "Post Goods Receipt" from the procure-to-pay process, version 3.2, in the German manufacturing variant. The SAP Signavio webhook delivers the payload at 14:07 UTC.

At the receiver, the payload is hashed; the SHA-256 over the canonicalized JSON yields a unique payload_hash. The webhook header carries an SAP Signavio change identifier that becomes correlation_id. The INGESTION_LOG insert succeeds with status RECEIVED. Stage 1 (tier 1, structural validation) passes: PROCESS_ID, ACTIVITY_ID, NAME, LEVEL, ROLE_REF, CONTROL_INTENT_REF, and APPLICATION_REF are all present and well-formed.

Stage 2 (tier 2, referential validation) checks that the activity's target application reference resolves to APPLICATION_ID 'S4HANA-PROD-EU' in APPLICATION; it does. It checks that ROLE_REF resolves to 'ROLE-WAREHOUSE-CLERK' in ROLE; it does. It checks that VARIANT_ID 'VAR-DE-MFG-01' has a valid PARENT_PROCESS_ID; it does. The activity's execution context is also checked for the CK_EC_DISTINCT_APPS constraint defined in Chapter 10, Section 10.2.3 (trigger and target applications must be distinct or explicitly marked single-app); the activity passes. Stage 2 passes.

Stage 3 (tier 3, semantic validation) checks that the activity has CONTROL_INTENT_REF 'CTRL-SOX-3WM' and that ROLE_REF is consistent with the SoD policy; it is. Stage 3 passes. Stage 4 (tier 4, AI-assisted) computes an anomaly score; the score is 0.4 (within normal cohort range, threshold 3 sigma not breached). Pattern detection finds the activity sequence has been seen before; novelty score is low. Embedding similarity finds no near-duplicate. Tier 4 produces an INFO row with explainability tag that reads "Routine activity in established variant" and does not gate publication.

The pipeline promotes the payload. PROCESS_VERSION is incremented from 3.1 to 3.2 with a new content_hash. ACTIVITY is upserted. EXECUTION_CONTEXT is upserted with the resolved trigger and target applications. PROCESS_VARIANT is upserted. A lineage edge is written linking ingestion_id, payload_hash, source_system 'SIGNAVIO-PROD-EU', and the canonical entities are updated. An AUDIT_EVENT row is appended to the hash chain. An event activity.published.v1 with the canonical activity ID is emitted on Event Mesh at 14:07:18 UTC. The end-to-end latency is eleven seconds. The INGESTION_LOG row updates the status to SUCCESS with the cached response payload, and the Ingestion API returns 200 OK to any subsequent retry of the same correlation_id and payload_hash.

If any single check had failed, the pipeline would have routed differently. A tier 1 failure would have moved the payload to status FAILED with class SCHEMA and notified the producer team. A tier 2 failure (for example, an APPLICATION_REF that did not resolve because SAP LeanIX had not yet ingested the application) would have moved the payload to status QUARANTINED and opened a governance work item. A tier 3 failure (for example, an SoD conflict) would have moved the payload to status QUARANTINED and routed to the SoD owner. A tier 4 finding alone would not have gated publication but appeared in the validation log as advisory and would have informed any downstream review.

This walkthrough is not unusual but routine. Hundreds of activities pass through the pipeline every day in a mature program, tens of payloads route to remediation each week, and a handful per quarter trigger bulk-drift halts that pause an entire domain. The discipline is what makes this process routine.

11.6 Summary

This chapter covered how to turn the canonical data spine from a structural design into a populated, governed, auditable operating asset. It specified how SAP Signavio process and activity data and SAP LeanIX application, interface, capability, and dependency data are ingested into the data spine on SAP BTP, how event-driven and batch patterns coexist, how synchronization is maintained through a daily reconciliation cycle, how errors are classified and retried, and how a four-tier validation pipeline gates every promotion to canonical state. The chapter routed every payload through the Ingestion API surface defined in Chapter 10, Section 10.4.1 and preserved Chapter 10's `AUDIT_EVENT` chain and lineage edges as the evidence layer.

11.6.1 Key Learning Points

- Identity is preserved across every hop. The immutable identifiers established in Chapter 9 and persisted by Chapter 10's schema are never rewritten by the pipeline. An apparent identity change in a source signals upstream corruption rather than a legitimate update.
- Idempotency is enforced at the API boundary, not in the iFlow. The Ingestion API's idempotency contract (200 OK with `replay: true` on duplicate and 409 Conflict on hash mismatch under the same correlation key) is the single source of truth for whether a payload has already been processed. The `INGESTION_LOG` that the API maintains is the durable record of that contract.
- Event-driven and batch patterns are additive, not exclusive. Event ingestion provides currency, batch ingestion provides completeness, and daily reconciliation closes the loop. A platform that runs only one pattern fails predictably in the way the omitted pattern would have caught.
- Validation is a precondition for consumption. No payload reaches an execution service or an AI grounding endpoint until `VALIDATION_RESULT` records a definitive verdict against it. The validation log and the `AUDIT_EVENT` hash chain are read together as the platform's evidence layer.
- AI participates as a pattern detector and explainer, never as an authoritative actor. The structural separation between gating tiers (1–3) and advisory tiers (4) is what allows AI to add value to validation without acquiring authority over canonical state. This is the structural basis on which the platform satisfies Section 2.2.2 of SAP API Policy v.4.2026a .
- The data spine is an FP&A-grade asset, not only an integration asset. Cost control, planning, and management reporting in SAP S/4HANA increasingly read the same canonical entities that AI grounding bundles read; the validation pipeline that gates AI also gates the financial reporting trace. Treating the platform as a joint deliverable across integration, data stewardship, and controlling produces a platform that supports both AI and audited financial reporting from the same logs.

11.6.2 Common Design Pitfalls

- Generating the correlation identifier on the receiving side. This destroys idempotency because every retry from the producer arrives with a fresh identifier. The correct pattern is to require the producer to supply a stable `correlation_id` and to reject any payload without one.

- Checking for duplicates by querying canonical entity tables instead of the `INGESTION_LOG`. The canonical tables answer the question "Does a row exist?" instead of "Have we already processed this payload?" Duplicate detection that uses the wrong source produces silent reprocessing and ambiguous lineage.
- Letting iFlows write directly to SAP HANA Cloud rather than calling the Ingestion API. Direct writes bypass the idempotency contract, the `AUDIT_EVENT` chain, and the role gating that Chapter 10 specifies. They create a parallel ingestion path that the rest of the platform cannot defend in audit.
- Auto-applying drift on identity attributes. An apparent change to `PROCESS_ID` or `ACTIVITY_ID` is almost always upstream corruption. Rewriting the data spine's keys to match destroys the very identity stability that the canonical model is built on.
- Treating bulk drift as ordinary drift. A daily diff that shows more than a few percent of a domain changed is almost certainly a symptom of an upstream incident, not legitimate change. Absorbing bulk drift contaminates the data spine; the architectural reflex is to halt the domain and investigate.
- Conflating Tier 4 advisory signals with gating verdicts. AI anomaly scores, embedding similarity, and pattern novelty are advisory inputs to human reviewers, not gates. Allowing them to gate publication transfers authority from the governance pathway to the model and breaks the boundary defined in Chapter 1.
- Letting the dead-letter queue become a parking lot. The dead-letter queue is a service-level instrument, not an archive. Aged dead-letter queue entries are evidence that the data spine and a source disagree about something unresolved, and they must be drained on an SLA, not accumulated.

11.6.3 Review Questions

1. Why must event-driven and batch ingestion patterns coexist, and what specifically goes wrong in a platform that runs only one of them?
2. How does the (`CORRELATION_ID`, `PAYLOAD_HASH`) composite key on `INGESTION_LOG` implement idempotency, and how does the Ingestion API's 200/409 contract from Chapter 10, Section 10.4.1 communicates that property to producers?
3. Explain the four-tier validation pipeline. Which tiers gate publication, which are advisory, and why is the boundary placed where it is?
4. In the daily reconciliation cycle, how is material drift distinguished from immaterial drift, and what specifically prevents the pipeline from absorbing identity drift automatically?
5. Trace the flow of a single SAP Signavio activity payload from webhook receipt to canonical promotion. Identify every place where identity is preserved, a foreign key is resolved, and a row is appended to the `AUDIT_EVENT` hash chain.

6. How does VALIDATION_RESULT participate in AI grounding eligibility through the AI grounding endpoints described in Chapter 10, Section 10.4.4, and how does that participation satisfy Section 2.2.2 of SAP API Policy v.4.2026a?
7. Describe the bulk-drift halt. Under what threshold does it activate, and why is halting safer than absorbing in this case?
8. How do hot restart, watermarks, and the INGESTION_LOG work together to produce a pipeline that tolerates restart without operator intervention?
9. Why does the validation discipline matter for FP&A and cost control reporting, not only for execution services and AI? Identify the specific lineage path that lets a controller trace a reported number back to the source payload that produced it.

11.6.4 Student Companion Notes

If you are studying this chapter within an SAP University Alliance program or an enterprise architecture or business transformation curriculum, you should focus on the structural reasoning behind the ingestion pipeline rather than the specific code or table definitions. The code and tables will evolve; the structural reasoning will not.

Three exercises are particularly valuable for this. First, take a real cross-system process from your own organization and diagram the ingestion lifecycle of one of its activities from SAP Signavio publication through canonical promotion. Mark every place where identity is preserved, a version state is added, a foreign key is resolved, and a validation rule is applied. Second, sketch the daily reconciliation outcome for that activity under three scenarios: a label correction, a control-intent change, and a silent deletion. Reason about which scenarios should auto-apply, which should quarantine, and why. Third, write the validation rule catalog for one custom semantic rule that applies to your organization (for example, a regulatory-specific control that must be present on all activities in a given jurisdiction) and decide which tier it belongs to and why.

A practical study sequence that has worked well in graduate transformation seminars is to read this chapter and then reread Chapter 9 and Chapter 10 again. On the first reading of Chapter 9, the canonical entities feel abstract; on the second reading, after Chapter 10 has implemented them and this chapter has filled them, they read as the precise targets of a real ingestion pipeline whose every stage you can now visualize. Reading the data spine as a static design and then as a populated and governed asset is a useful intellectual exercise; it is also how engineering teams reason about the platform they are building.

This chapter operationalizes the structural foundation laid in Chapter 9 and Chapter 10. The chapters that follow demonstrate how that operational asset becomes the basis for execution services, AI grounding, and ultimately the agentic AI capabilities that SAP's portfolio is now organized around.

Chapter 12
Process-Aware Execution Services

A data spine that is filled but not consumed is a museum of process truth. What turns the data spine into the runtime substrate of the enterprise is the discipline of the services that read it: governed, version-pinned, role-scoped, and audited on every call.

Chapter 9 and Chapter 10 taught you how to design and implement the canonical data spine on SAP Business Technology Platform (SAP BTP). Chapter 11 showed you how to fill it through governed ingestion from SAP Signavio and SAP LeanIX. This chapter makes it useful: it specifies the execution services through which the data spine becomes the runtime substrate of SAP Build Process Automation, AI grounding, and integrated execution across the SAP landscape.

The execution services are the layer at which the data spine meets the rest of the platform. SAP Build Process Automation workflows consume them as actions. SAP Build Work Zone tiles consume them as content application programming interfaces (APIs), SAP S/4HANA embedded analytics dashboards consume them as the source of process and architecture context for reported numbers, and Joule agents consume their AI grounding bundles. Every consumer reads the same canonical rows through the same Access API surface defined in Chapter 10, Section 10.4.2; no consumer reaches the data spine through a private path.

This chapter is intended for those of you who must design, deploy, and operate these services in production. It assumes the structural design of Chapter 9, the implementation of Chapter 10, and the ingestion discipline of Chapter 11. Where appropriate, it looks ahead to Chapter 13 (SAP AI architecture), Chapter 14 (data preparation rules for SAP AI), Chapter 16 (SAP Build Process Automation), and Chapter 18 (operating the integrated platform).

Three principles run through the chapter and should be carried into every service design decision. Identity and version are pinned on every read. Reads are role-scoped at the boundary, not in the consumer. Every call advances the `AUDIT_EVENT` hash chain, so that human users, automated services, and AI agents are visible on the same trace.

This chapter explains:

- How the execution service catalog is structured and how each service maps to the canonical data spine (see Figure 12.1)
- How process lookup and context enrichment services compose role-scoped, version-pinned reads
- How change impact resolution services walk the `EXECUTION_CONTEXT` and `INTERFACE` dependency graph

- How reusable integration services expose the data spine to SAP Build Process Automation as a governed action library
- How service composition, semantic versioning (semver), deprecation, and contract stability protect consumers across change
- How performance and scalability are achieved through tiered caching invalidated by Event Mesh
- How AI consumes the data spine through AI grounding endpoints only, never through API call sequencing
- How execution services preserve the financial planning and analysis (FP&A) trace from a reported number back to the source payload

12.1 Process Lookup Services

Process lookup is the first service in the execution catalog and the most heavily used. Almost every consumer of the data spine begins with a process lookup call: an SAP Build Work Zone tile that needs to display an activity, an SAP Build Process Automation workflow that needs to resolve the role responsible for a step, or an SAP S/4HANA embedded analytics report that needs to drill from a profit-center variance into the activity that produced it. The service is the read path through which the canonical data spine becomes visible to the rest of the platform, and the discipline of its contract is what keeps consumers stable across spine evolution.

Service consumers fall into four broad classes, each with characteristic latency tolerances and error-handling expectations. Workflow steps in SAP Build Process Automation expect single-digit to low-double-digit millisecond latency and tolerate transient errors through the action library's retry policy. Tile renderings in SAP Build Work Zone expect tens of milliseconds and surface latency degradations to the user as loading indicators. Drilldowns from SAP S/4HANA embedded analytics expect low-double-digit millisecond latency and treat errors as a hard failure of the drilldown path. Joule agents that consume AI grounding bundles expect single-second latency and tolerate occasional retries but cannot tolerate inconsistent bundles across retries; the version-pinning property is the structural guarantee that makes this expectation reasonable.

The process lookup service answers questions in the following the form: Given a `PROCESS_ID`, return the canonical `PROCESS` row, the most recent `PROCESS_VERSION` whose `PUBLISH_STATE` is `PUBLISHED`, and the `ACTIVITY` rows that belong to it. It is the entry point through which a consumer moves from an external identifier to a canonical artifact. The service guarantees four properties on every response. The version returned is the version that was current at the time of the request, the validation status of every returned activity is `PASS` in `VALIDATION_RESULT`, the role context of the caller has been applied as a row-level filter, and an `AUDIT_EVENT` row has been appended to the hash chain recording the read.

Figure 12.1 shows the catalog, the API surfaces it sits above, and the consumer ecosystem it serves. It serves as the structural reference for the rest of the chapter. Five service families are defined at the top of the catalog: process lookup (Section 12.1), context enrichment (Section 12.2), change impact resolution (Section 12.3), reusable integration for SAP Build (Section 12.4), and performance and scalability (Section 12.5). Below them sit the three governed API surfaces from Chapter 10 (Ingestion API, Access API, and Governance API), the AI grounding endpoints, and the Event Mesh integration. Below those sits the canonical data spine with its hash-chained AUDIT_EVENT log and VALIDATION_RESULT evidence rows.

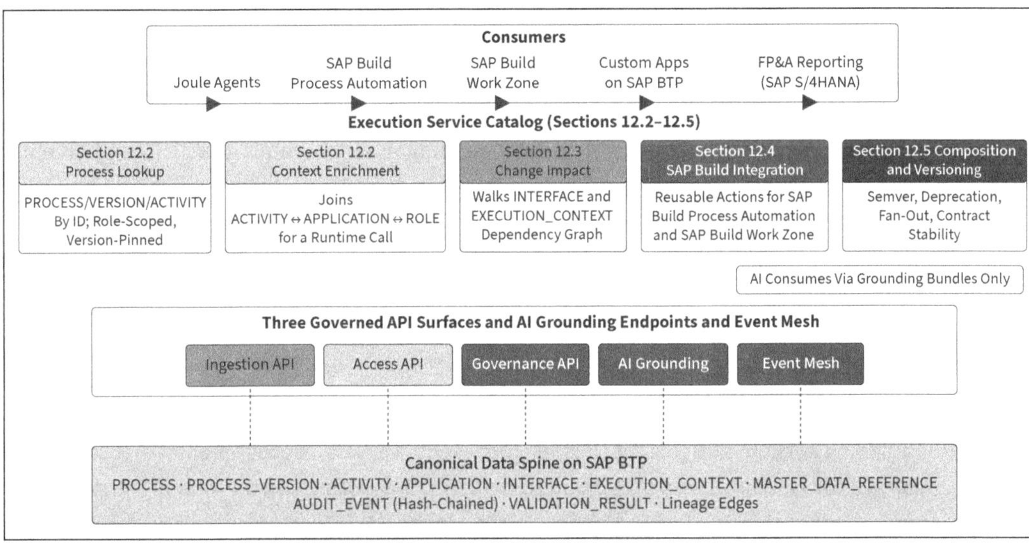

Figure 12.1: Execution Service Catalog and Architectural Boundaries

This section examines four aspects of the process lookup contract: the service's read-only boundary relative to canonical state, the latency budget and throughput targets the contract guarantees, a worked example tracing a single call from the runtime caller through to the canonical response, and the cost characteristics of the filter predicates the service exposes.

12.1.1 Service Boundaries and Read-Only Contract

Process lookup is a thin adapter over the Access API's process endpoint with three additional behaviors. It pins the version explicitly. The request to the Access API includes an AS-OF clause computed from the request timestamp, so that a subsequent request for the same canonical state can be served from the same row even if a publication has moved the canonical state forward in the meantime. It also filters the activity list against VALIDATION_RESULT/ An activity whose latest validation row is QUARANTINE or FAIL is excluded from the response, even if it appears in the underlying PROCESS_VERSION's activity list. Finally, it appends to the AUDIT_EVENT hash chain on every successful read.

The service does not have its own canonical state. It does not write into PROCESS, ACTIVITY, PROCESS_VERSION, or any other canonical table. Writes happen only through the Ingestion API (during ingestion, as described in Chapter 11) or the Governance API (during state transition; see Chapter 10, Section 10.4.3). The read-only contract is enforced by OAuth 2.0 scope assignment in SAP Authorization and Trust Management service. The process lookup service holds the access-api.read scope and nothing else. A service that has only the read scope cannot write to the data spine even if its developer mistakenly believed it could.

12.1.2 Latency Budget and Throughput Target

Process lookup carries the tightest latency budget in the catalog: p50 (50th percentile latency) of 12 ms, p95 (95th percentile latency) of 45 ms, and a throughput target of 5,000 requests per second per node. The budget assumes the Access API is reached through warm connections to SAP HANA Cloud and that the most frequently read rows are present in the SAP HANA buffer cache. It does not assume any external caching layer; caching is the responsibility of the action library that fronts the service (see Section 12.4). The autoscaler is configured to keep p95 within budget at peak load, with a minimum of two nodes per region for high availability.

Table 12.1 sets the budgets for the four core execution services together so that consumers can see the trade-offs as a single design decision. The pattern is consistent: The deeper the service walks into the canonical graph, the looser the latency budget and the lower the throughput target. Process lookup is a single-row read, context enrichment is a five-table join, change impact resolution is a multi-hop graph walk, and AI grounding bundle assembly is a graph walk with validation-lineage stamping. The budgets are not arbitrary; they are the operational expression of how much work each service does.

Service	p50	p95	Throughput Target	OAuth 2.0 Scope
Process lookup	12 ms	45 ms	5,000 req/s/node	access-api.process.read
Context enrichment	30 ms	120 ms	2,500 req/s/node	access-api.context.read
Change impact resolution	180 ms	850 ms	200 req/s/node	access-api.impact.read
AI grounding bundle	450 ms	1,800 ms	50 req/s/node	ai-grounding.bundle.read

Table 12.1: Latency Budgets, Throughput Targets, and OAuth 2.0 Scopes for the Four Core Execution Services

12.1.3 Worked Example: A Single Process Lookup Call

To better understand the contract, consider an SAP Build Work Zone tile that needs to display the current state of the procure-to-pay process for a procurement specialist in EU manufacturing. The tile calls the process lookup service with `PROCESS_ID = 'P2P'`. The service confirms the calling principal's role context against SAP Authorization and Trust Management service, computes the `AS-OF` timestamp, calls the Access API with the appropriate filter, and assembles the response. The response is the `PROCESS` row, the `PROCESS_VERSION` row pinned to version 3.2 (the current published version), and the activity list filtered to those whose validation status is `PASS` in `VALIDATION_RESULT`. The post goods receipt activity is among them; the quarantine inspection activity, whose latest `VALIDATION_RESULT` is `QUARANTINE` under rule REF-VAR-007, is not returned.

Before returning, the service appends an `AUDIT_EVENT` row recording the actor identity (the SAP Build Work Zone tile's service identity in SAP Authorization and Trust Management service), the requested `PROCESS_ID`, the returned `PROCESS_VERSION`, the response time, and a content hash of the row served. The audit row is what makes the read traceable. A controller who later questions a reported number that referenced this read can drill from the number to the `AUDIT_EVENT` row that recorded the read, from the audit row to the `PROCESS_VERSION` that was served, and from the version to the lineage edge that links the canonical row to the source payload that produced it. This trace is not a separate reporting feature; it is a side effect of every read served through the catalog.

AI Insight: Why Lookup Itself Is Audit-Recorded

An older generation of read services treated the read path as audit-free on the assumption that only writes need auditing. That assumption fails for AI grounding. When a Joule agent receives a grounding bundle assembled from process lookup outputs, the bundle is the basis on which the agent will produce its response; if a controller or auditor later questions the response, they need to know which version of the data spine the agent saw, under which role context, and at what moment in time. Audit-recording every read by both humans and AI is what makes that question answerable. The `AUDIT_EVENT` hash chain therefore advances on reads as well as writes, and the actor identity field distinguishes between human users, automated services, and AI agents without privileging any of them.

12.1.4 Filter Predicates and Their Cost

Process lookup is read-only, but read-only does not mean predicate-free. Consumers filter by jurisdiction, by application owner, by whether a process is currently AI-eligible, by deprecation state, and by a half-dozen other attributes. Each predicate has a cost, and the difference between a well-indexed predicate and a poorly indexed one is roughly two orders of magnitude in latency at scale. The service exposes a small, governed set of filter dimensions in the API contract, and the underlying tables carry composite indexes that match the most common filter combinations. Consumers cannot pass arbitrary SQL fragments and cannot construct filter combinations that the service has not pre-indexed. This is a design decision, not a limitation of the underlying database.

The reason is straightforward. The `PROCESS_VERSION` table is large—roughly 40,000 rows in a mature deployment, with `content_hash`, `publish_state`, jurisdiction, AI eligibility, and deprecation timestamp on every row. A query that filters on `publish_state` and jurisdiction returns in under thirty milliseconds because there is a composite index that exactly serves it. A query that filters on a column without an index walks the table and returns in two seconds. Allowing consumers to specify arbitrary predicates would expose the service to query patterns that the index strategy cannot anticipate, and the latency budget would be impossible to honor. Restricting predicates to the governed set keeps the contract honest: the latency budget the service publishes is the latency the consumer actually receives.

There is a softer benefit as well, and it matters for the AI grounding case. When the filter dimensions are governed and small, the cache hit rate is high. The lookup for all AI-eligible processes in EU jurisdiction returns the same payload to every consumer that asks the same question within the cache window, and one database query serves a thousand AI agent invocations. If consumers could ask arbitrary predicate combinations, the cache would fragment into millions of single-use entries, and the hit rate would collapse. The discipline of a small predicate surface is therefore not just an indexing convenience; it is what makes three-tier caching effective at AI-grounding scale.

There is one more aspect of predicate cost that affects FP&A consumers in particular. Drill-down queries from a controller's reporting tool—the kind of query that asks for every process touched by a specific cost center across a specific quarter—are not served by process lookup at all but by the analytical layer described in Chapter 14. Process lookup is a transactional service with sub-second latency expectations and a small predicate surface. Asking it to handle the controller's drilldown would either degrade the lookup latency for everyone else or force the service to refuse the query. The right architectural answer, and the one this platform follows, is to keep process lookup transactional and route analytical drilldowns to a separate service that reads from the same governed data spine but is tuned for scan-heavy workloads. This separation is the reason the controller's drilldown outlined in Section 12.5 traces through the analytical layer rather than through process lookup directly.

12.2 Context Enrichment APIs

Process lookup answers identifier-to-artifact questions. Context enrichment answers a related but different request: Given an `ACTIVITY` or `PROCESS_VERSION`, return the joined view that a runtime caller needs in order to act on it. The join is across ACTIVITY, APPLICATION, INTERFACE, EXECUTION_CONTEXT, MASTER_DATA_REFERENCE, and ACTIVITY_MASTERDATA_REF. The output is a single, normalized, role-scoped object that the consumer can render or pass to a downstream step without further joins.

Figure 12.2 traces a typical context enrichment call from the runtime caller through the API gateway, SAP Authorization and Trust Management service, the process lookup service, the context enrichment service, the Access API, and the canonical data spine. Figure 12.2 is the operational reference for what happens on every enrichment call. The flow is symmetrical with the process lookup flow described in the previous section, but its return path carries a richer object that joins across more entities. What the controller sees on the other end is as follows:

- A reported number in SAP S/4HANA points back to the canonical ACTIVITY served by context enrichment.
- From there, AUDIT_EVENT identifies the exact PROCESS_VERSION that was pinned and the lineage edge to the source payload.
- The trace from a number to its canonical artifact is a side effect of using the data spine through governed services, not a separate report.

Section 12.5.8 explains how this trace is exposed to FP&A and controlling users.

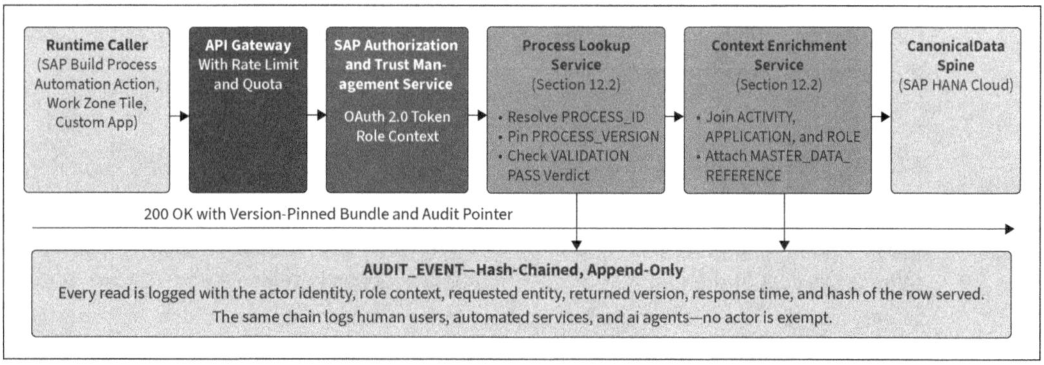

Figure 12.2: Process Lookup and Context Enrichment Flow

This section examines four facets of the context enrichment contract: the three properties that distinguish enrichment from a generic SQL join, the latency budget and caching behavior the contract guarantees, why context enrichment is the natural source for AI grounding bundles assembled at the AI grounding endpoints, and the schema of the enrichment output the consumer actually receives.

12.2.1 The Three Distinguishing Properties

Three properties distinguish the context enrichment service from a generic SQL join. First, the join is constrained by the EXECUTION_CONTEXT. An activity that participates in more than one execution context returns a different enrichment depending on which execution context the request specifies. The service rejects requests that do not specify an EXECUTION_CONTEXT_ID when the activity has multiple contexts, rather than returning a degraded result. The CK_EC_DISTINCT_APPS check constraint defined in Chapter 10, Section 10.2.3 applies on the read path as well as the write path. The trigger and target applications in the returned context must be distinct or the context must be explicitly marked as single-app. A request that would produce an inconsistent enrichment is rejected with HTTP 422.

Second, the join is version-pinned. Every entity in the returned object is the version that was current at the request timestamp, and the object carries the version timestamp so that a subsequent request can reproduce the same view. If APPLICATION rows have moved forward between two enrichment calls, the second call returns the new APPLICATION version even if the ACTIVITY row that referenced the application has not itself been republished. The ACTIVITY's content_hash anchors it, but the join always resolves to the current published version of each referenced entity unless an explicit AS-OF clause is supplied.

Third, the join is validation-gated. An activity whose latest VALIDATION_RESULT is not PASS does not appear in the enrichment, and an APPLICATION whose lifecycle state is RETIRED is excluded even if the underlying activity row still references it. The validation gate is applied at the context enrichment layer rather than in the consumer because the gate is a property of the data spine, not of any one consumer. Consumers that skip the enrichment and join the underlying canonical tables directly bypass the gate and are therefore not allowed in the data spine's read scopes.

12.2.2 Latency Budget and Caching

Context enrichment carries a looser latency budget than process lookup: p50 of 30 ms, p95 of 120 ms, and a throughput target of 2,500 requests per second per node. The looseness reflects the cost of the join: A typical enrichment touches between 4 and 12 rows across 5 tables, and the join is computed at request time, not precomputed. Caching at tier 1 (in-process, 60-second time to live [TTL], evicted by Event Mesh) and tier 2 (Redis, 600-second TTL, evicted by Event Mesh) absorbs most steady-state load; the hit rate target at tier 1 is at least 70 percent and at least 90 percent at tier 2.

Cache keys are versioned, so a hit on one of them (ACTIVITY_ID, ACTIVITY_VERSION, EXECUTION_CONTEXT_ID, ROLE) is always safe to serve because each canonical version is immutable. There is no cache coherence problem to solve because there is no shared mutable state. When a publication produces a new activity version, the Event Mesh subscriber in the action library evicts the relevant cache entries on the topic activity.published.v1. The next request for that activity resolves to the new version. The TTL is the safety net for cases where the event was missed. After it elapses, the cache is repopulated from the Access API with current state.

12.2.3 Why Context Enrichment Is the Source for AI Grounding

AI grounding bundles assembled at the AI grounding endpoints (see Chapter 10, Section 10.4.4) are typically composed of context enrichment outputs rather than raw canonical rows. The reason is that context enrichment already applies the validation filter, the lifecycle filter, the role-scope filter, and the version pin—the same four properties that will be required of any artifact eligible for AI consumption in Chapter 14. Bundles assembled from raw canonical rows would have to apply those filters again at the grounding endpoint. Assembling them from context enrichment outputs avoids the duplication and ensures that the AI consumer cannot, by accident or by design, request a less filtered view than a human consumer would receive.

This is the operational meaning of the principle from Chapter 1 that AI inherits the governance of the data spine, not the other way around. The validation discipline that we enforced in Chapter 11 at the write boundary is the same discipline that context enrichment enforces at the read boundary, and the same discipline that the AI grounding endpoint inherits when it composes its bundles. A single rule (`VALIDATION_RESULT` must be `PASS`) propagates through every layer of the read path.

12.2.4 Enrichment Output Schema

Table 12.2 documents the canonical output schema of the context enrichment service. Every field is sourced from a specific canonical entity, every field is filtered by the calling principal's role context, and every field is version-pinned to the response timestamp. Consumers that depend on this schema in their own contracts (such as a Joule agent's prompt template, an SAP Build Work Zone tile's rendering layer, or an FP&A drilldown query) can rely on the schema being stable within a major version of the Access API.

Output Field	Source Entity	Role-Scoped	Version-Pinned
`activity_id`	`ACTIVITY`	Yes	Yes (`ACTIVITY` `content_hash`)
`activity_name`	`ACTIVITY`	Yes	Yes
`execution_context_id`	`EXECUTION_CONTEXT`	Yes	Yes
`trigger_application`	`APPLICATION` (via `EXECUTION_CONTEXT`)	Yes	Yes (`APPLICATION` version)
`target_application`	`APPLICATION` (via `EXECUTION_CONTEXT`)	Yes	Yes

Table 12.2: Context Enrichment Service—Canonical Output Schema with Source Entity and Filter Properties

Output Field	Source Entity	Role-Scoped	Version-Pinned
interface_id	INTERFACE (via EXECUTION_CONTEXT)	Yes	Yes
role_required	ROLE (via ACTIVITY)	Yes	Yes
master_data_refs	MASTER_DATA_REFERENCE (via ACTIVITY_MASTER-DATA_REF)	Yes	Yes
validation_verdict	VALIDATION_RESULT (latest, tier <4)	No (always returned)	Yes (evaluation timestamp)
response_timestamp	Service-stamped	No	Yes (the AS-OF anchor)

Table 12.2: Context Enrichment Service—Canonical Output Schema with Source Entity and Filter Properties (Cont.)

AI Insight: The Validation Gate at Read Time

Chapter 11 introduced the four-tier validation pipeline as a write-time gate: a payload that fails tier 1, 2, or 3 does not reach canonical state. Context enrichment generalizes that gate to read time: An artifact whose latest VALIDATION_RESULT is QUARANTINE, GOVERNANCE, or FAIL is invisible to the read path even if it sits in the canonical tables. The two gates work together. A payload reaches the data spine only if it passes the write-time gate, and a consumer sees a canonical row only if it passes the read-time gate. AI consumes the same read path with the same gate, which is the structural foundation of the AI grounding contract that we will elaborate on in Chapter 14.

12.3 Change Impact Resolution Services

Change impact resolution answers a question that process lookup and context enrichment do not. Where those services move from an identifier to a canonical artifact, change impact resolution moves from a proposed change to the set of artifacts that the change would touch. It is the service that an architect calls before retiring an interface, a release manager calls before promoting a new APPLICATION version, and a controller calls before deciding whether a process change should be assessed for financial reporting impact. It is also the service that an SAP Build Process Automation exception handling workflow calls when a downstream system reports an unexpected error.

The service walks the dependency graph defined in Chapter 9 and persisted as edges in Chapter 10, Section 10.5.2. Walking the graph means following these relationships outward from a starting entity, one hop at a time; persisting them as edges means storing each relationship as an independent row in an edges table that the service can traverse with ordinary SQL joins, rather than collapsing the structure into a single denormalized `LINEAGE` table that would have to be recomputed on every change. Lineage is stored as edges, not as a single `LINEAGE` table. The graph is the union of edges from `PROCESS_VERSION` to `ACTIVITY`, from `ACTIVITY` to `APPLICATION`, from `APPLICATION` to `INTERFACE`, from `INTERFACE` to `APPLICATION` (the receiving side), from `EXECUTION_CONTEXT` to its participating applications, and from each canonical artifact to its source payload through the lineage edge written at publication. The service exposes the graph through three endpoints: forward impact (what depends on this?), reverse impact (what does this depend on?), and full neighborhood (everything within N hops).

The following sections walk through forward impact and reverse impact in turn, address the performance and caching considerations specific to graph walks, catalog the common traversal patterns by use case, work through the retirement of an interface as a concrete example, and close with the edge cases the service handles explicitly.

12.3.1 Forward Impact

Forward impact answers the question "What depends on this?" and is the most commonly requested query. Figure 12.3 illustrates a forward-impact query for an `APPLICATION` change: the upgrade of S4HANA-PROD-EU from the 2024 release to the 2025 release.

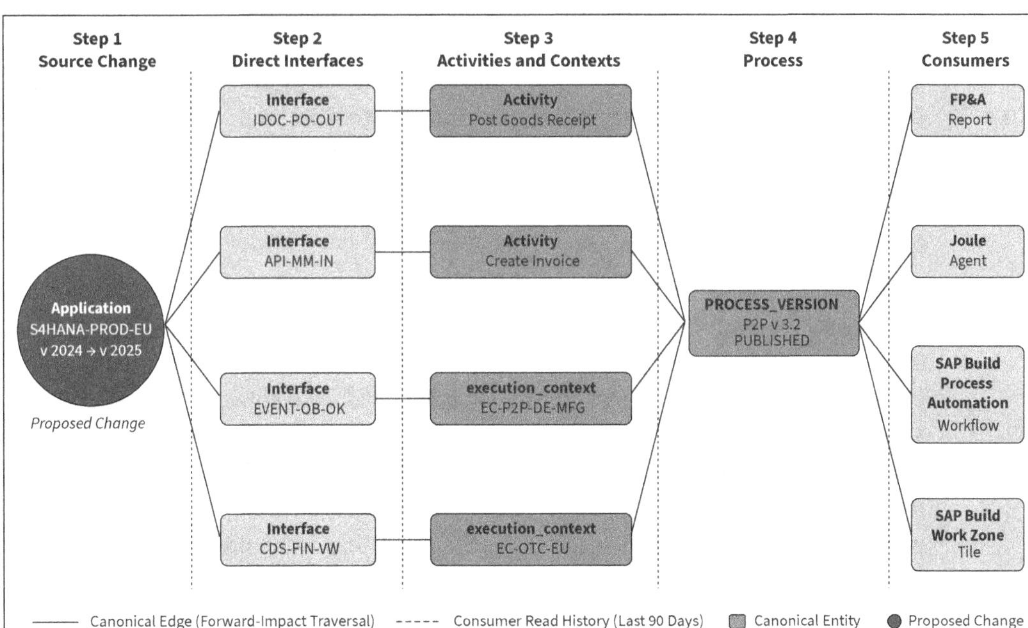

Figure 12.3: Change Impact Resolution—Walking the Dependency Graph

The service computes the immediate dependency set: the four INTERFACE rows whose source or target is S4HANA-PROD-EU. From each INTERFACE, the service expands to the EXECUTION_CONTEXT rows that include those interfaces, and from EXECUTION_CONTEXT to the PROCESS_VERSION rows that participate. From PROCESS_VERSION, the service expands to the consumer artifacts that have read those rows in the last ninety days: FP&A reports, Joule agents, SAP Build Process Automation workflows, and SAP Build Work Zone tiles. The result is a ranked impact set in which each row carries an estimate of read frequency, the role context under which it is most often read, and the date of the last read.

12.3.2 Reverse Impact

Reverse impact answers the question "What does this depend on?" and is less commonly requested but operationally critical. A reverse-impact query against a single ACTIVITY returns the EXECUTION_CONTEXT it participates in, the APPLICATION it executes against, the role it requires, and the MASTER_DATA_REFERENCE rows it depends on through ACTIVITY_MASTERDATA_REF. A reverse-impact query is what an architect runs when assessing whether a particular activity is safe to deprecate. If the activity depends on a master data reference that is itself deprecated, the deprecation is straightforward, but if it depends on a freshly published reference with active downstream consumers, the deprecation requires a coordination plan.

Both endpoints share the same depth and breadth controls. A request specifies the starting entity, the maximum depth (typically two or three hops), and the entity types to include in the traversal. Forward queries can be restricted to consumer artifacts; reverse queries can be restricted to canonical artifacts. The result set is bounded by the same role scope as process lookup and context enrichment. An architect with read scope only on the order-to-cash domain sees only order-to-cash artifacts in the result set, even if the underlying graph traversal touches artifacts from other domains.

12.3.3 Performance and Caching of Graph Walks

Graph walks are intrinsically more expensive than single-row reads. A typical forward-impact query against an APPLICATION with 12 interfaces and 25 participating execution contexts returns a result set of 2 to 300 hundred rows and takes between 150 ms and 300 ms against a warm spine. The latency budget for change impact resolution is therefore looser than for process lookup: p50 of 180 ms, p95 of 850 ms, and a throughput target of 200 requests per second per node. Section 12.5 describes how this budget is achieved through tier 2 caching of the graph adjacency lists in Redis, with eviction driven by Event Mesh on the topics interface.published.v1, application.published.v1, and execution_context.published.v1.

Graph adjacency lists are role-scoped. An architect with an order-to-cash scope sees a different graph than an architect with a procure-to-pay scope, even if the underlying canonical edges are the same. Caching graph walks blindly across roles would leak rows. The tier 2 cache key includes the calling role, not just the entity identifier, so that a hit returns the same view the role would have computed itself. This adds key cardinality to the cache but is the only safe way to cache role-scoped responses.

12.3.4 Graph Traversal Patterns by Use Case

Different consumers walk the graph in different ways. Table 12.3 catalogs the most common traversal patterns observed in production, the entry-point entity each pattern starts from, the typical depth and breadth of the traversal, and the consumer audience. The catalog is illustrative, not exhaustive. The operational platform typically carries a handfuls of additional bespoke traversals for specific business transformation programs. What matters for the architecture is that every traversal pattern is implemented as a parameterized call against the change impact resolution service rather than as a private query against the canonical edges; the service is the only legitimate read path for the graph.

Pattern	Entry Point	Depth	Primary Consumer	Typical Result Size
Application upgrade impact	APPLICATION	3 hops forward	Release manager	200-500 rows
Interface retirement impact	INTERFACE	2 hops forward	Architect	50-200 rows
Activity dependency review	ACTIVITY	2 hops reverse	Process owner	10-30 rows
Process variant comparison	PROCESS_VERSION	Full neighborhood, 1 hop	Variant analyst	50-150 rows
Master data trace	MASTER_DATA_REFERENCE	All ACTIVITY references	Data steward	100-500 rows
FP&A activity drilldown	ACTIVITY	1 hop reverse and lineage edge	Controller	5-15 rows
AI grounding scope check	PROCESS_VERSION	Full neighborhood, 2 hops	AI operations team	100-300 rows

Table 12.3: Graph Traversal Patterns Implemented by the Change Impact Resolution Service

12.3.5 Worked Example: Retiring an Interface

Imagine an architect plans to retire the interface IDOC-PO-OUT, which has been superseded by API-MM-IN as part of the S4HANA-PROD-EU 2025 upgrade. Before submitting the retirement request through the Governance API, the architect runs a forward-impact query on IDOC-PO-OUT. The result returns three EXECUTION_CONTEXT rows that reference

IDOC-PO-OUT as their source interface, eleven ACTIVITY rows that participate in those execution contexts, one PROCESS_VERSION (procure-to-pay version 3.2) that owns those activities, and four consumer artifacts that have read those activities in the last ninety days: an SAP Build Work Zone tile, an SAP Build Process Automation workflow, an FP&A drilldown report, and a Joule agent for procurement summarization. Each consumer artifact carries the date of last read, the average read frequency, and the role context of the most frequent reader.

The architect uses the impact set to construct the deprecation calendar required by the Governance API. The new interface API-MM-IN is published, the EXECUTION_CONTEXT rows are updated through governance workflows to use API-MM-IN, consumer artifacts are notified through Event Mesh on the topic interface.deprecation_warning.v1, and, after a ninety-day overlap, the IDOC-PO-OUT interface is marked RETIRED. The change impact resolution service is read again at each step to confirm that the consumer set is empty before the final retirement is committed. If a new consumer appears in the interim, the retirement is paused and the new consumer is contacted before the calendar resumes.

This is the discipline that change impact resolution makes feasible at scale. Without it, an architect would have to assemble the impact set by hand from architectural documentation that is typically out of date by the time it is consulted; with it, the impact set is a deterministic query against the canonical graph and is as current as the data spine itself. The service is therefore not only an architectural convenience but a control that prevents inadvertent breakage during change.

12.3.6 Edge Cases the Service Handles Explicitly

Three edge cases recur in production traversal queries, and each is handled by the service explicitly rather than by the consumer. The first is the cycle case: two execution contexts that reference each other through some chain of interfaces. Cycles are syntactically possible in the canonical graph but are semantically suspicious; the service detects cycles during traversal, breaks them at the cycle boundary, and includes a cycle marker in the response so that the consumer can decide whether to investigate the cycle as a modeling error. The marker is metadata on the response, not an error. The traversal still completes, but the consumer is warned.

The second is the orphan case: an ACTIVITY whose APPLICATION reference does not resolve because the APPLICATION row was retired without the activity being updated. Orphan rows are filtered out of the result set during the role-scope and validation-gate filtering, and a separate orphan_count field on the response tells the consumer how many rows were filtered. The orphan count is a useful operational metric. A rising orphan count is evidence that the deprecation discipline is breaking down, and the platform team uses the metric as an early warning that an APPLICATION was retired without all its dependent activities being updated.

The third is the burst case: a query whose result set is so large that the response would exceed the Access API's pagination limit. The service detects the burst at the first hop of the traversal, returns the first page with a continuation cursor, and logs the burst to SAP Cloud ALM so that the operations team can see whether bursts are concentrated on particular consumers or particular entities. The cursor is version-pinned: A consumer that paginates through a large result set sees a consistent view of the canonical state at the time of the first page, even if publications have moved canonical state forward in the meantime. Pagination is therefore as much a correctness property as a performance property.

AI Insight: Surfacing Anomaly Signals at Read Time

Tier 4 of the validation pipeline (see Chapter 11) produces anomaly signals that are advisory and do not gate publication. Change impact resolution is the natural consumer of those signals on the read path. When the service is asked for a forward-impact set on a freshly published artifact, it surfaces the anomaly score from Chapter 11's tier 4 results in the response so that the architect can see whether the artifact deviates from cohort norms before deciding to act on it. The signal is advisory at write time and read time, and it is never a gate; it is, however, a useful piece of context that has been carried forward across the pipeline. AI thus augments the architect's judgment at the moment the judgment is being made without ever replacing it. This is the operational meaning of Chapter 1's principle that AI is a pattern detector and explainer, never an authoritative actor.

12.4 Reusable Integration Services

SAP Build Process Automation is the principal consumer of the execution catalog. Citizen developers and pro-code developers alike build workflows that resolve process context, route work items by role, check publication state, subscribe to canonical events, and submit governance requests. The execution catalog exposes these capabilities to SAP Build Process Automation through a single, governed action library: a published, semver-pinned npm package that wraps the Access and Governance APIs in a small set of stable, contract-stable actions.

Figure 12.4 shows the architecture. SAP Build Process Automation workflows on the left consume actions from the library in the middle, the actions wrap the governed APIs on the right, and the APIs read from the canonical data spine on the far right. The action library is the only path through which an SAP Build Process Automation workflow reaches the data spine. Direct calls from a workflow to the Access API or the Governance API are technically possible but not allowed by code review and not supported by the platform team. The library is the canonical interface, and any need it does not satisfy is treated as a request for a new action.

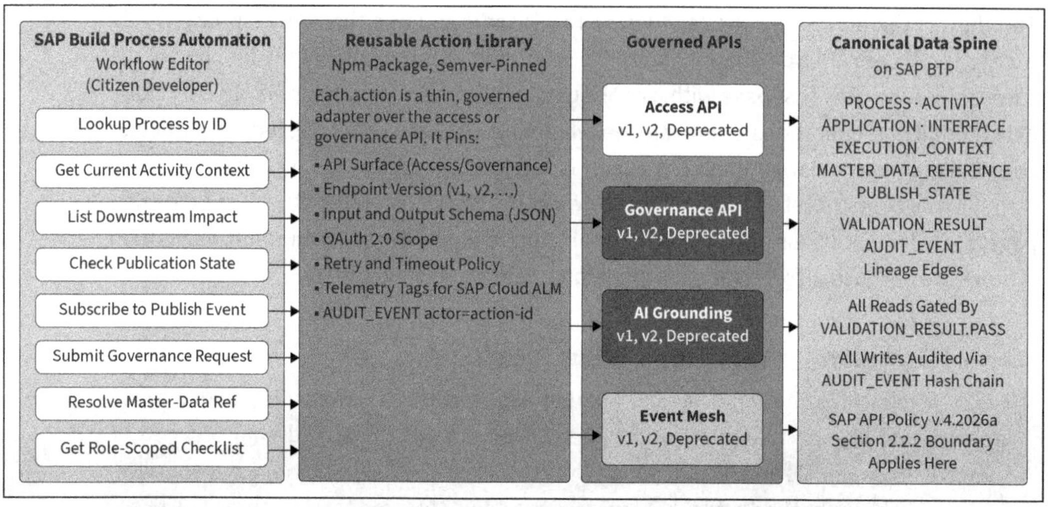

Figure 12.4: Reusable Integration Services for SAP Build Process Automation

This section examines five facets of reusable integration: the action library itself (the canonical interface between SAP Build Process Automation and the data spine), the versioning and deprecation discipline that protects long-lived workflows, the contract-stability commitments that make the library useful at scale, the action definition format the platform team uses to publish new actions, and the SQL view that backs the process lookup endpoint as a concrete example of a service contract.

12.4.1 The Action Library

The action library is a thin layer. Each action wraps a single endpoint in either the Access API or the Governance API and pins seven properties: the API surface, the endpoint version, the input schema, the output schema, the OAuth 2.0 scope, the retry and timeout policy, and the `AUDIT_EVENT` actor identity. None of these properties is the workflow developer's responsibility; all of them are encoded in the action definition. A workflow developer who drags the **Lookup process by ID** action into a workflow does not write any code, does not specify the API version, and does not configure the OAuth2 scope; the action handles all of those transparently.

Table 12.4 lists the canonical actions in the library at the time of writing. This list is not exhaustive; the library grows as new patterns emerge in production use. What matters for the architecture is that every action in the library wraps exactly one endpoint in exactly one governed API. An action that wrapped two endpoints would be a composition, and compositions are deliberately reserved for the composite actions layer described in Section 12.5. The single-wrap rule preserves the audit guarantee that every workflow step generates a single `AUDIT_EVENT` row, which keeps the audit chain linear and reconstructable.

Action Name	API Surface	Endpoint	Returns
Lookup process by ID	Access	`/process/{id}`	`PROCESS` and current `PROCESS_VERSION`
Get current activity context	Access	`/activity/{id}/context`	Enriched `ACTIVITY` view
List downstream impact	Access	`/impact/forward`	Ranked consumer set
Check publication state	Access	`/process/{id}/state`	`PUBLISH_STATE` row
Subscribe to publish event	Event Mesh	`process.published.v1`	Subscription handle
Submit governance request	Governance	`/governance/request`	Workflow instance ID (202)
Resolve master-data reference	Access	`/masterdata/{ref}`	`MASTER_DATA_REFERENCE` row
Get role-scoped checklist	Access	`/process/{id}/checklist`	Activity list filtered by role

Table 12.4: Canonical Actions in the SAP Build Process Automation Reusable Integration Library

12.4.2 Versioning and Deprecation of Actions

The action library is versioned by semver. A patch release fixes a bug in an existing action without changing its contract, a minor release adds a new action without changing any existing action, and a major release changes the contract of an existing action and therefore enters a deprecation calendar. The deprecation policy mirrors the API-version deprecation policy described in Section 12.5. A deprecated action is supported for a minimum of ninety days after its successor is released and carries a deprecation header that SAP Build Process Automation renders as a yellow banner in the workflow editor, and its removal requires a Governance API ticket signed off by a member of the platform team.

The library is published to the SAP BTP package registry as `@sap-spine/execution-actions`. Workflows pin a specific minor version in their dependency manifest, minor and patch updates are propagated through the standard SAP Build Process Automation patch process, and major updates require an explicit migration plan. The platform team operates a dashboard that tracks which version of the library each workflow is on and surfaces workflows that are pinned to a version entering its deprecation window.

12.4.3 Contract Stability

Contract stability is what allows the action library to be useful at scale. A workflow developer building a procurement approval flow today must be able to expect that the **Lookup process by ID** action will return the same shape next quarter. This is achieved through three commitments. The action's input schema is additive only within a major version, where new optional inputs are permitted but no input is renamed or removed. The action's output schema is additive only within a major version, where new fields are permitted but no field changes type or meaning. The action's OAuth 2.0 scope is fixed within a major version, where a new scope cannot be required without bumping the major version.

These commitments matter because workflows are long-lived. A procurement workflow that was deployed eighteen months ago is still the workflow that processes today's purchase orders. If its lookup action quietly began returning a different shape, the workflow would break in production. The library's commitments are therefore not stylistic; they are operational guarantees that a workflow deployed against a given library version will behave identically across the lifetime of that version.

12.4.4 Action Definition Format

Each action in the library is defined declaratively in JSON. The definition pins all seven contract properties so that every workflow that imports the action receives the same behavior. Listing 12.1 shows the definition of the **Lookup process by ID** action as it appears in version 1.4 of the `@sap-spine/execution-actions` package.

```
{
  "name": "lookup-process-by-id",
  "version": "1.4.0",
  "api_surface": "access",
  "endpoint": {
    "path": "/v1/process/{process_id}",
    "method": "GET",
    "version": "v1"
  },
  "input_schema": {
    "process_id": { "type": "string", "required": true, "pattern":
                    "^[A-Z0-9-]{3,64}$" }
  },
  "output_schema": {
    "process_id": "string",
    "process_version": "string",
    "publish_state": "PUBLISHED",
    "activities": "array<ActivitySummary>",
    "response_timestamp": "iso8601"
  },
  "oauth2_scope": "access-api.process.read",
```

```
  "retry_policy": {
    "max_attempts": 3,
    "backoff": "exponential",
    "initial_ms": 100
  },
  "timeout_ms": 2000,
  "audit_actor_template": "sbpa-action:{workflow_id}:lookup-process-by-id",
  "telemetry_tags": ["catalogue:execution", "service:process-lookup"]
}
```

Listing 12.1: Lookup Process by ID: The Declarative Action Definition That Pins the Seven Contract Properties of Every Action in the SAP Build Process Automation Library

AI Insight: The Action Library Is Also the AI Tool Catalog

When SAP Build Process Automation hosts agentic workflows that incorporate Joule agents, the action library doubles as the agent's tool catalog. The agent does not call the Access API directly; it invokes actions from the library through SAP Build Process Automation's standard tool-invocation pathway. Each invocation generates an `AUDIT_EVENT` row with the agent's identity as actor, exactly as a human-driven workflow step would. The agent therefore inherits the contract stability, the role scope, the deprecation discipline, and the audit guarantees of the library, without any additional governance scaffolding. This is the operational meaning of the SAP API Policy v.4.2026a Section 2.2.2 prohibition on AI-orchestrated API call sequences. The agent uses the same governed actions a citizen developer would use, and the data spine cannot tell the difference. The discipline that protects the citizen developer from contract drift is the same discipline that protects the data spine from agent drift.

Why Action Parameters Resolve Through the Same Lookup

When a composite action takes `ProcessID` as a parameter, the action does not validate the ID against a private list; it resolves the ID through process lookup like any other consumer. This means the action library has no parallel registry of valid processes, and a process that is retired through the Governance API immediately becomes unresolvable in every action that takes `ProcessID` as a parameter, without the action library having to do anything. The data spine is the single source of truth, and the action library trusts it.

The audit consequence is that every composite action invocation contributes the process lookup events for its parameters to the `AUDIT_EVENT` chain. A workflow that calls three actions, each taking a `ProcessID`, generates three lookup events plus the three action invocation events. The chain is busier than it would be in an unintegrated design, but every event is a hash-linked record of a real decision the platform made on the consumer's behalf. The verbose chain is the price of a verifiable platform.

The view defined in Listing 12.2 is what consumers query through the Access API; it is the shape they actually see, not the shape of the underlying tables. The columns are a deliberate subset: process identity, version `content_hash`, publish state, jurisdiction, AI eligibility flag, and the deprecation timestamp where applicable. Internal columns (the audit chain pointers, the validation cache identifiers, and the ingestion-batch markers from the load) are not part of the view, and consumers do not see them. This is the contract: The consumer sees what they need to see, the data spine retains what it needs to retain, and the boundary between the two is the view itself.

12.4.5 Service Contract: Process Lookup View

To help you understand the contract, we have defined the SQL view that backs the process lookup endpoint in Listing 12.2. The view is the operational specification of the service: It pins the joins, applies the validation filter, projects the role-scoped row set, and stamps the response with the version timestamp that the consumer can use to reproduce the read.

```
CREATE VIEW V_PROCESS_LOOKUP AS
SELECT
  p.PROCESS_ID,
  pv.VERSION_NO,
  pv.CONTENT_HASH,
  pv.PUBLISH_STATE,
  pv.OWNING_DOMAIN,
  pv.PUBLISHED_AT,
  a.ACTIVITY_ID,
  a.ACTIVITY_NAME,
  a.LEVEL,
  a.CONTROL_INTENT_REF,
  vr.RESULT AS VALIDATION_VERDICT,
  CURRENT_TIMESTAMP AS RESPONSE_TIMESTAMP
FROM PROCESS p
JOIN PROCESS_VERSION pv ON pv.PROCESS_ID = p.PROCESS_ID
JOIN ACTIVITY a ON a.PROCESS_VERSION_ID = pv.PROCESS_VERSION_ID
LEFT JOIN VALIDATION_RESULT vr ON vr.ENTITY_ID = a.ACTIVITY_ID
  AND vr.TIER < 4
  AND vr.EVALUATED_AT = (
    SELECT MAX(vr2.EVALUATED_AT) FROM VALIDATION_RESULT vr2
    WHERE vr2.ENTITY_ID = a.ACTIVITY_ID AND vr2.TIER = vr.TIER
  )
WHERE pv.PUBLISH_STATE = 'PUBLISHED'
  AND vr.RESULT = 'PASS'
  AND pv.OWNING_DOMAIN IN (SESSION_CONTEXT('AUTH_DOMAINS'));
```

Listing 12.2: V_PROCESS_LOOKUP

Three properties of this view deserve attention. The PUBLISH_STATE filter ensures that only PUBLISHED rows are returned, even if a draft of the same process exists in the data spine. The VALIDATION_RESULT join with TIER < 4 ensures that gating verdicts (tiers 1, 2, and 3 from Chapter 11) are enforced and tier 4 advisory signals do not gate the response. The SESSION_CONTEXT('AUTH_DOMAINS') predicate enforces the role scope at the database layer. A consumer with no authorized domains receives an empty result set, not an error. The view is the operational specification of what process lookup means, and changes to the view trigger a major-version bump on the service contract.

12.5 Performance and Scalability Considerations

An execution catalog that meets its functional requirements but cannot meet its performance budgets is not useful. Workflows time out, dashboards rendered slowly are abandoned, and AI grounding bundles assembled on the critical path of an agent's response delay every interaction. This section sets the latency budgets and throughput targets for the four core execution services, describes the three-tier caching pattern that achieves them under production load along with the cache-invalidation discipline that bounds stale reads, addresses service composition and versioning at scale, describes the capacity planning the platform team performs at each tier, explains the failure modes and graceful-degradation behavior the services exhibit under stress, summarizes the multi-region deployment topology, reviews the cost and resource considerations that affect every consumer, traces the FP&A drilldown that the layered architecture enables, and closes with the shape of an AI grounding bundle as it is assembled and consumed.

12.5.1 Latency Budgets and Throughput Targets

The latency budgets are codified in the action library's contract and enforced through service-level objectives (SLOs) on each service. Process lookup carries the tightest budget: p50 of 12 ms, p95 of 45 ms, and a throughput target of 5,000 requests per second per node. Context enrichment sits above it: p50 of 30 ms, p95 of 120 ms, and a throughput target of 2,500 requests per second per node. Change impact resolution is intrinsically more expensive because it walks the graph: p50 of 180 ms, p95 of 850 ms, and a throughput target of 200 requests per second per node. AI grounding bundle assembly is the most expensive: p50 of 450 ms, p95 of 1,800 ms, and a throughput target of 50 requests per second per node, because the bundle assembly walks the context enrichment outputs and the validation log for every artifact in the bundle.

These budgets assume the data spine is reached through warm connections to SAP HANA Cloud, that the most frequently read rows are present in the SAP HANA buffer cache, and that the caller's OAuth 2.0 token is cached at the gateway. Cold-start latencies are higher and are budgeted separately for autoscaling decisions. SLOs are computed over warm windows rather than cold starts because cold starts are an autoscaling property, not a service property.

12.5.2 Three-Tier Caching

Figure 12.5 and Table 12.5 show the three-tier caching pattern that achieves the performance budgets under production load. The pattern is intentional: Each tier is closer to the caller, has shorter TTL and tighter eviction, and is invalidated by Event Mesh on the topics that signal changes to the cached entities. The combination of TTL and event-driven eviction is the key to the design. TTL alone would either be too short (causing cache thrash) or too long (serving stale data after a publication). Event-driven eviction alone would be vulnerable to dropped events. The combination is self-healing in both scenarios.

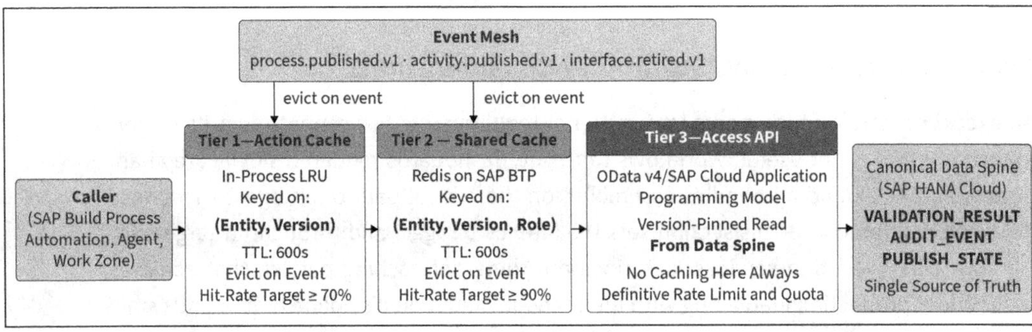

Figure 12.5: Performance and Scalability—Three-Tier Caching Pattern

Service	P50 Latency	P95 Latency	Throughput	Cache Class
Process lookup	12 ms	45 ms	5000 req/s	Tier 1
Context enrichment	30 ms	120 ms	2500 req/s	Tier 1 and 2
Change impact resolution	180 ms	850 ms	200 req/s	Tier 2 only
AI ground bundle	450 ms	1800 ms	50 req/s	Tier 2 and audit

Table 12.5: Performance Budgets per Service Tier

Tier 1 is an in-process least recently used (LRU) cache inside the action library. It is keyed on `entity`, `version` and has a TTL of 60 seconds. Cache entries are evicted on receipt of an Event Mesh message on the relevant topic (`process.published.v1`, `activity.published.v1`, and so on). The hit rate target is at least 70 percent for process lookup workloads; lower hit rates indicate either workload skew or a publication storm that should be investigated.

Tier 2 is a shared cache in Redis on SAP BTP. It is keyed on `entity`, `version`, `role` because graph adjacency lists are role-scoped; caching across roles would leak rows. The TTL is 600 seconds. Eviction is event-driven on the same topics as tier 1, plus the topics `interface.published.v1`, `application.published.v1`, and `execution_context.published.v1` for graph-walk caches. The hit rate target is at least 90 percent under steady-state load; lower hit rates indicate a publication storm or a misconfigured key.

Tier 3 is the Access API itself. It does not cache; every request is served from the data spine through SAP HANA Cloud. The role of tier 3 is to be the layer that is always definitive. Any request that misses both tier 1 and tier 2 is served from tier 3 against current canonical state. The Access API enforces rate limits and quotas; a consumer that exhausts its quota at tier 3 is throttled, not silently degraded.

12.5.3 Cache Invalidation and Stale Reads

Caching at the enrichment tier introduces a familiar problem in a more pronounced form: The cached payload is correct as of the moment the cache populated, but the underlying process can be republished a second later. The window between republication and cache eviction is the stale-read window, and consumers need to know how long it can be. The execution services answer this question explicitly. The default cache lifetime for an enrichment payload is sixty seconds; the default lifetime for a lookup result is fifteen minutes. The asymmetry reflects how the two payloads are used. Lookup results feed user interfaces and configuration screens where stale data is annoying but not consequential. Enrichment payloads feed AI agents and integration actions where stale data can drive a wrong decision.

Sixty seconds is not arbitrary. It is the answer to a specific question "What is the shortest interval over which a process owner can reasonably expect a republished version to take effect across all consumers?" Anything shorter forces every consumer to bypass the cache, and the cache stops being useful. Anything longer means a process owner who corrects an error has to wait minutes for the correction to propagate, and that delay erodes confidence in the publish action. The execution services also support explicit cache invalidation through Event Mesh. The `process.published.v1` event carries the affected process ID, and consumers that subscribe to the topic can flush their local caches immediately rather than waiting for the lifetime to expire. Subscribing is optional but recommended for any consumer where stale read consequences exceed the cost of subscription.

Stale reads at the AI grounding tier deserve a separate note. An AI agent that receives a sixty-second stale enrichment payload still gets a payload whose validation lineage is complete; the validations passed when the version was published, and that fact does not change retroactively. What can change is whether the version is still the current published version. The grounding payload therefore carries an `AS-OF` timestamp, and the agent's reasoning trace records both the `AS-OF` timestamp and the wall-clock time at which the agent acted. If a controller later asks why the agent used a particular version, the answer is in the trace: The agent acted on the version that was current at the `AS-OF` timestamp, and the `AS-OF` timestamp was within the agreed staleness budget. This is what makes AI grounding auditable in a way that an opaque cached read would not be.

Cache invalidation also interacts with the Governance API in a way that matters during deprecation. When a process is moved to deprecated state, the Governance API publishes a `process.deprecated.v1` event on Event Mesh, and consumers that subscribe receive the event within milliseconds. The recommended consumer behavior on receipt of this event is not to flush the cache—the deprecated version is still readable, and in-flight transactions still need to resolve it—but to flag the cached entry with the deprecation status so that subsequent reads know to choose the successor version where one exists. This soft-invalidation pattern is what allows deprecation to propagate without disrupting in-flight work, and it is what gives consumers the ninety-day deprecation window described in Section 12.3 to migrate at their own pace.

12.5.4 Service Composition and Versioning

Performance under change is as important as performance under steady state. The execution catalog is a layered composition in which higher-level services read from lower-level services and never the other way around. The discipline of the layering is what makes the catalog adaptable: A change to process lookup propagates upward through composite actions and through the action library in a controlled way, but no change at the top of the stack reaches into the canonical data spine.

Figure 12.6 shows the composition stack. The canonical data spine sits at the bottom, the three governed APIs and the AI grounding endpoints sit above it, the four core execution services (process lookup, context enrichment, change impact resolution, and the action library) sit above the APIs and the composite actions and domain services sit above the core services. Each layer reads only from the layer below, no layer reads from a layer above, and no layer reaches across to a peer at the same level.

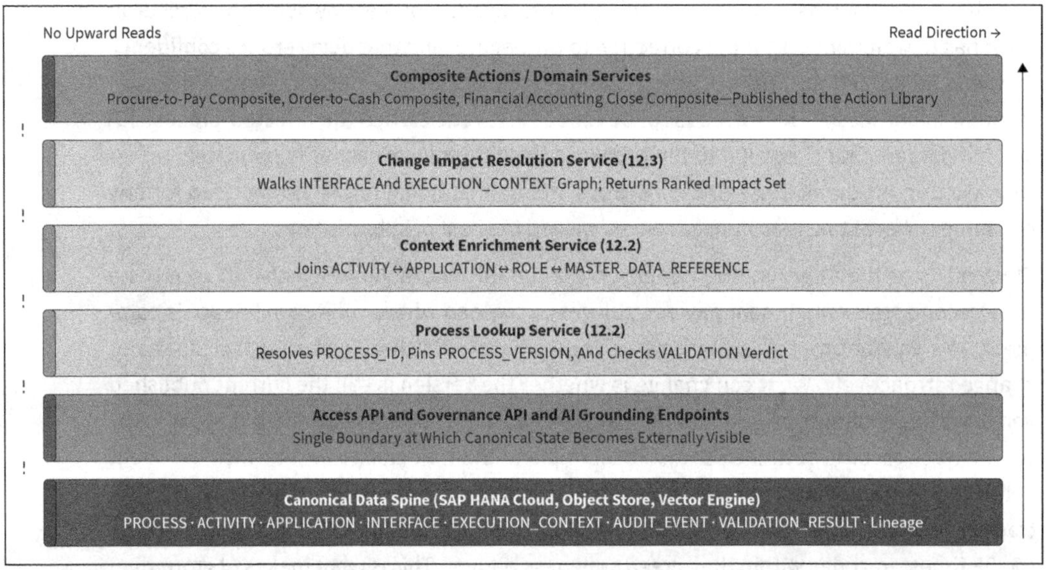

Figure 12.6: Layered Composition of Execution Services

Composite actions are the top of the catalog. Each composite action wraps a sequence of calls into the core execution services and exposes them as a single, semver-pinned action in the SAP Build Process Automation action library. Examples include **P2P close-period activity list, OTC variant comparison,** and **FI close pre-flight check**. Composite actions are themselves versioned artifacts published to the action library. A composite action that pins context enrichment v1 and change impact resolution v2 is a versioned artifact in its own right, and it follows the same semver and deprecation discipline as the actions below it.

Versioning is the mechanism that allows the catalog to evolve without breaking consumers. Every public surface in the catalog (API endpoints, actions, composite actions, and event topics) is versioned by semver. Patch and minor releases are additive, with new optional fields, new endpoints, new event topic versions, no removals, no contract changes to existing surfaces, and no `AUDIT_EVENT` actor change. Consumers can upgrade at their own pace. Major releases are breaking and trigger a deprecation calendar.

Figure 12.7 shows the deprecation lifecycle. A new major version is published; the previous major version enters a deprecation window of at least ninety days, extended to six months for AI consumers because Joule agents and Joule Studio agents have a slower upgrade cadence than human-driven workflows.

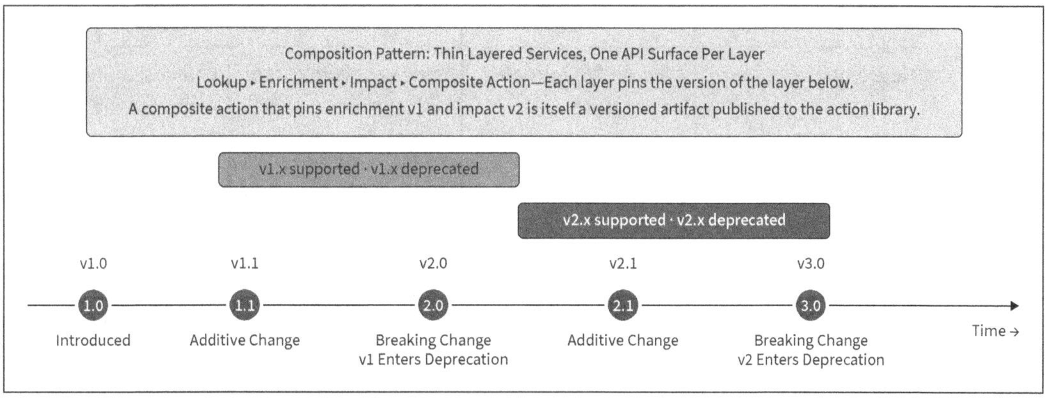

Figure 12.7: Service Composition, Versioning, and Deprecation

Consumers receive a deprecation header on every call to the deprecated surface. A Governance API ticket is required to retire the deprecated surface at the end of the window. The dashboard maintained by the platform team tracks the consumer count on every version and surfaces consumers that have not yet migrated. The retirement ticket cannot be approved if the consumer count is greater than zero.

Table 12.6 sets out the formal stages of a major version deprecation as the dashboard and the Governance API workflow understand them. The stages are not negotiable; they are the operational definition of what *deprecated* means at each point in the calendar. Consumers and platform team members can agree on what stage a surface is in without ambiguity, and the Governance API enforces the stage transitions through workflow approvals.

Stage	Duration	Consumer Experience	Platform Team Action
Active	Indefinite	Default; no banner	Monitor for new consumers
Successor published	0 days	Both versions visible; no banner yet	Begin migration outreach
Deprecation announced	Day 1 of window	Yellow banner in editor; deprecation header on every call	Track migration progress
Deprecation accelerated	Day 60 (humans), day 120 (AI)	Stronger banner; weekly outreach to remaining consumers	Direct contact with each consumer
Retirement window	Days 60-90 (humans), days 120-180 (AI)	Banner becomes red; calls log warning to SAP Cloud ALM	Final escalation to consumer owners
Retirement ticket open	Day 90 and beyond/day 180 and beyond	Calls return 410 Gone with migration pointer if attempted after retirement date	Governance workflow signs off
Retired	Permanent	Surface removed; calls fail closed	Surface archived in registry

Table 12.6: Major-Version Deprecation Stages for the Execution Catalog

12.5.5 Capacity Planning

Capacity is planned per service, not per consumer. A typical mid-sized SAP estate runs process lookup at 800 requests per second average and 4,000 requests per second peak, context enrichment at 400 average and 2,000 peak, change impact resolution at 5 average and 50 peak, and AI grounding at 2 average and 25 peak. The autoscaling targets are configured to keep p95 within budget at peak, with a minimum of two nodes per service for high availability. Cold start latencies are absorbed by the autoscaler's lookahead window and by the warm pool of two minimum nodes. The cost profile of the catalog is dominated by tier 2 Redis capacity and the SAP HANA Cloud query throughput; the four core services themselves are stateless and inexpensive to scale.

12.5.6 Failure Modes and Graceful Degradation

Three issues recur in production, and each has a defined graceful degradation behavior. A tier 2 cache outage—Redis becoming unavailable, a network partition, or a deployment of the cache cluster—forces every request through to tier 3, increasing p95 latency and reducing throughput. The action library detects the cache outage through circuit-breaker telemetry and routes traffic directly to tier 3, with explicit tier 2 bypass logging in SAP Cloud ALM so that the operations team can see the impact on the latency budget. Consumers continue to receive correct responses. The data spine itself is unaffected, and only performance is degraded.

An Access API rate-limit breach, meaning a consumer that exceeds its allocated request budget, returns HTTP 429 with a `Retry-After` header. The action library's retry policy honors the `Retry-After` value and pauses for the indicated interval before attempting again. Workflows that cannot tolerate the pause should be redesigned to batch their requests rather than to issue one request per record. The rate limit is calibrated to typical workload patterns, and a workflow that consistently exhausts its budget is a signal of either a workload design problem or a budget allocation that needs to be revisited through the Governance API.

An Event Mesh outage—dropped messages, broker partition, or a deployment of the Event Mesh service—prevents cache eviction events from reaching the subscribers in the action library. The TTL-based safety net then takes over: cache entries continue to expire on their TTL, the cache repopulates from tier 3 with current state, and consumers see slightly stale responses for at most TTL seconds after a publication. The platform's monitoring is configured to alert when Event Mesh subscriber lag exceeds thirty seconds, and the operations runbook prescribes a manual cache flush if the lag exceeds five minutes. The result is that the cache layer is self-healing in both scenarios (missed events fall back to TTL and missed TTL falls back to Tier 3) without operator intervention, except during sustained outages.

12.5.7 Multiregion Deployment

Mature programs deploy the execution catalog across multiple SAP BTP regions for resilience and for latency reasons. The deployment topology is asymmetric: SAP HANA Cloud is replicated across regions through standard SAP HANA system replication. Payloads from the Object Store service on SAP BTP are replicated through standard cross-region replication. Redis tier 2 caches are regional, not replicated, because replicated caches reintroduce the coherence problem that versioned keys eliminate. Consumers connect to the nearest region, and the action library's service discovery layer handles regional failover. Cross-region failover preserves the audit chain: `AUDIT_EVENT` rows written in the primary region are replicated to the secondary region as part of SAP HANA replication, so the chain remains intact across a failover.

Data residency obligations are honored at the Access API layer. A consumer in EU operates against the EU region, a consumer in US operates against the US region, and the data spine's role-scope filtering ensures that a user in one jurisdiction does not see rows that originated in another. The cross-region replication that supports failover is bounded to within-jurisdiction targets, meaning an EU primary fails over to an EU secondary, never to a US region. This is a property of the SAP BTP deployment configuration, not of the execution services themselves. The services see jurisdictional boundaries as a property of the role context, which is the right level at which to enforce them.

12.5.8 Cost and Resource Considerations

The cost profile of the catalog is dominated by three line items: SAP HANA Cloud compute and storage, Redis tier 2 capacity, and object store for the lineage payload archive. The four core execution services themselves are stateless and cost very little; their cost scales linearly with traffic and is typically less than 10 percent of the total. Programs that try to optimize cost by reducing the size of the tier 2 Redis cache typically discover within a quarter that the savings are illusory. The missed cache hits push more traffic to SAP HANA Cloud, the SAP HANA query cost rises faster than the Redis savings, and the consumer-facing latency budget begins to slip. The right trade-off is to size tier 2 generously and let the cache absorb steady-state load.

Object store cost is dominated by the lineage payload archive, which preserves every SAP Signavio and SAP LeanIX payload that has ever entered the data spine. The archive grows monotonically; programs that ingest 100,000 payloads per month accumulate roughly 50 GB per year. Compression is typical (the payloads are JSON and compress well); tiering to cold storage after twelve months reduces the cost of older archive entries while preserving their availability for audit. Deletion is not an option because the lineage payload is the structural foundation of the FP&A trace, and removing it would break the controller's drilldown at the bottom of the chain.

Capacity planning at the SAP Build Process Automation tier is a separate consideration. The action library itself is hosted as a Cloud Foundry application on SAP BTP and scales horizontally. The constraint at this tier is typically not computing capacity but the rate at which it can issue calls to the Access API. A workflow that fans out into a thousand parallel branches and calls lookup process by ID in each can exhaust the consumer's API quota within seconds. The right pattern is to batch the calls through a composite action that issues a single multirow request, not a thousand single-row requests. The action library's batch composite actions exist for exactly this reason.

12.5.9 Financial Planning and Analysis Drilldown and the Controller's Trace

The layered composition has an important consequence for financial reporting. A controller looking at a profit center variance in SAP S/4HANA embedded analytics can drill from the variance to the activities that produced it. The drill traverses the same execution services that an SAP Build Process Automation workflow or a Joule agent uses to read the data spine. The trace from a reported number to the source payload that produced it is therefore a side effect of using the catalog, not a separate reporting feature. Figure 12.8 shows the trace in detail.

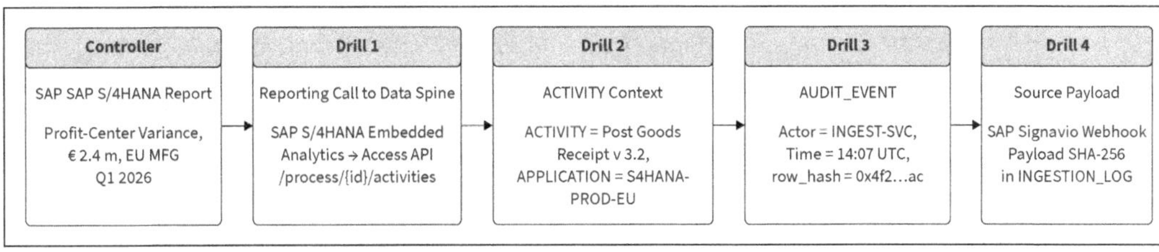

Figure 12.8: FP&A Drilldown—From Reported Number to Source Payload

The trace begins with a reported number in SAP S/4HANA embedded analytics: a profit-center variance of €2.4 million in EU manufacturing for Q1 2026. The embedded analytics drilldown calls the Access API for the activities that contributed to the variance. The response includes the post goods receipt v 3.2 activity in execution context EC-P2P-DE-MFG. The next drill reaches the `AUDIT_EVENT` row that recorded the activity read. The row identifies the actor (the embedded analytics service), the time of the read, the role context applied, and a content hash of the row served. The next drill reaches the lineage edge written when the activity was published. The edge identifies the source payload (the SAP Signavio webhook payload SHA-256 hash) and the `INGESTION_LOG` row that recorded the receipt. The trace terminates at the source payload, which is preserved indefinitely in Object Store as part of Chapter 11's ingestion discipline.

What this enables is unusual in enterprise reporting. A controller can defend any reported number by walking from the report to the canonical artifact and to the payload that produced it. An auditor can sample any `AUDIT_EVENT` row and reconstruct the read history, including the role context under which it was served. A Joule agent or SAP Business Process Automation workflow that referenced the same activity is visible on the same trace. Cost control decisions are tied to a stable, version-pinned object, not to a snapshot of a reporting cube that has since rolled forward. The same trace serves financial reporting, AI grounding evidence, and Sarbanes–Oxley Act (SOX) control testing without a separate evidence-collection layer for any of them. This is the architectural payoff of doing the catalog correctly.

For programs led from a finance, controlling, or business-transformation seat, this property is the most important deliverable of the platform. It changes the relationship between FP&A and IT: The platform is no longer something FP&A consumes through a reporting layer that obscures its lineage but something FP&A can defend, audit, and explain at the level of the canonical artifact. The execution services in this chapter are what make that property operational.

AI Insight: The FP&A Trace Is Also the AI Audit Trail

The drilldown trace described in this section is the same trace that an auditor uses when questioning an AI-driven decision. When a Joule agent recommends a remediation that touches a controlled process, an auditor can sample the AUDIT_EVENT row that recorded the agent's grounding bundle request, walk the bundle's content hashes back to the canonical artifacts, walk the lineage edges from those artifacts to the source payloads that produced them, and verify that the agent's recommendation was based on validated, role-appropriate, version-pinned canonical state. The trace is structurally identical to the FP&A trace; only the questioning party is different.

AI Insight: Why AI Grounding Has Its Own Latency Budget

AI grounding bundle assembly is by far the slowest service in the catalog, because the assembly walks the validation log for every artifact in the bundle and stamps each artifact with its validation lineage so that the AI consumer cannot, by accident or by design, ground on an artifact that has not passed validation. Caching at tier 2 helps, but the bundle is a versioned object, and a fresh bundle for a new entity must be assembled from scratch. The loose latency budget reflects the operational reality that AI grounding is a once-per-conversation operation, not a per-keystroke one. The cost is paid at the start of an agent's response and amortized across the rest of the interaction. The same trade-off would not be acceptable for process lookup, which is on the critical path of every workflow step. This is the structural basis on which SAP API Policy v.4.2026a Section 2.2.2 can be satisfied without making AI interactions feel sluggish: the bundle is heavy because it carries its full validation lineage, but the consumer pays the cost once.

12.5.10 AI Grounding Bundle Shape

The shape of an AI grounding bundle is the operational expression of the contract from Chapter 10, Section 10.4.4. As shown in Listing 12.3, the bundle is a single JSON object that carries the canonical artifacts the AI consumer is allowed to ground on, the validation lineage of each artifact, the role context under which the bundle was assembled, and a bundle identifier that anchors a row in the AUDIT_EVENT hash chain. An AI consumer that receives the bundle has everything it needs to produce a defensible response. An AI consumer that requested the bundle but did not yet receive it has no read access to the data spine through any other path.

```
{
  "bundle_id": "bnd-2026-01-14-7d3a1f",
  "audit_event_id": "ae-9c4f8e2b",
  "assembled_at": "2026-01-14T09:42:18.413Z",
  "role_context": {
```

```
    "principal": "joule-agent:procurement-summary",
    "domains": ["P2P-EU"],
    "scopes": ["ai-grounding.bundle.read"]
  },
  "entry_point": {
    "entity_type": "PROCESS",
    "entity_id": "P2P",
    "version_pinned_at": "2026-01-14T09:42:18.000Z"
  },
  "artifacts": [
    {
      "entity_type": "PROCESS_VERSION",
      "entity_id": "P2P-v3.2",
      "content_hash": "sha256:0x4f2ac9e1...",
      "validation": { "verdict": "PASS", "evaluated_at": "..." }
    },
    {
      "entity_type": "ACTIVITY",
      "entity_id": "POST-GOODS-RECEIPT",
      "content_hash": "sha256:0x8b1c7d3f...",
      "validation": { "verdict": "PASS", "evaluated_at": "..." }
    }
  ],
  "lineage": {
    "ingestion_log_ref": "ing-2026-01-12-a8c4e9",
    "source_payload_hash": "sha256:0x2d5e8f1c..."
  },
  "expires_at": "2026-01-14T10:42:18.413Z"
}
```

Listing 12.3: AI Grounding Bundle Response

There are three important properties in this bundle shape. First, the bundle is version-pinned. Every artifact carries its `content_hash`, and the bundle as a whole carries the `version_pinned_at` timestamp that the AI consumer can use to reproduce the bundle later if required. Second, every artifact carries its validation verdict. An AI consumer cannot ground on an artifact whose verdict is not present, and an artifact whose verdict is not `PASS` cannot appear in the artifacts array at all. Third, the bundle is anchored in `AUDIT_EVENT` through the `audit_event_id` field. An auditor questioning the bundle can sample the audit row, reconstruct the role context at assembly time, and verify the content hashes against the canonical data spine of record.

The `expires_at` field is the bundle's freshness contract. The AI consumer should not reuse a bundle older than its `expires_at` field; it should request a fresh bundle instead. The default expiry is one hour, configurable per consumer through the Governance API. The expiry is short enough that bundles do not accumulate as a parallel state store and long enough that an agent's multi-turn conversation can complete against a single bundle without paying the assembly cost on every turn. Consumers that need fresher bundles—for example, an agent that consumes change-impact results during a deprecation window—specify a shorter `expires_at`. Consumers that need long-lived bundles cannot specify a longer `expires_at`, because the freshness contract is one-way.

12.6 Summary

This chapter taught you how to turn the canonical data spine from a populated, governed asset into the runtime substrate of the integrated SAP estate. It defined the execution service catalog (process lookup, context enrichment, change impact resolution, the reusable integration library for SAP Build Process Automation, and the performance and composition layer above them) and showed how each service sits above the governed API surfaces from Chapter 10, reads from the canonical entities from Chapter 9 and Chapter 10, and depends on the validation discipline from Chapter 11. Throughout this chapter, we preserved the AI boundary established in Chapter 1: AI consumes the data spine through AI grounding endpoints only, governed by SAP API Policy v.4.2026a Section 2.2.2, and every read regardless of caller advances the `AUDIT_EVENT` hash chain.

12.6.1 Key Learning Points

- The execution catalog is a layered architecture, not a flat collection of microservices. Composite actions read from domain services, domain services read from the four core execution services, and the core services read from the Access API. No layer reads from a layer above it. The discipline of the layering is what allows the catalog to evolve without destabilizing consumers.
- Reads are version-pinned, role-scoped, and validation-gated at the boundary, not in the consumer. A process lookup response is identical for the same caller at the same `AS-OF` timestamp, regardless of intervening publications. A context enrichment response excludes artifacts whose validation status is not PASS even if the consumer requested them. An AI grounding bundle returns only artifacts that satisfy the validation contract.
- Every read advances the `AUDIT_EVENT` hash chain. The actor identity, role context, requested entity, returned version, and content hash of the row served are recorded for every call, regardless of whether the caller is a human user, an automated workflow, or an AI agent. The chain is the trace, and it is the same trace that an FP&A controller uses to defend a reported number, an auditor uses to test SOX controls, and an architect uses to demonstrate that an AI agent referenced only validated canonical artifacts.
- AI consumes the data spine through AI grounding endpoints only. An AI agent does not call process lookup or context enrichment directly and does not orchestrate sequences of API calls. It receives version-pinned, role-scoped bundles assembled by the data spine.

This is the structural realization of the boundary established in Chapter 1 and the operational compliance with SAP API Policy v.4.2026a Section 2.2.2. The action library that exposes the catalog to SAP Build Process Automation workflows is the same library that exposes governed actions to Joule agents; the agent inherits the contract stability and audit guarantees of the library, and the data spine cannot tell the difference between a human-driven workflow step and an agent-driven tool invocation.

AI Insight: The Citizen Developer's Quiet Compliance Win

A consequence of the action library is that citizen developers in SAP Build Process Automation (such as procurement specialists, FP&A analysts, controllers, and plant managers) build governance-compliant workflows by default, without having to know that they are doing so. Every action they drag into the editor is already role-scoped, validation-gated, version-pinned, and audit-recorded. A workflow that an FP&A analyst builds to summarize quarter-end activity exposure is the same kind of workflow, from the data spine's perspective, as a workflow that a platform engineer builds to drive ingestion reconciliation. The compliance discipline is in the substrate; it is not a cost the citizen developer pays. This is the practical answer to the question business-transformation leaders ask most often when proposing platforms of this kind: "How do we keep our citizen developers from circumventing governance?" The answer is to make compliance the cheapest path, and the action library is what makes it cheapest.

- The action library is the canonical interface between SAP Build Process Automation and the data spine. Direct calls from a workflow to the Access or Governance APIs are technically possible but disallowed by code review; the library is the only path. Each action wraps exactly one endpoint, pins the API version, the input and output schemas, the OAuth 2.0 scope, and the `AUDIT_EVENT` actor identity. Workflows pin a specific minor version of the library; major versions trigger a deprecation calendar.
- Performance is achieved through three-tier caching that is invalidated by Event Mesh, not by TTL alone. Versioned cache keys eliminate the cache coherence problem because canonical versions are immutable. Event-driven eviction closes the freshness gap, and TTL is the safety net for missed events. The combination is self-healing in both scenarios.
- The catalog produces an FP&A-grade trace as a side effect of being used. A controller can drill from a reported number through `AUDIT_EVENT` and lineage edges back to the canonical artifact and the source payload that produced it. The trace is not a separate reporting feature; it is the same trace that auditors and architects use, served from the same logs. This is the deliverable that makes the platform defensible for finance and controlling.

12.6.2 Common Design Pitfalls

- Allowing consumers to bypass the action library and call the Access or Governance APIs directly. This appears to save a layer but in practice introduces inconsistent retry policies, divergent audit actor identities, and contracts that drift away from the library's commitments. The library is the canonical interface; bypassing it produces operational artifacts that cannot be defended at audit.
- Caching across roles using a shared key. Graph adjacency lists, context enrichment outputs, and AI grounding bundles are all role-scoped; caching them across roles leaks rows that the consumer is not authorized to see. The tier 2 cache key must include the calling role, and the additional cardinality is the cost of correctness.
- Letting AI agents call execution services directly to assemble their own bundles. This violates SAP API Policy v.4.2026a Section 2.2.2 and removes the validation, role-scope, and version-pin guarantees that the AI grounding endpoint provides. The endpoint is the only AI consumption path; an agent that bypasses it is operating outside the endorsed architecture.
- Treating the SAP Build Process Automation action library as a documentation artifact rather than as a versioned deliverable. The library's contracts are operational guarantees, not stylistic choices. A workflow deployed against library v1.4 must continue to work after library v1.7 is released. Skipping the semver discipline produces silent breakage in long-lived workflows.
- Adding an action that wraps more than one endpoint. The single-wrap rule is what preserves the linear `AUDIT_EVENT` chain. An action that wraps two endpoints generates two audit rows under one actor identity and confuses the trace, and the platform's audit reconstruction tooling assumes one audit row per workflow step. Such actions belong in the composite actions layer, not in the core action library.
- Reading the data spine through a private path that bypasses the Access API. Direct database access from a service operator, even for diagnostic purposes, leaves no `AUDIT_EVENT` row and produces a read history that is structurally invisible to the platform's audit tooling. The Access API is the only legitimate read path. Diagnostic access uses a separate, audited break-glass procedure as outlined in Chapter 19.
- Confusing semver patch and minor releases of the action library with major releases. Patch and minor releases are additive and require no consumer action, while major releases are breaking and trigger a deprecation calendar. Treating a major release as a minor release silently breaks workflows; treating a minor release as a major release consumes deprecation budget that the library does not have. The discipline is to bump the version that matches what changed—nothing more and nothing less.

12.6.3 Review Questions

1. Why must process lookup, context enrichment, and change impact resolution be read-only relative to canonical state, and what specifically goes wrong if a service in the catalog is permitted to write to PROCESS, ACTIVITY, or EXECUTION_CONTEXT?
2. Trace a single process lookup call from the runtime caller through the gateway, SAP Authorization and Trust Management service, the service, the Access API, and the data spine. Identify every place where identity is preserved, the role context is applied, and a row is appended to the AUDIT_EVENT hash chain.
3. How does the context enrichment service satisfy the validation, lifecycle, role-scope, and version-pin requirements that Chapter 14 will impose on artifacts eligible for AI consumption? Why is it the natural source for AI grounding bundle assembly?
4. Explain the forward and reverse impact endpoints of the change impact resolution service. Trace what each returns when the input is an APPLICATION whose lifecycle state is moving from ACTIVE to RETIRED.
5. Describe the seven properties that each action in the SAP Build Process Automation reusable integration library pins, and explain why each property is the library's responsibility rather than the workflow developer's.
6. Explain the three-tier caching pattern. Why is caching by version safe? Why is event-driven eviction necessary, and why is TTL alone insufficient? What goes wrong if the tier 2 cache key omits the calling role?
7. Describe the deprecation calendar for a major version change in the action library. Why is the deprecation window for AI consumers longer than for human-driven workflows? Who must approve the retirement ticket, and on what condition?
8. Trace the FP&A drilldown from a reported profit-center variance in SAP S/4HANA's embedded analytics back to the source payload that produced it. Identify every layer of the trace and the structure that makes the trace possible (including the AUDIT_EVENT hash chain, lineage edges, INGESTION_LOG, and Object Store).
9. How do the AI grounding endpoints satisfy SAP API Policy v.4.2026a Section 2.2.2? Specifically, why is it correct that an AI agent does not call process lookup or context enrichment directly, and what would go wrong structurally if it did?

12.6.4 Student Companion Notes

If you are studying this chapter as part of the SAP University Alliances program, enterprise architecture, or business transformation curriculum, you should focus on the structural reasoning behind the execution catalog rather than the specific service signatures or table joins. The signatures and joins will evolve; the structural reasoning will not.

Three exercises are particularly valuable. First, take a real cross-system process from your own organization (for example, the procure-to-pay flow, the order-to-cash flow, or the financial close flow) and sketch the four core execution services that would be needed to expose it through SAP Build Process Automation. For each service, identify which canonical entities it reads, which role scope applies, which validation rules gate the response, and which AUDIT_EVENT actor identity the service would carry. Second, take a single proposed change (such as deprecating an interface, retiring an application, or publishing a new variant of an existing process) and walk through the forward-impact and reverse-impact queries that the change impact resolution service would return. Reason about which consumers would be affected, which would not, and what the deprecation calendar would look like. Third, trace an FP&A drilldown from a reported number in your organization's reporting platform back to the canonical artifact and the source payload that produced it. Identify which steps of that trace are currently available, which are missing, and what would have to change in your platform to make the full trace operational.

A practical study sequence that has worked well in graduate transformation seminars is to read this chapter a second time alongside Chapter 11. On your first reading, the execution services may feel like a different topic from the ingestion pipeline; by your second reading, after we have filled the data spine in Chapter 11 and made it useful in this chapter, the two chapters read as a single, coherent design in which the same validation discipline, audit chain, and role-scoping mechanism appear on both sides of the data spine. Reading the data spine as a write boundary, then as a read boundary, is a useful intellectual exercise. It is also how engineering teams reason about the platform they are building.

If you have a finance, controlling, or business transformation background, the FP&A trace described in Section 12.5 deserves your attention. The ability to drill from a reported number through AUDIT_EVENT and lineage edges back to the source payload that produced it is unusual in enterprise reporting and is the property that makes this architecture defensible for finance. If you are moving into a business transformation role from an FP&A or cost control background, you should expect to use this trace as the structural argument for why the platform is worth the investment of building it.

This chapter completes Part VI of the book. The chapters that follow demonstrate how the execution catalog becomes the basis for AI consumption.

PART VII

SAP AI Consumption of the Data Spine

Chapter 13

Introducing SAP AI Architecture

A data spine that is governed without being consumed is a museum. A data spine that is consumed without being governed is a liability. What turns the data spine into the substrate of enterprise AI is the discipline of an AI stack that consumes only through governed grounding, records every inference as an audit row, and treats the model as an executable artifact whose lifecycle is no less rigorous than the data it reads.

This chapter introduces the AI consumer of the execution catalog. It describes the SAP AI stack on SAP Business Technology Platform (SAP BTP)—SAP AI Core, SAP AI Launchpad, and Joule—and shows how each layer consumes the canonical data spine we designed in Chapter 9 and Chapter10, populated through the ingestion pipeline covered in Chapter 11, and exposed through the execution services from Chapter 12. The consumption is governed end to end. AI reaches the data spine only through AI grounding endpoints, the Access application programming interface (API) is closed to AI consumers by policy, and every inference advances the same `AUDIT_EVENT` hash chain that a controller relies on to defend any other reported number.

The structural commitment of this chapter is that the SAP AI stack is not a separate ecosystem bolted onto the data spine. It is the data spine's consumer of last resort, one consumer among many, and it inherits the same role scope, validation gates, version pinning, and audit obligations that we imposed on workflow steps, tile renderings, and analytical drill-downs in Chapter 12. The AI boundary we established in Chapter 1 (AI as pattern detector, anomaly scorer, explainer, and draft author, never a decider) is realized here as an engineering contract enforced at the SAP BTP gateway, not as a promise on a policy slide.

This chapter is for anyone who must take a useful canonical data spine and put SAP AI on top of it without breaking either side. It assumes the architectural model of Chapter 9, the implementation of Chapter 10, the ingestion discipline of Chapter 11, and the execution catalog of Chapter 12. It introduces the SAP AI stack at a level of detail sufficient to design integrations against it; the operational specifics of running the stack in production are taken up in Chapter 18, the data preparation rules that govern grounding bundles are specified in Chapter 14, the discipline of process suggestions with mandatory human validation is covered in Chapter 15, and the audit instrumentation that records `INFERENCE_EVENT` rows is described in Chapter 19.

Three principles run through this chapter and should be carried into every AI integration decision. First, the model is an executable artifact. It has a version, a content hash, a lifecycle state, and a governance verdict, and it is operated through SAP AI Core in the same way SAP Build Process Automation operates an action and the data spine operates a PROCESS_VERSION. Second, AI consumes only through AI grounding endpoints. There is no alternative path, no carve-out, no escape hatch, and the prohibition is enforced at the gateway. Third, every inference advances AUDIT_EVENT through an INFERENCE_EVENT row that joins on actor identity, so that the model's reasoning is reconstructible from the same chain a controller already reads for an ordinary transaction.

This chapter explains:

- How SAP AI Core organizes inference as scenarios, executables, deployments, and serving endpoints inside resource groups
- How SAP AI Launchpad catalogs SAP-built, partner, and open-source models, and how the generative AI hub abstracts foundation models behind a uniform interface
- How the AI lifecycle (REGISTERED, EVALUATED, DEPLOYED, DEPRECATED, and RETIRED) is governed by gates whose verdicts are recorded in AUDIT_EVENT
- How Joule and its agents consume the SAP Build Process Automation action library (see Chapter 12, Section 12.4) as their tool catalog, with no agent-side path to the Access API
- How AI consumption is policed by SAP API Policy v.4.2026a Section 2.2.2 at the SAP BTP gateway, and how INFERENCE_EVENT joins to the AUDIT_EVENT hash chain on actor identity
- How a single financial planning and analysis (FP&A) controller request ("Explain this variance") flows end to end through Joule, the AI grounding endpoint, the model, and back, with a reconstructible audit trace at every hop

Chapter 12 made the data spine useful. This chapter allows it to be used by AI without giving up any of the governance, version pinning, or audit guarantees that made it useful in the first place. By the end of this chapter, the AI tier should look not like a different kind of consumer but like the most disciplined consumer the platform exposes, precisely because the data spine refuses to be consumed any other way.

13.1 SAP AI Core Execution Model

SAP AI Core is the inference runtime of SAP BTP. It is the layer at which a trained model becomes a network-addressable endpoint that consumers can invoke, the layer at which scaling and isolation are applied, and the layer at which the runtime emits the telemetry and audit rows that make AI consumption observable. SAP AI Core is to a model what the SAP BTP, Cloud Foundry runtime is to a microservice: not the place where the artifact is built, but the place where it lives and serves traffic.

The execution model is organized around four objects whose distinction is sharper than it first appears: a scenario, an executable, a deployment, and a serving endpoint. A scenario is the logical AI task ("Explain a variance," "Classify a spend line," or "Summarize a contract") independent of any particular model or runtime. An executable is the runtime template that implements a scenario: a container image, environment variables, configuration secrets, and a declared scaling profile. A deployment is a running instance of an executable, bound to a resource group, allocated CPU and memory, configured with a specific model version, and addressable through a stable identifier. A serving endpoint is the network surface—REST, authenticated through the SAP Authorization and Trust Management service— through which consumers reach the deployment.

The four-object model is not a matter of taste; it is the structure that allows the same scenario to be served by different models in different jurisdictions, a deployment to be rolled forward through a model version change without consumer disruption, and the platform to record which scenario was served, by which executable, in which deployment, by which model version, at every inference. Figure 13.1 places the AI Core execution model in the context of the SAP AI stack and the canonical data spine; it serves as the structural reference for the rest of the section.

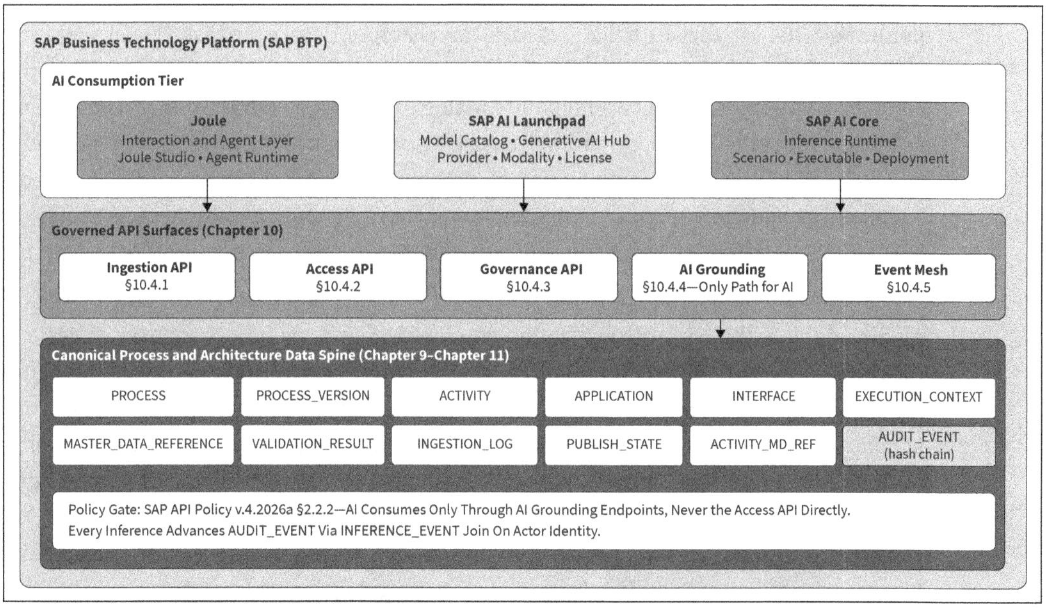

Figure 13.1: SAP AI Stack on SAP BTP: Joule, SAP AI Launchpad, and SAP AI Core Sit Above the Governed API Surfaces and the Canonical Data Spine

Figure 13.1 makes three observations explicit. The AI consumption tier (Joule, SAP AI Launchpad, and SAP AI Core) sits above the governed API surfaces covered in Chapter 10, not beside them. The API surfaces themselves are unchanged by the addition of an AI tier; the AI grounding endpoints defined in Chapter 10, Section 10.4.4 were designed from the outset to be the AI consumer's entry point, and the Access API remains the consumer surface for workflows, tiles, and analytical drilldowns that are not AI. The canonical data spine at the bottom of Figure 13.1 is consumed identically by all four consumer classes; it cannot tell the difference between a human-driven workflow read, a tile-rendered analytical read, and an AI-driven grounding read, except by the actor identity carried on the `AUDIT_EVENT` row each read appends.

This section works through the execution model in five steps: the four core objects (scenarios, executables, deployments, and serving endpoints), then how resource groups isolate deployments for tenancy and jurisdiction, how scaling policy turns capacity into a declared and budgetable property, how every served inference is recorded as an `INFERENCE_EVENT` joined to the audit chain, and finally how all of these come together in a worked deployment manifest.

13.1.1 Scenarios, Executables, Deployments, and Serving

A scenario is the platform's way of naming an AI task without committing to how the task is performed. The variance explainer scenario, for example, names the task "Given a reported financial variance, an FP&A role context, and a grounding bundle assembled from the canonical data spine, produce a human-readable explanation with citations." The scenario carries no model selection, no prompt template, and no scaling configuration. Instead, it carries an interface definition: the inputs it accepts, the outputs it produces, the error codes it can return, and the latency expectation a consumer can rely on. Multiple executables can implement the same scenario; the consumer addresses the scenario, not the executable, and the platform decides at deployment time which executable is bound.

An executable is the concrete realization of a scenario as a runnable container. It declares the container image, the environment variables and secrets the image needs at startup, the resource floor and ceiling it expects, and the scaling profile that governs how the platform should add or remove replicas under load. An executable can encode a fully self-contained model—a fine-tuned classifier, for example, packaged with its weights—or it can encode a thin orchestration layer that calls a foundation model through the generative AI hub. The choice is not architectural; both styles are valid, and the SAP AI Launchpad catalog (described in Section 13.2) records the choice as a catalog dimension.

A deployment is the running instance of an executable. It carries the bound model version explicitly. A deployment of the variance explainer scenario does not run "the current best model"; it runs "vex-3.2.1, content_hash 7f3a...d2c4, evaluated against test set T-2026-04, deployed at 2026-05-09 14:22:08 UTC." A consumer that calls the deployment cannot influence which model version answers. The model version is a property of the deployment, and the platform guarantees that all requests served by that deployment are served by that model version until the deployment is updated or replaced. As described in Section 13.5.3, this guarantee is what allows the FP&A controller of to ask, weeks after the fact, exactly which model produced the answer they are defending.

A serving endpoint is the network surface of a deployment. It is a REST URL of the form */v2/inference/{deployment_id}*, authenticated by an SAP Authorization and Trust Management service bearer token that carries the consumer's identity and the scopes the consumer has been granted. The endpoint is the only way to invoke the model; there is no back door, no console invocation in production, no shared service account that bypasses the consumer-identity contract. An inference call without a valid SAP Authorization and Trust Management service token is refused at the gateway, and a refused call generates an `AUDIT_EVENT(AUTH_REFUSED)` row that is structurally indistinguishable from a refused call against any other governed surface on SAP BTP.

Figure 13.2 puts the four objects in their characteristic flow and shows the resource group binding that wraps them. The resource group is the SAP AI Core construct that isolates deployments from one another for purposes of jurisdiction, role scope, billing, and operational fault containment. The variance explainer deployment in the FP&A example belongs to the rg-finance-prod resource group; a sibling deployment in the rg-marketing-prod resource group cannot reach its model weights, scaling configuration, or scaling history, and the two deployments compete for nothing because their capacity envelopes are separated at the platform layer. Jurisdiction-scoped resource groups are how a single multinational tenant separates EU data-residency-sensitive inference from US residency-sensitive inference without operating two SAP AI Core instances.

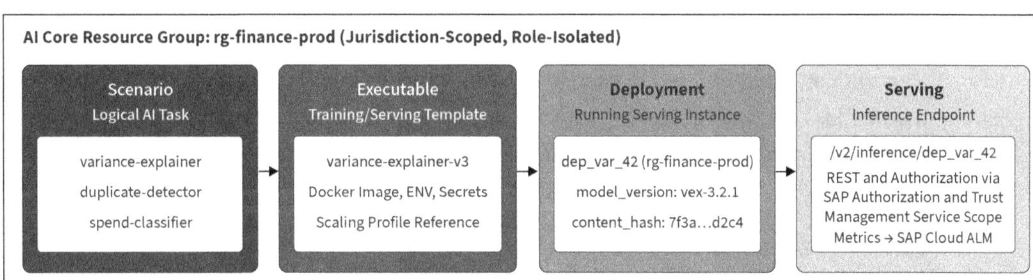

Figure 13.2: AI Core Execution Model—Scenario, Executable, Deployment, Serving with Resource Group Binding

The scaling policy includes minimum and maximum replicas, target inference latency, queue depth threshold, applied to the deployment, not the scenario. For the inference lineage, every served request emits `INFERENCE_EVENT(actor, model_version, content_hash, grounding_bundle_id, request_id)` and joins to `AUDIT_EVENT` on actor identity.

13.1.2 Resource Groups and Tenancy

A resource group in SAP AI Core is more than a budget envelope. It is the operational boundary inside which a set of deployments share a common configuration of secrets, a common bound SAP Authorization and Trust Management service scope assignment, a common audit-event prefix, and a common observability namespace in SAP Cloud ALM. Two deployments in the same resource group can be reasoned about together: Their inference latency dashboards aggregate, their cost views aggregate, their incident postmortems share a context. Two deployments in different resource groups cannot be reasoned about as a unit even if they implement the same scenario, because they are operationally distinct in every dimension that matters for production.

The choice of resource group structure follows the same logic as the choice of namespace structure in Kubernetes: It is a permanent decision that shapes everything that follows. The platform pattern that has worked across SAP customers is to use the cross product of jurisdiction and business function: rg-finance-eu-prod, rg-finance-us-prod, rg-supply-chain-apac-prod, and so on. The cross product is wider than it sounds (for example, a multinational with three jurisdictions and five business functions has fifteen production resource groups before staging) but the width pays for itself the first time a residency-restricted inference is accidentally routed to the wrong jurisdiction in a test environment and the resource group structure makes the mistake impossible to repeat in production.

Resource group assignment also governs which SAP AI Launchpad catalog entries are visible at deployment time. A resource group whose governance is tier 1 (finance-grade) can only deploy models whose catalog is tier 1; tier 2 (operational) deployments can use tier 1 or tier 2 models; tier 3 (sandbox) can use any model the catalog holds, but tier 3 resource groups cannot be addressed by production SAP Authorization and Trust Management service scopes. The visibility filter is enforced at the SAP AI Core control plane, not at the catalog, so a deployment manifest that references a non-permitted model is rejected at apply time with an audit row rather than at run time with an inference failure. The asymmetry is deliberate. Catch the governance error early, when a developer can fix it; do not let it surface as a runtime failure when a controller is waiting for an answer.

13.1.3 Scaling Policy and Capacity Reasoning

A deployment's scaling profile is declared on the executable but bound at the deployment. The profile names a minimum replica count, a maximum replica count, a target inference latency at the p95, and a queue depth threshold above which the autoscaler should add capacity. The values that work in practice differ widely across scenarios: a variance explainer scenario whose interactive consumer is a single controller in a finance close window may run with one or two replicas and a generous queue depth, while a duplicate detector scenario that consumes a stream of master data events may run with twenty replicas and a tight queue depth because backlog produces business-visible delay. The platform does not impose a default; the scenario owner specifies a profile that matches the consumer pattern.

Capacity reasoning for SAP AI Core deployments differs from capacity reasoning for ordinary services in one respect that you should appreciate if you're coming from an FP&A reader background. The dominant cost is rarely CPU or memory; it is model inference. A foundation model invocation through the generative AI hub is billed per input and output token; a fine-tuned classifier deployed in place is billed for the compute it consumes. The scaling profile's practical consequence is not just availability but cost. A deployment that scales aggressively in response to queue depth will incur cost in proportion to the scaling, and the cost shows up in the resource group's SAP Cloud ALM cost view in the same period the scaling occurred. Treating the scaling profile as a budget decision, not just an availability decision, is the right approach for an organization that wants to operate SAP AI Core deployments without surprises in the next quarter's controlling close.

13.1.4 Inference Lineage and INFERENCE_EVENT

Every served inference emits one `INFERENCE_EVENT` row. The row carries the actor identity (the SAP Authorization and Trust Management service principal that called the endpoint), the scenario identifier, the deployment identifier, the `model_version` and `content_hash` bound to the deployment, the `grounding_bundle_id` (when one was supplied), a request identifier, the token counts in and out (for billing reconciliation), and the inference latency. The row is the structural join point between the AI tier and the `AUDIT_EVENT` hash chain: `INFERENCE_EVENT.actor_identity` joins to `AUDIT_EVENT.actor_identity`, so that an auditor reading a `request_id` can pull both the `AUDIT_EVENT` chain and the `INFERENCE_EVENT` detail without correlating across separate stores.

From an FP&A and controlling perspective, the `INFERENCE_EVENT` row is what allows AI inference to be reasoned about as a cost element of a reported number. The variance explainer answer that a controller defends in front of an external auditor was produced by a specific deployment, consuming a specific bundle assembled at a specific instant, returning a specific token count for which a specific cost was incurred. All of those facts are in the `INFERENCE_EVENT` and `AUDIT_EVENT` rows together. The controller does not need a separate AI cost reconciliation tool; the same chain that defends the reported number defends the cost of producing the answer. This is the FP&A relevance of the inference lineage design choice, and it is the reason this section appears in a chapter written for an audience that includes the controlling community as much as the engineering community.

AI Insight: Why INFERENCE_EVENT Is Recorded for the AI, Not About the AI

An older generation of AI integrations recorded inference invocations as model-side log lines, separate from the audit trail used for ordinary transactional reads and writes. That separation produced two systems of record that had to be correlated by hand whenever an auditor asked what an AI agent saw, decided, or recommended, and it is a structural mistake. `INFERENCE_EVENT` is recorded in the same store as `AUDIT_EVENT`, joins to it on actor identity, and is queried with the same tooling. An auditor reading the `AUDIT_EVENT` chain for request req-9a1c sees the `INFERENCE_EVENT` row at the model hop alongside the `GROUNDING_ASSEMBLED` row at the data spine hop and the `RESPONSE_RETURNED` row at the consumer hop, without correlating across stores.

> The structural payoff is that the AI boundary established in Chapter 1 (AI is a pattern detector, anomaly scorer, explainer, and draft author, never a decider) becomes provable from the audit chain. An AUDIT_EVENT chain for a controller's reported number contains the INFERENCE_EVENT row that recorded the model that produced the explanation, the bundle the model consumed, and the human approval (if any) that promoted the model's suggestion into a state change. Auditors and controllers can prove from the chain itself that the model did not act unilaterally. The boundary is no longer a policy promise; it is an audit-readable fact.

13.1.5 Worked Example: Deployment Manifest

A deployment is created by submitting a deployment manifest to the SAP AI Core control plane. The manifest declares the scenario, the executable, the resource group, the model_version binding, and the scaling profile. The control plane validates the manifest against the SAP AI Launchpad catalog (Does this executable exist? Is the model version DEPLOYED rather than RETIRED?), against the resource group (Does the governance tier permit this model? Does the resource group have free capacity?), and against the consumer registry (Is this scenario already bound to a deployment in this resource group? If so, is this an in-place update?). A failed validation produces an AUDIT_EVENT(DEPLOYMENT_REFUSED) row with the specific failure code; a passed validation produces an AUDIT_EVENT(DEPLOYMENT_APPLIED) row, a new INFERENCE_EVENT subscription, and an Event Mesh announcement that downstream consumers can subscribe to. Listing 13.1 shows a complete manifest, expressed in JSON, for the variance explainer scenario in the rg-finance-prod resource group. You can see the model_version pinned, scaling profile declared, and the audit topic for INFERENCE_EVENT emission explicitly bound to the rg-finance-prod resource group.

```
{
  "apiVersion": "ai.sap.com/v1",
  "kind": "Deployment",
  "metadata": {
    "name": "dep-variance-explainer-v3",
    "resourceGroup": "rg-finance-prod",
    "labels": {
      "scenario": "variance-explainer",
      "tier": "T1-finance-grade",
      "consumer": "joule-agent:variance-explainer"
    }
  },
  "spec": {
    "scenario": "variance-explainer",
    "executable": "variance-explainer-runtime",
    "executableVersion": "3.2.1",
```

```
    "modelBinding": {
      "provider": "sap-genai-hub",
      "model": "anthropic-claude-3-7-sonnet",
      "modelVersion": "claude-3-7-sonnet-20251022",
      "contentHash": "7f3ac84d2c4a91e0",
      "lifecycleState": "DEPLOYED"
    },
    "scaling": {
      "minReplicas": 2,
      "maxReplicas": 12,
      "targetLatencyMsP95": 1800,
      "queueDepthThreshold": 8
    },
    "audit": {
      "inferenceEventTopic": "audit/inference/rg-finance-prod",
      "actorPrefix": "ai-core:dep-variance-explainer-v3:"
    }
  }
}
```

Listing 13.1: SAP AI Launchpad Deployment Manifest for the Variance Explainer Scenario

The manifest is the operational expression of the principles set out earlier in the section. The scenario is the consumer-facing identity. The executable and its version are pinned, not floating. The model binding carries both the catalog identifier and the content hash, so that a controller weeks later can prove which model produced an answer. The scaling profile commits to a latency target and a queue depth, both of which are operationally measurable; and the audit block names the topic on which `INFERENCE_EVENT` rows are emitted, so that downstream subscribers (the audit subsystem described in Chapter 19 and the cost-attribution view covered in Chapter 18) receive the rows without coupling to the deployment's internals. If you have worked through Chapter 10's API surfaces and Chapter 12's action library, you should recognize the contractual style: A manifest that pins every property a consumer needs to reason about and exposes them to audit as a single declarative object.

13.2 SAP AI Launchpad and Model Access

If SAP AI Core is the runtime that serves models, SAP AI Launchpad is the catalog that decides which models a runtime is allowed to serve. The catalog is not a passive index. It carries governance metadata about each entry (provider, modality, license terms, data residency constraints, governance tier, and lifecycle state) and the SAP AI Core control plane enforces that metadata at deployment time. A model that is not in the catalog cannot be deployed, a model whose catalog tier does not match the resource group's governance tier cannot be deployed in that resource group, and a model whose lifecycle state is `RETIRED` cannot be deployed at all. The catalog is the single source of truth about what the platform is permitted to run. Models are searchable, comparable, and accessible only through scope deployments in SAP AI Core.

The catalog spans three provider categories that should be reasoned about together rather than in isolation. SAP-built models—such as Document Information Extraction, Data Attribute Recommendation, Business Entity Recognition, and the SAP-built translation models—are services SAP has trained, evaluated, and operates inside its own SAP BTP boundary. Partner models—the foundation models offered by OpenAI (GPT family), Anthropic (Claude family), Google (Gemini family), and the Amazon Web Services (AWS) Bedrock-hosted catalog—are accessed through commercial agreements between SAP and the model provider, with traffic routed through the generative AI hub described in Section 13.2.2. Open-source models such as Mistral and Mixtral variants, Falcon, Llama 3, and the major embedding model families are deployed inside the platform's tenancy on SAP AI Core infrastructure rather than called out to a third party. The three categories carry different residency, latency, cost, and lifecycle properties, and the catalog records each model's category explicitly so that a deployment manifest cannot accidentally mix categories without the deployer noticing.

The generative AI hub abstracts foundation models behind a uniform interface. It is the layer at which a consumer that wants to invoke a foundation model, whether the model is partner-hosted or open-source, and does so through a single software development kit (SDK) surface rather than against the vendor SDK directly. As such, Joule and SAP Build Process Automation agents consume foundation models exclusively through this layer. The abstraction is operationally important: A consumer that calls the generative AI hub does not need to be rewritten when the underlying model is switched from one vendor to another, when a regional routing decision moves traffic from one foundation model deployment to another, or when content filters and output validators are inserted in the path. The same consumer code works across the catalog.

Every catalog entry carries six governance dimensions before it can be deployed: provider identity, modality (text, vision, embedding, multimodal), license terms and territorial restrictions, data residency, personally identifiable information (PII) handling capability, and governance tier (tier 1 finance-grade, tier 2 operational, and tier 3 sandbox). Models without all six dimensions cannot transition past `REGISTERED`, as the lifecycle gate refuses the `EVALUATE` request.

This section examines the catalog from three angles: the three provider categories and the governance properties that distinguish them, the generative AI hub through which foundation models are invoked behind a uniform interface, and the six catalog dimensions that together decide whether a given model may be deployed.

13.2.1 Provider Categories and Their Governance Properties

SAP-built models in the catalog carry the strongest residency and lineage properties because SAP has trained them, knows the training data's provenance, and operates them inside its own tenancy. The catalog entry for an SAP-built model includes the training corpus identifier, the evaluation suite identifier, the bias-scan verdict, and the model card produced by SAP's AI governance office. These properties are not optional. A model whose entry is missing any of them cannot transition past the `REGISTERED` state, and the lifecycle gate refuses the `EVALUATE` request with a specific reason code. This discipline is what allows a finance-grade deployment to bind to an SAP-built model with the same confidence that a transactional system binds to an SAP-built data type.

Partner models (the foundation models from major commercial providers) carry catalog entries assembled jointly by SAP and the provider. The entry records the contractual data-handling commitments the provider has made (no training on customer input, no log retention beyond the agreed window, or regional routing guarantees), the supported regions, the per-token pricing as exposed through the generative AI hub, and the model card the provider publishes. SAP does not own the training data of partner models and cannot represent the corpus the way it can for SAP-built models, but it can and does represent the contractual surface on which a customer deployment depends. The catalog is honest about the distinction; a deployment manifest reading a partner-model catalog entry sees fields whose values are contractual rather than audited, and the governance team that approves the deployment knows the difference.

Open-source models occupy the third category and behave differently from the other two in one important respect: They are deployed inside the platform tenancy, on SAP AI Core infrastructure, and their inference does not leave the customer's SAP BTP boundary. A Llama 3 70B deployment in the rg-finance-eu-prod resource group runs on SAP AI Core capacity allocated to that resource group, and inference traffic does not traverse a third-party boundary. The trade-off is operational. The customer is responsible for the capacity, the cost of running the inference workload at scale, and the lifecycle of the model artifact. The catalog records the model's license (Apache 2.0, the various Llama 3 community licenses, MIT, and the more restrictive variants), and the SAP AI Core control plane refuses to deploy a model whose license terms are not compatible with the resource group's intended use. License-incompatible deployment attempts produce `AUDIT_EVENT (LICENSE_REFUSED)` rows.

13.2.2 Generative AI Hub and the Foundation Model Interface

The generative AI hub is the catalog's consumer-facing surface for foundation models. It exposes a uniform SDK (chat completion, streaming responses, embedding generation, tool-use prompts, and structured output) that maps onto whichever underlying provider is bound at deployment time. A consumer that calls the generative AI hub does not write provider-specific code; the SDK call carries a model identifier from the catalog, and the hub routes to the appropriate provider, applies content filters, validates outputs against a schema if one is declared, and records the `INFERENCE_EVENT` row before returning the response. The abstraction is thin where it can afford to be—the prompt format and the tool-use semantics are exposed close to the underlying vendor's capabilities—and uniform where uniformity matters, principally on the audit, billing, and content filter surfaces.

Content filters and output validators warrant a moment of separate attention because they are the layer at which the platform applies safety constraints that the underlying model does not necessarily enforce on its own. The generative AI hub's content filters reject inputs that match known prompt-injection patterns, reject outputs that contain PII not present in the input, and reject outputs that violate a declared structural schema. The filters operate identically across providers, so a consumer cannot accidentally lower the safety floor by selecting a provider whose built-in filters happen to be more permissive. Filter outcomes are recorded as part of the `INFERENCE_EVENT` row, with explicit fields for `filter_input_verdict` and `filter_output_verdict`. A row in which either field is `REJECTED` is visible in the audit chain and can be reported on directly.

13.2.3 Catalog Dimensions and the Deployment Decision

A model catalog entry is described by six dimensions that decide whether a given deployment is permitted. Provider identity names the training and operating party (SAP, Anthropic, Mistral, and so on). Modality declares whether the model produces text, embeddings, vision, or multimodal output. License captures the terms under which the model may be used commercially, including the territorial restrictions some open-source licenses impose. Data residency states the regions in which inference may occur and the regions in which any retained data may be stored. PII-handling capability declares whether the model is approved to receive payloads containing personal data of customers or employees. Governance tier (tier 1 finance-grade, tier 2 operational, and tier 3 sandbox) names the operational rigor the model has been evaluated against and the deployment context in which it may be used.

Table 13.1 lists the dimensions, the values each can take, and the way each constrains the deployment decision. The table is the structural reference an engineer or governance officer consults when proposing a new model for the catalog; an entry that does not declare values for all six dimensions is incomplete and cannot transition past `REGISTERED`, regardless of who proposes it. The discipline of demanding all six dimensions before any model can be deployed is what makes the SAP AI Launchpad a governance instrument rather than a passive directory.

Dimension	Allowed Values	Deployment Constraint
Provider identity	SAP-built, partner (commercial agreement), and open-source (in-tenancy)	Records the operating party. SAP-built carries trained-corpus audit, partner carries contractual commitments, and open-source carries license terms.
Modality	Text, embedding, vision, and multimodal	Constrains the consumer scenarios that can bind to this model; a vision-only model cannot be bound to a text-completion scenario.
License	Apache 2.0, MIT, Llama 3 community, Mistral commercial, Anthropic commercial, SAP proprietary, OpenAI commercial, and restricted jurisdiction	Determines whether deployment is permitted in a given resource group; territorial restrictions on some open-source licenses are enforced here.
Data residency	EU, US, APAC, and multi-region (in tenancy only)	Must be compatible with the resource group's jurisdiction; an EU residency resource group cannot bind to a model with US-only residency.
PII handling	Approved and restricted (synthetic only); refused	Constrains the kind of grounding bundles that may be assembled for this model; refused models cannot bind to scenarios whose inputs include PII.
Governance tier	Tier 1 finance-grade, tier 2 operational, and tier 3 sandbox	Must be greater than or equal to the resource group's tier; tier 3 models cannot be deployed in tier 1 or tier 2 resource groups, regardless of other dimensions.

Table 13.1: SAP AI Launchpad Catalog Dimensions and the Deployment Constraints They Impose

The dimensions interact in ways that the deployment process should make visible to the deployer before the deployment is attempted. A scenario that consumes PII grounded from the canonical data spine cannot be bound to a model whose PII-handling dimension is `REFUSED`, regardless of how attractive the model's evaluation scores are. A resource group whose jurisdiction is EU cannot bind to a partner model whose data residency dimension is US-only. A finance-grade resource group in tier 1 cannot bind to a sandbox model whose in tier 3, even if a developer believes the sandbox model has been improving rapidly. These constraints fail closed at apply time, with a specific reason code in the `AUDIT_EVENT` row, so that the developer sees the constraint at the moment it is being violated rather than as a mysterious runtime failure.

AI Insight: Why Catalog Completeness Is the First Governance Gate

An incomplete catalog entry is the AI tier's equivalent of an under-validated PROCESS_VERSION row in the canonical data spine. It represents a structural commitment to operate something whose properties are not fully known. The discipline that requires all six dimensions to be present before any model can transition past REGISTERED is the same discipline that requires a PROCESS_VERSION to carry a content_hash, an activity list, and a publish_state before it can leave INGESTED. The form is a catalog field rather than a canonical column, but the structural intent is identical: refuse to operate on something whose declarable properties have not been declared.

From an FP&A perspective, the practical consequence of the discipline is that any AI-produced figure a controller defends is backed by a catalog entry whose six dimensions are known, recorded, and queryable. There is no AI model anywhere in the chain whose provider, license, residency, PII posture, or governance tier is undefined. When an external auditor asks where the model came from, the answer is one row in the SAP AI Launchpad catalog, joined to the INFERENCE_EVENT row, joined in turn to the AUDIT_EVENT chain that ends in the reported number. The whole defense is one query, and the query returns the same kind of answer for an AI-produced number as for any other number a controller has historically defended.

13.3 AI Lifecycle and Governance

An AI model is an artifact whose properties drift. The training data ages, the world the model was evaluated against changes, the consumer patterns evolve, and the model's real-world performance can degrade for reasons that have nothing to do with how the model was originally built. The lifecycle model SAP AI Launchpad imposes (REGISTERED, EVALUATED, DEPLOYED, DEPRECATED, and RETIRED) is a recognition of that drift. It is not a deployment workflow but a state machine that names what is true about a model at each stage of its life on the platform, and it carries gating verdicts at every transition that decide whether the model may move forward, must hold position, or must move backward. This state machine is shown in Figure 13.3.

Figure 13.3 shows the five states. There are four gates that separate them, as follows:

- PASS: State transition proceeds, AUDIT_EVENT row is recorded, and subscribers are notified through Event Mesh.
- CONDITIONAL: Transition is allowed with named conditions (e.g., usage cap, jurisdiction restriction) and conditions are stored on the MODEL_VERSION row.
- QUARANTINE: Transition is refused, the model returns to its prior state, QUARANTINE_REASON is recorded, and resubmission requires new evaluation evidence.
- WITHDRAW: Manually recalls a deployed model after the fact, flips DEPLOYED to DEPRECATED with reason code, and preserves the existing INFERENCE_EVENT rows.

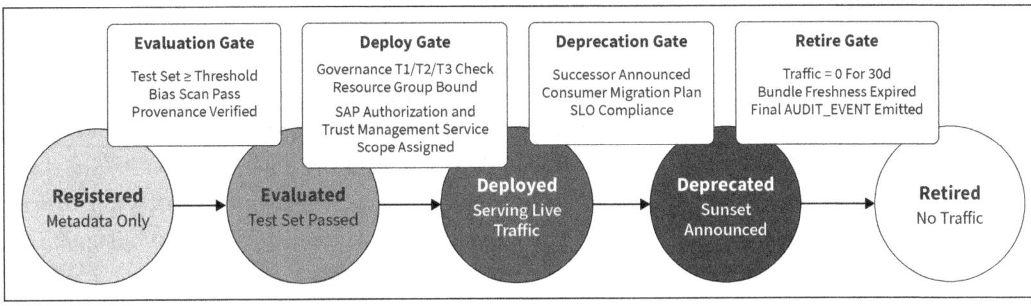

Figure 13.3: AI Lifecycle and Governance—Model State Machine with Evaluation, Deployment, Deprecation, and Retirement Gates

A REGISTERED model has a catalog entry but no evaluation evidence; it cannot be deployed. An EVALUATED model has a passed test set, a bias scan, and a verified provenance record. It can be considered for deployment but is not in production. A DEPLOYED model is serving traffic in at least one resource group and is subject to continuous evaluation. A DEPRECATED model has had its sunset announced, has a successor identified, and has a consumer migration plan in flight. It continues to serve while the migration is underway. A RETIRED model has no traffic for thirty consecutive days, has had its bundle freshness contract expire, and has had a final AUDIT_EVENT row emitted. It cannot be redeployed without a new evaluation cycle.

Evaluation is not a one-time gate. Daily drift checks, weekly bias scans, and quarterly reevaluation against an updated test set keep DEPLOYED models honest. A continuous evaluation failure triggers an automatic WITHDRAW verdict. Treating evaluation as a single upfront check is a governance anti-pattern. When it comes to FP&A, every gate verdict, with model_version and content_hash, is one row in AUDIT_EVENT, the same chain a controller reads for any other transaction.

This section follows the lifecycle through four lenses: the states a model occupies from registration to retirement and what each one promises, the gates that govern every transition and the verdicts they can return, the discipline of continuous evaluation that keeps a deployed model honest, and the way each lifecycle verdict becomes a row an FP&A controller can read in the same audit chain as any other transaction.

13.3.1 Lifecycle States and What They Promise

REGISTERED is the entry state for any model added to the catalog. The state captures that the catalog knows the model exists and has at least the six dimensions outlined in Section 13.2 recorded against it. REGISTERED is not permission to deploy but permission to consider deploying. A model can sit in REGISTERED indefinitely while the governance office evaluates whether it is worth the engineering investment of running it through the EVALUATE pipeline. Many models proposed for the catalog never leave REGISTERED, and the catalog is healthier for it.

EVALUATED is the state a model reaches after it has passed the evaluation pipeline. The pipeline runs the model against a curated test set, computes accuracy or relevant quality metrics, runs a bias scan against the demographic and process-context dimensions the governance office tracks, and verifies that the model's provenance is consistent with the catalog entry's claims. The verdict is recorded in three forms: a PASS or QUARANTINE flag, a score sheet documenting the metrics achieved, and a written governance opinion from the AI governance office. An EVALUATED model has a known quality profile that consumers and deployers can reason about; a REGISTERED model does not.

DEPLOYED is the state a model reaches after a deployment has been applied in a resource group. The same model version can be DEPLOYED in multiple resource groups simultaneously, and the state machine reasons about each deployment instance independently. Continuous evaluation runs on DEPLOYED models (drift checks against a moving reference distribution, periodic reevaluation against an updated test set, or ongoing scanning for content-filter rejection patterns that might indicate behavioral change) and a continuous evaluation failure can trigger an automatic transition to DEPRECATED with WITHDRAW reason.

DEPRECATED is the state a model enters when its sunset has been announced. The deprecation calendar carries a successor identifier (which model will replace this one), a consumer migration plan (which deployments must be repointed and by when), and a service-level objective (SLO) floor that the deprecated model is committed to maintaining during the migration. A DEPRECATED model continues to serve traffic during this window. Consumers that are not migrated by the migration deadline receive warning AUDIT_EVENT rows; consumers that are still bound after the deadline are migrated automatically to the successor model, with an AUDIT_EVENT(MIGRATION_FORCED) row recorded for each.

RETIRED is the terminal state. A RETIRED model has had no traffic for thirty consecutive days, has had its bundle freshness contract expire across all subscribed consumers, and has had a final AUDIT_EVENT(MODEL_RETIRED) row emitted. The catalog entry remains for historical query, as an auditor asking which model produced an inference from eighteen months ago must still be able to find the entry, but the model cannot be redeployed. Reintroducing a RETIRED model requires reevaluation against a current test set and a fresh transition through the gates; the catalog does not pretend that the model's previous EVALUATED verdict remains valid after thirty days of disuse.

13.3.2 Gates and Their Verdicts

Each transition between states is mediated by a gate. The evaluation gate decides whether REGISTERED may transition to EVALUATED. The deploy gate decides whether EVALUATED may transition to DEPLOYED. The deprecation gate decides whether DEPLOYED transitions to DEPRECATED. The retire gate decides whether DEPRECATED transitions to RETIRED. Each gate produces one of four verdicts (PASS, CONDITIONAL, QUARANTINE, or WITHDRAW) and each verdict produces a specific AUDIT_EVENT row. The verdicts are summarized in Table 13.2 with the conditions they impose on the resulting state.

Transition	Gate	PASS Conditions	QUARANTINE/WITHDRAW Conditions
REGISTERED to EVALUATED	Evaluation gate	■ Test set is greater than or equal to scenario threshold ■ Bias scan PASS ■ Provenance verified ■ Six catalog dimensions complete	Any condition fails → QUARANTINE with reason code; model returns to REGISTERED until evidence updated
EVALUATED to DEPLOYED	Deploy gate	■ Governance tier is greater than or equal to resource group tier ■ Resource group has capacity ■ SAP Authorization and Trust Management service scope assigned ■ License permits deployment in resource group jurisdiction	Tier or jurisdiction mismatch → QUARANTINE with TIER_MISMATCH or JURISDICTION_MISMATCH
DEPLOYED (continuous)	Continuous evaluation	■ Daily drift check within tolerance ■ Weekly bias scan PASS ■ Quarterly reevaluation PASS	Drift, bias, or reevaluation failure → Automatic WITHDRAW; model transitions to DEPRECATED with reason
DEPLOYED to DEPRECATED	Deprecation gate	■ Successor identified in catalog ■ Migration plan filed ■ SLO commitment for migration window declared	No successor or no plan → QUARANTINE; deployment must remain DEPLOYED until the governance office files plan
DEPRECATED to RETIRED	Retire gate	■ Traffic has remained at zero for 30 days or more ■ All subscriber bundles expired ■ Final AUDIT_EVENT emitted	Traffic still present → Transition deferred; CONDITIONAL until traffic confirmed at zero

Table 13.2: Lifecycle Gate Verdicts and the Conditions That Determine PASS, CONDITIONAL, QUARANTINE, or WITHDRAW

The four verdicts are not freely chosen by the governance officer who operates the gate; they are determined by whether the conditions named in the table are met. The officer's discretion is in the framing of CONDITIONAL (the named conditions under which a transition is allowed when the default would have been QUARANTINE) and in the framing of WITHDRAW, which is a deliberate revocation of a previous transition. PASS and QUARANTINE are mechanical: They follow from the evidence. Treating the gates as procedural rather than negotiable is what allows the lifecycle to be defended in front of an auditor as a controls regime; an officer who manually passed a model whose conditions were not met would generate an AUDIT_EVENT row whose verdict mismatch would be detectable on the next compliance review.

13.3.3 Continuous Evaluation: Why the Gate Is Not a One-Time Check

The most consequential design decision in the lifecycle model is that evaluation is continuous, not something you do once. A model that passes the evaluation gate on the day it is deployed is not permanently approved; it is approved subject to continuing to satisfy the evaluation conditions over time. The platform runs daily drift checks comparing the distribution of recent inputs and outputs against a reference distribution established at evaluation time, weekly bias scans against the same demographic and process-context dimensions used in the original evaluation, and quarterly reevaluations against an updated test set that the governance office maintains. A failure on any of these checks triggers an automatic WITHDRAW verdict, and the model transitions from DEPLOYED to DEPRECATED without waiting for a human review cycle.

The choice to make evaluation continuous reflects a specific operational reality: A foundation model whose underlying provider has rolled out a silent update is not the same model it was a month ago, even if its identifier in the catalog has not changed. Continuous evaluation is the instrument by which the platform detects such changes without depending on the provider to announce them. A continuous evaluation failure carries a reason code that names the specific check that failed and the magnitude of the deviation; the WITHDRAW row in AUDIT_EVENT carries the same reason code, so that a controller asking why a particular variance explanation looked different this quarter from last has a one-query answer.

An anti-pattern that organizations new to AI governance often reach for is treating evaluation as a one-time gate, on the reasoning that running reevaluations is expensive and the model has not been updated. While it is true the model has not been updated, the world it serves has, and the question evaluation answers is not "Is the model unchanged?" but "Does the model still satisfy the conditions under which we approved it?" The two questions have different answers, and the second is the one the platform has committed to. In Chapter 18, we will return to the continuous evaluation discipline as a recurring operational concern.

13.3.4 Lifecycle Verdicts and the Financial Planning and Analysis Trace

From an FP&A controller's point of view, the lifecycle gates are reportable events that share a structural property with the canonical data spine's PUBLISH_STATE transitions. Each verdict is one row in AUDIT_EVENT, with the same actor identity, the same hash-chain linkage, and the same queryable form as any other audited transition on the platform. A controller defending an AI-produced number can demonstrate that the model was in the DEPLOYED state on the day the number was produced, the most recent continuous evaluation check before that day was PASS, the model has not had a WITHDRAW verdict applied since, and the governance tier of the model was tier 1 (finance-grade), all from a single query against the audit chain.

The property that makes this defense available is the same one that the rest of the platform relies on: Every operationally significant event produces an AUDIT_EVENT row, and the rows are queryable as a graph. The AI lifecycle does not introduce a separate evidentiary regime for AI-specific events; it extends the existing regime with new event types (EVAL_PASS, EVAL_QUARANTINE, DEPLOY_APPLIED, WITHDRAW_FORCED, and MODEL_RETIRED) whose structure matches the rest. A controller does not need to learn a new tool to read AI events; the same audit tooling already in use for spine events surfaces them.

AI Insight: Lifecycle Verdicts as Reportable Events for the Controller

The Sarbanes–Oxley Act (SOX) controls that a US-listed customer applies to the canonical data spine are easily extendable to AI deployments because the lifecycle verdicts are AUDIT_EVENT rows. A SOX walkthrough for a finance-grade AI scenario looks the same as a SOX walkthrough for any other controlled process: name the process, identify the controls (the lifecycle gates), pull the AUDIT_EVENT rows that evidence the controls were operated, sample the rows for design and operating effectiveness, and document the conclusion. The platform did not have to invent an AI-specific control regime; it had to extend the AUDIT_EVENT event types and the rest of the regime worked unchanged.

The structural lesson, for an FP&A controller moving into business transformation, is that the value of imposing a single audit chain across the data spine, the execution catalog, and the AI tier compounds the longer the platform operates. Every additional event type that joins the chain becomes simultaneously a controllable, an auditable, and a reportable event without requiring new tooling. An AI lifecycle that lived outside the chain would require separate tooling for each of those properties and the controlling community to learn it. Inside the chain, it requires no new learning at all, only awareness that the AI verdicts are visible alongside everything else.

13.4 Joule as the Interaction Layer

Joule is the interaction layer of the SAP AI stack. It is the surface through which a human user (a controller asking why a variance moved, a buyer asking which supplier has the shortest lead time on a critical material, or an accounts payable specialist asking whether a supplier's recent invoices match the historical pattern) engages with SAP AI in natural language. It is the surface through which an agent (a variance explainer, duplicate detector, contract summarizer, or the hundreds of others a mature deployment will hold) engages with SAP AI as an automated consumer. Joule is not a model; it is the orchestration and tool invocation runtime that sits between human or agent intent and the AI tier below.

The architectural choice that defines Joule's place in the stack is that Joule does not reach the canonical data spine directly. It calls AI grounding endpoints to obtain version-pinned, role-scoped bundles, and it calls SAP Build Process Automation actions to invoke governed transactions or reads against the data spine. Both routes are governed by the same SAP Authorization and Trust Management service scope mechanism and produce the same `AUDIT_EVENT` rows that any other consumer produces. The user sees natural language; the underlying machinery sees governed API calls. There is no private path between Joule and the canonical data spine, and in Section 13.5, we will show that there is no allowed tool by which a Joule agent could construct one.

Figure 13.4 shows the five-stage pipeline through which a Joule interaction flows. The user prompt arrives carrying the user's role context. The Joule agent classifies the intent and decides which grounding bundles and tools are required. The AI grounding endpoint assembles the bundles from the data spine at a specific instant, the foundation model is invoked through the generative AI hub with the bundles attached as context. Finally, the response, with inline citations to the grounding bundle's artifacts, is returned to the user. Each of the five stages appends an AUDIT_EVENT row to the same hash chain, and the actor identity carries through end to end so that the chain reconstructs as a single linear trace.

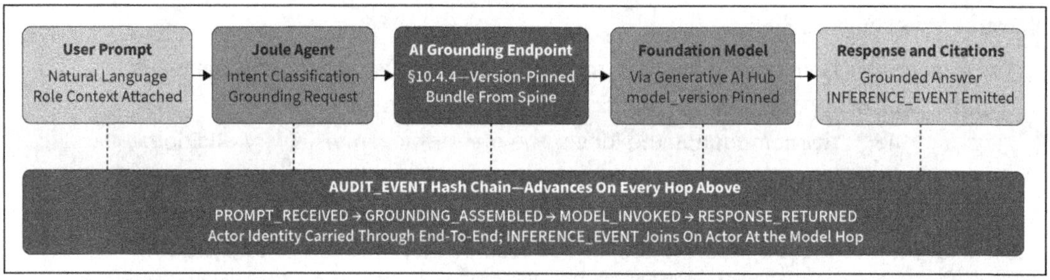

Figure 13.4: Joule Architecture—Prompt to Response Pipeline with AUDIT_EVENT Recorded at Every Stage

This section looks at Joule in three parts: how agents are authored in Joule Studio rather than written as code, the capability matrix that binds what each class of agent may do, and the way an agent's tool catalog is the same SAP Build Process Automation action library introduced in Chapter 12, Section 12.4, so that an agent inherits the same governance and audit guarantees as any other consumer of that library.

13.4.1 Joule Studio and Agent Authoring

A Joule agent is not written in code in the conventional sense. It is authored in Joule Studio, a low-code editor in which the agent author declares the agent's skills (the intents it is designed to handle), its tool catalog (the SAP Build Process Automation actions and grounding endpoints it is permitted to invoke), its prompt templates (the system-level framing the agent applies to each invocation of the foundation model), and its evaluation criteria (the test cases against which the agent is verified before promotion to `DEPLOYED`). The output of Joule Studio is an agent definition that the platform stores as a governed artifact with a version, a content hash, and a lifecycle state that follows the same five-state machine the rest of the AI tier uses.

The structural reason agent authoring is governed rather than ad hoc is that agents inherit privileges. An agent in production runs with a service principal that holds SAP Authorization and Trust Management service scopes, and those scopes determine which SAP Build Process Automation actions the agent can invoke and which AI grounding endpoints it can call. An agent definition that names a new tool the agent has not previously been permitted to use is, at the platform's level of reasoning, a privilege escalation request. The lifecycle gate that promotes the agent definition through `EVALUATED` to `DEPLOYED` is the same kind of governance gate that decides whether a new SAP Build Process Automation action can be added to the action library. Ad hoc prompting from a developer's laptop does not have this property; it is permitted in tier 3 sandbox resource groups for the development and exploration use case, but the resulting prompts cannot be promoted into production without passing through Joule Studio.

Joule Studio's evaluation harness is where the agent's behavioral contract is verified before deployment. The author declares a set of test interactions (a representative variance question, duplicate detection invocation, or contract summarization request), each with an expected outcome shape. This is not the exact wording the model will produce, which is not deterministic, but the structural shape: which grounding endpoint should be called, which tool invocation, which kind of citation should appear in the response. The evaluation runs the agent against the test set under controlled conditions, records the actual outcomes, and produces a `PASS` or `QUARANTINE` verdict on the agent definition. An agent that does not pass evaluation cannot be deployed.

13.4.2 The Agent Capability Matrix

Joule agents fall into three broad capability classes that govern what they can do without human intervention: read-only agents, grounded-write agents, and external-call agents. A read-only agent (a for example, a variance explainer) invokes grounding endpoints and read-side SAP Build Process Automation actions and returns information to the user. It cannot change state in any SAP system. A grounded-write agent (for example, a duplicate detector) can invoke a narrow set of write-side SAP Build Process Automation actions, but only on artifacts that have been retrieved through grounding and only with a human approval step interposed. An external-call agent can invoke SAP Build Process Automation actions that call out to external systems (such as a payment gateway, an external e-procurement platform, or a banking partner), again with mandatory human approval for any state-changing call.

Table 13.3 maps the three capability classes against the SAP Build Process Automation action library scopes from Chapter 12, Section 12.4, with the human-validation requirements for each class explicit. It is the structural reference an agent author consults when scoping a new agent; the choice of class is the most consequential design decision in agent authoring, because it determines which SAP Build Process Automation scopes the agent's service principal will hold and how many human-in-the-loop steps the agent's typical interaction will require. In Chapter 15, we return to the human validation requirement in detail; for the purposes of this chapter, the key point is that the requirement is enforced at the SAP Build Process Automation action library, not at the agent, and the platform refuses the action invocation if the required approval is not present.

Agent Class	SAP Build Process Automation Scopes Held	Permitted Invocations	Human Validation Requirement
Read-only	`access-api.read`; `grounding.assemble`	Lookup actions, enrichment actions, and grounding bundle assembly	None; reads do not require approval but generate `AUDIT_EVENT` rows
Grounded-write	`access-api.read`; `grounding.assemble`; `sbpa.write.bounded`	Narrow write actions on artifacts retrieved through grounding (e.g., flag duplicate or propose merge)	Mandatory human approval interposed before state-changing call; refusal generates `AUDIT_EVENT(APPROVAL_REFUSED)`
External-call	All read scopes; `sbpa.external.invoke`	External SAP Build Process Automation actions (payment, e-procurement, or banking partner integrations)	Mandatory human approval *and* four-eye second approver for high-value transactions

Table 13.3: Joule Agent Capability Matrix Mapped Against SAP Build Process Automation Action Library Scopes

Agent Class	SAP Build Process Automation Scopes Held	Permitted Invocations	Human Validation Requirement
Composite	Combinations of the above, declared per agent	Scope union of constituent classes; each invocation is subject to the rules of its class	Per-invocation rules apply; the agent does not aggregate approvals across invocations
Sandbox (tier 3 only)	Limited test scopes against tier 3 resource groups	Any test-tier invocation; no production data or state-changing actions	Not applicable; sandbox agents cannot reach production

Table 13.3: Joule Agent Capability Matrix Mapped Against SAP Build Process Automation Action Library Scopes (Cont.)

Every agent class except sandbox emits `AUDIT_EVENT` rows for every invocation, regardless of whether the invocation is a read or a write; the audit chain does not privilege state-changing operations. The human-validation step for grounded-write agents is interposed between the agent's decision and the platform's execution of the action, not between the agent and its tool catalog; the agent can propose any write within its scope, and the human approver decides whether the proposal is accepted, refused, or modified. The external-call agent class is genuinely narrower than it appears. The SAP Build Process Automation actions that constitute external invocations are themselves governed, the four-eye approval is a property of the SAP Build Process Automation action library rather than of the agent, and the agent has no mechanism for calling external endpoints outside the action library.

13.4.3 Joule's Tool Catalog Is the SAP Build Process Automation Action Library

The tool catalog available to a Joule agent is identical to the SAP Build Process Automation action library defined in Chapter 12, Section 12.4. There is no separate AI tool surface, no AI-only SDK, and no parallel library that exposes additional capabilities to agents that human-driven workflows do not have. A Joule agent that invokes the lookup-process-by-id action invokes the same action a human-authored SAP Build Process Automation workflow step invokes, with the same input schema, output schema, retry policy, audit emission, and SAP Authorization and Trust Management service-enforced scope check. The platform does not distinguish at the action layer between human-driven and agent-driven invocations except by the actor identity carried on the `AUDIT_EVENT` row.

The structural consequence is that the contract stability and audit guarantees of the SAP Build Process Automation action library extend automatically to Joule agents. An action whose semver discipline (see Chapter 12, Section 12.4) protects human-driven workflows against breaking changes protects agent-driven invocations with identical force. A deprecation calendar that runs over six months for human-driven consumers runs over twelve months for AI consumers, because agents reason from cached descriptions of the action surface more readily than human authors do and need longer to retrain. The library is the boundary, the contract is shared, and the audit chain is unified.

An immediate result is that growing the agent population grows the load on the SAP Build Process Automation action library proportionally, and the action library's operational planning (capacity, autoscaling, and cache sizing) must include the AI consumer population alongside the human workflow population. A deployment that adds twelve new Joule agents in a quarter without expanding the SAP Build Process Automation action library's capacity envelope will discover the omission as a latency degradation in workflows that were operating comfortably before. Chapter 18 returns to this operational concern in detail; we use it in this chapter so that agent authors do not reason about agent rollout in isolation from the action library's capacity plan.

AI Insight: Why a Shared Tool Catalog Is a Governance Property, Not an Engineering Convenience

When an AI tier is being added to a platform that already has a workflow tier, it is tempting to design a separate tool catalog for the AI consumer on the grounds that AI invocations have different latency profiles, retry behaviors, and error tolerances from human-driven workflows. This temptation should be resisted. A separate tool catalog creates two parallel governance surfaces—two semver disciplines, two deprecation calendars, and two audit emission policies—and the duplication is the source from which most AI-versus-workflow incidents originate in production deployments.

The structural reason a single tool catalog is the correct design is that it forces the AI tier's privileges to be visible and named. An agent that can invoke an SAP Build Process Automation action holds the scope the action requires, which is the same one a human-driven workflow would hold, and the scope is granted, recorded, and audited identically. There is no privilege the AI tier has that the workflow tier does not also have. The consequence for the FP&A controller and the auditor is that the AI tier is no harder to reason about than the workflow tier; it uses the same instruments and is equally traceable. That equivalence is the property a platform should hold itself to.

13.5 Integrating AI with SAP BTP APIs

The integration topology that connects the SAP AI stack to the canonical data spine is the most important architectural commitment in the chapter, and it is the commitment with the most direct policy backing. AI consumes the data spine through AI grounding endpoints only. The Access API, which we defined in Chapter 10, Section 10.4.2 as the consumer surface for the data spine, is closed to AI consumers by policy. The Ingestion API and the

Governance API are closed to AI consumers by both policy and OAuth 2.0 scope, since neither of those surfaces is in any agent's declared tool catalog. Event Mesh is available to agents as a notification surface, and an agent can subscribe to events the data spine emits, but it is not a path through which an agent can modify spine state.

The policy that articulates the prohibition is SAP API Policy v.4.2026a Section 2.2.2, which states that AI-orchestrated sequences of API calls outside SAP-endorsed architectures are prohibited at the platform level. The policy is enforced at the SAP BTP API gateway, not at the individual service. An agent that attempts to construct a sequence of Access API calls is refused at the gateway with an `AUDIT_EVENT(POLICY_REFUSED)` row, regardless of whether the agent's service principal happens to hold the underlying read scope for a legitimate non-AI reason. The enforcement is gateway-level because that is where the policy can be reasoned about as a single rule. A per-service enforcement would have to be reasoned about across hundreds of services and would be impossible to demonstrate as comprehensive in front of an auditor.

Figure 13.5 makes this prohibition visual. AI consumers (Joule and its named agents) sit above the SAP BTP API gateway, which enforces the SAP API Policy v.4.2026a Section 2.2.2 rule as a deny-by-default decision.

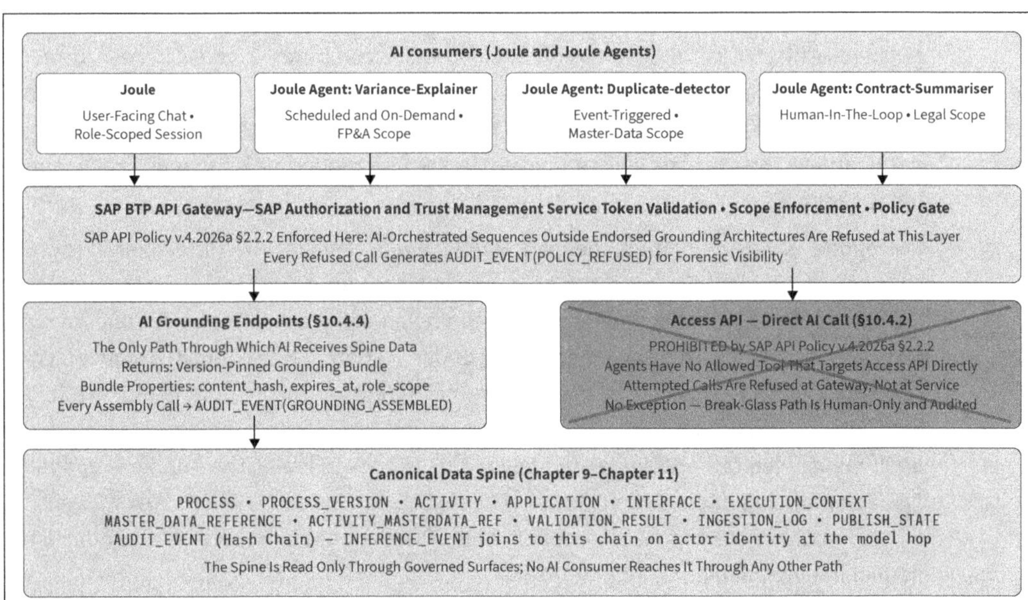

Figure 13.5: Integration Topology—Joule and Joule Agents Consume the Data Spine Only via AI Grounding Endpoints; Direct Access API Calls Are Prohibited

The allowed path leads to the AI grounding endpoints, which assemble version-pinned bundles from the data spine and return them to the agent. The prohibited path, a direct call from an AI consumer to the Access API, is struck through to note that no agent holds an allowed tool that targets the Access API directly and that any attempted call is refused at the gateway rather than at the service. The break-glass procedure for human-initiated direct Access API queries (a controller running an ad hoc analytical query, for example) operates through a separate human-authenticated path that does not flow through the AI stack at all.

This section ties the AI tier back to the governed API surfaces in four steps: the AI grounding endpoint as the only contract through which AI receives spine data, the consumption pathways an agent can and cannot use and the audit signature each one leaves, an end-to-end trace of a single controller question from prompt to grounded response, and the way inference cost itself becomes a controllable that the same audit chain can attribute to a reported number.

13.5.1 AI Grounding Endpoints: The Consumer Contract

An AI grounding endpoint is a thin assembler service that converts a grounding request from an agent into a version-pinned, role-scoped bundle of canonical artifacts. The request carries the agent's identity, the agent's role context, the scenario the bundle is intended to support, and the concrete inputs the scenario requires (a cost center identifier, a process identifier, or an application identifier, depending on the scenario). The endpoint resolves the request against the data spine, applies the role-scope filter, validates that every selected artifact has `VALIDATION_RESULT = PASS`, computes a content hash over the assembled bundle, and returns the bundle with metadata that names the assembly instant, the content hash, the role scope applied, the `expires_at` freshness window, and the bundle identifier. The agent receives a single object whose properties it can reason about without further calls.

The bundle is version-pinned, meaning every canonical artifact in it is a specific row at a specific instant, identified by `content_hash`, and it carries a `content_hash` that covers all artifacts together. It is also role-scoped, meaning an agent operating under one role context cannot share it with an agent operating under a different role context, because its role-scope metadata is checked at every consumer-side use. The bundle is also fresh: The `expires_at` field declares how long it may be reused, and a stale bundle must be discarded and reassembled rather than reused beyond its window. Finally, the bundle is auditable. The assembly call appended an `AUDIT_EVENT(GROUNDING_ASSEMBLED)` row whose `bundle_id` matches the bundle's identifier, so that the auditable history of its production is permanently available.

13.5.2 AI Consumption Pathways and Their Audit Signatures

Table 13.4 lists the AI consumption pathways available on the platform (the legitimate paths through which an AI consumer can interact with the data spine and the surrounding services) and names the `AUDIT_EVENT` signature each pathway produces. It is the operational reference for auditors and incident responders investigating an AI-related question;

reading the AUDIT_EVENT chain for an actor identity and recognizing the signatures listed in Table 13.4 is what allows the responder to reconstruct what the AI actually did in a specific interaction.

Pathway	What It Does	AUDIT_EVENT Signature
AI grounding endpoint	Agent submits a grounding request; endpoint assembles a version-pinned, role-scoped bundle from the data spine	GROUNDING_REQUESTED → GROUNDING_ASSEMBLED with bundle_id, content_hash, and expires_at
SAP Build Process Automation action (read class)	Agent invokes a read-class SAP Build Process Automation action (e.g., process lookup or context enrichment) through the action library	ACTION_INVOKED → READ_COMPLETED with action_id, scope_used, and returned content_hash
SAP Build Process Automation action (write class, approved)	Agent invokes a write-class SAP Build Process Automation action after human approval has been recorded	APPROVAL_RECEIVED → ACTION_INVOKED → WRITE_COMPLETED with action_id, approver identity, and target row
SAP Build Process Automation action (write class, refused)	Agent invokes a write-class SAP Build Process Automation action whose approval was refused or never received	ACTION_PROPOSED → APPROVAL_REFUSED or APPROVAL_TIMEOUT; no write occurs
Event Mesh subscription	Agent subscribes to a data spine event topic and is delivered events as they occur	SUBSCRIPTION_CREATED → EVENT_DELIVERED rows with topic, event_id, and agent identity
Foundation model inference	Agent invokes a foundation model through the generative AI hub with a grounding bundle attached	MODEL_INVOKED with bundle_id, model_version, content_hash, and token counts; joined to AUDIT_EVENT via INFERENCE_EVENT
Direct Access API call (prohibited)	Agent attempts to call Access API directly and is refused at gateway	POLICY_REFUSED with policy_id v.4.2026a §2.2.2, attempted endpoint, and attempted scope

Table 13.4: AI Consumption Pathways and the AUDIT_EVENT Signatures Each Pathway Produces

There are three important observations to note. First, the prohibited pathway is included in the table rather than omitted; a refused attempt produces an `AUDIT_EVENT` row whose forensic visibility is itself part of the platform's defense, and an auditor querying for `POLICY_REFUSED` rows can demonstrate not only that the policy holds but that attempted violations were detected and refused. Second, the SAP Build Process Automation write-class pathways are split into approved and refused variants because the platform records the approval state as an event in its own right. An `APPROVAL_REFUSED` row is as much a part of the audit chain as an `APPROVAL_RECEIVED` row, and the trace of refused proposals is a regular input to governance reviews. Third, the foundation model inference row is the only one in the table that joins to a second event type (`INFERENCE_EVENT`). The join is the AI tier's structural connection to the rest of the audit chain, and its existence is what makes the AI tier reconstructible from the same instruments as the data spine.

13.5.3 End-to-End Trace: The Controller's Variance Question

Figure 13.6 shows the end-to-end trace of a single FP&A controller interaction, which exercises every pathway in the previous table except the prohibited one. The controller asks Joule, in natural language, why operating expense in cost center CC-EU-4711 is up 8.3 percent against the prior period. The interaction flows through five stages (each a hop in the audit chain sense), and each stage appends an `AUDIT_EVENT` row to a single chain that ends in a response the controller can defend.

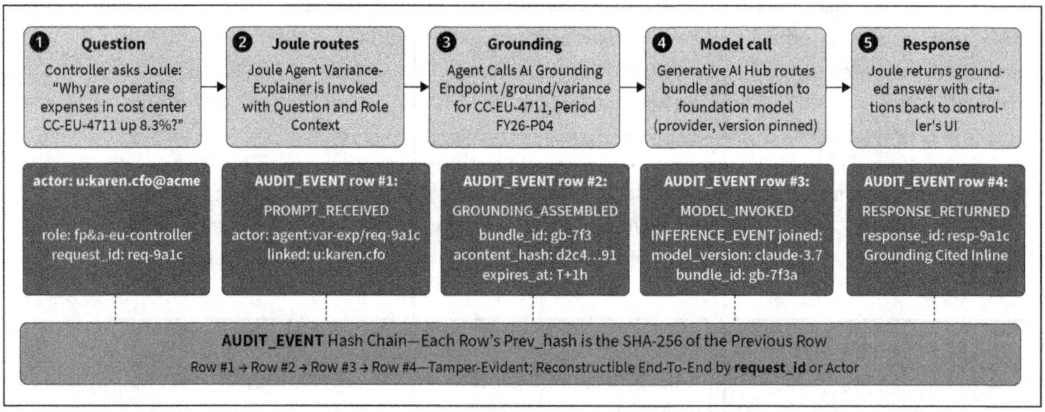

Figure 13.6: End-to-End Inference Trace—"Explain This Variance" Flows from Controller Request to Grounded Response with AUDIT_EVENT at Every Hop

In stage one, the controller's question arrives at Joule with the controller's identity (karen.cfo@acme) and role context (fp&a-eu-controller). Joule creates the interaction with `request_id` req-9a1c and routes the question to the variance explainer agent on the basis of intent classification. No `AUDIT_EVENT` row is appended at this stage because no governed surface has been called; the routing decision is internal to Joule and is recorded in Joule's interaction log, which is queryable but not part of the audit chain.

In stage two, the variance explainer agent receives the question and constructs the grounding request. The request names the scenario (variance explainer), the inputs (cost center CC-EU-4711 and period FY26-P04), the controller's role context (fp&a-eu-controller), and the agent's identity (agent:var-exp/req-9a1c, with the human controller's identity carried as the originating actor). The first `AUDIT_EVENT` row is appended with event type `PROMPT_RECEIVED`, carrying both the agent identity and the human identity so that the chain can be reconstructed either way.

In stage three, the agent calls the AI grounding endpoint `/ground/variance` with the request. The endpoint resolves CC-EU-4711 against the canonical data spine, retrieves the relevant `PROCESS` and `ACTIVITY` rows that touched the cost center during the period, filters against `VALIDATION_RESULT` to exclude any artifacts whose validation has not passed, applies the fp&a-eu-controller role scope to filter rows the controller is not entitled to see, computes the content hash over the assembled bundle (gb-7f3a, hash d2c4...91), and returns the bundle with an `expires_at` one hour into the future. The second `AUDIT_EVENT` row is appended, with event type `GROUNDING_ASSEMBLED`, carrying the `bundle_id`, `content_hash`, `expires_at`, and agent identity.

In stage four, the agent attaches the bundle to its prompt template and invokes the foundation model through the generative AI hub, which routes the call to the bound model (claude-3-7-sonnet at model_version claude-3-7-sonnet-20251022, as named in the deployment manifest in Listing 13.1). The model processes the bundle and the question and returns a grounded answer with inline citations to the bundle's artifacts. The third `AUDIT_EVENT` row is appended at this stage, with event type `MODEL_INVOKED`; an `INFERENCE_EVENT` row is emitted in parallel, carrying the `model_version`, the `bundle_id`, the input and output token counts, and the inference latency. The `INFERENCE_EVENT` joins to the `AUDIT_EVENT` chain on the agent identity so that an auditor reading the chain can pull both rows from a single query.

In stage five, Joule receives the grounded response, formats it for the controller's UI with inline citation links to the bundle's artifacts, and returns it to the controller's session. The fourth `AUDIT_EVENT` row is appended, with event type `RESPONSE_RETURNED`, carrying the `response_id` (resp-9a1c, matched to the `request_id` from stage one). The controller sees the answer, follows the citations to inspect the grounding bundle's artifacts directly, and accepts the explanation or asks a follow-up question that begins the trace again.

This results in four `AUDIT_EVENT` rows, one `INFERENCE_EVENT` row, and one chain linked end-to-end by `request_id` and by actor identity. The controller defending the variance explanation in front of an external auditor produces the chain as evidence, and the auditor can verify from the chain alone—without needing to interrogate Joule, the SAP AI Launchpad catalog, or the foundation model provider directly—exactly which bundle the model consumed, which model version produced the answer, which scope the controller was operating under, and which assembly instant the bundle represents. The trace is the defense.

For the cost lineage, the INFERENCE_EVENT also carries inference cost (tokens, model_version pricing, and resource_group budget) so AI consumption is one more cost element a controller can attribute to the reported number. No special instrument is needed because the same AUDIT_EVENT chain answers both the governance and the cost questions.

13.5.4 Cost Lineage: INFERENCE_EVENT as a Controllable

The INFERENCE_EVENT row carries one additional set of fields relevant to FP&A: the token counts and the model_version pricing. A foundation model inference through the generative AI hub is billed per input and output token at the provider's published rate for the model_version in question; the INFERENCE_EVENT row records both counts and is the platform's source of truth for the cost the inference incurred. A controller who wants to know what the variance explainer agent cost the organization last quarter queries INFERENCE_EVENT, filters by agent identity and date range, and sums the token-weighted cost. There is no separate cost reconciliation; the same chain that defends the answer attributes the cost of producing it.

The cost attribution becomes more interesting when the resource group dimension is brought in. An INFERENCE_EVENT row carries the resource group of the deployment that served the inference, and the resource group is bound to a business function and a jurisdiction by the resource-group convention outlined in Section 13.1. A controller can therefore break down the AI inference cost by business function, jurisdiction, or agent in a single query that operates on the audit chain rather than on a separate cost system. This breakdown is the kind of report a controller is asked to produce in any close cycle, and the AI tier's contribution to operating expense is presented in the same form as any other contribution. For an organization moving from FP&A into business transformation, the property that AI cost looks like every other cost in the same tools is what makes the AI tier defensible as an investment rather than a black box.

AI Insight: Why Policy Enforcement Belongs at the Gateway

The most consequential architectural choice in this section is that SAP API Policy v.4.2026a Section 2.2.2 is enforced at the SAP BTP API gateway rather than inside each service that the policy might otherwise govern. Gateway enforcement gives the platform a single, queryable, demonstrable point at which the policy is operationally instantiated. An auditor asking to see the policy in operation looks at the gateway, not at hundreds of services that each implement their fragment of the rule. The discipline that follows is that adding a new service to the platform does not require reimplementing the policy; the new service inherits the gateway's enforcement automatically.

From an FP&A and controlling perspective, the structural payoff is that the policy boundary is one row of evidence rather than many. An AUDIT_EVENT(POLICY_REFUSED) row carries the gateway as the actor, the policy identifier (v.4.2026a Section 2.2.2), the attempted endpoint, the agent that attempted the call, and the refusal reason. Querying for POLICY_REFUSED rows produces a forensic view of every attempted violation across the entire AI tier, regardless of which agent attempted what, which service was targeted, or which scope the agent happened to hold for unrelated reasons. The view is the controls evidence that an SOX walkthrough requires, and it is produced by a single query against the audit chain rather than by aggregating evidence across multiple subsystems.

13.6 Summary

This chapter introduced the SAP AI stack as it sits on top of the canonical data spine that we constructed in the preceding chapters. We began with SAP AI Core as the execution runtime, moved to SAP AI Launchpad as the model catalog and governance surface, examined the lifecycle through which a model moves from registered to retired, established Joule as the interaction layer through which humans and agents engage with AI, and closed by tying every AI consumption pathway back to the same governed API surfaces, AUDIT_EVENT chain, and SAP API Policy v.4.2026a Section 2.2.2 that the rest of the platform obeys. The architectural commitment is that AI does not get its own private path to the canonical data spine. It consumes through AI grounding endpoints for evidence, SAP Build Process Automation actions for transactions, and the generative AI hub for foundation model inference. Every one of those calls advances the audit chain in a way that a controller can reconstruct end to end.

We also made the FP&A case for why this architecture matters beyond compliance. INFERENCE_EVENT is a cost driver in its own right, tracked at the same level of granularity as any other operating expense. Lifecycle verdicts are reportable events that close on a defined remediation timeline, grounding bundles carry the version pins that make audit reproduction deterministic, and gateway-enforced policy turns the boundary between SAP and the rest of the AI ecosystem from many small enforcement points into one rule the auditor can read in an afternoon. Together, these properties make the AI tier a system that the controller can defend on the same terms as any other tier of the platform, by pointing at evidence rows in a chain instead of appealing to vendor assurances.

13.6.1 Key Learning Points

- SAP AI Core is the inference runtime on SAP BTP. It expresses workloads through a four-level object model (scenario, executable, deployment, and serving instance) and partitions runtime context across resource groups whose isolation guarantees are enforced by the runtime itself rather than by application-level convention. A deployment manifest pins the model version through content_hash, declares the resource group through which it will serve, and binds an SAP Authorization and Trust Management service scope under which its inferences will execute. Scaling policy is declared rather than imperative, which means that capacity planning becomes a question of policy parameters rather than of operational firefighting.

- SAP AI Launchpad is the catalog of models available for deployment. Models are organized by provider type (SAP-built, partner, and open-source) and every entry carries a set of governance dimensions that the deployment decision must consult: provider identity, modality, license, data-residency constraints, PII handling, and the governance tier through which the model must move. The generative AI hub is the foundation model interface within SAP AI Launchpad that normalizes the provider-specific contracts of GPT, Claude, or Gemini classes as well as open-source large models into a single SAP-mediated invocation surface. All foundation-model calls pass through the same SAP Authorization and Trust Management service-scoped, `AUDIT_EVENT`-instrumented path.
- The AI lifecycle moves a model through five terminal states: `REGISTERED`, `EVALUATED`, `DEPLOYED`, `DEPRECATED`, and `RETIRED`. Each transition is gated, and each gate produces a verdict that is recorded as an `AUDIT_EVENT` row. Continuous evaluation is the discipline that distinguishes the SAP lifecycle from a one-time certification model. The gates do not stop firing once a model is in production, and the same gate that licensed the initial deployment will demote the model if its evaluation metrics drift past the configured thresholds. The lifecycle is a closed loop, not a one-way street, and the controller's evidence of model fitness is the chain of gate verdicts rather than an approval from a single moment in time.
- Joule is the interaction layer of the SAP AI stack. It is the surface through which humans converse with the system in natural language and agents execute on behalf of those humans. Joule does not have direct access to the canonical data spine. It calls AI grounding endpoints for evidence and SAP Build Process Automation actions for transactions, and the agents it hosts are constrained to the tool catalog of those grounding endpoints and SAP Build Process Automation actions. The Joule agent capability matrix (read-only, grounded-write, external-call, composite, and sandbox) defines the boundary of what an agent of each capability tier is permitted to invoke, and the matrix is enforced at the gateway, not at the agent.
- Every AI consumption pathway lands on the same governed API surfaces that we cataloged in Chapter 10. Grounding endpoints assemble version-pinned, role-scoped bundles from the data spine. SAP Build Process Automation actions invoke governed transactions whose preconditions are checked against `PUBLISH_STATE`, `VALIDATION_RESULT`, and the consumer's SAP Authorization and Trust Management service scope. Foundation model calls pass through the generative AI hub and are recorded with the prompt fingerprint, the grounding bundle reference, the model and version, the token counts, and the response checksum. Direct calls from AI into the Access API are prohibited by SAP API Policy v.4.2026a Section 2.2.2, and the prohibition is enforced at the gateway through scope examination rather than at the agent through trust.
- `INFERENCE_EVENT` is the audit row that ties an AI invocation back to the rest of the audit chain. It joins to `AUDIT_EVENT` through actor identity, which means that a query that

reconstructs the chain for any business event will also reconstruct the AI events that influenced it. From an FP&A perspective, this property is what makes the AI tier defensible on the same terms as the rest of the platform. The variance the controller is asked to explain has a chain of evidence behind it that includes the model that wrote the explanation, the grounding bundle the model consulted, and the policy verdicts that licensed both. The chain is one query, not a federation across systems.

- The architectural payoff of the design is that the AI tier inherits the audit, identity, and policy properties of the rest of the platform rather than building its own. There is no parallel identity model for AI consumers, no parallel audit chain for inference events, and no parallel policy engine for AI-specific rules. The boundary between AI and the canonical data spine is a thin one (made up of the AI grounding endpoints, the SAP Build Process Automation action library, and the generative AI hub), and that thinness is what makes the platform's evidence story coherent regardless of which consumer is asking. Joule and a controller's spreadsheet both consume the same APIs, produce the same audit rows, and are governed by the same policy.

13.6.2 Common Design Pitfalls

- Letting Joule call the Access API directly. This anti-pattern attempts to use the Access API as a shortcut around the AI grounding endpoints on the reasoning that the Access API already returns the data the agent needs. The shortcut is precisely the construction that SAP API Policy v.4.2026a Section 2.2.2 prohibits, because the Access API returns raw spine reads without the version pin, role scope, and provenance metadata that a grounding bundle carries. The implementation effect is that the agent's evidence cannot be reproduced. The controller reading an AI-generated explanation cannot replay the same evidence the agent saw, because the Access API does not bind its responses to a reproducible read instant. The gateway refuses these calls and writes the refusal to the audit chain, which is the correct enforcement point, but teams that interpret the refusal as a friction to be worked around rather than a property to be respected will design themselves into a position from which they cannot defend their AI tier in an audit.
- Caching grounding bundles across roles. The temptation to cache a grounding bundle once and reuse it across multiple consumers is real, because grounding bundle assembly is the most expensive step in the inference pipeline and a cache hit eliminates the data spine read. This temptation is also wrong, because the grounding bundle is bound to the requesting actor's role scope. Reusing it for a different actor produces a bundle whose contents the new actor would not have been authorized to read on a fresh request. The audit row that the cached call writes will name the new actor, but the evidence the new actor saw will reflect the cache's original scope. This is the structural break that makes role-based access ineffective at the AI tier, and it is the kind of optimization that looks like performance engineering until the auditor asks why two agents with different scopes saw the same evidence, and the team has no answer.

- Treating AI evaluation as a one-time gate rather than a continuous lifecycle. A model that passes the `EVALUATED` gate is not certified for all time. The system the model serves changes—vocabularies shift, business rules evolve, prompt patterns drift, and the metrics that the gate measured at promotion time can decay below the configured thresholds without any change to the model itself. Teams that treat the gate as a one-time event will run models long past their effective expiry, and the `AUDIT_EVENT(GATE_PASSED)` row that licensed the deployment will be months or years old by the time the model produces its first defensibly wrong output. Continuous evaluation, with its scheduled reruns and its ability to trigger automatic demotion to `DEPRECATED`, is the discipline that prevents this drift, and it is the property that an audit will expect to find when it asks how the team knows its models are still fit for purpose.
- Treating the catalog as documentation rather than as control. The SAP AI Launchpad catalog is not a list of models the team finds interesting; it is the set of models that may be deployed under SAP-mediated governance. Adding a model to the catalog is a governance act, with provider identity verified, license terms recorded, data residency posture confirmed, and PII handling examined. Teams that treat the catalog as a wiki or a place to write down what is available will accumulate entries that have not passed the governance examination, and the deployment decision will appear to be informed by the catalog while in fact relying on entries the catalog's discipline has not vetted. The catalog must be the gate, and entries that have not passed the gate must not appear in it. Anything else turns the catalog into a source of false confidence.
- Permitting capability inflation across Joule agents. The Joule capability matrix exists to keep each agent at the lowest tier sufficient for its intent. The temptation is to promote a read-only agent to grounded-write because a single edge case requires a write or to promote a grounded-write agent to external-call because a single use case requires an outbound call. Each promotion enlarges the scope under which the agent operates and, with it, the scope of the audit trail the agent will generate. The right pattern is to compose two agents at the lower tiers rather than promote one agent to a higher tier, because composition preserves the audit boundary between the two operations whereas promotion merges them. The discipline matters more in retrospect than in the moment. An agent's capability tier is the first attribute an audit will ask about, and an agent whose tier was promoted for convenience is the kind of finding that surfaces months after the convenience has been forgotten.
- Reasoning about model cost without reasoning about inference frequency. The cost of an AI tier is the product of per-call cost and call volume, and teams that focus on per-call cost, typically by negotiating the foundation-model contract, without also instrumenting call volume will be surprised by the bill that arrives. `INFERENCE_EVENT` records the volume side of the equation at the same granularity as any other operational telemetry, which means that the controller has the data to attribute inference cost to the agent, the resource group, the scenario, and ultimately the business process. Teams that ignore this data will find that their AI tier's cost curve is unattributable, and they will be unable

to answer the FP&A question of which process is driving the spend. The controllable is the call volume, not just the call price, and it is visible through INFERENCE_EVENT.

- Building AI agents that reach outside SAP-endorsed integration paths. The temptation here is to write a Joule agent that calls an arbitrary external service (a third-party data feed, a non-SAP API, or a public web endpoint) on the grounds that the agent needs the data and the data is not in the data spine. SAP API Policy v.4.2026a Section 2.2.2 governs the boundary by requiring that external calls go through SAP-endorsed integration architectures, which in practice means SAP Build Process Automation external connectors or SAP Integration Suite flows that carry their own audit and policy properties. Agents that bypass this discipline produce inference outputs whose evidence chain has a gap precisely at the external call, and that gap is the kind that an audit cannot let pass. The agent will need to be rewritten to consume through the endorsed path before the deployment can be defended.

13.6.3 Review Questions

1. Describe the four-level object model that SAP AI Core uses to express an inference workload. What is the role of the resource group, and what isolation property does the runtime guarantee at the resource-group boundary?
2. The SAP AI Launchpad catalog records six governance dimensions for every model. Name them and explain why each is consulted at deployment time rather than at registration time.
3. Walk through the five lifecycle states a model can occupy. For each transition between states, identify the gate that licenses the transition and the AUDIT_EVENT verdict that the gate produces.
4. Why does the lifecycle treat evaluation as continuous rather than as a one-time gate? Describe a scenario in which a model that passed its initial EVALUATED gate is later demoted to DEPRECATED by a continuous rerun, and explain the AUDIT_EVENT chain the demotion produces.
5. The Joule capability matrix defines five capability tiers. For each tier, describe the scope the agent operates under, the kinds of tools the agent may invoke, and whether a human validation step is required before the agent's output produces a side-effect on the data spine.
6. SAP API Policy v.4.2026a Section 2.2.2 prohibits a specific class of AI consumption pathway. Identify the prohibited pathway, explain the gateway mechanism by which the prohibition is enforced, and describe the AUDIT_EVENT row that a refused call produces.
7. INFERENCE_EVENT joins to AUDIT_EVENT through actor identity. Construct an example query that reconstructs the end-to-end chain for an FP&A variance explanation, beginning with the controller's prompt and ending with the foundation model response. Identify each audit row that the query would return and the join key that connects it to the next row.

8. The chapter argues that grounding bundles must not be cached across roles. Explain the structural property of the grounding bundle that makes cross-role caching unsafe and describe the alternative caching pattern that preserves audit semantics while still minimizing the data spine–read cost of bundle assembly.
9. From the controller's perspective, `INFERENCE_EVENT` is a cost driver as well as an audit row. Describe how a controller would use `INFERENCE_EVENT` to attribute inference spend to a business process, and identify the three dimensions of attribution that the row supports.
10. This chapter closes by noting that the AI tier inherits the platform's audit, identity, and policy properties rather than building its own. Explain the architectural significance of this inheritance and contrast it with the alternative in which the AI tier maintains its own parallel control plane. What would be lost under the alternative, and what would the auditor see differently?

13.6.4 Student Companion Notes

The AI stack on the canonical data spine is the place at which our architectural argument earns its most demanding test. The previous chapters constructed a data spine whose properties (version pinning, role scoping, hash-chained audit, and gateway-enforced policy) were defined with human and conventional-software consumers in mind. In this chapter, we state that the same properties hold without modification when the consumer is an AI model, because the architectural commitment of the SAP design is that the consumer's identity does not change the discipline by which it consumes. If you have internalized this point, you have the framing you need for the remaining chapters; if you have not, later chapters will feel surprising in places they otherwise would not.

The introduction of `INFERENCE_EVENT` is the canonical entity that the chapter contributes to the data model. It is small in structure (consisting of the actor identity, model and version, grounding-bundle reference, prompt and response fingerprints, token counts, latency, and gateway verdict) but large in consequence, because it is the row that makes the AI tier auditable on the same terms as the rest of the platform. You should treat `INFERENCE_EVENT` as you treated `AUDIT_EVENT` in Chapter 9: as the entity that makes the rest of the architecture defensible, and as the first place to look when a question is asked about what the AI tier did and why.

If you are approaching this chapter from an FP&A and controlling background, you will find two structural payoffs that are worth holding onto. The first is that AI inference is a cost driver tracked at the same level of granularity as any other operational cost, which means that the AI tier can be reported on, budgeted for, and held to a variance discipline in exactly the way the rest of the cost base can be. The second is that the AI tier's evidence chain joins to the rest of the platform's evidence chain through a single key, which means that the controller defending a reported number does not have to reach across a federation of systems to assemble the evidence that the number is reproducible. Both payoffs depend on the discipline this chapter describes, and both fail quietly if it is allowed to slip.

Chapter 14 turns from the AI stack itself to the data that flows through it. We introduced AI grounding endpoints in this chapter as the consumer contract that AI inference must observe, but the contract is only as strong as the data that fills it. In Chapter 14, we specify the data preparation rules (eligibility, role scoping, grounding bundle assembly, and the discipline by which the platform prevents hallucination) that make the contract enforceable. It is the bridge between this chapter's architectural claims and the operational reality of running the AI tier, and it is the place where you will see why grounding bundles are the canonical unit of evidence rather than an implementation detail.

Chapter 14
Preparing Data for SAP AI

Enterprise data does not arrive at the AI tier in a form fit for AI consumption. Eligibility, role-based scoping, version control, grounding constraints, and the prevention of hallucination and drift are the disciplines that prepare it, and that discipline sits on the platform side of the boundary.

Chapter 13 introduced the SAP AI architecture and the consumer contract through which the AI tier reaches enterprise data. The contract is only as strong as the data that fills it. This chapter specifies the rules under which enterprise data is permitted to flow into SAP AI: what is eligible for AI consumption, how filtering and role-based scoping are applied at consumption time, how version control and approvals govern what reaches the model, what grounding rules and constraints bind the inference to evidence the platform recorded, and how hallucination and drift are prevented as structural properties of the platform rather than as algorithmic afterthoughts.

The structural commitment of this chapter is that data preparation for AI consumption is platform-side, not consumer-side. The Joule tool catalog introduced in Chapter 13 does not include filters, meaning the consumer cannot relax eligibility, widen role scope, reach behind a version pin, or fabricate evidence for a claim that has none. From the controller's point of view, the thinness of the consumer's surface is what makes the AI tier defensible: Every property the consumer might be tempted to manipulate is fixed by the platform before the model sees the prepared data, and every refusal lands in the same `AUDIT_EVENT` chain that defends the rest of the reported numbers.

This chapter takes the SAP AI architecture of Chapter 13 and makes the data flowing through it defensible on the same terms as the rest of the data spine. It assumes the canonical data model of Chapter 9 and Chapter10, the ingestion discipline of Chapter 11, and the execution catalog of Chapter 12. It looks ahead to Chapter 15, which covers process suggestions and the mandatory human validation that closes the loop on any AI-influenced state change; Chapter 18, which returns to the operational view of the AI tier; and Chapter 19, which closes the audit story with the full instrumentation that reconstructs an AI-influenced number from the data spine weeks after the fact.

Five principles run through the chapter and should be used in every preparation decision. First, every row of enterprise data is ineligible for AI consumption until a deliberate classification act promotes it, and the default-deny posture is the architectural correction of the permissive defaults earlier generations of customers wrote for analytical consumption. Second, filtering and role-based scoping are enforced at the platform tier under the requesting actor's role context, and the data set that reaches the model is never reused across actors. Third, every prepared data set is pinned to a specific PROCESS_VERSION and read at a single AS-OF instant of spine state, with approvals recorded as audit rows. Fourth, the AI consumption is bounded by grounding rules that name every source row the model received and require every factual claim in the response to cite a named row. Fifth, hallucination and drift are prevented as platform properties rather than as model behaviors, through structural constraints the platform enforces before, during, and after inference.

By the end of the chapter, every data set the AI tier consumes should be visibly the product of a platform discipline rather than the product of consumer behavior. If you're approaching this chapter from a financial planning and analysis (FP&A) perspective, you should see the five disciplines as evidence rows you can point to on the day an external auditor walks in.

14.1 Data Eligibility for AI Consumption

Data eligibility is the prerequisite for every other discipline in this chapter. A row that is INELIGIBLE for AI consumption cannot be filtered into a role-scoped data set, the subject of a grounding rule, or the source of a citation in an AI response. It does not enter the consumption pipeline at all. The idea is structurally simple but hard to execute, and it is anchored in the single eligibility policy that this section introduces.

14.1.1 The Eligibility Policy

The eligibility policy is the structured artifact that names an eligibility decision and joins to the data spine rows the decision applies to. Its substantive content is expressed through six fields. The policy_id is the primary key, referenced by foreign keys from PROCESS_VERSION, ACTIVITY, and MASTER_DATA_REFERENCE so that a row in any of those entities is permitted into an AI consumption path only if its referenced policy carries an ai_eligible_flag whose value resolves to TRUE for the consumer's actor context. The jurisdiction field ties the policy to the controlling jurisdiction of the data—EU, US, Asia-Pacific (APAC), or one of the more granular per-country values where the regulatory regime requires it. The sensitivity_class field places the data in one of three classes the rest of this section defines. The retention_window field expresses, in days, how long the row may remain eligible before the policy must be reevaluated. The ai_eligible_flag carries the runtime value the consumption gate reads. The last_review timestamp records when the policy was most recently affirmed.

The structural choice to make the eligibility policy a first-class artifact rather than a column on each consuming entity is deliberate. Policies change more slowly than the rows they govern. They are written by a smaller community than the community that writes the rows and require their own audit discipline. A finance-grade master data row whose policy changed from CONDITIONALLY_ELIGIBLE to INELIGIBLE because the underlying retention rule shifted—a quarter shorter, say, because a regulatory regime tightened—produces a single ELIGIBILITY_REVOKED audit row, joinable to every spine row that referenced the policy. The alternative, in which every consuming row carried its own eligibility column, would require either a costly bulk update or a lookup-time evaluation that defeated the purpose of having a column at all. The policy as a first-class artifact is the right shape for the operational reality.

14.1.2 Three Eligibility Classes

Eligibility has the following three classes:

- FREELY_ELIGIBLE rows may enter any AI consumption path for which they are otherwise in scope. Examples include public process documentation, published reference catalogs, and nonsensitive master data such as the chart of accounts as published in the corporate domain. Rows in this class do not need to be reevaluated at consumption time; the eligibility check is a row lookup, and a passing row falls straight through to the filtering and scoping stage.
- CONDITIONALLY_ELIGIBLE rows may enter an AI consumption path only when the conditions named on their eligibility policy hold. Conditions are expressed declaratively rather than as code, and the vocabulary is intentionally narrow: jurisdiction match (the requesting actor's role context must include the row's controlling jurisdiction), redaction required (a specific field on the row must be masked before the row enters the prepared data set), retention window not expired (the row's last update must fall within the policy's retention_window), and consent flag set (a referenced MASTER_DATA_REFERENCE row must carry an explicit consent attestation). A row whose conditions cannot be checked at consumption time is treated as INELIGIBLE; the platform never assumes a condition holds merely because the row arrived in the candidate set.
- INELIGIBLE rows may not enter any AI consumption path, regardless of the requesting actor's role context or the scenario's claimed need. The class includes personally identifiable information (PII) without a consent flag, contracts under litigation hold and therefore subject to legal counsel's approval before any AI consumption, regulated material outside its retention window, and any row whose eligibility policy has been revoked since its last review. The class is the discipline's safety net. A row in this class cannot be drawn into an AI consumption path by any combination of legitimate or illegitimate actor activity, and the platform's refusal to consume such a row is the audit row that demonstrates the discipline was applied.

14.1.3 The Ineligibility Default

Every new row of source data is INELIGIBLE for AI consumption until a deliberate classification act promotes it. The choice of default is the most consequential property of the eligibility discipline, and the section argues at some length that the inversion (default-deny) is the correct default and that the alternative is structurally untenable. An earlier generation of SAP customers, writing policies for analytical consumption rather than AI consumption, chose a permissive default: Data was eligible for analytical consumption unless excluded. The choice was justified at the time on the grounds that exclusions were rare and that requiring explicit inclusion would slow the analytical roadmap. The cost of the choice did not surface until the AI tier was introduced, at which point the permissive default had to be inverted row by row and policy by policy under conditions of regulatory pressure that did not exist when the default was originally chosen. The lesson the SAP AI tier has taken from that experience is that defaults are written when the consequences of being wrong are easy to imagine, and a default that fits analytical consumption fits AI consumption only by accident.

The default-deny posture has a second virtue beyond regulatory survivability: It makes eligibility visible. A team that must promote rows from `INELIGIBLE` to one of the other two classes is a team that has to think about which rows are AI-eligible and why, which prompts them to generate documentation an auditor will eventually ask for. A permissive default creates no such evidence because the default itself becomes the documentation, and that is exactly the kind of evidence an auditor cannot accept. From the controller's point of view, the default-deny posture is what turns eligibility from a vendor-assurance question into an evidence question, and the evidence is the `AUDIT_EVENT(ELIGIBILITY_CLASSIFIED)` row that the classifier wrote on each promotion.

14.1.4 Review Cadence and Reevaluation

Eligibility is not a one-time classification act. The policy that governs an eligible row must be reevaluated on a cadence proportional to the row's underlying volatility. `PROCESS` and `ACTIVITY` rows are reviewed annually because the substantive content of a process changes slowly and the review is principally a confirmation exercise. `MASTER_DATA_REFERENCE` rows are reviewed quarterly because the regulatory regime governing master data changes more frequently, and this is the slowest cadence that catches material regime changes before they reach an external audit. `INTERFACE` rows are reviewed annually as well as on contract amendment because an interface's eligibility is bounded by the data-processing agreement the controlling jurisdiction signed with the consuming party. `AUDIT_EVENT` and operational telemetry are reviewed annually and are only rarely promoted out of the `INELIGIBLE` default.

The reevaluation exercise is itself recorded in the audit chain. Each `policy_id` carries a `last_review` timestamp that the platform updates on every reevaluation, and the reevaluation produces an `AUDIT_EVENT(ELIGIBILITY_REAFFIRMED)` row carrying the reviewing authority, the date, and the outcome (reaffirmed, reclassified, or demoted). The platform treats a `policy_id` whose `last_review` is older than its cadence interval as expired. The rows referencing the policy are not automatically demoted to `INELIGIBLE`, but the platform writes a warning to the AI governance team's queue, and an external auditor reviewing the policy will read the stale `last_review` as a control weakness. The discipline of refreshing `last_review` on cadence is therefore not optional, and the governance team treats the expiry queue as a high-priority operational concern requiring prompt review. From the controller's point of view, the discipline is what lets the AI tier's eligibility classifications be defended as current rather than as historical—the `last_review` timestamp is the evidence the policy is still authoritative at the moment of consumption.

Figure 14.1 makes the state machine concrete. The default state `INELIGIBLE` sits on the left; promotion to `CONDITIONALLY_ELIGIBLE` or `FREELY_ELIGIBLE` is a forward transition recorded as `AUDIT_EVENT(ELIGIBILITY_CLASSIFIED)` with the classifier's identity, the `policy_id`, and the chosen class. Demotion in the reverse direction (from either eligible class back to `INELIGIBLE`) is the `AUDIT_EVENT(ELIGIBILITY_REVOKED)` event, which occurs when the jurisdiction the row sits in changes, the retention window expires, or the `sensitivity_class` is raised because of a downstream policy change. A third event type, `AUDIT_EVENT(ELIGIBILITY_REFUSED)`, is triggered by the platform itself when a candidate row that the consumption pipeline considered fails the eligibility check at runtime. This third event type is the runtime correlate of the classifier-side events and is the row an auditor reads when asking why a particular data set did not include a particular row.

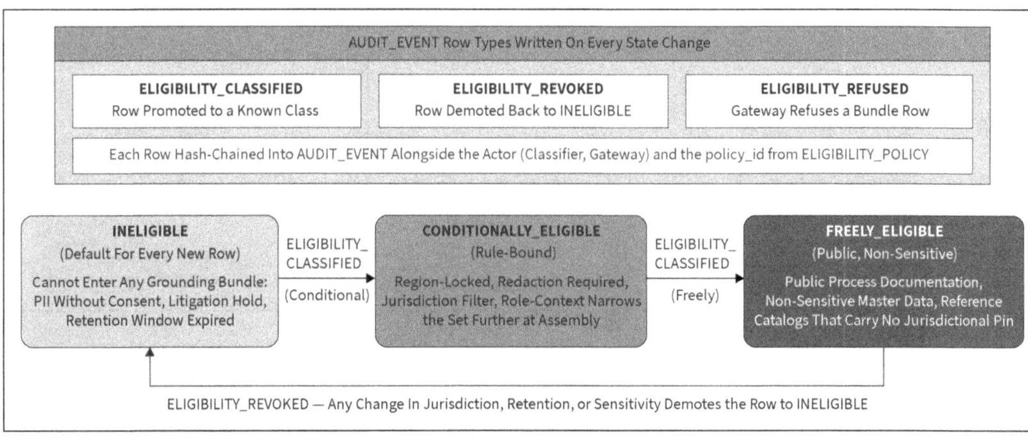

Figure 14.1: Eligibility State Diagram

14.1.5 The Eligibility Decision Matrix

Table 14.1 lists the decision matrix that every spine data source observes at eligibility classification time. The rows are the data sources, the columns are the four properties of the decision (default classification, classification authority, review cadence, and `AUDIT_EVENT` type on change), and the matrix is the structural reference a governance officer consults when writing a new eligibility policy entry. Every spine data source carries `INELIGIBLE` as its default classification. The classification authority differs by source, the review cadence is set by the source's underlying volatility, and every change writes one of two `AUDIT_EVENT` types. The platform's `ELIGIBILITY_REFUSED` event is independent of these and fires at consumption time when a row's classification cannot be honored for the scenario.

Data Source	Default Classification	Classification Authority	Review Cadence	AUDIT_EVENT Type on Change
PROCESS	INELIGIBLE	Process owner with governance sign-off	Annual or on republication	ELIGIBILITY_ CLASSIFIED/ ELIGIBILITY_ REVOKED
ACTIVITY	INELIGIBLE	Process owner inheriting from parent PROCESS	Annual or on parent change	ELIGIBILITY_ CLASSIFIED/ ELIGIBILITY_ REVOKED
MASTER_DATA_ REFERENCE	INELIGIBLE	Master data steward for the controlling jurisdiction	Quarterly or on regulatory change	ELIGIBILITY_ CLASSIFIED/ ELIGIBILITY_ REVOKED
INTERFACE	INELIGIBLE	Integration architect with security review	Annual or on contract amendment	ELIGIBILITY_ CLASSIFIED/ ELIGIBILITY_ REVOKED
AUDIT_EVENT	INELIGIBLE	Audit governance lead; rare promotion for control-evidence subsets	Annual	ELIGIBILITY_ CLASSIFIED/ ELIGIBILITY_ REVOKED
Operational telemetry	INELIGIBLE	Audit governance lead jointly with AI governance	Annual	ELIGIBILITY_ CLASSIFIED/ ELIGIBILITY_ REVOKED

Table 14.1: Eligibility Decision Matrix

The decision matrix exposes the breadth of the discipline. Every data source the data spine exposes is subject to it; there is no carve-out for internal sources or sources that the team considers low-risk. The classification authority differs by source because the substantive judgment differs. A master-data steward's judgment about a cost center row carrying PII is not the same as an audit governance lead's judgment about whether a subset of AUDIT_EVENT itself is AI-eligible, but the procedural shape is uniform. A row's eligibility moves only when an authoritative party writes an audited classification event.

14.1.6 Worked Example: A German Payroll Master-Data Row

Consider a `MASTER_DATA_REFERENCE` row representing a German payroll cost center with the identifier md-payroll-DE-CC-2057, attached to the eligibility policy pol-de-payroll-pii. The policy's jurisdiction is DE, its `sensitivity_class` is PII, its `retention_window` is 90 days, its `ai_eligible_flag` resolves to `TRUE` only for actors whose role context includes the controlling jurisdiction DE, and its `last_review` is was February 14, 2026. A Joule agent invoked by a controller whose role context is fp&a-eu-controller requests a variance-explanation against this row, and the consumption pipeline considers md-payroll-DE-CC-2057 as a candidate.

The eligibility check fires before the filtering and scoping check. The policy's `ai_eligible_flag` is evaluated against the actor's role context: fp&a-eu-controller is a composite scope that includes EU-wide read access but does not narrow to the controlling jurisdiction DE. The flag therefore resolves to `FALSE`, and the row fails the eligibility check. The platform writes an `AUDIT_EVENT(ELIGIBILITY_REFUSED)` row carrying the data set identifier under preparation, the actor identity, the `policy_id` pol-de-payroll-pii, the row primary key md-payroll-DE-CC-2057, and the refusal reason (`jurisdiction-mismatch-on-pii-row`). The prepared data set that is eventually emitted does not contain this row, the controller's variance explanation does not name this row, and the audit chain records the refusal as evidence the discipline was applied.

The same controller, asking the same question with a different scope —fp&a-de-controller, the scope that narrows to Germany—would produce a different prepared data set, because the eligibility check would resolve `TRUE` for the DE-controlling-jurisdiction case. The row would enter the candidate set, the preparation pipeline would proceed to the filtering and scoping check, and the resulting data set would include the row. From the controller's point of view, the gating discipline is visible at the granularity of the scope. An EU-wide question gets an EU-wide data set that excludes per-country PII; a DE-specific question with DE-specific scope gets a data set that includes it. The discipline is auditable in both cases.

AI Insight: The Cost of an Inversed Default

A permissive default for eligibility (meaning rows are eligible unless excluded) is the kind of choice that looks reasonable on the day it is made but is indefensible on the day an auditor walks in. The architect who chose the default did so when AI consumption was not yet a category and when the regulatory regime did not yet treat AI consumption of personal data as a separate concern. The controller who inherits the consequences is the one who must defend it. A permissive default exposes the controller because, at any moment, the question "What data did the AI see?" has the answer "All of it, minus the exclusions you wrote," and those exclusions cannot be guaranteed to be complete because they were never required to be.

The default-deny posture puts the discipline on the right side of the FP&A defense. Every row the AI consumes was promoted by an authoritative classifier, the promotion is in the audit chain, the promotion is queryable by `policy_id`, and the answer to "What data did the AI see?" is "The rows whose eligibility policies were promoted, and only those rows." The defense is one query against the data spine, not a survey of every team that ever wrote a configuration. From an FP&A perspective, the inversion is the property that makes the AI tier reportable on the same terms as the rest of the cost base.

14.2 Filtering and Role-Based Scoping

Filtering and role-based scoping is the next topic we'll discuss. Eligibility determines what can be shown to AI at all, while filtering and scoping determine who can see a given eligible row through an AI consumption path. The distinction matters because eligibility is a property of the row and the policy, while scope is a property of the requesting actor and the prepared data set. A row that is `FREELY_ELIGIBLE` can still be out of scope for a particular actor's request. A row that is `CONDITIONALLY_ELIGIBLE` for one role can be in scope for one actor and out of scope for another holding a narrower variant of the same role. The platform resolves the question per actor and per request every time, and the resolution is what the audit row records.

In the following sections, we will discuss why it is important to never reuse data sets across actors, the role-based scoping assembly algorithm, composite roles, and why it is better to have a refusal rather than a partial return.

14.2.1 Why Prepared Data Sets Are Never Reused Across Actors

Role-based scoping is enforced through SAP BTP's authorization service. The bearer token the consumer supplies carries a principal claim (the actor identity) and a scope set that enumerates the consumer's read authorities. The prepared data set is assembled under that scope set, and the scope set is stamped on the data set's metadata under a field named `role_scope_intersect`. A prepared data set carries the actor's role context as a first-class property, and the implication is that two actors with different scope sets receive different prepared data sets for the same logical question.

An anti-pattern frequently introduced as performance engineering is caching a prepared data set and reusing it across actors. This looks reasonable at first glance: Data preparation is the most expensive step in the inference pipeline, and a cache hit avoids the data spine read entirely. However, this is the wrong approach for a structural reason. The platform must foreclose at preparation time rather than diagnose afterwards. A prepared data set carries its `role_scope_intersect` field, and the data set was assembled under the assumption that its consumer's role context matches that intersect. A second consumer whose role context does not match the intersect would see rows that consumer was not authorized to read on a fresh request. The audit row that the cached call writes would name the second consumer correctly, but the evidence the second consumer saw would reflect the first consumer's scope. The audit chain would lie about what the second consumer was shown.

The platform avoids this anti-pattern by making data preparation mandatory on every request, regardless of how recently the same logical question was answered. The performance cost is real and is covered in Chapter 18, where the operational view of the AI tier describes the cache observability metrics and the runbook for high-frequency requests. The structural guarantee, however, is preserved. Every prepared data set is the product of an assembly act under the requesting actor's role context, and no actor ever receives a data set assembled under a different actor's context. From the controller's point of view, the guarantee is what makes the audit chain a faithful record of what each actor saw.

14.2.2 The Role-Scoped Assembly Algorithm

Listing 14.1 presents the assembly algorithm in pseudocode, offered at a level of detail sufficient for an engineer to implement the platform behavior and for a governance officer to read the implementation as a control. The five stages are linear, and each stage has a single outcome: Either the candidate set passes and assembly proceeds, or the candidate set fails and the prepared data set is refused entirely.

The function validates the actor token, resolves the role context (composite roles produce evidence subsets rather than a union), filters every candidate row through the role's read authorities, attaches the role identity and `AS-OF` pin to the data set metadata, and emits the data set only if every required row is in scope. Any row that fails the scope check causes the entire data set to be refused, with `AUDIT_EVENT(SCOPE_REFUSED)` naming the row.

```
function assembleRoleScopedDataSet(request, bearerToken):
    // Stage 1: validate the actor token
    actor = btpAuth.validate(bearerToken)
    if actor is null:
        emitAuditEvent(AUTH_REFUSED, request)
        refuse(reason: invalid_token)
        return

    // Stage 2: resolve the role context
    roleContext = btpAuth.resolveScopes(actor)
    // composite roles produce evidence subsets, not unions
    subsets = roleContext.partitionIntoEvidenceSubsets()

    // Stage 3: filter candidate rows through role scope
    candidates = spine.candidatesForScenario(request.scenario)
    dataSetRows = empty
```

```
for row in candidates:
    // Eligibility was checked upstream of this function
    if not roleContext.permitsRead(row):
        emitAuditEvent(SCOPE_REFUSED,
                       dataset_id, actor, row.pk)
        refuseDataSet(reason: row_out_of_scope, row: row.pk)
        return
    dataSetRows.append(row)

// Stage 4: attach role identity to data set metadata
dataSet = newDataSet(rows: dataSetRows)
dataSet.role_scope_intersect = subsets
dataSet.actor_principal = actor
dataSet.as_of_pin = spine.currentInstant()

// Stage 5: emit the data set or refuse
if any required row missing from dataSetRows:
    emitAuditEvent(SCOPE_REFUSED, dataset_id, actor, missing)
    refuseDataSet(reason: required_row_missing)
    return
emitAuditEvent(DATASET_ASSEMBLED, dataset_id, actor)
return dataSet
```

Listing 14.1: Role-Scoped Data Set Assembly

Figure 14.2 expands the pipeline visually. The five stages are arranged left to right, the two outcomes branch from stage 5, and the composite-role case is highlighted in the bottom strip because it is the part of the algorithm most often misunderstood.

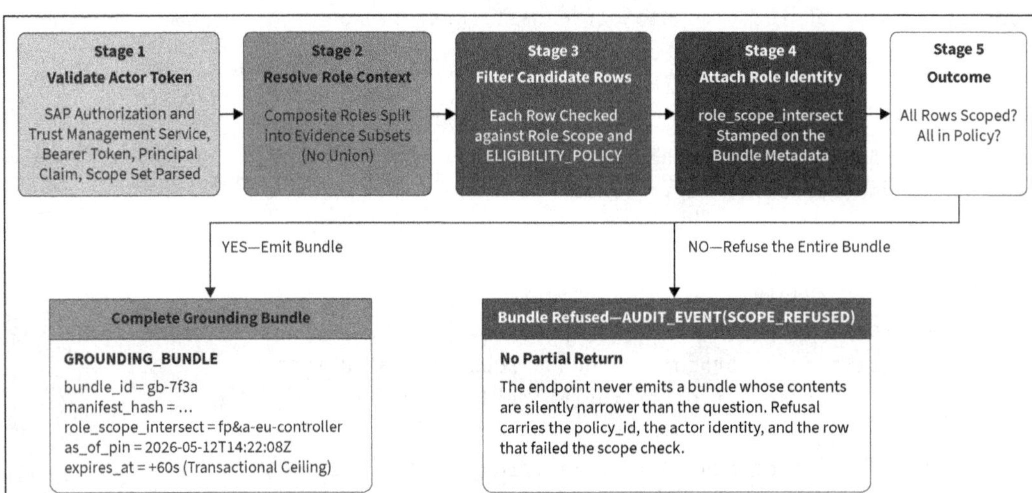

Figure 14.2: Role-Scoped Assembly Pipeline

Two outcomes shown explicitly: a complete data set stamped with the actor's role context or a refused data set with `AUDIT_EVENT(SCOPE_REFUSED)` and no silent partial return. A composite role is not a union of its components but a list of evidence subsets. We discuss the composite-role case in Section 14.2.3 because the discipline it expresses is the one that holds the SAP AI tier together for multinational controllers.

14.2.3 The Composite-Role Case

Consider a controller whose role context includes both fp&a-eu-controller and fp&a-na-controller. The controller is responsible for variance analysis across both jurisdictions, and the Joule agent serving the controller routinely receives questions whose scope spans both. A naive assembly logic would take the union of the two scopes (the rows readable under either one) and assemble a single data set from that union. The data set would be answerable, the model would produce a variance explanation, and the controller would receive the answer. The audit row would name the controller and the data set, and an external auditor reading the chain would have no way to know whether the answer's evidence subset for the EU sub-question was the EU subset or the North American (NA subset.

The SAP AI tier's assembly logic does not take the union. It produces a list of evidence subsets—in this case, two subsets, one per component role—and the downstream response must cite which subset answered which sub-question. The prepared data set carries the two subsets explicitly: rows tagged with the EU subset are answerable for EU sub-questions, rows tagged with the NA subset are answerable for NA sub-questions, and the variance explainer's response carries inline citations of the form `[ds-7f3a/row-12 (eu)]` or `[ds-7f3a/row-04 (na)]` that name the subset. The audit row carries the subset count and the per-subset row counts, so that the auditor reading the chain can ask the variance-explainer answer's questions one sub-question at a time without losing evidence boundaries.

The composite-role case is the clearest example of why the SAP AI tier insists on the discipline rather than the convenience. A union would be easier to assemble, cache, and explain to a developer reading the API for the first time. The union is wrong because it erases the evidence boundary that the multinational controller relies on when an auditor in one jurisdiction asks for the evidence subset relevant to that jurisdiction. This discipline preserves the boundary, while convenience would have erased it. The SAP AI tier chose the discipline, and that choice is also the architectural reason it could not have chosen otherwise.

A second example helps make this discipline concrete in a different jurisdictional shape. Consider a controller responsible for both group-level consolidation and country-level review for Germany, holding composite roles `group-consolidator` and `de-country-reviewer`. The composite is not a simple union of EU-wide and country-narrow scopes; `group-consolidator` brings the right to read consolidated entries across all jurisdictions but with PII redacted, while `de-country-reviewer` brings the right to read PII-bearing rows specifically for Germany. A single union scope would expose group-level numbers that include Germany-specific PII to the controller's queries about other countries, which is a privacy violation the role design is meant to prevent. The evidence subset assembly produces two distinct subsets: one for the `group-consolidator` scope (consolidated numbers with PII redacted, applicable to all jurisdictions) and one for the `de-country-reviewer` scope (PII bearing rows specifically for Germany). The variance explanation response carries citations tagged to the appropriate subset, and the controller's response to an auditor's question about the basis for any specific claim names the subset under which the claim was answered.

These two examples illustrate that the evidence subset discipline is not a narrow technical optimization but the architectural pattern that lets multinational and multi-role controllers consume AI responses without inheriting the privacy and audit complications that composite roles would otherwise produce. This discipline scales naturally to roles of any complexity—a controller holding four or five component roles produces four or five evidence subsets, each citation in the response names a specific subset, and the audit chain records the subset count and per-subset row counts at the level of the prepared data set. From an FP&A point of view, this discipline is what makes the AI tier deployable in multinational FP&A organizations rather than just in single jurisdiction prototypes.

14.2.4 Refusal Rather Than Partial Return

The pipeline emits a complete data set or no data set at all. A data set whose assembly encountered a row that failed the scope check is refused entirely, and the refusal is recorded as `AUDIT_EVENT(SCOPE_REFUSED)` naming the row that failed. The discipline is nonnegotiable. A partial-return pattern would assemble whatever rows passed the scope check and emit a data set containing only those rows. The argument for this pattern is that a partial data set is more useful than no data set, and the AI consumer can be told the data set is partial. The argument fails for the same structural reason that silent exclusion fails everywhere it appears in the SAP architecture: The consumer cannot tell what is missing from the data set's contents.

The model that consumes a partial data set will produce a confident answer based on the rows it received, and the controller defending the answer will not know that the answer was assembled from an evidence subset that the platform considered incomplete. The audit chain would correctly record that the data set was partial, but the controller would have to read the audit chain to know that the answer was unsafe, and the auditor would have to know to ask for it. The chain of inference between partial data set and unsafe answer is too long to be defensible, and the SAP AI tier refuses the data set outright to keep the chain short. From the controller's point of view, the refusal is the property that lets them treat a successfully returned data set as evidence of a question that was answerable in full.

AI Insight: Why Assembly Refuses Partial Returns

The discipline that prepared data sets are returned in full or not at all is the Sarbanes-Oxley Act (SOX) and SOC2 reasoning carried into the AI tier. A partial data set is the

AI-tier equivalent of a reported financial statement that omits material balances on the grounds that the auditor's questions about those balances could not be answered cleanly. The omission would not be tolerated in a financial report; it is not tolerated in a prepared data set. The platform refuses the data set outright so that the inability to answer is itself the evidence row, named in AUDIT_EVENT(SCOPE_REFUSED) with the actor identity, the data set identifier under assembly, and the specific row that failed the scope check.

From an FP&A perspective, the practical consequence is that a returned data set is a stronger evidence object than a partial data set could ever be. A returned data set is a complete answer to a complete question, assembled under a complete role scope. A controller defending an AI-influenced number from a returned data set has no ambiguity to manage; the question of "Was this answer assembled from the right evidence? " is settled by the existence of the data set itself. This discipline is costly but buys defensibility, and the SAP AI tier accepts the trade because the controller is the constituency for whom the trade matters.

14.3 Version Control and Approvals

Version control and approvals tie every prepared data set to a specific PROCESS_VERSION on the data spine and a specific moment of data spine state. It has three substantive components: the PROCESS_VERSION pin that names the approved version of the process the data set describes, the AS-OF pin that names the instant at which every row in the data set was read, and the approval chain that recorded the act of promoting the version and the data set into AI-consumable form. All three components are nonnegotiable and all three are recorded as audit rows.

14.3.1 The PROCESS_VERSION Pin

Every prepared data set names the PROCESS_VERSION it was assembled against. The PROCESS_VERSION is the unit at which a process is approved, published, and consumed: a PROCESS row may have many PROCESS_VERSION rows over its lifetime, but at any moment only one PROCESS_VERSION carries PUBLISH_STATE = PUBLISHED, and only rows linked to that PUBLISHED version are eligible for AI consumption. The data spine applies this same discipline to every consumer surface described in Chapter 9 through Chapter 12, and the AI tier inherits it directly. From the consumer's point of view, the PROCESS_VERSION pin is what makes the prepared data set semantically coherent: every row the model consumes describes the same approved version of the same process, and the model's response is bounded by that version's structural definition.

The pin is at the PROCESS_VERSION level rather than at the PROCESS level because a PROCESS row names a process in the abstract, such as order-to-cash, record-to-report, or procure-to-pay. The process is permanent in the sense that its identity persists across reorganizations and methodology changes. A PROCESS_VERSION row names the approved structural definition of the process at a moment in time: which activities it contains, what their order is, what controls apply, and which applications participate. Two PROCESS_VERSION rows of the same PROCESS may differ substantively, and an AI consumer that did not pin to a specific PROCESS_VERSION could conflate them. The pin at the PROCESS_VERSION level is what prevents the conflation.

14.3.2 The AS-OF Semantic

Every prepared data set is pinned to a single AS-OF timestamp, and the data spine resolves all reads at that instant. The pin is set when the request enters the preparation pipeline and stamped on the data set's metadata, and every row that enters the data set is read at that instant rather than at the wall-clock time of the row's lookup. The discipline is the AI-tier equivalent of the consistent-read semantics that any serious database offers for transactions, and it is prevents a particular failure point that would otherwise be invisible to the consumer.

Consider a data set assembled without an AS-OF pin. The assembly logic reads PROCESS_VERSION first, then ACTIVITY, then MASTER_DATA_REFERENCE, and finally AUDIT_EVENT. The reads are serial, and the data spine state changes between them. A republication of PROCESS_VERSION may complete between the first read and the third, and an audit-side correction may write a new AUDIT_EVENT row between the third read and the fourth. The data set the assembly logic emits would contain the PROCESS_VERSION read at pre-republication, ACTIVITY read at post-republication referring to the new PROCESS_VERSION, and MASTER_DATA_REFERENCE read at yet another time. The data set would be internally inconsistent. The model that consumes it would produce a confident answer based on evidence whose pieces never coexisted in the same data spine state. The audit chain would record each read individually, and the question of whether the answer was sound would require the auditor to reason about the temporal smear within the data set, which is an exercise that has no clean termination condition.

The AS-OF pin forecloses this failure mode by making every read happen at the same instant by construction. The data set is internally consistent because all of its rows were read from the same spine state. The model's confident answer is grounded in evidence the data spine never disagreed with itself about. The audit reconstruction is straightforward: Every row's content_hash is the value the data spine returns when read at as_of_pin, and the auditor's verification is a series of point lookups rather than an investigation of inter-read state drift. From the controller's point of view, the AS-OF pin is the property that makes the data set a coherent evidence object rather than a heterogeneous sample.

14.3.3 The Approval Chain

Every PROCESS_VERSION that enters AI consumption carries an approval chain that records the act of promoting the version from drafted to published, the parties who approved the promotion, and the constraints under which the approval was granted. The chain is anchored in the data spine's PUBLISH_STATE column on PROCESS_VERSION: a row transitions from DRAFT to REVIEWED to APPROVED to PUBLISHED, and each transition is an AUDIT_EVENT(VERSION_TRANSITIONED) row carrying the approver's identity and the transition rationale. Only rows whose current PUBLISH_STATE is PUBLISHED are eligible for AI consumption; rows in DRAFT, REVIEWED, or APPROVED states are visible to the internal governance flow but are foreclosed from AI consumption until they reach PUBLISHED.

The discipline that AI consumes only PUBLISHED rows is what makes the AI tier's outputs answerable in audit terms. A process owner asked to defend a model-generated answer can point to the PROCESS_VERSION the answer was based on, the PUBLISH_STATE of that version, and the approval chain that elevated it. The chain is queryable by version, the approvals are signed by their parties, and the auditor's question of "Was this version actually approved?" has a one-query answer. The alternative—letting AI consume rows in any state—would mean that the audit chain for an AI-influenced number could include unapproved intermediate versions, and the controller's defense would require explaining why the unapproved versions were consumable. Only allowing the AI tier to see PUBLISHED rows removes the explanation from the conversation.

The four states the chain recognizes deserve a brief description because the vocabulary recurs throughout Chapter 15 through Chapter 19:

- DRAFT is the initial state of every new PROCESS_VERSION row. The process owner has begun authoring the version but it is not yet ready for review.
- REVIEWED is the state a row enters when the process owner has completed authoring and the version has been submitted to peer review. The version is visible to the review community but not yet endorsed for production.
- APPROVED is the state a row enters when the peer review has concluded with a recommendation to publish. The version carries the authority of the review community but is not yet visible to consumers.
- PUBLISHED is the state a row enters when the controlling authority (typically a process governance committee or a designated senior process owner) has signed off on the version and authorized its consumption.

The four-stage chain is deliberately more elaborate than a simple draft-and-publish workflow because the additional states create audit-traceable evidence of the review and approval substantively occurring, rather than treating the entire chain as a single act.

14.3.4 Worked Example: Approval

Consider a revision to the variance-explanation process required because the chart of accounts has been restructured. The process owner authors the revision in SAP Signavio, generating a new PROCESS_VERSION row pv-vex-3.3.0 in PUBLISH_STATE = DRAFT. The row is invisible to AI consumption while it remains in DRAFT. Any preparation request that would have consumed pv-vex would still see pv-vex-3.2.1 as the PUBLISHED version, and pv-vex-3.3.0 does not enter the candidate set. The process owner completes the authoring work, submits the version for peer review, and an AUDIT_EVENT(VERSION_TRANSITIONED) row is written naming the submitter, the from state (DRAFT), and the to state (REVIEWED). The version is now visible to the review community but still invisible to AI consumption.

Peer review takes place over the following days, with two reviewers from the controlling community attesting that the revision correctly reflects the chart of accounts restructuring. Their attestations are recorded as additional AUDIT_EVENT rows, each naming the reviewer and the attestation rationale. When both reviewers have attested, the version transitions to APPROVED, with another VERSION_TRANSITIONED row naming the transition and the two attesting reviewers. The version is now endorsed by the review community but still invisible to AI consumption because the controlling authority has not yet signed off.

The process governance committee meets on its weekly cadence, reviews the queue of APPROVED versions awaiting publication, and signs off on pv-vex-3.3.0. The version transitions to PUBLISHED, with a final VERSION_TRANSITIONED row naming the committee chair as the publishing authority. The previous version pv-vex-3.2.1 transitions from PUBLISHED to SUPERSEDED, with its own VERSION_TRANSITIONED row recording the supersession. From the next AI preparation request onward, the candidate set for variance-explanation grounding refers to pv-vex-3.3.0 rather than pv-vex-3.2.1, and the manifest_hash of any data set assembled against the new version differs from any earlier data set's hash because the source rows differ. The controller defending an answer produced before the supersession can resolve the answer's as_of_pin to determine whether it was assembled against pv-vex-3.2.1 or pv-vex-3.3.0. The controller defending an answer produced after is similarly able to demonstrate that the version was approved and published before consumption.

This example illustrates the structural value of the four-stage chain. At every moment in the version's lifecycle, the audit chain holds an unambiguous answer to the question of which version was the AI-consumable one and who authorized the consumption. The controller's defense does not depend on the process owner's recollection or the governance committee's meeting minutes; it depends on AUDIT_EVENT rows the platform recorded automatically at each transition. This discipline is what lets the AI tier's outputs be defended through periods of process evolution. If you're approaching this from an FP&A perspective, you should recognize the discipline as the AI-tier analogue of the journal entry approval workflows that govern ordinary financial reporting.

Figure 14.3 provides a visualization of this discipline. Two requests are issued at different wall-clock times for the same logical question, a PROCESS_VERSION republication occurs between them, and the two prepared data sets emitted carry different content_hashes for the PROCESS_VERSION row. The difference is the consequence of the AS-OF discipline working correctly: The data sets' contents reflect the data spine state at their as_of_pin, the audit chain records both reads independently, and the controller asking which data set answered a given question has a single answer per question rather than an answer blurred across two points in time.

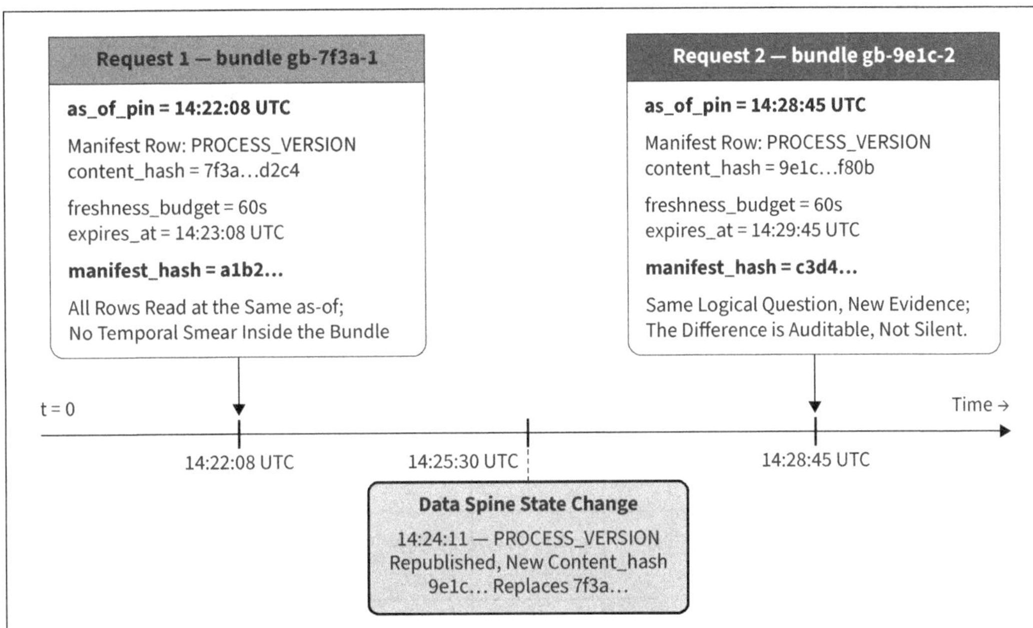

Figure 14.3: Version Control and AS-OF Timeline

14.3.5 Freshness Budgets and Their One-Way Property

Freshness budgets specify how much time may elapse between an event and the data being updated. The consumer can shorten the budget but not lengthen it. Default budgets are set by consumer class. Transactional grounding (Joule interactive agents responding to a controller's question in the user's session) carries a default budget of sixty seconds because an interactive conversation should reflect spine state recent enough to be intelligible as current. Analytical grounding (tile-rendered drill-downs, batch evaluations, and longer-horizon explanations) carries a default budget of one hour because an analytical view typically does not require sub-minute freshness. Workflow grounding (an SAP Build Process Automation step that invokes the AI to draft a recommendation) carries a default budget of five minutes. Third-party agent grounding through SAP Integration Suite carries the same five-minute default.

The platform enforces an upper bound of twenty-four hours on every prepared data set, regardless of the consumer's requested budget. A consumer may shorten its budget—for example, a regulatory scenario may require a freshness budget of five seconds for risk-sensitive reads—but no consumer may extend its budget beyond twenty-four hours. The ceiling is a structural property of the platform, not a configuration the platform's operations team can override. The reasoning is that a data set older than twenty-four hours is, in any conceivable consumer class, more useful as documentation than as evidence; a controller defending a number that was answered from spine state more than a day old is in a defense the controller would prefer not to have to construct.

Table 14.2 provides a summary of the freshness budget by consumer class. Every class has a default budget consistent with its conversational rhythm, a consumer floor below which the consumer may set the budget for risk-sensitive reads, a platform ceiling of twenty-four hours regardless of consumer request, and a one-way expiry semantic that forecloses cache-as-state-store reuse.

Consumer Class	Default Budget	Consumer Floor	Platform Ceiling	Expiry Semantic
Joule interactive (controller in session)	60 seconds	5 seconds	24 hours	One-way; rerequest required
Joule batch (overnight evaluation)	1 hour	60 seconds	24 hours	One-way; rerequest required
SAP Build Process Automation workflow step	5 minutes	30 seconds	24 hours	One-way; rerequest required
Embedded analytical drilldown	1 hour	60 seconds	24 hours	One-way; rerequest required
Third-party agent (SAP Integration Suite)	5 minutes	60 seconds	24 hours	One-way; rerequest required

Table 14.2: Freshness Budget by Consumer Class

Data set expiration is the runtime expression of the freshness budget. Every prepared data set carries `expires_at`, computed as `as_of_pin` plus the freshness budget. After `expires_at`, the data set is no longer authoritative; the consumer that wishes to ask the same logical question must rerequest, and the rerequest will produce a new data set with a new `AS-OF` pin and a new content set if the data spine state has changed in the interval. The expiration is one-way: A consumer may shorten its freshness budget on a per-request basis, but no consumer can pretend a data set is still fresh after `expires_at` has passed. This prevents the cache-as-state-store anti-pattern in which a consumer treats a data set's contents as durable evidence and reuses them across subsequent questions. A data set is not state but evidence-of a question at a single moment, and that moment is bounded.

AI Insight: Why Version Control and Approvals Are Not Tuned

A development team encountering this discipline for the first time will sometimes ask whether the freshness budgets or the PUBLISH_STATE filter can be relaxed for performance reasons, on the reasoning that a longer freshness budget or access to APPROVED but not yet PUBLISHED rows would amortize preparation costs across more requests. The question is reasonable on the throughput side but wrong on the audit side. Version control is a control variable, not a tuning knob. A consumer that relaxes either constraint is asking the platform to let it consume evidence the audit chain considers unfinished, and the controller defending an answer assembled from that evidence will have to explain why the answer's source was not actually approved.

From an FP&A and controlling perspective, the practical consequence is that the version and freshness pins are part of the evidence chain the controller defends, and the controller cannot inherit a posture that the AI consumer optimized away. The data set's PROCESS_VERSION is the version the controller will defend, the as_of_pin is the point in time the controller will defend, and the expires_at is the deadline by which the answer must have been emitted. All three are auditable, none is editable, and the controller's defense rests on their immutability rather than on the consumer's good faith.

14.4 Grounding Rules and Constraints

Grounding rules and constraints close the loop on the topics we covered in Section 14.1 through Section 14.3. Eligibility establishes what may be shown; scoping establishes to whom; version control establishes from which version and at what instant; grounding establishes from which source rows, with what content, and through what audit trail. Together, these four disciplines produce a prepared data set whose every property is structurally defensible, and grounding allows the data set to defend itself in the response the model returns.

14.4.1 Grounding as a Property of the Prepared Data Set

Every row in a prepared data set carries its source_entity, source_pk, content_hash, and the PUBLISH_STATE under which it was read. The data set carries a manifest enumerating all of its source rows together with each row's hash, and the manifest carries a hash that binds the data set as a single audit object. The manifest is recorded as a row in the data spine's audit subsystem, and its identifier (the manifest_hash) is the value the audit chain reads when an auditor wants to reconstruct what the model saw.

The properties on each row are mandatory. A row whose source_entity cannot be determined cannot enter a data set; a row whose source_pk is missing cannot enter a data set; a row whose content_hash cannot be computed at read time cannot enter a data set; a row whose PUBLISH_STATE is anything other than PUBLISHED at the as_of_pin cannot enter a data set. These four properties are the substantive content of the row's grounding, and a row that fails to provide any of them is treated identically to a row that failed the eligibility check: It is excluded, the platform writes an AUDIT_EVENT(GROUNDING_REFUSED) row naming the missing property, and the data set is either assembled without the row or refused entirely if the row was required for the scenario.

The manifest is the data set's own evidence object. A data set's manifest is the structured enumeration of its source rows, expressed as JSON, signed by the platform, and stored alongside the data set for the lifetime of the audit retention regime, which we will cover in Chapter 19. The manifest_hash is the SHA-256 of the manifest, computed at emission time and recorded on the inference record that fires when the consumer invokes the model with the data set. An auditor reading the inference record weeks later can resolve the manifest_hash to the manifest, the manifest to the source rows, the source rows to their spine reads at as_of_pin, and the data spine reads to their content_hashes. The reconstruction is deterministic at every step.

14.4.2 Manifest Signing and Immutability

The manifest is signed at emission by the platform's private key, and the signature is recorded as a column on the manifest row. The signing serves two purposes. First, it makes the manifest tamper-evident: An auditor who recomputes the manifest_hash against the recorded manifest and verifies the signature against the platform's public key has a deterministic answer to the question of whether the manifest is the manifest the platform emitted. Second, it makes the manifest unforgeable by any party other than the platform: A manifest that arrives at the audit subsystem without a valid platform signature is, by design, not a manifest the platform produced, and the audit subsystem refuses to record it. The combination of tamper evidence and unforgeability is what makes the manifest a legal-grade evidence object rather than merely an operational one.

Manifest immutability is a related but distinct property. Once a manifest row has been recorded, it cannot be amended, replaced, or deleted within its retention window; the only operation the audit subsystem permits is read. This is the same discipline the rest of the audit chain inherits from the SOX and SOC2 regimes: An evidence object is recorded once, observed thereafter, and never edited. The manifest's immutability is what lets an auditor's reconstruction of an AI-influenced number be repeated by a second auditor months later and produce the same answer. The manifest that the second auditor reads is the same one the first auditor read, and the audit's conclusions do not drift with time.

Figure 14.4 shows the manifest for a typical variance-explanation data set: five rows, five content_hashes, and one manifest_hash. The data set's role_scope_intersect, as_of_pin, and expires_at are stamped on the data set's header. The rows enumerate the source_entity in the first column, the source_pk in the second, the content_hash in the third, and the PUBLISH_STATE in the fourth. Figure 14.4 uses realistic identifiers that map to entities you encountered in previous chapters to illustrate how a controller reading the manifest would find rows whose identifiers map to other rows they have already seen in earlier evidence walkthroughs. A response that cites gb-7f3a/row-12 names a manifest row whose `content_hash` an auditor can resolve from the data spine at the `AS-OF` instant.

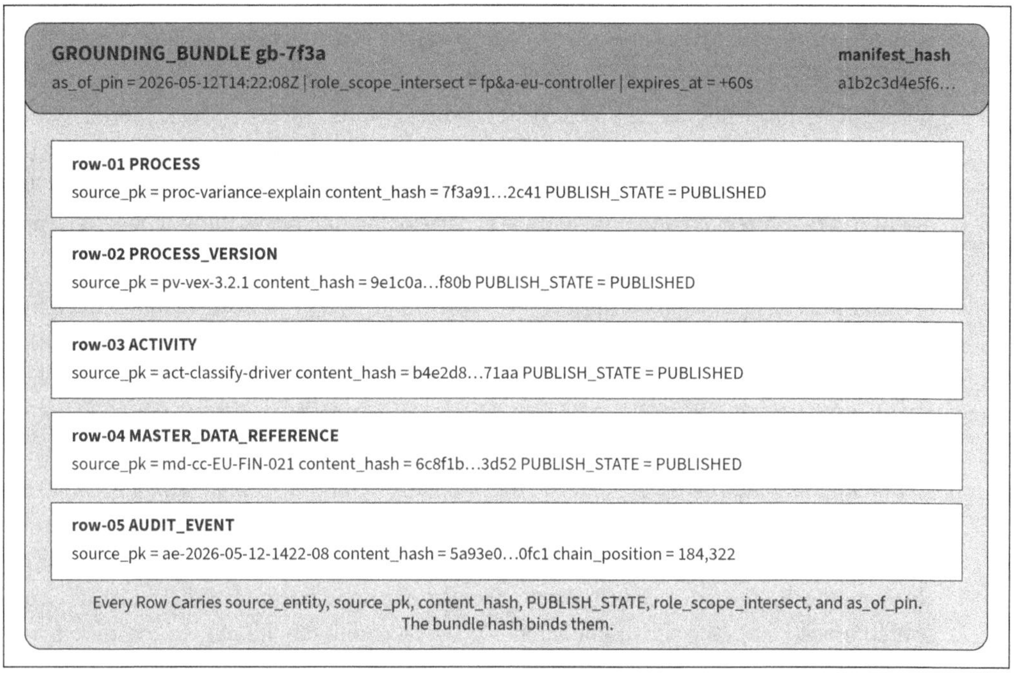

Figure 14.4: Grounding Rules and the Manifest

14.4.3 The Citation Contract

The citation contract is the runtime expression of the manifest discipline. A response that cites a data set row by its row identifier is a response an auditor can resolve. The auditor reads the data set identifier from the response, the row from the manifest's enumeration, the `source_entity` and `source_pk` from the manifest row, and the source row from the data spine at `as_of_pin`. It then computes the row's `content_hash`. The computed value matches the manifest's recorded value or the chain is broken; the auditor's investigation has a deterministic answer in either case. A response that cannot cite—whose claims do not name manifest rows—is refused by the validator before it reaches the user, and the refusal is an `AUDIT_EVENT(RESPONSE_REFUSED)`row that the controller can point at when asked what happens to ungrounded model output.

14.4.4 Grounding Constraints: What May and May Not Be Bound

Grounding rules govern not only how the data is bound to the model but what kinds of evidence are bindable in the first place. Three constraint classes operate at the grounding stage and collectively define the evidence types the AI tier accepts. Each constraint class is enforced at preparation time, with violations producing `AUDIT_EVENT(GROUNDING_REFUSED)` rows that name the specific constraint failure.

The first constraint class is the entity constraint: Only rows from entities the data spine recognizes as canonical may enter a manifest. The recognized entities are PROCESS, PROCESS_VERSION, ACTIVITY, APPLICATION, INTERFACE, EXECUTION_CONTEXT, MASTER_DATA_REFERENCE, ACTIVITY_MASTERDATA_REF, AUDIT_EVENT, VALIDATION_RESULT, INGESTION_LOG, and PUBLISH_STATE, as defined in Chapter 9. Rows from noncanonical sources—such as application-specific tables that have not been registered with the data spine, ad hoc CSV exports, screenshots, and external documents—cannot enter a manifest under any circumstance. The constraint is intentionally restrictive: The AI tier consumes only what the data spine has registered as canonical, and the constraint is what makes the manifest's evidence reproducible across reviewers.

The second constraint class is the freshness constraint: every row in the manifest must have been read at the data set's as_of_pin, and the read must have produced a deterministic content_hash. The constraint forecloses two failure modes that would otherwise be invisible in the manifest. The first is the case where a row was read at a different instant than the data set's pin (typically because of a preparation pipeline bug that read different rows at different moments in time), leaving the manifest temporally smeared; the constraint catches this by recomputing the row's content_hash at the pin and comparing against the value the manifest recorded. The second is the case where a row's content_hash could not be computed deterministically (typically because of a nondeterministic field such as a computed timestamp). The constraint catches this by treating the inability to compute a hash as a refusal rather than a soft failure.

The third constraint class is the consent constraint: Every row that carries PII must have an explicit consent attestation joinable through MASTER_DATA_REFERENCE, and the attestation must be valid at the data set's as_of_pin. The constraint is the runtime enforcement of the regulatory regime that governs PII consumption, which is applied regardless of the data set's role scope or the consumer's claimed need. A row whose consent evidence has expired between the original consent and the data set's pin cannot enter the manifest, even if the row would otherwise be eligible under its policy. The constraint is what lets the AI tier's PII consumption be defended as compliant with the regulatory regime in effect at the moment of consumption, rather than as compliant with the regime in effect at the moment of consent.

The three constraint classes together produce a grounding regime that is rich enough to defend against the failure modes that historically have caused AI consumption to be ruled inadmissible by external audit. The regime does not depend on the consumer's good faith or the model's behavior; it is enforced by the platform at preparation time and is auditable through the AUDIT_EVENT (GROUNDING_REFUSED) rows the constraint enforcement writes. From the controller's point of view, the regime is the property that lets them treat the manifest as evidence of a particular structural quality—where every row is canonical, every row is temporally consistent, and every PII row carries valid consent—rather than as evidence whose structural properties must be separately demonstrated.

Listing 14.2 presents a sample manifest in JSON at a level of detail sufficient for an engineer to implement the manifest emission logic and for an auditor to read the manifest as a self-describing object. Each row carries source_entity, source_pk, content_hash, and PUBLISH_STATE; the data set carries role_scope_intersect, as_of_pin, expires_at, process_version, and the manifest_hash that binds the rows. The manifest is the artifact an auditor resolves from the inference record when reconstructing what the model saw.

```
{
  "dataset_id": "ds-7f3a",
  "manifest_hash": "a1b2c3d4e5f6...",
  "as_of_pin": "2026-05-12T14:22:08Z",
  "role_scope_intersect": ["fp&a-eu-controller"],
  "expires_at": "2026-05-12T14:23:08Z",
  "process_version": "pv-vex-3.2.1",
  "publish_state": "PUBLISHED",
  "rows": [
    {
      "row_id": "row-01",
      "source_entity": "PROCESS",
      "source_pk": "proc-variance-explain",
      "content_hash": "7f3a91...2c41",
      "publish_state": "PUBLISHED"
    },
    {
      "row_id": "row-02",
      "source_entity": "PROCESS_VERSION",
      "source_pk": "pv-vex-3.2.1",
      "content_hash": "9e1c0a...f80b",
      "publish_state": "PUBLISHED"
    },
    {
      "row_id": "row-03",
      "source_entity": "ACTIVITY",
      "source_pk": "act-classify-driver",
      "content_hash": "b4e2d8...71aa",
      "publish_state": "PUBLISHED"
    },
    {
      "row_id": "row-04",
      "source_entity": "MASTER_DATA_REFERENCE",
      "source_pk": "md-cc-EU-FIN-021",
      "content_hash": "6c8f1b...3d52",
      "publish_state": "PUBLISHED"
    },
```

```
    {
      "row_id": "row-05",
      "source_entity": "AUDIT_EVENT",
      "source_pk": "ae-2026-05-12-1422-08",
      "content_hash": "5a93e0...0fc1",
      "chain_position": 184322
    }
  ]
}
```

Listing 14.2: Prepared Data Set Manifest, JSON Representation

AI Insight: Why Grounding Rules Are the Audit's First Move

When an external auditor reviews an AI-influenced number, their first move is not to ask about the model, the prompt, the resource group, or the deployment but to ask for the citations the response carried and to resolve those citations to source rows. This is because citations are the smallest, most direct evidence object the AI tier produces. A citation is one row identifier; a row identifier resolves to one manifest entry; the entry names one source row; the row's `content_hash` is one spine lookup. Four point queries answer the question of whether the model's claim is grounded in evidence the platform recorded, and the auditor can sustain the line of investigation across many claims without losing track.

From an FP&A perspective, the practical consequence is that the citation contract is the property that allows the AI tier to be audited at the same granularity as any other reported number. An ordinary financial number is defended at the level of the journal entry while an AI-influenced number is defended at the level of the citation. The shapes are different, but the audit posture is identical. A controller defending an AI-influenced quarterly variance walks the auditor through the citation list of the response, resolves each citation to a manifest row, and joins the manifest row to the data spine, completing the defense.

14.5 Preventing Hallucination and Drift

Preventing hallucination and drift is the final discipline. Where the previous disciplines prepare the data the model is allowed to consume, this one governs what the model is allowed to produce in response. It operates in two tenses: hallucination prevention as a runtime property of each individual inference, and drift detection as a longer-running property of the AI tier as a whole. The two are complementary, and both are platform-side properties rather than model-side behaviors.

14.5.1 Hallucination Prevention as Structural Property

In the SAP AI tier, hallucination prevention is the structural property that emerges from the manifest discipline described in Section 14.4 rather than an algorithmic technique applied to the model's output. The construction is direct: An AI consumer cannot answer a question for which it has not been given evidence, because the manifest enumerates the evidence the model received and the model's response must cite the manifest's rows for any factual claim. SAP considers a response that contains a factual claim with no manifest row to support to be hallucinated by definition, and the post-response validator that the generative AI hub applies refuses the response before it reaches the user.

The construction has three preconditions, each of which the previous sections have established. First, the manifest must enumerate every row of evidence the model received. A row in the data set but not on the manifest is a structural defect of the platform, not a runtime accident, and is treated as a priority 1 (P1) incident in Chapter 18's operational view. Second, the response must cite the manifest's rows by their identifiers—not by summary or paraphrase but by the literal row identifier in the form ds-7f3a/row-12. Third, the validator that the generative AI hub applies must refuse responses with uncited factual claims. The refusal is the runtime enforcement that completes the structural construction. Together, these three preconditions make hallucination prevention a property of the platform's contract rather than the model's behavior, and the controller's defense does not rely on trust in the model.

The construction claims that uncited content does not reach the user, not that the model never produces uncited content. This distinction matters because the model is, in the SAP AI tier, still a generative system whose outputs are not deterministic in their phrasing. The model may, on some invocations, produce phrasing that contains a factual claim the model could not support. The validator catches such phrasings before the response is returned and writes an `AUDIT_EVENT(RESPONSE_REFUSED)` row naming the data set, the model version, and the offending claim. The consumer then receives a refusal rather than an unsupported answer. The model is contained rather than corrected, and that containment is the SAP AI tier's architectural answer to the question of how to consume foundation-model output without inheriting foundation-model failure modes.

Figure 14.5 illustrates the disciplines in one end-to-end sequence. The controller asks the variance question; the preparation pipeline assembles data set ds-7f3a; the model produces a response in which every factual claim cites a manifest row by its `row_id`; the citation validator confirms the claims and the rows align; the controller receives the response.

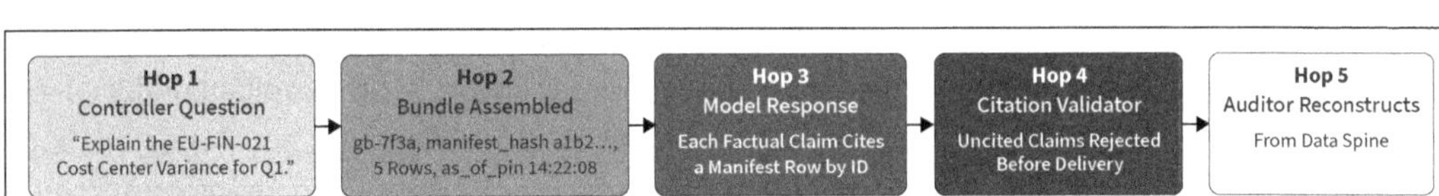

Figure 14.5: Citation Trace and Hallucination Refusal

Let's look at a sample response, where every factual claim resolves to an extant manifest row:

> *"The EU-FIN-021 cost center showed a 12.4 percent variance against budget (gb-7f3a/row-04), driven primarily by an external-services driver (gb-7f3a/row-03). The process version in effect at the AS-OF instant was vex-3.2.1 (gb-7f3a/row-02), published under* PROCESS *proc-variance-explain (gb-7f3a/row-01), with the read recorded in the* AUDIT_EVENT *chain at position 184,322 (gb-7f3a/row-05)."*

There are five citations and five manifest rows. Each row's content_hash is reproducible from the data spine at as_of_pin 14:22:08 UTC. A response with an uncited claim never reaches the controller; the validator at the fourth hop refuses it and writes the refusal to AUDIT_EVENT.

The auditor's reconstruction is as follows:

1. Resolve gb-7f3a from the BUNDLE_MANIFEST store and verify manifest_hash matches the value carried on the INFERENCE_EVENT row.
2. For each cited row, fetch the source row from the data spine AS-OF as_of_pin, recompute the content_hash, and compare it with the manifest entry.
3. Confirm the role_scope_intersect on the bundle is consistent with the actor identity carried on the INFERENCE_EVENT row.
4. Confirm the AUDIT_EVENT chain position of row-05 falls within the period under review and the chain hash is intact through that position.

The reconstruction is deterministic at every step, and the controller defending the answer is doing so from evidence the data spine already holds rather than from a federation across logs.

14.5.2 Drift as an Ongoing Concern

Drift is an ongoing concern that hallucination prevention does not address by itself. Drift takes three forms in the SAP AI tier, and each must be detected and corrected through its own mechanism. Source drift is the case where the source rows the AI was grounded against have changed since the grounding—a MASTER_DATA_REFERENCE row was corrected, a PROCESS_VERSION was republished, or an AUDIT_EVENT row was added that contradicts an earlier one—and the AI's previous answers may no longer be the answers the data spine would produce now. Policy drift is the case where the eligibility, scoping, or version-control policies under which earlier answers were prepared have changed, and answers that were compliant at production time may no longer be compliant under current rules. Model drift is the case where the foundation model itself has been updated, and a question that produced one answer last quarter may produce a different answer this quarter from the same prepared data set.

Source drift is addressed by the AS-OF discipline described in Section 14.3, which means that every AI-influenced answer is reconstructible against the data spine state at the answer's `as_of_pin`. An auditor asking whether the answer remains correct in light of a subsequent spine change is asking a question the discipline anticipates: The answer is correct as of the pin, and the question of whether the data spine has since changed is a separate audit question with a separate audit row. This discipline does not claim the answer is permanently correct but that the answer is correct as of a specific instant, and the instant is named.

Policy drift is addressed by periodic regrounding. The AI governance team, on a cadence aligned to the review cadence of the eligibility policies (typically quarterly), runs a replay against a representative sample of past inference records and compares the answers the replay produces under current policies to the answers the data spine originally produced. A material divergence is an audit finding rather than a correction, meaning the original answers were correct under the policies of the time. However, the divergence is the signal that the new policy regime would produce different answers going forward, and the FP&A team is alerted to the change before the change reaches their quarterly close.

Model drift is addressed by the deployment discipline covered in Chapter 13. The generative AI hub registers every model version, the deployment records every model invocation against its specific version, and the audit chain records the model version on every inference record. An auditor asking why a particular AI-influenced answer differs from a similar answer produced last quarter can resolve the model version on each inference record and determine whether the difference is attributable to spine state, policy change, or model update. The three are separable, and the separation is the property that lets the AI tier be defended through periods of model evolution.

14.5.3 Drift Detection in Practice

Drift detection is a periodic exercise that the AI governance team runs and that the FP&A team reviews. The exercise has four steps. First, a representative sample of past inference records is selected, typically the inference records that produced numbers material to a recent quarterly close, plus a random sample of inference records across the rest of the AI tier. Second, each inference record's prepared data set is reassembled under current policies and current spine state, and the model is reinvoked against the reassembled data set. Third, the replay's output is compared to the original output using a structural comparison: Do the same numbers appear, do the same citations resolve, and do the same conclusions still hold? Fourth, divergences are categorized as source-drift, policy-drift, or model-drift, and the categorization is the audit row that the FP&A team consults when assessing whether the AI tier's controls posture is intact.

This exercise's value is not in detecting individual incorrect answers, because the manifest discipline already makes individual answers reconstructible at their `as_of_pin`. Its value is in detecting systemic shifts in what the AI tier would say today about questions it answered differently yesterday and in giving the FP&A team advance warning of those shifts before they appear in quarterly results. From the controller's point of view, drift detection is the property that makes the AI tier's controls posture monitorable on a quarterly cadence rather than rediscoverable at each external audit.

14.5.4 A Worked Drift Detection Cycle

Consider the drift detection cycle that the AI governance team runs at the end of Q2. The team selects the 50 inference records material to the Q2 close (the AI-influenced answers that informed quarterly variance commentary, cost-allocation decisions, or accrual estimates), plus a random sample of 200 inference records from across the tier. Each record's `manifest_hash` is resolved, its prepared data set is reconstructed, and the data set is reassembled under current policies and current spine state. The reassembly produces a new manifest with a new `manifest_hash`, and the differences between the original manifest and the new manifest are categorized.

Of the 250 inference records replayed, 198 produce identical reassembled manifests. The source rows have not changed, the eligibility policies have not changed, the approval state of the consumed `PROCESS_VERSION` rows has not changed, and the model version remains the same. These 198 are recorded as no-drift findings and require no further action. Of the remaining 52, 31 show source drift. One or more source rows have been updated on the data spine, typically because of corrections or supplementations posted during Q2. The 31 findings are recorded with the specific row changes named, and the FP&A team reviews whether any of the 31 inference records were material to the close. Of those 31, 3 turn out to be material, because the source change shifts a variance commentary number by more than the materiality threshold. They are flagged for restatement consideration.

Of the remaining 21, 15 show policy drift: An eligibility policy has changed between Q1 and Q2 (typically because a regulatory regime has tightened a retention window or a sensitivity classification has been raised), and the original inference record consumed rows that would not be consumed today. The 15 findings are recorded with the specific policy change named, the FP&A team reviews whether the policy change should trigger restatement (typically not, because the answers were compliant at production time), and the governance team uses the finding to validate that the policy change is producing the intended downstream effect. The final 6 all show model drift: The foundation model was updated during Q2, and the model's response to the same prepared data set produces a different answer. These 6 findings are recorded with the model version names (original and current), the FP&A team reviews whether the model update is producing better or worse answers (typically both, with category-level distinctions), and the governance team feeds the findings back to the generative AI hub's model-version selection process.

The exercise produces, at quarter-end, a single drift report listing all 52 non-no-drift findings categorized by drift type, with the material findings flagged for FP&A action. The report is the artifact the controller reviews when preparing for the quarterly close, and the report's findings are joinable to the specific inference records through the `manifest_hash`. From the controller's point of view, the report is the AI tier's controls posture statement for the quarter, and the controller's review of the report is the AI-tier equivalent of the review of a control deficiency log in any other quarterly close exercise.

AI Insight: Why Hallucination and Drift Are Platform Concerns, Not Model Concerns

An older generation of AI integrations placed responsibility for hallucination prevention on the model's training or prompt engineering and responsibility for drift detection on the model vendor's release notes. Both placements are structurally indefensible for any system that needs to be auditable. A model whose hallucination is controlled by its training is a model whose controls posture varies with each retraining, and the controller defending an answer cannot reach into the training to verify. A drift posture that depends on vendor release notes varies with vendor disclosure practice, and the auditor cannot verify the vendor's claims.

The SAP AI tier's commitment is to make both hallucination and drift platform properties whose evidence the platform holds. Hallucination is prevented by the citation contract the platform enforces at response time; drift is detected by the replay the platform runs on its own cadence. Neither relies on trust in the model or in the model vendor. From an FP&A perspective, this is the property that makes the AI tier defensible across vendor changes. The controller's evidence is the same whether the underlying model is updated, replaced, or supplemented by a different vendor's model through the SAP AI Agent Hub, because the evidence is the platform's manifest and the platform's audit chain rather than the model's behavior.

14.6 Summary

This chapter specified the five disciplines that govern the data flowing into SAP AI and the responses flowing back out. The chapter began with eligibility as the default-deny precondition, covered filtering and role-based scoping as the discipline that refuses partial returns and produces evidence subsets rather than unions, discussed version control and approvals through the `PROCESS_VERSION` pin and the `AS-OF` semantic, established grounding through the manifest and the citation contract, and closed with the platform-side prevention of hallucination and drift. The architectural commitment is that data preparation is platform-side rather than consumer-side, and the SAP AI tier therefore consumes only what the five disciplines together permit.

The chapter also made the FP&A case for why these disciplines matter beyond compliance. Every discipline produces an evidence row the controller can point at when defending an AI-influenced number, every row is queryable by a single key, and the five together produce a manifest that joins to the rest of the audit chain through the inference record introduced in Chapter 13. The controllability of the AI tier rests on the immutability of these five disciplines, and the discipline-by-construction posture is what lets the controller defend an AI-influenced quarterly number on the same terms as any other quarterly number.

14.6.1 Key Learning Points

- Data eligibility defaults to INELIGIBLE for every new row of source data, and promotion to one of the two eligible classes is a deliberate, audited classification act. The default-deny posture is the architectural correction of the permissive defaults earlier SAP customers wrote for analytical consumption. The eligibility policy joins to PROCESS_VERSION, ACTIVITY, and MASTER_DATA_REFERENCE through policy_id, and its six fields (jurisdiction, sensitivity_class, retention_window, ai_eligible_flag, last_review, and policy_id) are the substantive content of the eligibility judgment.
- Filtering and role-based scoping is the discipline that prepared data sets are never reused across actors and that the assembly logic produces a complete data set or refuses the data set entirely. Composite roles produce evidence subsets rather than unions, the manifest carries the subset tagging, and the downstream citation contract names the subset under which each claim was answered. The no-partial-return discipline is nonnegotiable, and the AUDIT_EVENT(SCOPE_REFUSED) row that a refused assembly writes is the evidence the controller produces when asked what the platform does with out-of-scope data.
- Version control and approvals pin every prepared data set to a specific PROCESS_VERSION at PUBLISH_STATE = PUBLISHED and to a specific AS-OF instant of spine state. The freshness budget is a one-way controllable: The consumer may shorten its budget but never lengthen it. Default budgets are sixty seconds for transactional grounding, one hour for analytical grounding, five minutes for workflow and third-party agent grounding, and the platform ceiling is twenty-four hours regardless of consumer request.
- Grounding rules and constraints bind every prepared data set to a manifest of named source rows. Each row carries source_entity, source_pk, content_hash, and PUBLISH_STATE; the data set carries role_scope_intersect, as_of_pin, expires_at, process_version, and manifest_hash; the manifest is signed and immutable. The manifest is the auditor's primary artifact, joined to the inference record through manifest_hash and resolvable at every step from the auditor's first query to its last.
- Hallucination prevention is the structural property that emerges from the citation contract: A response is permitted to reach the user only if every factual claim cites a manifest row, and uncited claims are refused by the platform's validator. Drift detection runs on the AI governance team's cadence and categorizes divergences as source-drift, policy-drift, or model-drift, giving the FP&A team advance warning before drift reaches quarterly results. Both are platform concerns rather than model concerns, and an FP&A defense does not depend on the model's cooperation.

- FP&A controllability emerges from all five disciplines simultaneously and from none of them in isolation. The controller defending an AI-influenced number does so through a single chain of evidence: the eligibility policy for what the AI could see, the `role_scope_intersect` for whom the data set was assembled under, the `process_version` and `as_of_pin` for at which version and instant, the `manifest_hash` for from which source rows, and the citation list for through which response claims. The chain is one query against the data spine, the answer is the same shape as the answer for any non-AI number, and the controller's defense does not require any AI-specific evidence apparatus that the rest of the platform does not already provide.

14.6.2 Common Design Pitfalls

- Allowing permissive default for AI eligibility. This carries over the permissive default from earlier analytical-consumption regimes on the reasoning that rows are AI-eligible unless excluded. The pattern fails because the default itself becomes the documentation, and the auditor cannot verify completeness of the exclusions. The discipline is the default-deny posture, with every promotion recorded as `AUDIT_EVENT(ELIGIBILITY_CLASSIFIED)` and every demotion as `ELIGIBILITY_REVOKED`.
- Returning partial prepared data sets when some required rows fail role-scope checks. This may seem helpful because it emits the rows that pass the scope check and marks the data set as partial in its metadata. However, the AI consumer cannot tell from the contents what is missing, and the model that consumes the partial data set produces a confident answer based on incomplete evidence. The discipline is to refuse the data set outright and write `AUDIT_EVENT(SCOPE_REFUSED)` naming the row that failed.
- Letting a freshness budget extend through caller-side overrides. This allows a consumer to specify a freshness budget longer than the platform default on the reasoning that the consumer knows its tolerance better than the platform does. This reasoning fails because the freshness budget is not a tuning knob; it is the contract between the consumer and the audit chain about how old the evidence may be. The discipline is that consumers may shorten their freshness budget but may not extend it, and the platform ceiling of 24 hours is enforced regardless.
- Letting AI consume rows in `PUBLISH_STATE` other than `PUBLISHED`. This assumes that draft, reviewed, or approved (but not published) rows are visible to internal governance and therefore ought to be visible to AI. However, the controller's defense of an AI-influenced number depends on the auditor's ability to confirm the version was actually approved. The discipline is that AI consumes only `PUBLISHED` rows, and the `AUDIT_EVENT(VERSION_TRANSITIONED)` row chain is the evidence the consumer was allowed to see the version.
- Allowing an AI response to assert a fact for which the manifest contains no row. This accepts model output that contains factual claims the manifest does not support on the

reasoning that the claim is probably right and the user wants a fluent answer. This reasoning fails because the citation contract is the only mechanism by which the AI tier's responses are auditable, and an uncited claim is a claim no auditor can resolve. The discipline is that the platform's validator refuses uncited claims before the response reaches the user.

- Caching prepared data sets across actors with different role scopes. This approach reuses a previously assembled data set for a second consumer on the reasoning that the data set has already paid the cost of spine reads. The discipline is that every assembly is a fresh act under the requesting actor's role context, and the audit row written on the cached call would lie about what the second consumer was shown if reuse were permitted.

14.6.3 Review Questions

1. Describe the five disciplines that govern every prepared data set for AI consumption. For each discipline, identify the source entity from the data spine that supplies the property and the `AUDIT_EVENT` row type that records changes to the property.
2. The eligibility policy carries six fields. Name them, explain what each field governs, and identify the consuming entities (`PROCESS_VERSION`, `ACTIVITY`, `MASTER_DATA_REFERENCE`) that reference `policy_id` as a foreign key.
3. Walk through the three eligibility classes. For each class, describe the default conditions under which a row enters the class, the conditions under which it is demoted, and the `AUDIT_EVENT` row type that records the demotion.
4. A controller holds two component roles, fp&a-eu-controller and fp&a-na-controller. The controller asks a variance explanation question whose evidence spans both jurisdictions. Describe the assembly logic's output, the structure of the data set's manifest, and the form the model's response citations take.
5. Explain the `AS-OF` semantic and contrast it with a latest read assembly. Identify the issue that approach exposes, the audit-side consequence of that issue, and the property the `AS-OF` pin restores.
6. A citation contract has three preconditions. Identify them, explain how each precondition is enforced (at which stage of the request pipeline by which component), and describe the `AUDIT_EVENT` row that fires when a precondition is not met.
7. Drift takes three forms in the SAP AI tier. Name them, identify the mechanism that addresses each, and construct a query that retrieves a quarterly drift report's findings categorized by drift type.
8. The freshness budget is a one-way controllable. Explain what one-way means in practice, identify two consumer classes and their default budgets, and describe the platform ceiling.

9. A row's eligibility policy changes from CONDITIONALLY_ELIGIBLE to INELIGIBLE after a regulatory regime tightens its retention window. Describe the AUDIT_EVENT chain the change produces, the assembly behavior the next time the row is considered as a candidate, and the controller's reconstruction query.
10. From the controller's perspective, an AI-influenced quarterly number must be defended to an external auditor. Construct the reconstruction query the controller writes, naming each evidence object (eligibility policy, role_scope_intersect, process_version, as_of_pin, manifest_hash, and citation list) and the join keys between them.

14.6.4 Student Companion Notes

This chapter is where the architectural argument transitions from describing properties of the data spine to describing properties of the data that flows through it into SAP AI. The transition is not a change of subject; it is the next layer of the same subject. The five disciplines described in this chapter are the properties the AI consumption tier of Chapter 13 was designed to enforce, and the citation contract is the runtime payoff of the audit chain established in Chapter 9 and elaborated on in Chapter 10 through Chapter 13. If you have internalized the audit-chain reasoning of the earlier chapters, you should read this chapter as its extension into the AI tier rather than as an unrelated set of new disciplines.

If you come from an FP&A and controlling background, this chapter offers a single structural payoff worth holding onto. The five disciplines do not exist for the AI tier's sake; they exist for the controller's sake. Eligibility is the controller's evidence that the AI was not shown ineligible data. Scoping is the controller's evidence that the AI was not shown data the consumer was not authorized to see. Version control is the controller's evidence that the answer's vintage and approval state are bounded. Grounding is the controller's evidence that every claim resolves to source. Hallucination prevention and drift detection are the controller's evidence that the model did not invent and has not silently changed. The AI tier's discipline is the controller's defense kit.

From the controller's point of view, the most important exercise we recommend is writing a reconstruction query for a recent AI-influenced number. Take an AI-influenced figure from the most recent quarterly report, find the inference record that produced it, join it to the manifest through manifest_hash, join it to the citation list through inference_event_id, and reconstruct the source rows from the data spine at as_of_pin. The exercise will either complete cleanly, in which case the AI tier's discipline is intact, or expose a gap, in which case the gap is the conversation the controller should have with the AI governance team before the next quarterly close. Either outcome is valuable, and this exercise is the cheapest internal audit you can run.

Chapter 15 explores the next stage of the discipline: process suggestions generated using pattern detection and similarity analysis, with scoring, explainability, and the mandatory human-in-the-loop validation that closes the loop on any AI-generated proposal before it becomes a process change.

Chapter 15

Generating Process Suggestions Using AI

A model can notice what a thousand process reviews would miss. It can rank what it noticed and explain why it ranked it that way. What it should never be allowed to do is decide. The model can make suggestions, but a human should make the final decision. The platform provides the audit row that joins them.

Chapter 14 specified the evidence side of the AI conversation. It defined what enterprise data is eligible for AI consumption, how filtering and role-based scoping bound the data set to the requesting actor, how version control and approvals pin the data set to a single instant of data spine state, how grounding rules bind every inference to a manifest of named source rows, and how hallucination and drift are prevented as platform properties rather than model behaviors. This chapter covers the decision side. It specifies how the AI tier turns the prepared data set into process suggestions, how those suggestions are scored and prioritized, how they are explained and made traceable, and how the mandatory human-in-the-loop validation closes the loop before any suggestion is allowed to become a change to the data spine.

The structural commitment of this chapter was first stated in Chapter 1, Section 1.5, where we set the boundary on what AI should not automate. That boundary is not a policy reminder pinned to the top of the chapter; it is the shape of the machinery the chapter specifies. An AI-generated process suggestion is a proposal, never a decision. It is produced from evidence the platform prepared, scored by a model whose weights are governed configuration, explained in terms a controller can read, routed to a human whose role context covers the affected process, and applied to the data spine only after that human records an explicit verdict. The verdict is itself an audit row that is joinable to the suggestion, the evidence the suggestion was drawn from, and the inference record that produced it.

This chapter takes the prepared data sets of Chapter 14 and builds a suggestion pipeline that is defensible on the same terms as the rest of the data spine. It assumes the canonical data model of Chapters 9 and 10, the ingestion discipline of Chapter 11, the execution catalog of Chapter 12, the SAP AI architecture of Chapter 13, and the data preparation discipline of Chapter 14. It looks ahead to Chapter 16, which shows how SAP Build Process Automation turns an accepted suggestion into a governed workflow; Chapter 18, which returns to the operational view of the AI tier including the queue-depth and verdict-latency metrics the suggestion pipeline emits; and Chapter 19, which closes the audit story with the full instrumentation that reconstructs a complete suggestion-to-change chain weeks after the fact.

Five principles run through the chapter and should be carried into every suggestion-pipeline design decision. First, a suggestion is a first-class data spine artifact, not an application-side notification; it carries its evidence trace, its score breakdown, and its inference reference as durable columns. Second, pattern detection is bounded. The analysis is permitted to surface only four families of pattern (deviation, redundancy, harmonization, and control gap) and a candidate that cannot be tied to named data spine rows is discarded before it is scored. Third, the score routes but never decides. Priority bands change the service-level agreement (SLA), the validator pool, and the signoff requirement, but they never change whether a human validates. Fourth, every suggestion carries an explanation record that resolves to the evidence and the inference event, so a controller can always answer the question "Why was this suggested?" with a row-level answer. Fifth, the human verdict is mandatory, explicit, and immutable; the absence of a verdict is never an implicit acceptance, and an amendment is a new verdict rather than a mutation of the recorded one.

This chapter helps controllers specify the row they will eventually point to when an external auditor asks what changed a process and why. They do not point at the model's output but at the validation verdict as well as joins to the suggestion for the proposal, the explanation record for the evidence and the score, the inference event for the model version, and the audit chain for the data spine effect. The defense of an AI-influenced process change is the same as the defense of any other governed change: a row plus its join graph.

By the end of this chapter, you will see how every process change that an AI suggestion influenced should be visibly identifiable as the product of a human verdict recorded against a traceable proposal. If you're approaching this chapter from a financial planning and analysis (FP&A) background, you should see the suggestion pipeline as the decision half of the defense kit whose evidence half Chapter 14 specified.

15.1 Definition of Process Suggestions

A process suggestion is a proposal, produced by the AI tier from a prepared data set, that some part of the modeled enterprise should change. The proposal might be that an activity be realigned to its approved variant, two activities performing the same control be consolidated, a family of regional process variants be harmonized under a common pattern, or an activity an audited process expects to carry a control be flagged because it does not. In every case, the suggestion is a proposal and nothing more. It does not change the data spine, instruct a workflow, or notify a user that something has already happened. It is a first-class data spine artifact whose entire purpose is to be reviewed by a human who will accept, reject, or modify it.

The distinction between a suggestion and an application notification is what this section covers. An application notification is a message. It tells a user that an event occurred, lives in the application's store, and carries no obligation to be acted upon or to be auditable weeks later. A process suggestion is a governed proposal. It lives on the data spine as a row, carries the evidence it was drawn from and the inference event that produced it, and cannot be dismissed into silence because every suggestion is resolved by an explicit verdict that becomes part of the audit chain. The architectural choice to make the suggestion a data spine artifact rather than an application message is the choice that makes the activities described in the rest of this chapter possible.

In the following sections, we will walk through the suggestion artifact, its table, its lifecycle, how it helps enforce the stated AI boundary, and where suggestions sit in the overall pipeline.

15.1.1 The Suggestion Artifact

The suggestion is anchored in a single canonical artifact this chapter introduces. Its substantive content is expressed through a small set of durable columns. The `suggestion_id` is the primary key. The `bundle_id` is a foreign key to the grounding bundle covered in Chapter 14: the prepared, eligible, role-scoped, version-pinned data set the suggestion was drawn from. The `inference_event_id` is a foreign key to the `INFERENCE_EVENT` row covered in Chapter 13: the model invocation that produced the suggestion. The `target_entity` and `target_pk` name the data spine row the suggestion would change, by entity and primary key. The `pattern_family` records which of the four permitted families the suggestion belongs to. The `proposed_state` is the declarative description of the change being proposed. The `risk_class` and the composite score govern routing. The `lifecycle_state` records where the suggestion is in its journey from draft to resolved. The `created_at` is the server-side timestamp at which the suggestion became a data spine row.

The structural choice to carry `bundle_id` and `inference_event_id` as foreign keys rather than as descriptive text is deliberate, and it is the same choice made in Chapter 14 for the eligibility policy. A suggestion that names its grounding bundle by key can be reconstructed: The controller resolves the bundle, the bundle's manifest names the source rows, the source rows are reads against the data spine at the bundle's `AS-OF` pin, and the chain is deterministic from the suggestion to the evidence. A suggestion that described its evidence in prose would be a suggestion the controller could read but not verify. The artifact is the right shape because the controller's question is a join, not a reading exercise.

Figure 15.1 places the suggestion in the loop we have been building since Chapter 13. The evidence side of the loop (the prepared data set in Chapter 14) feeds pattern detection, which produces a suggestion. The suggestion is scored, prioritized, and explained. The explained suggestion is routed to a human. The human records a verdict. The verdict becomes an audit row, and only an accepted verdict becomes a change to the data spine.

The accepted change then becomes part of the state the next prepared data set is built on, and the loop closes—not by the model and not by the human alone, but by the discipline that joins them. No AI-generated suggestion becomes a process change without an explicit human verdict.

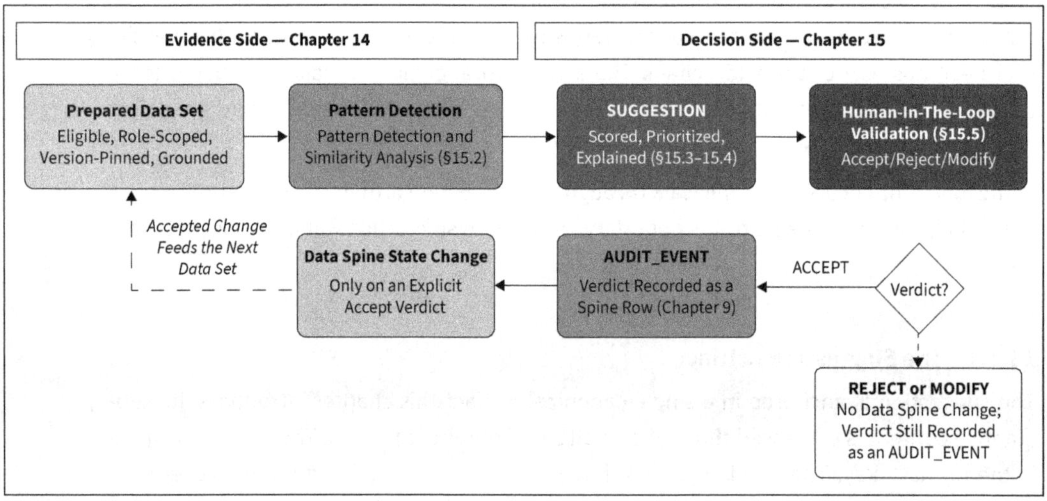

Figure 15.1: The Process Suggestion Loop

The discipline of every subsequent section rests on the property the figure asserts: There is no path from the model's inference to a change on the data spine that does not pass through a human verdict. The absence of that path is enforced by the data spine's commit gateway, not by the goodwill of the application that surfaces the suggestion. The only commit handler that exists for a process change influenced by a suggestion requires a validation verdict row with `verdict_value = ACCEPT` as a foreign-key precondition, and the handler refuses to commit if no such row resolves. When you review Figure 15.1, you should be looking for the path it does *not* draw, because that path is both this chapter's architectural commitment and the operational expression of the AI boundary stated in Chapter 1.

15.1.2 The Suggestion Table

The suggestion artifact is persisted on the data spine as a governed table in the same persistence layer and under the same lifecycle discipline as the canonical entities described in Chapter 9. The table is deliberately narrow. It carries the keys that make the suggestion reconstructible, the columns that govern routing, and the lifecycle state. It does not carry anything that belongs to the application that exposes it. The data definition in Listing 15.1 is the shape an engineer implements, which a controller will query when defending a change.

```
CREATE TABLE SUGGESTION (
  Suggestion_ID         NVARCHAR(50)  PRIMARY KEY,
  Bundle_ID             NVARCHAR(50)  NOT NULL,
```

```
  Inference_Event_ID   NVARCHAR(50)  NOT NULL,
  Target_Entity        NVARCHAR(50)  NOT NULL,
  Target_PK            NVARCHAR(50)  NOT NULL,
  Pattern_Family       NVARCHAR(20)  NOT NULL,
  Proposed_State       NCLOB         NOT NULL,
  Risk_Class           NVARCHAR(10)  NOT NULL,
  Composite_Score      DECIMAL(5,4)  NOT NULL,
  Priority_Band        NVARCHAR(4)   NOT NULL,
  Lifecycle_State      NVARCHAR(20)  NOT NULL,
  Created_At           TIMESTAMP     NOT NULL,
  CONSTRAINT fk_sg_bundle     FOREIGN KEY (Bundle_ID)
      REFERENCES BUNDLE_MANIFEST (Bundle_ID),
  CONSTRAINT fk_sg_inference FOREIGN KEY (Inference_Event_ID)
      REFERENCES INFERENCE_EVENT (Inference_Event_ID),
  CONSTRAINT ck_sg_family     CHECK (Pattern_Family IN
      ('DEVIATION','REDUNDANCY','HARMONIZATION','CONTROL_GAP')),
  CONSTRAINT ck_sg_state      CHECK (Lifecycle_State IN
      ('DRAFT','ROUTED','PENDING_VALIDATION','ACCEPTED',
       'MODIFIED','REJECTED','COMMITTED','CLOSED'))
);
```

Listing 15.1: Create Table Suggestion

Two constraints carry this chapter's discipline into the schema itself. The check constraint on Pattern_Family is the database expression of the four-family boundedness we cover in Section 15.2: The table will not store a suggestion whose pattern is outside the governed vocabulary, so an ungoverned pattern cannot become a suggestion even through a defective code path. The foreign key on Bundle_ID is the database expression of the traceability requirement. A suggestion cannot exist without a grounding bundle it can be reconstructed from, and the referential integrity is enforced by the persistence layer rather than left to the application's good behavior.

15.1.3 The Suggestion Lifecycle

A suggestion moves through a defined lifecycle, and every transition is an AUDIT_EVENT row. A new suggestion enters DRAFT when pattern detection drafts it. It moves to ROUTED when the prioritization logic has assigned it a priority band and a validator pool. It moves to PENDING_VALIDATION when a specific validator has been selected and the SLA clock has started. It moves to one of ACCEPTED, MODIFIED, or REJECTED when a validation verdict is recorded. It moves to COMMITTED when an accepted or modified change has been applied to the data spine through the governed commit chain, and it moves to CLOSED when a rejected suggestion has been archived. There is no transition from any pre-verdict state directly to COMMITTED, and the absence of that transition is the lifecycle expression of the AI boundary.

The lifecycle has one property worth dwelling on because it is the property an auditor tests first: monotonicity. A suggestion never moves backward through the lifecycle. A `REJECTED` suggestion is not reopened; if the same pattern is detected again on a later bundle, the pipeline drafts a new suggestion with a new identity rather than resurrecting the closed one. A `COMMITTED` suggestion is not uncommitted; a change that must be undone is undone by a new suggestion and a new verdict that propose and approve the compensating change. The monotone property is what lets the controller reason about a suggestion's history as an append-only sequence rather than as a mutable record whose current state might not reflect what actually happened.

From the controller's point of view, the lifecycle is the column the controller filters on when scoping an internal review. A query for every suggestion in `lifecycle_state` `COMMITTED` over a reporting period returns exactly the set of AI-influenced changes that reached the data spine; a query for `REJECTED` returns the set the validators declined; a query for `PENDING_VALIDATION` at quarter-end returns the work-in-progress the controller must account for in the close. The lifecycle is the controller's index into the AI tier's decisions, and it is governed and audited for exactly that reason.

15.1.4 Enforcing the Stated AI Boundary

Chapter 1, Section 1.5 established the non-goals of the integration: the parts of the enterprise that AI must not be permitted to automate even when the underlying technology becomes capable of it. Chapter 1's argument was about policy; this chapter's argument is about architecture. AI should not make decisions about any policies that can be circumvented by a defective code path, an over-eager integration, or a future maintainer who does not know the policies exist. An architecture in which there is no code path from an inference to a data spine change without a human verdict cannot be circumvented by any of those, because the path the policy forbids is a path the architecture does not contain.

This is the single most important idea in the chapter. The AI boundary is not a rule the platform checks; it is a property of the platform's shape. The commit gateway described in Section 15.5 has exactly one entry condition that produces an AI-influenced data spine change, a verdict row whose value is `ACCEPT` or `MODIFY`, and no other function in the platform writes such a change. An engineer auditing the codebase for the boundary does not search for a policy document; they enumerate the callers of the data spine write path on an AI-influenced target and confirm that the only caller is the commit gateway and that the gateway cannot be reached without a verdict. The boundary lives in the call graph, and a boundary that lives in the call graph cannot accidentally be removed by a maintainer without the removal being visible in code review.

If you're approaching this chapter from an FP&A and controlling background, you should understand that the difference between a policy boundary and an architectural boundary is the difference between assurance and evidence. A policy boundary is something the controller is told to be true; an architectural boundary is something the controller can demonstrate to be true by pointing at the absence of a code path the way the controller points at the presence of an audit row.

15.1.5 Where the Suggestion Pipeline Sits Operationally

The location of the suggestion pipeline matters, because the architecture only holds if it cannot be bypassed by running it outside the platform's governance. The pipeline runs as a governed workload on the platform's AI execution layer, the same layer Chapter 13 specified for inference. It consumes grounding bundles produced by the preparation discipline outlined in Chapter 14. It writes `SUGGESTION`, `SUGGESTION_EXPLANATION`, and `VALIDATION_VERDICT` rows to the data spine through the same governed persistence layer as every other canonical entity, and it emits `AUDIT_EVENT` rows through the audit chain Chapter 9 established. There is no out-of-band pipeline, no shadow datastore, and no path by which a suggestion exists anywhere other than as a data spine row.

This matters because the most common way an AI boundary erodes in practice is not a deliberate violation but an expedient one: a team under delivery pressure stands up a quick suggestion service of its own, reading the data spine directly and writing changes through a side path. The architecture forecloses the expedient path the same way it forecloses the deliberate one. The data spine's write path for an AI-influenced target rejects any caller that is not the governed commit gateway, and the gateway cannot be reached without a verdict row. A team that stands up its own service can produce suggestions, but it cannot make them into data spine changes, because AI-influenced changes enter the data spine only through the governed path defined in this chapter.

From the controller's point of view, the operational placement is what makes the controller's reconstruction query complete. Because every suggestion is a data spine row and there is no out-of-band pipeline, the query the controller writes against the data spine returns every AI-influenced change without exception. Completeness is not a property the controller has to verify; it is a property the architecture guarantees, because there is nowhere else for an AI-influenced change to be.

From an FP&A and controlling perspective, the loop shown in Figure 15.1 is the controller's defense kit in one image. Asked how a process came to change, the controller traces the loop from right to left: the audit row names the verdict, the verdict names the suggestion, the suggestion names the inference event and the grounding bundle, and the bundle names the source rows under the four disciplines described in Chapter 14. The defense is a single chain of joins, the answer is the same shape as the answer for a manual process change, and the controller's habit of walking from the change backwards through the audit chain works identically whether the change originated with a process architect or with an AI suggestion a human accepted. The shared shape is what lets the controller defend AI-influenced changes the same way they defend manual ones.

AI Insight: Why a Suggestion Is a Data Spine Artifact, Not an Application Message

An older generation of AI integrations treated the model's output as a notification: A panel in the application showed a recommendation, the user clicked a button, and the application called its own write path with the chosen value. That arrangement is structurally indefensible for any system that must be auditable. The recommendation lived only in the application's memory, the click lived only in the application's log, and the row that finally landed on the data spine carried the user's identity but no trace of the suggestion the user was reviewing. An auditor walking the chain backwards found a manual change and had no structural way to discover that a model proposed it.

The discipline that places the suggestion on the data spine as a first-class artifact is the architectural correction. The model's proposal is a data spine row, the human's verdict is a data spine row, the join between them is a data spine relationship, and the audit event that fires on accept is a data spine row joinable to both. The application that surfaces the suggestion becomes a presentation layer over a chain it does not own, and the chain remains intact whether the human acted through Joule, the SAP Build Process Automation inbox, or a process governance review.

If you are approaching this from an FP&A perspective, the practical consequence is that the controller's reconstruction query never reaches into application logs. Every artifact the controller might want lives on the data spine, every relationship lives on the data spine, and the controller's habit of defending a change by walking the audit chain works the same way for an AI-influenced change as for any other governed change. The structural commitment outlined in this chapter is what allows the habit continue to work in the AI era.

15.2 Pattern Detection and Similarity Analysis

Pattern detection is the stage that turns a prepared data set into candidate suggestions. The input is the grounding bundle covered in Chapter 14: a set of `ACTIVITY`, `VARIANT`, `EXECUTION_CONTEXT`, `APPLICATION`, and `INTERFACE` rows that are already eligible, already role-scoped, already pinned to a single `AS-OF` instant, and already named in a manifest. The output is a set of candidate suggestions, each carrying the data spine rows the pattern was detected from and the reference-corpus rows the candidate was compared against. The stage in between is a pipeline of feature extraction, similarity scoring, and pattern classification, and the discipline of the stage is that the pipeline is bounded: It is permitted to produce only four families of pattern, and a candidate it cannot tie to named data spine rows is discarded before it is scored.

Figure 15.2 illustrates this pipeline. The candidate set is the bundle's rows. Feature extraction turns each row into a feature vector capturing the row's structural shape, role graph, control intent, and application and interface references, the same semantic properties the canonical data model in Chapter 9 and Chapter 10 made first class. Similarity analysis measures the distance between each candidate's feature vector and a reference corpus using a fuzzy-logic evaluator and a semantic graph distance, the Python-based components we introduce as the reasoning library. Pattern classification places each strong match into one of the four permitted families. The final stage drafts a candidate suggestion with its evidence trace attached.

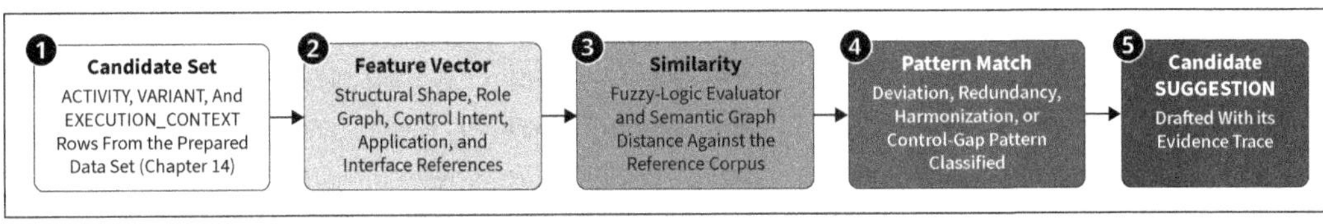

Figure 15.2: Pattern Detection and Similarity Analysis

The four pattern families are the entire vocabulary the analysis is permitted to use, and the boundedness is deliberate:

- A deviation pattern is detected when an activity diverges from the approved variant for its process; the suggestion proposes realignment to the approved variant, not the creation of a new one.
- A redundancy pattern is detected when two activities perform the same control function under different names; the suggestion proposes a consolidation candidate for human review, never an automatic merge.
- A harmonization pattern is detected when comparable processes in different jurisdictions differ only in incidental ways; the suggestion proposes a common pattern with the jurisdictional differences made explicit rather than erased.
- A control-gap pattern is detected when an activity that an audited process expects to carry a control does not reference one; the suggestion flags the gap for a controller and never invents a control value to fill it.

The vocabulary is bounded to four families for the same reason the eligibility default of Chapter 14 was default-deny. An open-ended pattern detector's outputs cannot be governed because the governance regime would have to anticipate every pattern the detector might surface. Conversely, every output from a four-family detector falls into a category the governance regime already understands and has a defined validator pool and remediation path for. The boundedness is not a limitation of the model's capability; it is a property of the platform's governance, imposed on the model's output before that output is allowed to become a suggestion.

The discipline that a candidate with no resolvable evidence trace is discarded before scoring is the structural reason a controller can always ask why a suggestion was made and receive a row-level answer. The model is not permitted to propose a change it cannot ground in rows the data spine already holds. A candidate whose similarity was strong but whose evidence rows cannot be resolved against the bundle's manifest is not a weak suggestion to be shown with a caveat; it is not a suggestion at all, and the pipeline writes an AUDIT_EVENT(CANDIDATE_DISCARDED) row recording why. The discarded candidate audit row is the runtime correlate of the ELIGIBILITY_REFUSED row described in Chapter 14: The platform's refusal to propose is itself the evidence that the discipline was applied.

15.2.1 Worked Example: A Harmonization Suggestion

Imagine a multinational manufacturer runs a goods-receipt process in two jurisdictions. The European variant and the North American variant were modeled independently in SAP Signavio, ingested into the data spine through the flows described in Chapter 11, and prepared into a grounding bundle under the disciplines covered in Chapter 14 for a process-harmonization review requested by a transformation lead whose role context covers both jurisdictions. The bundle contains the ACTIVITY rows of both variants, their EXECUTION_CONTEXT rows, the APPLICATION and INTERFACE references each activity carries, and the PROCESS_VERSION rows that own them.

Pattern detection extracts a feature vector for each activity. The two variants' feature vectors are close on every axis except one: The European variant carries an additional approval activity that references a control the North American variant does not. The fuzzy-logic evaluator scores the structural similarity high; the semantic graph distance confirms that the role graphs and the control intents are congruent everywhere except at the extra approval step. The classifier places the match in the HARMONIZATION family rather than the REDUNDANCY family, because the difference is a jurisdictional control rather than a duplicated activity. This distinction matters: A redundancy suggestion would propose consolidation, which would erase a control the European jurisdiction requires. A harmonization suggestion proposes a common pattern with the jurisdictional difference made explicit.

The evidence trace resolves cleanly. The suggestion names the ACTIVITY rows of both variants by primary key, names the PROCESS_VERSION rows that own them, names the reference-corpus rows the similarity was measured against, and resolves every one of them against the bundle's manifest. Because the trace fully resolves, the pipeline drafts a suggestion whose pattern_family is HARMONIZATION and whose proposed_state is a common goods receipt pattern, with the European approval step modeled as a jurisdictional variant point rather than a structural divergence. An AUDIT_EVENT(SUGGESTION_GENERATED) row is written, and the suggestion enters DRAFT.

15.2.2 The Reference Corpus Is Itself a Governed Artifact

Pattern detection measures similarity against a reference corpus, and the reference corpus is not an opaque training set. It is a governed artifact on the data spine: a versioned, role-scoped collection of approved process patterns, each one traceable to the PROCESS_VERSION rows it was derived from and each one carrying the eligibility and scoping properties Chapter 14 specified. The corpus is governed for the same reason the scoring weights are governed: A similarity measured against an ungoverned corpus would be one the controller could not reconstruct, because they would have no way to know what the candidate was compared against.

The corpus is scoped the same way the grounding bundle is scoped. A similarity measured for a transformation lead whose role context covers only one jurisdiction is measured against the slice of the corpus that lead is authorized to see, not against the whole corpus. The scoping discipline is the reason a harmonization suggestion cannot leak a pattern from a jurisdiction the requesting actor is not entitled to: The pattern is not in the actor's segment of the corpus, so the similarity that would have surfaced it never runs. This is the same evidence-subset discipline Chapter 14 imposed on composite roles, applied to the corpus rather than to the bundle.

The corpus is maintained through the same governed configuration regime that owns the scoring weights and the priority-band thresholds. Adding a pattern to the corpus is an audited configuration change; retiring one is another. A controller reconstructing a historic suggestion resolves the corpus version in force at the suggestion's `created_at` timestamp rather than the corpus as it stands today.

15.2.3 Failure Modes the Discipline Is Designed Against

A discipline is best understood through what it refuses. The first failure mode is the confident hallucination: A model proposes a plausible-sounding change that no data spine evidence supports. The evidence trace requirement refuses it before scoring. The second is the scope-leak: A similarity measured against a pattern the requesting actor is not entitled to see surfaces a suggestion that exposes that pattern. The corpus scoping discipline refuses it. The third is the ungovernable pattern: A detector generates a type of change the governance regime has no validator pool for. The four-family bound refuses it.

The fourth failure mode is the most subtle: the erasure of a deliberate difference. Two regional process variants differ because a jurisdiction requires the difference; an uninformed consolidation detector reads the difference as redundancy and proposes to remove it. The discipline that separates the `REDUNDANCY` family from the `HARMONIZATION` family is designed precisely against this, and the classifier is biased toward harmonization whenever a difference correlates with a jurisdiction boundary in the role graph. The bias is deliberate: The cost of a missed consolidation is an inefficiency a later review will catch, while the cost of an erased control is a finding an external auditor will catch, and the discipline is calibrated to fail toward the cheaper error.

Each failure mode the discipline refuses is recorded as an audit row rather than silently dropped. The refusals are evidence, exactly as the eligibility refusals were evidence in Chapter 14, and the controller who wants to know what the pipeline declined to propose has a row-level answer for every category of refusal.

Listing 15.2 shows the pattern-detection pipeline in pseudocode. The function takes a grounding bundle, extracts a feature vector for each candidate row, measures similarity against the reference corpus, classifies strong matches into one of the four families, and drafts a suggestion only for a candidate whose evidence trace fully resolves against the bundle's manifest. The function is deterministic given the bundle and the corpus, and its only side effects are the audit rows the lifecycle requires. A candidate whose trace does not resolve is discarded with an AUDIT_EVENT row rather than surfaced as a low-confidence suggestion.

```
function detectPatterns(bundle):
  suggestions = []
  corpus = REFERENCE_CORPUS.for_scope(bundle.role_scope_intersect)

  for row in bundle.candidate_rows:
      # 1. Feature extraction from the canonical semantics (Ch. 9/10)
      fv = featureVector(row.structural_shape,
                         row.role_graph,
                         row.control_intent,
                         row.application_refs,
                         row.interface_refs)

      # 2. Similarity against the role-scoped corpus
      match = fuzzySimilarity(fv, corpus) +
              semanticGraphDistance(fv, corpus)
      if match.strength < SIMILARITY_FLOOR:
          continue

      # 3. Classify into exactly one permitted family
      family = classify(match)    # DEVIATION | REDUNDANCY |
                                  # HARMONIZATION | CONTROL_GAP
      if family is None:
          continue

      # 4. Resolve the evidence trace against the bundle manifest
      trace = resolveEvidence(row, match.reference_rows, bundle.manifest)
      if not trace.fully_resolved:
          emit AUDIT_EVENT(CANDIDATE_DISCARDED,
                           target = row.pk,
                           reason = 'evidence_trace_unresolved')
          continue

      # 5. Draft the suggestion as a first-class spine artifact
      s = SUGGESTION(bundle_id = bundle.id,
                     inference_event_id = bundle.inference_event_id,
                     target_entity = row.entity,
```

```
                    target_pk = row.pk,
                    pattern_family = family,
                    proposed_state = match.proposed_state,
                    evidence_trace = trace,
                    lifecycle_state = 'DRAFT')
        emit AUDIT_EVENT(SUGGESTION_GENERATED, suggestion_id = s.id)
        suggestions.append(s)

    return suggestions
```

Listing 15.2: Pattern Detection Pipeline

From the controller's point of view, the value of bounding the detector to four families and discarding ungrounded candidates is that the universe of suggestions is finite and explicable. A controller preparing for an audit can enumerate every suggestion in the reporting period, group them by pattern family, and know in advance which validator pool reviewed each family and which remediation path each accepted suggestion followed. The discarded candidates are as informative as the surfaced ones. A sudden rise in `CANDIDATE_DISCARDED` rows for a particular bundle is a signal that the reference corpus or the bundle's scoping has drifted, and that signal is the kind of evidence a controller takes to the AI governance team before the next quarterly close.

15.3 Scoring and Prioritization Logic

A candidate suggestion that has survived pattern detection carries a similarity strength, but similarity strength is not a priority. A weak similarity on a high-risk control activity may deserve a controller's attention before a strong similarity on a low-impact workflow variable. Scoring is the stage that turns the suggestion's properties into a composite score, and prioritization is the stage that turns the composite score into a routing decision. The discipline of both stages is that the score routes but never decides. A higher score changes the SLA, the validator pool, and the signoff requirement, and it never changes whether a human validates the suggestion.

Figure 15.3 illustrates the scoring model. Four factors are combined into a composite score. Confidence captures the similarity strength and the density of the evidence trace, as a suggestion drawn from many corroborating rows scores higher than one drawn from a single weak match. Impact captures the breadth of the change, meaning how many activities and interfaces the proposed change would touch. Risk captures the control sensitivity of the target rows, as a change to an activity that carries a financial control scores higher than a change to a presentational variable. Effort captures the inverse of the estimated change cost, because a cheap change should, all else equal, be proposed sooner than an expensive one. The four factors are combined under governed weights into a single composite score.

Priority bands change the SLA, the validator pool, and the signoff requirement. They never change whether a human validates. A priority 3 (P3) suggestion is reviewed more slowly than a priority 1 (P1) suggestion; neither is applied without an explicit verdict. The weights and thresholds are stored as governed configuration rows so a controller can reconstruct the score after the fact.

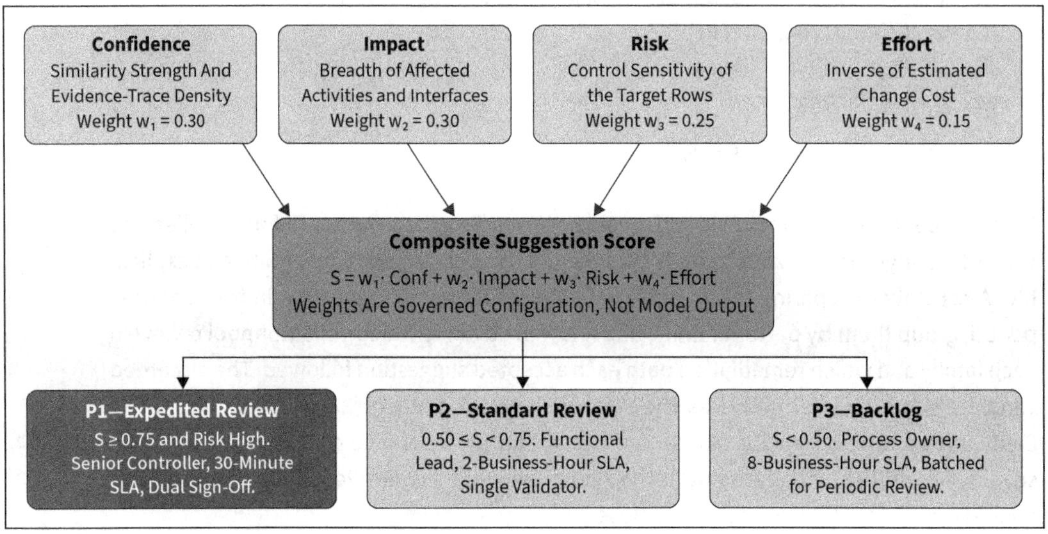

Figure 15.3: Scoring and Prioritization

15.3.1 The Four Factors in Detail

Each of the four factors is computed from data spine evidence rather than from the model's self-assessment, and the distinction is the reason the score is defensible. Confidence is not the model's stated confidence; it is a function of the similarity strength the fuzzy-logic evaluator returned and the cardinality of the resolved evidence trace. Impact is computed by walking the data spine's relationship graph from the target rows outward and counting the activities, interfaces, and downstream configuration objects a change would touch. Risk is read from the control sensitivity already recorded on the target rows. Effort is the only factor that is an estimate. It is bounded and carries the lowest weight precisely because it is the least evidential of the four.

The choice to derive three of the four factors from data spine evidence and to give the one estimated factor the lowest weight keeps the score honest. A scoring model that lets the model grade its own confidence could inflate its way to the front of the queue; a scoring model that reads confidence from the evidence trace cannot. The score is a property the platform computes about the suggestion, not a property the suggestion asserts about itself.

15.3.2 Worked Example: Scoring a Control-Gap Suggestion

To continue our manufacturer example, imagine that a separate suggestion is raised in the same review. An activity in the North American good-receipt variant posts a goods movement that the audited process expects to carry a three-way-match control, but the activity references no control object. Pattern detection classifies this as a `CONTROL_GAP` suggestion. Confidence is high because the gap is unambiguous and the evidence trace is dense. Impact is moderate because the change touches one activity and the two interfaces it posts through. Risk is high because the target activity carries a financial posting and the missing control is a three-way match. Effort is moderate because adding the control reference is a contained change.

Under the governed weights, confidence 0.30, impact 0.30, risk 0.25, effort 0.15, the factor values of approximately 0.88, 0.55, 0.95, and 0.50 combine to a composite score of about 0.72. A separate escalation rule promotes any `CONTROL_GAP` suggestion whose risk factor is at the top of the scale to P1 regardless of composite score, because a missing financial control is the kind of finding a controller must see within thirty minutes. The escalation rule is itself governed configuration, versioned and audited, so the controller reconstructing the routing decision can show exactly which rule promoted the suggestion and when that rule took effect.

15.3.3 Weights Are Configuration, Not Learning

The decision to make the scoring weights governed configuration rather than learned parameters is the decision an external auditor probes hardest. A learned weighting, one the model adjusts as it observes which suggestions validators accept, would produce a ranking the controller could not reconstruct because the weights at the moment of any historic suggestion would be a function of every validation that preceded it. A configured weighting is reconstructible by a single lookup. In this chapter, we choose reconstructibility over self-tuning accuracy deliberately—the same choice we made in Chapter 14 when we chose a default-deny eligibility posture over a permissive one. The auditable form is preferred to the convenient form.

This does not mean the weights never change. They change through the same governed configuration regime that owns the reference corpus and the priority-band thresholds. A governance review that concludes the impact factor is underweighted records a new weight set, the change is an audit row, and from the effective date forward new suggestions are scored under the new weights while historic suggestions retain the weights they were scored under. This is the version-pinning discipline we covered in Chapter 14, now applied to the scoring configuration.

The observation that validators accept some kinds of suggestion and reject others is not discarded; it is fed back through the governance review rather than directly into the weights. A quarterly analysis of verdict outcomes by pattern family and score band is the input to the governance review that may adjust the weights. The model does not learn its own importance; the governance regime decides it based on evidence the controller can audit.

15.3.4 Rescoring and the Stability Guarantee

A suggestion is scored once, at the moment it is drafted, and that score is durable. It is not rescored while it waits in the validation queue, for the same reason we pinned a grounding bundle to a single `AS-OF` instant in Chapter 14. A suggestion whose score changed while a validator was reviewing it would be one whose priority band could shift underneath the review in progress. The stability guarantee is that the score, the band, and the routing are fixed at draft time and are reconstructible as fixed.

There is one governed exception, and it proves the rule. If the underlying data spine evidence changes materially while the suggestion is still pending, the suggestion is invalidated rather than rescored. The pipeline writes an `AUDIT_EVENT(SUGGESTION_STALE)` row, the suggestion transitions to `CLOSED` with a stale reason, and if the pattern still holds against the changed evidence a fresh suggestion is drafted with a new identity and a new score. The discipline never mutates a live suggestion's score; it either preserves it unchanged or closes the suggestion and starts a new one.

From an FP&A perspective, the stability guarantee is what makes a quarter-end reconciliation tractable. The set of suggestions pending at the close is a fixed set with fixed scores and fixed bands; it does not shift while the controller is counting it. The controller's close procedure for AI-influenced changes is the same procedure as for any other governed change: Count the committed, account for the pending, and note the closed, on a set that does not move while it is being counted.

The composite score is a weighted sum, and the choice to make it a weighted sum rather than a learned black-box ranking makes it defensible. A controller asked why a suggestion was prioritized the way it was does not have to interrogate a model; the controller reads the score breakdown and the governed weights, and reproduces the arithmetic. The weights are not the model's output; they are configuration rows written by the governance regime, versioned like every other governed artifact, and changed only through an audited configuration change. A controller reconstructing a historic suggestion resolves the weights that were in force at the suggestion's `created_at`, not the weights in force today.

Table 15.1 sets out how the composite score maps to priority bands and what each band changes. Given a composite score and a risk level, the band determines the validator pool, the SLA budget, and whether a dual signoff is required. The mapping is configurable per customer through the data spine's policy entities, and changes to the mapping are themselves audit rows so a controller can reconstruct which mapping was in force at the time of any historic validation.

Priority Band	Composite Score	Validator Pool	SLA Budget	Dual Signoff
P1 —expedited	$S \geq 0.75$ and high risk	Senior controller whose role context covers the target process	30 minutes	Required
P2—standard	$0.50 \leq S < 0.75$	Functional lead, jurisdiction-bound to the target rows	2 business hours	Not required
P3—backlog	$S < 0.50$	Process owner; suggestions batched for periodic review	8 business hours	Not required

Table 15.1: Composite Score to Priority Band Mapping

The property the table is built to defend is that the band changes the speed and the rigor of the review, never its existence. A P3 suggestion may be reviewed more slowly than a P1 suggestion, by a process owner rather than a senior controller, and in a periodic batch rather than within thirty minutes, but it is still reviewed. Neither a P1 nor a P3 suggestion is applied to the data spine without an explicit verdict, and the SLA budget is the maximum time a suggestion may wait for that verdict, not a timer after which the suggestion is auto-accepted. A suggestion whose SLA expires without a verdict does not become accepted; it escalates, and the escalation is its own audit row.

AI Insight; Why an SLA Timeout Is Never an Implicit Acceptance

An improper implementation of suggestion routing treats an SLA timeout as an implicit acceptance on the reasoning that the validator's silence is consent. This reasoning collapses on the day the auditor walks in. An SLA timeout means the platform did not receive a verdict; treating the absence of a verdict as an acceptance means the audit row the data spine writes claims a human acted when none did. The chain the controller defends collapses at the first row, because the row asserts an act that did not occur.

The discipline of the platform is that an SLA timeout is its own transition, recorded as `AUDIT_EVENT(SLA_BREACHED)`, and the suggestion remains unresolved while the routing logic escalates it to the next tier. A second timeout escalates again. The escalation path is finite; at its top is the governance officer, and beyond that is the option to record a deliberate `REJECT` verdict with a rationale that names the operational failure. None of those paths is an implicit acceptance; every one of them is an explicit human act recorded as such on the data spine.

From an FP&A perspective, the implication is that an AI-influenced process change can never appear because nobody got around to reviewing the suggestion in time. If the pipeline reached the SLA without a verdict, the suggestion is still unresolved, the data spine still holds its prior state, and the controller defending the process has the same answer the validator would have given: The suggestion was not accepted, no change occurred, and no defense is required.

15.4 Explainability and Traceability

A suggestion that cannot be explained cannot be validated, because the human asked to accept or reject it has nothing to consider except the model's assertion that the change is a good one. Explainability is the discipline that attaches an *explanation record* to every suggestion, a structured artifact that names the evidence rows the pattern was detected from, the reference corpus rows the similarity was measured against, the score breakdown, the governed weights, and the model version and deployment that produced the suggestion in a natural language rationale that a controller can read. Traceability is the discipline that the explanation record resolves. Every reference in it is a foreign key to a data spine row, not a description, so the explanation can be verified rather than merely read.

Figure 15.4 sets out the explanation record and the verdict chain it anchors. The left panel is the explanation record itself, with its evidence rows, reference-corpus rows, score breakdown, model and deployment references, and natural-language rationale.

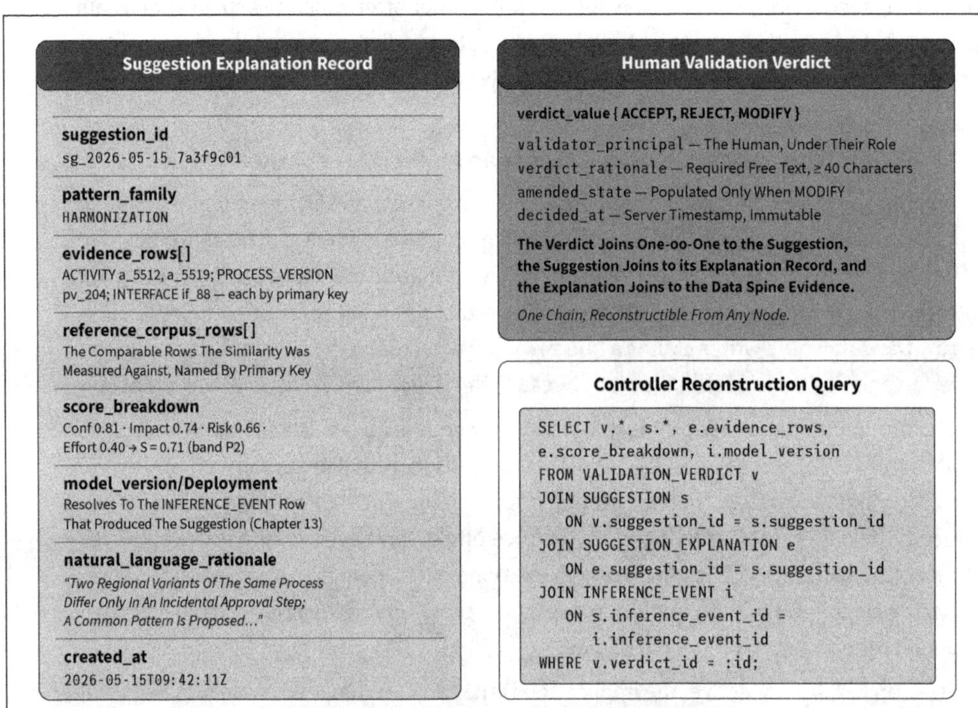

Figure 15.4: Explainability, Traceability, and the Verdict Chain

The right panel is the validation verdict, joined one-to-one to the suggestion, and the reconstruction query the controller writes to walk the entire chain from the verdict back to the data spine evidence in a single statement. The explanation record names the evidence rows, the reference-corpus rows, the score breakdown, and the inference reference, with a natural-language rationale for the human. The verdict joins one-to-one to the suggestion, the suggestion joins to its explanation record, and the explanation joins to the data spine evidence and the inference event. The chain is reconstructible from any node by a single join query.

15.4.1 The SUGGESTION_EXPLANATION Table

The explanation record is persisted as its own table, joined one-to-one to the suggestion. Separating the explanation from the suggestion is deliberate. The suggestion is a small routing artifact queried constantly by the pipeline, while the explanation is a larger evidential artifact read mainly by validators and controllers, and the two have different access patterns and different retention requirements.

As shown in Listing 15.3, the unique constraint on `Suggestion_ID` enforces the one-to-one join, and each suggestion has exactly one explanation. The `Rationale_Cited_Rows` column is the schema expression of the citation discipline. Every sentence of the natural-language rationale that makes a factual claim names the evidence row it rests on, and the cited-rows column is the machine-checkable list the platform validates the rationale against before the explanation is persisted. A rationale whose cited rows do not all resolve against the bundle manifest is refused, and the refusal is an `AUDIT_EVENT(RATIONALE_REFUSED)` row.

```
CREATE TABLE SUGGESTION_EXPLANATION (
  Explanation_ID        NVARCHAR(50)  PRIMARY KEY,
  Suggestion_ID         NVARCHAR(50)  NOT NULL,
  Evidence_Rows         NCLOB         NOT NULL,
  Reference_Corpus_Rows NCLOB         NOT NULL,
  Score_Breakdown       NCLOB         NOT NULL,
  Governed_Weight_Set   NVARCHAR(50)  NOT NULL,
  Model_Version         NVARCHAR(50)  NOT NULL,
  Deployment_ID         NVARCHAR(50)  NOT NULL,
  NL_Rationale          NCLOB         NOT NULL,
  Rationale_Cited_Rows  NCLOB         NOT NULL,
  Created_At            TIMESTAMP     NOT NULL,
  CONSTRAINT fk_se_sg FOREIGN KEY (Suggestion_ID)
      REFERENCES SUGGESTION (Suggestion_ID),
  CONSTRAINT uq_se_sg UNIQUE (Suggestion_ID)
);
```

Listing 15.3: Create Table Suggestion Explanation

15.4.2 What the Validator Sees

Table 15.2 lists the fields the validator is shown when a suggestion reaches them, and the source of each field. The table is the contract between the explanation record and the validation interface. Every field the validator relies on resolves to a data spine row, so a validator who doubts any line can resolve the key and read the underlying row rather than trusting the presentation.

Field Shown to the Validator	Source	Resolvable To
Proposed change	`SUGGESTION.proposed_state`	The target data spine row by entity and primary key
Why it was raised	`SUGGESTION_EXPLANATION.nl_rationale`	The cited evidence rows, each by primary key
Evidence it rests on	`SUGGESTION_EXPLANATION.evidence_rows`	`ACTIVITY`, `PROCESS_VERSION`, and `INTERFACE` rows in the bundle
What it was compared to	`reference_corpus_rows`	The reference corpus rows the similarity used
Score and why	`score_breakdown` and `governed_weight_set`	The versioned weight configuration in force
Which model produced it	`model_version` and `deployment_id`	The `INFERENCE_EVENT` row from Chapter 13

Table 15.2: What the Validator Sees, and What Each Field Resolves To

The table is built to defend one property: The validator never decides on the basis of something the controller cannot later see. Every field in the validation interface resolves to a durable data spine row, the explanation record that assembled those fields is immutable once persisted, and the verdict the validator records joins to the explanation the validator actually saw.

15.4.3 The Explanation Under Adversarial Review

The real test of an explanation record is not whether a cooperative validator finds it helpful but whether a hostile external auditor finds it sufficient. A hostile auditor assumes the rationale is wrong until the evidence proves it right and reads the rationale only as a set of claims to be checked against the cited rows. The explanation record is designed a hostile auditor. A rationale sentence that cannot be tied to a row is not a weak point the auditor must argue about; it is a sentence the platform would have refused at persistence time, so it cannot be in the record at all.

This is the operational meaning of the citation discipline. The natural-language rationale is not the model's argument for the change but a human-readable index into the evidence, and the evidence is the argument. An auditor who distrusts the prose can ignore it entirely and read only the cited rows, and the cited rows are sufficient on their own. Because the prose is a constrained restatement of the rows rather than an independent claim, the two cannot diverge.

There is a second adversarial reading the record is designed for: the auditor who asks not whether the explanation is true but whether it is the explanation the validator actually saw. The explanation record is written once, before the suggestion is routed, and is never updated. The explanation's `created_at` precedes the verdict's `decided_at` by construction. An auditor can therefore demonstrate not only that an explanation exists but that this specific explanation existed before the human decided and was not edited afterward to fit the decision. The temporal ordering is part of the evidence, enforced by the schema rather than asserted by the platform.

The natural-language rationale is the part of the explanation record a validator reads first, and the discipline that governs it is the same citation discipline we imposed on AI responses in Chapter 14. The rationale is permitted to assert only what the evidence rows support; a sentence in the rationale that makes a claim no evidence row substantiates is a rationale the platform refuses, and the refusal is an `AUDIT_EVENT(RATIONALE_REFUSED)` row. The rationale is not the model's free-form opinion of the change; it is a constrained restatement of what the evidence rows already say, phrased so a controller can read it without resolving every foreign key by hand. A controller who wants to verify the rationale resolves the keys; a controller who wants to triage quickly reads the prose. Both are reading the same governed artifact.

Traceability is what makes the natural-language rationale safe to rely on. Because every reference in the explanation record is a foreign key, the rationale's claims are verifiable. A validator who doubts a sentence can resolve the evidence row it rests on and read the row directly. The explanation record is a claim accompanied by its proof. That proof is not a separate document the validator must locate but a join the validator can run from the suggestion itself. The architectural choice that every explanation reference is a key rather than a description is the choice that turns explainability from a vendor assurance into an evidence relationship.

From an FP&A and controlling perspective, the explanation record is the answer to the question an external auditor always asks about an AI-influenced change—not "Did a human approve it?" but "What did the human see when they approved it?" The explanation record is exactly what the human saw, preserved as a data spine row, joined to the verdict the human recorded. The controller does not reconstruct the human's reasoning from memory or from an application log. They resolve the explanation record the human reviewed, and the record is the same record weeks later than it was at the moment of the verdict. The immutability of the explanation record is what makes the controller's defense of the human's judgment possible.

15.5 Human-in-the-Loop Validation

Human-in-the-loop validation is the stage at which the suggestion stops being a proposal and becomes a resolved artifact. The human assigned by the priority band reviews the suggestion and its explanation record and records exactly one verdict: ACCEPT, REJECT, or MODIFY. The verdict is mandatory. Every suggestion is resolved by an explicit verdict, and there is no lifecycle path by which a suggestion becomes a change without one. The verdict is also explicit. It is recorded as a data spine row carrying the validator's identity, their role context, a required rationale, and a server-side timestamp. Finally, the verdict is immutable. Once recorded, the row is append-only, and an amendment is a new verdict on a new suggestion rather than a mutation of the recorded one.

In the following sections, we will show you the VALIDATION_VERDICT table, the commit chain, and the path agents submit through, and then briefly cover what the verdict does not do.

15.5.1 The VALIDATION_VERDICT Table

The verdict is the row this entire chapter converges on, and it is persisted with the same first-class discipline as the suggestion and the explanation. Its columns are the minimum required for the verdict to be attributable, reconstructible, and idempotent.

As shown in Listing 15.4, there are three important constraints in this context. The check constraint on Verdict_Value confines the verdict to the governed vocabulary. The unique constraint on Idempotency_Key is the schema expression of the idempotence property: a second commit attempt carrying the same key fails the uniqueness check rather than applying a second change. The nullable Amended_State and Revokes_Verdict_ID columns carry the MODIFY and REVOKE semantics, and the original verdict row is never updated in either case because the schema offers no path to mutate it.

```
CREATE TABLE VALIDATION_VERDICT (
  Verdict_ID           NVARCHAR(50)  PRIMARY KEY,
  Suggestion_ID        NVARCHAR(50)  NOT NULL,
  Verdict_Value        NVARCHAR(10)  NOT NULL,
  Validator_Principal NVARCHAR(100) NOT NULL,
  Validator_Role       NVARCHAR(100) NOT NULL,
  Verdict_Rationale    NCLOB         NOT NULL,
  Amended_State        NCLOB,
  Revokes_Verdict_ID   NVARCHAR(50),
  Idempotency_Key      NVARCHAR(100) NOT NULL,
  Decided_At           TIMESTAMP     NOT NULL,
  CONSTRAINT fk_vv_sg FOREIGN KEY (Suggestion_ID)
      REFERENCES SUGGESTION (Suggestion_ID),
  CONSTRAINT ck_vv_val CHECK (Verdict_Value IN
```

```
      ('ACCEPT','REJECT','MODIFY','REVOKE')),
  CONSTRAINT uq_vv_idem UNIQUE (Idempotency_Key)
);
```

Listing 15.4: Create Table Validation Verdict

15.5.2 The Governed Commit Chain

An accepted or modified verdict becomes a change to the data spine through a governed commit chain, not through a direct write. Listing 15.5 shows the chain. The gateway resolves the verdict, confirms it is ACCEPT or MODIFY, selects the state to apply, checks the idempotency key against prior commits, applies the change inside a transaction, writes the AUDIT_EVENT row that records the change with the verdict as a foreign key, and advances the suggestion lifecycle to COMMITTED. Any failure rolls the transaction back and writes an AUDIT_EVENT(COMMIT_FAILED) row.

```
function commitVerdict(verdict):
    if verdict.value not in ('ACCEPT','MODIFY'):
        return            # REJECT/REVOKE handled elsewhere

    # 1. Idempotence: a repeated commit is a no-op
    prior = AUDIT_EVENT.find(idempotency_key =
                             verdict.idempotency_key)
    if prior is not None:
        return prior.audit_event_id

    # 2. Select the state to apply
    state = verdict.amended_state if verdict.value == 'MODIFY'
                                  else suggestion.proposed_state

    # 3. Apply inside a transaction
    begin_transaction()
    try:
        applyToSpine(suggestion.target_entity,
                     suggestion.target_pk, state)
        ae = emit AUDIT_EVENT(STATE_CHANGED,
                   target = suggestion.target_pk,
                   verdict_id = verdict.id,
                   idempotency_key = verdict.idempotency_key)
        suggestion.lifecycle_state = 'COMMITTED'
        commit_transaction()
        return ae.audit_event_id
    except:
```

```
        rollback_transaction()
        emit AUDIT_EVENT(COMMIT_FAILED, verdict_id = verdict.id)
        raise
```

Listing 15.5: The Governed Commit Chain

Listing 15.5 makes this chapter's structural commitment executable. The function has exactly one entry condition that produces a data spine change, a verdict whose value is ACCEPT or MODIFY, and there is no other function in the platform that writes an AI-influenced change to the data spine. The boundary is a property of the call graph, and that is the strongest place it can live.

15.5.3 Hybrid and Non-SAP Agents Submit Through the Same Path

The suggestion pipeline does not assume the suggesting agent is an SAP agent. A Joule agent operating in SAP S/4HANA, a Joule agent reaching into SAP ERP through a governed connector, an SAP Build Process Automation workflow agent, a third-party large language-model agent registered through the SAP AI agent governance surface, and an in-house custom agent built on the platform's reasoning library all submit suggestions through the same path. The SUGGESTION row carries an agent_principal that records which agent emitted it, but the column that records the agent's identity does not change the pipeline the suggestion flows through.

This is the cross-vendor governance property the multiagent estate depends on. As the number of agents that can propose process changes grows, the risk is not that any one agent is malicious but that the governance surface fragments. The single-path property is the architectural answer. Agent sprawl does not become agent risk, because every agent, whatever its origin, enters the same funnel: the same SUGGESTION artifact, the same scoring, the same explanation, the same mandatory human verdict, and the same audit chain. The funnel is the governance, and it does not care which vendor built the agent.

From the controller's point of view, the single-path property means the reconstruction query does not branch on agent origin. The controller defending an AI-influenced change writes the same join whether the suggestion came from Joule or from a third-party agent, and the answer is the same in both cases.

15.5.4 What the Verdict Does Not Settle

An ACCEPT verdict settles that a human, under a named role context, agreed that the model's proposed change should be applied. It does not settle that the change is correct in some absolute sense. The verdict is an attributable human decision recorded at a moment in time, not a guarantee of correctness. This is what makes the REVOKE path coherent: A later determination that an accepted change was wrong does not contradict the original verdict, because the original verdict never claimed the change was right. It claimed a human accepted it then, which remains true.

Nor does a verdict settle the question of the model's quality. A pattern of REJECT verdicts against a particular pattern family is data the governance review covered in Section 15.3 consumes, not a verdict on the model. Conflating the two would let an individual validator's judgment silently retrain the model, which is exactly the learned-weighting failure mode that Section 15.3 forbids. The verdict closes a suggestion; it does not tune a model. The separation is deliberate and structural.

From an FP&A and controlling perspective, the precision about what the verdict does not settle is what keeps the controller's defense honest. The controller does not defend an AI-influenced change by claiming it was correct; they defend it by demonstrating that an authorized human, shown reconstructible evidence, made an attributable decision under a governed process, and that the decision and its basis are preserved on the data spine. That is the same defense the controller mounts for a hand-made change.

The three verdict values have precisely defined semantics. ACCEPT means the validator agrees with the proposed_state and instructs the platform to apply it as a governed change to the data spine. REJECT means the validator disagrees; no data spine change occurs, the rationale is recorded, and the suggestion is closed. MODIFY means the validator agrees that a change is warranted but not the change the model proposed; the validator's amended_state is recorded as the change to be applied, and it is the validator's amended state, not the model's proposed state, that the data spine receives. In every case, the rationale is required and is subject to the same citation discipline as the explanation record's rationale. A verdict rationale that asserts what no evidence supports is refused.

Table 15.3 lists the verdict semantics and the data spine consequence of each one. It represents the contract the commit gateway enforces: The only input that produces a data spine change is an ACCEPT or a MODIFY verdict, and the change applied is the model's proposed state for an ACCEPT and the validator's amended state for a MODIFY. A REJECT produces no change and closes the suggestion. Table 15.3 deliberately omits any path by which a suggestion becomes a change without an explicit, attributable human act.

Verdict	Meaning	State Applied to the Data Spine	Suggestion Outcome
ACCEPT	Validator agrees with the model's proposed change	The model's proposed_state, via the governed commit chain	Closed as accepted
MODIFY	Validator agrees a change is warranted but not the proposed one	The validator's amended_state, not the model's proposed_state	Closed as modified
REJECT	Validator disagrees that the change should be made	No change; the prior data spine state is preserved	Closed as rejected

Table 15.3: Validation Verdict Semantics

The verdict immutability property deserves the same emphasis the `AS-OF` pin immutability received in Chapter 14. A recorded verdict is append-only. If the validator concludes after recording an `ACCEPT` that the acceptance was a mistake, the remedy is not to edit the verdict but to issue a new verdict: a `REVOKE` verdict that names the original, carries its own rationale, and triggers a compensating change to the data spine through the same governed commit chain that applied the original. The original verdict row is never modified, because a modified verdict row would make the audit chain lie about what the validator decided and when. The controller defending the process change reads both verdict rows (the original and the revocation) and the pair is a truthful record of what happened.

Idempotence is the second structural property of the verdict. Every commit attempt carries the verdict's identity as an idempotency key, and the commit gateway checks for a prior commit with the same key before applying a change. A network retry, a queue redrive, or an operator who clicks the accept control twice does not produce two changes; the gateway returns the existing audit event without reapplying the side effect. This is the same discipline we imposed on ingestion in Chapter 11: The safe outcome of a repeated operation is the outcome of the operation performed once, and that safety is a property of the platform rather than a discipline the operator must remember to observe.

From the controller's point of view, the verdict is the row the entire chapter has been building toward. It is the artifact the controller points at when an external auditor asks who changed the process and on what basis. It joins to the suggestion for the proposal, the suggestion joins to the explanation record for the evidence and the score, the explanation joins to the inference event for the model version and to the grounding bundle for the prepared data set, and the bundle resolves to the source rows under the four disciplines outlined in Chapter 14. The controller's defense of an AI-influenced process change is a single reconstruction query, the answer is the same as the answer for any other governed change, and the defense requires no AI-specific evidence apparatus the rest of the platform does not already provide. The decision side of the conversation, like the evidence side Chapter 14 specified, reduces to a row and its join graph.

15.6 Summary

This chapter covered the decision side of the AI conversation. It defined the process suggestion as a first-class data spine artifact rather than an application notification, established pattern detection and similarity analysis as a bounded pipeline that may surface only four families of pattern, defined a transparent composite scoring model whose weights are governed configuration, showed how to make every suggestion traceable through an explanation record that resolves to data spine rows, and closed the loop with a mandatory, explicit, immutable human verdict. The architectural commitment is that no AI-generated suggestion becomes a change to the data spine without an explicit human verdict, and the boundary stated in Chapter 1, Section 1.5 was enforced here as the structural shape of the loop rather than as a policy reminder.

This chapter also made the FP&A case for why the suggestion pipeline matters beyond compliance. Every stage produces an evidence row the controller can point at when defending an AI-influenced process change: the suggestion for the proposal, the explanation record for the evidence and the score, the inference event for the model version, the grounding bundle for the prepared data set, and the verdict for the human decision. The five rows join into a single chain, the chain resolves with one query, and the answer is the same as what the controller has produced for governed process changes for decades.

15.6.1 Key Learning Points

- A process suggestion is a first-class data spine artifact, not an application notification. It carries its grounding bundle, its inference event, its target rows, its pattern family, its proposed state, and its lifecycle state as durable columns, and it cannot be dismissed into silence because every suggestion is resolved by an explicit verdict that becomes part of the audit chain. The artifact property is what makes the decision side of the AI conversation defensible on the same terms as the evidence side.
- Pattern detection is bounded to four families: deviation, redundancy, harmonization, and control gap. A candidate that cannot be tied to named data spine rows is discarded before scoring with an `AUDIT_EVENT(CANDIDATE_DISCARDED)`row, and the model is never permitted to propose a change it cannot ground in rows the data spine already holds. The boundedness is a property of the platform's governance imposed on the model's output, not a limitation of the model's capability.
- The composite score is a transparent weighted sum of confidence, impact, risk, and effort, computed under governed weights that are versioned configuration rows rather than model output. The score routes the suggestion into one of three priority bands, and the band changes the SLA, the validator pool, and the signoff requirement. It never changes whether a human validates the decision. An SLA timeout is never an implicit acceptance; it is an escalation recorded as its own audit row.
- Every suggestion carries an explanation record whose every reference is a foreign key to a data spine row rather than a description. The natural-language rationale is a constrained restatement of what the evidence rows support, subject to the same citation discipline we imposed on AI responses in Chapter 14, and a rationale that asserts what no evidence supports is refused. Traceability is what makes the rationale safe to rely on. The claim is accompanied by its proof, and the proof is a join the controller can run.
- The human verdict is mandatory, explicit, and immutable. `ACCEPT` applies the model's proposed state through the governed commit chain, `MODIFY` applies the validator's amended state instead, and `REJECT` preserves the prior state. A recorded verdict is append-only, and a mistaken acceptance is remedied by a new `REVOKE` verdict rather than a mutation. Idempotence makes a repeated commit safe by construction.
- FP&A controllability emerges from the chain as a whole. The controller defending an AI-influenced process change writes a single reconstruction query that joins the verdict

to the suggestion, the suggestion to the explanation record, the explanation to the inference event and the grounding bundle, and the bundle to the source rows under the four disciplines outlined in Chapter 14. The result is a single query that yields a single answer, in the same shape as any other governed change.

15.6.2 Common Design Pitfalls

- Surfacing the model's output as an application notification rather than a data spine artifact. This shows the recommendation in a panel and calls the application's own write path on the user's click, but it is wrong because the recommendation and the click live only in the application's log. The audit chain records a manual change with no structural trace that a model proposed it. The discipline is that the suggestion is a data spine row and the verdict is a data spine row, joined to each other and to the evidence, so the chain is intact regardless of which application surfaced the suggestion.
- Treating an SLA timeout as an implicit acceptance. This allows a suggestion to become a change because the validator did not respond in time. It fails because the audit row then claims a human acted when none did, and the chain collapses at the first row. The discipline is that an SLA timeout is an escalation recorded as `AUDIT_EVENT(SLA_BREACHED)` and the suggestion remains unresolved until an explicit verdict is recorded.
- Surfacing a low-confidence candidate with a caveat instead of discarding it. This shows a candidate whose evidence trace does not resolve, annotated as uncertain, on the reasoning that the human can judge it. It fails because a candidate the platform cannot ground is a candidate the controller cannot defend, and the caveat does not survive into the audit chain. The discipline is that an ungrounded candidate is discarded with an audit row, not surfaced with a footnote.
- Letting the model's proposed state apply on a `MODIFY` verdict. This applies the model's original proposal when the validator chose `MODIFY`, treating the validator's amendment as advisory. However, the change the data spine receives must be the change the human decided, not the change the model proposed. The discipline is that a `MODIFY` verdict applies the validator's `amended_state` and the commit gateway enforces it.
- Editing a recorded verdict to correct a mistaken acceptance. This mutates the verdict row when the validator concludes the acceptance was wrong. However, a mutated verdict makes the audit chain lie about what was decided and when. The discipline is that a mistaken acceptance is remedied by a new `REVOKE` verdict naming the original, with its own rationale and a compensating change through the governed commit chain.
- Treating the suggestion score as a decision threshold. This auto-applies suggestions above a score cutoff and routes only the rest to humans on the reasoning that high-confidence suggestions are safe. This wrongly assumes that confidence is authority. A high score changes how fast and how rigorously a human reviews, never whether a human reviews. The discipline is that every suggestion, at every score, is resolved by an explicit human verdict.

15.6.3 Review Questions

1. Distinguish a process suggestion from an application notification. Identify the columns that make the suggestion a first-class data spine artifact and explain why `bundle_id` and `inference_event_id` are foreign keys rather than descriptive text.
2. Pattern detection is bounded to four families. Name them, give the remediation each suggestion proposes, and explain why an open-ended detector would be ungovernable.
3. Walk through the pattern-detection pipeline shown in Listing 15.2. Identify the point at which a candidate with an unresolved evidence trace is discarded and name the audit row the discard produces.
4. The composite score combines four factors under governed weights. Name the factors, explain why the weights are configuration rows rather than model output, and describe how a controller reconstructs a historic suggestion's score.
5. Explain the difference between the three priority bands. For each band, identify the validator pool, the SLA budget, and the signoff requirement, and explain what the band never changes.
6. An SLA budget expires before a verdict is recorded. Describe the transition the platform takes, the audit row it writes, and why the timeout is not an implicit acceptance.
7. Describe the explanation record. Identify the references it carries, explain why each is a foreign key, and describe the citation discipline the natural-language rationale is subject to.
8. The three verdict values have precisely defined semantics. For each, state the meaning, the state applied to the data spine, and the suggestion outcome, and explain why a `MODIFY` applies the validator's amended state rather than the model's proposed state.
9. A validator concludes after recording an `ACCEPT` that the acceptance was a mistake. Describe the remedy, the audit rows it produces, and why the original verdict row should never be modified.
10. From the controller's perspective, an AI-influenced process change must be defended to an external auditor. Construct the reconstruction query, naming each artifact (verdict, suggestion, explanation record, inference event, grounding bundle, and data spine source rows) and the join keys between them.

15.6.4 Student Companion Notes

This chapter is where the architectural argument we have been building since Chapter 13 completes its arc. Chapter 13 specified the AI architecture and the consumer contract, and Chapter 14 specified the data preparation that makes the contract substantive. This chapter specified the decision discipline that makes the AI tier's output safe to act on. If you have internalized the audit-chain reasoning of Chapter 9 through Chapter 14, you should treat Chapter 15 not as a new subject but as the closing of the loop those chapters opened: the evidence side and the decision side of the same conversation, joined by the same audit chain the canonical data spine established.

The single most important idea to take from this chapter is that the AI boundary outlined in Chapter 1, Section 1.5 is not enforced by a rule the platform reminds itself of but by the shape of the machinery. There is no commit handler that turns an inference into a data spine change without a verdict row, and the absence of that handler is the boundary. If you can explain why the boundary is structural rather than procedural, you have understood the chapter, and you have understood why we treat AI as a consumer and a proposer but never a decider.

If you're approaching this chapter from an FP&A and controlling perspective, you should understand that it completes the defense kit that Chapter 14 began. Chapter 14 gave the controller the evidence side, meaning what the AI was permitted to see. Chapter 15 gives the controller the decision side, meaning what a human chose to do about what the AI proposed. The two together let the controller defend an AI-influenced process change with the same one-query reconstruction the controller has always used for governed changes, and this chapter's practical recommendation is the same as Chapter 14's: Take a recent AI-influenced change, find the verdict, and walk the chain back to the source rows. This exercise either completes cleanly or exposes a gap, and either outcome is the cheapest internal audit you can run.

Chapter 16 turns from the suggestion itself to what happens after it is accepted. SAP Build Process Automation is the layer that turns an accepted verdict into an approval, remediation, or AI-assisted workflow on top of the data spine, with the audit and governance controls the verdict chain requires.

PART VIII
SAP Build Consumption

Chapter 16
SAP Build Process Automation: Workflow on Top of the Data Spine

A suggestion is a sentence the model wrote. A verdict is a sentence a human signed. A workflow is the machinery that makes the signed sentence happen—once, in order, with every hand that touched it named. The data spine remembers what was decided, SAP Build Process Automation remembers how it was carried out, and the join between them is the only audit you should ever need.

In Chapter 15, we closed our discussion of AI by specifying its decision side. An AI-generated process suggestion is produced from a platform-prepared data set, scored by governed configuration, explained in terms a controller can read, and routed to a human who records an explicit verdict. The verdict is an audit row, joinable to the suggestion, the evidence the suggestion was drawn from, and the inference that produced it. Chapter 15 ended at the moment the verdict is recorded. This chapter begins one instant later, and answers the question Chapter 15 deliberately left open: Once a human has made a decision, what carries out that decision?

The answer is SAP Build Process Automation. It is the execution layer that turns an accepted verdict into an approval, remediation, or AI-assisted human task on top of the canonical data spine, with the audit and governance controls the verdict chain requires. The structural commitment of this chapter is the boundary we first stated in Chapter 1, Section 1.5 and enforced through Chapter 13 to Chapter 15: AI may propose, a human must decide, and only a governed handler may write. This chapter shows that this boundary is not a policy the workflow engine is asked to respect but the shape of the workflow machinery itself. There is no path through SAP Build Process Automation that turns a proposal into a spine change without a recorded human verdict, and the absence of that path is the boundary.

This chapter is for those who decide where workflow state lives, the governance officer who must defend a control to an auditor, and the controller or financial planning and analysis (FP&A) professional who must explain, weeks later, why a process changed and on whose authority. This chapter completes the FP&A defense kit we began in Chapter 14 and Chapter 15. Chapter 14 gave the evidence side: what the AI was permitted to see. Chapter 15 gave the decision side: what a human chose to do about it. This chapter gives the execution side: how the chosen action was carried out, by whom, in what order, and against which version of the spine. The three together let the controller reconstruct an AI-influenced change with the same one-query walk they have always used for any governed change.

Five principles run through this chapter and should be carried into every workflow built on the spine. First, routing is derived from process context, never hard-coded into the workflow. Second, an exception is a typed event, and remediation is a governed workflow rather than an ad hoc fix. Third, an AI recommendation inside a workflow is advisory, and the human task is the only decision point. Fourth, every workflow instance writes its own audit row, and that row is joinable to the spine without reading a log. Fifth, the workflow engine holds workflow state but never owns canonical truth. It reads context from the data spine and writes back only through the commit gateway we reserved for it in Chapter 10.

This chapter explains:

- How an accepted verdict becomes a governed workflow trigger, and why the trigger is keyed to the verdict row rather than to the suggestion
- How approval workflows derive their approver chain from `EXECUTION_CONTEXT` and `PROCESS_VERSION` rather than from a form, and how separation of duties (SoD) is enforced structurally
- How exceptions are classified into typed, severity-banded events, and how auto-remediation, guided remediation, and escalation are chosen and governed
- How an AI recommendation is surfaced inside a human task without ever becoming the decision, and why the commit boundary is the absence of a handler rather than the presence of a rule
- How governance and audit handling make every workflow instance reconstructible from the data spine, and what an FP&A-focused one-query defense looks like for an AI-influenced change

Figure 16.1 places this chapter in the arc we have been building since Chapter 9.

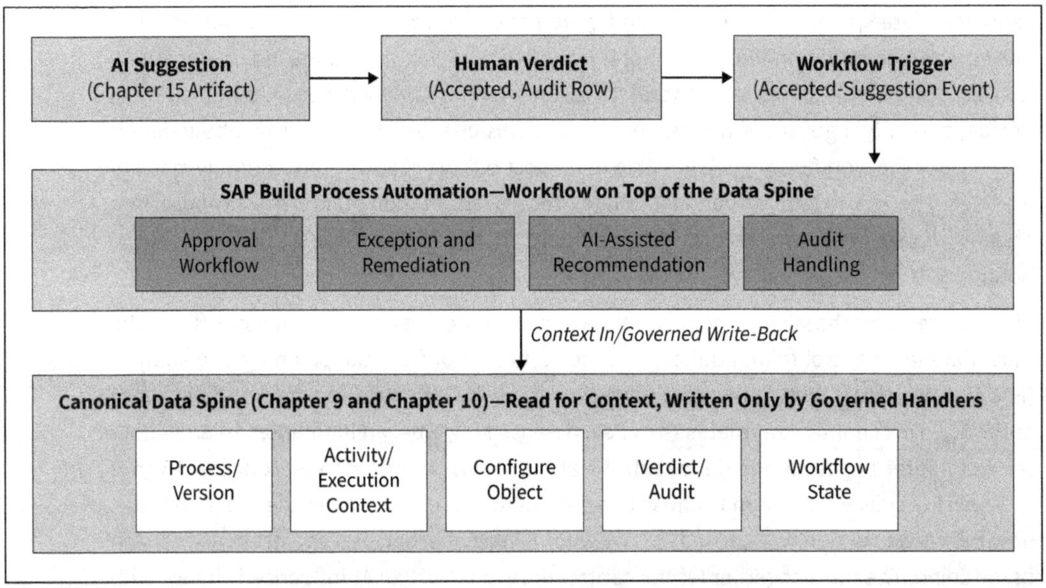

Figure 16.1: From Accepted Verdict to Governed Workflow

The suggestion is the proposal, the verdict is the decision, the workflow is the governed execution of that decision, and the data spine is the memory that makes all three reconstructible. SAP Build Process Automation sits between the verdict and the data spine, never beside the decision.

The remainder of the chapter describes the four workflow families (approval, exception and remediation, AI-assisted recommendation, and governance and audit handling) and shows how each is the same governed contract with the data spine seen from a different operational angle. Before we get to the families, we will discus SAP Build Process Automation itself, because the families only make sense once you know what the product is and, more importantly, what it should not be allowed to become.

16.1 Positioning SAP Build Process Automation on the Data Spine

SAP Build Process Automation is a low-code workflow and automation service on SAP Business Technology Platform (SAP BTP). Its building blocks (processes, forms, decisions, automations, and visibility scenarios) are designed to let a business expert assemble a working workflow quickly. That speed is a virtue and, in a governed landscape, a hazard. The same low-code surface that lets a team build an approval in an afternoon also lets a team encode an approver in a form, a routing rule in a decision table, and a write to a downstream system in an automation, all without ever consulting the canonical data spine. The product does not automatically enforce the architecture of this book. The architecture must be imposed on the product, and this section show you how.

16.1.1 Building Block and What Each Is Allowed to Do

A process in SAP Build Process Automation is the orchestration: the sequence of triggers, tasks, and branches. In this architecture, a process is allowed to orchestrate, but it is never the source of a routing fact. A form is the human-facing surface of a task. It is allowed to display context and collect a decision, but it is never the place an approver, a risk band, or a control classification is chosen. Those are read from the data spine and shown on the form, not entered into it. A decision is a rules artifact. It can compute a presentation or a routing branch from spine-supplied inputs, but it is never the authority for who can approve; it is a switch operated by spine facts, not a register of them. An automation is an unattended sequence, often a user interface (UI) or application programming interface (API) interaction. It can read and call the commit gateway but never write canonical truth by any other path. A visibility scenario is the analytics surface. It reports on workflow state and never becomes a second, ungoverned copy of the audit chain.

Table 16.1 shows the whole governance of the product in one. Nothing in SAP Build Process Automation is banned; everything is bounded. A review of any workflow built in the product reduces to checking that no building block has taken on its forbidden role, and that check is mechanical rather than a matter of taste.

SAP Build Process Automation Building Blocks	Permitted Role on the Data Spine	Forbidden Role
Process	Orchestrate tasks and branches	Hold a routing or approver fact
Form	Display context; collect a verdict	Capture an approver or risk band
Decision	Compute a branch from spine inputs	Be the authority for who may approve
Automation	Read context; call the handler	Write canonical truth by any other path
Visibility	Report on workflow state	Become a parallel ungoverned audit copy

Table 16.1: Every SAP Build Process Automation Building Block has a Permitted Role and a Forbidden Role

16.1.2 Projects, Releases, and Why Versioning Lives in the Data Spine

SAP Build Process Automation organizes work into projects that are versioned and released. It is tempting to think of that release versioning as the version of record for the process the workflow serves. It is not, and this distinction is one you will recognize immediately if you have read Chapter 9. Releasing an SAP Build Process Automation project versions the workflow: the orchestration, forms, and decision logic. The `PROCESS_VERSION` in the data spine versions the business process: its activities, execution context, and control gates. These are different things changing for different reasons on different cadences, and conflating them reintroduces exactly the version ambiguity we built the data spine to eliminate in Chapter 9. The workflow project release is operational metadata; the `PROCESS_VERSION` is canonical truth. The audit chain joins on the latter, and the workflow instance records both so that an auditor can see which workflow build acted on which process version without the two ever being assumed equal.

16.1.3 Destinations, Identity, and the Single Write Path

SAP Build Process Automation reaches the data spine through SAP BTP destinations and consumes identity through the same Identity Authentication or SAP Authorization and Trust Management service role collections the rest of the landscape uses. Two disciplines from earlier chapters are important here. First, the destination the workflow uses to write is not a general data spine API; it is the commit gateway and only that handler. The workflow has no destination that permits a direct table write, because a capability that does not exist cannot be misused. Second, the identity the workflow acts under is never a generic service account for approvals; the deciding human's identity is carried on the verdict, and the handler records that identity, not the workflow's technical user, as the author of the change. The workflow's technical identity moves the work, the human's identity owns the decision, and the data spine records the second, never the first, as the accountable party.

Note

This is the concrete realization of a key principle we outlined in Chapter 10: SAP Build Process Automation is the home of the workflow state, but that workflow state is not canonical truth. The destination discipline is what makes that separation unbreakable: the workflow cannot reach canonical truth except through the one handler, so no amount of low-code freedom inside the product can erode the governed boundary outside it.

16.2 Approval Workflows

An approval workflow is the most familiar object in any SAP landscape and, for that reason, the one most often built wrong. The common mistake is to treat the approver as a property of the form: a field on the request, a value chosen by the requester, a routing table maintained by hand. In a single-system process that mistake is survivable. In the integrated, AI-influenced landscape this book has been constructing it is fatal, because the approver of a change to a process is a fact about the process, not a fact about the request. The canonical data spine already knows that fact. The approval workflow's only job is to ask the data spine and act on the answer.

This section describes the approval workflow family that specifies a governed approval contract end to end. First, we establish the structural foundations: why the workflow trigger is keyed to the accepted suggestion rather than the raw one, how the approver chain is derived entirely from process context held in the data spine, and why SoD is enforced as a structural property rather than a procedural check. Next, we turn to execution: what the approval writes back through the commit gateway, an annotated reference approval process showing every governed obligation, and the discipline an approval form must follow when it displays context and collects only a verdict. Finally, we take the pattern deeper: the anatomy of the verdict-keyed trigger event, how risk bands derive the depth of the approver chain, an example of a cross-region sequencing change, and the time-and-escalation behaviors that handle real-world problems, including the standing approver problem.

16.2.1 Why the Trigger Is Keyed to the Accepted Suggestion, Not the Raw Suggestion

In Chapter 15, we emphasized that a suggestion is not a decision. That distinction has a concrete consequence for how an approval workflow is started. The workflow is never triggered by a SUGGESTION in DRAFT, ROUTED, or PENDING_VALIDATION. It is triggered by a SUGGESTION whose lifecycle_state has reached ACCEPTED (in Chapter 15, we defined this state as the one a SUGGESTION enters only when a VALIDATION_VERDICT row with verdict_value = ACCEPT is recorded against it) and whose subject is a proposed change to a PUBLISHED process version. The trigger carries the suggestion_id together with the VALIDATION_VERDICT that moved it to ACCEPTED, because that verdict row is the only artifact that already encodes a human author, a timestamp, and an explicit decision. A workflow keyed to a pre-verdict SUGGESTION could, in principle, start before anyone decided anything. A workflow keyed to lifecycle_state = ACCEPTED cannot, by design, start before a human has signed, because that state is unreachable without the verdict.

This is the first place our Chapter 1 boundary becomes machinery rather than principle. The event subscription that starts an approval workflow filters on SUGGESTION.lifecycle_state = 'ACCEPTED' (and equivalently 'MODIFIED', the other commit-eligible value). There is no subscription on a pre-verdict state. The boundary is not enforced by a check inside the workflow but by the fact that the workflow has no way to hear about a SUGGESTION that has not yet been accepted through an explicit VALIDATION_VERDICT.

Note

The accepted suggestion trigger is also what makes the FP&A reconstruction in Section 16.5 a single join. Because the workflow instance stores the suggestion_id and the VALIDATION_VERDICT it was started from, and the SUGGESTION already carries bundle_id and inference_event_id as foreign keys (FKs), the controller never has to correlate timestamps across systems. The FKs that the data spine already holds do the correlation.

16.2.2 Deriving the Approver Chain from Process Context

Once the workflow holds the accepted suggestion_id and its VALIDATION_VERDICT, it resolves context before it resolves an approver. It reads the PROCESS_VERSION the suggestion targets (through target_entity and target_pk), the ACTIVITY records affected by the proposed_state, and the EXECUTION_CONTEXT rows that name the trigger and target applications for those activities. Only then does it determine who must approve. The approver is not a name on a form. It is the owning role of the affected execution context, looked up against the role model the data spine already carries and resolved to people through the identity provider rather than through a workflow-local table.

Figure 16.2 shows the reference pattern. A single-approver low-risk change and a multi-level high-risk change have the same shape; only the SoD check in the middle differs. The pattern is deliberately uniform because uniformity is what makes the audit trail comparable across thousands of instances.

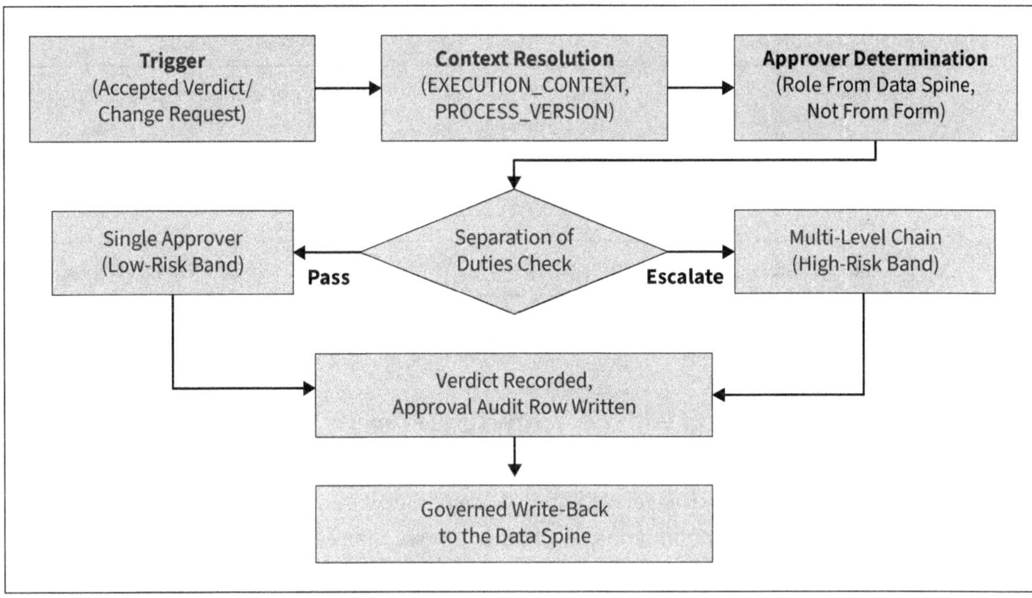

Figure 16.2: The Approval Workflow Reference Pattern

Consider an example aligned with the cost center to position scenario we introduced in Chapter 9. An AI suggestion proposed adding a cost center category validation gate to the published position creation process. A finance process owner reviewed it and recorded an accepted verdict. The approval workflow now starts. It reads the targeted PROCESS_VERSION (pv-pos, V1.3), finds the affected activity (ACT_VALIDATE_CC_CATEGORY), and reads its execution context, which names S4_CORE as the target application and the finance controlling role as the owning role. Because the change touches a control gate, the SoD check escalates: the proposer (the finance process owner who accepted the verdict) is structurally barred from being the approver of the same change. The workflow therefore routes to the controlling authority and, because the risk band is high, to a second approver in governance. None of this routing is configured in the workflow; all of it is read from the data spine. Approver resolution proceeds as an ordered walk through the data spine rather than a lookup against a form. Table 16.2 lists each step, the data spine entity it reads, and the routing fact that step establishes.

Resolution Step	Data Spine Entity Read	What It Determines
Target version	PROCESS_VERSION	Which definition the change applies to
Affected activities	ACTIVITY	Scope of the change
Ownership	EXECUTION_CONTEXT	Which role owns the affected step

Table 16.2: Approver Resolution Is Entirely Derived from the Data Spine

Resolution Step	Data Spine Entity Read	What It Determines
Risk band	The EXECUTION_CONTEXT and CONFIG_OBJECT links	Whether a control gate is touched
Approver identity	Role model and identity provider (IdP)	Which people may approve
SoD constraint	VERDICT (proposer)	Who is barred from approving

Table 16.2: Approver Resolution Is Entirely Derived from the Data Spine (Cont.)

16.2.3 Separation of Duties as a Structural Property

SoD is the structural protection that prevents a single actor from both proposing and authorizing the same change. In Chapter 1, we named it as a non-goal of AI. In Chapter 9, we encoded ownership boundaries. In Chapter 15, we made the proposer an audit row. We close the loop in this chapter by making the proposer's identity an input to the approver resolution rather than a manual reviewer instruction. The workflow reads the author of the accepted verdict and removes that identity, and any role that identity holds in the relevant scope, from the set of eligible approvers. If the removal empties the eligible set (for example, at a small organization where the only qualified approver is the proposer), the workflow does not silently proceed. It raises a governance exception, which we treat as a first-class typed event rather than an error, as we will describe in Section 16.5.

Tip

When a reviewer asks "How do you know an approver did not approve their own change?" the strongest answer is not a policy document. It is the event subscription and the approver-resolution query. Showing that the proposer's identity is subtracted from the eligible set by the same code path on every instance is a structural answer, and structural answers survive audits that procedural answers do not.

16.2.4 What the Approval Writes Back, and How

When the final approver records an approval, the workflow does not write to the canonical data spine directly. It calls the commit gateway we specified as the only writer of an AI-influenced governed change in Chapter 15, the gateway whose single foreign-key precondition is a VALIDATION_VERDICT row with verdict_value of ACCEPT or MODIFY. This gateway is idempotent on the suggestion_id and replays the same approval produces the same single lifecycle transition to COMMITTED, never a double write. It verifies, at write time, that the PROCESS_VERSION is still in the PUBLISH_STATE the workflow read it in; if the version moved underneath the workflow, the gateway refuses and the workflow reresolves rather than writing a stale decision. The AUDIT_EVENT row it writes records the suggestion_id, the VALIDATION_VERDICT, the approver identities in order, the manifest_hash at write time, and the workflow instance identifier. That row is the join target for everything in Section 16.5.

Listing 16.1 is intentionally not in SAP Build Process Automation syntax. It is the contract the workflow must satisfy regardless of how the steps are drawn in the modeler, and it is the same commit gateway Chapter 15, Section 15.5 specified, reached now from a workflow rather than from the validation UI. The four assertions are the entire boundary: act only on an accepted or modified SUGGESTION, act only with an ACCEPT or MODIFY verdict, refuse on PUBLISH_STATE drift, and subtract the proposer. A workflow that satisfies these is a governed approval; one that omits any of them is a liability no diagram can disguise.

```
FUNCTION apply_approval(suggestion_id, approver_chain, instance_id):
    sg = spine.read_suggestion(suggestion_id)
    v  = spine.read_validation_verdict(sg.suggestion_id)
    ASSERT sg.lifecycle_state IN ('ACCEPTED','MODIFIED')
    ASSERT v.verdict_value    IN ('ACCEPT','MODIFY') – never pre-verdict
    pv = spine.read_process_version(sg.target_pk)
    ASSERT pv.publish_state == sg.observed_publish_state – refuse drift
    ASSERT proposer(v) NOT IN approver_chain         – SoD, structural
    WITH GOVERNED TRANSACTION:                       – the commit gateway
        spine.transition(sg, to='COMMITTED',
                         keyed_by=suggestion_id)     – idempotent
        spine.write_audit_event(
            kind='APPROVAL_COMMITTED',
            suggestion_id=suggestion_id,
            verdict=v.verdict_value,
            approvers=approver_chain,
            manifest_hash=sg.bundle.manifest_hash,
            workflow_instance=instance_id)
    RETURN 'COMMITTED'
```

Listing 16.1: The Approval Commit Is a Guarded, Idempotent Transition Through the Gateway, Never a Direct Table Write

16.2.5 An Annotated Reference Approval Process

We will now look at an abstract contract beside a concrete reference process and annotate exactly where each governed obligation lives. Listing 16.2 is the skeleton of an approval process as it would be assembled in SAP Build Process Automation, written in a deliberately product-neutral pseudo-definition so you can see the obligations rather than the modeler's visual grammar. Every numbered comment marks a point where this chapter's architecture constrains what the low-code surface is otherwise free to do.

```
PROCESS approval_on_accepted_suggestion:

  TRIGGER event = spine.suggestion.accepted      # (1) producer-filtered
      bind  payload -> instance.context          #     stored once, frozen

  STEP resolve_context:                          # (2) read, never ask
      pv   = spine.read_process_version(
                 instance.context.target_pk)
      acts = spine.read_activities(pv)
      exec = spine.read_execution_context(acts)
      band = spine.classify_risk(acts, exec)     #     band is derived

  STEP resolve_approvers (DECISION):             # (3) switch, not register
      eligible = role_model.holders(exec.owning_role, band)
      eligible = eligible MINUS proposer(
                 instance.context.validation_verdict_id)
      IF eligible IS EMPTY:                      # (4) refuse, do not dilute
          RAISE governance_exception('SOD_EMPTY_SET')
      chain = build_chain(band, eligible)        #     depth from band

  STEP human_approval (FORM) for each role in chain:
      display  context, change, chain, band, sod_result, ai_panel?
      collect  verdict { verdict_value, rationale? } # (5) one field only
      timer    escalate -> resolve_approvers         #     reresolve, audit

  STEP commit (AUTOMATION):                          # (6) one door only
      result = spine.commit_gateway.apply_approval(
                   suggestion_id  = instance.context.suggestion_id,
                   approver_chain = collected_chain,
                   instance_id    = instance.id) #     idempotent, guarded
      ON drift: re-run resolve_context           #     refuse stale write

  END  # no path reaches the spine except the commit gateway in step (6)
```

Listing 16.2: A Reference Approval Process, Annotated with Its Governed Obligations

You can read the six annotations as the entire governance of the process. The first annotation is the accepted-suggestion trigger: The process cannot start from a pre-verdict `SUGGESTION` because the event it subscribes to is emitted only by the data spine and only when `lifecycle_state` has reached `ACCEPTED` through an explicit `VALIDATION_VERDICT`, and the payload is frozen onto the instance so nothing is silently reread mid-flight. The second annotation is the read-not-ask discipline: Process version, activities, and execution context are read from the data spine, never collected from a requester. The third annotation is the decision-as-switch principle: The decision artifact computes a chain from data spine-supplied inputs but is never itself the register of who may approve. The fourth annotation is the structural refusal: An emptied eligible set raises a governance exception rather than diluting SoD to make the process complete. The fifth annotation is the single-field form: The only thing the human supplies is a decision and an optional rationale. The sixth annotation is the single path: The only way to change canonical truth is the idempotent, drift-guarded handler.

A review of any real approval workflow built in the product reduces to checking that these six annotations are honored. The check does not require reading the workflow's logic in depth; it requires confirming that the trigger is the data spine event and not a manual start, no form field carries a routing fact, the decision reads rather than registers authority, the empty-set branch raises rather than proceeds, and the only write is the handler call. A workflow that passes this six-point inspection is governed by design; one that fails any point has a named, fixable defect rather than a vague governance concern. This is implementation guidance in the truest sense: the small set of invariants a reviewer can verify mechanically on any instance, in any region, regardless of how the workflow was drawn.

Tip

A useful exercise for an implementation team is to take an existing, pre-architecture approval workflow from the landscape and map it against these six annotations. The gaps are almost always the same two: The approver is captured on a form rather than resolved from the data spine, and the write is a direct downstream call rather than the commit gateway. Both are mechanical to remediate once named, and naming them against a reference is faster than discovering them in an audit.

16.2.6 Approval Forms: What They Display and What They Must Never Collect

The form is where the architecture is most often quietly defeated, because a form is the easiest place to add a field, and a field is the easiest place to hide a decision. We therefore state the correct discipline as a rule rather than a guideline. An approval form displays the change, the affected process version, the resolved approver chain, the risk band, the SoD result, and, where present, the AI recommendation with its cited evidence. It collects exactly one thing: the approver's verdict, with an optional rationale. It never collects the approver's identity, the risk band, the next approver, or whether SoD applies. Every one of those is a fact the data spine already holds, computed by context resolution, and shown on the form for the human to see, not asked of the human to supply. A form field that captures a routing fact has moved authority from the data spine into the form, and an auditor who finds such a field has found the exact point at which the governed architecture stopped being governed.

There is a practical reviewer test you can apply without reading any code. Look at the approval form and imagine removing every field except the verdict control and its rationale. If the workflow can still route, escalate, and enforce SoD with those fields gone, the form is a display surface, and the architecture holds. If removing a field breaks routing, that field was carrying authority it should never have carried, and that break is your finding. The strongest forms in this architecture are the ones that would still produce a correct, governed outcome if every field but the decision were hidden, because everything else came from the data spine.

Figure 16.3 makes this rule concrete. Every field above the verdict is displayed, read from the data spine by context resolution; the single bordered control at the bottom is the only thing the human supplies. The annotation on the left-hand side is the reviewer test stated as a picture: Hide every displayed field and the form must still route correctly, because none of those fields was ever the source of a routing fact.

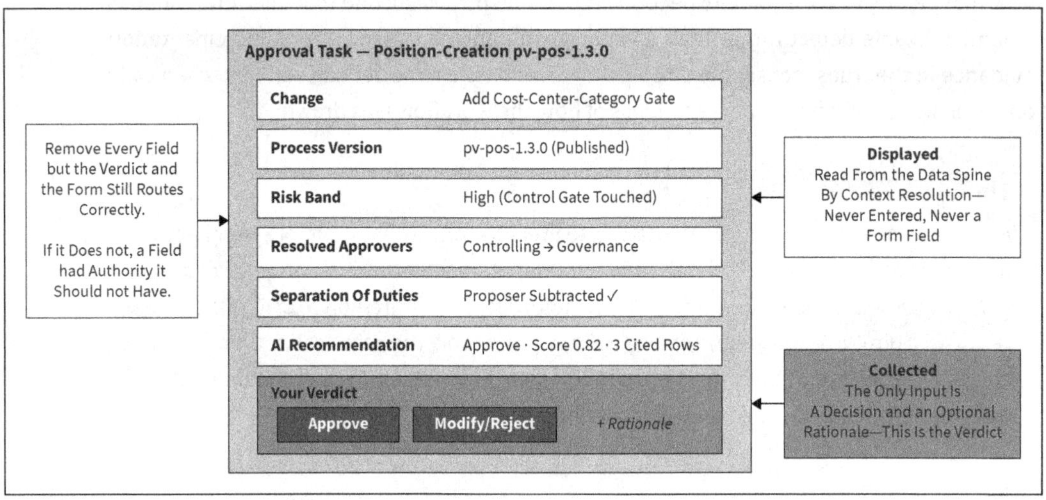

Figure 16.3: Anatomy of a Governed Approval Form

16.2.7 The Anatomy of a Verdict-Keyed Trigger

The trigger is where this chapter's entire boundary is first realized in running software. In SAP Build Process Automation, the approval workflow is started by an event subscription, and the subscription is not a generic listener that tells you a verdict changed. It is a filtered subscription whose predicate is evaluated by the data spine before the event is ever emitted. The event is published only when a verdict transitions into the accepted state and the verdict's subject is a proposed change to a published process version. A draft verdict, a rejected verdict, or a verdict whose subject is not a governed change never produces an event, so the workflow engine never sees it. The filter is on the producing side, not the consuming side. This matters because a consumer-side filter can be bypassed by a second consumer; a producer-side filter cannot because there is nothing to consume.

As shown in Listing 16.3, every field in the payload is an FK or an immutable fact drawn from the SUGGESTION and its VALIDATION_VERDICT exactly as we defined them in Chapter 15. suggestion_id, bundle_id, and inference_event_id are the same FKs the SUGGESTION row already carries, not new identifiers we invented for this chapter. The workflow stores the whole payload on the instance at start time and never rederives any of it. This is deliberate: A workflow that rereads the SUGGESTION halfway through could observe a different value if anything changed, and an audit that depends on reread values is an audit that cannot be trusted. The instance is anchored to the payload it started with, and the commit gateway rechecks the data spine state at commit time rather than the workflow rereading it mid-flight. If you are a controller, you should recognize this pattern from financial close: You reconcile against the snapshot you opened the period with, not against a moving balance.

```
{
  "event": "spine.suggestion.accepted",
  "emitted_when": {
    "suggestion.lifecycle_state": "ACCEPTED",
    "validation_verdict.verdict_value": "ACCEPT",
    "process_version.publish_state": "PUBLISHED"
  },
  "payload": {
    "suggestion_id": "sg-77120",
    "validation_verdict_id": "vv-2026-031847",
    "verdict_value": "ACCEPT",
    "decided_by": "u.finance.controller.02",
    "decided_at": "2026-03-18T09:14:22Z",
    "bundle_id": "gb-2026-0312-77120",
    "inference_event_id": "inf-2026-0312-44910",
    "target_entity": "PROCESS_VERSION",
    "target_pk": "pv-pos-1.3.0",
    "observed_publish_state": "PUBLISHED",
    "manifest_hash": "9f2c...e71a"
  }
}
```

Listing 16.3: The Trigger Event Is a Narrow, Producer-Filtered Contract, Not a Generic Change Notification

16.2.8 Risk Bands and the Shape of the Approver Chain

The reference pattern shown earlier in Figure 16.2 branches on an SoD check, but the depth of the approver chain is governed by a risk band, and the risk band is not a workflow constant (see Table 16.3). It is derived from what the change touches in the data spine. A change that modifies only descriptive metadata of an activity is low band. A change that alters sequencing or execution context is medium band. A change that adds, removes, or modifies a control gate—anything mapped through the `EXECUTION_CONTEXT` and `CONFIG_OBJECT` links to a configuration object with a control classification—is high band. The band is computed by the same context-resolution step that finds the approver, from the same rows, so the band and the approver are always consistent. A high band change cannot be routed as if it were low band, because the band is read, not asserted.

Risk Band	What the Change Touches	Approver Chain	SoD Rule
Low	Descriptive metadata only	Single owning-role approver	Proposer subtracted
Medium	Sequencing or execution context	Owning role and architecture authority	Proposer and proposer role subtracted
High	A control gate or master data constraint	Owning role, architecture, and governance	Proposer, role, and dual control
Governance	Spine-governed configuration (weights, bands)	Governance authority only	No self-approval of governance configuration

Table 16.3: The Approver Chain Depth Is Derived from the Data Spine, Not Configured in the Workflow

The scoring weights and priority-band thresholds we treated as versioned configuration rows in Chapter 15 are themselves changeable, and a change to them is the most sensitive change in the entire system, because it silently alters how every future suggestion is scored. Such a change is therefore its own highest band: It can be approved only by the governance authority, never by the team that benefits from the new weights, and the approval workflow for it is the same machinery with the eligible set reduced to governance. The architecture does not need a special workflow for its most dangerous change; it needs the same workflow with the data spine supplying a narrower answer to the question "Who may approve?"

16.2.9 Worked Example: A Cross-Region Sequencing Change

In our first example, we added a control gate. Consider a different shape for this example. An AI suggestion proposes resequencing two activities in the published position-creation process so that a cost center availability check runs before, rather than after, the organization assignment step, a change that reduces a known issue with the telemetry we described in Chapter 8. A regional process owner for the EU variant accepts the verdict. The trigger fires. Context resolution reads pv-pos-1.3.0-eu, the branched version we modeled as a lineage child of V1.3 in Chapter 9. Because the change alters sequencing, the band is medium, and the approver chain is the owning role plus the architecture authority. However, the resolution also discovers, from the lineage graph, that the global parent V1.3 does not carry the same resequencing. The workflow does not silently apply a divergence. It records, on the instance, that the change creates a branch divergence, and the approval task presented to the architecture authority states the divergence explicitly. The architect approves the EU branch change knowing it diverges, and the divergence is now an approved, recorded fact rather than an accident discovered later during an audit of why EU and global behave differently.

This is the kind of outcome for which the canonical data spine exists to make ordinary. Without lineage in the data spine, a regional resequencing is a local edit nobody notices until two regions disagree in production. With lineage in the data spine and a workflow that reads it, the divergence is surfaced at approval time, decided deliberately, and recorded as part of the same audit chain. If you have ever had to explain why two subsidiaries closed the same process differently, you will recognize the value of having that explanation be a row rather than an investigation.

16.2.10 Time, Escalation, and the Standing Approver Problem

Approval workflows live in real time, and real time introduces two failure modes the architecture must answer rather than ignore. The first is the unattended task: an approver who does not act. The workflow carries an escalation timer, but the escalation target is resolved the same way the original approver was—from the data spine's role model, never from a hard-coded copy. When a task escalates, the new approver is the next eligible role-holder with the proposer still subtracted, and the escalation itself is written as an audit event so that the question "Why did a different person approve this?" has a recorded answer. The second failure mode is the standing approver problem: an organization small enough that the only eligible approver is also the proposer. The architecture refuses to dilute SoD to make the workflow complete. It raises a governance exception, and in Section 16.5, we explain why that refusal is itself a first-class, auditable outcome rather than an error to be suppressed. A platform that quietly lets the proposer approve their own change when nobody else is available has not solved the standing approver problem; it has hidden it until an auditor finds it.

16.3 Exception and Remediation Flows

Approval workflows handle the planned path: a change is proposed, decided, and authorized. Exception and remediation flows handle the unplanned one: something the data spine expected to be true is not. A validation rule fails. A WalkMe telemetry signal shows users abandoning a step the process says is mandatory. An interface SAP LeanIX listed as active stops answering. A configuration object a published activity depends on is missing in the target system. Each of these is a deviation between modeled intent and operational reality, and each is the kind of thing that, when left to informal handling, becomes the silent drift we warned the data spine exists to prevent in Chapter 9.

The discipline of this section can be summarized in one sentence: An exception is a typed event, and remediation is a governed workflow. The first half of that sentence stops exceptions from being log lines nobody joins. The second half stops remediation from being an engineer's ad hoc fix that no audit can reconstruct. This section covers exception and remediation handling in seven steps: how a deviation becomes a typed, severity-banded event; the schema that makes all exceptions queryable as one population; the three remediation paths (auto, guided, and escalation); how auto-remediation rules stay governed and idempotent; the remediation case as a durable spine object; how workflow state reconciles with spine truth; and a worked example from friction signal to governed change.

16.3.1 An Exception Is a Typed, Severity-Banded Event

When a deviation is detected by a validation run, a telemetry threshold, an interface health probe, or a dependency check, it is not written to a log and surfaced on a dashboard. It is recorded as a typed exception event linked to the canonical entity it concerns. A failed cost center category validation links to the `ACTIVITY` and the `CONFIG_OBJECT` it tested. A telemetry friction exception links to the `ACTIVITY` and the `WALKME_ASSET`. An interface-down exception links to the `INTERFACE` and the activities that depend on it. The link is what makes the exception joinable to the same audit chain that everything else in Part IV shares. The type and the severity band together decide what happens next. Figure 16.4 traces this path: Detection produces a typed event, classification selects the remediation path from the type and severity band, and the path either auto-remediates, opens a guided case, or escalates.

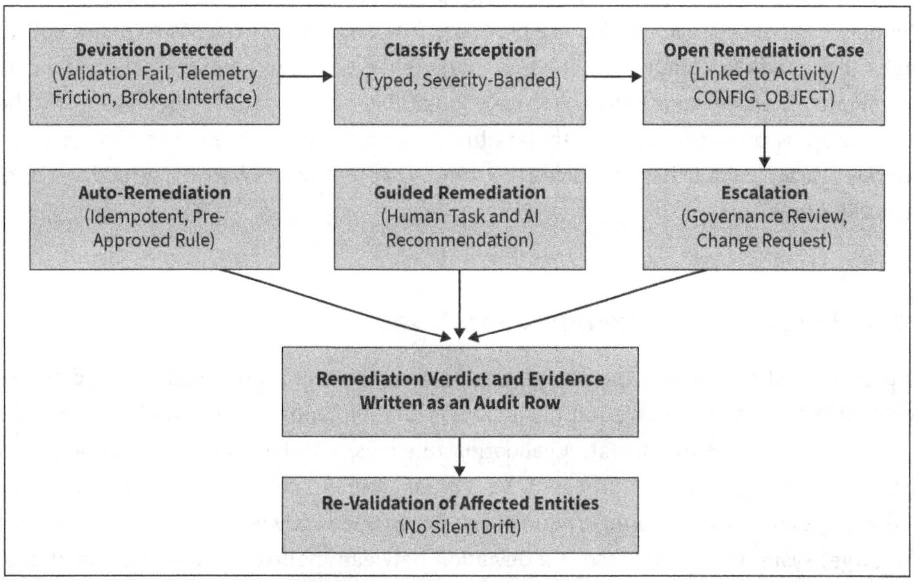

Figure 16.4: Exception and Remediation Flow

Table 16.4 summarizes this mapping and lists each exception type with its typical source, default severity band, and default remediation path.

Exception Type	Typical Source	Default Band	Default Path
Validation failure	Data spine validation run	High	Guided or escalate
Telemetry friction	WalkMe signal	Medium	Guided
Interface unavailable	Health probe	High	Escalate
Missing dependency	Configuration check	High	Escalate
Stale version reference	Drift check	Medium	Auto or guided
Idempotent retry failure	Execution service	Low	Auto-remediate

Table 16.4: Exception Type and Band Select the Remediation Path

16.3.2 Three Governed Remediation Paths

Auto-remediation is reserved for the narrow set of exceptions whose fix is idempotent, pre-approved as a governed rule, and incapable of changing canonical meaning. A transient retry failure on an execution service that is safe to replay is the archetype. The auto-remediation rule is itself a versioned, governed artifact. It has an owner, an approval state, and an audit row each time it fires. *Auto* does not mean *unwitnessed*. It means the witness was recorded once, at rule-approval time, and again, briefly, each time the rule acts.

Guided remediation is the common path. It opens a human task that carries the exception, the affected spine context, and, where the AI tier has something useful to say, a recommendation. The human decides, the decision is a verdict, and the verdict drives the write-back. Guided remediation is, structurally, the same machinery as an approval workflow with a different trigger. That sameness is intentional. The fewer distinct shapes the platform has, the fewer distinct audit stories a controller must learn.

Escalation is the path for exceptions whose remediation is itself a change to a published process or architecture fact. An interface that is permanently gone is not remediated by a human task; it is remediated by a change request that reenters the approval workflow outlined in Section 16.2. Escalation is therefore not a dead end. It is the point at which an operational exception becomes a governed change proposal, and from there the approval family takes over again.

Note

The reason escalation routes back into the approval family rather than into a separate change process is the fifth principle of this chapter: The platform should have as few governed write paths as possible. Every remediation that changes canonical truth converges on the same commit gateway. A controller who has audited one governed change has, in effect, audited them all.

16.3.3 Worked Example: A Failed Control-Gate Validation

Let's return to the position to creation process. After V1.3 published with the new cost center category validation gate, a nightly validation run finds that for one company code the `CONFIG_OBJECT` the gate depends on, `CFG_CC_CATEGORY`, is not present in the target SAP S/4HANA system. This is a high-band validation-failure exception. It is recorded as a typed event linked to `ACT_VALIDATE_CC_CATEGORY` and `CFG_CC_CATEGORY`. Auto-remediation is not available, as creating a configuration object changes canonical meaning and is explicitly outside the auto path. Guided remediation opens a human task for the controlling authority that owns the activity. The task carries the exception, the affected version, the missing configuration object's IMG path, and an AI recommendation that the gate be made conditional on company-code scope until the configuration is delivered. The controller does not have to accept the recommendation; Section 16.4 explains why the recommendation can never make a decision. The controller records a verdict to escalate, which raises a change request, which reenters the approval workflow as a proposed amendment to V1.3. The exception, the verdict, the change request, and the eventual approval are one joinable chain.

If you have an FP&A background, you will understand how this example is the whole point of the discipline. Imagine a control gate failed in production, and three weeks later an auditor asks what happened. The answer is not an anecdote assembled from chat logs. It is a single query: The exception row carries the verdict, the verdict carries the change request, the change request carries the approval, and the approval carries the `manifest_hash`. The cheapest internal audit a controller can run is to take one exception and walk that chain to its source. Either it completes cleanly or it exposes a gap, and either outcome is worth more than spending an entire quarter checking dashboards.

16.3.4 The Exception Event Schema

The claim that an exception is a typed event is only meaningful if the type is a schema rather than a free-text category. Every exception, regardless of source, is written in the same shape, and that shape is what makes 10,000 heterogeneous deviations queryable as one population. The schema names the canonical entities the exception concerns by their stable identifiers, carries the type and band, and carries an evidence reference rather than an evidence copy. This is the same discipline we applied to master data in Chapter 9, now applied to exception evidence. The remediation path is not in the event; it is decided by the classifier from the type and band, so that the same exception type always enters the same path and the path is auditable as policy rather than as an engineer's judgment. Listing 16.4

shows the schema. Every exception, regardless of source, is recorded in this shape: the typed identifier, the canonical entities it concerns, the evidence reference, and the classifier's decision.

```
{
  "exception_id": "ex-2026-0319-00482",
  "type": "VALIDATION_FAILURE",
  "severity_band": "High",
  "detected_at": "2026-03-19T02:11:07Z",
  "detected_by": "spine.validation.nightly",
  "concerns": {
    "activity_id": "ACT_VALIDATE_CC_CATEGORY",
    "config_object_id": "CFG_CC_CATEGORY",
    "process_version_id": "pv-pos-1.3.0",
    "scope": { "company_code": "DE01" }
  },
  "evidence_ref": "vr-2026-0319-55217",
  "rule_id": "VR.CONFIG.PRESENT",
  "remediation_path": null
}
```

Listing 16.4: Every Exception Is the Same Typed, Data Spine-Linked Shape Regardless of Its Source

The `remediation_path` is null at detection time on purpose. It is filled by the classifier, not the detector, because the detector knows only that something is wrong, while the classifier knows the governed policy that maps a type and a band to a path. Separating detection from classification is the same separation the architecture applies everywhere: The component that observes is not the component that decides what observation means, and the decision is governed configuration rather than embedded code.

16.3.5 Auto-Remediation Is a Governed Rule, Not a Code Path

Auto-remediation is the most dangerous of the three paths because it acts without a human in the loop, so the architecture is at its strictest here. An auto-remediation rule is a versioned data spine artifact with an owner, an approval state, an explicit idempotency guarantee, and a meaning-preservation assertion. It is not a branch in a workflow diagram. The rule names exactly which exception type and band it applies to, the precondition it checks before acting, the idempotent action it performs, and the audit row it writes every time it fires. A rule that cannot state its idempotency guarantee cannot be approved, and a rule that could change canonical meaning cannot be an auto-remediation rule at all; by definition, it belongs in the guided or escalation path. Listing 16.5 shows the structure of one such rule, written deliberately as a versioned, governed artifact rather than as embedded code.

```
RULE AR.RETRY.IDEMPOTENT_EXEC_SERVICE  (version 3, state=Approved)
  applies_to:    type=IDEMPOTENT_RETRY_FAILURE, band=Low
  precondition: target_service.idempotent == TRUE
                AND attempts < max_attempts
                AND no_canonical_write_in_action == TRUE
  action:        replay(original_request)        - safe by precondition
  on_success:    write_audit(kind='AUTO_REMEDIATION',
                             rule='AR.RETRY.IDEMPOTENT_EXEC_SERVICE',
                             exception_id=:ex, outcome='RESOLVED')
  on_exhausted: escalate(to='GUIDED', exception_id=:ex)
  owner:         platform.operations
  approved_by:  governance.authority   approved_at: 2026-02-04
```

Listing 16.5: An Auto-Remediation Rule Is a Governed, Idempotent, Meaning-Preserving Artifact

The `on_exhausted` clause is the safety net that keeps auto-remediation honest. An auto rule never loops forever and never fails silently. When its bounded attempts are exhausted, it converts itself into a guided remediation, which requires a human and a verdict. Auto-remediation is therefore not an escape from the governed path. It is a fast lane that merges back into it the moment a simple case stops being simple. The controller can audit the entire auto-remediation population with one query against the audit rows it writes, and they can prove that every exhausted auto rule produced a human-decided follow-up rather than an unreported error.

16.3.6 The Remediation Case as a Data Spine Object

When remediation is guided or escalated, it is not merely a workflow instance; it opens a remediation case, and the case is a first-class data spine object linked to the exception and to the canonical entities the exception concerned. This matters because cases outlive workflow instances. A workflow instance is operational and may be retried, replayed, or migrated; a remediation case is the governed record that says "This deviation was opened, worked, decided, and closed." It persists in the data spine with the same immutability discipline as a verdict. The case carries its own lifecycle (open, in progress, decided, and closed) and the transitions are audit rows. An auditor who says "Show me every control gate exception in Q1 and how each was resolved" is asking for a single query over the case population, not a forensic exercise across an operational engine's internal state.

16.3.7 Reconciliation Between Workflow State and Data Spine Truth

The separation between workflow state and canonical truth, asserted in the positioning section and revisited later for audit, has one operational consequence that must be engineered rather than assumed: The two can disagree, and the architecture needs a defined answer for when they do. A workflow instance can crash after a human recorded a verdict but before the commit gateway committed. A handler can commit and the workflow can fail to record that it committed. A retry can replay a step whose effect already landed. These are not out of the ordinary; they are common problems in any distributed system, and in Chapter 11, we named the same class of problem for ingestion. The answer here is the same answer we gave there: The data spine is authoritative, the workflow reconciles to it, and reconciliation is a defined cycle rather than a manual rescue.

Reconciliation is possible only because of two earlier design choices. First, the commit gateway is idempotent on the `suggestion_id`, so a replay cannot apply twice; a second call with the same key observes the transition already made and returns the same result. Second, the workflow instance stores the `suggestion_id` it is acting on, so a reconciliation pass can ask the data spine, authoritatively, whether the change for that verdict was applied. If the data spine says it is applied and the workflow believes it is not, the workflow advances to its post-commit state without recalling the handler. If the data spine says it is not applied and the workflow believes it is, the workflow presents the commit step again. The reconciliation never guesses and never writes; it reads the authoritative side and moves the operational side to match. If you are a controller, you might recognize this pattern as the same one a subledger uses against a general ledger: One side is truth, the other reconciles to it, and the reconciliation itself is logged.

This is also why the audit chain stays trustworthy under failure. A crashed or replayed workflow can make operational state messy, but it cannot make the governed record wrong, because the governed record was written once, idempotently, by the only path that may write it, and the reconciliation only ever moves the workflow toward what the data spine already recorded. Operational mess is recoverable, and governed truth is never corrupted. The boundary between them is a defined cycle that we will instrument and monitor at production volume in Chapter 18.

16.3.8 Worked Example: Telemetry-Driven Remediation

Not every exception is a hard failure. The most valuable ones are often soft. WalkMe telemetry, linked to the data spine through the adoption domain we specified in Chapter 8, shows that the position-creation process for a particular activity (entering the cost center reference) is being abandoned and restarted by a high fraction of users in one region. No validation failed, and no interface is down. The process works exactly as modeled; users simply cannot complete the step reliably. This is a medium-band telemetry-friction exception, linked to the `ACTIVITY` and the `WALKME_ASSET`. Auto-remediation is not available, as there is no idempotent fix for a usability problem. The exception opens a remediation case and a guided task for the process owner, carrying the telemetry evidence and an AI recommendation: the suggestion pipeline (covered in Chapter 15). The AI notices that a similar friction pattern was resolved elsewhere by adding an inline validation hint and proposes the same change here. The process owner reviews the recommendation, modifies it after discovering the regional data shows the real problem is a defaulting rule and not a missing hint, and records a verdict to escalate a scoped change. The escalation reenters the approval family. The friction signal, the case, the verdict, the modified recommendation, and the eventual approved change are one chain.

This example closes a loop we opened in Chapter 1 and again in Chapter 8: Design produces a process, delivery executes it, telemetry observes how it actually behaves, and improvement feeds back into the process through a governed decision, never as an automatic reaction to a metric. The telemetry did not change the process. It opened a case. A human changed the process, and the data spine recorded the whole arc. From an FP&A perspective, this is the difference between a dashboard that says a step is slow and an audit trail that explains what the organization decided to do about it and why.

Note

The recurring shape across both worked examples in this section is that escalation always reenters the approval family from Section 16.2. The platform deliberately has only one way for canonical truth to change—an accepted verdict driving the commit gateway—and every remediation that needs to change canonical truth converges on it. Auditing one governed change teaches the controller how to audit all of them, regardless of whether the change began as a suggestion, a validation failure, or a friction signal.

16.4 AI-Assisted Recommendations

The third workflow family is the one where Chapter 1's AI boundary is most tempting to cross and therefore the one where it must be the most visibly uncrossable. An AI-assisted recommendation is the appearance of the AI tier inside a running workflow, such as in an approval task, a guided remediation task, or a standalone review. The recommendation is produced by the same governed pipeline we specified in Chapter 13 to Chapter 15: grounded on a grounding bundle, scored by governed configuration, and explained with cited evidence. The only thing that changes in this chapter is where it appears. What does not change, and what must be demonstrably unchanged, is that it never decides.

This section covers AI-assisted recommendations in seven steps: how the recommendation is advisory by design, what the recommendation panel must show to be auditable rather than a black box, why the panel must never prefill the human's decision, a worked example of a recommendation inside an approval task, the disposition analytics that signal whether the model is well calibrated, how the verdict records the human's three-way response to the advice, and which metric to use first when creating an AI-assisted program.

16.4.1 The Recommendation Is Advisory by Design

Figure 16.5 shows an AI recommendation inside a human task. The important feature of the diagram is what is missing from it. There is no arrow from the recommendation panel to the data spine. Every path from the recommendation passes through the accept or the modify-or-reject control, and both of those write a human verdict before anything reaches the write-back handler. The AI tier sits outside the task boundary and connects to it with a dashed, advisory-only line. The boundary is not the absence of permission. It is the absence of a direct connection.

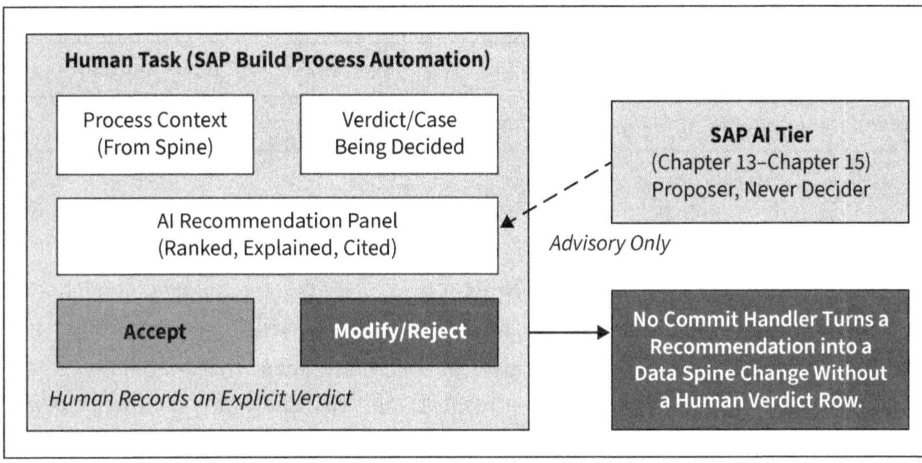

Figure 16.5: AI-Assisted Recommendation Inside a Human Task

This is the same argument we made in Chapter 15 about the suggestion pipeline. We are restating it at the workflow layer because it must hold at every layer, or it holds at none. In Chapter 15, the boundary was the absence of a commit handler that turned an inference into a spine change without a verdict. In this chapter, the boundary is the absence of a workflow transition that turns a recommendation into a write-back without passing through the human task's verdict control. A reviewer can be shown the workflow model and asked to find the transition that bypasses the human task. There is none to find, and the inability to find it is the control.

16.4.2 What the Recommendation Panel Must Carry

An advisory recommendation is only useful if the human can interrogate it as fast as they can read it. The panel inside the task therefore carries four things, and a workflow that omits any of them has built an oracle, not an assistant. It carries the ranked recommendation in plain language. It carries the score and the band, so the human sees how confident the pipeline was and against what threshold. It carries the cited evidence (the named data spine rows the recommendation was grounded on, not a summary of them) so the human can check the reasoning against the source. It also carries the `Inference_Event_ID`, so the recommendation the human saw is reproducible long after the model behind it has been redeployed. Table 16.5 lists the four elements the panel must carry, the data spine source of each one, and why the human reader needs it.

Panel Element	Source	Why the Human Needs It
Ranked recommendation	SUGGESTION artifact	Shows the proposal in readable form
Score and band	Scoring configuration	Calibrates trust against threshold
Cited evidence rows	Grounding manifest	Lets the human verify, not just believe
Inference record ID	INFERENCE_EVENT	Makes what was shown reproducible
Boundary statement	Workflow contract	States plainly that this is advisory

Table 16.5: The Advisory Panel Is a Tool, Not an Answer

16.4.3 An Annotated Human Task Contract

The reference approval process showed where governance lives in an orchestration. The human task is where governance lives in the one moment that matters most—the decision—and it deserves the same treatment. Listing 16.6 shows the contract a human task must satisfy whenever an AI recommendation is present, written again as a product-neutral definition so the obligations are visible rather than the modeler's form designer. The annotations mark the points where the advisory boundary becomes machinery.

```
HUMAN_TASK decide_with_ai_assist:

  INPUTS (read-only, from spine):                  # (1) display, not edit
      change, process_version, case_or_verdict

  AI_PANEL (advisory region, beside the control): # (2) beside, never in
      recommendation_text
      score, band
      cited_evidence  = grounding.named_rows()     #     rows, not summary
      inference_id                                 #     reproducible
      boundary_note   = 'Advisory. You decide.'

  DECISION_CONTROL:                                # (3) nothing preselected
      options      = [ACCEPT, MODIFY, REJECT]
      default      = NONE                          #     no pre-fill
      cost(ACCEPT) == cost(MODIFY) == cost(REJECT) #     equal-effort rule

  COLLECT verdict:                                 # (4) verdict_value is data
      decision      in {ACCEPT, MODIFY, REJECT}
      verdict_value in {ACCEPT,MODIFY,REJECT}      #     3-way, recorded
      rationale?    free text
```

```
ON SUBMIT:                                           # (5) verdict, then exit
    write spine.verdict(decided_by = actor,
                        verdict_value = verdict_value,
                        inference_id = AI_PANEL.inference_id)
    # task ends here; no path from panel to spine

END  # the only output is a human verdict row
```

Listing 16.6: The Human Task Contract When an AI Recommendation Is Present, Annotated

The five annotations are the advisory boundary made literal. The first annotation establishes that task inputs are read-only context drawn from the data spine, not editable fields. The second annotation places the AI panel beside the decision control and never inside it, and insists the cited evidence is the named rows rather than a summary the human cannot verify. The third annotation is the anti-coercion rule: No option is preselected and agreeing costs the human the same as disagreeing, so the recorded decision is a choice rather than the path of least resistance. The fourth annotation captures the three-way disposition against the recommendation, which is the data the disposition analytics of the previous subsection depends on. The fifth annotation is the exit discipline: Submission writes a human verdict carrying the inference identifier, and the task ends there because there is no defined path from the panel to the data spine.

If you look at this listing beside Figure 16.5, you will notice the same fact is visible in both representations: Figure 16.5 shows there is no arrow from the recommendation panel to the data spine, and the contract shows there is no statement in the definition that writes from the panel. A reviewer can be handed either and asked to find the bypass. Neither contains one, and the consistent absence across the diagram, the contract, and the running workflow is what makes the advisory boundary a structural property rather than a claim. For the controller, the practical consequence is the one we keep returning to throughout this chapter: The question is never "Did the AI decide?" but "Show me the verdict," and the contract guarantees the verdict exists, names a human, and carries the inference it was shown, which are three things an audit of an AI-influenced decision actually needs.

16.4.4 The Verdict Records What the Human Did with the Advice

When the human acts, the `VALIDATION_VERDICT` they record is richer than a yes or a no. Its `verdict_value` is one of the three values we defined in Chapter 15 (`ACCEPT`, `MODIFY`, or `REJECT`) and the `SUGGESTION`'s `lifecycle_state` moves to the matching `ACCEPTED`, `MODIFIED`, or `REJECTED`. In this chapter, we call that three-way outcome the verdict's disposition relative to the recommendation, not because disposition is the word the controller already uses for what was decided about the proposal, but because disposition is nothing more than the canonical `verdict_value` read in the AI-assist context. `ACCEPT` is accepted as proposed, `MODIFY` is accepted with change, and `REJECT` is declined. This three-way disposition is not a bureaucratic detail. It is the data that lets the organization answer, a year later, the only question that matters about an AI assist program: When humans disagreed with the model, were they right? A landscape that records only the final action cannot answer that question. A landscape that records the `verdict_value` can. If you have an FP&A background, you will recognize that value immediately, because it is the same value a variance analysis has. The gap between what was proposed and what was decided is where the learning is.

> **Tip**
>
> A practical recommendation for any team setting up an AI-assisted family: Use the modification disposition first. The rate at which humans accept with modification, and the nature of the modifications, is the earliest and cheapest signal of whether the scoring configuration (covered in Chapter 15) is calibrated for the process in question. It is a controller's metric, and it is available before any model-quality metric stabilizes.

16.4.5 Why the Panel Cannot Prefill the Decision

There is a subtle and tempting error that violates the boundary while appearing to respect it: the prefilled decision. The human task still exists and the human still clicks a button, but the decision control arrives with the AI's recommended choice already selected, so that approval requires one click and disagreement requires several. This passes a superficial inspection, as there is a human task and a verdict, but it quietly converts the human from a decider into a rubber stamp. The architecture should forbid it structurally. The decision control in the task must render with no option selected. The recommendation should be displayed beside the control, never inside it. Selecting the recommended option and selecting any other option should cost the human exactly the same number of actions. This is not a usability nicety; it is the boundary. A panel that preselects the model's answer has built the auto-decider we prohibited in Chapter 1, Section 1.5, disguised as a human task.

The test a reviewer should apply is simple: If accepting the AI's recommendation is cheaper for the human than rejecting it, the panel is not advisory, it is coercive, and the verdict it produces cannot be trusted as an independent decision. Equal effort to agree and to disagree is what makes the recorded disposition real data rather than the path of least resistance.

16.4.6 Worked Example: A Recommendation Inside an Approval Task

The AI-assisted family is not a separate workflow but a capability that appears inside the other families. Return to the first worked example of Section 16.2, the proposed cost center category control gate. When the approval task reaches the governance approver, the task carries more than the change. It carries an AI recommendation produced by pipeline covered in the Chapter 15, grounded on the grounding bundle from Chapter 14: The model observes that three peer processes already carry an equivalent gate and that the proposed gate's configuration matches the established pattern, and it recommends approval with a noted caveat that one peer implements the gate as a warning rather than a hard stop. The governance approver reads the recommendation, reads the cited evidence (the named peer process versions, not a summary of them), and decides. Crucially, the approver disagrees with the caveat: In this control context, a warning is insufficient, and the hard stop is correct. The approver approves the change and records the `VALIDATION_VERDICT` with `verdict_value MODIFY`, with the modification being the explicit rejection of the model's caveat. The model was useful and partly wrong, and the system recorded exactly that. A year later, when someone asks whether the AI's control-design advice has been reliable, this row is one data point in an answerable question rather than an anecdote.

16.4.7 Disposition Analytics: The Controller's Earliest Signal

Because every AI-assisted verdict records a three-way disposition against the recommendation, the organization can ask a question no model-quality metric answers directly: Where, and for which processes, do experienced humans most often overrule the model? That question is a single aggregate over the verdict population. It is particularly useful in an FP&A context, because it is a variance analysis in everything but name—the systematic gap between what was proposed and what was decided, grouped by where it occurs. Listing 16.7 shows how to count, for each domain, how often humans accepted, modified, or rejected the AI's recommendation.

```
SELECT  pv.domain,
        vv.verdict_value,
        COUNT(*)                        AS n,
        ROUND(100.0 * COUNT(*) /
              SUM(COUNT(*)) OVER (PARTITION BY pv.domain), 1) AS pct
FROM    VALIDATION_VERDICT vv
JOIN    SUGGESTION s        ON s.suggestion_id = vv.suggestion_id
JOIN    PROCESS_VERSION pv  ON pv.version_id   = s.target_pk
WHERE   vv.decided_at >= :period_start
  AND   s.target_entity = 'PROCESS_VERSION'
GROUP BY pv.domain, vv.verdict_value
ORDER BY pv.domain, vv.verdict_value;
```

Listing 16.7: Disposition Analytics Is One Aggregate over the Verdict Population

A domain in which humans accept the model's recommendation unmodified almost every time is either a domain where the model is genuinely strong or a domain where the human task has quietly become a rubber stamp, and the disposition rate alone does not distinguish the two, which is exactly why the rule against prefilling matters. A domain with a high accept-with-modification rate is where the scoring configuration from Chapter 15 most needs recalibration, and it is actionable long before any offline model quality evaluation completes. The controller does not need to understand the model to read this table; they only need to recognize a variance, which is exactly what they are trained to do.

16.5 Governance and Audit Handling

The first three families describe what the workflows do. This one describes why any of it can be trusted. Governance and audit handling is not a step appended to the end of a workflow. It is a property of every workflow in this chapter: Each instance writes an audit row, each row is joinable to the data spine without reading a log, and the join reconstructs the full chain from change back to the evidence the data spine already holds. This section specifies that property and shows the one-query defense it makes possible. This section covers governance and audit handling in nine steps: how every workflow instance writes a governed audit row, the one-query reconstruction that traces a change back to its source evidence, how governance exceptions are recorded when the platform refuses to proceed, the refusal ledger that captures what the platform declined to do and why, the control catalog that maps each control to an audit objective, the population-level evidence this produces, how audit handling scales across programs and regions, how workflow state and canonical truth stay separated, and a worked example tracing one change end to end through all four families.

16.5.1 Every Workflow Instance Is an Audit Row

A workflow instance in this architecture is not an opaque runtime object whose history lives in the engine's internal store. At the moments that matter (such as trigger, human decision, and write-back), it writes a row into the `AUDIT_EVENT` hash chain we defined in Chapter 9 and implemented in Chapter 10. The row carries the `suggestion_id` and `VALIDATION_VERDICT` it was started from, the workflow instance identifier, the participating identities in order, the `manifest_hash` at write time, and the disposition. Because the row lives in the same governed store as the verdict, the suggestion, the `INFERENCE_EVENT` row, and the source rows, the entire history is one connected graph of FKs rather than a set of correlated logs.

Figure 16.6 shows the governance and audit chain Part IV has been driving toward. Every arrow is an FK the canonical data spine already stores. Nothing is reconstructed from logs. Source rows feed the grounding bundle, and the grounding bundle feeds the inference, which produces the suggestion. That suggestion is decided into a verdict, the verdict starts a workflow instance, and the workflow instance produces a governed spine change. Nothing in the chain is reconstructed from a log. The reconstruction is a join.

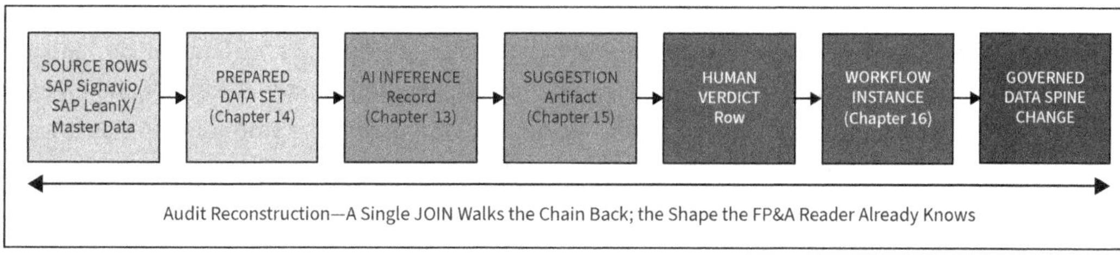

Figure 16.6: The Governance and Audit Chain

16.5.2 The One-Query Reconstruction

Since Chapter 13, we have promised that you can defend an AI-influenced change with the same one-query walk a controller has always used for a governed change. This is where that promise is delivered, because the workflow instance is the last link the chain was missing. Listing 16.8 demonstrates the shape of the reconstruction. It is deliberately written as the join a controller would recognize, not as engine-specific API calls.

```
SELECT  ae.audit_event_id,
        ae.applied_at,
        ae.workflow_instance,
        vv.validation_verdict_id,
        vv.decided_by,
        vv.verdict_value,
        s.suggestion_id,
        s.inference_event_id,
        s.bundle_id,
        src.source_entity,
        src.source_pk
FROM    AUDIT_EVENT           ae
JOIN    VALIDATION_VERDICT    vv  ON vv.validation_verdict_id
                                     = ae.validation_verdict_id
JOIN    SUGGESTION            s   ON s.suggestion_id = vv.suggestion_id
JOIN    INFERENCE_EVENT       i   ON i.inference_event_id
                                     = s.inference_event_id
JOIN    GROUNDING_BUNDLE      gb  ON gb.bundle_id     = s.bundle_id
JOIN    MANIFEST_ROW          src ON src.manifest_hash
                                     = gb.manifest_hash
WHERE   ae.kind = 'APPROVAL_COMMITTED'
  AND   ae.audit_event_id = :event_under_audit;
```

Listing 16.8: The Audit Reconstruction Is One Join over Rows the Data Spine Already Holds

The query returns, for one committed change under audit, the workflow instance that applied it, the verdict that authorized it, the human who decided it, the disposition relative to the AI recommendation, the suggestion the verdict answered, the inference that produced the suggestion, the grounding bundle the inference was grounded on, and the named source rows that the bundle manifest enumerated. That is the entire chain Figure 16.6 shows, returned by one statement, because the data spine stored the FKs as it went. No log was parsed, and no timestamp was correlated. The controller runs the query, reads one row outward in both directions, and either the chain is whole or the gap is named.

Note

This is the reconstruction we told you to expect in Chapter 15 and that we will use for production scale in Chapter 19. The single most valuable exercise a controller can perform after reading this chapter is to take one recent AI-influenced change, run this join, and walk it. The exercise either completes cleanly or exposes exactly one missing FK, and a named missing key is a finding a team can close in a sprint. An unreconstructible change is a finding a team closes in a crisis.

16.5.3 Governance Exceptions Are First Class, Not Errors

In Section 16.2, we noted that an emptied approver set raises a governance exception rather than a silent proceed. In this section, we make that rule general. Any condition that would force a workflow to violate the boundary (a drifted version at write time, an emptied eligible-approver set, a recommendation arriving without an `INFERENCE_EVENT` row, or a verdict whose author cannot be resolved to an identity) is not a runtime error to be retried. It is a governance exception, recorded with the same typed, joinable discipline as the operational exceptions covered in Section 16.3, and routed to the governance authority. The platform's refusal to proceed in these cases is not a fault. It is the control doing exactly what it was built to do, and it is recorded as such so that the refusal itself is auditable.

This is the deepest expression of this chapter's thesis. A governed platform is not one that always proceeds. It is one whose refusals are as reconstructible as its actions. The controller who can show an auditor not only every change that was made but every change the platform declined to make, and why, has a defense kit no dashboard can assemble and no after-the-fact narrative can fake.

16.5.4 What an Auditor Actually Asks, and the Answer's Shape

It is worth leaving the architecture for a moment and sitting on the other side of the table, because the value of everything in this chapter is measured by how an audit conversation goes. An auditor examining an AI-influenced process change does not ask abstract questions about model governance. They asks four concrete questions, in roughly this order, and the architecture is designed so that each one can be answered with a single row, not a project.

The first question is "Who decided this?" The answer is the `decided_by` identity on the verdict the workflow instance was started from: a named human, not a service account, because the handler recorded the human carried on the verdict and never the workflow's technical user. The second question is "What were they looking at when they decided?" The answer is the suggestion the verdict answered, the cited evidence manifest behind it, and the `INFERENCE_EVENT` row that produced it: the exact rows, not a summary, reproducible because the inference identifier was stored. The third question is "Who else could have stopped this and did not?" or in audit language, "Where was SoD?" The answer is the approver chain on the approval audit row together with the subtracted-proposer fact, which shows not only who approved but that the proposer was structurally excluded from approving. The fourth question is "What would have happened if a control had failed?" This is where the refusal ledger earns its place: The auditor can be shown comparable instances where a control did fire, such as a drift refusal or an emptied approver set, demonstrating that the controls are live, not decorative.

Four questions, four single-row answers, all reachable from one change identifier through FKs the data spine already holds. If you have an FP&A background, you will recognize that this is exactly the structure of a well-run financial audit. The auditor does not want a narrative; they want to pull one transaction and trace it to source and to authorization. A control environment is judged by whether that trace completes without manual reconstruction. This chapter's purpose is to make an AI-influenced process change trace exactly the way a controller already expects a journal entry to trace. AI is a novelty; the audit shape is deliberately not novel at all, because an audit shape a controller already trusts is worth more than a new one a controller must be taught to trust.

16.5.5 Where Workflow State Lives, and Why It Matters for Audit

In Chapter 10, we named SAP Build Process Automation as the store for workflow state and made the point that workflow state is not canonical truth. We will now explain why that separation matters for audit. Workflow state (such as which step an instance is on, which task is open, or which timer is pending) is operational and variable. Canonical truth (such as which version is published, which verdict authorized it, or which change was applied) is governed and, once published, immutable. If the two shared a store, a workflow retry or a step replay could appear to modify canonical history. Because they are separate, and because the only bridge between them is the commit gateway, no amount of workflow churn can rewrite what the data spine recorded. The audit chain is stable precisely because the workflow engine cannot reach past the handler. Operational state is allowed to be messy; governed truth is not allowed to be touched except through one path.

Figure 16.7 shows this chapter's structure in one view. Four workflow families sit on SAP Build Process Automation. They serve four kinds of actor, including the controller and FP&A professionals this chapter has kept in view. They share one governed contract with the data spine: Read context, propose, route to a human, and write only through the commit gateway. Below the contract is the canonical data spine covered in Chapter 9 and Chapter 10, holding process and version, activity and execution context, configuration, verdict and audit, workflow state, and AI artifact. Figure 16.7 does not show new architecture. It combines every previous chapter of Part IV, as viewed from the execution layer.

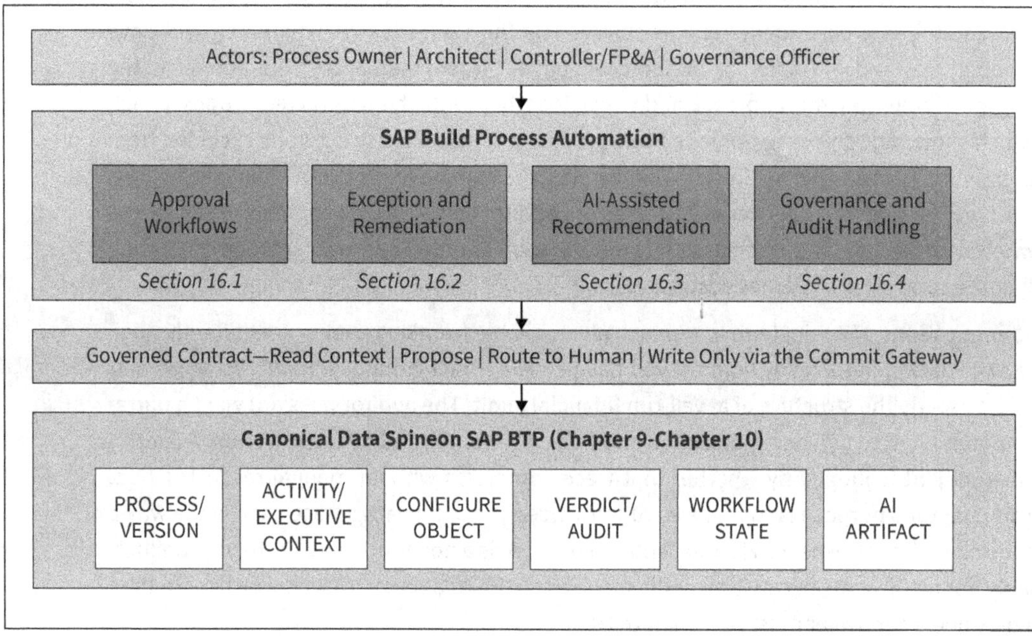

Figure 16.7: SAP Build Process Automation on the Data Spine—Reference Model

16.5.6 The Refusal Ledger

Most audit thinking concerns itself with what a system did. The deepest control in this architecture concerns what it declined to do. Every governance exception of Section 16.5.3—the emptied approver set, the drifted version refused at write time, the recommendation that arrived without an `INFERENCE_EVENT` row, the verdict whose author could not be resolved—is written to the same governed store as the actions, with the same joinable discipline. Together, these rows form the *refusal ledger*: the auditable record of every point at which the platform was asked, implicitly or explicitly, to cross its own boundary and did not.

The refusal ledger is the single most powerful artifact a governance officer can put in front of an auditor, and the reason is psychological as much as technical. Any system can show the changes it made. A system that can also show, with the same fidelity, the changes it refused to make and the precise rule each refusal enforced has demonstrated that its controls are live rather than documented. To a skeptical auditor, a control that has never visibly fired is indistinguishable from a control that does not work. The refusal ledger is the evidence that the SoD subtraction, the drift check, and the verdict requirement are not paragraphs in a design document but code paths that have stopped real attempts and recorded that they did. Table 16.6 lists each refusal condition, the rule that triggers it, and how the platform records it, so that every refusal is itself an auditable event.

Refusal Condition	Rule Enforced	Recorded As
Eligible approver set emptied	SoD	Governance exception, routed
Version drifted at write time	No stale governed write	Refused write, reresolve event
Recommendation lacks `INFERENCE_EVENT` row	AI traceability	Governance exception, task held
Verdict author unresolvable	Accountable decision	Governance exception, escalated
Auto-rule attempts exhausted	No silent failure	Converted to guided remediation

Table 16.6: The Refusal Ledger Records What the Platform Declined to Do and Why

16.5.7 The Control Catalog: Mapping Workflow Controls to Audit Frameworks

So far, everything in this chapter has been described as architecture. The controller and the internal auditor live in a different vocabulary (control objectives, control activities, and evidence of operating effectiveness) and this chapter's value to them depends on the architecture mapping cleanly onto that vocabulary. That mapping is the bridge between the engineer who builds the workflow and the auditor who must assess it. Each structural property this chapter specified is, in audit language, a control activity with a named control objective and a standing source of evidence. The evidence is not a screenshot taken for the audit; it is a query over rows the data spine produced as a byproduct of operating, which is precisely the kind of evidence an auditor weighs most heavily because it cannot be staged.

In Table 16.7, the right-hand column is the one that changes an audit conversation. In a conventional control environment, evidence of operating effectiveness is a sample: The auditor selects a number of changes and asks for the approval records, and the control is judged on whether the sample is clean. In this architecture, the population is the evidence, because every change without exception carries its verdict, its approval chain, and its inference reference as joinable rows. The auditor does not need to sample to gain assurance that the control operated on the items sampled; they can test the assertion that no change exists without a verdict at all, which is a single query that returns the empty set or returns the exact exceptions. A control whose failure mode is a queryable empty-set assertion is a stronger control than one whose assurance rests on a sample. If you have ever sat through sampling-based testing, you will recognize immediately how much cheaper and more conclusive population-level evidence is.

Control Objective	Structural Control Activity	Standing Evidence
Only authorized changes reach production	Verdict-keyed trigger; one write handler	Join from change to accepted verdict
Proposer cannot approve own change	Proposer subtracted from eligible set	Approval row with subtracted-proposer fact
No stale or out-of-sequence change	Drift recheck at commit; idempotent key	Refused-write rows in the refusal ledger
AI does not decide	Advisory panel with no write path	Every change has a human verdict row
Decisions are reproducible	Inference ID carried on the verdict	Verdict joins to inference and evidence
Controls are operating, not documented	Governance exceptions recorded as first class	Refusal ledger shows controls firing

Table 16.7: Each Structural Control Maps to an Audit Control Objective with Standing, Unstaged Evidence

The Sarbanes–Oxley Act (SOX), SOC2, and SAP audit-chain semantics those references describe are satisfied here not by adding a control layer on top of the workflow but by the workflow being shaped so that its ordinary operation produces the evidence those frameworks require. The control catalog is not a compliance artifact maintained alongside the system; it is a reading of the system's structure in the auditor's language. A team that builds the four workflow families to the contracts this chapter specified has, without any additional compliance project, built a control environment whose operating effectiveness is continuously and queryably evidenced.

16.5.8 Audit Handling Across Programs and Regions

A single chapter's worked examples can make the reconstruction look easy because they concern one process in one landscape. Real programs run many processes across many regions, often with the branched and regulatory variants we modeled as lineage in Chapter 9. The audit discipline covered in this chapter scales to that reality precisely because it never depended on locality. The reconstruction join of the previous subsection does not care how many regions exist; it walks FKs, and a FK is region-agnostic. A controller auditing a German payroll process change and a controller auditing a global procurement process change run the same query with a different change identifier. The shape is invariant, which is the property that lets a central governance function supervise a federated program without learning a new audit story per region.

This is also where we will point you to Chapter 18 and Chapter 19. In this chapter, we established that every workflow instance is an audit row and that the reconstruction is a join. In Chapter 18, we establish the monitoring that keeps the workflow tier healthy at the volume a multi-region program generates (queue depth, verdict latency, and escalation rates) so that the audit discipline does not silently degrade under load. In Chapter 19, we establish the full instrumentation that makes the one-query reconstruction dependable not just the day after a change but quarters later, when the auditor actually arrives. If you are a controller, you should read the three chapters as one promise made in installments: Chapter 16 proves the chain exists, Chapter 18 proves it stays healthy, and Chapter 19 proves it survives time.

16.5.9 One Change, End to End

We will now trace a single change through all four families once, because they operate as a single contract, and you can see that in how the change moves through them without ever leaving it. A nightly validation run detects that a published control gate activity depends on a configuration object missing in one company code. The exception family records a typed, high-band data spine-linked exception, and the classifier routes it to guided remediation because creating configuration changes canonical meaning and is barred from the auto path. The AI-assisted family surfaces a recommendation inside the guided task (grounded, scored, and cited) that the gate be scoped to company code until the configuration is delivered. The human, an owning controller, reads the cited evidence, modifies the recommendation to add an expiry condition the model did not propose, and records a `VALIDATION_VERDICT` with `verdict_value` `MODIFY` and an escalate disposition. The approval family receives the accepted-suggestion trigger, resolves the approver chain from the data spine at high band because a control gate is touched, subtracts the proposing controller from the eligible set, routes to the architecture and governance authorities, and, on final approval, calls the commit gateway, which rechecks version state, refuses if it drifted, and otherwise writes the change and its audit row. The governance and audit family was present the entire time. The exception, the case, the recommendation's `INFERENCE_EVENT` row, the modified verdict, the approval chain, and the `manifest_hash` are one connected graph of FKs.

Weeks later, an auditor asks why that company code's position-creation behavior differs from the global default. The controller runs one join. It returns the change, the workflow instance that applied it, the verdict and the human who decided it, the disposition that records the model was modified and how, the suggestion and the inference behind it, the grounding bundle, and the named source rows. The controller reads one row outward in both directions and answers the auditor in a sentence, with evidence. That sentence—grounded, decided by a named human, governed, and reconstructible—is the entire purpose of every structural choice we have made since Chapter 9. The workflow did not make it possible to change the process. The data spine and the workflow together made it impossible to change the process without being able to explain it.

Figure 16.8 traces exactly that change as a swimlane diagram, with one lane per family, and it shows you how the four families function together as one path. Read left to right, the exception family detects and classifies the gap, the AI-assisted family offers a scoped recommendation inside the guided task, and the human verdict carries the decision down into the approval family, which resolves the chain at high band with the proposer subtracted and calls the one handler. The governance and audit lane spans the whole width, because the reconstruction join touches every step.

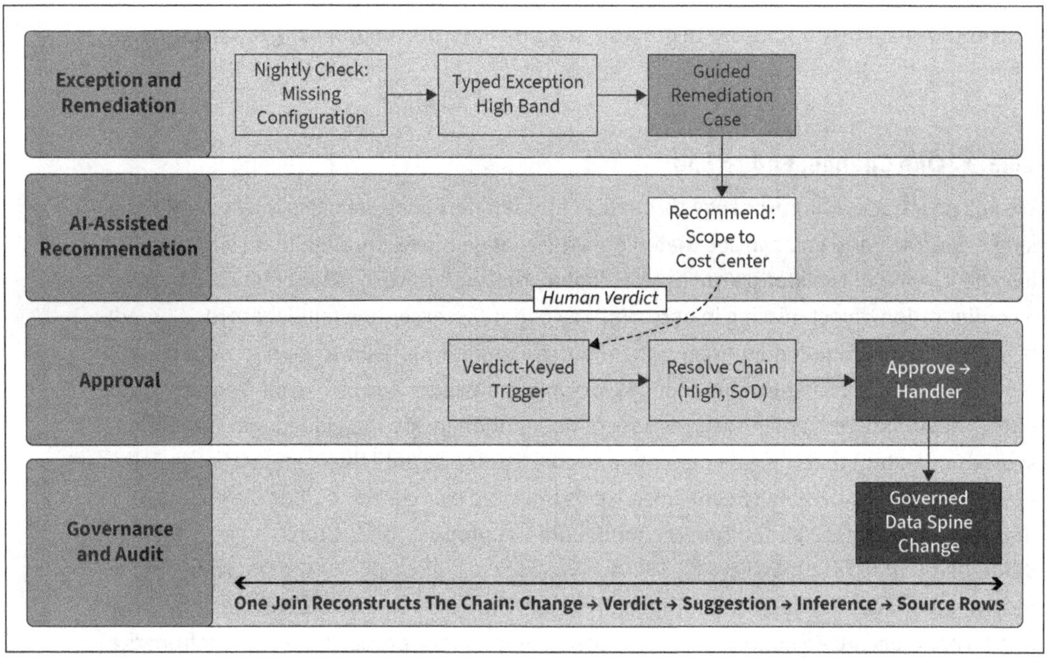

Figure 16.8: One Change, End to End, Across Four Families

16.6 Summary

This chapter specified the execution side of the AI conversation. SAP Build Process Automation is the layer that turns an accepted verdict into a governed workflow on top of the canonical data spine. We described the four workflow families (approval, exception and remediation, AI-assisted recommendation, and governance and audit handling) and showed that each one is the same governed contract with the data spine seen from a different operational angle: Read context, propose, route to a human, and write only through the commit gateway.

This chapter's structural claim is the one we have made since Chapter 1, Section 1.5, now repeated at the workflow layer. The AI boundary is not a rule the workflow engine is asked to honor. It is the shape of the workflow machinery: a trigger that can only hear a verdict, an approver chain derived from the data spine rather than a form, a SoD enforced by subtracting the proposer in code, a recommendation with no direct connection to the data spine, and a write path that exists only through a single guarded handler. Together with the evidence from Chapter 14 and the decision from Chapter 15, this chapter's execution completes the picture of an AI tier whose every input is platform prepared, whose every output is human decided, and whose every action and every refusal is auditably reconstructible from the data spine.

16.6.1 Key Learning Points

- An approval workflow is triggered by a verdict, never by a suggestion. The event subscription filters on an accepted verdict, so the workflow has no way to start before a human has signed.
- Approver resolution is derived entirely from process context (`PROCESS_VERSION`, `ACTIVITY`, `EXECUTION_CONTEXT`, and the role model) not from a field on a form. Routing is a fact about the process, not the request.
- SoD is structural. The proposer's identity is subtracted from the eligible approver set by the same code path on every instance, and an emptied set raises a governance exception rather than a silent proceed.
- An exception is a typed, severity-banded event linked to the canonical entity it concerns. Auto-remediation is reserved for idempotent, preapproved, meaning-preserving fixes. Guided remediation is the common path, and escalation reenters the approval family.
- An AI recommendation inside a workflow is advisory by design. There is no transition from the recommendation to the data spine that does not pass through a human verdict, and the verdict records the disposition relative to the recommendation.
- Every workflow instance writes a governed audit row joinable to the verdict, the suggestion, the inference, and the source rows. An AI-influenced change is reconstructed by one join, not by correlating logs.
- Workflow state and canonical truth live in separate stores bridged only by the commit gateway, so no workflow churn can rewrite governed history. The audit chain is stable because the engine cannot reach past the handler.

16.6.2 Common Design Pitfalls

- Treating the approver as a property of the form. A requester-chosen or table-maintained approver cannot survive an audit in an integrated landscape, because the approver of a process change is a fact about the process the data spine already holds.
- Triggering a workflow from a suggestion to save a step. This does not actually save a step; it removes the human decision the entire architecture exists to guarantee, and it is invisible until an auditor asks who decided.
- Building auto-remediation for convenience rather than for narrow, idempotent, meaning-preserving cases. Auto-remediating anything that can change canonical meaning reintroduces exactly the ungoverned drift we aimed to prevent in Chapter 9.

- Surfacing an AI recommendation without its cited evidence and inference record. This turns an assistant into an oracle. The human can then only believe or ignore it, and neither is the interrogation the boundary requires.
- Letting workflow state and canonical truth share a store. This is a subtle error that works until a retry or replay appears to mutate governed history. By then, the audit chain is already untrustworthy, and the failure is structural rather than incidental.

16.6.3 Review Questions

1. Why is an approval workflow triggered by a verdict rather than by a suggestion, and what would be lost if it were triggered by the suggestion?
2. How is the approver chain derived from process context, and which canonical entities are read in the resolution?
3. In what sense is SoD a structural property in this architecture rather than a procedural one?
4. What three remediation paths follow from an exception, and what distinguishes the cases in which auto-remediation is permitted?
5. Where is the AI boundary located in the AI-assisted recommendation family, and why is the absence of a direct connection a stronger control than the absence of a permission?
6. What does the one-query audit reconstruction return, and why is it a join rather than a log correlation?
7. Why must workflow state and canonical truth live in separate stores, and what role does the commit gateway play in keeping the audit chain stable?

16.6.4 Student Companion Notes

If you are a student approaching this chapter through an SAP University Alliance program or an enterprise architecture curriculum, you should think of it as closing the execution loop we opened in Chapter 1 and built structurally from Chapter 9 onward. The goal is not to memorize SAP Build Process Automation syntax but to understand why each workflow family is the same governed contract seen from a different angle.

First, take the cost center to position scenario from Chapter 9 and draw the approval workflow for a proposed change to the published version, labeling which spine entity supplies each routing decision. Do not draw an approver box with a name in it; draw the query that resolves the name. Second, take one exception type from Table 16.4 and trace it through classification, the chosen remediation path, the verdict, and the write-back, naming the audit row written at each step. Third, take the one-query reconstruction from Listing 16.8 and, for a hypothetical change, write out the chain of FKs it walks. This exercise is complete when every arrow in Figure 16.5 corresponds to a join in your query.

If you have an FP&A and controlling background, our practical recommendation is the same exercise from Chapter 14 and Chapter 15, now completable end to end. Take a recent AI-influenced process change, run the reconstruction join, and walk the chain from the change back to the source rows. The exercise either completes cleanly or exposes a single named missing key, and either outcome is the cheapest and most honest internal audit the function can run.

A note on vocabulary: In this chapter, we deliberately reuse the nouns defined in earlier chapters rather than coining new ones. The `SUGGESTION`, its `lifecycle_state`, the `VALIDATION_VERDICT` and its `verdict_value`, the grounding bundle and its `manifest_hash`, the `INFERENCE_EVENT`, the commit gateway, and the `AUDIT_EVENT` hash chain were all introduced in Chapter 13 through Chapter 15. This chapter adds no new canonical entity. Where we say a workflow is triggered by an accepted suggestion in this chapter, we mean `lifecycle_state = ACCEPTED` reached through a `VALIDATION_VERDICT` with `verdict_value = ACCEPT`, and where we say the disposition relative to a recommendation, we mean `verdict_value` read in the AI-assist context. The continuity is intentional. If you learned the vocabulary to defend an evidence question in Chapter 14 or a decision question in Chapter 15, you can defend an execution question in this chapter with the same words and the same joins.

This chapter specified the execution side of the AI conversation. Chapter 17 turns to SAP Build Work Zone and the process-aware user experiences that surface these workflows, their tasks, and their AI insights to the people who act on them, with the same role-based discipline the data spine enforces everywhere else.

Chapter 17

SAP Build Work Zone: Process-Aware User Experiences

Evidence is what the platform was allowed to see. A decision is what a human chose. Execution is how the choice was carried out. Experience is the only part of the process a person actually interacts with—and if that surface can show them something the data spine would have rejected, then every guarantee underneath it was decoration. The experience layer either upholds the architecture or it quietly throws it away.

Chapter 16 specified the execution side of the AI conversation. An accepted verdict becomes a governed workflow, the workflow derives its routing from process context, and the only write reaches the canonical data spine through the single commit gateway Chapter 10 reserved for it. Chapter 16 ended with the workflow running and the audit row written. It deliberately left one question for this chapter: a workflow that no one can see, a task no one can find, and an AI insight no one can interrogate are, from the point of view of the person who has to act, indistinguishable from a workflow that does not exist. This chapter solves that riddle. It is where the governed spine, the four workflow families, and the AI tier finally meet the human being who must do something about them.

The tool for this problem is SAP Build Work Zone. It is the experience layer that projects the spine, the workflows, and the AI insights to people—a controller defending a number, an approver clearing a task, a process owner inspecting a journey, an operator watching a region—with the same role-based, version-pinned, audit-consistent discipline the data spine enforces everywhere else. The structural commitment of this chapter is the one we have carried since Chapter 1, Section 1.5, now stated for the interface a user touches: The experience layer may show only what the data spine would already permit that user to read, may collect only a decision an SAP Build Process Automation human task is entitled to record, and may itself write nothing. There is no path through SAP Build Work Zone that turns a click into a data spine change without a recorded human verdict carried by a governed workflow, and the absence of that path is the boundary.

This chapter builds on Chapter 14's evidence (what the AI was permitted to see), Chapter 15's decision (what a human chose to do about it), and Chapter 16's execution (how the chosen action was carried out). It walks through how a user sees and acts on all three. The stages are not independent. An experience layer that adds a capability the evidence, decision, or execution stages did not sanction does not extend the architecture; it punctures it. The entirety of this chapter argues that the experience layer adds nothing, that it is the data spine, the workflows, and the AI tier seen by a human, and that everything which makes it trustworthy was decided before the user ever opened a browser.

This chapter is for those who assemble the work pages and spaces, who decide what a tile is allowed to query, who must defend a control to an auditor, and for those who, weeks later, must explain why a process changed, on whose authority, and on what evidence. Part of the promise of this chapter is the guarantee that what the controller saw on screen when they made the call is itself a reconstructible fact, joinable to the verdict, the inference, and the grounding bundle by the same one-query walk.

Five principles run through the chapter and should be carried into every space, page, tile, and Joule surface built on the data spine. First, navigation is derived from the process model, never manually built as a menu. Second, a task view is a role-scoped assembly produced per actor, per request, and never reused across actors. Third, an AI insight in the experience layer is advisory by construction; it is the recommendation panel introduced in Chapter 16, Section 16.4 rendered for a human, and the human task remains the only decision point. Fourth, every interaction boundary the platform enforces structurally has an experience-layer expression, and a control the user is not entitled to is absent, not disabled. Fifth, SAP Build Work Zone holds no canonical data. It reads context through the governed read path and writes only by routing a human decision into SAP Build Process Automation.

This chapter explains:

- What SAP Build Work Zone is on the canonical data spine, what each of its components is allowed to do, and, more importantly, what each must never be allowed to become
- How process-driven navigation is resolved from `PROCESS`, `PROCESS_VERSION`, `ACTIVITY`, and `EXECUTION_CONTEXT` rather than from a stored menu and how the published version pins what the user sees
- How a role-based task view is assembled under the requesting actor's role scope, why separation of duties (SoD) is visible and not merely enforced, and why a view that cannot be assembled in full is refused rather than trimmed
- How an AI insight is surfaced as a cited card that carries its grounding bundle and inference identity, why the card cannot prefill a decision, and how the controller's earliest signal is one aggregate over the verdict population
- How every structural boundary has an experience-layer expression, why the absent control is the strongest control, and what a one-query reconstruction returns for an AI-influenced change seen end to end

The remainder of the chapter develops the four experience surfaces (process-driven navigation, role-based task views, surfaced AI insights, and user interaction boundaries) and shows how each one is the same governed contract with the data spine seen from a different human angle. Before we get to the surfaces, we will begin by positioning SAP Build Work Zone itself, because the surfaces only make sense once you know what the product is and what discipline must be imposed on it.

17.1 Positioning SAP Build Work Zone on the Data Spine

SAP Build Work Zone is the unified entry point on SAP BTP through which users reach the applications, tasks, content, and conversational assistance of an SAP landscape. Its components—spaces, pages, sections, cards and tiles, content widgets, the launchpad and search, and the embedded Joule surface—are designed to let an experience team assemble a coherent digital workplace quickly. That speed is a virtue and, in a governed landscape, the same hazard it was for SAP Build Process Automation in Chapter 16. The same low-code surface that lets a team build a role's home page in an afternoon also lets a team hard-code a navigation tree, cache a task list, paste a target URL that bypasses the execution services, and wire a tile straight to a database view. The product does not automatically enforce the architecture we describe in this book. You must impose the architecture on the product, and this section shows you how to do so.

The imposition rests on one structural claim that we elaborate on in the rest of the chapter: SAP Build Work Zone is a projection of the canonical data spine, not a second copy of it. A projection has no data of its own. It shows, for a named user at a named instant, a role-scoped, version-pinned view of facts the data spine already holds, and it reflects state changes only after a governed workflow has written them. The moment any part of the experience layer holds a fact the data spine does not (a route, an approver, a task list, or a status) that fact is ungoverned, and the audit chain we have been building since Chapter 9 has a hole in the exact place a user can see it.

17.1.1 Components and What Each Is Allowed to Do

A space in SAP Build Work Zone is the top-level grouping that organizes a user's digital workplace. In this architecture, a space is allowed to group pages by process or role, but it is never the place a process-to-role relationship is defined; that relationship is read from the data spine, not maintained in the space designer. A page is the arrangement of sections and cards that a user sees. It is allowed to lay out content, but it is never the authority for which content a given user may see; the page requests content and the read path decides what comes back. A card or tile is the unit of content and navigation. A tile is allowed to display a value and carry a navigation target, but the value is the output of a governed read and the target is resolved from the process model, never typed into the tile by a designer. A content widget is an embedded interactive surface. A widget may render context and collect an interaction that becomes a workflow trigger, but it is never the place a decision is recorded; the decision is recorded by an SAP Build Process Automation human task. The launchpad and search are the discovery surface. They expose only the spaces, pages, and apps the user's role scope already permits, and a search can never return a target the user could not have navigated to deliberately. The embedded Joule surface is the conversational entry point. Joule is the surface, governed exactly as we governed it in Chapter 13, and it acquires no additional privilege by being embedded in a work page.

Table 17.1 contains the structural reference an experience architect consults before assembling a space. Every component has a permitted role that serves the user and a forbidden role that, if allowed, silently relocates a governed fact into an ungoverned surface. The reviewer's test is to hide every value on the page and still have the navigation resolve correctly, because no value on the page was ever the source of a routing or authorization fact.

Component	Permitted Role	Forbidden Role
Space	Group pages by process or role read from the data spine	Define a process-to-role relationship
Page	Lay out sections and request governed content	Decide which content a user may see
Card/tile	Display a governed value; carry a resolved target	Store a navigation target or a cached value
Content widget	Render context; raise a workflow trigger	Record a decision or write canonical data
Launchpad/search	Expose only role-scoped, permitted destinations	Surface a target outside the user's scope
Joule surface	Converse via grounding endpoints and SAP Build Process Automation actions	Acquire privilege by being embedded

Table 17.1: Every SAP Build Work Zone Component Has a Permitted Role and a Forbidden Role on the Data Spine

17.1.2 Why the Experience Layer Holds No Data

The reason SAP Build Work Zone must not hold any canonical data is the reason Chapter 14 gave for never reusing a prepared data set across actors, restated for the surface a user looks at. A fact stored in the experience layer—such as a navigation tree built in the space designer, a task list cached in a tile, or an approver name rendered into a page—is a fact that was true for some user at one moment and is now detached from the data spine that governs it. The moment the data spine's version advances, the process is republished, the role model changes, or a workflow reassigns a task, the stored fact is stale, and nothing on the page tells the user it is stale. The user acts on the stale fact with complete confidence, and the audit chain, which correctly records what the data spine held, cannot explain what the user actually saw, because what the user saw never came from the data spine.

Figure 17.1 illustrates this projection. People reach the experience layer at the top. SAP Build Work Zone, in the middle, is a projection surface: spaces, pages, tiles, widgets, the Joule surface, and search. Every value it shows is read (version-pinned and role-scoped) through the governed read path: context enrichment and the process-aware execution services covered in Chapter 12, Section 12.2, and the Access API read surface in Chapter 10, Section 10.4. Below that is the canonical data spine, covered in Chapter 9 and Chapter 10. Note that there is no arrow from the experience layer to the data spine because it lacks a direct connection. The only way a user interaction becomes a data spine change is by leaving the experience layer entirely, becoming a human task in SAP Build Process Automation, and reaching the data spine through the commit gateway. The absence of the direct write connection is the boundary, exactly as the absence of a direct connection from the recommendation panel was the boundary in Chapter 16, Section 16.3.

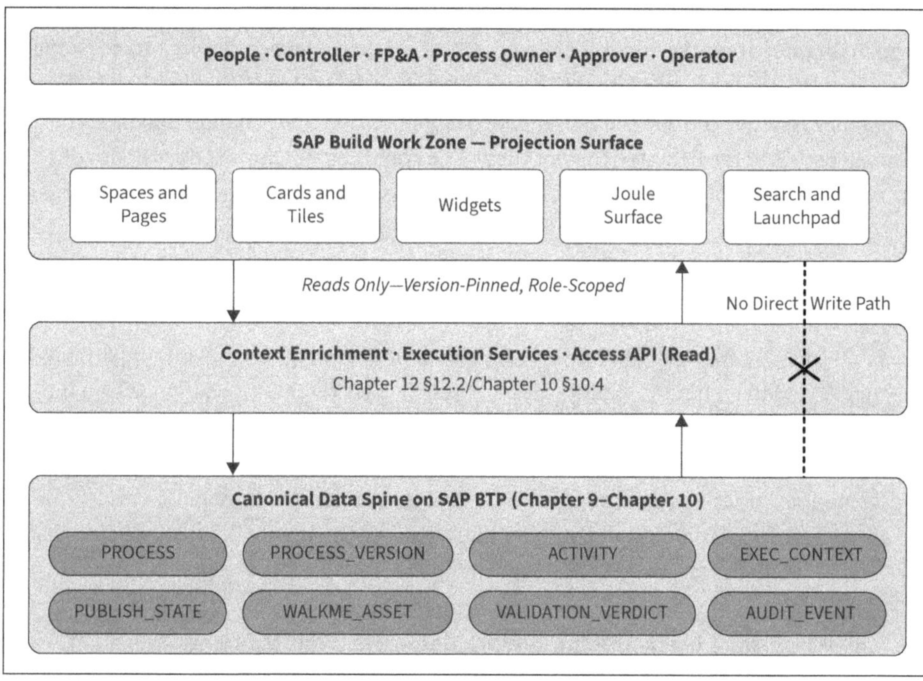

Figure 17.1: SAP Build Work Zone Is a Projection over the Canonical Data Spine, with a Single Governed Read Path and No Direct Write Path

17.1.3 The Single Read Path and the Absent Write Path

The experience layer reads through one path and writes through none. The read path is the context enrichment service and the process-aware execution services of Chapter 12. This is a deliberate reuse, not a coincidence of convenience. Chapter 12, Section 12.2 already established that context enrichment applies the validation filter, the lifecycle filter, the role-scope filter, and the version pin before any consumer sees a row, and Chapter 12 named the SAP Build Work Zone tile's rendering layer as one of the consumers that depends on the enrichment output schema. An SAP Build Work Zone tile that renders an activity name, an execution context, a trigger and target application, or a validation verdict is consuming the same schema (introduced in Table 12.2 in Chapter 12) a Joule agent's prompt template and a financial planning and analysis (FP&A) drilldown query consume, with the same four filters already applied. The experience layer does not get a less-filtered view than any other consumer; it gets the same view, which is the structural property that makes a screenshot of an SAP Build Work Zone page defensible evidence rather than an unverifiable artifact.

The write path does not exist. SAP Build Work Zone has no credential, no destination, and no API surface that targets the data spine's write side. When a user does something that must change state—such as approving a change, recording a remediation decision, or accepting or rejecting an AI suggestion—the interaction is handed to SAP Build Process Automation as a human task, and the task carries the decision through the machinery outlined in Chapter 16 to the commit gateway. The experience layer's role in a state change is to present the context and route the human's decision; it is never the writer. This is the concrete realization, at the surface a user touches, of the principle Chapter 10 stated for persistence and Chapter 16 stated for workflow: The commit gateway is the only path of entry, and no amount of low-code freedom in the space designer can create an alternate one.

Note

SAP Build Work Zone also offers a standalone site mode and federated content from SAP S/4HANA and SAP SuccessFactors launchpads. Federation does not change the discipline of this chapter. A federated tile reaches its target application through the same role-scoped resolution; a federated launchpad exposes only what the user's scope already permits. The rule is invariant under federation: The experience layer projects permitted, governed reads and routes decisions into governed workflows, regardless of how many launchpads it federates.

17.1.4 What May Live in SAP Build Work Zone and What May Not

The projection discipline is easy to state and easy to violate by degrees, so it is worth drawing the line precisely between what an experience team should legitimately keep in SAP Build Work Zone and what it should not. The test is not whether an item is data; it is whether the item is a fact about the enterprise or a fact about the presentation. A fact about the enterprise (which process a role performs, which activity an approval belongs to, what a cost center's status is, or who proposed a change) is canonical and lives only on the data spine. A fact about the presentation (a user prefers the dense table layout, a tile is pinned to the top of a page, or a chart defaults to a twelve-month window) is not a fact about the enterprise, and keeping it in the experience layer creates no governance exposure because nothing about the audit chain depends on it.

The boundary matters because the two are easy to conflate in a design review. A saved filter looks innocuous until the filter encodes which cost centers a user works with, at which point it has quietly become a scope assertion living outside the role model, and a defect in how it is applied is a scope leak. A pinned tile is harmless until the pin encodes which approvals the user considers theirs, at which point it is a task-ownership fact that belongs to the workflow engine, not to a layout preference. The reviewer's question is therefore sharper than "Is this data?" Their true question is "If this stored item was wrong or stale, could a user act incorrectly and could the audit chain fail to explain why?? If the answer is yes, the item is canonical and does not belong in the experience layer. If the answer is no, it is presentation and is safe to keep. The roles outlined in Table 17.1 govern the component; this test governs the temptation to let a component hold a little data because it is convenient.

17.2 Process-Driven Navigation

The first experience surface is navigation. This section focuses on the structural claim that navigation is a query against the process model, not a menu maintained by a designer. Everything that makes SAP Build Work Zone navigation trustworthy follows from refusing to store the navigation tree. We'll explore this topic alongside process-to-space mapping, search, deep links, and a continuation of our cost center to position example.

17.2.1 Navigation Is a Query, Not a Menu

The conventional way to build a digital workplace is to design a navigation tree: a hierarchy of menu entries, each pointing at a target, maintained by an experience team as the landscape evolves. The conventional way is wrong in an integrated landscape for the same reason a hard-coded approver was wrong in Chapter 16. A stored menu is a copy of a relationship the data spine already owns (which process a role performs, which activities a process contains, or which application executes an activity) and the copy drifts the instant the data spine changes. For example, a process is republished with a renamed activity, but the menu still points at the old label. An application is rehosted from on-premise to cloud in SAP LeanIX, but the menu still targets the old endpoint. A role's process assignment changes, but the menu still shows the old workspace. None of these drifts is visible on the page, and each of them puts a user one click away from acting on a model the enterprise no longer considers authoritative.

The discipline that forbids this drift is resolving navigation at request time from the process model, exactly as Chapter 16 resolved an approver chain at trigger time from process context. The navigation a user sees is not stored; it is the answer to a question posed against `PROCESS`, `PROCESS_VERSION`, `ACTIVITY`, and `EXECUTION_CONTEXT`, filtered by the user's role scope and pinned to the published version. Listing 17.1 shows the resolution as pseudocode, written at the level a developer can implement and a governance officer can read as a control.

```
FUNCTION resolve_navigation(actor_token, space_id):
    # Stage 1 - actor and scope
    actor = authz.principal(actor_token)
    scope = authz.scope_set(actor_token)
```

```
# Stage 2 - the processes this role performs, published only
procs = enrich.processes_for_role(
            actor.role_context,
            publish_state = 'PUBLISHED',
            as_of = now())

nav = []
FOR p IN procs:
    pv = enrich.published_version(p.process_id)   # PUBLISH_STATE gate
    FOR act IN enrich.activities(pv.version_id, scope):
        ec = enrich.execution_context(act.activity_id)
        target = resolve_target(ec.target_application,
                                act.fiori_app_id)
        ASSERT scope.permits(target)              # deny-by-default
        nav.append(node(process = p.process_id,
                        version = pv.version_id,
                        activity = act.activity_id,
                        label   = act.activity_name,
                        target  = target))
# Stage 3 - audit and return; nothing is persisted in Work Zone
audit.emit('NAV_RESOLVED', actor, space_id,
           version_set = versions(nav))
RETURN nav
```

Listing 17.1: Navigation Is Resolved from Context Enrichment at Request Time, Never Read from a Stored Menu

Three properties of the listing carry the section's weight. First, every label and every target is read, not stored: `act.activity_name` and the resolved target come from the context enrichment schema outlined in Table 12.2 in Chapter 12, so a republish that renames an activity changes the navigation on the next request with no designer action. Second, the `scope.permits(target)` assertion is the deny-by-default rule rendered as navigation. A target the user's scope does not permit is not greyed out and is never appended, so it cannot be seen, hovered, or guessed. Third, the function persists nothing in SAP Build Work Zone and emits a `NAV_RESOLVED` audit row naming the version set it resolved against, so the question "What navigation did this user see and against which version of the model?" is answerable by a join, not by interviewing the experience team.

AI Insight: Why a Resolved Tree Is Also the Cheapest Thing to Make Defensible

It is tempting to treat a stored navigation tree as a performance optimization. Resolving the tree on every request costs spine reads, and a cached tree is free. This temptation is the same one Chapter 14 identified for cached prepared data sets, and it fails for the same structural reason. A cached tree is a fact detached from the role context and version that produced it; a second user whose scope differs, or the same user after a republish, sees a tree that no longer reflects what the data spine would authorize. The audit row would name the user correctly while the tree they saw reflected a different scope or an older model, and the chain would lie about what the user was shown.

The structural payoff from an FP&A perspective is the one this book returns to in every chapter. Because navigation is resolved and audited rather than stored, the question an auditor asks about an AI-influenced change ("Could the person who made this call even have reached the screen they say they used?") is answered by the NAV_RESOLVED row joined to the verdict, not by a screenshot whose provenance no one can establish. Resolution becomes the less expensive option once the cost of an indefensible audit is priced in.

17.2.2 The Process-to-Space Mapping

Process-driven navigation rests on a mapping that the data spine owns and the experience layer reads. A PROCESS maps to a space, a PROCESS_VERSION pins what that space shows, an ACTIVITY maps to a page or a navigation node within the space, and the activity's SAP Fiori app identifier, carried on the data spine as part of the WalkMe and adoption model of Chapter 7, maps to the concrete navigation target. This mapping is the same one Chapter 7 established when it positioned WalkMe content against L4/L5 process steps and SAP Build Work Zone. Table 17.2 states the mapping and names the data spine entity each navigation element resolves from so that an experience architect can see at a glance that no element of the navigation is invented in the space designer.

Navigation Element	Resolves From	Pinned/Scoped By
Space (the workplace grouping)	PROCESS	Role context; PUBLISH_ STATE = PUBLISHED
What the space shows	PROCESS_VERSION	Published version, AS-OF the request
Page/navigation node	ACTIVITY	Role scope on the activity
Navigation target	SAP Fiori app ID via EXECUTION_CONTEXT	scope.permits(target), deny-by-default
In-app guidance on the page	WALKME_ASSET	Mapped to the same ACTIVITY/SAP Fiori app ID
Order and grouping	Process structure (L4/L5)	Read (not maintained) in SAP Build Work Zone

Table 17.2: Each Navigation Element Resolves from a Data Spine Entity, Never from the Space Designer

Because the space is a projection of a PROCESS and the page a projection of an ACTIVITY, the digital workplace is automatically as correct as the process model and never more correct than it. This is the desired property. An experience layer that could be more current than the published process model would be a second source of truth; an experience layer that is exactly as current as the published model is a faithful window onto it. The discipline does not make the workplace better than the model. It makes it honest about the model, which is the only property an auditor and a controller can rely on.

17.2.3 Search Is a Scoped Projection, Not an Index

Search deserves its own treatment because it is the navigation surface that most quietly defeats the others. A user who cannot find an unauthorized target in the menu can often find it in search, if search is built the conventional way, as an index of all content, queried and then filtered. This reintroduces every leak Listing 17.1 closed in the menu to the search box. A global index is a stored set of targets, the filter is an after-the-fact guard, and a defect in the guard surfaces a destination the user could never have navigated to deliberately. Worse, search results often carry enough metadata such as titles, snippets, and breadcrumb paths to disclose the existence of elements the user is not entitled to even when the link itself is withheld.

The same discipline applies here as at every surface: Search returns a projection scoped before results exist, not a global result set scoped on the way out. A query is resolved against the same role-scoped, version-pinned process model navigation resolves against, so a result can only ever be a destination the user could have reached deliberately. The invariant is exact, and it is a test an architect can apply: Search should never return a target that navigation would not have offered the same user at the same instant. If the two can disagree, search has become a second, less-governed navigation surface, and the weaker one is the one a determined user will use. Holding search to the navigation invariant means there is only one navigation surface with one set of rules, regardless of whether the user clicked a tile or typed a query.

Navigation is pinned to the published version through the same PUBLISH_STATE gate Chapter 9 introduced and Chapter 14 reused for AI grounding. A process may have many versions, such as drafts under revision, a version in approval, or the version currently published. The experience layer shows the published version and only the published version. A draft activity, however far advanced, does not appear in any user's navigation, because the resolution in Listing 17.1 reads PUBLISH_STATE = PUBLISHED and pins the version AS-OF the request, exactly as Chapter 14, Section 14.3 pinned a prepared data set AS-OF its assembly instant. The user navigates the model the enterprise has approved, not the model someone is still editing, and the NAV_RESOLVED audit row records which version that was.

The AS-OF anchor matters for the controller specifically. When a process is republished, such as a new version promoted through the approval family outlined in Chapter 16, the navigation changes on the next request. The change is itself an audited transition of PUBLISH_STATE, not a silent edit to a menu. A controller reconstructing why a user navigated to a particular activity on a particular date does not have to establish what the menu looked like then; the menu was never stored. They read the published version AS-OF that date from PUBLISH_STATE and reresolve the navigation deterministically. The experience layer has no history of its own because it needs none. Its history is the data spine's version history, which the controller already knows how to read.

17.2.4 Worked Example: The Cost Center to Position Journey

This cost center to position example has served as a running example since Chapter 9. A cost center request originates in SAP Master Data Governance and is activated in controlling in SAP S/4HANA. Replication is confirmed, and a position is created in SAP SuccessFactors referencing that cost center. Chapter 9 used the scenario to motivate EXECUTION_CONTEXT; Chapter 16 used it to draw an approval workflow. This chapter uses it to show what process-driven navigation looks like to the person who performs the journey, because it is exactly the kind of scenario a manually built menu cannot represent correctly.

Figure 17.2 draws the journey as it is navigated, one lane per owning system. The user does not choose a menu entry labeled **SAP SuccessFactors** and hope it is the right one. Each step is an ACTIVITY whose EXECUTION_CONTEXT names the trigger and target application, and the navigation target for each step is resolved from that execution context. The journey crosses three ownership boundaries (SAP Master Data Governance, controlling in SAP S/4HANA, and SAP SuccessFactors) and the user is never asked to know where a step executes, because the data spine already knows and the experience layer reads it. A manually built menu would have encoded these three targets as three static links; the first rehost, rename, or resequence would have broken one of them silently. The resolved navigation cannot break silently, because there is nothing static to break. Every target is the current answer to a question about EXECUTION_CONTEXT, asked at the moment the user looks at it.

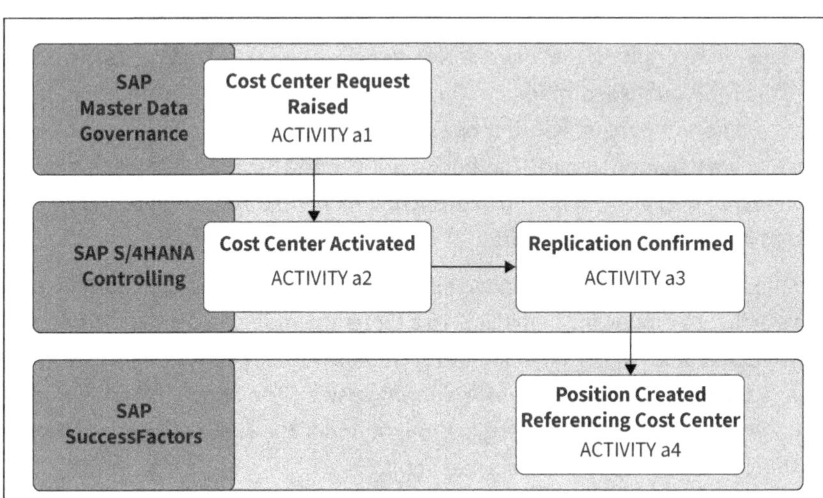

Figure 17.2: Process-Driven Navigation for the Cost Center to Position Journey, Derived from ACTIVITY and EXECUTION_CONTEXT

The journey also demonstrates the separation Chapter 9 insisted on. The navigation expresses sequence and ownership; it does not express authorization. Whether this user may raise the cost center request, activate it, or create the position is decided by the role scope on each `ACTIVITY`, not by the presence of a navigation node. The two are deliberately distinct. The journey is a property of the process; the right to perform a step is a property of the actor against the process. The experience layer renders both faithfully and invents neither.

17.2.5 Deep Links, Bookmarks, and the Stale URL Problem

The most common way an experience layer leaks a stale fact is one no design review catches because no designer built it: the bookmark. A user navigates to an activity, the browser shows a URL, the user bookmarks it or pastes it into a runbook, and months later they follow that URL again. The straightforward way to build a deep link—and the one most implementations reach for first—is to encode the resolved target (a concrete application or object) directly in the URL. It works immediately, which is what makes the problem easy to miss: The result is a stored navigation fact carrying all the drift modes Listing 17.1 was designed to eliminate, smuggled back in through the address bar. The bookmarked URL keeps working long after the process was republished, the activity renamed, or the role reassigned, but it creates a new problem: It lands the user on a target the current model would no longer resolve to, with nothing on the page admitting the discrepancy.

The discipline that closes the address bar is to make a deep link name the process intent, not the resolved target. A governed deep link carries the `PROCESS` and `ACTIVITY` identifiers, which identify what the user meant to do, and the target is resolved on arrival by the same Listing 17.1 function that resolves a fresh navigation, under the arriving user's scope and against the published version `AS-OF` arrival. A bookmark then behaves correctly after a change. If the activity still exists and the arriving user is in scope, it resolves to the current target. If the activity was retired, the link resolves to with a message that reads "This step no longer exists in the published model" rather than a silently wrong destination. If the arriving user is a different person whose scope does not permit the activity, it resolves to nothing, exactly as navigation would. The bookmark is not a stored route; it is a stored question, reanswered every time it is followed, and the `NAV_RESOLVED` audit row is emitted on arrival just as it is for ordinary navigation.

For the controller this turns a notorious blind spot into a nonevent. The question "Did this user reach this activity by following an old link to a target the model had moved?" does not need to be asked, because an old link cannot reach a moved target. It carries intent, not destination, and the destination is always the current answer. The address bar, historically the place experience-layer drift hides best, is brought under the same resolve-don't-store rule as everything else, and the audit chain reaches it.

As explained in Chapter 7, a `WALKME_ASSET` is mapped to the same `ACTIVITY` a navigation node resolves from, so the in-app guidance a user sees on a page is guidance for the activity the page projects, not generic help bolted onto a screen. When the process is republished and an activity changes, the guidance follows the activity because both are keyed to the same spine identifier, which is the entire point of Chapter 7's discipline and the reason WalkMe content does not drift away from the process the way free-text documentation does.

The return path closes the loop we have described since Chapter 1. A user's interaction with WalkMe-guided navigation emits a `TELEMETRY_EVENT`, linked to the same `ACTIVITY`, that flows back through the adoption model of Chapter 8 into the data spine. Friction at any step—hesitation, repeated guidance invocation, or abandonment—becomes a signal the process owner can see, and the signal is joinable to the activity, the process version, and the navigation that produced it. The experience layer is therefore not only a window onto the process; it is the instrument through which the process learns how it is actually performed. It does so without becoming a separate store, because the telemetry is keyed to spine identifiers the moment it is captured.

17.3 Role-Based Task Views

The second experience surface is the task view: the inbox, the worklist, and the tile that shows pending tasks. It is the single most-used surface in any operational landscape. This section restates Chapter 14's discipline for the surface a person works from: A task view is a role-scoped assembly produced for one actor at one instant, never a shared list filtered on the way out. The difference between assembling under a scope and filtering after the fact is invisible on screen and decisive in audit, and this section explains why.

17.3.1 A Task View Is a Role-Scoped Assembly

The wrong way to build an inbox is to query the full set of open tasks and then hide the rows the current user may not see. This is due to the same reason Chapter 14 gave for never preparing one data set and reusing it across actors: A list that exists in full before the scope is applied is a list that can leak through any defect in the filter, such as a missed predicate, a cached page, a count that includes hidden rows, or an export that runs before the hide. The right way was specified in Chapter 14, Section 14.2: Resolve the actor's role scope first, then assemble the view from only what that scope admits, so that the rows the actor may not see are never materialized in the actor's view at all. The absent row, like the absent navigation target and the absent wire, is the control.

Figure 17.3 deliberately takes the same shape as the prepared-data assembly described in Chapter 14 because it is the same operation. The actor's token is validated to an identity and a scope set. The role scope is resolved with the `role_scope_intersect` from Chapter 14, the intersection of what the role grants and what the data demands. The view is assembled from an SAP Build Process Automation task state joined to the context enrichment projection, under that intersection. The intersection is then stamped onto the assembled view so that anything done with it later (such as a drill-down, an export, or a follow-on navigation) inherits the same scope and cannot widen it. If the view cannot be assembled in full under one actor's scope, it is refused. It is never returned partial. The figure has two terminal states: a full view or VIEW_REFUSED. There is no state that lists *most* of your tasks, because a partial inbox permits a user to act confidently on an incomplete picture.

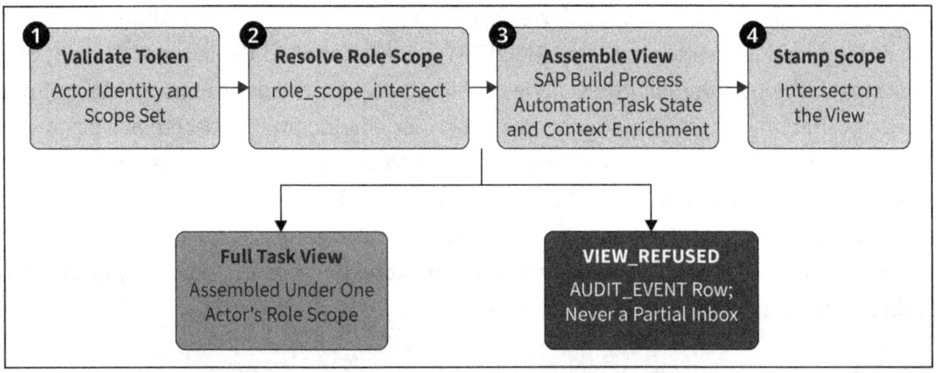

Figure 17.3: The Role-Scoped Task-View Assembly, the Same Discipline We Applied at the Experience Layer in Chapter 14

17.3.2 Where the Tasks Come From: SAP Build Process Automation

Every row in a governed task view is an SAP Build Process Automation human task, and the four workflow families of Chapter 16 are the four kinds of row a user sees. An approval task is a row from the approval family, present only because a verdict from Chapter 15 was accepted and the Chapter 16 subscription started a workflow. A remediation task is a row from the remediation family. An exception task is a row from the exception family. A review task is a row from the periodic-review family. The task view is therefore not a new construct; it is the Chapter 16 workflow population, projected per actor. This is the structural reason a task view holds no data. The task's existence, owner, due state, and disposition are all facts on the data spine, written by the workflow engine through the commit gateway, and the experience layer renders them without owning any of them.

The keying matters for the controller. Because each task row carries the identifiers Chapter 16 stamped on the workflow (the PROCESS_VERSION, the ACTIVITY, the EXECUTION_CONTEXT, and, for an AI-influenced task, the VALIDATION_VERDICT and the INFERENCE_EVENT behind it), a task in the inbox is joinable to the evidence and the decision that produced it without leaving the data spine. The inbox is not a dead end that requires a separate system to explain; it is one more place the one-query reconstruction described in Chapter 16, Section 16.4 reaches.

17.3.3 Separation of Duties Is Visible, Not Just Enforced

Chapter 16 made SoD structural by subtracting the proposer's identity from the eligible approver set, so the person who proposed a change cannot be the person the workflow routes the approval to. The experience layer's obligation is to make that subtraction visible rather than merely enforced. The proposer of a change does not see an approve control on their own change that rejects them when they click it. They do not see the approve control at all, because the task was never routed to them, so it was never assembled into their view. The user's experience of SoD is the absence of the control, which is the strongest possible expression of it: There is nothing to attempt, nothing to deny, and nothing to explain.

Listing 17.2 is the operational heart of this section. The proposer subtraction is a `CONTINUE`, not a post-hoc filter. The row is never added, so it cannot leak. The deny-by-default check is the same rule navigation used in Listing 17.1. The `assembly_incomplete` branch is the refusal-rather-than-partial discipline made literal. If any source the view depends on is unavailable, the view is refused with an audited reason, because a task list that is missing rows the user cannot know are missing is more dangerous than no task list at all. The `stamp_scope` call is the guarantee made in Chapter 14 that nothing done with the view later can widen its scope.

```
FUNCTION assemble_task_view(actor_token):
    actor = authz.principal(actor_token)
    scope = authz.scope_set(actor_token)
    rsi   = role_scope_intersect(actor.role_context, scope)

    tasks = []
    FOR t IN sbpa.open_human_tasks(rsi):          # assembled under scope
        ctx = enrich.task_context(t.task_id)      # Table 12.2 schema
        IF t.kind == 'APPROVAL'
           AND ctx.proposer_id == actor.id:
            CONTINUE                              # SoD: never your own
        IF NOT scope.permits(ctx.activity_id):
            CONTINUE                              # deny-by-default
        tasks.append(row(task = t.task_id,
                         version   = ctx.process_version,
                         activity  = ctx.activity_id,
                         verdict   = ctx.validation_verdict,
                         inference= ctx.inference_event))

    IF assembly_incomplete(tasks):                # any source unavailable
        audit.emit('VIEW_REFUSED', actor, reason)
        RAISE view_unavailable                    # never a partial inbox

    stamp_scope(tasks, rsi)                       # inherited downstream
    audit.emit('VIEW_ASSEMBLED', actor, count(tasks))
    RETURN tasks
```

Listing 17.2: The Task-View Assembly Is Role-Scoped and Subtracts the Proposer, so SoD Is an Absence, Not a Denial

17.3.4 Delegation and Substitution Without Scope Escalation

Every operational landscape needs substitution. It's inevitable that a user will go on leave and someone must clear their tasks in the meantime. Substitution is where task views most often leak scope, because the naive implementation copies the absent user's worklist to the substitute, which is precisely the prepared-and-reused list we forbade in Chapter 14, now reused across users rather than across requests. A copied worklist gives the substitute a set of rows assembled under the absent user's scope, not the substitute's, and the substitute can then act on tasks their own role would never have admitted. The breach is invisible because the substitute sees a normal inbox and has no way to know the rows were assembled under someone else's entitlements.

The governed form of substitution is a delegation fact on the data spine, not a copied list in the experience layer. Delegation is modeled as an explicit, time-bounded, audited relationship (actor A's tasks for activities in set S may be acted on by actor B from date to date) and it is carried into the routing from Chapter 16, not into the SAP Build Work Zone inbox. When the substitute opens their task view, the assembly of Listing 17.2 runs under the substitute's own identity and scope, and the delegated tasks appear only if the delegation relationship admits them and the substitute's scope also permits the underlying activity. The intersection still governs: Delegation can widen whose tasks B sees, but it can never grant B an entitlement B's own role lacks because `role_scope_intersect` is evaluated against B's scope regardless of the delegation. SoD also survives substitution. The proposer subtraction is evaluated against the acting identity, so a substitute who happens to have proposed a change still cannot approve it, even while acting for someone who could.

If you're approaching this chapter from an FP&A perspective, the audit consequence is the one you care about. A delegated action records the acting identity, the delegating identity, and the delegation relationship that authorized the substitution, as three distinct audited facts. The reconstruction question "Who actually approved this, and under whose authority were they allowed to?" has a precise answer that names both people and the time-bounded grant between them. A copied worklist could never answer that question because the copy erased the distinction between the two identities at the moment it was made.

Chapter 14 treated the composite-role case explicitly, and the experience layer inherits the treatment without amendment. A user who holds more than one role does not get a task view assembled by aggregating the roles into a superuser scope. The view is assembled under the union of the roles' scopes evaluated as a scope, with the `role_scope_intersect` applied so that the intersection of grant and demand still governs every row. This distinction is the same one we drew in Chapter 14: A union of scopes is not the same as a sum of privileges. Holding a cost center approver role and a position maintenance role lets a user see the approval tasks one role admits and the maintenance tasks the other admits; it never lets the user approve a change they proposed under the other role because the proposer subtraction in Listing 17.2 is evaluated against the actor identity, not the role.

Table 17.3's third column is the one a governance officer reads first. The structurally absent rows are absent because of the assembly, not because of a setting on the view that a well-meaning administrator could change. There is no configuration flag whose flip turns a maintainer's view into an approver's view because the rows are a function of the actor's scope evaluated at assembly time, not a property of the view object. This is the experience-layer form of the invariant we defend throughout the book: The boundary is a structural property of how something is built, not a checkbox someone remembered to tick.

Role Context	Rows the View Assembles	Rows That Are Structurally Absent
Cost center approver	Approval tasks for cost center changes in scope	Changes this actor proposed; out-of-scope activities
Position maintainer	Remediation and maintenance tasks in scope	Approval tasks; activities outside the role
Process owner (read)	Review tasks; periodic-review worklist	Approve/remediate controls (no write privileges)
Composite (approver and maintainer)	Union of the in-scope rows of both roles	Own proposals; anything neither role includes
Operator	Exception tasks for monitored regions in scope	Approval and remediation outside operations scope

Table 17.3: The Task-View Rows Differ by Role, Never by Configuration of the View

AI Insight: Refusal Rather Than a Partial Inbox Is an AI-Safety Property, Not Just a Hygiene One

Refusing a task view that cannot be assembled in full looks like ordinary defensive engineering until an AI agent is the consumer. Chapter 13 established that a Joule agent reaches governed data only through the SAP Build Process Automation action library and the grounding endpoints, never through a private path. When a user asks Joule "What is waiting for me?" Joule answers from the same role-scoped task assembly Listing 17.2 produces. If that assembly could return a partial list, the agent would summarize an incomplete picture with the same fluent confidence it summarizes a complete one, and the user would have no way to tell the difference. The failure would be laundered through natural language into certainty.

Because the assembly refuses rather than returns partial, the agent's worst case is the response "I cannot retrieve your task list right now," which is a recoverable, auditable, honest failure. The refusal-rather-than-partial discipline is therefore not only inbox hygiene; it is the property that stops an AI mediated summary from quietly becoming a confident lie. For the controller, this is the same one-query guarantee in a different costume. The system never presents an incomplete view as a complete one, whether the consumer is a screen or an agent.

17.4 Surfacing AI Insights

The third experience surface is the one this part of the book has been converging on: the place a user actually meets an AI insight. Chapter 13 through Chapter 16 governed how an insight is produced, grounded, decided, and executed. This section governs how it is shown. This section is particularly important if you're approaching this chapter from an FP&A perspective, because an AI insight in the experience layer is advisory by construction. It is the recommendation panel from Chapter 16, Section 16.3 rendered for a human. It carries its own evidence, and it cannot be the element that changes the data spine. The human task must do that.

17.4.1 An Insight Is a Citation, Not an Answer

The defining property of a surfaced insight is that it arrives with its evidence attached. An insight card is not a sentence the model produced on its own. It is a sentence the model produced with the grounding bundle `manifest_hash` that fixes exactly what the model was allowed to see, the `INFERENCE_EVENT` identifier that records the invocation, the `SUGGESTION` identifier and its `lifecycle_state`, and, once a human has acted, the `VALIDATION_VERDICT` that records what was decided. The card is a citation in the literal sense: Every claim it makes is traceable to the evidence that was in scope when the claim was generated, by the same `manifest_hash` we defined in Chapter 14 and carried into the suggestion in Chapter 15. An insight without its bundle is not a weaker insight; it is simply not surfaced, because an uncited claim that cannot be defended cannot be shown.

Table 17.4 demonstrates the contract a card must satisfy to be rendered. Five of the six elements are evidence; the sixth is a route to a human task. There is no element on the card that is a write control, because the card has no write capability—a point the next section makes structural.

Element on the Card	Data Spine Entity It Cites	Why It Must Be Present
The claim itself	`SUGGESTION` (text, `lifecycle_state`)	What is being advised, and its current state
Evidence fingerprint	Grounding bundle `manifest_hash`	Fixes exactly what the model could see
Inference identity	`INFERENCE_EVENT`	Ties the claim to one audited invocation
Explanation	`SUGGESTION_EXPLANATION`	Why the model advised this, in scope terms

Table 17.4: What an Insight Card Must Carry Before the Experience Layer Will Surface It

Element on the Card	Data Spine Entity It Cites	Why It Must Be Present
Disposition (after action)	VALIDATION_VERDICT (verdict_value)	What a human decided relative to the claim
Route to act	SAP Build Process Automation human task handle	The only control that can change the data spine

Table 17.4: What an Insight Card Must Carry Before the Experience Layer Will Surface It (Cont.)

17.4.2 Joule in SAP Build Work Zone

Joule appears inside SAP Build Work Zone as the conversational surface, and its embedding changes none of the Chapter 13 governance. In a work page, Joule reaches process and master data only through the AI grounding endpoints. It performs governed reads only through the SAP Build Process Automation action library we covered in Chapter 12, Section 12.4. It carries the user's SAP Authorization and Trust Management service scopes rather than a service identity, and every invocation writes an INFERENCE_EVENT and an AUDIT_EVENT. Embedding Joule in a work page gives it proximity to the user, not privilege over the data spine. The grounding endpoints remain the Access API closed to AI, as covered in Chapter 10, Section 10.4. The action library remains the only governed-read surface, and the scopes remain the user's own.

Figure 17.4 draws the path and, as with Figure 17.1, its meaning is in the direction connection that is not present. A user prompt with role context enters Joule, which orchestrates and routes tools.

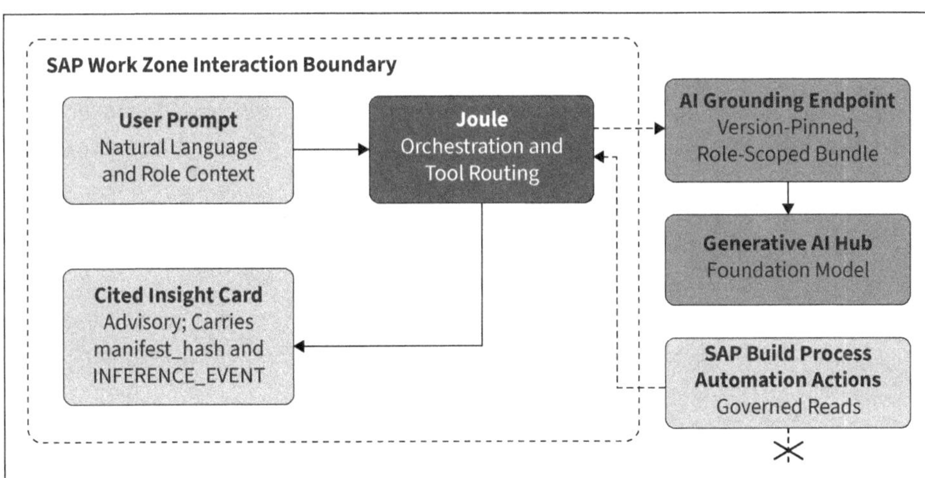

Figure 17.4: Joule Inside SAP Build Work Zone: A Prompt Becomes a Cited Insight, and There Is No Private Path from Joule to the Data Spine

It calls an AI grounding endpoint for a version-pinned, role-scoped bundle, the generative AI hub for the foundation model, and SAP Build Process Automation actions for governed reads. It returns a cited insight card carrying the `manifest_hash` and the `INFERENCE_EVENT`. What Figure 17.4 does not show is a line from Joule to the canonical data spine, because there is no private path. Joule cannot read the data spine except through the grounding endpoints and cannot change it except by handing a decision to an SAP Build Process Automation human task, exactly as the recommendation panel could not as covered in Chapter 16, Section 16.3. The boundary is the absence of a direction connection, not a permission someone set to deny.

17.4.3 The Insight Card Contract, Annotated

Listing 17.3 shows the insight card contract as the experience layer enforces it. It is deliberately short, because the discipline is not in what the card does but in what it structurally cannot do.

```
FUNCTION render_insight_card(suggestion_id, actor_token):
    scope = authz.scope_set(actor_token)
    sg = enrich.suggestion(suggestion_id, scope)    # role-scoped read
    ASSERT sg.grounding.manifest_hash IS NOT NULL   # no bundle, no card
    ASSERT sg.inference_event IS NOT NULL           # no event, no card

    card = {
       claim       : sg.text,
       state       : sg.lifecycle_state,
       evidence    : sg.grounding.manifest_hash,
       inference   : sg.inference_event,
       explanation : sg.explanation,
       verdict     : sg.validation_verdict,      # null until decided
       act         : sbpa.human_task_handle(sg) # the only actuator
    }
    # NOTE: there is no card.commit, card.write, or card.apply.
    # 'act' opens an SBPA human task; the task, not the card,
    # carries any decision to the commit gateway.
    audit.emit('INSIGHT_SURFACED', actor, suggestion_id,
               evidence = card.evidence)
    RETURN card
```

Listing 17.3: The Insight Card Carries Its Evidence and Writes Nothing; Acting Routes to a Human Task

The two assertions are the gate: A suggestion without a grounding `manifest_hash` or without an inference event is never rendered, so an uncited claim cannot reach a user's eyes. The comment is the specification, not documentation. There is no commit, write, or apply method on the card because the card is a projection, and a projection that could write would violate the architecture we've outlined in this book. The act handle opens an SAP Build Process Automation human task, the task is the actuator, and the card is the evidence.

17.4.4 Multi-Suggestion Cards and the Comparison Trap

A model often returns not one suggestion but several, such as three candidate cost center mappings, two remediation options, or a ranked list of probable causes. Rendering several suggestions on one card introduces a problem the single-suggestion case does not: The card can imply a recommendation through order, emphasis, or a confidence number, and the human's role degrades from deciding to ratifying the top of a list. This is the prefill problem from Section 17.3.4 wearing a different costume, and the architecture forecloses it the same way—structurally, not by visual restraint.

Each suggestion in a multi-suggestion result is a distinct `SUGGESTION` with its own `lifecycle_state`, its own `SUGGESTION_EXPLANATION`, and its own line in the grounding bundle. The card surfaces them as peers, not as a ranked list with a winner. The human task that the card routes to does not preselect one; it requires the human to record a verdict that names which suggestion, if any, was accepted, and the `verdict_value` carries that selection as the human's act, not the model's ordering echoed back. A confidence score, where the model produces one, is shown as an attribute of each suggestion's evidence, never as a sort key the card applies on the user's behalf, because a sort the card performs is a recommendation the card makes, and the card does not get to make recommendations. Only the human task records a decision. The reconstruction is unambiguous as a result. The audit chain records which of the peer suggestions the human chose and which the model offered, as separable facts, so a later review can always distinguish places where the model ranked the suggestion first from places where a human made a decision, which are the two things an automation-bias inquiry most needs to tell apart.

A subtle problem remains even when the card cannot write: a card that prefills the human task with the model's preferred answer, so the human's role degrades to clicking **Confirm**. Chapter 16, Section 16.3 forbade this for the recommendation panel, and the same must be done here. The card displays the claim and its evidence, the human task presents the decision unfilled, and the human supplies the verdict. The `verdict_value` records the human's disposition relative to the insight (accepted, rejected, or modified), not the model's suggestion echoed back. The distinction is the difference between a tool that informs a judgment and a tool that manufactures the appearance of one, and the architecture is built so that only the first is possible.

The reason this is structural rather than a matter of user interface (UI) taste is that the human task is a Chapter 16 construct with its own audited input. The verdict the task records is the human's, captured by the task, written by the workflow through the commit gateway, and joined to the `INFERENCE_EVENT` for the model's contribution. Even if a card visually emphasized the model's preferred option, the verdict is still a separately captured human fact, and the audit chain records the human's choice and the model's suggestion as two distinct events with two distinct identities. The architecture does not merely discourage automation bias; it makes the human's act and the model's act separately accountable, so that no decision can later be mischaracterized as the other.

17.4.5 When the Bundle Is Stale: Insight Expiration and Regrounding

An insight card carries the grounding bundle `manifest_hash` that fixes exactly what the model could see when it produced the claim. That fingerprint is also the card's expiry condition. The bundle was assembled `AS-OF` an instant against a published version, exactly as Chapter 14 specified; the world the bundle described can move on. The cost center the suggestion reasoned about can be revalidated, the process can be republished, and the rows the bundle fingerprinted can change. A card that keeps presenting a claim after its grounding has gone stale is no longer a citation but an assertion whose evidence no longer exists, which is exactly the type of uncited claim the architecture refuses to surface.

The experience layer therefore treats the `manifest_hash` as live, not decorative. When a card is rendered, the current grounding fingerprint for the suggestion's scope is compared to the one the card carries. If they match, the card is current and is shown with its evidence. If they diverge, the card is not silently refreshed nor shown stale. It is marked as expired, and acting on it requires regrounding—a fresh inference against a fresh bundle, producing a new `SUGGESTION` with a new `manifest_hash` and a new `INFERENCE_EVENT`. The expired card and the regrounded card are distinct audited objects, not a singular modified one, so the record never shows a claim whose evidence changed underneath it. This is the freshness discipline from Chapter 14 applied to the surface a person reads: A claim is only ever as current as the evidence it cites, and when the evidence moves the claim must be reearned, not refreshed in place.

The controller's benefit is the elimination of a subtle reconstruction error. Without expiration, an auditor reviewing a decision could find a card whose displayed claim no longer matched the evidence on the data spine and not be able to tell whether the human had acted on the old claim or the new state. With expiration, the verdict is always joined to a specific, non-stale `manifest_hash`, so the reconstruction states without ambiguity which evidence set the human actually saw when they decided. This is the same guarantee we attached to the suggestion in Chapter 15, now preserved all the way to the user's screen.

The earliest signal a controller gets that AI is influencing outcomes is not a quarterly report but an SAP Build Work Zone analytics tile that reads the `VALIDATION_VERDICT` population. The tile shows how many AI-originated suggestions were accepted, rejected, or modified, over which processes, in which period. This is the acceptance posture we promised the controller would be able to see in Chapter 16. The structurally important property is that the tile is one aggregate over rows the data spine already holds. It is not a new store, a copy, or a metrics pipeline that could disagree with the system of record. It is a read, governed exactly as every other read in this chapter, returning a count of verdicts that already exist.

Figure 17.5 traces the path from card to spine and back to the controller's tile. The insight card is advisory. The human task is the only decision point. The VALIDATION_VERDICT records the verdict_value and the disposition. The SAP Build Process Automation workflow carries it to the commit gateway and the data spine. The disposition tile reads one aggregate over the verdict population. The loop closes without a new store anywhere in it, because the controller's earliest signal is a read over rows that the act of deciding already wrote. This is the same architectural economy we emphasize throughout the book: The cheap option and the correct option are the same once you refuse to keep a second copy.

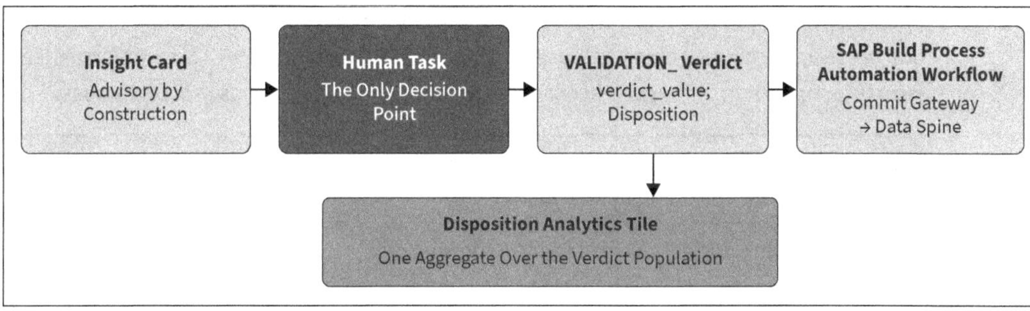

Figure 17.5: The Insight-to-Verdict Path in the Experience Layer: The Card Displays, the Human Task Decides, and the Verdict Records the Disposition

Tip

When an experience team is asked to add an AI adoption dashboard, your instinct might be to stand up an analytics store that ingests inference logs. Resist it. Every number such a dashboard would show is already a query over INFERENCE_EVENT and VALIDATION_VERDICT on the data spine. A tile that reads those entities through the governed read path is current by construction, reconcilable to the system of record by definition, and free of the divergence risk a parallel store always carries. The dashboard is a read, not a system.

17.5 User Interaction Boundaries

The fourth experience surface is the set of boundaries themselves. The preceding sections each enforced a boundary as a side effect of doing something useful. This section focuses on the boundaries. We state the thesis of this chapter as a rule the experience layer must obey without exception: Every boundary the platform enforces structurally has an experience layer expression, and the experience layer can never offer an action the data spine would refuse.

17.5.1 The User Interface Offers Only What the Data Spine Would Permit

The deny-by-default principle outlined in Chapter 10 has a precise experience layer form: A control the user is not entitled to is absent, not present and disabled. A greyed-out **Approve** button is an information leak. It tells the user a control exists, that someone may use it, and why they may not (often through a tooltip or an error). An absent control tells the user nothing because there is nothing to tell. The experience layer renders the deny-by-default decision by not assembling the control into the view, the same way navigation does not append an unauthorized target and the task view does not materialize an out-of-scope row. This is one pattern applied at every surface. The boundary is realized as an absence the user cannot perceive, not a presence the user is told not to touch. Table 17.5 summarizes each structural boundary, its experience layer expression, and what the user actually sees.

Structural Boundary (Where Defined)	Experience Layer Expression	What the User Perceives
Deny-by-default authorization (Chapter 10, Section 10.4)	Unentitled control not assembled	No such control on the page
Role scope on activity (Chapter 14, Section 14.2)	Out-of-scope navigation node not appended	The destination is simply not there
SoD (Chapter 16)	Own-proposal approval row not assembled	Nothing to approve on your own change
AI is advisory (Chapter 16, Section 16.3)	Card has no write method; routes to a task	Acting always opens a human decision
Single commit gateway (Chapter 10)	No write capability anywhere in SAP Build Work Zone	Every change goes through a workflow
Refusal rather than partial (Chapter 14)	View refused if it cannot be assembled whole	An honest message that the view is unavailable, never a partial return

Table 17.5: Each Structural Boundary Has an Experience Layer Expression, and the Expression Is an Absence

17.5.2 The Absent Control Is the Strongest Control

Chapter 16 closed its central argument with the observation that the boundary is the absence of a wire. The experience layer's version of that sentence is that the boundary is the absence of a control. The two are the same structural idea seen from two sides. A direct connection that does not exist cannot carry an unauthorized write; a control that is not assembled cannot be clicked, scripted, screen-scraped, or discovered by a curious user with developer tools open. Every alternative (a disabled control, a hidden control, or a control that validates upon submission) is a presence that depends on enforcement holding everywhere it is reached. An absence depends on nothing because there is nothing to enforce. This is why we have insisted, at every surface, that the right expression of a boundary is to not build the element at all, rather than to build it and guard it.

Listing 17.4 shows control assembly the way we have stated navigation and task assembly: as a resolution that produces only what the actor is entitled to, so that the boundary is the set the loop never appends to rather than a check a later handler must remember to perform.

```
FUNCTION assemble_controls(actor_token, object_ctx):
    scope = authz.scope_set(actor_token)
    actor = authz.principal(actor_token)
    controls = []
    FOR c IN candidate_controls(object_ctx):
        # entitlement is evaluated BEFORE the control exists
        IF NOT scope.permits(c.required_scope):
            CONTINUE                     # not disabled - absent
        IF c.kind == 'APPROVE'
           AND object_ctx.proposer_id == actor.id:
            CONTINUE                     # SoD - never your own
        IF c.writes_state:
            c = route_to_human_task(c)   # never a direct write
        controls.append(c)
    # the page renders 'controls'; whatever is not here cannot
    # be clicked, scripted, or discovered - there is no node
    audit.emit('CONTROLS_ASSEMBLED', actor, object_ctx,
               rendered = ids(controls))
    RETURN controls
```

Listing 17.4: A Control the Actor Is Not Entitled to Is Never Assembled, so the Boundary Is an Absence, Not a Guarded Presence

The listing has no `else` branch that disables a control and no handler that rejects a forbidden click because there is no forbidden control to click. The unentitled candidate was skipped before it became a node on the page. The `writes_state` branch is the single-commit-gateway rule rendered at the experience layer. A control that would change state is not rendered as a writer but rewritten into a route to a human task, so even a permitted, in-scope control cannot itself touch the data spine. The `CONTROLS_ASSEMBLED` audit row records exactly what the actor was offered, which is the experience-layer half of the reconstruction: the auditor can establish not only what the user did but what the user could have done and confirm the two are consistent with the actor's scope at that instant.

AI Insight: An Absent Control Is the Only One an Agent Cannot Talk Its Way Around

The absent control discipline reads as classical UI security until the actor is an AI agent rather than a person with a mouse. A disabled or hidden control still exists in the page model and the API surface. A person is stopped by the disabled state, but an agent that enumerates the interface, reads the accessibility tree, or calls the underlying endpoint directly is not stopped by a visual state at all. Every enforcement that lives in the rendering (greying, hiding, and validate-on-submit) is precisely the enforcement an agent is structurally positioned to ignore.

An absent control has no page node and no endpoint the agent is entitled to reach because the entitlement was evaluated in Listing 17.4 before the control existed, and the `writes_state` branch rewrote even permitted controls into routes to a human task. There is nothing for the agent to enumerate, call, or reason about. The boundary is not a state on an element but the absence of the element. This is why our insistence on absence over guarding is not stylistic. In a landscape where some actors are agents, the absent control is the only expression of a boundary an agent cannot talk, enumerate, or call its way around. It is the same property we used to protect the data spine in Chapter 16 (the boundary is the absence of a direct connection), now applied at the surface a user, or an agent acting for one, actually touches.

17.5.3 Break-Glass Is a Separate, Human-Authenticated Path

There are operational moments—such as a severity 1 incident, a regulator deadline, or a payroll run that must complete—when an authorized human must act outside the normal flow. Chapter 13 specified break-glass as a separate, strongly authenticated, heavily audited path that is never reachable through the AI surface. The experience layer honors that specification literally. A break-glass action in SAP Build Work Zone is a distinct, separately authenticated journey (a different tile, an explicit reauthentication, or an explicit reason) that writes an `AUDIT_EVENT` of type `BREAK_GLASS` before it does anything else. It is never an option on an insight card, a fallback Joule can invoke, or a control that appears because a normal one was refused. The point of break-glass is that it is loud, deliberate, and unmistakable in the audit log; an experience layer that made it convenient or reachable from the AI surface would have destroyed the only property that makes it acceptable.

17.5.4 The Mobile and Offline Case

Everything we have argued so far in this chapter assumes that the experience layer reads the data spine at the moment the user looks at it. Mobile and offline scenarios break that assumption deliberately. An offline client, by definition, is shown facts captured earlier and acts on facts that will be reconciled later. Offline is therefore the clearest example of the stale-fact hazard, and it needs an explicit rule instead of optimistic hope that it rarely matters.

The rule is that an offline cache can hold a projection for reading but can never hold a decision as though it were recorded. A user can take a role-scoped, version-pinned snapshot offline and read it; the snapshot carries the version it was pinned to and the instant it was taken, so it is honest about its own staleness rather than pretending to be live. What the offline client cannot do is record a verdict locally and present it to the user as complete. An offline decision is a captured intent that becomes a verdict only when the client reconnects and the decision is routed through the same SAP Build Process Automation human task and commit gateway every online decision uses, with the assembly and the SoD checks from Listing 17.2 evaluated at reconnection against the then-current model, not against the model that was current when the snapshot was taken. If the task was reassigned, the process republished, or the user's scope changed while the user was offline, the captured intent can be refused on reconnection loudly, rather than committed silently against a model that no longer holds.

Table 17.6 shows what an offline client is and is not permitted to do. Being offline does not earn the experience layer an exception; it earns it an obligation to be even more explicit about staleness, because the user cannot see the data spine to notice the gap themselves. An offline client that committed from the device, or that showed a locally recorded decision as final, would have cut exactly the alternate path to the data spine that the entire architecture exists to prevent—and that path would simply be on a mobile device in someone's pocket. The one-query reconstruction still holds across an offline decision precisely because the decision did not become a verdict until it came back through the governed path, with its captured-offline provenance recorded as part of the audit fact.

Offline Operation	Permitted?	Why?
Hold a role-scoped, version-pinned read snapshot	Yes	A projection that is honest about its pin and instant
Show the snapshot's version and capture time	Yes	The snapshot must not pretend to be live
Capture a decision as an intent to route later	Yes	Intent is not a verdict; it carries no authority yet
Record a verdict locally and show it as done	No	A verdict exists only when the workflow writes it
Commit any change from the device	No	The commit gateway is the only writer, online or not
Reevaluate scope and SoD at reconnection	Required	The current model governs, not the snapshot

Table 17.6: What an Offline Client Can and Cannot Do, by the Same Projection Rule Applied Online

Chapter 16 introduced the refusal ledger, the record of what the platform declined to do and why: a workflow not started because no accepted verdict existed, a write rejected because it did not arrive through the commit gateway, or an action refused because scope did not permit it. The experience layer surfaces the refusal ledger as a governance-space tile, so that what the platform refused is as visible to an auditor as what it did. This is a deliberate inversion of the usual instinct to hide failures. In a governed landscape a refusal is not a failure; it is the control working, and a control whose operation cannot be seen cannot be audited. Surfacing the refusals turns the boundary from an assertion the architecture makes about itself into a fact a governance officer can read off a tile.

From an FP&A perspective, the refusal ledger is the other half of the defense kit. The audit chain proves what happened and on whose authority. The refusal ledger proves what was prevented and why. An auditor asking "Could someone have pushed an unapproved change through the experience layer?" is answered not by an assurance but by a query: The refusal ledger shows the attempts the boundary turned away, each with its reason and its actor, and the absence of any successful out-of-band write is then a demonstrable property rather than a claim.

17.5.5 One Interaction, End to End: Across the Boundaries

The four surfaces are not four products; they are four faces of one governed contract, and a single user interaction touches all of them in sequence. Figure 17.6 follows one journey end to end so that you can see the boundaries not as a list but as a path a real person walks.

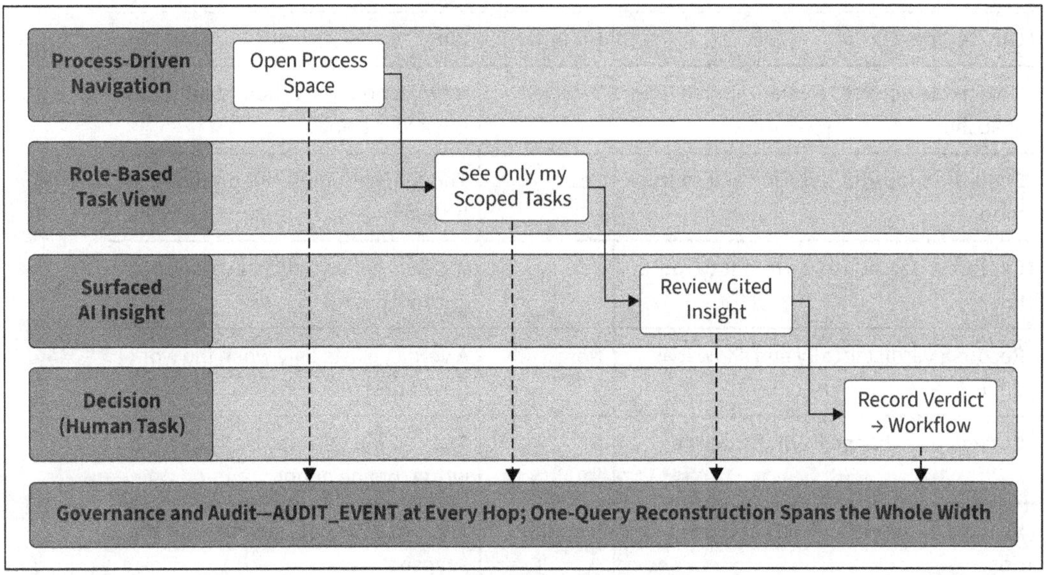

Figure 17.6: One Interaction End to End: Navigation, Role Task View, AI Insight, and Decision, Audit-Joined Across the Entire Width

Figure 17.6 reads left to right as one journey. The user opens a process space (navigation resolved from the model in Listing 17.1). They see only their scoped tasks, the role-scoped assembly from Listing 17.2. They review a cited insight from the card contract in Listing 17.3. They record a verdict, which routes to a workflow from the machinery outlined in Chapter 16. The lane spanning the full width is governance and audit: Every hop emits an AUDIT_EVENT, and the one-query reconstruction from Chapter 16, Section 16.4 runs the length of the journey, not just at its end. Figure 17.6 emphasizes that the experience layer is not a place where governance is added but one where governance that already exists becomes visible to a person, hop by hop, with the audit join holding the whole way across.

Let's now walk through the figure using our running example of a cost center that we have carried through from Chapter 9. A controller opens the cost center governance space. Navigation resolves from PROCESS and the published PROCESS_VERSION under the controller's role scope; the NAV_RESOLVED row records the version set. The space presents the controller's task view, assembled under their scope by Listing 17.2. It contains the review tasks for cost centers they own and excludes, structurally, any approval for a change they themselves proposed. One review task carries an AI insight (a suggested reclassification the model produced, as covered in Chapter 15) surfaced as a cited card carrying the grounding manifest_hash, the INFERENCE_EVENT, and the SUGGESTION_EXPLANATION, with its bundle confirmed non-stale by the expiry check. The controller reads the evidence, disagrees with part of it, and records a modified verdict in the human task; verdict_value captures the modification as the controller's act, distinct from the model's suggestion.

The verdict starts a workflow as outlined in Chapter 16, which then carries the change to the commit gateway and the data spine. Weeks later, an auditor asks what happened. One query, joined on the data spine identifiers, returns the navigation the controller saw and its version, the task view they worked from, the insight and its exact evidence, the modified verdict and that it was the human's, the workflow that executed it, and the audit event at every hop. The whole journey is reconstructed without interviewing anyone or trusting a screenshot.

Now imagine how Figure 17.6 would look like if the architecture were wrong. A manually built menu would break the first lane silently. A filtered-after-the-fact inbox would leak in the second lane. An insight card with a write method would short-circuit the third lane straight to the data spine. A greyed control would advertise the fourth lane. In each failure, the governance lane at the bottom would still be drawn, but it would no longer span the width. There would be a hop it could not join, and the one-query reconstruction would return a gap exactly where the user acted. The architecture we cover in this chapter exists to keep that lane unbroken.

17.6 Summary

SAP Build Work Zone is the experience layer of the integrated platform, and its only legitimate role is to be a faithful, role-scoped, version-pinned, and audit-consistent projection of a data spine it does not own. Navigation is resolved from the process model, never stored as a menu, so the workplace is exactly as current as the published model and honest about it. A task view is assembled under one actor's role scope and refused rather than returned partial, so a user never acts confidently on an incomplete picture and an agent never launders a partial list into a fluent certainty. An AI insight is surfaced as a cited card that carries its grounding bundle and inference identity and has no method that writes, so the human task remains the only decision point and the model's contribution and the human's choice stay separately accountable. Every structural boundary is expressed as an absence the user cannot perceive rather than a presence the user is told not to touch, and what the platform refused is surfaced to governance alongside what it did.

Nothing in this chapter was a new capability. Every guarantee the experience layer offers was decided in an earlier chapter: the data spine in Chapter 9 and Chapter 10, the governed read path in Chapter 12, the AI tier in Chapter 13 through Chapter 15, and the workflow machinery in Chapter 16. This chapter focused on how the surface a person touches introduces nothing of its own and removes nothing that the governed layers provide. That is this chapter's central discipline: The experience layer earns the architecture by being a window, and a window is trustworthy precisely because it manufactures nothing it shows.

17.6.1 Key Learning Points

- Navigation is a query against `PROCESS`, `PROCESS_VERSION`, `ACTIVITY`, and `EXECUTION_CONTEXT` resolved at request time and pinned to the published version, never a menu maintained in the space designer. A stored navigation tree is a copy of a data spine relationship and drifts the instant the model changes, with the drift invisible on the page.
- A task view is a role-scoped assembly produced for one actor at one instant, with the proposer subtracted so SoD is an absence rather than a denial. A view that cannot be assembled in full is refused, never trimmed, because a partial inbox lets a person (or an agent summarizing for them) act confidently on an incomplete picture.
- An AI insight is advisory by construction. It is surfaced as a cited card carrying the grounding bundle `manifest_hash`, the `INFERENCE_EVENT`, the `SUGGESTION` and its `lifecycle_state`, and the `SUGGESTION_EXPLANATION`. It has no write method, and acting on it opens an SAP Build Process Automation human task, which is the only decision point and the only one that reaches the commit gateway.
- Every structural boundary has an experience-layer expression, and the correct expression is an absence: An unentitled control is not assembled, an out-of-scope target is not appended, an own-proposal approval is not materialized. The absent control is the strongest control because it depends on nothing being enforced.

- SAP Build Work Zone holds no canonical data. It reads through context enrichment and the execution services and writes through nothing; the only path from an interaction to a data spine change is a recorded human verdict carried by a governed workflow, and the one-query reconstruction covered in Chapter 16, Section 16.4 here reaches every hop of the journey, including what the user saw.

17.6.2 Common Design Pitfalls

- Hard-coding the navigation tree. A hard-coded navigation tree substitutes a stored menu for a resolved query and drifts silently when the model changes; resolve navigation at request time and persist nothing in the experience layer.
- Filtering the inbox after the fact. This materializes every open task and hides the unauthorized rows, leaking through any defect in the filter. Instead, assemble the view under the actor's scope and refuse if it cannot be completed.
- Prefilling the insight card. This seeds the human task with the model's preferred answer and manufactures the appearance of a judgment. Instead, present the task unfilled and capture the human verdict as a separately audited fact.
- Greying out a control. This advertises a capability and depends on enforcement holding wherever it is reached. Instead, express the boundary as the absence of the control, not its disablement.
- Writing directly from a tile. A writing tile creates a secondary path to the data spine. The experience layer should read through the governed path and route decisions into workflows. It never writes anything under any circumstance.

17.6.3 Review Questions

1. Explain why a navigation tree maintained in the SAP Build Work Zone space designer is a governance defect rather than a performance optimization. Reference the drift modes a republish and an application rehost would introduce, and state what the `NAV_RESOLVED` audit row lets a controller reconstruct that a stored menu would not.
2. State the difference between assembling a task view under an actor's role scope from filtering a complete task list down to the actor's rows. Give one way the filtered approach can leak and explain why the leak is invisible on screen but decisive in audit.
3. The chapter argues that refusal rather than a partial inbox is an AI-safety property and not only inbox hygiene. Reconstruct that argument using the role of Joule as a consumer of the task assembly.
4. An insight card must carry six elements before the experience layer will surface it. List them, name the data spine entity each cites, and explain why a card with no write method is a structural property rather than a UI choice.

5. Restate the rule "The boundary is the absence of a direct connection" in the experience layer's terms, and explain why an absent control is stronger than a disabled, hidden, or validate-on-submit control.
6. Trace the one-query reconstruction across Figure 17.6 for an AI-influenced change. Hop by hop, name the audit events emitted and the data spine identifiers that join navigation, the task view, the insight, the verdict, and the workflow into a single walk.

17.6.4 Student Companion Notes

You can build the chapter's argument on your own by rederiving each surface from a boundary you met earlier in the book rather than memorizing four independent designs. Process-driven navigation is Chapter 9's `EXECUTION_CONTEXT` and Chapter 16's resolve-don't-store discipline applied to menus. The role-scoped task view is Chapter 14's assemble-under-scope and refusal-rather-than-partial disciplines applied to inboxes. The cited insight card is Chapter 16, Section 16.3's advisory recommendation applied to the user's screen. The interaction boundaries are Chapter 10's deny-by-default and single commit gateway disciplines applied to controls. If you can state which earlier boundary each surface expresses, you understand the chapter; if you are memorizing the surfaces separately, you have missed its point.

For an additional exercise, carry the following FP&A-focused question through every section: weeks later, can the controller reconstruct what the person saw, what the AI contributed, what the human decided, and on whose authority, with one query and no interview? Test each design decision in the chapter against that question. A resolved navigation tree passes it; a stored menu fails it. An assembled task view passes it; a filtered list fails it. A cited card passes it; an uncited claim fails it. This question is the same instrument the whole book has used, now turned on the surface a human actually touches.

With the experience layer specified, the platform is complete as something a person can see and act on. What remains is to run it. Chapter 18 turns to operating the integrated platform, covering observability, incident response, capacity, and the operational model that keeps the data spine, the workflows, the AI tier, and the experience layer healthy under real load.

PART IX

Operations, Security, and Extensibility

Chapter 18
Operating the Integrated Platform

A platform is always not what you designed; it is what survives an average workday. Every guarantee we've outlined in this book was a promise made at design time. Operations is where the promises are kept or quietly broken, and the only honest measure of an architecture is whether, weeks later, a controller can still reconstruct what happened, on whose authority, and on what evidence, without an interview.

Chapter 17 established the platform as something a person can see and act on. Chapter 9 and Chapter 10 explained how the canonical data spine holds governed truth. Chapter 11 demonstrated how the ingestion and reconciliation flows keep it current, and Chapter 12 described how the execution services read it under contract. In Chapter 13 through Chapter 15, we covered how the SAP AI tier reasons over it without authority to change it, and in Chapter 16 and Chapter 17, we showed how the SAP Build experience projects all of it to approvers, process owners, operators, and the controller. Each of those chapters ended with a guarantee, but none of them proved that the guarantee still holds at 3 a.m., in the second region, during a bulk upstream refresh, while an auditor is asking a question. That proof is the subject of this chapter.

Operating the integrated platform is not a maintenance afterthought. It is the discipline that converts a designed architecture into a defensible one. A platform that cannot show it was healthy when a number was produced has not earned the trust the earlier chapters assumed; it has only asserted it. This chapter turns the structural promises into operated properties: signals that are watched, quality that is scored, drift that is detected and resolved without silent mutation, and a footprint that scales across programs and jurisdictions without losing the audit chain. We'll also cover in depth the operational dashboards we first introduced in Chapter 11.

The chapter is written for the platform engineer who runs the landscape, the architect who sets its service-level objectives (SLOs), the governance officer who must defend a control, and—as in every chapter of this part—financial planning and analysis (FP&A) professionals who, weeks later, must explain why a reported figure can be trusted. For the latter, operations is the final piece of the defense kit: the evidence that the platform which produced the number was itself in a known-good state, reconstructible by the same one-query walk defined in Chapter 16, Section 16.4 and extended to the surface a person touched in Chapter 17.

Five principles run through the chapter and should be carried into every dashboard, runbook, and on-call rotation built on the platform. First, every operational signal is correlated to the same identity the audit chain uses, so health and evidence are the same query asked twice. Second, quality is an operated property, scored continuously against SLOs, not a one-time gate at ingestion. Third, drift is expected, classified, and resolved through governed transitions; the data spine never silently mutates and never loses history. Fourth, scaling adds programs, regions, and tenants without weakening identity, isolation, or the one-query reconstruction. Fifth, operations does not grant AI new authority (the principle this book has carried since Chapter 1, Section 1.5). The AI tier remains an advisor under the same governance the rest of the platform enforces, and operations is one more place that boundary is proven rather than assumed.

This chapter explains:

- How the platform is observed, including the three signal classes, the golden signals per layer, the SLOs that govern them, the burn-rate alerting that protects the error budget, and the severity model that turns a signal into a paged human, all correlated through the same `CORRELATION_ID` and `AUDIT_EVENT` the data spine already records
- How data quality is managed as a continuously operated property, including the quality scorecard by domain, the quality SLOs, the steward operating model, the quarantine economy, and the bounded role AI plays as a quality pattern detector that proposes but never approves
- How version drift is detected and resolved without silent mutation, including the drift taxonomy carried forward from Chapter 11, the resynchronization decision flow, forensic replay, hot-restart, and the regrounding of the AI tier when a pinned version moves underneath it
- How the platform scales across programs and jurisdictions, including program tenancy on resource groups, the multi-region topology, jurisdiction-bounded replication, the federated platform-team operating model, and the per-program cost lineage that makes the AI tier a controllable rather than an unbounded expense
- What a one-query reconstruction returns when the question is not "Who approved this?" but "Was the platform that produced this healthy at the time?" and why operations makes that question answerable

The remainder of this chapter describes the four operational disciplines (monitoring and observability, data quality management, version drift and resynchronization, and scaling across programs) and shows how each is the same governed contract with the data spine seen under operational load rather than at design time. Before we get into the disciplines, we will look at the operating model itself, because the disciplines only make sense once you know what is being operated and to what standard.

18.1 The Operating Model of the Integrated Platform

The integrated platform is a coordinated set of SAP Business Technology Platform (SAP BTP) services: SAP HANA Cloud as the canonical relational core, the Object Store for the lineage payload archive, the SAP HANA Cloud vector engine for grounding embeddings, Event Mesh for change propagation, SAP Build Process Automation for governed workflow, SAP AI Core for inference, and SAP Build Work Zone for the experience surface. Each of these were introduced in an earlier chapter as a design element and we will cover them now as something that must run. Operating them well rests on one claim that the rest of this chapter elaborates on: The platform is operated as a single system with one identity model, one correlation key, and one severity language, even though it is assembled from many services. The moment any service is operated in isolation—its own dashboard, its own on-call, its own definition of "healthy"—the platform loses the property that makes it defensible, that a question about a number and a question about the platform's health resolve to the same evidence.

We separated the three roles of SAP BTP in Chapter 1, Section 1.4 (the data spine, integration layer, and extension layer) and that separation was justified by their different governance regimes. Operations inherits that same separation. The data spine is operated for integrity and lineage, the integration layer is operated for flow and reconciliation, and the extension layer is operated for latency, availability, and cost. Collapsing all three operating regimes into a single "Is the platform up?" question is the operational equivalent of the architectural mistake Chapter 1 warned against, and it fails in the same way: Symptoms are visible long before the structure is understood, and the team debugs those symptoms rather than the underlying cause.

The platform is operated is not operated to a standard of zero incidents. Incidents are certain. The standard is that every incident is detected by a signal before it is reported by a user, classified by a severity model rather than by improvisation, resolved through a governed transition rather than an ad hoc fix, and (most importantly from an FP&A perspective) reconstructible after the fact with the same one-query walk that reconstructs an approval. An operations practice that cannot answer the question "Was the platform healthy when this figure was produced, and how do you know?" has not met the standard, however few incidents it has had.

This standard rules out the heroic operator whose knowledge lives in their head rather than in a runbook, because that knowledge cannot be audited and does not survive their absence. It rules out the silent fix that restores service without leaving a record, because a fix that cannot be reconstructed is, from the auditor's chair, indistinguishable from an undetected corruption. It rules out the parallel monitoring estate that a well-meaning team stands up beside the governed platform just to keep an eye on things, because a second source of operational truth that does not share the audit chain's identity is precisely the ungoverned side-channel the canonical layer was built to eliminate. Operations, in this architecture, is not a set of tools bolted onto a finished platform; it is the platform's own discipline, expressed in the platform's own vocabulary.

If you're approaching this chapter from a finance, costing, or planning background, you will recognize the shape of this argument immediately, because it is the shape of internal control. A financial control is not the absence of error; it is a designed, owned, evidenced procedure that detects error, routes it to an accountable owner, and leaves a trail an auditor can follow. The four operational disciplines of this chapter are that same idea applied to a software platform: Monitoring is the detective control, the severity model is the escalation procedure, resynchronization is the corrective control, and the correlation invariant is the evidence. We use the controller's perspective as a recurring test throughout the book; in this chapter, the perspective is not a test applied from outside but the design principle itself.

18.2 Monitoring and Observability

Monitoring is the practice of watching known signals against known thresholds. Observability is the stronger property that, when a signal moves, the platform carries enough structured context to explain why without a new deployment to add instrumentation. The integrated platform requires both, and it requires them to be expressed in the same identity vocabulary the data spine and the audit chain already use. This section defines the three signal classes the platform emits, the golden signals within each, the SLOs that govern them, the burn-rate alerting that protects the error budget, the severity model that routes a signal to a human, and the correlation discipline that makes a health question and an audit question the same query.

18.2.1 Three Signal Classes

Every observable signal in the platform belongs to exactly one of three classes, and the classification is not cosmetic. Ot determines who owns the signal, what SLO governs it, and which runbook responds to it. Figure 18.1 shows the three classes feeding a single operational backbone.

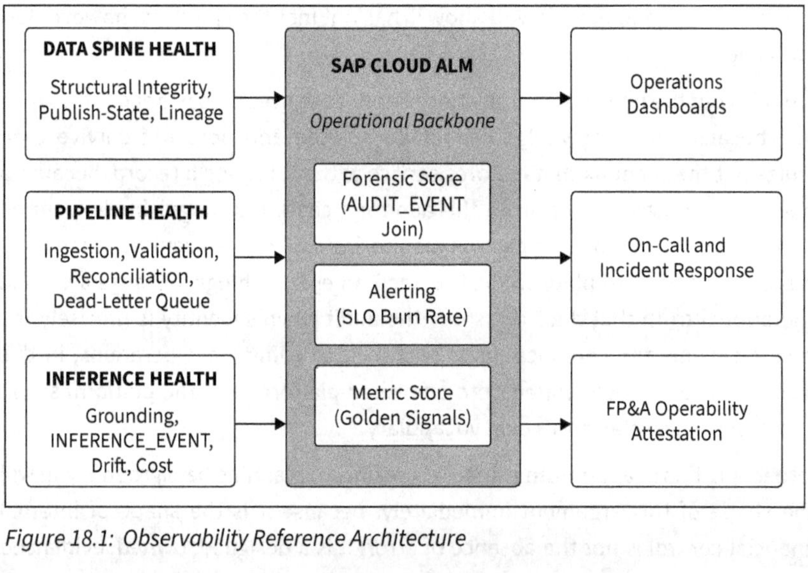

Figure 18.1: Observability Reference Architecture

Every signal carries the CORRELATION_ID set at ingestion, so a controller can ask "Was the platform healthy when this number was produced?" and answer with one join.

Spine health is the structural-integrity class. It answers whether the canonical data spine itself is internally consistent: whether every PROCESS_VERSION in PUBLISHED state has a resolvable PROCESS, whether every ACTIVITY resolves its EXECUTION_CONTEXT, whether the PUBLISH_STATE overlay agrees with the entities it gates, and whether the hash-chained AUDIT_EVENT sequence (introduced in Chapter 10, Section 10.5) is unbroken. Spine-health signals are owned by the data platform team and are, by design, low-volume and high-severity. A broken audit chain or an orphaned published version is never a warning. It is an incident, because it undermines the property every other process depends on.

Pipeline health is the flow class. It is the operational expression of the metrics introduced in Chapter 11: ingestion success ratio per source, validation tier pass and fail rates, reconciliation drift counts by classification, dead-letter-queue depth and residency, watermark lag, and idempotency-violation counts. These signals are owned by the integration team and are higher-volume and more graduated than spine-health signals; a single transient failure is noise, a sustained drop in the success ratio is a leading indicator, and an idempotency violation is, as Chapter 11 stated, treated as data corruption and therefore as a severity 1 incident.

Inference health is the AI class. It is the operational expression of the metrics covered in Chapter 13 through Chapter 15: grounding-bundle assembly latency and failure, the rate at which prepared data sets are refused rather than returned partial, INFERENCE_EVENT volume and cost, suggestion acceptance and rejection ratios, and grounding drift signals from the detection cycle of Chapter 14. Inference health signals are owned jointly by the AI-platform team and the governance function because an inference health regression is rarely only a performance problem. A rising refusal rate or a falling acceptance ratio is frequently the earliest visible symptom of drift in the data underneath the AI, which is precisely the condition the architecture in Chapter 14 was designed to prevent.

18.2.2 The Golden Signals per Layer

Within each class, the platform watches a small, fixed set of golden signals rather than an unbounded forest of metrics. This discipline is deliberate, because a dashboard that shows everything shows nothing, and an on-call engineer paged at 3 a.m. needs four numbers to solve the problem, not four hundred. Table 18.1 sets out the golden signals by layer, the SLO that governs each, and the team that owns the response. The objectives restate, rather than invent, the targets Chapter 11 and Chapter 12 already established, so that an operator and an architect mean the same thing when they talk about being within budget.

Layer	Golden Signal	SLO	Owner
Spine	Audit-chain continuity; orphaned published entities	Zero unbroken-chain or orphan findings	Data-platform team
Ingestion	Per-source success ratio (24-hour rolling)	≥ 99% per source	Integration team
Reconciliation	Completeness drift at daily cycle	≤ 1% of domain artifacts	Integration team
Ingestion	Freshness: publication to spine update	≤ 30 minutes p95	Integration team
Execution	Process lookup latency	p50 12 ms/p95 45 ms	Extension team
Execution	Context enrichment latency	p50 30 ms/p95 120 ms	Extension team
AI	Grounding-bundle assembly latency	p50 450 ms/p95 1,800 ms	AI-platform team
AI	Audit readiness: time to a definitive verdict	≤ 10 minutes	Governance function

Table 18.1: Golden Signals, Governing SLOs, and Ownership by Layer

Two properties of this table deserve emphasis. First, every objective is a restatement of a target an earlier chapter set, not a new number introduced by operations; the freshness, completeness, and audit-readiness objectives are Chapter 11's SLOs verbatim, and the latency budgets are Chapter 12's contract verbatim. Operations does not get to redefine what healthy means; it must watch whether the design's definition is being met. Second, the audit-readiness objective—the maximum time any artifact can sit without a definitive validation verdict—is an operational signal that is simultaneously a governance control, which is why its owner is the governance function and not the integration team. It is the first place in this chapter where a health signal and an audit obligation are the same measurement.

18.2.3 Service-Level Objectives and the Error Budget

An SLO is a target; an error budget is the permission to fail that the target implies. A 99 percent success ratio objective for a source is, read correctly, a statement that 1 percent of payloads may fail within the window without the objective being breached. The error budget is what makes alerting rational: An alert should fire not when an error occurs but when the rate of errors would, if sustained, exhaust the budget before the window closes. This is the discipline Chapter 11 introduced as burn-rate alerting, and this chapter covers how it applies to operations.

Figure 18.2 shows the two-window discipline. A short-window alert goes off when the recent error rate is high enough to exhaust the daily budget within hours. It is a fast incident detector and it pages a human. A long-window alert goes off when the rolling error rate, though individually unremarkable, would exhaust the monthly budget; it is a slow-regression detector and it opens a tracked ticket. The combination is not redundant. The short window catches the authentication change that breaks a source at 2 a.m.; the long window catches the modeling team and the architecture team drifting one percent further apart every week until, a quarter later, the completeness objective is structurally unreachable. A platform that alerts only on the short window is blind to the regression that will define its next audit finding.

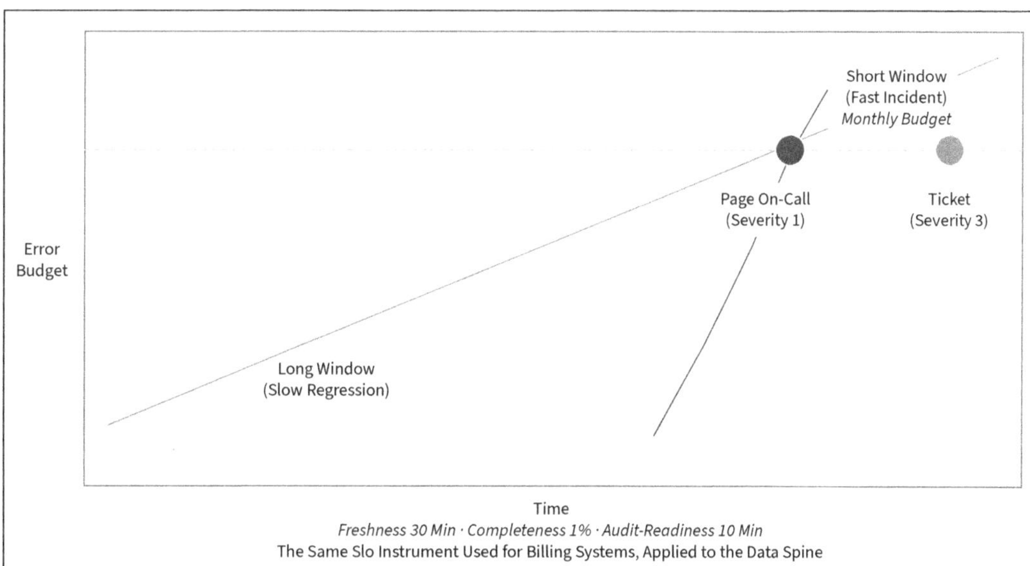

Figure 18.2: SLO Burn-Rate Alerting—The Two-Window Discipline

The error-budget framing has a consequence operators must internalize: A budget that is never spent is a sign of an objective set too loosely, not of operational excellence. If a source has run at 100 percent for a year, the 99 percent objective is not protecting anything and should be tightened until it constrains behavior. If your objective is so loose that the system always meets it, it will not warn you about problems. This is the same reasoning Chapter 14 applied to freshness budgets: A budget exists to be a meaningful constraint, and an operations team that treats an unspent budget as a trophy has misunderstood what the budget is for.

18.2.4 The Severity Model and On-Call Posture

A signal that breaches an objective must page a human in bounded time, and that boundedness is governed by a severity model rather than by the on-call engineer's judgment at the moment of the page. The model is the one Chapter 11 introduced for the ingestion pipeline, generalized to the whole platform so that an operator means the same thing by *severity 1* whether the failing component is a source adapter, an SAP HANA replica, or a Joule grounding endpoint. Let's look at each severity level in more detail:

- Severity 1 is reserved for conditions that undermine a structural guarantee: a broken audit chain, an orphaned published entity, an idempotency violation, a bulk-drift halt, an identity-drift quarantine, an authorization failure that halts a source, or a region-level loss of the canonical core. Severity 1 pages immediately and requires acknowledgment within fifteen minutes, because every one of these conditions, left unattended, corrupts the property the FP&A reconstruction depends on.
- Severity 2 is reserved for conditions that degrade a SLO without yet breaking a guarantee: a per-source success ratio below 99 percent for a single source, a dead-letter-queue entry past its retry budget, a sustained latency-budget breach on an execution service, or a grounding-assembly failure rate above tolerance. Severity 2 pages within the working day and requires resolution within twenty-four hours, after which an unresolved severity 2 escalates to severity 1 on the same reasoning Chapter 11 applied to aged dead-letter entries.
- Severity 3 is reserved for governed work that has an owner and a service-level agreement but no immediacy: an open governance work item from material drift, an advisory tier 4 validation finding, a deprecation-calendar action, or a quality-scorecard number that has crossed its target but not its critical threshold—for instance a domain's completeness slipping from 99.6 to 98.8 percent. Severity 3 is routed by ticket and tracked against a domain service-level agreement, never paged.

On-call posture layers three horizons on top of the severity model, exactly as Chapter 11 prescribed for reconciliation. At the daily horizon, the on-call engineer reviews the operational summary at handover, confirms that every severity 2 is owned, and acknowledges any severity 1. At the weekly horizon, the platform lead reviews objective-burn trends per layer and tunes thresholds whose noise floor has shifted. At the monthly horizon, the architecture council reviews aggregate platform health (burn-rate history, drift trends, cost lineage, and deprecation progress) and decides on platform-level investment. This discipline is what allows a platform that ingests thousands of payloads a day, serves thousands of execution requests a second, and runs a governed AI tier to be operated by a small, federated team rather than become a perpetual war room.

18.2.5 Correlation: One Identity for Health and Evidence

The property that distinguishes this observability model from a conventional one is that every signal carries the same correlation identity the audit chain carries. When a payload enters the platform at the SAP BTP boundary it is assigned the CORRELATION_ID described in Chapter 11 as the data spine of idempotency and used in Chapter 10 as part of the hash-chained AUDIT_EVENT. Every metric increment, alert, log line, and forensic record the platform emits about that payload carries the same identifier. The consequence is the one we have been building toward since Chapter 9: A health question and an audit question are the same query asked with a different projection.

When the question "Was the platform healthy when this figure was produced?" inevitably arises in FP&A environments, the answer is not assembled from a separate monitoring system that must be time-correlated by hand against the audit log. It is the same one-query reconstruction defined in Chapter 16, Section 16.4 and extended to the experience layer in Chapter 17, projected onto the operational signals that share the CORRELATION_ID: the ingestion that produced the source data, the validation verdict that admitted it, the reconciliation state at the time, the execution-service latency and error signals on the read path, and the inference health if AI was in the path. The platform does not have a monitoring story and an audit story. It has one story, told twice.

Operational Rule: The Correlation Invariant

No signal may be emitted by any platform component without the CORRELATION_ID of the payload, request, or inference it concerns. A signal that cannot carry the correlation identity is, by definition, one that cannot be joined to evidence, and an operations practice that admits such signals has reintroduced the gap the canonical layer was built to close. The invariant is enforced at the instrumentation library, not by convention. A component that attempts to emit an uncorrelated signal fails its own startup health check.

18.2.6 The Observability Data Model

Observability is only as good as the structure of what is recorded. The platform records three correlated stores, each a first-class part of the canonical architecture rather than a side effect of a logging framework. The metric store holds the golden signals as time series, each point tagged with the layer, the source or service, the resource group, and, where the signal concerns a specific artifact, the entity identity. The event store holds the discrete operational events: an INGESTION_LOG row, a validation verdict, a reconciliation classification, an INFERENCE_EVENT, and a state transition. The forensic store holds the immutable, dated material the platform may need to reconstruct a past condition: the daily snapshots described in Chapter 11, the lineage payload archive outlined in Chapter 12, and the hash-chained AUDIT_EVENT sequence from Chapter 10. The three stores are not independent. They share the CORRELATION_ID, and they are designed so that a single join across them answers both the operational question and the audit question without a second system in the path.

This is the structural reason we insist on the correlation invariant in this chapter. A conventional observability stack treats metrics, logs, and traces as three loosely coupled telemetry types correlated after the fact by timestamp and best-effort trace propagation. That is adequate when the only consumer is an engineer diagnosing a latency spike. It is inadequate when the consumer is a controller who must establish, to an auditor's standard, that a specific reported figure was produced by a platform in a known-good state. Timestamp correlation degrades under clock skew, retries, and asynchronous processing exactly where the audit question is hardest. Identity correlation does not degrade because the identity is assigned once at the boundary and carried, not reconstructed. The observability data model is therefore a governance decision before it is an engineering one.

18.2.7 Dashboards as Operated Artifacts, Not Decoration

We first introduced operational dashboards in Chapter 11, and we will take a closer look at them here. Those dashboards are governed artifacts with owners and a change history, not screens a team improvises. The platform maintains a fixed set of dashboard surfaces, each scoped to one signal class and one audience, so that the on-call engineer, the integration lead, the AI-platform team, and the controller each have a view calibrated to the decision they must make rather than a single wall of numbers no one reads. Table 18.2 names the standard surfaces, the question each answers, and the audience it serves.

Surface	Signal Class	Question It Answers	Primary Audience
Platform health	Data spine	Is canonical truth structurally intact right now?	On-call; data-platform team
Ingestion flow	Pipeline	Are sources current and reconciled within budget?	Integration team
Drift and quality	Pipeline	Where is the data spine and real-world facts disagreeing?	Stewards; integration lead
Execution SLO	Execution	Are read-path latency budgets being met?	Extension team
Inference health	AI	Is the AI tier grounded, bounded, and trusted?	AI-platform team; governance
Operability attestation	All	Was the platform healthy when this ran?	Controller; FP&A; audit

Table 18.2: Standard Dashboard Surfaces, the Operational Question Each Answers, and Its Audience

The last surface is the one we have been building toward and the one that distinguishes this observability model from a conventional one. The operability attestation surface is not a real-time display; it is a query interface keyed on a CORRELATION_ID or a reported figure that returns the platform's health state for the window in which that figure was produced. It is the operational projection of the one-query reconstruction, and it exists so that the controller never has to ask the platform team to pull the logs. The attestation is a self-service artifact, governed and audited like every other read path, because a control whose evidence requires a favor from an engineer is not one an auditor will accept.

18.2.8 Worked Example: The Incident Response Lifecycle

Detection is necessary but not sufficient; an unowned alert is noise with a timestamp. The platform operates a defined incident lifecycle so that every severity-bearing signal moves deterministically from detection to closure with an owner at every stage. The lifecycle has the following five stages, and each stage writes a correlated event so the incident itself is reconstructible by the same path as the data it concerned:

1. **Detect**
   A golden signal breaches its objective; the burn-rate logic shown in Figure 18.2 classifies it as a fast incident or a slow regression and assigns a provisional severity from the model of Section 18.1.
2. **Page or ticket**
   Severity 1 and 2 alerts page the on-call engineer within the bounded acknowledgment window, while severity 3 opens a tracked work item against a domain service-level agreement. The page carries the CORRELATION_ID and a link to the relevant dashboard surface, never a bare metric.
3. **Triage and classify**
   The engineer confirms or revises the severity, identifies the signal class and the owning layer, and, critically, records whether any structural guarantee is at risk, because that determination, not the size of the spike, drives the response path.
4. **Contain and resolve**
   Containment uses only governed mechanisms: a domain freeze, a circuit breaker, a quarantine, or a controlled failover. Resolution is a governed transition, never a manual edit of canonical truth. If the data spine must change, it changes through resynchronization, not through an operator's hand.
5. **Reconstruct and close**
   The incident is closed only when its timeline is reconstructible from the correlated stores and a post-incident record is written. For any incident that touched the read path of a reported figure, the closure record explicitly states the operability impact, so the controller's later attestation query returns a complete answer rather than a gap.

The lifecycle deliberately forbids two shortcuts that production teams reach for under pressure. The first is the undocumented containment—the operator who restarts a service and clears the alert without recording what they did—because an unrecorded containment is a hole in the exact place an auditor will look. The second is the severity downgrade of convenience, in which a severity 1 alert is quietly relabeled severity 3 to stop the pager rather than because the structural risk has passed. Both are caught for the same reason quality-rule weakening is caught: The severity assignment and every change to it are correlated, audited events, so the convenient downgrade leaves precisely the trail that makes it indefensible.

Let's now jump into the worked example. At 2:14 a.m., the password the platform uses to pull process data from SAP Signavio is changed on the SAP Signavio side without anyone telling the platform team. The platform can no longer authenticate, and incoming process records start to fail.

The platform automatically recognizes these as authentication failures, not as bad data. Rather than retrying blindly or letting partial data through, it stops accepting from that one source and pages the on-call engineer within the fifteen-minute window. The alert names the source, the reason, and the exact set of affected records by their correlation identifiers, not just a red light.

By 2:21 a.m., the engineer confirms the important point: the data spine itself is untouched, so nothing is actually broken. Every consumer is still reading the last-known-good version of the process data, and because the platform refused the failing records rather than half-importing them, nothing incorrect was written. A source being unavailable is an inconvenience, not a corruption.

By 2:40 a.m., the new credential is supplied through the proper secured channel. Operators should never type credentials directly into the platform. Ingestion picks up exactly where it left off, with no records lost and none duplicated, because the platform tracks precisely how far it had got.

By 3:05 a.m., the overnight reconciliation confirms the outage introduced no drift. The incident is closed with a complete, reconstructable timeline. Crucially, because no reported figure depended on data read during the outage, the recorded operability impact is **None**—and that **None** label, not a silent gap, is what a controller's later query returns. The platform can prove it was healthy, not merely fail to show that it wasn't.

18.3 Data Quality Management

Chapter 11 defined the four-tier validation pipeline as the boundary between received data and canonical truth. That pipeline is a gate. It decides, at ingestion, whether a payload may update the data spine. Data quality management is the discipline that recognizes the gate is necessary but not sufficient. Data that was valid when it entered can become outdated while it sits in storage, because the system it describes keeps moving: an application is rehosted, a role is reorganized, a cost center is blocked, or a control is retired. Quality is therefore not a property a payload has at the gate; it is a property the data spine has continuously, and one that must be operated.

18.3.1 Quality Is an Operated Property

The architectural claim of this section is that data quality is measured, scored, and owned on the same cadence as availability, not audited once a quarter. The instrument is the data-quality scorecard shown in Table 18.3. If you come from a financial-planning background, you already know exactly what this is, because it is the same instrument as a monthly operating review: a short report with one row per business area, a small number of measured indicators per row, a target for each indicator, and a named owner accountable for the number. The scorecard simply applies that familiar discipline to the canonical data spine instead of to revenue or headcount. Its rows are the six data domains covered in Chapter 9 (process, architecture, the master-data domains, and so on) and for each domain it reports four measured numbers, refreshed continuously rather than monthly.

	Freshness	Completeness	Validity	Lineage
Process	Meets SLO	Meets SLO	Within tolerance	Meets SLO
Architecture	Meets SLO	Within tolerance	Meets SLO	Meets SLO
Execution context	Within tolerance	Within tolerance	Meets SLO	Within tolerance
Master data	Meets SLO	Meets SLO	Within tolerance	Meets SLO
Adoption and telemetry	Within tolerance	Within tolerance	Meets SLO	Within tolerance
Governance	Meets SLO	Meets SLO	Meets SLO	Meets SLO

Table 18.3: The Data Quality Scorecard by Domain

The four measured numbers are deliberately chosen so that no one has to form an opinion to produce them; a quality indicator that requires a judgment call is one that will quietly stop being produced. Freshness is how far behind its source the data spine's view of a domain is, in minutes, computed from the same watermark maintained in Chapter 11, which is the operational equivalent of how many days old a financial close is. Completeness is the percentage of expected records that are actually present, computed from the daily reconciliation's four-bucket comparison, which is the operational equivalent of a coverage or population-completeness check. Validity is the percentage of published records that still pass the deterministic tiers 1 through 3 of the validation covered in Chapter 11 when checked against today's referential state, not the state they entered in. This is the operational equivalent of rerunning your control checks against current master data rather than trusting last quarter's. Lineage integrity is the percentage of records whose hash-chained audit trail is unbroken and whose lineage row resolves, which is the operational equivalent of whether every figure still traces cleanly back to its source document. The steward's job is to act on these numbers, never to invent them. Table 18.4 shows quality dimensions, the metric that computes them, the governing objective, and the responding action.

Dimension	Computed From	Quality Objective	Action on Breach
Freshness	Watermark lag vs. system of record	≤ 30 min p95 per domain	Investigate source channel; severity 2
Completeness	Daily reconciliation four-bucket diff	≥ 99% of expected artifacts	Reconciliation review; severity 2
Validity	Reevaluated tiers 1–3 pass rate	≥ 99.5% of published artifacts	Quarantine failing rows; steward work item
Lineage integrity	Audit-chain continuity and lineage resolve	100% (no tolerance)	Severity 1; freeze affected domain

Table 18.4: Quality Dimensions, the Metric That Computes Them, the Governing Objective, and the Responding Action

The asymmetry in the last row is intentional, and it is the single most important line in the table. Freshness, completeness, and validity have tolerances because the world legitimately moves faster than any pipeline and a small, bounded lag is a normal operating condition. Lineage integrity has no tolerance because a break in the audit chain is not a degraded condition, it is the loss of the property that lets every other guarantee be defended. A platform may run for a day at 99.2 percent completeness and remain trustworthy; it may not run for a minute with a broken lineage chain and remain trustworthy. Operations must encode that asymmetry, or it has not understood what it is protecting.

18.3.2 The Steward Operating Model

Every number on the scorecard has exactly one accountable steward, and that steward is a role in the system of record's own governance, not a member of the platform team. The process domain numbers are owned by the process governance authority that owns SAP Signavio publication, the architecture domain numbers by the architecture lead who owns the SAP LeanIX portfolio, and the master data numbers by the relevant master data governance owner in SAP Master Data Governance. This is a direct application of Chapter 1's principle that authority lives with the system of record: The platform team operates the tool that detects a quality breach, but it does not own the decision about how to resolve one, because resolving it is a change to governed truth and that authority was never the platform's to hold. It is the same separation a controller relies on every close: The reporting system flags the variance, but the business owner explains and corrects it.

When a measured number crosses its threshold—for example, the architecture domain's completeness falling from 99.6 percent to 98.7 percent, dropping below its 99 percent target—the scorecard automatically opens a steward work item in SAP Build Process Automation, routed to the accountable owner by the same domain-ownership derivation used for approval chains in Chapter 16. The work item carries the current number, the trend over recent cycles, the specific records responsible, and, where the AI tier has produced one, a drafted impact summary. It does not carry a decision. The steward decides; the platform records the decision and remeasures on the next cycle. The discipline is identical to the one enforced for remediation in Chapter 16: A quality breach is a typed, severity-banded, owned event with a deadline, not an error someone happens to notice.

18.3.3 The Quarantine Economy

When validity or lineage breaches at the artifact level, the affected artifacts are quarantined rather than deleted or autocorrected, exactly as Chapter 11 prescribed for failed validation and suspected identity drift. Quarantine has an economy that operations must manage deliberately. A quarantined artifact is invisible to execution services and to the AI tier (it is, by design, not in any `PUBLISHED` set) but it is not gone. Consumers continue to read the last known good, published version under its current `VERSION_NO` while the quarantine holds. The cost of quarantine is therefore staleness, not data loss. It is explicitly bounded and explicitly visible, which is precisely what the architecture prefers over silent correctness.

This economy fails in two characteristic ways, and operations must watch for both. The first is a growing quarantine population that no steward is draining, which is the quality equivalent of a growing dead-letter queue and escalates on the same rule: A quarantined artifact that has not been adjudicated within its service-level agreement is escalated because an unresolved quarantine is a place where the data spine and the world disagree and no one has decided who is right. The second is the temptation, under delivery pressure, to release a quarantined artifact by lowering a validation rule rather than by fixing the data. That is not a remediation; it is a governance bypass, and it is caught because validation rules are themselves versioned, governed artifacts whose change is an audited event. A rule weakened to release a quarantine leaves exactly the evidence trail that makes the shortcut indefensible.

18.3.4 AI as a Quality Pattern Detector and Its Boundary

Quality scoring produces a stream of breaches and trends. Many are routine; some, taken together, are signals a human will not see at the speed they accumulate. The AI tier, configured with read-only access to the scorecard, the reconciliation log, and the data spine, is well suited to detecting these patterns: the domain whose completeness erodes a fraction every week because two upstream surveys disagree, the activity that fails validity every Tuesday because of a recurring batch job, or the correlation between a freshness regression in one domain and an acceptance-ratio drop in the AI tier that depends on it. This is the same pattern-detector role we assigned AI in reconciliation in Chapter 11 and formalized as the `SUGGESTION` pipeline in Chapter 15, now applied to quality.

The boundary is the same one we drew in Chapter 1, Section 1.5 and have enforced in every chapter since. The AI tier surfaces a quality pattern as a scored, explainable `SUGGESTION` carrying its `SUGGESTION_EXPLANATION`; it does not change a validation rule, release a quarantine, republish an artifact, or adjudicate a steward work item. It proposes, the steward (with the authority of the system of record) decides, and the platform records the decision as a `VALIDATION_VERDICT` and scores it again. Operations does not relax this boundary under load. The temptation to let the AI tier autocorrect the obvious ones to clear a backlog is precisely the temptation Chapter 1 named as an audit failure waiting to occur, and it is rejected here for the same reason: An autocorrected quality breach is an ungoverned modification of canonical truth, no matter how obvious the correction looked at the time.

18.3.5 The Validation Rule Lifecycle

The scorecard is only as trustworthy as the rules behind its validity dimension, and those rules are not static. New control requirements appear, a source begins emitting a field that must now be checked, or a rule proves too strict and produces false quarantines. Operations must change rules without making the scorecard incomparable across time or opening a path to silently weaken control. This discipline is the same one introduced in Chapter 11 for the rule registry, operated here as a governed lifecycle. Every `VALIDATION_RULE` is a versioned artifact with its own `PUBLISH_STATE`. A rule change is proposed, reviewed by the control owner, and published through the same workflow that governs any other canonical change. Every validation result records the rule version it was evaluated against. The consequence is that a validity number is always interpretable. It is not a bare percentage but one measured against a named, governed rule set, so a movement in the number caused by a rule change is distinguishable from a movement caused by a change in the data itself. Without that distinction, a quality trend is uninterpretable, and an uninterpretable trend is worse than no trend because it invites false confidence. The controller's instinct here is exactly right: A metric whose definition silently changed is not a metric but a trap.

This is also where the most consequential anti-pattern in data quality operations is structurally blocked. Under delivery pressure, the path of least resistance to a failing validity number is to relax the rule that is failing rather than to fix the data that is wrong. Because the rule is a governed, versioned artifact whose change is an audited event routed to the control owner, the relaxation is not a quiet configuration tweak; it is a recorded decision with a named approver, visible in the same one-query reconstruction as everything else. The architecture does not prevent a team from weakening a rule, because sometimes a rule is genuinely wrong, but it makes the weakening a governed act with an owner and a trail, which is the difference between a control change and a control bypass.

18.3.6 Worked Example: Reconciliation Is the Engine of the Scorecard

The scorecard is not a separate measurement apparatus but a projection of the daily reconciliation cycle covered in Chapter 11. Completeness is the reconciliation's four-bucket diff aggregated by domain. Freshness is the watermark lag the same cycle already tracks. Validity is the pass rate of tiers 1 through 3 when the cycle reevaluates published artifacts against present referential state rather than the state they entered in. Lineage integrity is the audit-chain continuity check the cycle performs as it walks the data spine. This matters operationally because the scorecard cannot drift away from the reconciliation if it has no independent source of truth. A team that builds a quality dashboard from a separate extract has built a second opinion that will, inevitably, disagree with the reconciliation at the worst possible moment. The scorecard is trustworthy precisely because it manufactures nothing; it reads what reconciliation already established and presents it by domain and dimension.

In our worked example, the architecture domain holds the platform's record of which applications and interfaces support which business processes. In week 1, its completeness number reads 99.6 percent, which is above the 99 percent target and requires no action. Nothing looks wrong.

By week 4, the number has drifted to 99.1 percent. It is still above target, so a simple threshold check would stay silent. But it has fallen every week for four weeks, and the slow-regression alert (covered in Section 18.1) watches the trend, not just the threshold. It opens a low-urgency work item for the architecture lead before the number ever goes red.

The AI tier, which can read the reconciliation log but cannot change anything, has already spotted the pattern a human would have taken weeks to see: A particular group of interface records keeps arriving from SAP LeanIX but failing the referential check because an application was rehosted upstream and given a new identifier while a set of interfaces still point at its old one. The AI writes this up as a plain-language explanation with a recommended fix and attaches it to the work item. It does not apply the fix.

The architecture lead (the one who owns this data, not the platform team lead) reviews it and decides the interfaces are real and must be corrected at the source in SAP LeanIX instead of quietly forced through in the data spine. The affected interface records stay quarantined. Every consumer keeps reading the last-known-good version in the meantime, so nothing silently changes downstream.

As such, SAP LeanIX is corrected upstream. The next overnight reconciliation sees the source and the data spine agree and clears the quarantine, returning completeness to 99.7 percent. Every step—the trend alert, the AI explanation, the human decision, the upstream fix, and the remeasurement—carries the same correlation identifier, so months later a controller asking "Was this data sound when the number was reported?" gets the whole story from one query. The problem was caught a quarter before it would have surfaced as an audit finding. That early catch is the entire reason quality is operated continuously instead of inspected once.

18.4 Version Drift and Resynchronization

Drift is the central operational risk of an integration platform, and Chapter 11 introduced its taxonomy and daily detection cycle. This section covers where drift is operated: where the taxonomy becomes a runbook, resynchronization is performed without silently mutating the data spine, forensic replay answers a historical question, and a moved version is regrounded into the AI tier so that an agent never reasons over a definition that has shifted underneath it.

18.4.1 The Drift Taxonomy, Carried Forward

The taxonomy from Chapter 11 is repeated here unchanged because the value of a taxonomy is that it does not change between the process that defines it and the process that operates it. Drift is classified from the daily reconciliation's four-bucket diff (ADDED, CHANGED, REMOVED, and UNCHANGED) and every CHANGED or REMOVED row is assigned exactly one class, which determines whether it can be auto-applied or must be routed to governance. Table 18.5 reproduces the classification so that an operator running the runbook and an architect working through the processes described in Chapter 11 are working from the same rule, not two descriptions of it.

Drift Class	Example	Auto-Apply?	Resynchronization Action
Cosmetic	Label spelling, description text	Yes	Auto-apply; increment patch VERSION_NO; LINEAGE row
Lifecycle (advance)	Application PLAN to ACTIVE	Yes (policy)	Auto-apply; emit lifecycle event
Lifecycle (retire)	Application to PHASE_OUT/END_OF_LIFE	No	Governance review; downstream impact analysis
Identity	PROCESS_ID or ACTIVITY_ID change	No	Quarantine; suspect upstream corruption; severity 1
Control intent	Activity gains/loses control tagging	No	Governance review by control owner
Role assignment	Activity ROLE_REF change	No	Governance review by separation of duties (SoD) owner
Cross-reference	Activity APPLICATION_ID change	No	Governance review by architecture lead

Table 18.5: Drift Classification and Resynchronization Action

Drift Class	Example	Auto-Apply?	Resynchronization Action
Bulk	> 5% of a domain changed in one cycle	No	Freeze domain; page on-call; severity 1
Silent deletion	Artifact present in spine, absent from snapshot	No	Quarantine; require explicit retire decision

Table 18.5: Drift Classification and Resynchronization Action (Cont.)

The two rows that carry the most architectural weight are identity drift and silent deletion, and operations must treat both as explained in Chapter 9. Identity is immutable by construction. An apparent identity change in a source therefore signals corruption in the source, not a key the data spine should rewrite. Silent deletion is not inferred retirement. An artifact missing from a snapshot is quarantined and an explicit retire decision is demanded through governance. An operations team that, under pressure to clear a noisy reconciliation, relaxes either rule has not made the platform more efficient; it has made it untrustworthy in a way that remains invisible until an audit.

18.4.2 The Resynchronization Decision Flow

Figure 18.3 shows the daily cycle as the runbook an on-call engineer actually executes. It begins with the 02:00 UTC snapshot pull specified in Chapter 11, writing an immutable dated snapshot table. It proceeds to the hash-based diff against the live spine using the `SOURCE_HASH` attribute computed at ingestion and then branches on the classification. The first branch is the bulk-drift guard. If more than the configured fraction of a domain shows `CHANGED` in a single cycle, no change is applied, not even cosmetic ones. The domain is frozen and on-call is paged at severity 1, because in production bulk drift is almost always a symptom of an upstream incident and absorbing it corrupts the data spine. Identity drift and silent deletion never auto-apply. They quarantine and demand an explicit decision, since identity is immutable by construction, as described in Chapter 9.

Below the bulk guard the flow separates immaterial from material drift. Immaterial drift (cosmetic and policy-permitted lifecycle advances) is auto-applied. The `VERSION_NO` is incremented, the `LINEAGE` row is written, and the corresponding published event is emitted so the caches outlined in Chapter 12 evict and the AI tier regrounds. Material drift opens a governance work item in SAP Build Process Automation routed to the steward and the relevant authority by domain, and the artifact remains in its previous `PUBLISHED` state until the work item resolves. Consumers continue to read the prior version. The AI tier may draft an impact summary and a recommended resolution for the work item, but it does not approve any changes, because approval is authority and the AI tier holds none. Every path converges on the same terminal step: a LINEAGE row and, where state changed, a new VERSION_NO, so that the reconstruction defined in Chapter 16 reaches the resynchronization exactly as it reaches an approval.

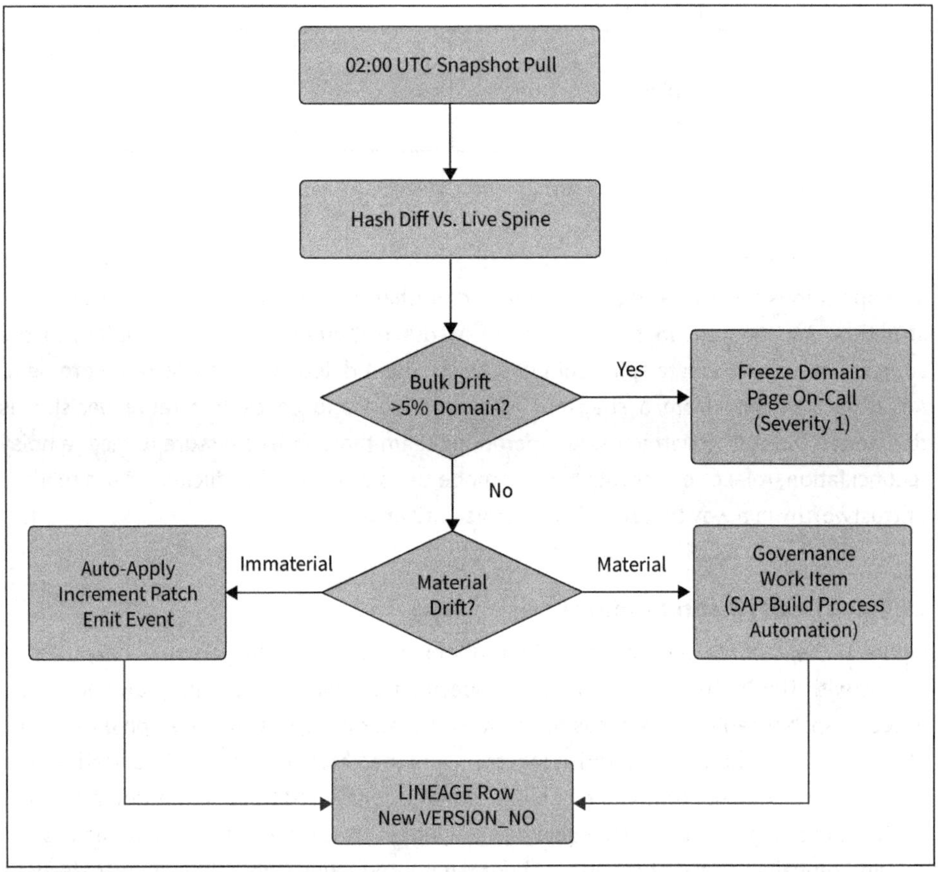

Figure 18.3: Resynchronization Decision Flow: The Daily Reconciliation Cycle, Read as an Operator Runbook

18.4.3 Drift Never Mutates in Place

Figure 18.4 shows the formal state machine, and its single most important property is that the data spine never silently mutates and never loses history. An artifact moving from PUBLISHED to DRIFT DETECTED does not overwrite the published version; the published version remains available to consumers under its current VERSION_NO until a new version is approved. Revalidation, governance review, and republication are explicit, logged transitions, not background corrections. While drift is held in DRIFT DETECTED, QUARANTINED, or GOVERNANCE REVIEW, consumers read the last known good version and the platform is honest about which version that is. This is the operational expression of commitment that identity is stable and version is state, outlined in Chapter 9. Resynchronization changes state through governed transitions, but it never edits identity, and it never makes a change that the one-query reconstruction cannot see.

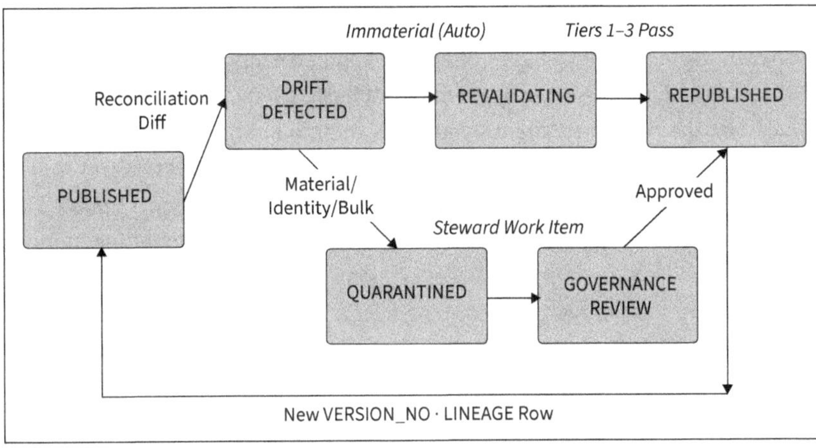

Figure 18.4: Drift Lifecycle and Resynchronization State Machine

18.4.4 Forensic Replay

Operations must be able to answer a historical question without disturbing the live platform: What did the data spine show on the day this configuration was approved, or this number was reported? The capability is forensic replay, and it is operated exactly as specified in Chapter 11. Daily snapshots are retained for thirty days on hot storage and archived thereafter. Replay selects the snapshot for the date in question, materializes it through the same deterministic validation pipeline with reconciliation suppressed so no present-day comparison contaminates the result, and writes the output to a distinct `RECON_REPLAY_yyyy_mm_dd` schema so the live data spine is never touched. The replay is itself a correlated, audited operation. It carries a `CORRELATION_ID`, it writes an `AUDIT_EVENT`, and its output is joinable to the original reconstruction. Forensic replay is the mechanism that allows the FP&A request ("Show me the data spine as it was when this figure was produced") to be answered with reconstructed fact rather than recollection, which is the whole point of the lineage discipline we have carried since Chapter 9.

18.4.5 Hot Restart and Regrounding

Two operational properties keep resynchronization safe under the failures that are routine at scale. Hot restart is the property, established in Chapter 11, that an integration flow restart, a node failure, or a controlled deployment loses no data and produces no duplicates, because the watermark and the idempotency log are committed in the same transaction as the canonical update. Operations relies on this absolutely. At scale, restarts are not incidents, they are normal, routine tasks, and a platform that needed operator intervention on every restart could not be operated by a small team. The on-call runbook therefore treats a clean restart as a nonevent and reserves human attention for the watermark that does not advance, which is the real signal.

Regrounding is the property that when a pinned version moves, the AI tier does not continue to reason over the version it last saw. Chapter 14 pinned every prepared data set to a PROCESS_VERSION and bound it with a signed manifest hash. When resynchronization publishes a new version, the published event invalidates the grounding bundles that pinned the prior one, and the AI tier must reassemble and resign before it answers again. Operationally, this means a resynchronization is not complete when the data spine is updated but when the caches described in Chapter 12 have evicted and the grounding bundles described in Chapter 14 have regrounded. An operations practice that declares victory at the data spine update has left the AI tier reasoning over a stale pin, which is the exact drift condition the architecture in Chapter 14 was built to prevent, reintroduced at the operational seam.

Operational Rule: Resynchronization Is Not Done at the Data Spine

A resynchronization is operationally complete only when three conditions hold: the new VERSION_NO is published with its LINEAGE row, the execution-service caches have evicted on the published event, and every grounding bundle that pinned the prior version has been invalidated and is eligible for reassembly. Declaring completion at the first condition is the most common operational error. It is invisible until an agent answers from a stale pin and the answer cannot be defended in audit.

18.4.6 The Reconciliation Cadence, Operated

Reconciliation is a continuous practice, not a quarterly project, and the cadence Chapter 11 prescribed is operated here on three horizons. At the daily horizon, the on-call engineer reviews the reconciliation summary at handover, confirms that every open drift work item is routed to the correct owner, and acknowledges any severity 2 drift alert. At the weekly horizon, the integration lead reviews drift trends per domain, identifies recurring patterns (an activity that drifts every release because two teams are out of step or an application whose lifecycle attribute oscillates because two surveys disagree) and tunes the bulk-drift threshold for any domain whose noise floor has shifted. At the monthly horizon, the architecture council reviews aggregate drift health and decides whether the pattern of recurring drift is a tooling problem, a process problem, or a source-system service-level problem, and invests accordingly. The cadence is what converts a stream of individually unremarkable drift events into the early signal of a structural divergence, which is the difference between operating a platform and merely watching it.

The cadence has an explicit relationship to the AI tier that operations must respect. In Chapter 11, we assigned AI a pattern-detector role over the reconciliation log, and in Chapter 15, we formalized it as the SUGGESTION pipeline. Operationally, the AI tier reads the drift stream and surfaces recurring patterns the weekly review would otherwise discover late or not at all. This is genuinely useful and genuinely bounded because the AI tier accelerates the detection of a drift pattern. It does not decide that the pattern warrants a change, and it does not make the change. The architecture council decides, and the AI tier informs the decision. Operations preserves this boundary under exactly the pressure that tempts teams to cross it (a noisy quarter, a backlog of drift work items, or a deadline) because the boundary is not a performance setting but the property that keeps the platform defensible.

18.4.7 Snapshot Retention Economics

Forensic replay depends on retained snapshots, and retention is a deliberate economic decision, not a default. Daily snapshots are held on hot SAP HANA Cloud tables for thirty days and archived to cold storage thereafter, exactly as specified in Chapter 11. The thirty-day hot window is chosen to cover the typical incident lifecycle (detection, investigation, remediation, and post-mortem) with enough margin for a regulator to request a point-in-time view that predates a contested change. The cold archive is bounded but never deleted, because the lineage payload it preserves is the structural foundation of the FP&A drilldown we built in Chapter 12. Deleting it would not save meaningful cost but would break the controller's trace at exactly the point an auditor reaches for it. Operations sizes the hot window for investigation speed and the cold archive for evidentiary completeness, and it treats a proposal to shorten either as a governance change requiring the same review as any other control change—because that is what it is.

18.4.8 Worked Example: A Drift Event, End to End

To help you understand this discipline, we will follow one real drift event from the moment it is detected to the moment a controller is satisfied.

Let's start with the situation: One step in the period-close process is marked, in the process-governance system, as a step that carries a financial control—the kind of step an auditor cares about. Overnight, someone editing the process in SAP Signavio removes that control marking, almost certainly by mistake.

At 2 a.m., the platform takes its nightly picture of every source and compares it to what the data spine holds. The comparison flags exactly one changed item: This period-close step has lost its control marking. Nothing else in the domain changed, so this is not a mass upset. The identifier did not change, so this is not corruption. However, losing a control marking is a change to control intent, which the platform is required to treat as material.

Because it is material, the change is not absorbed. The step keeps its previous, control-marked version. Every downstream consumer (the execution services, the AI tier, and the screens people use) continues to see the step exactly as it was, control intact. The data spine has not been edited and nothing downstream has silently shifted.

The platform opens a work item and routes it to the person who owns this control, not to the platform team. The AI tier attaches a plain summary that describes which later steps, which AI answers, and which reported figures depend on this step so that the owner can judge the impact quickly. The AI does not decide.

The control owner recognizes the upstream edit as a mistake, not a deliberate removal of the control. The right fix is to restore the marking in SAP Signavio at the source, not to accept the change in the data spine. A correction is raised upstream; the data spine keeps holding the prior, correct version while that happens.

SAP Signavio is corrected. The next night's comparison sees source and spine agree again. The platform records a new version with a full history entry, refreshes its caches, and regrounds the AI tier so no agent is left reasoning from the old picture. Only now is the change considered fully resolved—not when the data spine was updated, but when everything that reads from it has caught up.

Weeks later, a controller questions a period-close figure that ran through this step. One query returns the whole story: The change was detected, the step was held at its correct version, a named owner judged it, the source was corrected, and throughout the entire episode every consumer read the version with the control still in place. The figure is defensible not because drift never happened but because the drift was handled in the open, with evidence, from end to end.

18.5 Scaling Across Programs

The platform described so far is operated for one program. Mature enterprises run many platforms concurrently—a finance transformation, a supply-chain redesign, a human resources (HR) harmonization—often in different regions and under different data-residency obligations. Scaling across programs is the discipline of adding programs, regions, and tenants without weakening identity, isolation, the SLOs, or the one-query reconstruction. This section defines the program-tenancy model, the multi-region topology, the federated operating model, and the cost lineage that makes the AI tier a controllable rather than an unbounded expense.

18.5.1 Program Tenancy on Resource Groups

A program is not a copy of the platform; it is a tenant of it. The canonical data spine remains singular—there is one `PROCESS` namespace, one identity model, and one audit chain—because the entire value of the data spine is that it is the place where meaning is reconciled, and a per-program spine would reintroduce the fragmentation we argued against in Chapter 1 and Chapter 9. Isolation between programs is achieved through the resource group construct for SAP AI Core introduced in Chapter 13 and the role-scoped exposure we enforced in Chapter 10 through SAP Authorization and Trust Management service scopes. A program operates within its resource group, sees the canonical entities its scope entitles it to see, and runs its own AI workloads under its own tenancy, but it does not get its own copy of the data.

This model has the direct operational consequence that capacity is planned per service, not per program, exactly as Chapter 12 prescribed. A new program does not double the platform; it adds load to shared, horizontally scaled services whose autoscaling targets are already calibrated to keep the p95 latency budgets within contract at peak. The four core execution services are stateless and inexpensive to scale. The cost that actually scales with programs is the shared tier 2 cache capacity and the SAP HANA Cloud query throughput, which is why we warned against under-sizing the cache in Chapter 12 and why operations sizes it generously and lets it absorb steady-state load rather than discovering the false economy a quarter later.

18.5.2 Multiregion Topology and Jurisdiction-Bounded Replication

Programs that span regions deploy the platform across multiple SAP BTP regions for resilience and latency, and the topology is the same asymmetric one we defined in Chapter 12. The canonical core is replicated across regions through SAP HANA system replication. The object store lineage archive is replicated through standard cross-region replication. The Tier 2 caches are regional and deliberately not replicated, because a replicated cache reintroduces the coherence problem that versioned keys were designed to eliminate. Consumers connect to the nearest region and the service-discovery layer handles failover, and the hash-chained `AUDIT_EVENT` (the property that matters most for audit) is replicated as part of SAP HANA replication. The chain survives a failover intact, and the one-query reconstruction still resolves after a region has moved. Figure 18.5 demonstrates multiregion topology and scaling across programs.

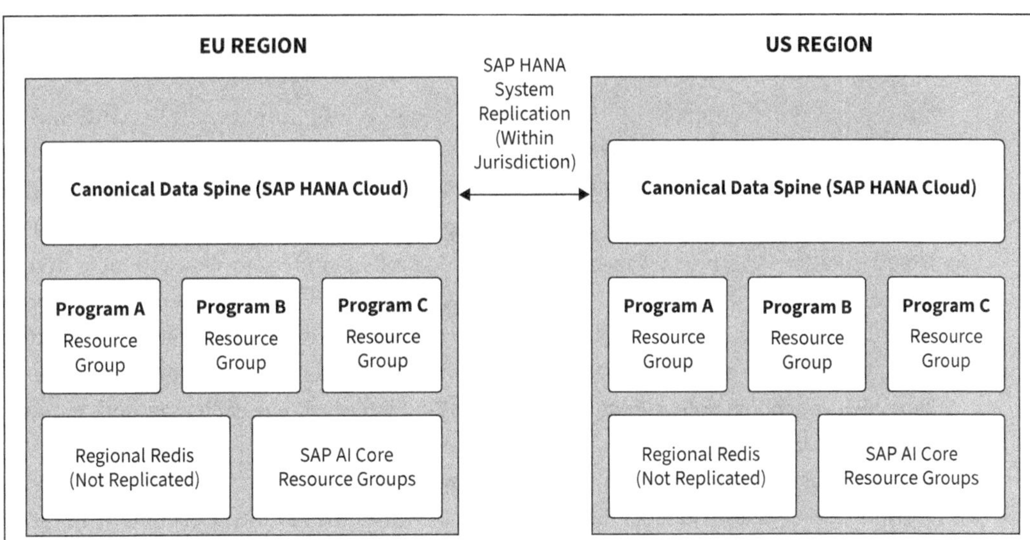

Figure 18.5: Scaling Across Programs—Multiregion Topology

Data residency is operated at the Access API boundary, not bolted on. A consumer in one jurisdiction operates against that jurisdiction's region. The data spine's role-scope filtering ensures a user in one jurisdiction does not see rows that originated in another. Cross-region replication for failover is bounded within-jurisdiction, meaning that an EU primary fails over to an EU secondary, never across a residency boundary. Operations must treat a cross-jurisdiction replication path as a severity 1 misconfiguration because it is a compliance breach, not a performance defect, and it is exactly the kind of condition the processes described in Chapter 19 will hold the platform accountable for.

18.5.3 The Program Onboarding Lifecycle

Adding a program is a governed lifecycle, not a configuration change, because a new tenant inherits access to a shared canonical truth and that access must be provisioned with the same discipline the rest of the platform enforces. The onboarding lifecycle has four stages. First, the program is provisioned a resource group and the SAP Authorization and Trust Management service scopes defined in Chapter 10, scoped to exactly the canonical entities its role context entitles it to read. Nothing is granted by default because deny-by-default is the property the entire access model rests on. Second, the program's stewards and governance authorities are registered, so that a drift or quality work item arising from the program's domains routes to a named owner from day one rather than to an unowned queue. Third, the program's SLOs are set as the platform's standard objectives, not negotiated, because divergent objectives across tenants make the aggregate platform unobservable, as we explained in Section 18.4. Fourth, the program's onboarding is itself recorded as a correlated, audited event, so that the question "When did this program gain access to this domain, and on whose authority?" is answerable by the same one-query walk as every other governance question. A program that cannot answer that question about its own onboarding is one whose every subsequent action inherits an unprovable origin.

18.5.4 Verifying Tenant Isolation

Isolation that is asserted but not verified will fail silently. Operations continuously verifies three isolation properties rather than assuming the resource group and scope model holds. Read isolation is verified by a standing synthetic probe per program that attempts to read an entity outside the program's scope and asserts the read is refused, not merely empty. The distinction we drew in Chapter 17 between an absent result and a denied one matters here too, because a silently empty result is indistinguishable from data that does not exist and hides a scoping defect. Cost isolation is verified by reconciling each program's metered consumption against its resource group, so that no program's load is silently charged to the shared pool or to another tenant. Residency isolation is verified by asserting that no replication path crosses a jurisdiction boundary, treated as the severity 1 condition we established in Section 18.4. The probes are correlated and audited. A failed isolation probe is a platform incident, not a test failure, because in a multitenant governed platform, an isolation breach is a governance breach.

18.5.5 Scaling the AI Tier

The AI tier scales differently from the rest of the platform and operations must account for the difference explicitly. The execution services are stateless and scale horizontally at low marginal cost; the AI tier scales along three axes that are each separately bounded. The first is inference capacity, managed through SAP AI Core resource groups per program as described in Chapter 13, so that one program's inference load cannot starve another's. The second is grounding throughput, bounded by the assembly latency budget covered in Chapter 12 and the freshness budget covered in Chapter 14. Adding programs adds grounding load, and operations sizes the SAP HANA Cloud vector engine and the context enrichment tier so that the p95 grounding budget holds at the aggregate peak rather than per program in isolation.

The third is the model and contract surface, which scales not with load but with change. Every model deployment, every grounding endpoint, and every action is a versioned artifact under the deprecation calendar described in Chapter 12, and operations runs that calendar as a standing obligation, with the longer deprecation window reserved for AI consumers because Joule and Joule Studio agents upgrade on a slower cadence than human-driven workflows. A platform that scales the AI tier's capacity but neglects its deprecation calendar accumulates a population of agents pinned to surfaces no one dares retire, which is its own kind of unmanaged drift. Table 18.6 summarizes the scaling dimensions, what bounds each one, and the operational control.

Dimension	What It Scales With	Bound	Operational Control
Execution services	Request volume	p95 latency budget (Chapter 12)	Per-service autoscaling; minimum two nodes
Canonical core	Programs × artifacts	SAP HANA Cloud throughput	Generous tier 2 cache; query governance
Inference capacity	Inference volume	SAP AI Core resource-group quota	Per-program resource groups (Chapter 13)
Grounding throughput	Programs × bundles	Assembly p95 budget (Chapter 12/ Chapter 14)	SAP HANA Cloud vector engine sizing at aggregate peak
Contract surface	Rate of change	Deprecation calendar (Chapter 12)	Standing calendar; extended AI window
Cost	All of the above	Per-program budget	`INFERENCE_EVENT` and Access API cost lineage

Table 18.6: Scaling Dimensions, What Bounds Each, and the Operational Control

The table's final row is the one that makes the rest operable. Every other dimension scales a resource; the cost dimension makes that scaling visible to the people who must answer for it. Because every INFERENCE_EVENT and every Access API call carries the CORRELATION_ID and resolves to a resource group, the cost of scaling is not an aggregate platform overhead that lands as an unexplained line in a shared budget; it is attributable to the program, the activity, and ultimately the reported figure that incurred it. This is the single most important consequence of holding the AI tier to the same correlated, governed discipline as everything else. If you're approaching this chapter with a background in planning and cost control, it is important to know that a capability you cannot attribute is a capability you cannot forecast, allocate, or defend in a budget review. The governance discipline we have argued throughout this book is, read from the controller's chair, also the only thing that makes the AI tier a plannable cost rather than an unbounded one.

18.5.6 The Federated Team and Its Accountabilities

One platform team operating many programs works only if the boundary between what the platform team owns and what a program owns is as explicit operationally as it is architecturally. The platform team owns the tool: the data spine's integrity, the ingestion and reconciliation flows, the execution and AI services, the observability backbone, the SLOs, and the runbooks. A program owns the authority: its process, architecture, master data governance, stewards, and control owners, as well as the decisions on its drift and quality work items. The platform team detects a breach and routes it. The program decides how a breach to its governed truth is resolved. This is not a division of labor that can be renegotiated per program for convenience because the moment a program is allowed to operate its own objectives or adjudicate its own structural drift, the federation loses the single property that justifies it: that the aggregate platform is observable and defensible as one system. The federated model is the operational expression of the separation of the three SAP BTP roles we described in Chapter 1, and it fails in exactly the way we predicted when the separation is collapsed for expedience.

18.5.7 Work Example: Cost Lineage and the AI Tier as a Controllable

Scaling across programs makes cost an operational signal, and the AI tier is where cost is least bounded by default. Chapter 13 introduced INFERENCE_EVENT as the record of every inference, and Chapter 12 detailed the cost profile of the execution catalog, dominated by SAP HANA Cloud, tier 2 Redis, and the object store lineage archive, with the stateless services themselves a minor line item. Operations joins these into per-program cost lineage. Because every INFERENCE_EVENT and every Access API call carries the CORRELATION_ID and resolves to a resource group, the cost of the AI tier and the execution catalog is attributable to the program that incurred it, to the activity that triggered it, and ultimately to the reported number that consumed it.

From an FP&A perspective, this closes a loop we have been building since Chapter 12. The drilldown from a reported figure to the activities that produced it is not a separate reporting feature; it is a side effect of every consumer using the same correlated catalog. Cost lineage extends that drilldown by one step, from the activity to what it costs to compute, including the AI inference in the path. The AI tier becomes a controllable in the financial sense (a line item with an owner, a trend, and a variance that a controller can question) rather than an unattributable platform overhead. Operations is what makes that possible, and it is the clearest answer to the question of why the AI tier was held to the same governed, correlated discipline as everything else. A capability you cannot attribute is a capability you cannot control, and a capability you cannot control does not belong in a regulated execution path.

Let's look at an example. Imagine it is quarter-end. A profit-center variance looks wrong and the controller has to explain it. Using the drilldown we built in Chapter 12, they trace the figure back to the specific business activities that produced it. So far, this is an ordinary investigation. Then they ask the question we posed in this chapter: When those activities ran, was the platform underneath them actually healthy?

Because every operational signal carries the same identifier as the audit trail, a single query (not a request to the platform team and not an afternoon of cross-referencing logs) returns the complete picture for exactly that time: that the source data was admitted and passed validation; that the relevant data domains were reconciled, with anything uncertain explicitly held rather than silently applied; that the services answering the requests were within their performance limits; that any AI in the path was properly grounded and what it cost; and, if needed, a faithful reconstruction of exactly what the data spine showed on that day.

This is the whole point. There's no interview and no second monitoring system stitched to the audit log by hand. It is the same one-query reconstruction we have used since Chapter 16, now focused on the platform itself. The controller does not have to take the platform's health on trust; they can see it, for that specific value, on that day. That is precisely what it means to operate to an evidentiary standard, and it is the standard Chapter 19 will turn into a formal control framework.

18.6 Summary

Operating the integrated platform is the discipline that converts the designed architecture covered in Chapter 9 through Chapter17 into a defensible one. This chapter argued that the platform is operated as a single system under one identity model, one correlation key, and one severity language, even though it is assembled from many SAP BTP services, because an incident in any layer is visible to a user and consequential to a financial statement through all of them. Monitoring and observability watch a small set of golden signals against SLOs that restate, rather than reinvent, the targets we set in Chapter 11 and Chapter 12, with burn-rate alerting that protects the error budget against both fast incidents and slow regressions. Data quality is operated as a continuously scored property by domain, owned by stewards who hold the authority of the system of record, with a quarantine economy that prefers bounded, visible staleness over silent correctness. Version drift is detected and resolved through governed transitions that never modify the data spine in place and never lose history, with forensic replay to answer historical questions and regrounding to keep the AI tier from reasoning over a moved pin. Scaling across programs adds tenants, regions, and jurisdictions without weakening identity, isolation, or the one-query reconstruction, and makes the AI tier a cost controllable rather than an unattributable overhead.

Nothing in this chapter was a new capability. Every guarantee operations protects was decided in an earlier chapter: the data spine in Chapter 9 and Chapter 10, the ingestion and reconciliation discipline in Chapter 11, the execution contract in Chapter 12, the AI tier in Chapter 13 through Chapter 15, the workflow machinery in Chapter 16, and the experience layer in Chapter 17. Operations only insisted that every one of those guarantees still holds under real load, and that the proof it holds is the same one-query reconstruction the book has used throughout, now focused on the platform itself.

With the platform operated to an evidentiary standard, what remains is to formalize the standard itself. Chapter 19 closes the book by covering security, controls, and audit, where the one-query reconstruction this chapter used for operations becomes the formal evidentiary backbone an auditor tests, the severity model becomes a control framework, and the correlation invariant becomes the property on which the entire compliance posture of the integrated platform rests. The experience layer was the last place governance had to reach. Operations is where the reach was proven to hold, and audit is where the proof is made formal.

18.6.1 Key Learning Points

- The integrated platform is operated as one system with one identity model, one correlation key, and one severity language. A health question and an audit question resolve to the same one-query reconstruction because every operational signal carries the `CORRELATION_ID` the hash-chained `AUDIT_EVENT` already records.
- SLOs are restated from the design chapters, never reinvented by operations. Freshness, completeness, and audit-readiness are the objectives from Chapter 11, and the latency budgets are the contract from Chapter 12. Operations watches whether the design's definition of healthy is met; it does not get to redefine it.
- Quality is an operated property, not an ingestion gate. It is scored continuously by domain across freshness, completeness, validity, and lineage integrity. Lineage integrity alone has zero tolerance, because a broken audit chain is the loss of the property every other guarantee depends on.
- Drift never mutates the data spine in place. Identity is immutable, version is state, and resynchronization changes state through explicit, logged transitions; identity drift and silent deletion are quarantined and adjudicated, never absorbed, and bulk drift freezes the domain rather than being applied.

- A resynchronization is complete only when the new version is published with its LINEAGE row, the execution caches have evicted, and the grounding bundles that pinned the prior version have been invalidated. Declaring completion at the data spine update leaves the AI tier reasoning over a stale pin.
- Scaling adds programs as tenants of one spine, not copies of it. Isolation is by resource group and role scope, capacity is planned per service, replication is jurisdiction-bounded, and per-program cost lineage makes the AI tier a controllable with an owner, a trend, and a questionable variance.

18.6.2 Common Design Pitfalls

- Running per-service operations as independent silos. This gives each SAP BTP component its own dashboard, on-call, and definition of healthy, and loses the property that a number question and a platform question are the same query. You should instead operate the platform as one system with one correlation key.
- Framing the unspent budget as a trophy. This turns an SLO that is never approached as a sign of operational excellence. An objective that never constrains is one that will not warn you, and it should be tightened until it does.
- Letting auto-fix backlog clearance proceed unchecked. This allows the AI tier to silently correct "obvious" quality breaches to drain a queue. However, it is an ungoverned modification of canonical truth and an audit failure waiting to occur, no matter how obvious the fix looked.
- Treating the rule-weakening as a legitimate quarantine release. This lowers a validation rule to release quarantined data under delivery pressure. Validation rules are versioned governed artifacts and the weakening is itself audited, so this shortcut leaves exactly the evidence that makes it indefensible.
- Conflating the data spine update completion with a resynchronization completion. This leaves caches and grounding bundles stale. True completion requires eviction, regrounding, or an agent who answers from a moved pin.
- Allowing the cross-jurisdiction failover path to treat residency as a configuration detail. A replication path across a residency boundary is a severity 1 compliance breach, not a performance tuning choice.

18.6.3 Review Questions

1. Explain why every operational signal in the integrated platform must carry the CORRELATION_ID. Reference the one-query reconstruction of Chapter 16, Section 16.4 and state what an FP&A professional can reconstruct that a separate, time-correlated monitoring system could not.

2. Distinguish the short-window and long-window burn-rate alerts. Give one incident each is designed to catch that the other would miss, and explain why an unspent error budget is a defect rather than an achievement.
3. The quality scorecard tolerates a small shortfall in completeness but none in lineage integrity. Justify that difference in terms of what each number protects, and state the severity and the action each breach triggers.
4. Trace a single material drift event through Figure 18.3 from the 02:00 UTC snapshot to the terminal `LINEAGE` row. Name every governed transition and explain why the artifact remains readable in its prior `PUBLISHED` version throughout.
5. This chapter argues a resynchronization is not complete at the data spine update. Reconstruct that argument using the grounding pin from Chapter 14 and explain why the defect is invisible until an agent answers from a stale bundle.
6. Explain why scaling across programs keeps a single canonical spine rather than a per-program copy and how per-program cost lineage turns the AI tier into a financial controllable. Reference resource groups, role scope, and `INFERENCE_EVENT`.

18.6.4 Student Companion Notes

You can build this chapter's argument by rederiving each operational discipline from a guarantee you saw in earlier chapters, rather than by memorizing four independent practices. Monitoring is watching the SLOs from Chapter 11 and the hash-chained audit chain from Chapter 12. Data quality is the validation pipeline from Chapter 11 operated continuously instead of only at the gate. Drift and resynchronization are the reconciliation cycle from Chapter 11 and the identity-is-immutable principle from Chapter 9 run as a runbook. Scaling is the separation of the three SAP BTP roles from Chapter 1 and the capacity model from Chapter 12 extended to many tenants. If you can state which earlier guarantee each discipline protects, you understand the chapter; if you are memorizing the four disciplines separately, you have missed its point.

Carry the FP&A operability question through every section: Weeks later, can a controller establish that the platform which produced a number was healthy at the time, with one query and no interview? Test each design decision against it. A correlated signal passes it; an uncorrelated dashboard fails it. A governed resynchronization passes it; a silent in-place mutation fails it. A jurisdiction-bounded replication passes it; a cross-residency failover fails it. The question is the same tool we have used throughout the book, now focused on the platform itself rather than on a single decision within it.

With the platform operated to an evidentiary standard, Chapter 19 is the natural next step. It formalizes security, controls, and audit, making the one-query reconstruction this chapter used for operations into the evidentiary backbone an auditor tests. It is the point at which everything we have built throughout the book is finally examined rather than merely asserted.

Chapter 19
Security, Controls, and Audits

A platform earns trust not by being clever but by being accountable. Every read can be authorized, every change can be traced to a named decision, and every refusal is itself a record. Security is not the wall around the system; it is the grain that runs through it.

This is the final chapter of the book, and it carries two responsibilities. The first is to consolidate the security, control, and audit architecture that we have introduced incrementally across the previous chapters into a single, coherent picture. The second is to close the argument we opened with in Chapter 1: that process and architecture data, integrated through a governed canonical layer on SAP Business Technology Platform (SAP BTP), is what makes execution-grade and AI-grade automation safe.

Security and audits are not a bolt-on to that argument. They are the proof that the argument holds.

This chapter comes last rather than first because security chapters that open a book are often read as constraints: a list of things the architecture must not do, written before the architecture exists. When security is covered in the closing chapter, it is read as a property: a demonstration that the architecture already does the right thing, by construction, and can prove it.

We designed the previous chapters to make this one straightforward to argue and difficult to dispute. If the consolidation that follows feels anticlimactic, that is the result of eighteen chapters of structural discipline.

The earlier chapters each contributed a structural control without always naming it as such. Chapter 2 established identity, access segmentation, and auditability as first-class architectural concerns rather than operational afterthoughts. Chapter 9 separated identity from version and made publication a governed boundary.

Chapter 10 introduced hash-chained audit events, lineage as a first-class query path, and separation of duties (SoD) through SAP Authorization and Trust Management service scopes. Chapter 11 and Chapter 12 made the validation log the audit spine of the ingestion and execution tiers. Chapter 13 through Chapter 15 constrained what AI may consume and required every suggestion to be explainable and grounded.

Chapter 16 and Chapter17 introduced the refusal ledger and the control catalog. Chapter 18 operated all of it under monitoring, data-quality, and drift discipline. This chapter assembles those pieces and shows that, taken together, they form a defensible whole rather than a collection of locally sensible mechanisms.

This chapter covers:

- How data access and authorization are enforced structurally rather than procedurally
- What it means for a control to be effective and how that effectiveness is evidenced rather than asserted
- How audit trails are made tamper-evident and how evidence is reconstructed on demand
- How regulatory compliance is satisfied as a query against existing controls, not a separate program

Throughout the chapter, the recurring principle of the book is restated in its final and strongest form: AI proposes, governance approves, and the system of record records. Every mechanism in this chapter exists to make that sentence enforceable, observable, and auditable. You should leave this chapter able to answer, for any control in the platform, three questions an auditor will ask without rehearsal: What is the control, what structural mechanism enforces it, and what query produces the evidence that it operated? A practitioner who can answer those three questions for every control has, in effect, already passed the audit; the formal audit merely confirms it.

Before we begin, a note on audience: This chapter addresses three roles at once. It demonstrates to architects that the controls compose, to control owners (often from a finance, internal-audit, or risk function) that effectiveness is demonstrable without a reconstruction project, and to AI governance leads that an autonomous agent is bounded by construction rather than by hope. The same mechanisms serve all three roles because our central claim is that these are not three separate problems requiring three separate solutions. They are one problem, and the integrated data spine is the one solution.

19.1 Data Access and Authorization

Access control in a fragmented landscape is a negotiation. Each system has its own roles, its own administrators, and its own interpretation of who should see what. The negotiation is never finished, because every new project adds a system, every system adds roles, and the union of those roles is a surface no single person can reason about.

The integrated platform replaces that negotiation with a single structural model. Because every consumer reaches process and architecture truth through the canonical data spine, authorization can be enforced at one boundary, expressed in one vocabulary, and audited from one place. This section describes that model: the layered security architecture, the authorization vocabulary, the segmentation of the data spine, the treatment of AI agents as named actors, and the way authentication and identity feed the whole.

Before describing the model, it helps to name the failure it replaces, because it's so common that it often feels inevitable. In a fragmented model, the question "Who can change this configuration?" never has a single answer. Each system has its own answer, maintained by a different team, and they only reconcile when an audit demands it.

That reconciliation is the expensive part, and it is expensive precisely because there is no single place where the answer lives. The integrated platform does not make the question easier to answer. It gives it one answer, which is a different and stronger property.

19.1.1 Identity Is the First Control, Not a Prerequisite to It

It is tempting to treat authentication as plumbing that precedes the interesting controls. In this architecture, it is the first control. Every principal that touches the platform, whether a human through SAP Build Work Zone, an execution service, or an AI agent, authenticates against the corporate identity provider federated through Identity Authentication using SAML or OpenID Connect (OIDC).

There are no anonymous callers and no shared technical accounts whose actions cannot be attributed to a single responsible identity. This is not a stylistic preference. Every downstream control we cover in this chapter, from SoD to the hash-chained audit trail, depends on the actor being a specific, named identity. An audit trail that records that the integration user made a change is not an audit trail at all; it is a record that something happened and an admission that no one knows who caused it.

This discipline extends to machine identities, and that is where most landscapes quietly fail. A service that ingests from SAP Signavio authenticates as itself, with its own credential, on its own rotation schedule, as described in Chapter 11. An AI agent authoring a suggestion authenticates as that agent.

The credential rotation is operated, monitored, and audited exactly as described in Chapter 18 for the ingestion adapters, because a machine identity whose credential never rotates is a standing risk that an auditor will (correctly) treat as a finding. The common anti-pattern is a single powerful technical account shared by several integrations because issuing distinct identities felt like overhead during the project. That account becomes, within a year, the single identity that can do anything and that no log can disambiguate. The architecture forbids it not by policy but by making distinct service identities the only supported integration pattern.

There is a second-order benefit to named machine identities that is easy to overlook. When every agent and every service has its own identity, the audit trail does not merely record that an action occurred; it records which specific component took it. This turns the audit trail into an operational debugging instrument as well as an assurance artifact.

The team investigating why a suggestion was generated, as well as the auditor investigating whether an agent acted within its mandate, are reading the same record for different purposes, and the record serves them both without modification. A control that also pays for itself operationally is one that survives the budget cycle.

19.1.2 Defense in Depth as an Architectural Property

The security architecture is layered deliberately so that no single failure exposes canonical truth or the audit trail. Identity and authentication establish who is calling. Authorization and segmentation establish what that caller can see and do.

The canonical data spine enforces the published-only read path and the identity-version separation introduced in Chapter 9. The execution and AI tier enforces the single write path through SAP Build Process Automation and the human-in-the-loop verdict introduced in Chapter 15 and Chapter 16. The audit layer records everything the other layers did.

An attacker, or more commonly a misconfigured integration or a malformed agent call, must defeat each of these layers independently. They are not arranged in series so that one breach cascades; they are arranged so that each is a separate, evidenced control whose failure is observable. Let's look at each of the five layers in turn:

1. **Identity and authentication**
   Consists of Identity Authentication/corporate identity provider (IdP), SAML and OIDC, and agents modeled as named service identities, never anonymous callers
2. **Authorization and segmentation**
   Consists of SAP Authorization and Trust Management service scope grants, role-based exposure of the data spine, and SoD enforced at the scope boundary
3. **Canonical data spine (system of trust)**
   Consists of published-only read path and identity/version separation; AI may read and propose but never mutate canonical truth
4. **Execution and AI services**
   Consists of a single write path through SAP Build Process Automation and human-in-the-loop verdicts; the refusal ledger records every denied call
5. **Audit and evidence**
   Consists of hash-chained, append-only audit events, lineage as a first-class query, and cryptographic anchoring of retained evidence

When it comes to these layers, there is one point that is easy to lose in implementation. Each band is owned by a different discipline and governed by a different regime, exactly as Chapter 1 argued for the three roles of SAP BTP. Identity is owned by the corporate identity provider and Identity Authentication.

Authorization is owned by the platform security model. The canonical layer is owned by the data spine governance team. The execution and AI tier is owned by the integration and AI platform teams.

Audit is owned by the control function, which must be organizationally independent of the teams it audits. Collapsing these into one administrator account in the name of operational convenience is the single most common way the architecture is undermined after go-live, and it almost never appears in a design document but in an incident, six months later, when the convenience account turns out to hold every scope.

The phrase *defense in depth* is sometimes misread as redundancy, as though the layers are copies of one control for safety. They are not. Each layer addresses a different class of failure.

Identity addresses impersonation. Authorization addresses overreach by a correctly identified actor. The published-only read path addresses contamination of downstream consumers by unstable state.

The single write path addresses unauthorized mutation. The audit layer addresses the question of what actually happened when one of the others is tested. Because the layers are independent in what they defend against, a weakness in one does not imply a weakness in the others, and an auditor can reason about each separately. Redundant controls, by contrast, tend to fail together, because they usually fail for the same reason.

The architectural consequence of independence is that the security posture degrades gracefully rather than catastrophically. If the authorization layer is misconfigured for a particular scope, the published-only read path still prevents draft contamination, the single write path still requires a verdict, and the audit layer still records the anomaly so that the misconfiguration is detected rather than exploited silently. A landscape whose controls are entangled has no graceful degradation; the first failure is often the last warning. The investment in keeping the layers independent is, in practice, an investment in having early warnings rather than late incidents.

19.1.3 The Authorization Vocabulary: Scopes Bound to Canonical Operations

Authorization in the integrated platform can be expressed in one sentence: SAP Authorization and Trust Management service scopes bound to canonical operations. A scope is not a job title or an organizational role but a precise statement of a permitted operation against a precise canonical target.

The four operations that matter for governance are read, propose, approve, and publish. Every consumer, whether human, service, or AI agent, resolves to a set of scopes, and every call resolves to the scope it requires before any canonical state is touched. The resolution is server-side and deterministic; it is not a client assertion that the platform trusts.

A scope is not a job title, and mapping scopes to job titles is the most common way this model is weakened in practice. A job title is an organizational convenience that aggregates many unrelated permissions for administrative ease. The moment a scope is defined as everything a senior process architect can do, the scope has become a role. That role accretes permissions over time as the job changes, and SoD becomes unanalyzable because no one can specify what the role now contains.

Binding scopes to canonical operations keeps each scope small enough to reason about. The organizational mapping from people to scopes is then a separate, reviewable concern rather than a property baked into the scope itself.

The SoD constraint shown in Table 19.1 is the structural heart of the access model. No principal may simultaneously hold the scope that proposes a change and the scope that approves it (`spine.Propose` and `gov.Approve`), and no principal may hold both the scope that approves and the scope that publishes (`gov.Approve` and `gov.Publish`). The constraint is evaluated when scopes are granted, not when calls are made.

SAP Authorization and Trust Management Service Scope	Permitted Operation	Canonical Target	Held By
`spine.Read`	Read published entities and lineage	`PROCESS_VERSION (PUBLISHED)`	Execution and AI services
`spine.Propose`	Write a `SUGGESTION` (advisory only)	`SUGGESTION` (advisory store)	AI grounding service
`gov.Approve`	Record a `VALIDATION_VERDICT`	`VALIDATION_VERDICT`	Authorized human approver
`gov.Publish`	Transition entity to `PUBLISHED`	`PUBLISH_STATE`	Publication authority (distinct role)

Table 19.1: The Authorization Model: SAP Authorization and Trust Management Service Scopes Bound to Canonical Operations

A forbidden combination cannot be assembled in the first place, which means the control cannot be defeated by a caller who simply waits for the right moment, nor by an emergency exception that temporarily widens a role. This is the difference between a procedural SoD, which depends on people behaving and on someone noticing when they do not, and a structural one, which depends on the authorization model refusing to express the conflict at all.

The distinction matters because procedural SoD is the control most often cited in audit findings across SAP landscapes. It fails not because the policy is wrong but because the enforcement is human and the landscape is large. A grant-time structural constraint removes human error from the enforcement path.

The reviewer still decides the policy, but the policy, once decided, is enforced by a service that cannot be persuaded or rushed and records its own decisions. The reviewer's judgment is preserved exactly where judgment belongs, in deciding the policy, and removed exactly where judgment is a liability, in enforcing it consistently a thousand times a day under operational pressure.

Consider the emergency change scenario, which is the case procedural controls handle worst and the one auditors probe hardest. A production incident is in progress, a configuration change is needed urgently, and the person available to make it is also, by organizational accident, the person who would normally approve it. In a procedural model, the pressure of the incident is the force that defeats the control: The separation is waived, often verbally, and the waiver is reconstructed afterward from memory, if it is reconstructed at all.

In the structural model, the person simply cannot hold both scopes, so the emergency change still requires a second identity to approve, and the incident record shows that it did. The control does not bend under pressure because there is no human in the enforcement path to apply the pressure to.

Implementation

The scope-grant evaluation is itself a governed, audited change. Granting the approval scope to a principal who already holds the propose scope is not merely discouraged by policy. The grant is rejected by the authorization service and the rejection is written to the same hash-chained log that records the operations the scopes authorize. The conflict matrix is configuration, reviewed under change governance, not code that an administrator can quietly amend. A change to the conflict matrix is therefore as auditable as a change to a process version, which is the property that lets an auditor trust the matrix rather than merely the people who maintain it.

19.1.4 The Four Canonical Operations in Detail

The authorization vocabulary reduces to four operations. We define each term precisely in the following list, because that precision is what makes the SoD constraint analyzable rather than merely aspirational:

- Read returns a role-scoped, `AS-OF` pinned projection of published canonical state. It never returns draft state, and it never returns an attribute the requesting identity is not entitled to. The request is refused as a whole rather than partially populated.
- Propose creates a `SUGGESTION` record. A suggestion is advisory only. It carries its grounding citations and its explanation, and it has no effect on canonical state until a verdict is issued by a different identity.
- Approve issues a `VALIDATION_VERDICT` against a specific proposed change. The verdict records the approver identity, the rationale, and the canonical version the verdict authorizes. It is the only operation that can turn a proposal into an authorized change.
- Publish transitions an approved version to `PUBLISHED_STATE`, making it visible to execution services and AI grounding. Publication is the structural boundary; nothing reaches a runtime consumer that has not passed it.

The constraint that no identity may hold both propose and approve operations, and none may hold both approve and publish operations, is now stated clearly rather than against vague role names. That precision is exactly what lets the authorization service reject a conflicting grant deterministically.

19.1.5 Worked Example: An AI Agent Reaches Its Boundary

Consider an AI grounding agent that has analyzed a recurring cost-center allocation pattern and produced a strong suggestion to change an allocation rule. We will trace what the agent can and cannot do, step by step:

1. The agent authenticates as its own service identity. The authentication event is recorded; there is no anonymous or shared-account path.
2. The agent issues a read for the relevant published process version. The projection it receives is eligibility-filtered and `AS-OF` pinned. Draft variants are invisible to it by design.
3. The agent issues a propose. A `SUGGESTION` record is created with the agent identity, the grounding citations, and the explanation. No canonical state has changed.
4. The agent attempts an approve because its reasoning concluded the change is correct. The authorization service rejects the call because the agent identity does not hold the approve scope and cannot acquire it.
5. The rejected attempt is written to the refusal ledger with the agent identity, the attempted operation, the canonical target, and the timestamp. The boundary did not merely hold; it produced positive evidence that it held.

The outcome is the architectural thesis in miniature. The agent proposes, governance must approve, and the system of record will record. The agent's confidence in its own suggestion is irrelevant to the boundary, because the boundary is enforced at authorization time and does not consult the agent's reasoning at all.

19.1.6 Role-Based Segmentation of the Data Spine

A single canonical record is not exposed identically to every consumer. The same `PROCESS_VERSION` row appears differently depending on who sees it: a process modeler sees draft-plus-published semantics; an execution service sees a resolved, published execution context; an AI grounding bundle sees a published, eligible, role-scoped, `AS-OF` pinned projection; and an auditor sees full lineage with no write path at all. Segmentation is computed from one record set, not configured per consumer and then maintained by hand, because manually maintained per-consumer copies are exactly the drift surface that we aimed to eliminate in Chapter 9. Figure 19.1 shows data access segmentation in action, with role-scope exposure of the data spine.

The property that makes this trustworthy is the same one we insisted on in Chapter 9: Because every projection derives from a single governed record set, segmentation cannot drift between consumers the way duplicated extracts inevitably do. A consumer never receives an attribute it is not entitled to. Critically, the platform refuses rather than partially returns.

If a grounding request would require an attribute the requesting actor is not entitled to, the request is refused as a whole and the refusal is recorded. A half-populated response is more dangerous than no response, because a downstream model, and frequently a downstream human, treats absence as fact rather than as a withheld value. The controller who sees a blank where a cost center owner should be does not conclude that the value was withheld; they conclude there is no owner, and they act on that incorrect conclusion.

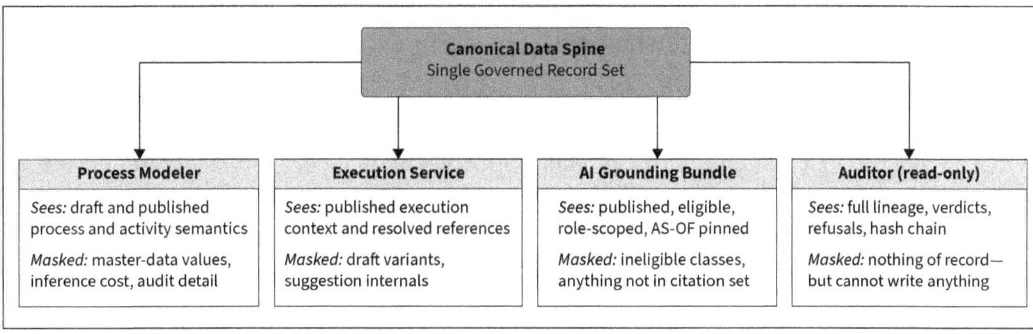

Figure 19.1: Data Access Segmentation: Role-Scoped Exposure of the Data Spine

This refusal-over-partial-return discipline was introduced for AI data preparation in Chapter 14 and for the experience layer in Chapter 17. We restate it here as a general access principle because it applies uniformly: Execution services, AI services, and user interfaces all consume through projections, and all of them are subject to the same refusal semantics. The uniformity is the point.

A control that behaves differently depending on which consumer triggered it is one an auditor cannot reason about because the auditor would have to test every consumer path separately rather than testing the one projection rule once. Uniformity is not an aesthetic preference in this regard. It is what makes the control testable in finite time.

Segmentation also has a temporal dimension that is easy to neglect. An auditor examining a decision taken two quarters ago needs to see the data as it was scoped then, not as it would be scoped under today's entitlements. Because projections are computed from the governed record set with an `AS-OF` pin rather than from a live, modifiable copy, the segmentation that applied at the moment of a decision is reconstructable.

This matters in practice because entitlements change, people move roles, and the question "Who could see this?" is meaningless without the qualifier of "*When* did they see it?" The architecture makes the qualifier answerable rather than rhetorical.

19.1.7 AI Agents as Named Actors

We have argued from Chapter 1 onward that an AI agent is a named actor in the SAP landscape, subject to the same authorization model as a human user. This section makes that argument concrete. An agent authenticates as a service identity, never anonymously and never under a shared technical account that obscures which agent acted.

It may hold the read and propose scopes. It is structurally barred from the approve and publish scopes, in exactly the same way a human proposer is barred from approving their own proposal. The agent therefore inherits SoD for free. It cannot be the actor that approves what it suggested because it cannot hold the scope that approves. It also cannot directly execute a configuration change because execution flows through the single write path that requires a verdict the agent cannot issue.

This is the architectural answer to a question that recurs in every AI governance review: How do you stop the agent from doing something it should not? The answer is that you do not stop it at runtime with a prompt instruction or a model guardrail, both of which are probabilistic and degrade under adversarial input. You prevent it at authorization time, where the prevention is deterministic and the attempt, if made, is recorded.

The agent that tries to publish does not produce a bad outcome that must later be detected but a refusal ledger entry that proves the boundary held. This reframing, from behavioral guardrail to structural authorization, is the single most important idea this chapter contributes to AI governance, and it is the direct application of the five non-goals we stated in Chapter 1.

This reframing inverts the usual conversation about AI safety in the enterprise, which asks how confident we are that the model will behave. That question has no satisfying answer, because model behavior is statistical and adversarial inputs are unbounded.

The architecture refuses to ask it. It asks instead what the model is permitted to do, and it answers that question with scopes that the model does not hold and cannot acquire. A model that is not granted the publish scope cannot publish regardless of how it is prompted, how it is jailbroken, or how it fails. The safety property does not depend on the model's behavior at all, which is the only kind of safety property worth attesting to an auditor.

It follows that increasing the capability of the model does not increase the risk surface of the platform, which is a counterintuitive but important consequence. A more capable model proposes better suggestions; it does not acquire new scopes by being more capable. The boundary is defined by what the agent identity is granted, not by what the model can reason about. This is precisely why the architecture is described as AI-ready rather than AI-dependent. It is built to absorb better models as improvements to the proposal quality without any change to the control surface, and that decoupling is the whole point of treating the agent as a named actor rather than as a special case.

Table 19.2 is intentionally sparse, and that sparseness enforces the control. Most cells are negative. The few positive cells never line up in a way that lets a single principal carry a change from proposal to publication unaccompanied.

Principal	May Read	May Propose	May Approve	May Publish
Process modeler	Published and draft semantics	No	No	No
Execution service	Published execution context	No	No	No
AI grounding agent	Published, eligible projection	Yes (advisory only)	No	No
Governance approver	Published and lineage	No	Yes	No
Publication authority	Published and lineage	No	No	Yes
Auditor	Full lineage, read-only	No	No	No

Table 19.2: Granted Scopes

An auditor reviewing this table is not reviewing intentions but the scope grants that the authorization service will actually enforce, because the table is generated from those grants rather than maintained alongside them. The distinction between a generated control register and a maintained one will recur in the next section. It is the difference between assurance and the appearance of assurance, and it is the hinge on which the credibility of the entire control environment turns.

19.1.8 Worked Example: A Data Residency Question

A privacy regulator asks a direct question: "Did any personal data about residents of a given jurisdiction leave that jurisdiction during the last reporting period?" In a fragmented landscape this is answered with a network diagram and a series of assurances. In the integrated platform, it is answered structurally, as follows:

- The jurisdiction runs the data spine as a self-contained instance, with replication bounded by jurisdiction rather than by storage convenience, as described in Chapter 18.
- Every canonical access and replication event is recorded in the same hash-chained lineage that evidences every other control, scoped to the region.
- The question is answered by querying that lineage for any replication or access event whose source is in-jurisdiction and whose destination is out-of-jurisdiction over the period.
- The answer is the query result, not an attestation about intended design. If the result is empty, residency held, demonstrably; if it is not, the exact events are identified.

The difference is the same this chapter keeps returning to, now applied to geography. A diagram states what the system was designed to do. The lineage states what the system actually did, and only the second is evidence.

19.2 Control Effectiveness

Access control answers the question of who may act. Control effectiveness answers a harder question: Given that controls exist, do they actually work, and can their working be demonstrated to an independent party who is not inclined to take the organization's word for it? This distinction is the one auditors care about most, and it is the one that fragmented landscapes fail at most often.

A control that is well designed but produces no evidence of operation is, for assurance purposes, indistinguishable from no control at all. This section defines effectiveness as the product of design adequacy and operating evidence, walks through the three classes of control the platform uses, and shows why the refusal ledger and the generated control catalog are what make the difference between assurance and its appearance.

19.2.1 Effectiveness Is Design Adequacy Multiplied by Operating Evidence

It is worth stating the definition of control precisely because the imprecision of everyday usage is where control programs go wrong. A control is effective when it is adequately designed for the risk it addresses and when there is evidence that it operated as designed over the period in question. These are multiplied, not added. A perfectly designed control with no operating evidence has zero effectiveness, because nothing can be demonstrated, and a well-evidenced control that does not actually address the risk also has zero effectiveness, because the evidence points to the wrong thing.

Most audit findings are not design failures but evidence failures. The control was probably operating, but the organization cannot prove it without a reconstruction project, and a control that requires a project to evidence is one that fails the next cycle too.

The multiplicative framing has a practical consequence that control owners frequently resist at first. It means that effort spent improving a control's design while its operating evidence remains weak produces no improvement in effectiveness, because zero multiplied by a larger number is still zero. The highest-leverage work in most control environments is therefore not designing better controls but making the operation of existing, adequate controls capable of providing evidence. From this angle, the integrated platform is primarily an evidence architecture. The controls it implements are not novel in their design intent, but they are unusual in that every one of them emits evidence of its own operation as a side effect of operating.

The phrase *as a side effect of operating* is the load-bearing part of that sentence. Evidence that must be deliberately produced is evidence that competes with operational priorities and loses. Evidence that is produced automatically because the control cannot operate without producing it is evidence that survives operational pressure, staff turnover, and the slow entropy that degrades every manually maintained artifact. The architecture is designed so that the evidence is not a deliverable the team owes the auditor but a residue the system cannot help leaving behind.

19.2.2 The Three Control Classes

Controls in the integrated platform fall into three classes, and each class is evidenced differently. Preventive controls stop the wrong outcome from occurring. Detective controls surface a deviation that has occurred.

Corrective controls return the system to a known good state. A mature control environment uses all three, because no single class is sufficient: Prevention can be circumvented by an unforeseen path, detection without correction merely documents decline, and correction without prevention is an endless cycle of cleanup that never reduces the underlying rate of error.

Table 19.3 deliberately pairs each class with its evidence of operation, because the evidence is what distinguishes a real control from an aspirational one.

	Preventive	**Detective**	**Corrective**
What It Does	Stops the wrong outcome occurring: ■ Scope boundary blocks unauthorized canonical writes ■ Published-only read path keeps draft state out of AI grounding ■ SoD constraint prevents conflicting scope combinations ■ Single write path through SAP Build Process Automation	Surfaces a deviation that occurred: ■ Daily reconciliation classifies material vs. immaterial drift ■ Validation log flags structural integrity failures ■ Continuous AI evaluation gates detect model drift ■ Golden-signal SLO breach alerts	Returns the system to a good state: ■ Governed remediation cases with a typed severity band Resynchronization never modifies drift in place ■ Hot restart and regrounding after a version correction ■ Break-glass as a separate audited path
Evidence of Operation	The refusal ledger, where every blocked call is a positive record that the preventive control fired	Drift snapshots and validation verdicts retained for forensic replay	The remediation case is a data spine object: opened, worked, and closed with a verdict

Table 19.3: Control Effectiveness Model: Preventive, Detective, and Corrective

The preventive controls should be familiar from previous chapters: the scope boundary that blocks unauthorized canonical writes, the published-only read path that keeps draft state out of AI grounding, the SoD constraint that prevents conflicting scope combinations, and the single write path through SAP Build Process Automation. The detective controls are the daily reconciliation and drift classification from Chapter 18, the continuous AI evaluation gates from Chapter 15, and the hash-chain recomputation described in Section 19.3. The corrective controls are the severity-banded remediation case from Chapter 18 and the resynchronization discipline that repairs a divergence without mutating canonical history in place.

The three classes are not of equal cost to evidence, and the rest of this section will acknowledge this asymmetry. Detective controls are the cheapest to evidence because the event they surface is already recorded; the finding exists whether or not anyone planned to evidence it. Corrective controls are moderately expensive because the correction must be linked to the deviation it corrected, and the link is the evidence rather than either event alone.

Preventive controls are the most expensive to evidence and, paradoxically, the most valuable to have because a deviation prevented is a deviation that never became an incident. The next section is about how the architecture resolves this paradox, because resolving it is the central trick on which the entire assurance argument depends.

19.2.3 Why the Refusal Ledger Is the Most Valuable Evidence

Most control frameworks struggle to evidence preventive controls. It is straightforward to show that a detective control found something because the finding is itself a record. It is much harder to show that a preventive control stopped something because the event that was stopped did not happen, and a nonevent leaves no natural trace.

This is the structural reason preventive controls dominate audit findings: not because they fail more often, but because their success is invisible. The refusal ledger, introduced for AI boundaries in Chapter 16 and Chapter 17, solves this generally. When a principal attempts an operation it lacks the scope for, the attempt and its denial are written to an append-only, hash-chained ledger.

The denial is the evidence. An auditor asking whether the AI can publish is not answered with a policy document; they are answered with the ledger, which shows every attempt and its refusal, and equally shows the absence of any successful publish event attributed to an AI identity.

Figure 19.2 shows the anatomy of a refusal entry and, more importantly, the inversion it achieves: A control whose entire purpose is that nothing happens is made to produce a record every time it succeeds. Additionally, the absence of refusals is also evidence.

A well-behaved agent that never attempts an action outside its mandate produces an empty refusal record for those operations, and that emptiness, combined with the positive audit trail of what it did do, demonstrates that the agent operated within its boundary by design rather than by luck. Auditors are trained to distrust the absence of records because in most systems, absence means nothing was captured. The integrated platform makes the absence meaningful by ensuring that any boundary violation would have produced a record, so that here, and only here, the absence of a record is positive evidence of the absence of a violation.

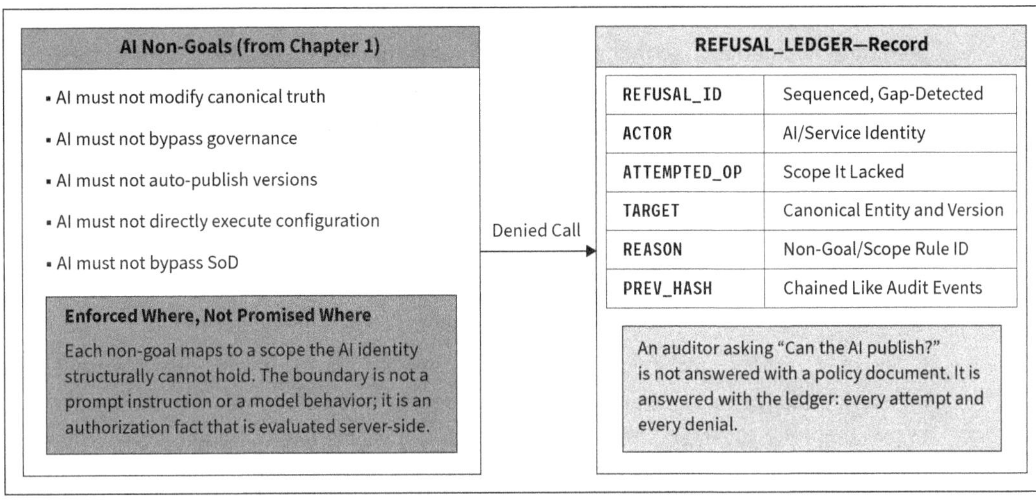

Figure 19.2: The Refusal Ledger: A Preventive Control That Produces Positive Evidence

This argument is the one auditors find most unfamiliar and, once they accept it, most powerful. In an ordinary system, the statement "We found no evidence of the agent overstepping" is weak, because there being no evidence of X is consistent with both X not happening and with X happening, but not being logged. The refusal ledger collapses that ambiguity.

Because every attempt to overstep would, by design, write a refusal entry, the absence of refusal entries is logically equivalent to the absence of attempts. The auditor is no longer reasoning from absence of evidence to a hopeful conclusion; they are reasoning from a guaranteed-complete log to a certain one. Completeness of the log is the property that does all the work, and that completeness is guaranteed because the refusal is written by the same authorization decision that performs the denial, not by a separate logging step that could be skipped.

The refusal ledger also reframes a class of conversations that, in fragmented environments, are unresolvable arguments about intent. When something goes wrong, the fragmented-environment question is "Did someone try to do this and get away with it?" That question generates suspicion that cannot be settled, because the absence of a trace proves nothing.

The integrated environment question is "What does the refusal ledger show?" and it has an answer. When repeated across an entire control environment, replacing an unresolvable argument with a query is the difference between an assurance function that adjudicates blame and one that reports facts.

19.2.4 The Control Catalog as a Generated Register

In Chapter 16, we introduced the control catalog as the mapping between workflow controls and audit frameworks. This chapter generalizes it. The control catalog is the authoritative register of every control in the integrated platform: its class, the structural mechanism that implements it, the canonical objects it governs, and the query that produces its operating evidence.

It is not a spreadsheet maintained by the control function alongside the system. It is generated from the platform, so that a control cannot exist in the catalog without a corresponding enforcement mechanism, and a mechanism cannot be added without appearing in the catalog. This generated property is the entire difference between a catalog an auditor can rely on and one they must independently verify, because a maintained catalog is an assertion about the system while a generated one is a projection of it.

The problem with a maintained catalog is so common that it often goes unnoticed. A maintained control register is accurate on the day it is written and decays from that moment. A control is decommissioned but stays in the register because no one removed it.

A new enforcement mechanism is added but does not appear in the register because the register update was a separate task that was deprioritized. Within a year, the register and the system have diverged, and that divergence is invisible until an auditor tests a control the register claims exists and either finds nothing or finds an enforcement mechanism the register never mentioned. A generated catalog cannot diverge because it is recomputed from the enforcement mechanisms every time it is read; there is no separate artifact to fall out of date. Table 19.4 shows a summary of controls and their mechanisms.

Control	Class	Structural Mechanism	Operating Evidence
Unauthorized write prevention	Preventive	Scope boundary and SoD constraint	Refusal ledger entries
Draft isolation from AI	Preventive	Published-only read path	Bundle manifests and eligibility log
Change authorization	Preventive	Verdict-keyed single write path	Validation verdict records
Drift detection	Detective	Daily reconciliation classification	Drift snapshots (retained)
Model degradation detection	Detective	Continuous evaluation gates	Gate verdicts over time
Integrity breach detection	Detective	Hash-chain recomputation	Chain divergence reports

Table 19.4: Catalog of Controls

Control	Class	Structural Mechanism	Operating Evidence
Governed remediation	Corrective	Severity-banded remediation case	Remediation case lifecycle
Version correction	Corrective	Resynchronization without mutation	Snapshot and replay lineage

Table 19.4: Catalog of Controls (Cont.)

The catalog is the artifact an external auditor is handed first because it lets them choose which controls to test and tells them exactly which query produces the evidence for each. In a fragmented landscape, this conversation takes weeks, because each control is evidenced by a different team in a different system in a different format, and a meaningful fraction of the audit budget is consumed simply establishing what the controls are and where their evidence lives. In the integrated platform, the conversation is bounded, because the evidence for every control is a query against the same retained lineage, and the auditor can run a sample of those queries themselves rather than receiving prepared extracts whose provenance they would otherwise have to challenge.

There is a governance subtlety in letting auditors run the queries themselves that is sometimes raised as an objection. Allowing an auditor to execute evidence queries does not grant them any write capability; it grants them the auditor scope from the table in the previous section, which is read-only over full lineage. The independence of the audit is therefore strengthened rather than compromised because the auditor no longer depends on the audited organization to prepare and hand over evidence that they had the opportunity to shape. The architecture makes auditor independence a technical property rather than a procedural promise, and technical properties are the only kind an auditor is professionally obliged to trust.

19.2.5 The Dimensions of a Catalog Entry

Each control catalog entry is more than a name. It records, for every control, a fixed set of the following dimensions so that an auditor can move from the control to its evidence without asking anyone:

- **Class**
  Preventive, detective, or corrective. This sets the expectation for what evidence of operation looks like.
- **Structural mechanism**
  The specific platform construct that enforces the control, named precisely enough to be located in the architecture.

- **Governed objects**
  The canonical entities and versions the control acts upon, so the scope of the control is unambiguous.
- **Evidence query**
  The exact query that returns the control's operating evidence over an arbitrary period. This is the dimension that makes the catalog auditable rather than descriptive.
- **Owner**
  The discipline accountable for the control, which must be organizationally distinct from the discipline the control constrains where SoD requires it.

Because every entry carries its own evidence query, an auditor selecting a sample of controls to test is also, in the same action, selecting the queries that will test them. There is no intervening step in which someone decides what evidence to assemble, and removing that step is what removes the opportunity for the evidence to be shaped.

19.2.6 Worked Example: Evidencing a Preventive Control Over a Quarter

A control owner must attest that unauthorized writes to canonical state were prevented over the last quarter. We can trace the attestation as follows:

- The owner locates the control in the catalog: unauthorized write prevention, preventive, enforced by the scope boundary and SoD constraint.
- The catalog entry names the evidence query: refusal ledger entries for write operations over the period, plus the absence of any successful unauthorized write event.
- The owner runs the query over the quarter. It returns every blocked attempt and confirms no successful unauthorized write occurred.
- The owner attests based on that result. The auditor independently runs the same query against the same lineage and obtains the same result.

There is no binder, reconciliation, or gap between what the owner attested and what the auditor found, because both read the same evidence through the same query. The attestation and the audit have become the same act performed by two parties, which is the strongest form an attestation can take.

19.2.7 Effectiveness Is a Property of the Architecture, Not the Effort

The most important point in this section is that control effectiveness in the integrated platform is structural rather than effortful. In the traditional model, effectiveness is something an organization works to demonstrate at audit time. Evidence is collected, reconciled, formatted, and explained, often by the same people who operate the controls, which is itself a SoD weakness rarely acknowledged. In the integrated model, effectiveness is a property the architecture exhibits continuously, because every control either prevents and records its prevention, detects and retains its detection, or corrects and tracks its correction. The audit does not cause the evidence to come into existence; it queries evidence that already exists and that the organization itself relies on operationally between audits.

The phrase *relies on operationally between audits* is the proof of effectiveness. Evidence that the organization uses to run the platform day to day is evidence that is continuously tested by reality, not merely by an annual inspection. If the refusal ledger were inaccurate, operations would notice, because operations depends on it to understand why a legitimate call was denied.

If the drift snapshots were unreliable, the reconciliation function would notice, because it acts on them daily. The audit, in this model, is not the first time the evidence is scrutinized; it is the formal confirmation of evidence that has been under continuous informal scrutiny by the people whose work depends on its being correct. That is a far stronger basis for assurance than evidence assembled specifically for, and seen only by, the auditor.

Binders No More

A practical consequence for finance and controlling functions is that the controller who must attest to the effectiveness of a control over a reporting period no longer assembles a binder. They run the catalog query for that control over that period and read the result. The attestation rests on the same evidence the auditor will independently query, which removes the reconciliation gap between what the business attests and what the auditor finds. That gap, in fragmented environments, is where most late audit findings originate, because the binder was assembled from one set of extracts and the auditor sampled another, and the two no longer agree by the time the difference is noticed.

19.3 Audit Trails and Evidence

An audit trail is only as valuable as it is trustworthy, and only as usable as it is queryable. A trail that can be silently altered proves nothing, because an adversary, or an embarrassed operator, can make it say whatever is convenient. A trail that is tamper-evident but takes weeks to reconstruct into an answer is technically sound and operationally useless, because by the time the answer arrives the decision it was meant to inform has already been taken without it. This section describes how the integrated platform makes its audit trail both tamper-evident and immediately queryable as well as how a single business question can be reconstructed end to end from it.

19.3.1 The Hash-Chained Audit Event

Every consequential operation in the integrated platform writes an audit event. The event records the actor, the scope-resolved operation, the canonical target and version, and a cryptographic hash of the before-and-after canonical state. Crucially, each event also records the hash of the immediately preceding event.

The audit log is therefore a chain. Altering any historical record changes its hash, which invalidates the predecessor reference in every event that follows it. Tampering is not made impossible by this construction, because no construction makes tampering impossible, but it is made detectable, and for assurance purposes detectability is the property that matters. An undetectable alteration is a forgery; a detectable one is merely an incident, and incidents can be investigated.

Figure 19.3 shows the record anatomy and the chain. Two operational details make the construction robust in practice rather than merely on paper. First, event identifiers are monotonic and gap-detected, so the removal of an event, not its alteration, is visible as a sequence break rather than as a silent shortening of history.

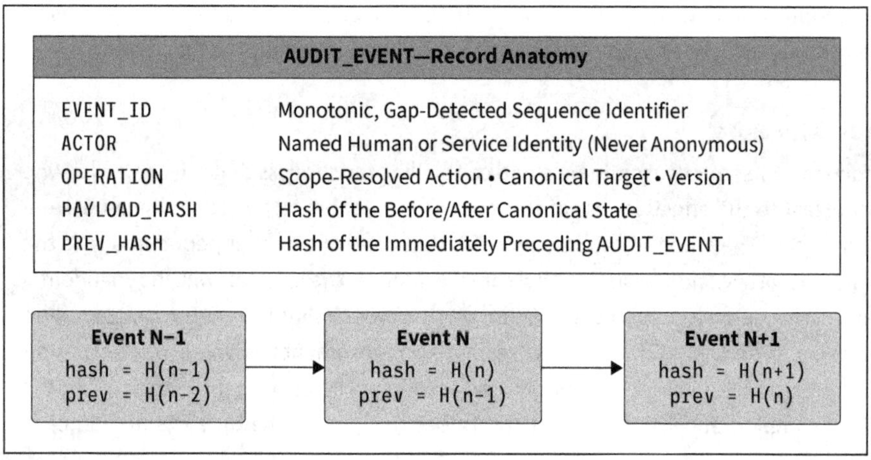

Figure 19.3: The Hash-Chained Audit Event: Tamper-Evident by Construction

Second, chain heads are periodically anchored to an external, independent time source, so the window during which silent alteration is even theoretically possible is bounded to the interval between anchors rather than being open-ended. The combination means that an investigator can recompute the chain over retained events and identify the first divergent link with precision, rather than being left to argue in the abstract about whether the trail can be trusted. The shift from a trust argument to a recomputation is the whole value of the construction.

Precision about what the chain does and does not promise matters here, because overclaiming is how cryptographic controls lose credibility with auditors who understand them. The chain does not prevent a sufficiently privileged actor from altering history; nothing short of write-once media does, and even that has problems. What the chain guarantees is that any alteration of a historical event invalidates every subsequent link, so that the alteration cannot be localized and hidden.

An attacker who wishes to change one past event must rewrite the entire chain from that point forward and reanchor it, and the external anchors, being independent, will not match. The honest claim, and the one the architecture makes, is not that the past cannot be changed but that the past cannot be changed quietly. That is a weaker claim than tamper-proof but a far more defensible one.

19.3.2 Lineage as a First-Class Query Path

A tamper-evident trail that cannot be queried efficiently is a forensic resource, not an operational one, and the difference determines whether the trail is used or merely retained. The integrated platform treats lineage as a first-class query path, a discipline we introduced in Chapter 10 and operated in Chapter 18. Every canonical change is addressable by a stable identifier, and the relationships between a trigger, a suggestion, a verdict, a published version, and the audit events that record them are queryable links rather than facts that must be reassembled from separate logs by someone who remembers how the systems connect.

Figure 19.4 traces the question that the book has used as its recurring touchstone, the controller's variance question: Why did this cost-center allocation change between quarters, who approved it, and on what evidence? In a fragmented landscape, this question initiates a project. Someone exports the process history, someone else reconciles it against the configuration transport log, a third person reconstructs the approval from email, and the answer is a narrative assembled from fragments, defensible only to the extent that the people who assembled it are trusted and available.

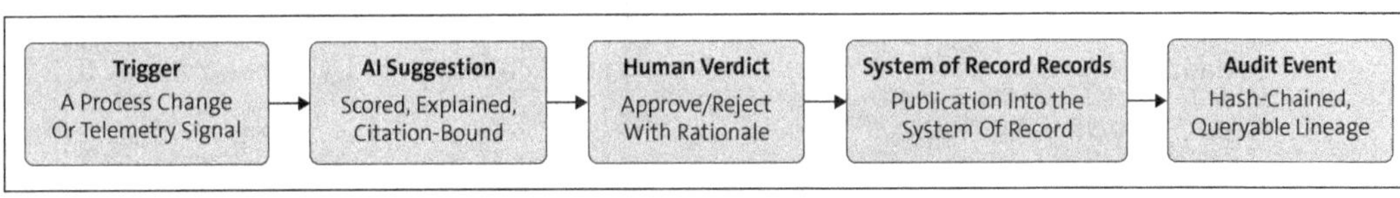

Figure 19.4: The Evidence Chain: From a Single Change to an Audit-Ready Answer

In the integrated platform the same question is a single traversal: the triggering signal, the suggestion and its citations, the approver identity and rationale, the published version, and the hash-chained audit events, with each link addressable and none reconstructed from memory. The answer is not prepared; it is retrieved, and it is identical whether the controller retrieves it or the auditor does.

The phrase *identical whether the controller retrieves it or the auditor does* is the property that finance and controlling functions value most once they have experienced its absence. In a fragmented model, the controller's reconstructed answer and the auditor's independently reconstructed answer are two narratives built from overlapping but not identical fragments by people with different incentives, and the gap between them is precisely what audit meetings are spent arguing about. When both parties run the same traversal against the same lineage, there is no gap to argue about. The conversation moves from whose reconstruction is correct to what the single retrievable answer implies, which is the conversation an assurance function should be having but rarely gets to.

19.3.3 What the Auditor Actually Asks

Auditors bring specific questions, and the architecture should be judged against those questions rather than an abstract notion of auditability. The questions are remarkably consistent across regulatory regimes and across audit firms, which is useful because it means an architecture that answers them well answers most of what will ever be asked of it.

Each row of Table 19.5 pairs a question with the query that answers it. The question "Who could have changed this, and who actually did?" is answered by the scope grants at the period combined with the actor on every audit event touching the entity. For example, the demand "Show me the approval for this production change" is answered by the validation verdict, the approver identity, the rationale, and the released version.

The Question	The Shape of the Answer
"Who could have changed this, and who actually did?" *An SoD and access question*	Scope grants at the period and the actor on every `AUDIT_EVENT` touching the entity
"Show me the approval for this production change." *A change-governance question*	The `VALIDATION_VERDICT`, the approver identity, the rationale, and the version it released
"Prove the AI did not act outside its mandate." *An AI-accountability question*	The refusal ledger for the AI identity, plus the absence of any `gov.*` event attributed to it
"Has the audit trail itself been altered?" *An audit-integrity question*	Recomputed hash chain versus the anchored chain heads—divergence, if any, is pinpointed
"Why did this number move between quarters?" *The controller's variance question*	The lineage from trigger to suggestion to verdict to published version, as one trace

Table 19.5: What an Auditor Actually Asks, and the Shape of the Answer

The demand "Prove the AI did not act outside its mandate" is answered by the refusal ledger for the AI identity together with the absence of any governance event attributed to it. "Has the trail itself been altered?" is answered by recomputing the hash chain against the anchored heads. The question "Why did this number move between quarters?" is answered by the lineage traversal. None of these answers is a document that someone prepared in advance; each is a query the platform was built to serve, which means the audit does not perturb the system it inspects.

There is a consistency between these answers that is itself a control, and it is the consistency that experienced auditors probe for hardest. Every answer derives from the same retained lineage, so the answers cannot contradict one another the way evidence drawn from separate systems routinely does. In fragmented environments, a great deal of audit effort goes into resolving contradictions between sources that should agree but do not: the change log says one thing, the approval email says another, and the configuration transport says a third.

Those contradictions are not usually evidence of wrongdoing but of fragmentation. Removing the fragmentation removes the contradictions, and removing the contradictions is what shortens the audit far more than any fast query does.

19.3.4 Worked Example: The Controller Variance Question, Step by Step

This book's recurring touchstone is the controller's question: Why did this cost-center allocation change between quarters, who approved it, and on what evidence? Here is the exact traversal that answers it in the integrated platform:

1. Start from the published canonical version of the allocation rule in the current quarter and retrieve its stable identifier.
2. Follow the version lineage backward to the immediately prior published version and identify the change between them.
3. From the change, follow the link to the `VALIDATION_VERDICT` that authorized it, retrieving the approver identity, the rationale, and the authorized version.
4. From the verdict, follow the link to the originating `SUGGESTION`, retrieving the proposing identity, the grounding citations, and the explanation.
5. From the suggestion, follow the link to the triggering signal that prompted it, establishing why the change was considered at all.
6. At every node, retrieve the associated hash-chained `AUDIT_EVENT` and confirm the chain is intact across the span.

The answer is the assembled traversal: a triggering signal, a grounded suggestion, a named approver with a rationale, a published version, and an unbroken audit chain. None of it was prepared in advance. It is the same answer whether the controller retrieves it for a management review or the auditor retrieves it for an assurance opinion, and that identity of answers is precisely what eliminates the reconciliation argument that consumes fragmented-environment audits.

19.3.5 Worked Example: Forensic Replay of a Past Decision

Suppose a dispute arises over a decision taken two quarters ago, and the question is not what the state is now but what it was when the decision was made. Forensic replay answers it, as follows:

1. Identify the decision point: the verdict and the canonical version it authorized, with its timestamp.
2. Reconstruct the state as of that timestamp from retained snapshots and the audit chain, rather than from the live system, which has since moved on.
3. Reconstruct the role-scoped projection that the deciding identity would have seen then, using the `AS-OF` pin so that entitlements as they were, not as they are, govern the reconstruction.
4. Confirm the audit chain between the snapshot and the decision is intact, so the reconstruction is demonstrably faithful rather than merely plausible.

Replay answers the question disputes actually rest on, because a decision can only be fairly judged against the information available when it was taken. An architecture that can only show the present state cannot conduct that judgment honestly; one that can replay the past can.

19.3.6 Retention, Anchoring, and Forensic Replay

Evidence has a lifecycle, and an architecture that ignores that lifecycle eventually finds that the evidence it needs has aged out or become unverifiable. The platform tiers retention so that hot evidence is immediately queryable, warm evidence is retained for the full audit and regulatory horizon, and cold evidence is cryptographically anchored before archival so that its integrity remains demonstrable even after it leaves primary storage. Forensic replay, which we introduced in Chapter 11 and operated in Chapter 18, is the mechanism by which a historical state is reconstructed from retained snapshots and the audit chain rather than from a live system that has since moved on. Replay is what allows an investigator to answer not only what the state is now but what it was at the moment a decision was taken, which is almost always the question that matters in a dispute, because disputes are about decisions and decisions are made against the state that existed then, not the state that exists now.

The discipline that ties retention to assurance is that the retention policy is itself a control in the catalog, with its own evidence: The platform can demonstrate, by query, that evidence required for a given obligation over a given horizon was in fact retained and remained verifiable. An organization that cannot prove its evidence was retained essentially has no evidence, and that problem is unnoticeable until the evidence is needed and found missing. Making retention a cataloged, evidenced control is what removes that silent problem, which is this chapter's recurring objective. A control that can fail without anyone noticing is the only kind of control an auditor is right to fear.

19.4 Compliance Considerations

Compliance is often run as a separate program with its own staff, evidence repository, and annual rhythm. That separation is the source of most of its cost and most of its friction, and it is rarely questioned because it is the way compliance has always been done. The argument of this section is that, in an integrated platform, compliance should not be a separate program at all.

This involves the same controls described in the Section 19.2 and Section 19.3, queried through a regulatory lens. Once obligations map to controls, and controls map to evidence queries, a compliance assessment becomes a traversal rather than a collection exercise, and the staff who ran the collection exercise are free to do the analysis that actually reduces risk.

19.4.1 Obligation, Control, Evidence

Every compliance obligation, regardless of the regime it comes from, has the same three-part structure: an obligation that states what must be true, a control that makes it true, and evidence that demonstrates the control operated. The integrated platform is built so that the second and third parts already exist for every obligation that maps onto access, change governance, data protection, audit integrity, or AI accountability, which together cover the majority of obligations that bear on a process-and-AI platform. The obligations that are not structured this way are usually organizational rather than technical, and the platform does not pretend to satisfy those; it satisfies the technical core and makes the organizational remainder visible rather than hidden.

This is particularly important because compliance architectures lose credibility when they overclaim. There are obligations, particularly around governance, accountability, and human oversight, whose satisfaction is partly organizational and cannot be reduced to a query. The architecture does not assert otherwise.

What it does is draw a clear line between the obligations it satisfies structurally, where the evidence is a query, and the obligations that require organizational attestation, where the evidence is a documented human process. Drawing that line explicitly is a compliance asset, because the most damaging audit findings are not the obligations an organization failed to meet but the obligations it did not realize it had not addressed because the boundary was never drawn.

Table 19.6 maps representative obligations to the controls that satisfy them and the queries that evidence them. An access-control obligation, whether it arrives as an IT general control under a financial reporting regime or as an information security management requirement, is satisfied by the scope grants and the SoD constraint, and evidenced by the scope-grant audit events and the refusal ledger. A change governance obligation is satisfied by the verdict-keyed single write path and evidenced by the validation verdicts and publication events.

Obligation	Data Spine Control that Satisfies It	Evidence (A Query, Not a Document)
Access control (Sarbanes–Oxley Act Information Technology General Controls [SOX ITGC], ISO 27001 A.9)	SAP Authorization and Trust Management service scope grants and SoD constraint: unauthorized canonical write is structurally impossible	Scope-grant audit events and the refusal ledger: replayable per actor and per period
Change governance (SOX change management)	Verdict-keyed approval and single write path: no change reaches the system of record without an approver	`VALIDATION_VERDICT` and publication audit events: approver identity and rationale on every change
Data protection (General Data Protection Regulation [GDPR] Art. 5, 25, 32)	Eligibility classes and role-scoped projection: ineligible data never enters a grounding bundle	Eligibility decisions and bundle manifests (signed): what AI saw, and what it was barred from seeing
Audit trail integrity (SOX, ISO 27001 A.12.4)	Hash-chained, append-only audit events: tamper is detectable, not merely discouraged	Chain recomputation and cryptographic anchoring: first divergent link is identifiable
AI accountability (EU AI Act, internal AI policy)	Non-goal scopes and citation contract: every AI output is grounded and explainable	`SUGGESTION_EXPLANATION` and `INFERENCE_EVENT`: what was suggested, on what evidence, at what cost

Table 19.6: The Compliance Mapping Matrix: Obligation to Control to Evidence

A data protection obligation under a privacy regime is satisfied by the eligibility classes and role-scoped projection and evidenced by the eligibility decisions and the signed bundle manifests. An audit-integrity obligation is satisfied by the hash chain and evidenced by chain recomputation. An AI-accountability obligation, of the kind now appearing in AI-specific regulation and in internal AI policy, is satisfied by the non-goal scopes and the citation contract and evidenced by the suggestion explanations and inference events.

Table 19.6 shows a pattern that becomes visible only when the obligations are laid side by side. The same small set of structural mechanisms satisfies obligations drawn from regimes that were written independently, by different bodies, for different purposes, in different decades. This is not a coincidence, and it is not the architecture being clever. It is a consequence of the fact that the regimes, despite their surface differences, are all asking variations of the same underlying questions about who could act, who did act, what they were permitted to see, and whether the record of all that can be trusted. An architecture that answers those underlying questions well answers the regimes as a byproduct, which is why mapping a newly introduced regime onto the platform is usually a labeling exercise rather than an engineering one.

19.4.2 Why Mapping to Queries Rather Than Binders Matters

The practical difference between mapping obligations to queries and mapping them to binders is the difference between an audit cycle measured in hours and one measured in weeks. A binder is a point-in-time artifact that begins to age the moment it is assembled, must be reassembled for every cycle, and is only as trustworthy as the assembly process that produced it, which is itself rarely audited. A query is evaluated against current retained lineage every time it is run, returns the same answer to the business and to the auditor because both run the same query against the same lineage, and carries its own provenance because the lineage it traverses is itself audited. The reduction in cost is real and is usually what attracts attention, but the more important gain is the elimination of the reconciliation gap between what the organization believes its compliance posture to be and what an independent party finds when they test it.

One particular problem with the binder model is rarely discussed openly: The binder is assembled by the people most motivated for it to look complete. This is not an accusation of bad faith but an observation about incentives. The team that assembles compliance evidence under deadline pressure makes a thousand small judgment calls about what to include, how to present it, and which gaps to characterize as immaterial, and every one of those calls bends, even slightly, toward the appearance of compliance.

The query model removes the judgment calls. The query returns what the lineage contains, the auditor runs the same query, and there is no assembly step in which incentives can operate. Removing the assembly step is removing the place where the appearance of compliance and the fact of it diverge.

19.4.3 Worked Example: Decomposing a Compliance Obligation

Consider a typical obligation raised by an internal audit function: Changes to financially material configuration must be authorized by someone other than the person who proposed them, and the authorization must be evidenced. It can be broken down as follows:

- **Obligation**
  A proposer may not authorize their own change to financially material configuration, and the authorization must be demonstrable.
- **Control**
  The SoD constraint that prevents any identity from holding both the propose and approve scopes, evaluated at scope-grant time.

- **Evidence**
  The scope grants in force over the period, the `VALIDATION_VERDICT` records showing a distinct approver identity for each material change, and the refusal ledger showing that any attempt to self-approve was blocked.

The obligation initially reads as something requiring its own compliance sub-program. The decomposition reveals that the control and its evidence already exist structurally; nothing new must be built, and nothing must be assembled at audit time. The obligation is satisfied by a query, and the work that remains is labeling, not engineering.

19.4.4 Why the Same Mechanisms Satisfy Independent Regimes

A small set of mechanisms can satisfy obligations from regimes when written independently and for different purposes. The regimes differ on the surface but converge on a few underlying questions:

- Who was permitted to act, and was that permission appropriate and separated where it needed to be?
- Who actually acted, and can that action be attributed to a specific, named identity?
- What was the actor permitted to see, and was anything exposed that should not have been?
- Can the record of all of the above be trusted not to have been altered?

An architecture that answers these four questions well answers the regimes as a byproduct, because the regimes are, beneath their differing vocabularies, asking these same four questions. This is why onboarding a newly introduced regulation onto the platform is usually an exercise in mapping its language onto existing evidence queries rather than an exercise in building new controls.

19.4.5 Jurisdiction, Residency, and Multiregion Compliance

Compliance is not uniform across jurisdictions, and the platform does not pretend otherwise. In Chapter 18, we described jurisdiction-bounded replication and program tenancy on resource groups. The compliance consequence is that data residency and sovereignty obligations are satisfied structurally: a region runs the entire spine as a self-contained instance, replication is bounded by jurisdiction rather than by storage convenience, and the evidence that residency held is the same lineage that evidences every other control, scoped to the region.

An obligation that data about a given population never left a given jurisdiction is answered by querying the lineage of that data, not by attesting to a network diagram that may or may not still describe the running system. The diagram describes intent; the lineage describes what happened, and an auditor cares only about the second.

The distinction between a diagram and lineage is the same distinction we have been making throughout this chapter, bow applied to geography. A residency attestation backed by an architecture diagram is one that the system was designed not to move data across a boundary. A residency attestation backed by lineage is confirms that the data did not, in fact, cross the boundary, over the period in question, demonstrable by query.

The first is a statement about intent and is the kind of attestation that produces findings when the running system has quietly diverged from the design. The second is a statement about reality and is the kind of system that does not diverge, because it is recomputed from what happened rather than asserted from what was planned.

19.4.6 The Compliance Posture Is Continuous, Not Periodic

Because every control evidences its own operation continuously, the platform has a compliance posture at every moment, not only at audit time. This changes the nature of the relationship with assurance functions. Continuous control monitoring, which in fragmented environments is an expensive overlay that samples and extrapolates and is itself a frequent source of findings, becomes a direct read of evidence that already exists.

An organization can answer the question "Are we now compliant with a query rather than with a project?" and the answer is grounded in the same lineage an external party would independently traverse. The strategic value is not the speed of the answer but its honesty: The organization that can see that its posture continuously cannot maintain the comfortable truth common in periodic models that everything is fine between audits because no one is looking.

That last point is the one executives find most uncomfortable and most valuable. A periodic compliance model creates an incentive not to look between audits, because looking creates knowledge, and knowledge of a gap creates an obligation to remediate it on a schedule the organization did not choose. A continuous model removes the option of not looking.

The posture is visible whether or not anyone wants it to be. This is uncomfortable precisely because it is honest, and it is valuable for exactly the same reason. The gap that is found continuously and remediated quietly is far cheaper to fix than the same gap found by an external party who was looking when the organization had arranged not to.

19.5 Summary

This chapter consolidated the security, control, and audit architecture of the integrated platform and used it to close the book's central argument. We enforced data access and authorization through a single vocabulary of SAP Authorization and Trust Management service scopes bound to canonical operations, with SoD evaluated structurally at scope-grant time rather than procedurally at call time, and with AI agents treated as named actors that are barred from governance scopes by construction rather than by instruction. We defined control effectiveness as the product of design adequacy and operating evidence, establishing the refusal ledger as the mechanism that finally makes preventive controls capable of providing evidence and the generated control catalog as the register that binds every control to its enforcement mechanism and its evidence query, so that the catalog is a projection of the system rather than an assertion about it.

We showed how audit trails can be tamper-evident through hash-chaining and external anchoring and usable through lineage as a first-class query path, so that the recurring controller's variance question is answered by a single traversal rather than a forensic project, and answered identically whoever asks it. We reframed compliance not as a separate program but as the same controls queried through a regulatory lens, with obligations mapping to controls and controls mapping to evidence queries, which collapses the audit cycle from weeks to hours and eliminates the reconciliation gap between attested and observed posture. The unifying theme, which we stated in every section, is that security and audit in the integrated platform are structural properties of the architecture, exhibited continuously, rather than efforts undertaken in anticipation of inspection. A control that cannot fail silently, evidenced by a query that cannot be shaped, registered in a catalog that cannot drift, is the standard we have held every mechanism to in this chapter.

19.5.1 Key Learning Points

- Security in the integrated platform is structural, not procedural. Authorization is one set of SAP Authorization and Trust Management service scopes bound to canonical operations, and SoD is enforced at scope-grant time so that a conflicting combination cannot be assembled, rather than at call time, where it can be defeated by timing or by emergency exception.
- An AI agent is a named actor. It may read and propose, and it is barred from approve and publish operations by construction, which means it inherits SoD for free and cannot approve what it suggested. The boundary is enforced at authorization time, where it is deterministic, not at runtime, where it would be probabilistic. Increasing model capability improves proposal quality without expanding the control surface.
- Control effectiveness is design adequacy multiplied by operating evidence. A control with no evidence trail is, for assurance purposes, not a control at all. The refusal ledger is what makes preventive controls evidenceable, because it turns a nonevent into a positive record, and the guaranteed completeness of the ledger makes the absence of refusals into proof of the absence of attempts.
- The audit trail is tamper-evident by hash-chaining and external anchoring, and it is usable because lineage is a first-class query path. The honest claim is not that the past cannot be changed but that it cannot be changed quietly. The controller's variance question is answered by one traversal, identical whether the business or the auditor retrieves it.

- Compliance is not a separate program. It is the same controls queried through a regulatory lens. The same small set of structural mechanisms satisfies obligations from independently written regimes because the regimes all ask variations of the same underlying questions. Mapping obligations to queries rather than binders removes the assembly step where appearance and fact diverge.
- Our closing principle was enforced at every layer: AI proposes, governance approves, and the system of record records. Each step is bound to a structural mechanism, not to an intention. The architecture is not built once; it is sustained.

19.5.2 Common Design Pitfalls

- Collapsing the layered ownership of the security architecture into a single administrator account for operational convenience. This silently undoes defense in depth and is the most common post-go-live regression, because it never appears in a design document, only in an incident six months later.
- Defining scopes as job titles rather than canonical operations. A scope that means everything a senior architect can do accretes permissions over time until SoD is unanalyzable. Keep scopes small enough to reason about and treat the people-to-scopes mapping as a separate, reviewable concern.
- Treating SoD as a runtime check rather than a scope-grant constraint. A runtime check can be defeated by timing and by emergency exceptions; a grant-time constraint cannot express the conflict in the first place and therefore has nothing to defeat under pressure.
- Evidencing preventive controls with policy documents instead of refusal records. An auditor cannot test a policy; they can only test a ledger. If prevention produces no record, it cannot be demonstrated to have operated, regardless of how well it was designed.
- Allowing the control catalog to be a spreadsheet maintained alongside the platform rather than generated from it. A catalog that can drift from the enforcement mechanisms manufactures false assurance that no one discovers is false until it matters.
- Running compliance as an annual collection exercise on top of the platform rather than as a continuous query into it. The collection model reintroduces exactly the reconciliation gap the integrated architecture was built to eliminate, and the binder always looks complete because the people assembling it are the people most motivated for it to appear that way.
- Treating retention as infrastructure rather than as a cataloged, evidenced control. Evidence that cannot be proven to have been retained and to have remained verifiable is absent in practice, and that absence is silent until the evidence is needed and found missing.

19.5.3 Review Questions

1. Why is SoD described as a structural property in this architecture, and how does evaluating it at scope-grant time differ, in both enforcement and auditability, from evaluating it at call time? Use the emergency change scenario in your answer.
2. Explain why the refusal ledger is described as the most valuable evidence for preventive controls, and reconstruct the argument under which the absence of refusals becomes positive evidence rather than a gap. What property of the ledger does the argument depend on?
3. What two construction details make the hash-chained audit trail robust in practice rather than only on paper, and state precisely what the chain does and does not promise. Why is the weaker claim the more defensible one to an auditor?
4. How does mapping a compliance obligation to an evidence query rather than to a binder eliminate the reconciliation gap between attested and observed compliance posture, and why is the binder assembly step the place where appearance and fact diverge?
5. Restate the book's closing principle in one sentence, and for each of its three verbs, name the structural mechanism in this chapter that enforces it. Then explain why increasing the capability of the AI model does not expand the control surface.

19.5.4 Student Companion Notes

You should read this final chapter as the convergence of the whole book rather than as a standalone treatment of security. Every control named here was introduced earlier for a local reason; the value of this chapter is in seeing that the local controls compose into a defensible global architecture. The exercises below are designed to make that composition explicit rather than to test recall.

First, take any one control from the control catalog table and trace it backward through the book to the chapter that introduced it. Identify the local problem it originally solved and explain how that same local control also serves the global assurance argument made here. This exercise succeeds when you can state the control's local purpose and its global purpose in one sentence each and see that they are the same mechanism viewed from two distances.

Second, take the controller's variance question used throughout the book and write out, step by step, the lineage traversal that answers it in the integrated platform. Then write out, honestly, how the same question would be answered in a fragmented environment you have personally seen, and compare the degree of trust you would rationally place in each answer and why. Pay particular attention to where in the fragmented case two sources that should agree do not.

Third, choose a compliance obligation relevant to your own organization and break it down into the obligation, control, and evidence structure. Determine whether the control already exists structurally in the architecture described here, and if so, state the query that would evidence it. This exercise is most valuable when the obligation initially appears to require a separate program but breaking down its structure reveals that it does not, because that is the moment this chapter's central claim becomes concrete rather than rhetorical.

Fourth, and as a capstone for the whole book, write a single page that traces one change from a triggering signal through suggestion, verdict, publication, and audit, naming the structural control that applies at each transition and the query that would evidence it. If you can write that page without referring back to the chapters, you have internalized the architecture; if you cannot, the gaps in your page tell you which chapters are worth rereading.

This chapter is the end of the book, but it is the beginning of your work as a practitioner. The architecture described here is not a diagram to admire but a discipline to operate, and the enterprises that operate it well will be the ones for whom each increase in AI capability becomes an asset on a trustworthy base rather than a liability amplified by fragmentation.

19.6 Conclusion: The Integrated, AI-Ready Enterprise

We began this book with a claim and have spent eighteen chapters establishing it. The claim was that transformation programs do not fail because architects lack vision or because process teams lack discipline, but because the artifacts those teams produce never meet in a place that machines, auditors, and AI systems can trust. Every chapter has answered that single problem.

We configured SAP Signavio so that process intent is modeled at execution depth rather than as documentation. We configured SAP LeanIX so that architecture is bound to that intent rather than describing systems in isolation. We positioned WalkMe so that adoption telemetry rejoins the loop instead of being lost the moment a project closes.

We described how SAP BTP is not middleware but the canonical data spine where identity, version, and ownership are preserved. We placed SAP AI on top of that spine under strict grounding, eligibility, and explainability discipline. We used SAP Build to turn governed suggestions into governed action. We showed how operations, security, and audit are the proof that the whole construction holds rather than an overhead imposed upon it.

The synthesis is best seen as a closed loop. Design becomes delivery, delivery generates telemetry, telemetry drives improvement, and improvement reenters design, with the canonical data spine at the center reconciling all four and the security and audit layer wrapped around the whole so that every transition is authorized, every change is evidenced, and every refusal is recorded. None of the individual SAP tools is sufficient on its own. The integration discipline that connects them, and the governance that holds that discipline accountable, is what determines whether an enterprise has built a platform it can trust or merely a collection of tools that appear to be connected because they share a vendor.

This architecture does not promise that AI will run the enterprise. It promises the opposite: that AI will propose, that governance will approve, and that the system of record will record, and that each of those three actions is enforced by a structural mechanism rather than by good intentions or by a model that behaves well in testing.

It does not promise that integration becomes easy. It promises that integration, done as a governed canonical discipline rather than as point-to-point connectivity, becomes durable, auditable, and safe to expose to increasingly capable AI. The enterprises that internalize this distinction will find that the same foundation that makes their automation trustworthy also makes their audits shorter, their changes traceable, and their operating model legible to the people who are accountable for it but did not build it.

The frontier will keep moving. Foundation models will improve, agents will become more capable, and the pressure to expose more of the operating model to autonomous reasoning will intensify rather than relent. Our argument is that the correct response to that pressure is not a better model but a better foundation.

An enterprise that has built the integrated, governed, auditable spine described here can adopt each new AI capability as an increment on a trustworthy base, testing it against the same eligibility, grounding, and refusal discipline that already governs everything else. An enterprise that has not will find that every increase in AI capability amplifies the fragmentation it never resolved, because a more capable agent pointed at incoherent data produces more confident incoherence, faster. The choice between those two trajectories is architectural. It is available now, and it is the reason we wrote this book.

A closing word to all of you who have read this far. The architecture in this book is not a destination that a program reaches and then maintains. It is a discipline that an organization practices, every time a new process is modeled, a new integration is built, a new agent is granted a scope, or a new obligation arrives.

Each chapter can be read in an afternoon, but the disciplines they describe take years to make habitual and a single inattentive quarter to erode. The enterprises that succeed with this material will not be the ones that understood it fastest but the ones that kept practicing it after the program that introduced it had closed and the attention had moved on. That is the unglamorous truth underneath every diagram in this book, and it is the right note on which to end: The integrated, AI-ready enterprise is not built once. It is sustained.

The Author

Kumar Piraviperumal has worked with SAP technologies since 1997, building a career that spans enterprise delivery, architecture leadership, and professional education. Over nearly three decades, his work has focused on how large organizations design, integrate, and operate complex SAP landscapes under real-world constraints.

He currently leads a global SAP business transformation practice at Cognitus, an IBM Company, with revenues exceeding USD 100 million, operating across more than 19 countries. In this capacity, Kumar is responsible for shaping transformation strategy, enterprise and solution architecture standards, delivery governance, and execution models for organizations undertaking large-scale SAP S/4HANA–centric programs. His work is grounded in environments where scale, regulatory requirements, and operational continuity are non-negotiable.

Kumar's industry experience is concentrated in utilities, automotive, and large-scale information technology and services organizations. Across these sectors, he has led and advised initiatives involving end-to-end process transformation, enterprise integration, financial consolidation, planning, and operational modernization. He has supported global rollouts, regulatory-driven change programs, and post-merger integration efforts, often in situations where SAP business transformation management tools, SAP BTP, and analytics platforms were introduced concurrently, without an established integration model.

Alongside his practitioner work, Kumar has maintained a strong commitment to academic engagement and professional enablement. He has written for SAPinsider magazine and the MIT Technology Journal, and has taught and contributed to courses at MIT Open University, focusing on enterprise systems, transformation frameworks, and the application of emerging technologies in large organizations. He has also been actively involved with SAP University Alliances and SAP Education Services, delivering enablement programs for consultants, architects, and client teams worldwide.

Kumar has directly taught and supported core SAP curricula, including S4H100 and S4H200, SAP BPC for consolidation and planning, SAP Analytics Cloud, and SAP Cloud ERP, group reporting. This sustained involvement in SAP education has shaped his perspective on the gap between formal product training and the realities of execution in complex enterprise environments (a gap that strongly informs the content of this book).

He is a frequent speaker and sponsor at SAP-focused events such as SAP enterprise architecture conferences, SAP Sapphire, SAPinsider events, SAP Transformation Excellence Summit, and SAP TechEd, and regularly collaborates with SAP product teams, partners, and enterprise architecture leaders. His current work increasingly focuses on execution-grade transformation and the responsible application of AI in SAP landscapes.

This book reflects Kumar's combined experience as a practitioner, architect, and educator. It presents a field-tested, technically rigorous view of how SAP Signavio, SAP LeanIX, SAP BTP, WalkMe, SAP Build, SAP Build Work Zone, and SAP AI can be integrated into a coherent, governed execution platform, grounded not in abstract theory, but in sustained delivery and academic inquiry.

Index

C

D

E

K

L

M

N

O

P

Q

R

S

T

U

V

W